Dispersity, Structure and Phase Changes of Proteins and Bio Agglomerates in Biotechnological Processes

Arno Kwade · Ingo Kampen
Editors

Dispersity, Structure and Phase Changes of Proteins and Bio Agglomerates in Biotechnological Processes

Springer

Editors
Arno Kwade
Institut für Partikeltechnik
Technische Universität Braunschweig
Braunschweig, Niedersachsen, Germany

Ingo Kampen
Institut für Partikeltechnik
Technische Universität Braunschweig
Braunschweig, Niedersachsen, Germany

ISBN 978-3-031-63163-4 ISBN 978-3-031-63164-1 (eBook)
https://doi.org/10.1007/978-3-031-63164-1

This Springer imprint is published by the registered company Springer Nature Switzerland AG
The registered company address is: Gewerbestrasse 11, 6330 Cham, Switzerland

Preface

On the way to a bio-based economy and society, a wide variety of products are already being manufactured using biotechnological processes. An important challenge of such process chains is the targeted control of the interactions of proteins and bioagglomerates with the environment. During the process chain, which includes cultivation, downstream processing and formulation, mechanical, thermal, and material stresses affect the biotechnological products. Since these products are much more sensitive compared to chemically produced low-molecular molecules and inorganic particles, material changes such as structure, surface properties, or even changes in the productivity of cell assemblies occur. Due to these effects, the yield of biotechnological processes can be reduced severely.

In order to overcome these challenges, scientists from process engineering and life science worked together in the DFG priority program No. 1934 "Dispersity, Structure and Phase Changes of Proteins and Bio Agglomerates in Biotechnological Processes" (DiSPBiotech). They investigated the effect of various stress events on proteins and bioagglomerates on three levels: The micro-level studied the change of single proteins, including the molecular level. At the meso-level, protein aggregates like protein complexes, amyloid fibrils, virus-like particles and crosslinked protein crystals were investigated. The macro-level focused on cell aggregates like pellets from filamentous microorganisms. For the investigation, various measurement methods were developed or adapted from inorganic materials to biological materials from the multidisciplinary consortium, leading to new insights in this field. This book presents the results of the projects within the priority program No. 1934 DiSPBiotech.

As scientific speaker and general manager of the priority program, we would like to thank all involved partners for their intensive and highly motivated work within the last years and the preparation of many papers and especially the different book chapters showing state-of-the-art results about the influence of different kinds of stress on proteins and bioagglomerates.

Braunschweig, Germany

Arno Kwade
Scientific Speaker
DFG priority program No. 1934

Ingo Kampen
General Manager

Contents

Correlation of Structure and Properties of Enzyme Crystals in Consideration of Downstream Processing and Formulation

Jan Wichmann, Marcel Staar, Marta Kubiak, Jennifer Solarczek, Janine Mayer, Dieter Jahn, Ingo Kampen, Carsten Schilde, Anett Schallmey, and Rebekka Biedendieck

1 Introduction

Enzymes are biocatalysts that catalyze a wide range of reactions with usually high chemo-, regio-, and stereoselectivity [1]. The moderate conditions, under which they often operate in their natural environment, can differ from optimal reaction conditions during their industrial application, as the applied higher reaction temperatures, organic solvents, or extreme pH values can offer the chance to run reactions more effectively. However, this can also challenge the stability of the biocatalysts itself. One possibility to overcome these limitations and additionally enable the reuse of

J. Wichmann, M. Staar, M. Kubiak: These authors equally contributed to this work.

J. Wichmann · J. Mayer · D. Jahn · R. Biedendieck (✉)
Institute of Microbiology and Braunschweig Integrated Centre of Systems Biology (BRICS), Technische Universität Braunschweig, Braunschweig, Germany
e-mail: r.biedendieck@tu-braunschweig.de

J. Wichmann
e-mail: jan.wichmann@tu-braunschweig.de

J. Mayer
e-mail: janine-1993@gmx.de

D. Jahn
e-mail: d.jahn@tu-braunschweig.de

M. Staar · J. Solarczek
Institute of Biochemistry, Biotechnology and Bioinformatics, Technische Universität Braunschweig, Braunschweig, Germany
e-mail: m.staar@tu-braunschweig.de

J. Solarczek
e-mail: jennifer_solarczek@web.de

A. Kwade and I. Kampen (eds.), *Dispersity, Structure and Phase Changes of Proteins and Bio Agglomerates in Biotechnological Processes*,
https://doi.org/10.1007/978-3-031-63164-1_1

enzymes for multiple reactions is enzyme immobilization [2]. Available strategies can be divided into carrier-dependent immobilization, where a soluble enzyme is entrapped in or attached to a non-catalytic matrix or support, and carrier-free immobilization, where enzyme molecules are covalently connected to each other via cross-linking [3]. Although carrier-dependent immobilization has been predominantly used in the past, it comes with some disadvantages, such as the high mass of the non-catalytic carrier (up to 90%), which reduces volumetric activity, as well as high costs of the carrier itself [4]. One example of carrier-free immobilization is the generation of cross-linked enzyme crystals (CLECs), which are obtained upon crystallization of an enzyme and subsequent chemical cross-linking of enzyme molecules within a crystal via functional groups of amino acid side chains on the enzyme surface. In this way, intermolecular connections are formed, protecting the crystal lattice from disassembling under conditions that would otherwise lead to dissolution of non-cross-linked crystals [5].

Typically, enzyme crystallization is known in connection with X-ray structure analysis, for which high-purity enzyme crystals are required. On technical scale, the production of enzyme crystals is mainly used as a purification method during downstream processing and for final formulation of a product [6]. Crystallized enzymes offer several advantages over their soluble form, such as higher stability [7] as well as easier processability in industrial processes. However, as native enzyme crystals are mechanically unstable, their industrial use is usually limited [8]. Next to an easier separation from liquids and, thus, a facilitated reuse for multiple reactions [9], the cross-linking of enzyme crystals further stabilizes enzymes for application at high temperatures, at extreme pH values, or in organic solvents and additionally protects them from proteolytic digestion [9, 10]. Due to such benefits, CLECs have been widely used in different reactor systems such as packed-bed reactors or on microfluidic chips [9–11]. In order to obtain a CLEC preparation with high activity recovery, specific requirements have to be met. Firstly, the active sites of the corresponding enzyme molecules within the CLEC have to be accessible for substrates and products by diffusion through the solvent channels of the CLEC. Secondly, the cross-linker should not react with amino acid residues that are essential for catalysis and thirdly,

A. Schallmey
Institute of Biochemistry, Biotechnology and Bioinformatics, Center of Pharmaceutical Engineering (PVZ) and Braunschweig Integrated Centre of Systems Biology (BRICS), Technische Universität Braunschweig, Braunschweig, Germany
e-mail: a.schallmey@tu-braunschweig.de

M. Kubiak · I. Kampen · C. Schilde
Institute of Particle Technology and Center of Pharmaceutical Engineering (PVZ), Technische Universität Braunschweig, Braunschweig, Germany
e-mail: marta-kubiak@web.de

I. Kampen
e-mail: i.kampen@tu-braunschweig.de

C. Schilde
e-mail: c.schilde@tu-braunschweig.de

the cross-linked crystal lattice must allow the enzyme to adopt its active conformation in case the enzyme can exist in several conformations or if it requires a certain flexibility for catalysis [12, 13]. Moreover, the catalytic activity of CLECs correlates with particle size [12] and cross-linking time [14]. With increasing crystal size, the resistance of diffusion also increases, which leads to a decrease in biocatalytic activity. On the other hand, the handling of bigger particles in a process is easier. Also a longer cross-linking time leads to a decreased biocatalytic activity, while on the other hand, enzymes are exposed to a wash out in case of short and thus incomplete cross-linking [15].

Due to the increasing demands on the tailored properties of CLECs in terms of enzymatic activity as well as mechanical stability during technical applications, the elucidation of the relationship between enzyme structure and the resulting particle properties is of particular interest [16] as our current understanding of such relationships is very limited. In this regard, the mechanical stability of these enzyme crystals is not only closely related to the crystal structure of the corresponding enzyme but also to the enzyme itself. Since the structure of an enzyme depends on the underlying protein sequence, the introduction of amino acid exchanges in that protein sequence by means of protein engineering will enable a tailoring of enzyme structure and thus also crystallization and crystal properties. The challenge of protein engineering, however, lies in the correct prediction of suitable amino acid exchanges to obtain enzyme structures that then fulfill the specified functions. In order to produce customized CLECs with desired properties, all relationships, interactions, and influences along the entire manufacturing chain, starting from the protein sequence to the final CLEC, have to be investigated and elucidated (Fig. 1).

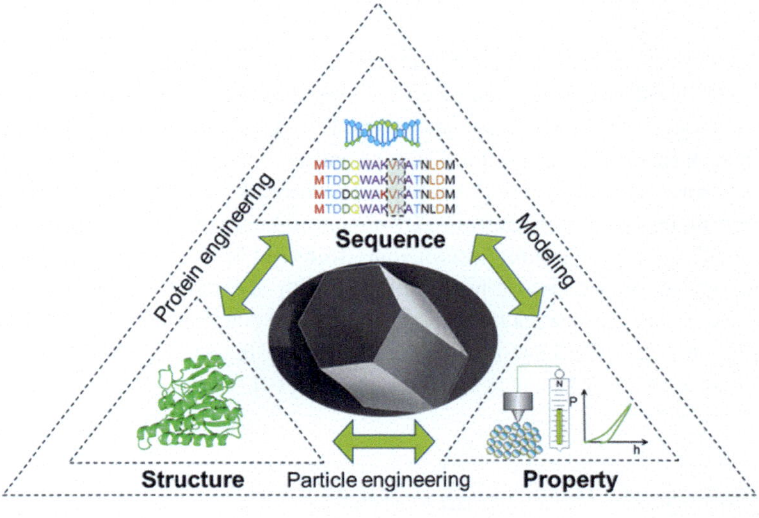

Fig. 1 Structure–property, sequence-structure, and sequence-property relationships to produce tailored cross-linked enzyme crystals (CLECs). Figure taken from [16], licensed under the Creative Commons Attribution 4.0 International license (CC BY 4.0)

With regard to the **sequence-structure** relationship (Fig. 1), various enzyme engineering methods were employed in our project to exchange amino acids of the wild-type enzymes in order to enhance crystallization as well as cross-linking within the produced protein crystals with the aim to improve the mechanical properties of resulting CLECs. To elucidate the **structure–property** interactions, a parametric study of the influence of process and formulation parameters was performed during protein crystallization, using purified wild-type enzyme as an example, and methods for reproducible sample preparation were established. Moreover, a structure-based model was developed to represent the cross-linking within the crystal. With the aid of this model, the anisotropic crystal properties were correlated with the potential cross-linking sites based on the crystal structures of wild-type enzyme and variants, and the influence of specific exchanges on the resulting cross-linking and, thus, on the mechanical crystal properties was elucidated (**sequence-property** relationship).

The three relationships were studied at the example of penicillin G acylases (PGA) and the halohydrin dehalogenase HheG and corresponding variants of both. In this project, a detailed process chain was followed for the generation and characterization of the individual CLECs. It started with the genetic engineering, production, and purification of the given enzyme variant in sufficient yields, followed by enzyme characterization, and, if relevant, crystal structure determination. The next step included the crystallization of an adequate amount of the given enzyme followed by cross-linking to obtain respective CLECs. Finally, the CLECs were characterized regarding their biochemical, thermal, and mechanical behavior.

1.1 Penicillin G Acylases (PGA)

Penicillin G acylases (PGA; EC 3.5.1.11) are heterodimers with a smaller α-subunit and a larger β-subunit. They belong to the superfamily of N-terminal nucleophile hydrolases found in bacteria, yeast, and fungi [17]. PGAs are industrially used for the production of semisynthetic antibiotics, especially for penicillins and cephalosporins [18]. In general, the semisynthetic antibiotics are synthesized by first splitting a β-lactam antibiotic that originates from a fermentation-based production into an organic acid and a β-lactam-containing intermediate followed by coupling the latter with a different acyl group to yield a new antibiotic with altered properties. During this process, PGAs can catalyze both, the splitting and the coupling [19, 20] (Fig. 2). Due to the high demand of β-lactam antibiotics including the semisynthetic substances, the production of PGA has grown to a large scale [19].

1.2 Halohydrin Dehalogenase HheG

Halohydrin dehalogenases (HHDH; EC 4.5.1.-) are bacterial lyases belonging to the superfamily of short-chain dehydrogenases and reductases (SDR) and catalyze the reversible dehalogenation of β-haloalcohols with formation of the corresponding

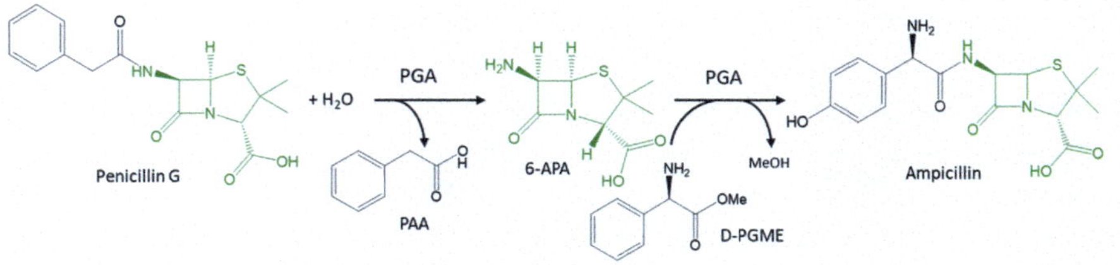

Fig. 2 Enzymatic synthesis of ampicillin from penicillin G catalyzed by PGA. PAA: phenylacetic acid; 6-APA: 6-aminopenicillanic acid; D-PGME: D-phenylglycine methyl ester

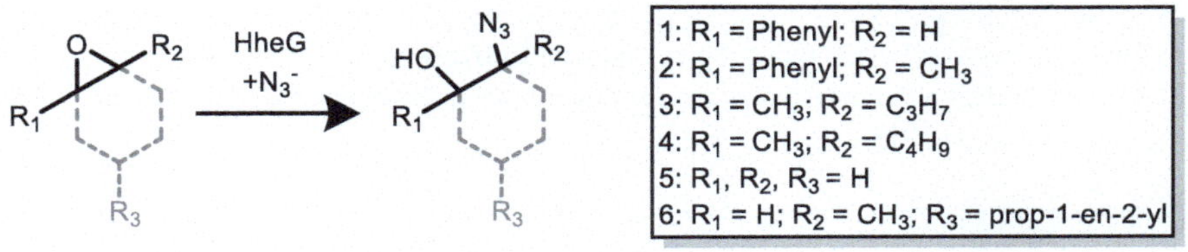

Scheme 1 Azidolysis of various epoxides catalyzed by HheG. Nomenclature as followed: **1**: styrene oxide, **2**: 1-methylstyrene oxide, **3**: *trans*-2,3-hexene oxide, **4**: *trans*-2,3-heptene oxide, **5**: cyclohexene oxide, **6**: limonene oxide. Cyclohexane ring and chemical rest R_3 only present for substrates **5** and **6**

epoxides [21]. In the reverse reaction, i.e. epoxide ring opening, they also accept a range of different *N*-, *C*-, *O*-, and *S*-nucleophiles giving access to a plethora of possible β-substituted alcohol products that can be used as building blocks in the pharmaceutical and fine chemicals industry [22]. Industrially, HHDHs are applied for the synthesis of a chiral statin side chain precursor, used in the production of atorvastatin (Lipitor™), or the removal of a toxic haloalcohol in food applications [23, 24]. Among all currently known HHDHs, G-type halohydrin dehalogenases such as HheG from *Ilumatobacter coccineus* are of special interest as they feature much broader substrate spectra including also sterically demanding cyclic as well as acyclic internal epoxides (Scheme 1) [25–27].

2 Materials and Methods

2.1 Engineering of HheG and PGA

Amino acid positions for mutagenesis were selected as described in the corresponding chapters. Site-directed mutagenesis was performed using the PfuUltra II Hotstart PCR Mastermix (Agilent Technologies) with the corresponding primers as described in [28–30] with 250 nM (engineering of HheG) or 500 nM (engineering of PGA) concentration each and 100 ng template plasmid. The mutagenesis protocol was in agreement with the manufacturer's instructions with an additional DpnI digestion (37 °C, overnight) step.

2.2 Generation of Cross-Linked PGA Crystals

PGAs were recombinantly produced and secreted with *Priestia megaterium* MS941 as described previously [30, 31]. They were purified directly from the cell-free culture supernatant by different ion exchange chromatography protocols, suitable for the respective PGAs [31]. Depending on the type and variant of the PGA, different further purification techniques including buffer exchange and gel permeation chromatography were required for crystallization [30, 31]. Crystallization of PGAs based on the addition of PEG_{8000} (PGA from *Bacillus* sp. FJAT-27231), calcium chloride, sodium acetate, 2-methyl-2,4-pentanediol (PGA from *Bacillus thermotolerans*), and potassium iodide and PEG_{3350} (PGA from *P. megaterium*) [30, 32]. Furthermore, seeding with crushed crystals of the respective PGAs was used in several cases [30, 32]. Cross-linking was carried out with glutaraldehyde, directly added to a suspension of washed crystals [32].

2.3 Generation of Cross-Linked HheG Crystals

HheG wild-type and variants were produced and purified as described before [28]. Small-scale crystallization in 2 μL volume using vapor-diffusion was performed in 96-well CrystalQuick SW plates (Greiner Bio-one). Protein solution and crystallization buffer were mixed 1:1 and analyzed microscopically after incubation for 72 h at 8 °C. The composition of respective crystallization buffers has been described earlier [28]. 2 mL batch crystallization was performed for 24 h at 8 °C under gentle shaking using a 1:1 ratio of protein solution and crystallization buffer. Cross-linking was performed for 24 h at 8 °C under gentle shaking using 2 mM bis-maleimidoethane (BMOE) as cross-linker for cysteine-variants and 4–40 mM disuccinimidyl tartrate (DST), dimethylpimelimidate (DMP), dimethylsuberimidat (DMS), dithiobis(succinimidyl)propionate (DSP), or ethylene glycol bis(sulfosuccinimidyl succinate) (Sulfo-EGS) for lysine-variants as well as wild-type HheG. Cross-linking of HheG variant V46K for biocatalytic applications was performed with 4 mM DSP. All cross-linkers were purchased from Thermo Fisher Scientific.

2.4 Activity and Thermal Stability Determination of Cross-Linked HheG Crystals

Biocatalytic reactions were generally performed using 100 μg biocatalyst, 20 mM cyclohexene oxide (Thermo Fisher Scientific), and 40 mM azide in 1 mL Tris·SO_4, pH 7.0, at 22 °C and 900 rpm. Reactions were extracted with an equal volume of *tert*-butylmethylether (TBME) and dried over $MgSO_4$. Samples were analyzed via gas chromatography (GC) on a GC2010 plus (Shimadzu). Achiral analysis was carried out using an OPTIMA 5 MS column (30 m length, 0.25 mm inner diameter, 0.25 μm film thickness, Macherey Nagel), while chiral analysis was performed using a HYDRODEX γ-DIMOM column (25 m length, 0.25 mm inner diameter, 0.25 μm film thickness, Macherey Nagel) with 40 cm s^{-1} hydrogen carrier gas flow. Details regarding the determination of temperature and pH profiles, solvent stability and half-life as well as respective GC temperature programs and corresponding retention times of substrate and products have been described previously [28].

A thermal shift assay in combination with SYPRO orange dye (Thermo Fisher Scientific) has been applied to determine melting temperatures of CLECs and soluble enzyme as described earlier [28].

2.5 *Assay for Determination of PGA Activity*

Instead of the natural substrate penicillin G, the synthetic substrate 2-nitro-5-[(phenylacetyl)amino] benzoic acid (NIPAB) was used in a kinetic enzyme assay for determination of PGA activities. Cleavage of NIPAB by PGAs releases the dye 2-nitro-5-aminobenzoic acid (NABA), which absorbs at 405 nm. For activity determination, 90 μL of substrate solution (600 mg L^{-1} NIPAB in 9.4 mM NaH_2PO_4 and 40.6 mM Na_2HPO_4, pH 7.5) were placed in flat-bottomed microtiter plates at 37 °C. To start the reaction, 10 μL of pre-warmed cell-free supernatant or solution of purified enzyme were added, mixed, and incubated at 37 °C. Absorbance (A) at 405 nm was determined every 6 s over a period of 3 min in the microplate reader (NanoQuant infinite M200 pro, Tecan). Volumetric enzymatic activity EA [U mL^{-1}] was calculated using the time-dependent slope of absorbance dA/dt [min^{-1}] according to the following formula.

$$EA = \frac{\frac{dA}{dt} \cdot V_R}{\varepsilon \cdot d \cdot V_E} \tag{1}$$

where V_R is the volume of the total reaction mixture (0.1 cm^3), V_E is the volume of the enzyme solution (0.01 cm^3), ε is the molar extinction coefficient (8.98 cm^2 μmol^{-1}), and d is the layer thickness of the reaction solution (0.315 cm).

2.6 *Diffraction Data Collection and Structure Determination*

Diffraction data of crystals were collected on beamline P11 (PETRA III synchrotron, DESY, Hamburg, Germany) or on beamline PXIII of the Swiss Light Source (SLS, Paul Scherrer Institute, Villigen, Switzerland). Crystals of the HheG variant D114C, which were used for structure determination, were grown in 46 mM HEPES, pH 7.34,

5.33% (w/v) PEG_{8000}. Details regarding crystallization screening conditions, data integration, and structure determination have been described previously [28]. Details regarding structure determination of PGAs, including the crystallization conditions and crystallization techniques used to obtain the corresponding crystals, have been described previously for the PGA from *P. megaterium* and *B. thermotolerans* and the wild-type PGA from *Bacillus* sp. FJAT-27231 [31] as well as for variants of the latter [30].

2.7 *Micromechanical Characterization of Cross-Linked Enzyme Crystals*

An atomic force microscope (AFM) (Bruker NanoWizard3, former JPK) and a classical nanoindenter (Hysitron TriboIndenter Ti900) were used to characterize mechanical properties at different scales [16, 32–35]. The AFM tip with a radius of 150 nm allows the investigation of mechanical properties of the surface at low forces by generating small penetration depths in nanometer range. The nanoindenter (Berkovich tip) allows the use of much higher indentation forces, so that mechanical behavior at larger penetration depths (up to 5 μm) can be investigated. A mechanical investigation in both depth ranges was of particular interest from a modeling view, which was based on the molecular crystal structure.

3 Results and Discussions

For both studied enzyme types, PGA and HheG, all required steps along the process chain for the generation of stable and active CLECs, i.e. genetic modification of underlying enzymes, protein production and purification, crystallization and cross-linking as well as biochemical and mechanical characterization of resulting CLECs, were investigated. For each enzyme, however, the focus was set differently depending on the encountered enzyme-specific challenges. For PGA enzymes, the production and crystallization had to be improved, while the cross-linking of PGA crystals with glutaraldehyde worked well. In contrast, for HheG the cross-linking of respective enzyme crystals had to be optimized specifically, as glutaraldehyde as cross-linker resulted in an inactivation of HheG at longer incubation times. HheG production and crystallization, instead, did not represent challenges in the project. In the following, first the results regarding the identification and engineering of PGAs regarding their production and crystallizability are presented, followed by the engineering of HheG towards higher thermal stability as well as improved cross-linking, and the detailed characterization of resulting HheG variants and CLECs. Finally, mechanical properties of both PGA and HheG CLECs are presented and the impact of molecular engineering and cross-linking is discussed.

3.1 *Properties and Structures of PGAs from Gram-Positive Bacteria*

At present, industrially used PGAs mainly derive from Gram-negative bacteria, in particular from *Escherichia coli* and *Achromobacter sp.* CCM 4824 [18]. PGAs from Gram-positive bacteria are less characterized. They have around 30% amino acid sequence identity with PGAs from Gram-negative bacteria and can be secreted directly into the culture medium by Gram-positive production hosts simplifying the downstream processing for these enzymes [36]. In order to further explore and analyze PGAs from Gram-positive bacteria, the amino acid sequence of the PGA from Gram-positive *P. megaterium* (formerly *Bacillus megaterium*) strain ATCC14945 (BmPGA) was used for a BLAST search (NCBI) to identify homologs from related Gram-positive species. PGAs from six different *Bacillus* species were identified revealing amino acid identities of 52–72% with the PGA from *P. megaterium.* All six PGAs were recombinantly produced and secreted carrying their native signal peptides with *P. megaterium* strain MS941, lacking an extracellular protease which reduces the extracellular protease activity by 98.5% [37]. In addition to the PGA from *P. megaterium*, the PGAs from *Bacillus sp.* FJAT-27231 (FJAT-PGA) and *B. thermotolerans* (BtPGA) could be successfully produced and secreted. All three PGAs could be purified by ion exchange chromatography, followed by gel permeation chromatography. Compared to BmPGA, the newly identified PGAs showed increased thermal stability by up to 18.3 °C (BmPGA: 56.0 °C; BtPGA: 64.5 °C; FJAT-PGA: 74.3 °C). The specific activities regarding the hydrolysis of NIPAB were 14.0 U mg_{PGA}^{-1} (BmPGA), 12.6 U mg_{PGA}^{-1} (BtPGA), and 14.4 U mg_{PGA}^{-1} (FJAT-PGA), respectively [31].

The three PGAs could be successfully crystallized using different buffer conditions. The crystals scattered to resolutions of 2.30 Å (BmPGA), 2.07 Å (BtPGA), and 1.75 Å (FJAT-PGA), respectively, in X-ray structural analysis. The elucidation of the protein structure succeeded for all three PGAs by molecular replacement using known PGA structures. The newly elucidated structures were assigned the PDB IDs 6NVX (FJAT-PGA), 6NVY (BtPGA), and 6NVW (BmPGA), respectively. All three PGAs showed a triangular shape with a central deep trough, containing the positively charged active site, which is also typical for PGAs form Gram-negative bacteria. The new PGA models superimposed with rmsd (root-mean-square deviation) between 0.62 and 0.84 Å over all C-α positions of the structures, which shows a high structural similarity between them. However, the charges of the surfaces differed among the three enzymes (Fig. 3).

Major structural differences were observed in comparison to PGAs from Gram-negative bacteria. Structural alignment of the FJAT-PGA with the PGA from *Providencia rettgeri* [39] and the PGA from *E. coli* (EcPGA) [40] resulted in overall rmsds of 1.84 Å and 1.87 Å respectively. A striking structural difference was an insertion of the β-subunit into the entrance of the binding pocket with a length of about ten amino acids, which is present in PGAs from Gram-negatives but was not observed for the PGAs from Gram-positives that hence have larger binding pockets (Fig. 4).

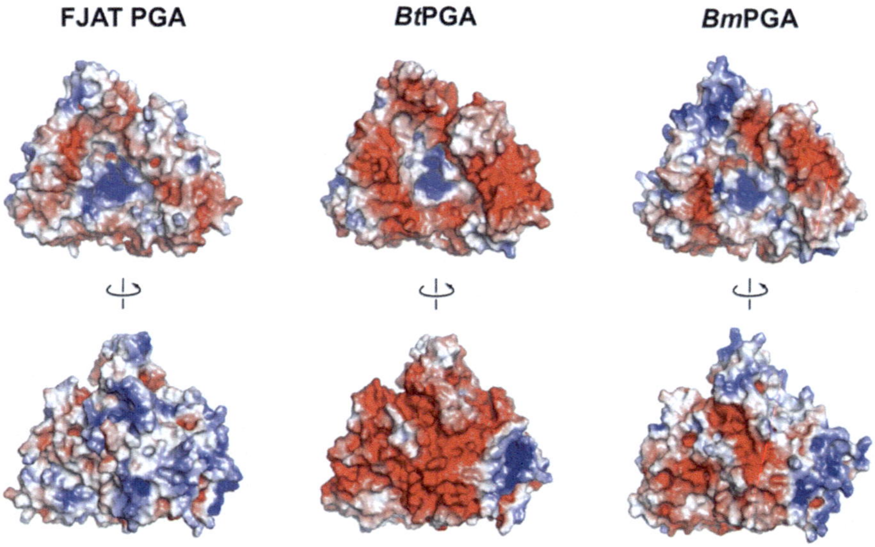

Fig. 3 Front and backside view of the electrostatic potential mapped to the molecular surfaces of FJAT-PGA, BtPGA, and BmPGA displayed at ± 5 kb Te^{-1} (blue: positive, red: negative). The electrostatic potential was calculated with the software APBS [38]. This figure and all other molecular representations have been prepared with PyMol (The PyMOL Molecular Graphics System, Version 2.0 Schrödinger, LLC). Reprint by permission from Springer Nature: Ref. [31]

It was further found, that a Penicillin G molecule taken out of a EcPGA structure in complex with this substrate did not fit well into the binding side of the FJAT-PGA structure (Fig. 4) and that some residues in the vicinity of conserved residues, required for enzyme activity, differed between the EcPGA and the FJAT-PGA. These three observations indicate that there could be differences in the accessibility of the binding site and in the substrate binding itself between PGAs from Gram-negatives and Gram-positives.

In summary, two new PGAs (FJAT-PGA and BtPGA) were identified and could be recombinantly secreted with *P. megaterium* like the known BmPGA [31, 41]. Their high thermostabilities and activities justify choosing them as model enzymes for further studies, aiming at realizing their immobilization in CLECs. The obtained crystal structures allowed for structure-based protein engineering and structure–property–correlation analysis of their crystal lattices.

3.2 Improvement of Recombinant PGA Production and Secretion

In contrast to the intracellular production and the accumulation of a recombinant protein in Gram-negative organisms such as *E. coli*, its secretion offers advantages such as simplified and faster product recovery or the possibility of a continuous cultivation without cell disruption [42–44]. For its successful secretion, a protein

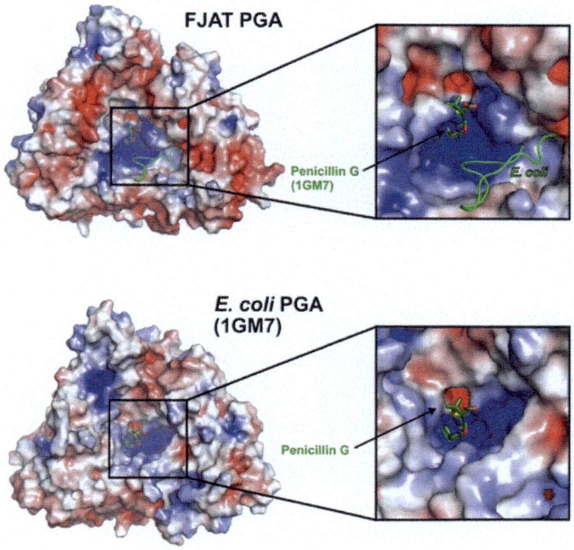

Fig. 4 Comparison of the molecular surfaces of FJAT (top) and *E. coli* PGA (bottom), colored according to their electrostatic potential at ± 5 kb Te^{-1}. The inserts show a magnification of the active sites. Penicillin G, present in the *E. coli* protein (PDB entry 1GM7) [40], is shown in stick representation and has been modeled into FJAT-PGA to indicate that the active site of this enzyme is not ideally suited to accommodate this substrate. The green loop shows an insertion in the β-subunit of PGAs from Gram-negative bacteria that reduces the active site pocket of these enzymes. Reprint by permission from Springer Nature: Ref. [31]

must carry an N-terminal signal peptide responsible for protein translocation. The so-called Secretion (Sec) pathway, by which proteins are transported in an unfolded manner across the cell membrane, is the most commonly used secretion mechanism in *Bacillus* spp. Studies in *Bacillus subtilis* have shown that more than 90% of its secreted proteins are transported via this pathway [45]. However, the amount of protein secreted depends on the combination of signal peptide and protein. As the best suited combination of signal peptide and given protein cannot be predicted bioinformatically, an optimal combination must be found individually [43]. Therefore, a system that allows easy and rapid cloning and screening is needed to identify the best signal peptide for a given recombinant protein in a given production organism [43].

Six newly identified Gram-positive PGAs were recombinantly secreted with *P. megaterium* using their native Sec-dependent signal peptide with widely varying quantities [31]. While the strongly secreted PGAs could be purified and further characterized, different Sec-dependent native signal peptides from *P. megaterium* should be tested for the only poorly secreted ones to enhance their secretion. Using the native signal peptide of the *Bacillus massiliogorillae* PGA (BmasPGA) for guiding the enzyme out of the cell, a volumetric PGA activity of 0.015 U mL^{-1} could be measured which was insufficient for purification and characterization [31]. Although the BmasPGA was recombinantly produced in higher amounts, due to an inefficient secretion the majority of the enzyme stayed as insoluble protein inside the cell [46].

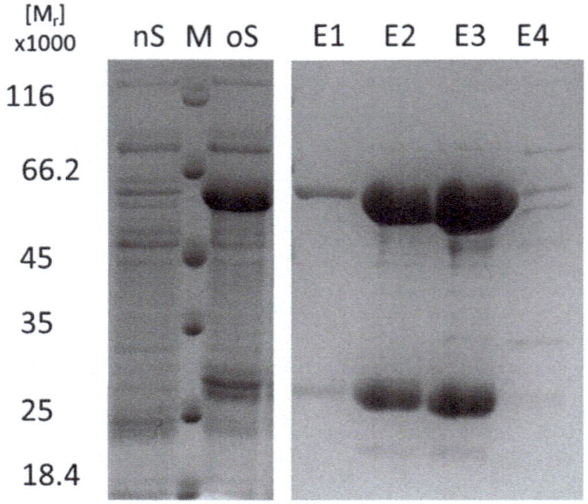

Fig. 5 Extracellular production and purification of recombinant BmasPGA. *P. megaterium* carrying a recombinant plasmid for BmasPGA secretion (nS–native signal peptide, oS–optimal signal peptide) was grown in full medium at 37 °C for 22 h after induction of recombinant protein production. Proteins in the cell-free supernatant were precipitated by ammonium sulphate (lines nS and oS). For the purification, the cell-free culture supernatant was obtained by centrifugation and sterile filtration. BmasPGA was purified by cation exchange chromatography. For elution, 50 mM sodium phosphate buffer (pH 5.5) containing 100 (E1) to 400 mM NaCl (E4) was used. M, Pierce™ Unstained Protein MW Marker, Thermo Scientific. Parts of the Figure taken from [46], licensed under the Creative Commons Attribution 4.0 International license (CC BY 4.0)

Based on the genome sequence of *P. megaterium* strain DSM319 [47], 182 Sec-dependent signal peptides were predicted for this organism. The sequences were used for the construction of a plasmid-based signal peptide library for *P. megaterium*. This allowed for multiple combinations of different *P. megaterium* signal peptides and the BmasPGA via simple cloning steps. After transformation of *P. megaterium* with the different constructs, a screening using PGA-activity in the cell-free supernatant as a read-out was performed in mL-scale. Replacing the native signal peptide with individual different *P. megaterium* own signal peptides, the amount of recombinantly secreted BmasPGA could be increased almost 30-fold (0.432 U mL^{-1}) [46] which now allowed for the purification directly from the cell-free supernatant in sufficient amounts (Fig. 5). With this, a deeper characterization of this new enzyme will be possible.

3.3 Optimization of PGA Crystallization by Protein Engineering

The crystallization conditions identified for the structural elucidation of the wild-type FJAT-PGA (FJAT WT) [31] allowed successful crystallization, but however within multiple days up to weeks. For establishing a reliable production of FJAT-PGA CLECs, the crystallizability of the enzyme needed to be enhanced. In detail, the crystallization should reach a high crystal yield after a maximum time of 48 h with a low amount of precipitating agent. Successful crystallization at low concentrations of the PGA and despite organic or saline impurities in the mother liquor should furthermore reduce the effort regarding purification and concentration prior to crystallization [30].

To enhance the crystallizability, multiple protein engineering strategies were combined. In a first step, surface entropy reduction (SER) was carried out [48]. The properties of the surface of proteins are crucial for their crystallizability, since the formation of a crystal lattice is based on interactions of the surface amino acids of the individual molecules [49]. Amino acid residues with high conformational entropy, such as lysine and glutamic acid, located on the surface of proteins can hinder their crystallization. These flexible residues are more difficult to arrange into a crystal lattice due to a large number of conformational possibilities, often restricted during crystallization. Thus, an entropic barrier has first to be overcome. Size and charge of lysine or glutamic acid residues can furthermore cause steric or electrostatic repulsions. Small, uncharged amino acid residues with low entropy, such as alanine, therefore thermodynamically favor the formation of crystal contacts [48, 50].

For the FJAT-PGA, three clusters of three amino acids each where calculated by the Surface Entropy Reduction prediction (SERp) server [48] to be suitable targets. By exchanging all predicted nine amino acids with alanine residues, a new nine-fold modified PGA variant termed FJAT SER (FJAT-PGA α_K201A, α_K202A, α_E203A, β_K135A, β_E136A, β_K137A, β_K403A, β_E404A, β_E405A) was created without any loss of activity compared to the wild-type enzyme (FJAT WT). After its recombinant production and secretion with *P. megaterium*, its purification by cation exchange chromatography and a buffer exchange with a salt-free HEPES buffer, the new variant was successfully crystallized under conditions suitable for crystallization of FJAT WT. But while FJAT WT crystallized in a compact crystal form after some weeks (Fig. 6a), FJAT SER reliably spontaneously formed bigger rod-like crystals within less than 24 h (Fig. 6b). The crystallization process could be further accelerated by seeding with crushed crystals for secondary nucleation. While FJAT WT requires at least two days to crystallize in a seeded batch approach, crystals of FJAT SER filled the crystallization drop after less than 30 min (Fig. 6c) [30]. Its crystal structure was solved with a resolution of 1.63 Å (PDB-ID: 8BRQ), revealing an altered crystal packing compared to crystals of FJAT WT with new crystal contacts enabled by some of the exchanged patches on the surface [30]. FJAT SER was chosen for further structure-based crystallizability enhancement.

Based on the FJAT SER crystal structure, entire flexible regions were identified and shortened. Similar to large amino acid residues, also whole flexible regions can cause an entropic hurdle for crystallization as the arrangement in a crystal lattice may restrict their conformational freedom [51, 52]. Thus, regions in the FJAT SER model with high B-factors compared to other sections were shortened by deletions in the FJAT SER variant. The resulting variants were termed FJAT flex variants and had deletions at the C-terminus of the α- (flex 1) or β-subunit (flex 5 and 7) or in

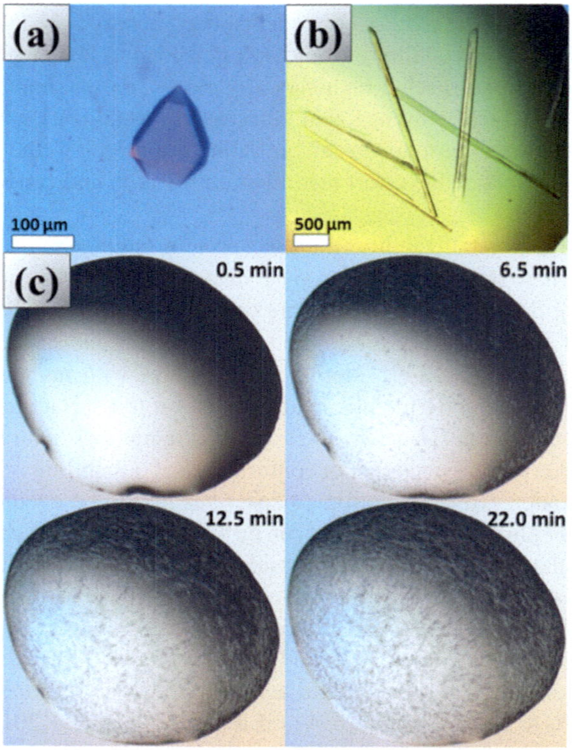

Fig. 6 Forms of FJAT WT (**a**) and FJAT SER (**b**) crystals and crystals of FJAT SER after seeded batch crystallization (**c**). **a**–**b** are from hanging drop approaches. The reservoir buffer (R) contained 100 mM HEPES (pH 7.5) and PEG_{8000}. The PGA solution (P) contained 10 mM HEPES (pH 7.0). The concentration of PGA and PEG_{8000} in the solutions and the mixing ratio differed: (a): R: 40 mg mL^{-1} PEG_{8000}, P: 8 mg mL^{-1} FJAT WT, R:P = 6:4, 26 d of incubation **b** R: 100 mg mL^{-1} PEG_{8000}, P: 10 mg mL^{-1} FJAT SER, R:P = 5:5, 19 h of incubation. **c** Crystallization occured without concentration by vapor diffusion. 4 µL of 100 mM HEPES (pH 7.5) containing 80 mg mL^{-1} PEG_{8000} were mixed with 4 µL of 8 mg mL^{-1} FJAT SER in 10 mM HEPES (pH 7.0) and 2 µL of a seed suspension of crushed FJAT SER crystals in 100 mM HEPES (pH 7.5) containing 80 mg mL^{-1} PEG_{8000}. The drop was incubated in a small closed petri dish. Figure taken from [30], licensed under the Creative Commons Attribution 4.0 International license (CC BY 4.0)

a flexible loop in the β-subunit (flex 3) (Fig. 7). The new variants were screened for improved crystallizability by determining the lowest concentration of purified PGA variant on the one hand and PEG_{8000} as precipitant on the other hand at which reliable crystallization of a FJAT flex variant occurred after 2, 24, and 48 h and were compared to the crystallizability of FJAT SER [30]. While the truncation of the C-terminus of one subunit (FJAT flex1, FJAT flex 5, FJAT flex 7) in most cases led to higher concentration limits of PGA or PEG_{8000} compared to FJAT SER, a deletion of β_K114 in the flexible loop (FJAT flex 3) caused reliably crystallization at a 60% decreased PGA concentration and a 25% decreased PEG_{8000} concentration after 2 h (Fig. 7).

The deletion in FJAT flex 3 did not influence the enzymatic activity compared to FJAT SER. Moreover the crystals of this new variant had the same rod-like form as

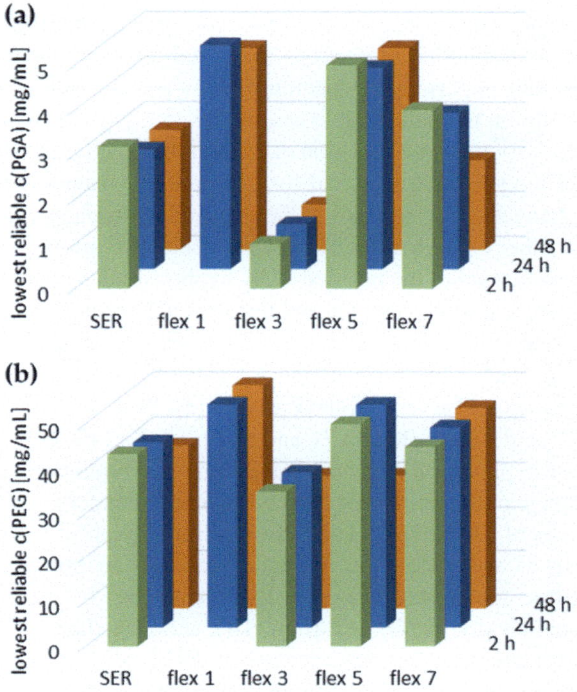

Fig. 7 Crystallizability screening assay of FJAT flex variants. The variants based on FJAT SER with the following exchanges: FJAT flex 1: α_D199del, FJAT flex 3: β_K114del, FJAT flex 5: β_S528stop, FJAT flex 7: β_N535stop. A solution of PGA in 10 mM HEPES (pH 7.0) and a solution of PEG_{8000} in 100 mM HEPES (pH 7.5) were mixed in equal proportions, creating conditions with **a** 50 mg mL^{-1} PEG_{8000} and 0.5–5 mg mL^{-1} PGA and **b** 5 mg mL^{-1} PGA and 5–50 mg mL^{-1} PEG_{8000}. For each condition, six drops with a volume of 4 μL were incubated under paraffin oil and observed after 2 h, 24 h, and 48 h. A crystallization under a certain condition was rated as reliable, if visible crystals grew in more than four of the respective drops. Figure taken from [30], licensed under the Creative Commons Attribution 4.0 International license (CC BY 4.0)

crystals of FJAT SER [30]. However it was observed that 95% of initially dissolved FJAT flex 3 had moved to the crystalline phase after 24 h, while FJAT SER reached a crystal yield of 71% after 24 h and 76% after 48 h, which was close to the equilibrium of this variant under the given conditions. Besides a locally reduced flexibility, there could be an additional explanation for the enhanced crystallizability of FJAT flex 3. Since FJAT SER and FJAT flex 3 shared the same crystal form, it is speculated that their crystal packing is also comparable. Following this assumption, analysis of the location of β_K114 in the FJAT SER model revealed a spatial proximity to a crystal contact and to β_K308 on the interacting PGA with which this contact is formed. Steric obstacles or electrostatic repulsion between β_K114 and β_K308 could hamper the formation of the adjacent crystal contact. Thus, the deletion of β_K114 might have lowered this hurdle, if the contact actually formed in more or less the same way as observed for FJAT SER [30].

A second structure-based approach to increase crystallizability of FJAT SER was to intensify hydrophobic interactions in crystal contacts. Hydrophobic interactions are important in permanent specific protein–protein–interaction interfaces, optimized by evolution to enable stable assemblies [53, 54]. Therefore, they could also contribute to favor the crystallization of PGA enthalpically. The crystal contacts in the FJAT SER model where searched for hydrophobic patches, which than were extended by exchanges of amino acids to have several hydrophobic residues located opposite of each other on both sides of a crystal contact resulting in different FJAT hypho variants. As several of the purified FJAT hypho variants could not be concentrated without forming a solid phase the addition of NaCl was necessary. The addition of same NaCl concentrations completely prevented the crystallization of FJAT SER within 48 h, while four of the FJAT hypho variants showed reliable crystallization carrying the exchanges SER + β_K408V (hypho 3), SER + β_K408L (hypho 4), SER + α_T167V + β_G14A (hypho 7), and SER + β_K114V + β_K308L (hypho 8). Of these, FJAT hypho 4 and 7 reached minimum concentration limits of PGA and PEG_{8000} in the range of the limits that FJAT SER reached under salt-free conditions. The concentrations of FJAT hypho 3 and PEG_{8000} for a reliably crystallization were even decreased by 81 and 86%, respectively, compared to the limits reached by FJAT SER under salt-free conditions. The crystal yields reached by these three variants within 24 h were 86% (hypho 7), 92% (hypho 4), and 96% (hypho 3), respectively, which were all higher than the 76% reached by FJAT SER in the absent of salt. Furthermore, the activity of all three hypho variants was not changed significantly compared to FJAT SER. Crystal structures of the three new variants (PDB-IDs: FJAT hypho 3: 8BRR; FJAT hypho 4: 8BRS; FJAT hypho 7: 8BRT) verified, that the changed crystal contacts formed as intended (Fig. 8). FJAT hypho 7 (α_T167V + β_G14A) might have a higher crystallizability due to hydrophobic interactions between the two changed residues and β_I369. FJAT hypho 4 (β_K408L) should also have a strengthened hydrophobic interaction at the engineered contact, that additionally reduced the conformational entropy of the contact residue at position β_408. This entropy reduction was even higher in FJAT hypho 3 having a different exchange at the same position (β_K408V). Replacing β_K408 with a valine residue also appears to allow a closer contact in the engineered region (Fig. 8c and d), which could be the reason for a slight deformation of the crystal lattice compared to FJAT SER, recognizable in the space group and the unit cell parameters [30]. In consequence, FJAT hypho 3 had 25% less contact residues, defined as residues with a maximum distance of 4 Å to another PGA molecule in the crystal lattice, than FJAT SER. Nevertheless, more than 85% of the remaining contact residues where identical with the contact residues of FJAT SER, indicating that the overall packing strongly resembles FJAT SER crystals. Hence, the reduced number of contact residues might be an additional factor, contributing to the improved crystallizability of FJAT hypho 3.

Finally, exchanges from FJAT flex and FJAT hypho variants were combined resulting in the so far best crystallizing variant FJAT com 2. This variant carried 12 amino acid exchanges and one deletion compared to FJAT WT combining the changes of FJAT SER, flex 3, hypho 3, and hypho 7 [30].

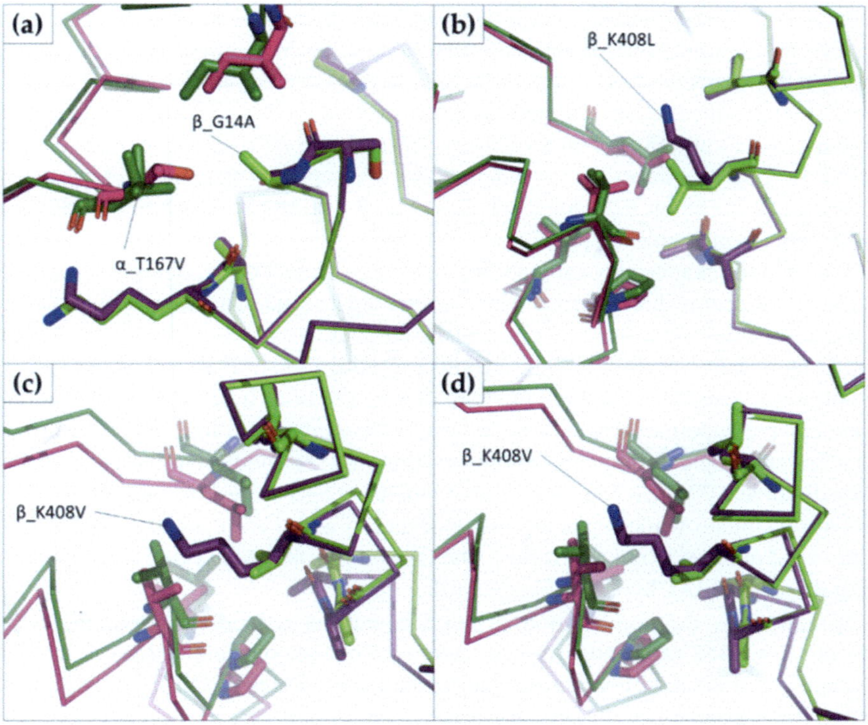

Fig. 8 Changed positions of FJAT (**a**) hypho 7, (**b**) hypho 4, (**c**) and (**d**) hypho 3. The right side of each picture shows a structure alignment of FJAT SER (purple) to the given FJAT hypho variant (light green); the left side shows one interacting symmetry mate of the aligned PGA (FJAT SER: pink; FJAT hypho: dark green). Residues with a maximum distance of 4 Å to another PGA in the crystal lattice where considered as crystal contact residues and are shown as stick models, if they are in the vicinity of the changed sides. The FJAT hypho 3 crystal structure had two PGAs in the asymmetric unit, which were analyzed individually ((**c**): dimer 1, (**d**): dimer 2). Changed positions are labeled accordingly. Figure taken from [30], licensed under the Creative Commons Attribution 4.0 International license (CC BY 4.0)

Next, to further simplify and fasten the production of PGA CLECs, shortening the purification process of PGA from the cell-free supernatant down to one step was addressed. So far, the established crystallization of PGA variants based on cation exchange chromatography followed by a concentration step, gel permeation chromatography for desalting, and a final concentration step. After the enhancement of the crystallizability for FJAT SER, flex 3 (SER + deletion of β_K114), hypho 3 (SER + β_K408V), and com 2 (SER + α_T167V + β_G14A + deletion of β_K114 + β_K408V) compared to FJAT WT all variants were recombinantly produced and purified via cation exchange chromatography from the cell-free supernatant resulting in purities of 94.4% for FJAT SER down to 85.7% for FJAT com 2. These protein solutions were directly used for crystallization with PEG_{8000} at pH 7.5, both with and without corresponding seeds in the form of crushed crystals. Based on the protein concentration in the mother liquor, the crystallization process was finished after 24 h. While FJAT WT precipitated under all conditions, FJAT SER and FJAT flex 3 precipitated in the absence but crystallized in the presence of seeds, and FJAT hypho 3 and FJAT com 2 even crystallized independently of seed addition. The decrease of the protein concentration in the mother liquor was between 78% (FJAT flex 3) and 87% (FJAT SER). If the values are considered after correction by the non-PGA proteins, crystal yields of about 92% for FJAT SER, flex 3, and hypho 3 as well as 98% for FJAT com 2 were achieved [30].

Finally, the enhanced crystallizability of the engineered PGA variants was used for PGA crystallization directly in the concentrated supernatant of a *P. megaterium* PGA-secreting culture. As organic compounds from complex medium like yeast extract and tryptone/peptone strongly inhibited the crystallization of PGA, the variant FJAT hypho 3 was secreted to defined medium during a fed-batch bioreactor cultivation. The cell-free supernatant was concentrated 1000-fold with a molecular cut-off of 50 kDa, reaching PGA concentrations of 10 mg mL^{-1}. By mixing the concentrate with dissolved PEG_{8000} low amounts of small crystals were obtained along with brown precipitate while replacing the concentrate with pure defined medium as a control did not lead to crystallization. The form of the obtained crystals matched the crystal form of FJAT hypho 3 (Fig. 9) [30].

In summary, the combination of surface entropy reduction, shortening of a flexible loop, and especially strengthening of hydrophobic interactions at crystal contacts could successfully improve the crystallizability of the FJAT-PGA. Although hydrophobic patches on the protein-surface have been discussed to hinder crystallization by promoting amorphous aggregation [55], the presented results suggest that considering hydrophobic interactions in crystal contact engineering can be highly beneficial for protein crystallization. It was further demonstrated that the engineered PGA variants not only enable a more reliable and faster crystallization, but also allow

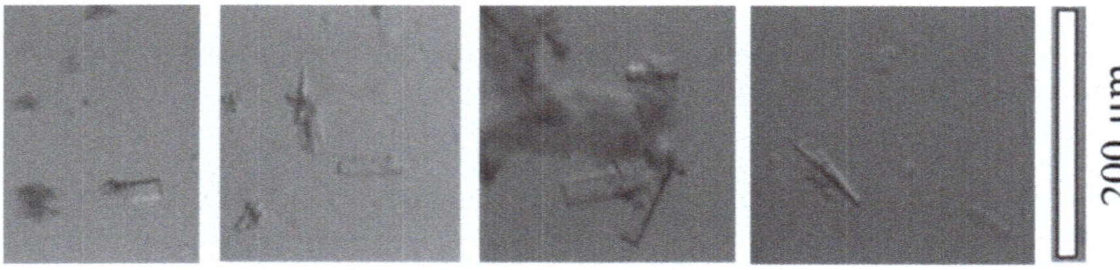

Fig. 9 Crystals obtained from concentrated cell-free culture supernatant. FJAT hypho 3 (SER + β_K408V) was recombinantly produced and secreted during a fed-batch cultivation using defined medium in a bioreactor. The cell-free supernatant was concentrated 1000-fold by cross-flow filtration and subsequent centrifugal filtration (cutoff: 50 kDa). The concentrate was diluted 3 : 4 with 100 mM HEPES (pH 7.5) and mixed with 300 mg mL^{-1} PEG_{8000} in 100 mM HEPES (pH 7.5) and seeds, suspended in the same solution of PEG_{8000}. The ratio of mixing was 2 : 1 : 2 (diluted concentrate : seed suspension : PEG_{8000} solution). The drops were incubated under paraffin oil for six days at 22 °C. Figure taken and modified from [30], licensed under the Creative Commons Attribution 4.0 International license (CC BY 4.0)

a massive shortening of the purification process prior to crystallization, while maintaining high crystal yields. Since crystallization (and the purification effort which is often associated with it) represents the major bottleneck of CLEC technology [9, 56], protein engineering proved to be a tool to overcome this limitation and could make CLEC technology applicable for a variety of enzymes. The crystallization of the PGA-variant FJAT hypho 3 in concentrated culture supernatant further provides a step towards realizing preparative crystallization of the FJAT-PGA as an alternative purification method, which could enable a higher throughput than chromatography and avoid consumption of column materials [6].

3.4 Protein Engineering of HheG for Improved Thermal Stability

During initial characterization of HheG, it became evident that this enzyme exhibits a very low stability regarding temperature and the use of different organic co-solvents, which are required to solubilize higher substrate amounts, hampering a possible industrial application of this enzyme. The half-life time of HheG was later determined to be only around 21 h at 22 °C. Though enzyme immobilization has been demonstrated to increase enzyme stability significantly, we first aimed for a thermostabilization of HheG by semi-rational protein engineering, to later investigate the impact of that thermostabilization on the mechanical properties of respective HheG CLECs. Based on the crystal structure of HheG (PDB: 5O30), amino acid residues were selected that either (i) exhibited high flexibility (high B-factor values) in the crystal structure, (ii) would enable the introduction of additional intra- and intermolecular interactions (e.g. hydrogen bonds, hydrophobic interactions), (iii) would increase surface hydrophilicity by introduction of additional lysine and arginine residues, or (iv) that were predicted by the FireProt webserver to influence the thermal stability of HheG. Respective positions were randomized, i.e. exchanged by all other 19 proteinogenic amino acids, via site-saturation mutagenesis and resulting HheG variants were evaluated regarding their apparent melting temperatures T_m, as a measure of thermal stability, using the thermal shift assay. Additionally, the activity of all generated variants in the azidolysis of cyclohexene oxide was determined and compared to wild-type HheG. In total, 865 mutants were generated, produced, and tested, of which 14 displayed an increase in apparent melting temperature (Fig. 10). At the same time, enzyme activity was also affected, either positively or negatively, by the respective mutations. The most thermostable HheG variants, exhibiting a ΔT_m of +14 K, carried amino acid exchanges at position 123. At the same time, those T123 variants displayed also up to threefold higher specific activity. Thus, a single amino acid exchange at this position did not only improve thermostability but also activity of HheG. This improvement in thermal stability was also accompanied by a slight increase in solvent stability [57].

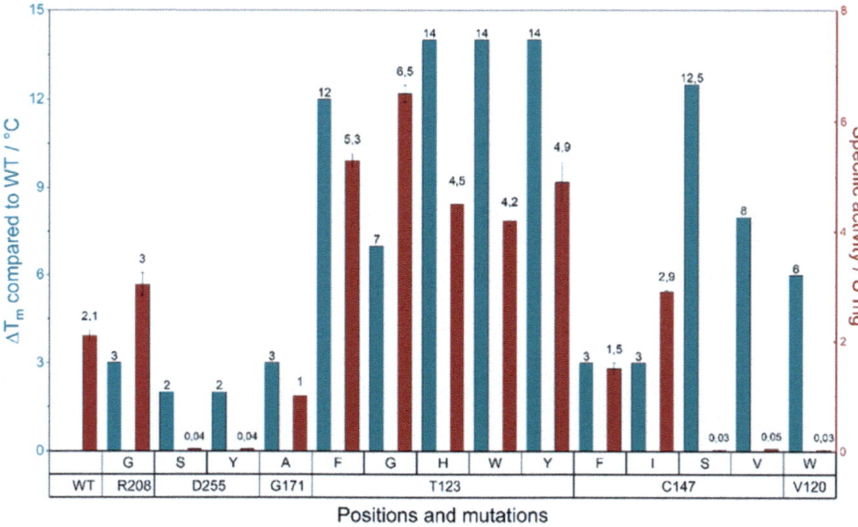

Fig. 10 Melting temperature deviations (ΔT_m) compared to wild-type HheG (WT) and specific activities of thermostabilized HheG variants. Melting temperatures were determined via thermal shift assay and specific activities were obtained for epoxide ring opening reactions of cyclohexene oxide using azide as the nucleophile. Figure taken from [57], licensed under the Creative Commons Attribution 4.0 International license (CC BY 4.0)

To gain molecular insights on the impact of those mutations on the enzyme structure and respective enzyme properties, the crystal structures of two selected HheG variants, T123W (PDB: 6I9U) and T123G (PDB: 6I9V), were determined and, based on those structures, molecular dynamics (MD) simulations were performed. This revealed a highly flexible loop (residues 39–47) near the active site of HheG, adopting either an open or closed conformation depending on the amino acid at position 123 (Fig. 11). In the closed conformation, an increase in buried surface area and a denser packing (volume decrease) of the protein were observed, both contributing to the improved thermal stability. Additionally, new interactions between the aromatic residue at position 123 and residues on the flexible loop are established, as revealed from the crystal structure of HheG T123W, further increasing thermal stability of this variant compared to HheG T123G. Later, it was demonstrated by Estévez-Gay et al. through extensive MD simulations that the flexible loop of HheG also impacts on the presence and size of an additional substrate entrance channel to the enzyme active site, which is not found in most other studied HHDHs. Thus, the conformation of this loop, which is influenced by the amino acid at position 123, will also affect enzyme activity, as seen in our T123 variants.

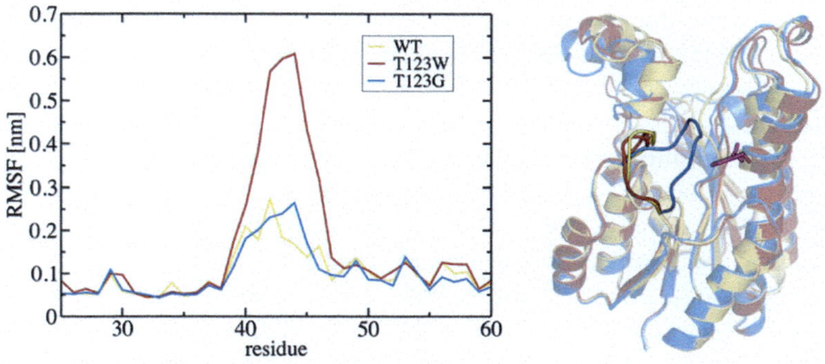

Fig. 11 Left: RMSF plots derived from MD simulations for residues 25 to 60 from HheG wild-type (yellow), as well as variants T123W (red) and T123G (blue). Right: Superposition of average structures of one monomer taken from dimer MD simulations with the flexible loop highlighted in foreground. Colors as on the left; residue 123 is shown in magenta. Figure taken from [57], licensed under the Creative Commons Attribution 4.0 International license (CC BY 4.0)

3.5 HheG Engineering for the Use of Alternative Cross-Linkers in CLEC Preparation

For cross-linking of enzyme crystals during CLEC preparation, glutaraldehyde (GA) is typically used as chemical cross-linker, as it is not only able to cross-link lysine, arginine and tyrosine side chains but also forms oligomers in solution, which enables to cross-link residues of various distances with each other [58]. However, HheG CLECs cross-linked with GA displayed only low or no residual activity [28], probably due to a cross-linking of catalytic residues Tyr165 and Arg169 in the active side of HheG. To overcome this limitation, a crystal contact engineering approach was developed to incorporate new cross-linking sites at desired positions in the crystal interface with the aim to use alternative cross-linkers for CLEC generation. In this context, two different strategies were applied: (a) the insertion of cysteines as new cross-linking sites for thiol-specific cross-linking and (b) the insertion of additional lysine residues for lysine (i.e. primary amine)-specific cross-linking. Strategy (a) was implemented using the PRODIGY crystal webserver [59]. Here, residues facing each other within a cut-off distance of 5 Å in the axial and tangential crystal contact of neighboring HheG molecules were determined first. This revealed three residues, M45, D114, and A221, facing each other in the crystal interface (Fig. 12). Those residues were subsequently exchanged by cysteine residues resulting in three HheG single variants: M45C, D114C, and A221C. In contrast, strategy (b) was adapted from Abdul Wahab et al. [60]. Here, the relative surface accessibility tool from the Swiss PDB Viewer was used to identify surface-exposed residues in HheG using a cut-off value of 40% [61]. Resulting residues were afterwards analyzed using the PDBePISA web server to select suitable amino acid positions within the crystal contact for subsequent replacement by lysine [62]. This resulted in HheG single variants V46K, D115K, A141K, R185K, and A217K (Fig. 12). In most cases, the different mutations did not influence activity or thermal stability of HheG variants. Only variants M45C, V46K, and D115K displayed lower specific activity compared to HheG wild-type.

In the next step, the crystallization behavior of the generated HheG variants was studied, as successful crystallization is a prerequisite for later CLEC generation. Wild-type HheG crystallized well via sitting drop vapor diffusion using 10 mM HEPES (pH 7.3) with 10% (w/v) PEG_{4000} as precipitant [34]. Starting from this buffer composition, a grid screening in microtiter plate format was performed varying the buffer pH, PEG_{4000} and protein concentrations to identify optimal crystallization conditions for each HheG variant (Table 1) [28, 29] Additionally, a crystallizability rate was determined as the ratio of the number of wells with crystallization event and the total number of wells per variant (Table 1). Thus, the higher the crystallizability rate, the higher the ability of a variant to crystallize under different conditions.

Except for three variants (A221C, D115K, A217K), all other designed HheG variants were able to crystallize in 2 μL scale. Crystals of variant M45C were initially not found in our grid screening (0% crystallizability rate), but successful crystal formation of HheG M45C could be demonstrated in later crystallization experiments. All other variants displayed an even improved crystallizability rate (at least by a factor of two) compared to wild-type HheG [28, 29].

As vapor-diffusion crystallization cannot be used at larger scales, e.g. for industrial applications, batch crystallization via precipitation was tested in 2 mL scale for all crystallizing variants using in each case the optimal buffer conditions listed in Table 1. Due to this change in crystallization method, variants M45C and R185K did not crystallize anymore, but adjustments in e.g. precipitant or protein concentration

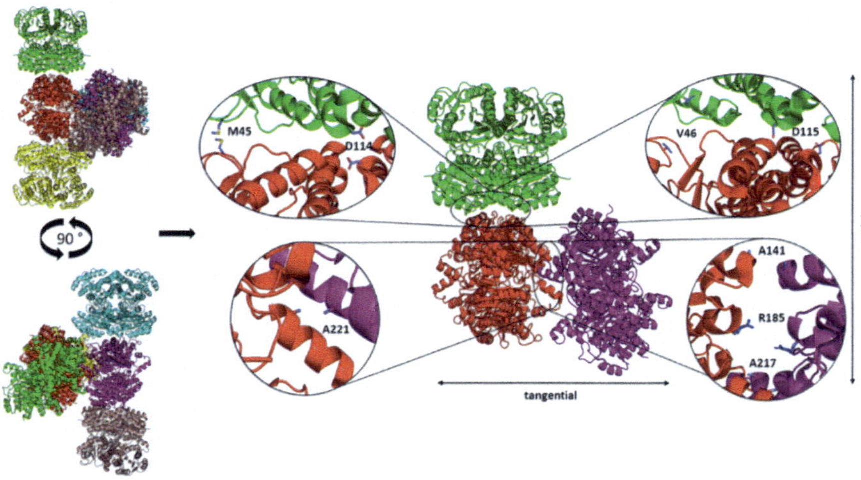

Fig. 12 Selected amino acid positions in the axial and tangential crystal contacts of HheG for replacement by cysteine (M45, D114, A221) or lysine (V46, D115, A141, R185, A217) residues. Parts of the figure taken from [28, 29], licensed under the Creative Commons Attribution 4.0 International license (CC BY 4.0)

Table 1 Crystallization characteristics of different HheG variants based on sitting drop vapor diffusion crystallization in 2 μL scale

Variant	Crystallizability rate[a] [%]	Best crystallization conditions
WT	17.4	20 mg mL^{-1} protein solution; 10 mM HEPES, pH 7.0, 12% (w/v) PEG_{4000}
M45C	0	24 mg mL^{-1} protein solution; 10 mM HEPES, pH 7.0, 12% (w/v) PEG_{4000}
D114C	41.7	24 mg mL^{-1} protein solution; 10 mM HEPES, pH 7.0, 8% (w/v) PEG_{4000}
A221C	0	–
V46K	49.3	24 mg mL^{-1} protein solution 10 mM HEPES, pH 7.3, 6% (w/v) PEG_{4000}
D115K	0	–
A141K	62.5	24 mg mL^{-1} protein solution 10 mM HEPES, pH 7.3, 8% (w/v) PEG_{4000}
R185K	38.2	24 mg mL^{-1} protein solution 10 mM HEPES, pH 7.3, 12% (w/v) PEG_{4000}
A217K	0	–

[a]Calculated based on the number of wells with crystals divided by the total number of wells per variant

might solve this problem in future trials [63]. In contrast, variants D114C, V46K, A141K as well as wild-type HheG were able to crystallize in 2 mL batch scale and respective crystallization kinetics were determined using the Avrami equation, which describes phase changes under isothermal conditions (Fig. 13, Table 2) [64].

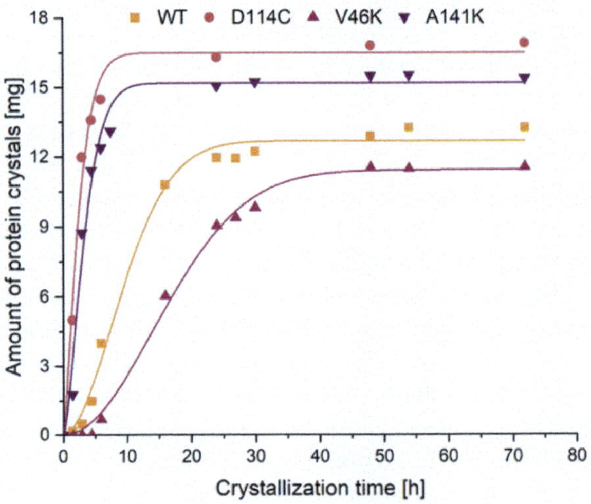

Fig. 13 Avrami-based crystallization kinetics of different HheG variants. Crystallization was performed under gentle shaking at 8 °C. A total amount of 24 mg protein was used for crystallization. Figure taken from [29], licensed under the Creative Commons Attribution 4.0 International license (CC BY 4.0)

Table 2 Crystallization kinetic parameters of different HheG variants based on the theory of Avrami of isothermal phase change

Variant	S_{max} [mg]	k [h^{-1}]	n	$t_{0.5}$ [h]
WT	12.7	0.011	1.9	9.4
D114C	16.5	0.249	1.3	2.2
V46K	11.4	0.003	2.0	16.6
A141K	15.2	0.133	1.5	3.0

Variants D114C and A141K displayed improved crystallization kinetics compared to HheG wild type as the velocity constant (k) was at least tenfold higher, the half-time of the crystallization process ($t_{0.5}$) was at least threefold lower and also the maximal amount of crystals (S_{max}) obtained at the end of the crystallization process was increased. In contrast, the crystallization behavior of variant V46K in batch crystallization was inferior to that of HheG wild type. Resulting Avrami exponents n for 3-dimensional, kinetically-controlled crystal growth are all in the expected range of typically 1.5–2.5 [65]. As lower Avrami exponents are usually connected to crystallization processes exhibiting faster nucleation rates [66], the improved crystallization of variants D114C and A141K could be explained by faster nucleation.

Afterwards, the cross-linking of HheG wild type and variants D114C, V46K, and A141K was studied using either cysteine- or lysine-specific cross-linkers. In case of variant D114C, which carried newly incorporated cysteine residues facing each other in the crystal contact with a distance of C_α atoms of 8.5 Å, bis-maleimidoethane (BMOE) with a spacer length of 8 Å was selected as cross-linker (Table 3) [67]. After successful cross-linking of D114C crystals, respective cross-linking sites could be even resolved in the crystal structure of HheG D114C CLECs cross-linked with BMOE (Fig. 14).

Additionally, the apparent melting temperature of respective CLECs was increased by 8 K compared to soluble HheG D114C and the CLECs were still catalytically active, even though a slight activity decrease compared to soluble enzyme was observed (Fig. 15). This is probably explained by a diffusion limitation of substrate within the CLEC. For comparison, when HheG D114C crystals were cross-linked with GA instead, respective CLECs were either not active anymore (when GA was applied via soaking) or not fully stabilized (when GA was applied via vapor diffusion) as indicated by a low apparent melting temperature (Fig. 15). The latter could be increased by additional cross-linking with BMOE. Nevertheless, HheG D114C CLECs cross-linked only with BMOE displayed even higher activity, while the apparent melting temperature of both CLEC preparations was the same.

The difference in apparent melting temperature between successfully cross-linked crystals and soluble enzyme was also used to determine cross-linking efficiency in strategy (b). Here, additional lysine residues were incorporated in the crystal contact to assist already existing lysine residues in cross-linking via lysine-specific cross-linkers. In the crystal contact of HheG wild type, four lysine residues are naturally present with distances of ε-amino groups between 9.5 and 17.6 Å. Generated HheG

Table 3 Structure and spacer length of different cross-linkers used for unspecific (GA), cysteine-specific (BMOE) and lysine-specific cross-linking (DST, DMP, DMS, DSP and Sulfo-EGS) of HheG CLECs in this study

Cross-linker	Cross-linker type	Spacer length [Å]
Glutaraldehyde (GA) 		5.0 and longer
Bis-maleimidoethane (BMOE) 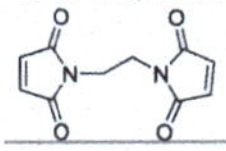	Sulfhydryl	8.0
Disuccinimidyl tartrate (DST) 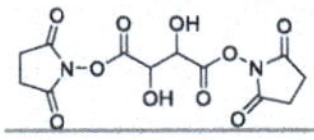	*N*-hydroxy-succinimid	6.4
Dimethylpimelimidate (DMP) 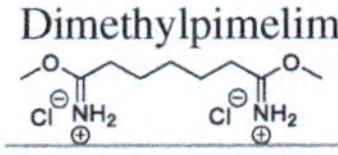	Imidoester	9.2
Dimethylsuberimidat (DMS) 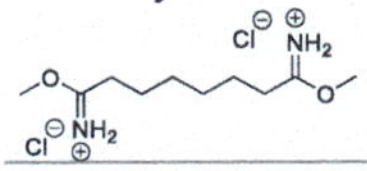	Imidoester	11.0
Dithiobis (succinimidyl) propionate (DSP) 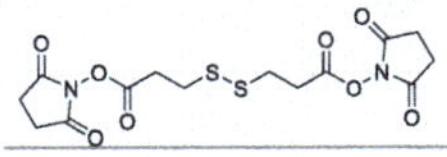	*N*-hydroxy-succinimid	12.0
Ethylene glycol bis (sulfosuccinimidyl succinate) (Sulfo-EGS) 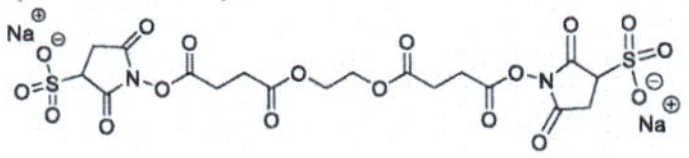	*N*-hydroxy-succinimid	16.1

variants V46K and A141K featured additional lysine residues with distances between C_α atoms of 14.3 Å and 11.7 Å, respectively. Thus, lysine-specific cross-linkers of varying spacer length were tested to cover different distances between lysine residues during cross-linking (Table 3).

Crystals of HheG wild-type as well as HheG variants V46K and A141K were cross-linked using either disuccinimidyl tartrate (DST), dimethylpimelimidate (DMP), dimethylsuberimidate (DMS), dithiobis(succinimidyl)propionate (DSP) or ethylene glycol bis(sulfosuccinimidyl succinate) (Sulfo-EGS) (Table 3) in different concentrations (4, 10, 20, and 40 mM) and apparent melting temperatures of resulting CLECs were determined via thermal shift assay (Fig. 16). Among all tested variants, cross-linkers DST and DMS had no positive effect on the thermal stability of the formed CLECs, indicating a low degree of cross-linking. For CLECs of HheG wild type (Fig. 16A), a high concentration ($\geq$20 mM) of cross-linkers DMP, DSP, and Sulfo-EGS yielded already stable CLECs, as indicated by an increase in apparent melting temperature, without incorporation of additional lysine residues. In contrast, for V46K CLECs cross-linked with DMP, DSP and Sulfo-EGS even higher apparent melting temperatures compared to HheG wild-type CLECs were

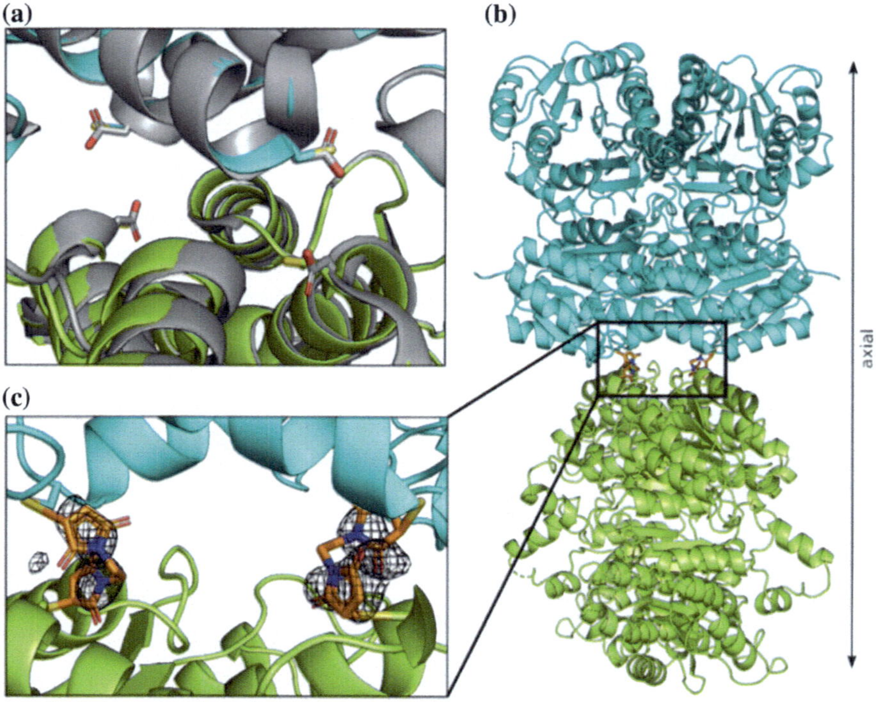

Fig. 14 Crystal structure of HheG D114C (PDB: 7QY3). (a) Overlay of residue 114 of variant HheG D114C (colored) and the wild-type structure (gray). (b) Cross-linked chains of neighboring tetramers in axial orientation, BMOE shown in orange. (c) Magnified view of the bound cross-linker BMOE fitting into the respective electron density ((Fo-Fc) at 3σ). Figure taken from [28], licensed under the Creative Commons Attribution 4.0 International license (CC BY 4.0)

obtained (Fig. 16B). Moreover, cross-linking of V46K with DSP yielded the most stable CLECs already at a cross-linker concentration of only 4 mM. For variant A141K (Fig. 16C), a similar trend was observed as for V46K with the difference that 10 mM DSP were required for efficient cross-linking. The direct comparison of variants V46K and A141K with HheG wild-type CLECs cross-linked with DSP and Sulfo-EGS (Fig. 16D) illustrates that the incorporation of the additional lysine residues reduced the required amount of cross-linker for effective cross-linking and yielded CLECs of higher stability compared to HheG wild type. Overall, our data also indicate that *N*-hydroxysuccinimid (NHS)-type cross-linkers (DSP, Sulfo-EGS) were more effective for CLEC generation than imidoester-type cross-linkers (DMP, DMS). This might be caused by the higher stability of NHS-type cross-linkers as well as our used buffer system with a physiological pH value of 7.0–7.5 [68–71]. Moreover, also the spacer length of the cross-linker is indeed important for effective cross-linking as DST (the shortest NHS-type cross-linker tested) was not able to yield stable CLECs.

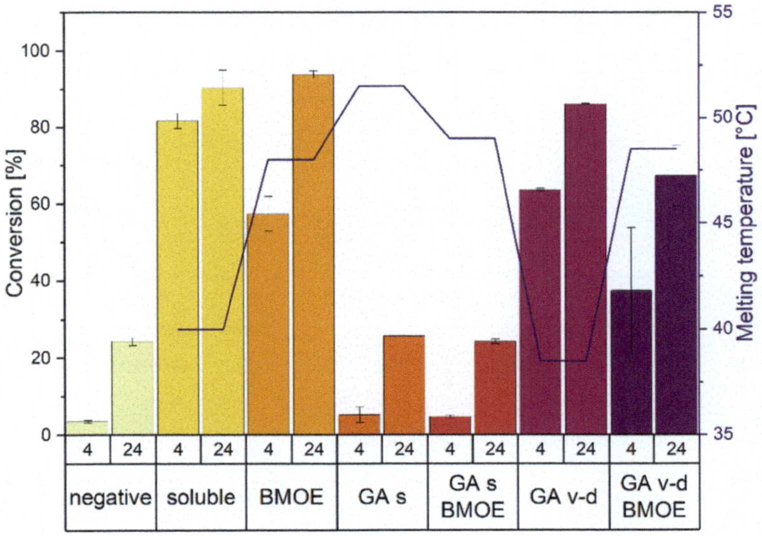

Fig. 15 Activity and stability of HheG D114C CLECs prepared using different cross-linkers and cross-linking methods [negative: negative control reaction without enzyme addition; soluble: soluble HheG D114C; BMOE: CLECs cross-linked with BMOE via soaking; GA s: CLECs cross-linked with GA via soaking; GA s/BMOE: CLECs cross-linked with GA and BMOE via soaking; GA v-d: CLECs cross-linked with GA via vapor diffusion; GA v-d/BMOE: CLECs cross-linked with GA via vapor diffusion and BMOE via soaking]. Activity is represented by the obtained conversion in the azidolysis of cyclohexene oxide after 4 and 24 h in reactions using 20 mM cyclohexene oxide, 40 mM azide and 100 μg biocatalyst (soluble enzyme or CLECs) in 1 mL Tris·SO_4, pH 7.0. Conversion was measured by achiral GC. Reactions were performed in quintuplicate. Stability is represented by apparent melting temperature, which was determined in triplicate using the thermal shift assay

When looking at the activity of formed CLECs of HheG wild type as well as variants V46K and A141K cross-linked with either DSP or Sulfo-EGS, a clear difference between the resulting CLECs is observed (Fig. 17). Thus, only V46K crystals cross-linked with DSP still yielded highly active CLECs, while the activity after cross-linking with Sulfo-EGS was almost abolished. In case of A141K CLECs, activity was drastically reduced for both cross-linkers. In contrast, HheG wild-type crystals cross-linked with Sulfo-EGS displayed higher residual activity than respective CLECs prepared with DSP as cross-linker, although a significant reduction in activity was observed in both cases. This illustrates that the right choice of cross-linker does not only play a crucial role for the (thermal) stability of resulting CLECs but also their activity.

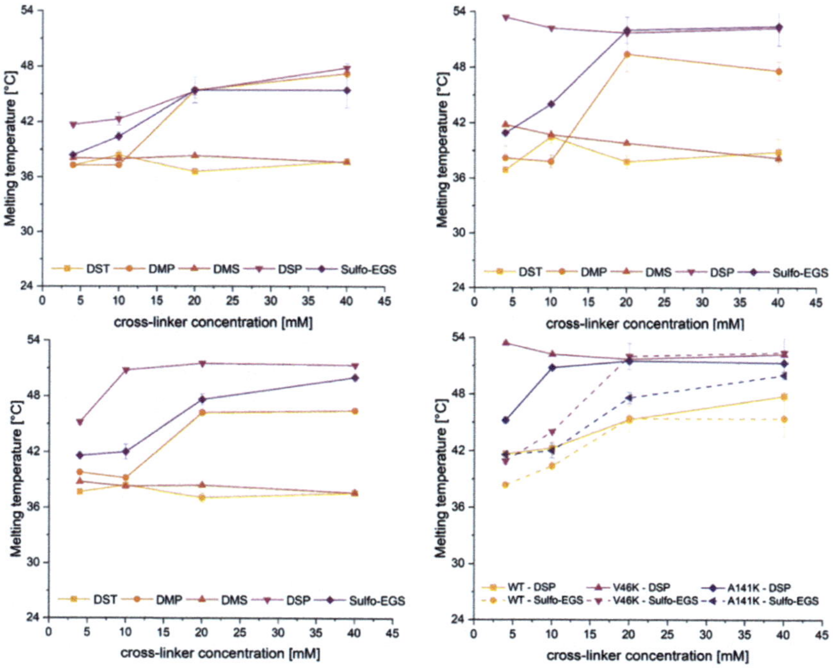

Fig. 16 Influence of the concentration of different cross-linkers on CLEC stability. Apparent melting temperatures of CLECs of wild-type HheG (top left) and variants V46K (top right) and A141K (bottom left) were determined after cross-linking with five different cross-linkers as well as increasing cross-linker concentrations. Bottom right: Comparison of the different CLEC variants cross-linked with either DSP or Sulfo-EGS. Melting temperatures were determined in triplicate via thermal shift assay. Figure taken from [29], licensed under the Creative Commons Attribution 4.0 International license (CC BY 4.0)

3.6 *Characterization of Cross-Linked Enzyme Crystals of Different HheG Variants*

Two types of HheG CLECs were studied further regarding their biochemical characteristics and biocatalytic applicability: variant D114C cross-linked with BMOE and variant V46K cross-linked with DSP. Both variants yielded stable and active CLECs during cross-linking tests and required only low cross-linker concentrations for CLEC preparation.

First, a particle size distribution analysis was performed, as the CLEC activity will also depend on the actual crystal size. Both CLECs, D114C and V46K, displayed a median value of around 2 µm, which is significantly below 10 µm. Thus, a possible diffusional limitation within the CLECs during catalysis is not to be expected [72].

Additionally, CLEC activity was investigated under varying process conditions such as different temperatures and pH values, as well as the addition of organic co-solvent. Respective temperature and pH profiles for CLECs of HheG variants

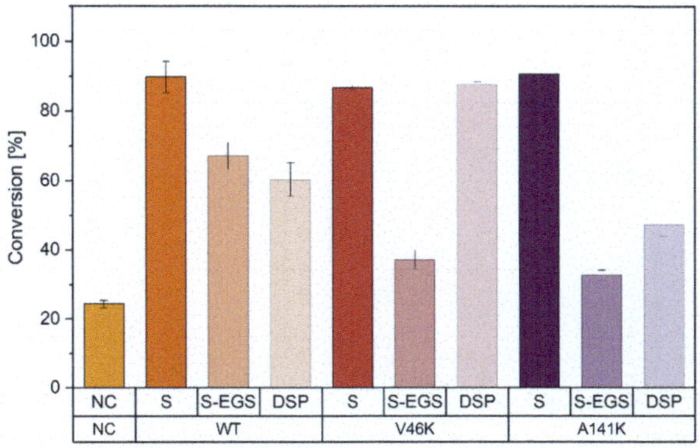

Fig. 17 Conversion obtained in reactions of 20 mM cyclohexene oxide with 40 mM azide using 100 μg biocatalyst (soluble or CLECs) after 24 h. 20 mM cross-linker concentration was used during cross-linking with Sulfo-EGS. In case of DSP, 4 mM, 10 mM or 20 mM cross-linker concentration was applied for cross-linking of HheG variants V46K, A141K and wild-type, respectively. Measurements were performed in quintuplicate. 'NC' indicates negative control and 'S' soluble enzyme. Parts of the figure taken from [29], licensed under the Creative Commons Attribution 4.0 International license (CC BY 4.0)

D114C and V46K were determined and compared to soluble enzyme (Fig. 18). In both cases, the CLECs displayed higher optimal reaction temperatures (by at least 10 K) and were still highly active above 30 °C, at which the corresponding soluble enzymes were already inactivated (Fig. 18A and C). Thus, the thermal stability of both CLEC variants is enhanced, which is in line with the observed higher apparent melting temperatures. Moreover, also the pH profiles of the CLECs were broadened compared to soluble enzyme with a shift of the apparent pH optima towards more acidic conditions (Fig. 18 B and D). This shift can be explained with an increased pH stability of the CLECs and the catalytic mechanism of HHDHs, where the catalytic tyrosine residue (Tyr165) of HheG donates a proton to the negatively charged oxygen during epoxide ring opening [21]. Here, re-protonation of this tyrosine will be facilitated at more acidic pH. Analysis of the apparent melting temperatures of both CLEC preparations under acidic pH values indeed revealed a drastic gain in stability for both CLECs compared to soluble enzyme (+11.0 K for D114C CLECs and +20.8 K for V46K CLECs at a pH of 6.0 using citrate buffer).

Organic solvents are commonly used in biocatalysis to increase substrate solubility, which is also required in epoxide ring opening reactions using halohydrin dehalogenases. Therefore, the activity of D114C and V46K CLECs in reactions with different water-miscible organic solvents was investigated. As shown in Fig. 19, both CLEC preparations were still significantly active in the presence of 25% (v/v) dimethylformamide, methanol, ethanol and isopropanol, while respective reactions with soluble enzymes only displayed conversions in the range of the negative control reaction without enzyme addition. Dimethylsulfoxide as co-solvent did not have a significant negative effect on neither CLECs nor soluble D114C and V46K. In contrast, 25% (v/v) acetonitrile fully inactivated both soluble enzymes

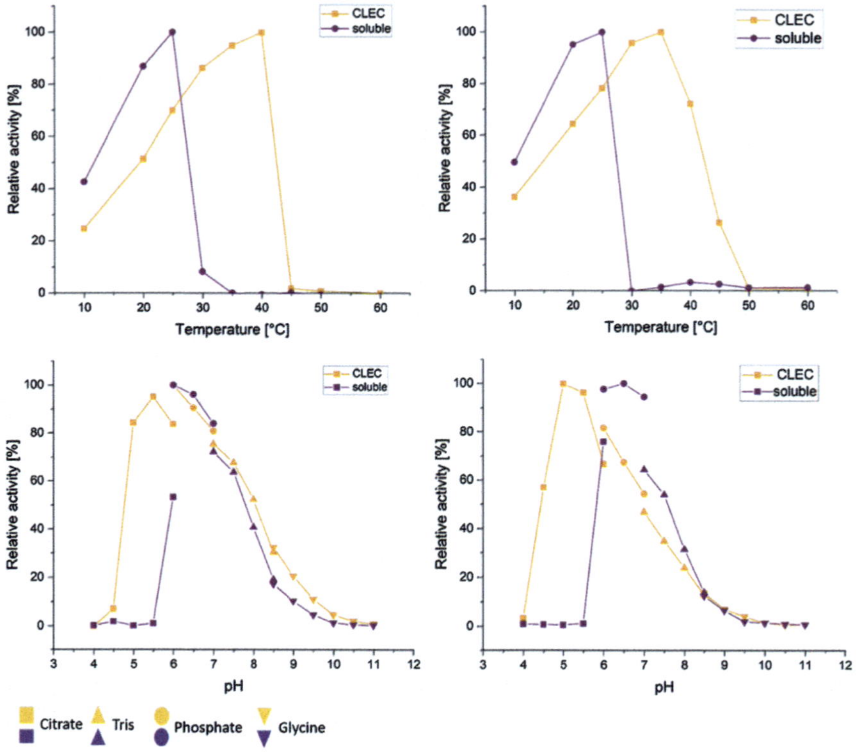

Fig. 18 Temperature (top) and pH (bottom) profiles of HheG D114C CLECs (left) and HheG V46K CLECs (right) in comparison to respective soluble enzymes in the epoxide ring opening of 20 mM cyclohexene oxide with 40 mM azide using 100 μg biocatalyst. Reactions were performed for 2 h in duplicate and the highest determined conversion was set to 100% relative activity Parts of the figure taken from [28, 29], licensed under the Creative Commons Attribution 4.0 International license (CC BY 4.0)

and corresponding CLEC preparations. This inactivation might actually be caused by a possible inhibition of HheG by acetonitrile, as previously reported for DMSO and the halohydrin dehalogenase HheC from *Agrobacterium radiobacter* AD1 [73], since the stability of V46K and D114C CLECs in the presence of 10% (v/v) acetonitrile was not much more reduced than in the presence of 10% (v/v) of the other co-solvents [28, 29]. In summary, both HheG CLEC preparations exhibited significantly higher stabilities toward temperature, pH and co-solvent concentration than respective soluble counterparts.

Furthermore, long-term stability of HheG D114C and V46K CLECs was evaluated by determination of enzymatic half-life times ($t_{1/2}$). First, deactivation rate constants at temperatures above 30 °C were measured experimentally and from this, deactivation rate constants as well as half-life times at 22 °C were approximated using the Eyring equation [74]. CLECs of variants D114C and V46K displayed impressive

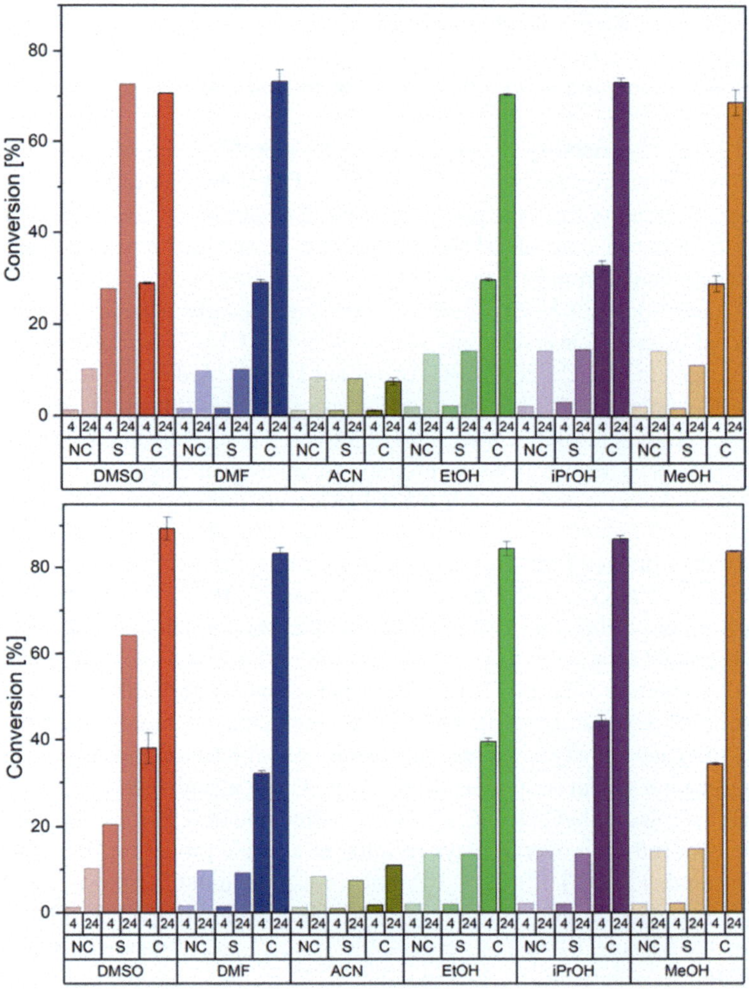

Fig. 19 Conversion of 100 μg HheG_D114C (top) and HheG_V46K (bottom) CLECs as well as respective soluble enzymes in reactions of 20 mM cyclohexene oxide and 40 mM azide in the presence of 25% (v/v) water-miscible organic solvents. Investigated solvents were dimethyl sulfoxide (DMSO), dimethyl formamide (DMF), acetonitrile (ACN), ethanol (EtOH), isopropanol (iPrOH), and methanol (MeOH). Samples were taken after 4 and 24 h. Reactions were performed in duplicate. "NC "indicates negative control without enzyme, "S" soluble enzyme, and "C" CLECs. Parts of the figure taken from [28, 29], licensed under the Creative Commons Attribution 4.0 International license (CC BY 4.0)

increases in enzymatic half-life to 64 and 82 days, respectively, at 22 °C compared to only 21 and 23 h, respectively, for their soluble counterparts. This result is in line with previous literature reports that CLEC formation enhances enzymatic half-life [13, 75].

This long-term stability of our HheG CLECs at 22 °C enabled also their use in repetitive batch reactions in the azidolysis of cyclohexene oxide over 21 days with daily separation from the product stream and addition of fresh reaction media (Fig. 20). While in case of D114C CLECs the determined conversion per cycle decreased gradually from 94% in cycle 1 to 65% in the last cycle, conversion in repetitive batch reactions with V46K CLECs dropped significantly in the first 4 cycles. This unexpected drop in conversion only in case of V46K CLECs is probably explained by differences in CLEC size. CLECs of variant D114C used in this repetitive batch reaction had microscopically determined sizes of around 100 μm, whereas CLECs of V46K were much smaller with sizes of around 2 μm. Thus, the bigger particles are much easier to recover from the product stream by centrifugation. Instead, much smaller particles as in the case of V46K CLECs were probably not effectively separated from the product stream under the used centrifugation conditions and hence, also partially removed from the reaction. Despite this, the activity recovery of V46K CLECs after 21 days was still above 40%, in case of D114C CLECs even 70%. In comparison, cross-linked enzyme aggregates (CLEAs) of halohydrin dehalogenase HheC displayed an activity recovery of 74% after 10 reaction cycles of only 30 min each (total reaction time of 5 h) in the dehalogenation of 5 mM 1,3-dichloro-2-propanol [76]. Overall, both HheG CLEC preparations exhibited formidable long-term stability and activity under reaction conditions, which prompted us to investigate their application also in continuous-flow setups. Using a fluidized-bed reactor (FBR) containing 4 mg of HheG D114C CLECs at a flow rate of 0.25 mL min^{-1} at 20 mM epoxide concentration over 21 days resulted in a stable process with a cumulative turnover number (TON) of 1.22×10^5 mol product per mol of enzyme after 21 days (Fig. 21). At an industrially more relevant epoxide concentration of 150 mM, a space–time yield (STY) of 11.2 $kg_{product}\ g_{CLECs}^{-1}\ h^{-1}\ L_{reactor\ volume}^{-1}$ was achieved with the same FBR. In comparison, the corresponding STY in a packed-bed reactor containing 50 mg HheG D114C CLECs increased to 23.9 $kg_{product}\ g_{CLECs}^{-1}\ h^{-1}\ L_{reactor\ volume}^{-1}$, demonstrating the high industrial applicability of our CLECs [77].

3.7 *Mechanical Characterization of PGA CLECs by AFM-Based Nanoindentation*

To investigate the possibility of using the newly identified PGAs from *Bacillus* species (BmPGA, BtPGA, FJAT-PGA [31]) in the form of CLECs, crystallization of the three wild-type enzymes on siliconized vials was established. Further, using variations of the concentration and pH-value of the buffer, the concentration of the high-purity PGAs, and the concentration ratio of corresponding PGA and buffer,

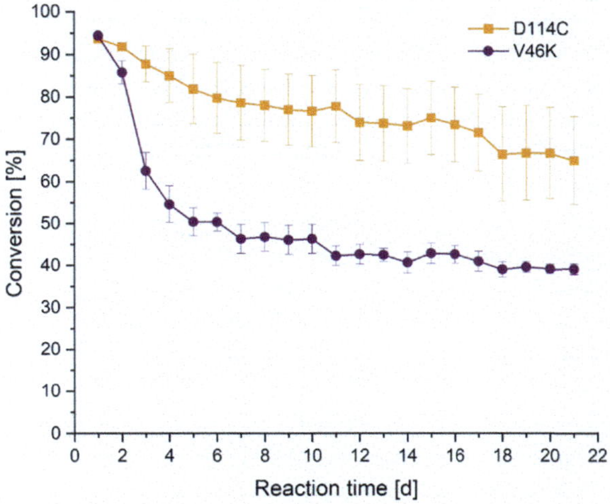

Fig. 20 Repetitive batch reactions of cyclohexene oxide and azide at 22 °C with daily reuse of CLECs of HheG variants D114C and V46K. Each reaction cycle (1 d) contained 100 μg CLECs, 20 mM cyclohexene oxide and 40 mM azide in a total of 1 mL Tris·SO_4, pH 7.0. After 24 h of reaction, the CLECs were centrifuged, the supernatant was removed and fresh reaction media including the substrates was added to start a new cycle. Standard deviations are based on quintuple measurements. For comparison, chemical background conversion in this reaction (one cycle) without enzyme addition reached 24%. Parts of the figure taken from [28, 29], licensed under the Creative Commons Attribution 4.0 International license (CC BY 4.0)

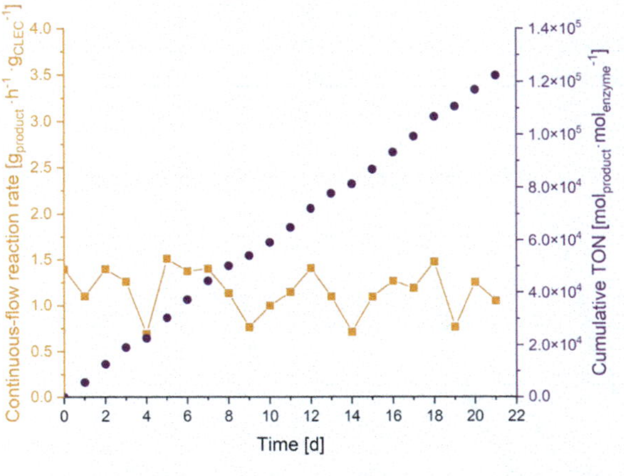

Fig. 21 Continuous-flow reaction rate and cumulative turnover number (TON) over 21 days in the fluidized-bed reactor (FBR) achieved at a flow rate of 0.25 mL min^{-1}. The reaction flow contained 20 mM cyclohexene oxide and 40 mM azide in 50 mM Tris·SO_4, pH 7.0. The FBR contained 4 mg D114C CLECs, and the continuous reaction was performed at room temperature. Sampling was carried out every 24 h and the corresponding continuous-flow reaction rate was determined by achiral GC via the attained conversion. Figure taken from [77], licensed under the Creative Commons Attribution 4.0 International license (CC BY 4.0)

the PGA crystals were optimized to have a consistent morphology, be single, and have a flat surface. Seeding further increased the reproducibility of PGA crystallization [78]. After the individual crystallization of the PGAs, the mother liquor was removed, crystals were washed and cross-linked in a solution with 5% (w/v) of GA (Table 3) for 24 h, washed again, and could finally be stored in H_2O for activity and mechanical measurement [34].

For cross-linked EcPGA crystals, Margolin and co-workers showed activities of up to 71% compared to the free enzyme [12]. CLECs of BmPGA, BtPGA, and FJAT-PGA were also found to be active, however BmPGA-CLECs only showed 5% activity compared to CLECs of the other PGAs, which could be due to the higher packing of this PGA. So far the specific activity of the here mentioned CLECs could not be determined [34].

The mechanical properties of the PGA-CLECs were investigated by nanoindentation using an atomic force microscope (AFM). It was found that the CLECs of the FJAT-PGA exhibited the highest hardness with a median of 25.7 MPa, followed by the BmPGA CLECs (22.6 MPa), and the BtPGA CLECs (11.4 MPa). For the Young's modulus, the same order was found. The FJAT-PGA-CLECs had the largest Young's modulus of the PGAs studied (2050.5 MPa), making them the stiffest, least deformable PGA CLECs, followed by the BmPGA (1591.9 MPa), and the BtPGA CLECs (868.2 MPa) (Fig. 22).

Overall, nearly all of the FJAT-PGA-CLECs were harder and stiffer than all of the measured BtPGA CLECs. Furthermore, the distributions of hardness and Young's modulus for the BtPGA CLECs were very narrow, while the distributions for the BmPGA CLECs were much broader and thus showed greater variability. As for the structure elucidation [31], anisotropy was also observed in the mechanical measurements of the BmPGA CLECs as different mechanical strengths of the BmPGA

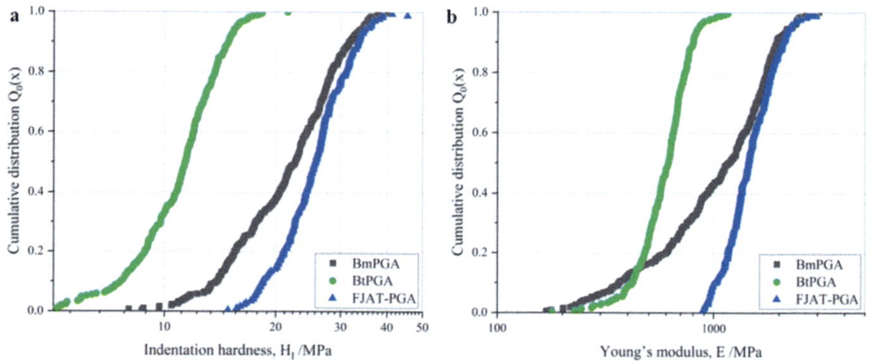

Fig. 22 Cumulative distribution of hardness (**a**) and Young's modulus (**b**) for BmPGA, BtPGA, and FJAT-PGA CLECs determined by AFM nanoindentation. All crystals were cross-linked by GA. The AFM nanoindentation measurements were performed using a spherical cantilever tip with a radius of 300 nm in a force-controlled mode. The indentations were carried out with an indentation speed of 2 $\mu m\ s^{-1}$ until an indentation force of 250 nN was reached. Figure taken from [32], licensed under the Creative Commons Attribution 4.0 International license (CC BY 4.0)

CLECs were measured depending on the side from which stress was applied. This is possible because the orientation in which molecules are arranged can differ depending on the direction which was also the case for CLECs of the HheG due to two different crystal faces [35]. As no such clear step could be observed in the cumulative distribution of the BmPGA CLECs, their anisotropy was more likely due to a variety of factors or multiple different surfaces. Furthermore, in HheG CLECs, anisotropy was found to occur more frequently at short cross-linking times [35]. Hence, a longer cross-linking time could reduce anisotropy in BmPGA CLECs.

The mechanical stability of the CLECs cannot be correlated with the thermal stability of the native PGAs [31]. Although the FJAT-PGA was found to be the most thermally and micromechanically stable PGA, the BmPGA CLECs showed only slightly lower mechanical hardness, although it was significantly less thermostable than the FJAT-PGA. In contrast, the BtPGA showed intermediate thermostability but the lowest mechanical stability of the measured PGA CLECs. For thermal stability, intramolecular interactions such as hydrogen bonding or salt bridges within a molecule are mainly responsible [79]. For mechanical stability, however, beside intramolecular also intermolecular interactions, especially at crystal contacts, and the arrangement of molecules in the crystal lattice play a role. Moreover, the mechanical strength is strongly determined by the morphology of the crystals and by the cross-linking with GA [33].

Overall, the determination of the micromechanical properties showed that the FJAT and BmPGA CLECs were significantly more stable than the BtPGA CLECs. The arrangement of the molecules in the crystal lattice showed that in the BmPGA crystals, the individual molecules are arranged very close to each other and show only small channels (water content of 43.5%), while in the BtPGA crystals relatively large free spaces between the BtPGA molecules are evident (water content of 64.8%). In the FJAT-PGA crystal, the packing was denser than in the BtPGA crystal, but the water channels were slightly larger than in the BmPGA crystals (water content of 54.6%) (Fig. 23).

As lysine residues are mainly cross-linked by GA [80], their number was determined in the three PGAs. The least stable BtPGA heterodimer has only 49 lysine residues, which is significantly less than FJAT- (78) and BmPGA (74). Due to the lower number of lysine residues, the cross-linking of the BtPGA CLECs was likely to be less pronounced, possibly making the corresponding CLECs less stable than FJAT and BmPGA CLECs.

A comparison of the micromechanical properties of *Bacillus* spp. PGA CLECs with the micromechanical properties of HheG and lysozyme CLECs showed that FJAT-PGA and BmPGA CLECs were significantly more stable [33–35]. FJAT-PGA CLECs at 25.7 MPa and BmPGA CLECs at 22.6 MPa were about twice as hard as lysozyme and HheG CLECs. Hardness and Young's modulus of BtPGA, HheG and lysozyme CLECs were about the same. The crystal water content of HheG crystals was 65%, similar to that of BtPGA crystals, and together with the low number of lysines of only 0.17 kDa^{-1} may have resulted in low mechanical stability. The lysozyme crystals had the lowest value for crystal water content at 40.5%, but did not show correspondingly high strength. Possibly, the lower number of

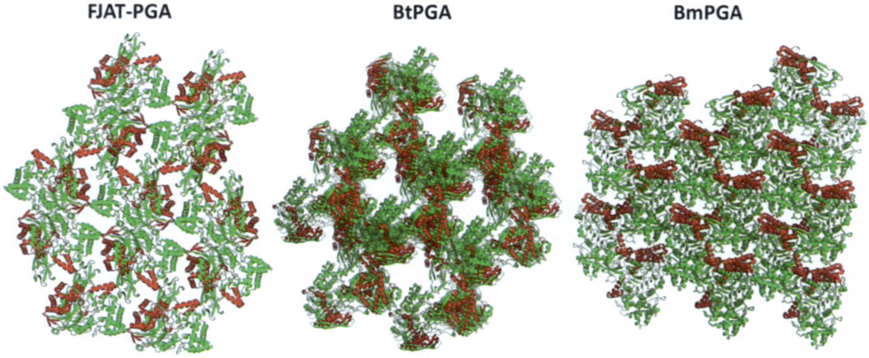

Fig. 23 Arrangement of symmetry mates 50 Å around the asymmetric unit of FJAT-, Bt-, and BmPGA (6NVX, 6NVY and 6NVW from [31]). The α-subunits are shown in red and the β-subunits in green. Figure taken from [32], licensed under the Creative Commons Attribution 4.0 International license (CC BY 4.0)

lysines (0.42 kDa^{-1}) compared to those of FJAT-PGA and BmPGA (0.90 kDa^{-1} and 0.86 kDa^{-1}, respectively) additionally contributed to this. However, the lysozyme CLECs were cross-linked for less time (~10 h). It is likely that a greater hardness would have been observed with a cross-linking time of 24 h. For HheG crystals, increasing the cross-linking time from 10 to 24 h resulted in a threefold increase in hardness [35].

Furthermore, it was noticed that all five proteins crystallized in other space groups and crystal systems. When compared with the evaluation of all PDB structures published up to February 2007, it was found that the average crystal water contents of crystals of different crystal systems vary. The average crystal water contents for orthorhombic and monoclinic crystals such as FJAT-PGA and BmPGA crystals, 49% and 48%, respectively, are significantly smaller than the average crystal water contents of hexagonal (57%), trigonal (55%), and tetragonal crystals (56%) [81]. This was also confirmed for the crystals studied here, with the exception of lysozyme (Table 4). Possibly, the packing density of lysozyme is particularly high due to its low molecular weight of 14.3 kDa and thus the water content is lower. By testing more crystals, it could be investigated whether there is a significant dependence between water content, crystal shape, and mechanical properties. For this purpose, lysozyme could be studied as a model protein, since lysozyme crystals have already been obtained in different crystal systems [82]. In addition, attempts could be made to crystallize proteins in the orthorhombic or monoclinic form, for example, by genetic methods, to examine them mechanically, and to compare them with CLECs of the same enzyme but different crystal form.

Table 4 Micromechanical and crystal properties of FJAT-, Bt-, and BmPGA CLECs as well as HheG and lysozyme CLECs. Table taken from [32], licensed under the Creative Commons Attribution 4.0 International license

	FJAT-PGA (6NVX)[a]	BtPGA (6NVY)[a]	BmPGA (6NVW)[a]	HheG (5O30)[a]	Lysozyme (3LYZ)[a]
Median hardness [MPa]	26	11	23	Prismatic face: 11[b] basal face: 14.5[b]	12[b]
Median Young´s modulus [MPa]	1452	614	1170	Prismatic face: 500[b] basal face: not shown[b]	1073[b]
Solvent content [%]	54.6	64.8	43.5	65.0	40.5
Number of lysine residues per molecular weight [kDa^{-1}]	0.90	0.57	0.86	0.17	0.42
Space group	$P2_12_12_1$	H3	$P2_1$	$P3_121$	$P4_32_12$
Crystal system	Orthorhombic	Hexagonal	Monoclinic	Trigonal	Tetragonal

[a]The PDB codes of the structures are given in parentheses [31, 83, 84]
[b]Median hardness and Young´s modulus of HheG CLECs from [35] and of lysozyme CLECs of [33]

3.8 *Structure–Property Correlation of CLECs of Different HheG-Variants*

As mentioned above protein crystals contain on average about 50% solvent (Table 4), usually in large channels between the packed molecules [85], which in combination with the weak forces holding the molecules within the crystal lattice, implies a very high sensitivity to mechanical stress. To stabilize the crystal lattices, bifunctional linkers are usually used, which form strong covalent bonds between the amino acid residues, e.g., the ε-amino groups of lysine residues [86] Cross-linking fixes the enzyme in its crystalline form, improving both thermal and proteolytic stability, as well as achieving insolubility in aqueous and organic solvents [87]. Due to protein engineering methods amino acids on the surface of the folded wild-type enzyme structure can be genetically exchanged, to incorporate new potential cross-linking sites within the subsequently fabricated protein crystals and thus improve the mechanical crystal properties.

In order to reliably measure the mechanical properties, certain requirements with regard to crystallization must be met: (i) the crystals adhere to the surface of the support, (ii) are present individually, (iii) their shape is fully formed, and (iv) one

Fig. 24 Cross-linked HheG crystals produced according to the set requirements. Figure taken from [16], licensed under the Creative Commons Attribution 4.0 International license (CC BY 4.0)

crystal face is orthogonal to the direction of the measurement. Furthermore, they must have (v) a minimum particle size (about 70 μm) and (vi) have a reproducible production. In the search for crystallization conditions that would allow the crystals to form with the requirements, a parametric study was carried out, in which the crystallization methods (hanging and sitting drop) and the crystallization conditions (type of precipitant, concentration, pH, temperature, ionic strength) were varied. The optimal conditions were then tested for their reproducibility [16].

Analogous to crystallization, the subsequent cross-linking also had to meet certain criteria, such as (i) complete and homogeneous cross-linking within the crystal, (ii) no negative influence on its catalytic activity, (iii) no change in surface properties of the crystals, (iv) no crystal damage after cross-linking, and (v) no detachment of the crystals from the surface of the carrier. In Fig. 24 a sample is shown, which completely fulfills the requirements [16].

The subsequent mechanical characterization of the material properties was carried out via instrumented indentation tests, including nanoindentation. The method of nanoindentation was derived from the classical hardness testing methods [88], where a hard tip is pressed into the surface of the specimen until the specified force or deformation is reached. In nanoindentation, the force curve and the correlating penetration depth are continuously recorded in the form of a force–displacement curve (Fig. 25), which subsequently allows conclusions to be drawn about the mechanical behavior of the material. The indentation tests can be performed with a classical nanoindenter or with AFM force spectroscopy. The atomic force microscope allows the use of very low forces in nanonewton range and thus enables much finer resolution and small motion steps in depth direction from pico- to micrometer.

The hardness and reduced Young's modulus (Er) are calculated according to Eqs. (2) and (3), respectively, in which S is the slope of the unloading and Ac is the contact area of the indent.

$$H = \frac{P}{A_c} \tag{2}$$

$$E_r = \frac{\sqrt{\pi}}{2} \cdot \frac{S}{\sqrt{A_c}} \tag{3}$$

Additionally, elastic and plastic indentation work were calculated according to formula (4) [89]:

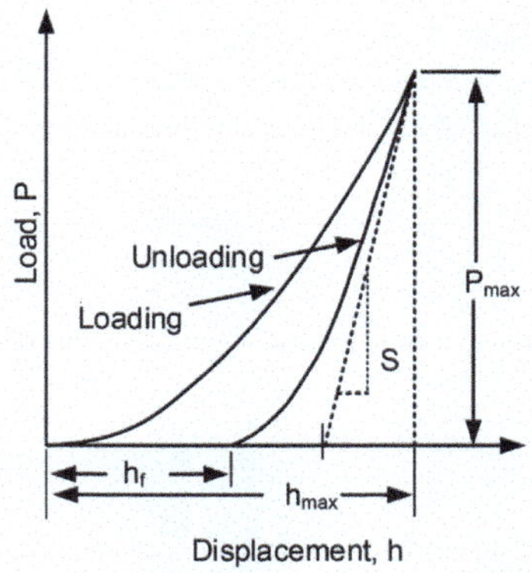

Fig. 25 Schematic force–displacement curve of a nanoindentation including the relevant measurement parameters for calculating the hardness and the Young's modulus of a material sample. Reprinted with permission from [33]. Copyright 2018 American Chemical Society

$$E_{elastic} = \int_0^{h_{max}} F_{unload}(h)dh;\ E_{plastic} = \int_0^{h_{max}} F_{load}(h)dh - E_{elastic} \tag{4}$$

where h_{max} is the maximum displacement during indentation, F_{load} and F_{unload} are the indentation forces during loading and unloading of the sample, elastic deformation is defined as the area under the unloading curve, and plastic deformation is defined as the area enclosed between loading and unloading curves.

It is necessary to describe crystal deformations and all mechanical properties using a statistical or cumulative distribution, because of the high variability on the crystal surface, failures within the inner crystalline structure, and additional influencing factors (e.g., temperature drift) [33]. Since the modeling of the mechanical behavior of crystals can only be done using a very small section of the crystal structure, relative to the whole crystal, the mechanical properties of wild-type, E85K and T123W crystals are first investigated using atomic force microscopy. Here, very small forces in the nanonewton range can be applied, allowing measurement at low indentation depths. In this order of magnitude, the mechanically investigated region is comparable to the crystal section derived and modeled from crystal structure analysis in this work. However, the question arises as to whether a small crystal section is representative of the entire crystal and thus also represents the mechanical properties in deeper crystal regions. For this reason, additional mechanical properties of wild-type and D114C crystals were measured down to the micrometer range with the aid of a nanoindenter and used to validate the model.

The investigation of mechanical properties as a function of different formulation parameters of the crystallization process with innumerable crystal variants would go beyond the experimental scope, which is why crystallization conditions resulting in comparable crystal morphologies were sought for all protein mutants within the scope of this work. For this purpose, process and formulation parameters, such as precipitant, temperature, ionic strength, substrate concentration, pH, as well as process variations, such as those caused by the protein batch or the crystallization method, were screened and found to have different effects on the formation of the HheG

Table 5 Formulation parameters for crystallization of HheG wild-type proteins

Parameter	Wild-type
Precipitation solution	10% (w/v) PEG_{4000}
Buffer	10 mM HEPES, pH 7.3
$C_{protein\ in\ drop}$	11–18 g L^{-1}
$C_{precipitation\ in\ drop}$	40–60 g L^{-1} PEG_{4000}
Crystallization time	72 h

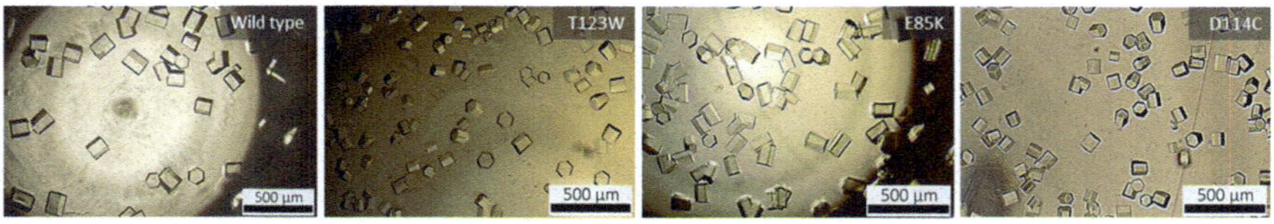

Fig. 26 Hexagonal crystals of the model proteins HheG WT, HheG T123W, HheG E85K, and HheG D114C

crystals. Considering these effects and using a systematic approach, it was possible to identify suitable formulation parameters for crystallization of isometric hexagonal prisms with an average size of 70–100 μm (Table 5).

The model proteins T123W, E85K, and D114C are derived from the HheG wild-type and their crystal structures thus show a very high structural similarity. Even though they differ strongly in terms of their catalytic activity and melting temperature, the same conditions could be used for crystallization except for the mutant T123W. Hexagonal crystal prism of the model proteins are shown in Fig. 26.

Following crystallization, the crystals of the model proteins were cross-linked. For the cross-linking, following cross-linker according to Table 3 were used: GA, DMP, DST, Sulfo-EGS, and BMOE.

As already mentioned above, GA is a homobifunctional linker, which has been most commonly used as a cross-linker due to its low cost, ease of use, and high effectiveness [33, 58]. This reactive linker forms a mixture of diverse oligomers in aqueous solutions, and therefore the resulting cross-linking is rather unpredictable [87, 90]. Its tendency to self-polymerize increases with increasing linker concentration [91] and pH [58]. Therefore, the onset of semi-optimal cross-linking parameters may involve either incomplete or excessive cross-linking [87]. In the first case, the crystal would completely (or only partially) redissolve. Excessive cross-linking, on the other hand, may result in distortion of the protein structure and loss of activity [92, 93]. To resolve this conflict of interest, studies were carried out to allow homogeneous cross-linking without affecting the catalytic activity. For this purpose, firstly, the influence of cross-linking duration on mechanical properties of CLECs was investigated.

As can be seen from Fig. 27, the cross-linking time during the first 24 h is decisive for mechanical stability. During this time, the hardness of the prismatic crystal faces increases almost threefold, from about 4.2 to 11 MPa (median values), the elastic modulus from 140 MPa to about 360 MPa. Even longer cross-linking does

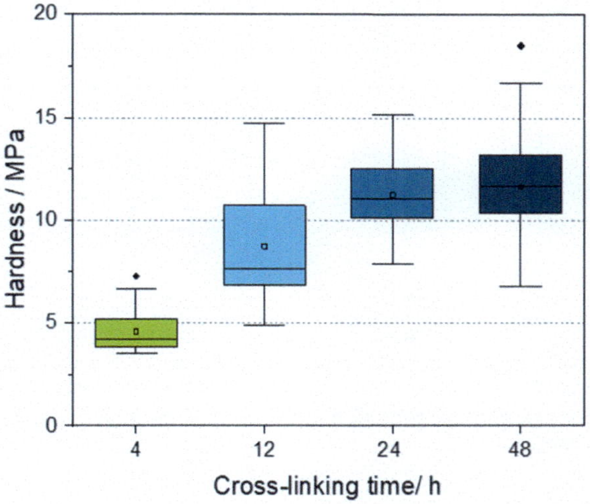

Fig. 27 Hardness of prismatic faces of HheG crystals as a function of the cross-linking time. Each boxplot takes into account more than 150 indentation measurements. According to Kubiak et al. approximately 150 indents performing with the AFM measurements are needed for accurate describing of cumulative distribution. Reprinted adapted with permission from [35]. Copyright 2018 American Chemical Society

not lead to any significant change in mechanical properties. Only a higher distribution of mechanical properties on the crystal surfaces is observed. In comparison to the measured values after 24 h cross-linking time, the distribution increases by 37% for the hardness, and by 55% for the Young's modulus. At this cross-linking time the mechanical properties are homogeneously distributed between all investigated surfaces, so that further cross-linking do not change the mechanical properties significantly. Another aspect that can be seen in Fig. 27 is a broad and asymmetric distribution of mechanical properties on the crystal surface at a cross-linking time of 12 h. A detailed examination of the cumulative distribution of the crystal faces again showed that 1/3 of prismatic crystal faces differs from the another 2/3 of that faces [35]. At longer cross-linking times (~24 h), the anisotropy disappears and the mechanical properties are homogeneously distributed between all investigated surfaces [35]. This indicates that the nature of the anisotropy is due to the complete cross-linking of the prismatic surfaces.

Mechanical investigation of different crystal faces (prismatic and face) at a cross-linking time of 24 h showed that they exhibit structurally based anisotropic properties, which are shown in Fig. 28.

It can be seen from Fig. 28 that the median value of hardness (14.5 MPa) and modulus of elasticity (460 MPa) from the prismatic face is about 23% lower than from the basal face. A similar width of the distribution (see whiskers of the boxplot) indicates a comparable biovariability of the two faces. Thus, both crystal anisotropy and anisotropy of the prismatic faces due to insufficient cross-linking time were found to have a structurally based origin. On the one hand, the molecular packing must yield different densities depending on the direction, which results in different mechanical properties after cross-linking. On the other hand, the accessibility of the functional groups must be different, which is why one part of the crystal faces can

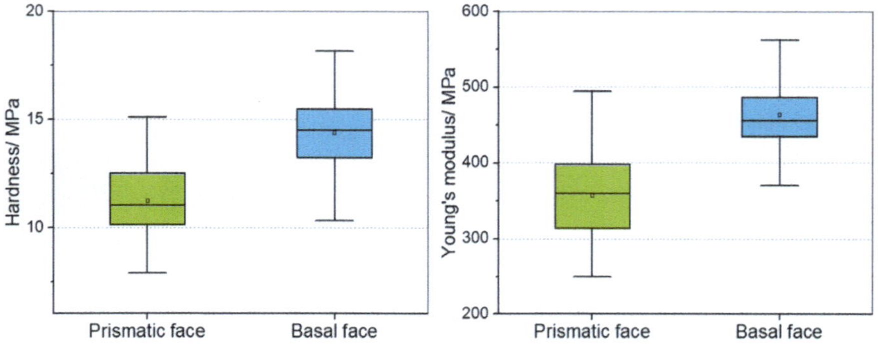

Fig. 28 Comparison of hardness [33] and elastic modulus of different crystal faces cross-linked over 24 h. Reprinted adapted with permission from [35]. Copyright 2018 American Chemical Society

be cross-linked faster than the other [35]. This phenomenon was discussed in detail in [35] and is explained in more detail below ("Property-sequence correlation").

In the results presented so far for "structure–property correlation", the HheG crystals were cross-linked using GA and the soaking method. It was already shown that cross-linking by the soaking method with GA results in a loss of the catalytic activity of the HheG wild-type crystals (Fig. 15). This can be avoided by using either a vapor diffusion method or another linker. As an alternative to GA a mixture of three linkers (DST, DMP, and Sulfo-EGS) with all different lengths was used (~6, ~9, and ~16 Å), which specifically react with lysine residues that are in sufficiently close proximity to each other (Table 3). Figure 29 compares the mechanical properties of the wild-type crystals cross-linked with different linkers.

The results showed that, using the alternative lysine cross-linker, the hardness of the anisotropic crystal faces decreased by about 70% compared to the cross-linking

Fig. 29 Comparison of the hardness of wild-type crystals cross-linked using the different linkers for 24 h. Figure taken from [16], licensed under the Creative Commons Attribution 4.0 International license (CC BY 4.0)

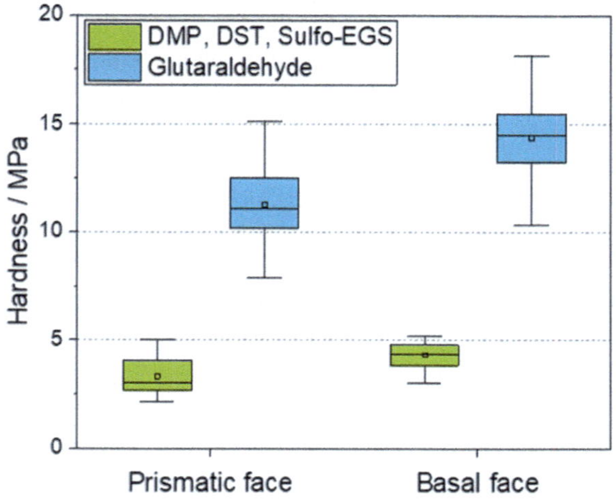

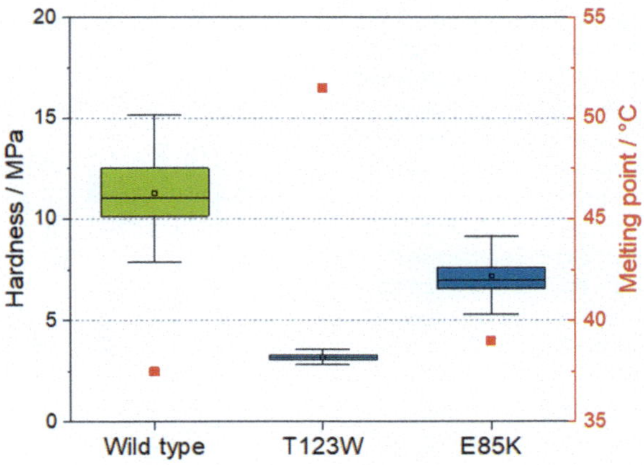

Fig. 30 Comparison of the hardness of model proteins wild-type, T123W, E85K cross-linked with GA for 24 h. The results represent the micromechanics of the prismatic crystal faces. On the secondary axis, the corresponding melting point was added from the solubilized proteins

with the GA. Besides the high stability of CLECs cross-linked with GA, its low price is of great advantage. Because of its negative impact on the activity of HheG crystals, this advantage cannot be taken in this case.

One aim of this work was to produce customized CLECs that can be used in process engineering processes and also reused. This can be achieved by different cross-linking (time, linker) and/or by the use of protein engineering. However, based on measurements from Fig. 30, it could be shown that the protein engineering does not necessarily improve the mechanical properties of the protein crystals, the measured hardness of model protein variants decreases in case of E85K CLECs around 36% up to 73% by T123W CLECs.

From Fig. 30 it can also be seen, that there is no correlation between the thermal stability of the proteins and the mechanical properties of the crystals built from these proteins. This finding could also be applied to the PGA CLECs [32].

For the previously presented mutant model proteins E85K and T123W, it could be clearly shown by means of AFM force spectroscopy, that these have a significantly lower hardness than the wild-type crystals. Since the crystals are held together by crystal contacts, a mutation can lead to an increase in the spacing and therewith to a weakening of crystal contacts and thus of their strength.

Now, the influence of the genetic modification of position D114 on the mechanical properties of CLECs is considered in more detail. By replacing it with cysteine, a cross-linking site was incorporated, which can be cross-linked by means of a suitable linker–in this case a BMOE linker–in addition to the lysines. In this series of experiments, crystals were cross-linked partly by the glutaraldehyde and vapor diffusion method and partly by the BMOE linker via the soaking method. In addition, some of the crystal samples were cross-linked first with GA and then with the BMEO linker. All three samples were mechanically examined and compared with the wild-type CLECs. In this experimental study, fractions of elastic energy were considered.

In Fig. 31, the influence of the cross-linker on the fraction of elastic energy of the prismatic faces of D114C crystals is shown. It can be seen, that the fraction of elastic energy increases with the degree of cross-linking and is the lowest for cross-linking

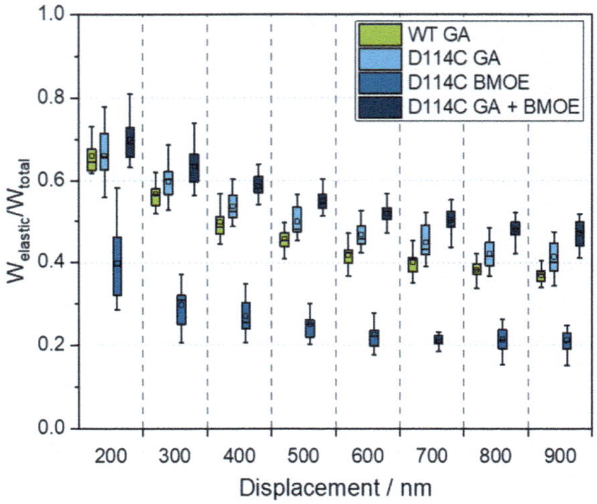

Fig. 31 Influence of cross-linking on the fraction of elastic energy of prismatic faces of D114C crystals as a function of penetration depth, as well as the comparison with the wild-type crystals. Figure taken from [16], licensed under the Creative Commons Attribution 4.0 International license (CC BY 4.0)

with BMEO and highest for cross-linking with both linkers. At the penetration depth of 200 nm, the fraction of elastic energy is 39% (BMEO), 66% (GA), and 69% (GA + BMEO). In comparison, the fraction of elastic energy of wild-type crystals cross-linked with GA is 64% and is thus slightly lower than the median value from the D114C crystals, which were also cross-linked with GA. The fraction of elastic energy decreases with increasing depth according to the so called indentation size effect (ISE), which was discussed in detail by Kubiak et al. [34]. At a depth of 900 nm, the fraction of elastic energy of the crystals cross-linked with BMOE is the lowest (around 20%) and increases to 40% when cross-linked with GA. By additional cross-linking with BMEO the fraction of elastic energy rises by additional 7–47%, which is 20% higher than the elastic fraction of the wild-type crystals [16].

3.9 Property-Sequence Correlation of CLECs of Different HheG-Variants

Based on the crystal structure, a structural model was developed to represent the mechanical behavior of cross-linked crystals at the structural level. MATLAB version R2017a software was used for this purpose. The developed toolbox allows for reading the Protein Data Bank file format, the database (.pdb), which represents the structure of three-dimensional molecules based on the atomic coordinates of amino acids of the protein. A.pdb file format contains coordinates from the smallest unit of the crystal (the so-called asymmetric unit), which can be multiplied to a larger crystal structure according to the given symmetry operators, which are also present in the file. The developed model allows for rebuilding the whole crystal using the symmetry operators, searching for atoms of relevant amino acids residues, e.g., sulfur atoms of the

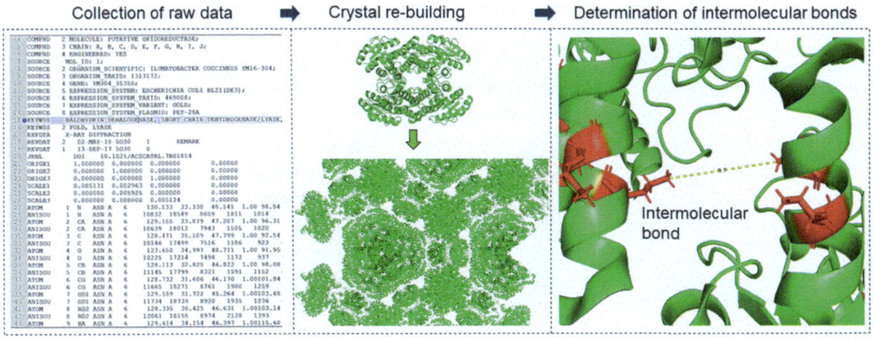

Fig. 32 Overview approach to modeling mechanical crystal behavior based on crystal structure: investigation and classification of inter- and intramolecular cross-linking bonds. Figure taken from [16], licensed under the Creative Commons Attribution 4.0 International license (CC BY 4.0)

cysteine residues, and calculating the distances between them (Fig. 32). The amino acid residues that would be cross-linked by glutaraldehyde were determined based on the literature [14, 15, 94, 95]. According to those studies, cross-linking bonds can be expected between three residual pairs: the ε-amines of lysine residues (Lys–Lys), two neighbored arginine residues (Arg–Arg), and arginine and lysine residues (Lys–Arg) having a maximal distance of 10 Å to each other [35]. For alternative cross-linkers (e.g., BMOE) the distance and the amino acids residues were adapted according to Table 3. For the crystals, all bonds within a unit cell and its surrounding of 100 Å (50 asymmetric units) were considered [16, 35].

Then, the possible cross-linking bonds were analyzed regarding their distance and direction. Finally, the directions of the bonds in relation to the mechanically stressed crystal face were considered in order to calculate the direction-dependent crystal strength. For the investigation of the anisotropic crystal strength, the intra- and intermolecular bonds were considered separately, assuming that only the intermolecular bonds contribute to the improvement of the stability of the crystal structure in an environment other than the crystallization cocktail [96]. Furthermore, the strength of these bonds depends on their orientation with respect to the applied force during indentation, according to Fig. 33.

The first approach for evaluating the direction-dependent crystal strength (dimensionless equivalent value for a mechanical parameter) is based on the assumption that mainly intermolecular cross-linked bonds influence the mechanical crystal properties. In addition, the strength of these bonds are depending on their orientation with respect to the applied force. Similar studies on the anisotropic particle strength as a function of porosity, particle structure or size have already been published in recent Literature [89, 97–99]. Rumpf investigated a continuum-based model for the tensile strength of aggregates. It is based on all partial stress contributions in a cross-sectional area of the structure in the direction of the applied force [97]. Schilde et al. reported that this model can also be applied to the compressive strength measured by nanoindentation of aggregates [89, 98, 99]. Based on these results the fraction of each

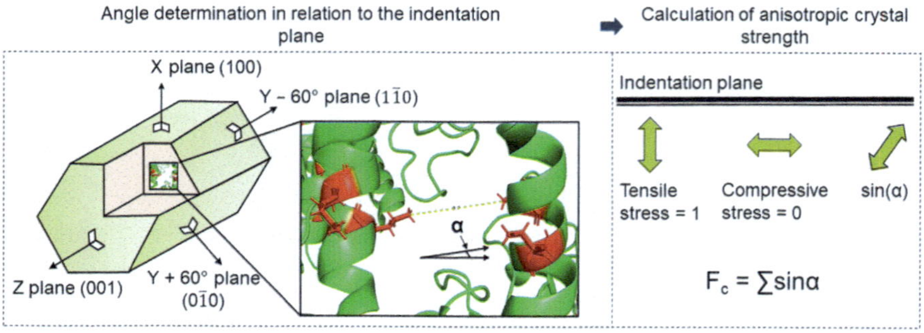

Fig. 33 Overview approach to modeling mechanical crystal behavior based on crystal structure: calculation of anisotropic crystal strength using a direction of cross-linking bonds in relation to the indentation plane. Figure taken from [16], licensed under the Creative Commons Attribution 4.0 International license (CC BY 4.0)

bond in the loading direction is calculated for each measured crystal surface. For this purpose, the sine of the angle between the normal of the crystal surface ($\perp$) and the direction of the bond is used as a factor (Fig. 30, right). The determination of the directional crystal faces (the planes), which were used as surface normals in the calculation, was done by means of the geometry data of the unit cell and the plane equations. As long as the mechanical binding properties of the possible amino acid pairs connected via cross-linkers are unknown, it is assumed that they contribute equally to the mechanical properties of the crystal. Under this assumption, by summing up all bond proportions orthogonal to the respective plane of indentation, a representative value is determined for the total strength of the crystal surface (resistance of the cross-link bridges to the normal force = dimensionless equivalent value of crystal strength).

Table 6 summarizes the modeled direction-dependent crystal strength of all intermolecular bonds smaller than 10 Å (cross-linking with GA), or Lys–Lys* bonds with a well-defined distance for alternative cross-linker (6, 9, and 16 Å, Table 3).

In Fig. 34, a comparison of the experimental and modeled results is shown. The experimental and modeled data were interpreted by comparing the percentage differences between the anisotropic faces. For instance, the difference in the hardness of the prismatic face to the basal face of crystals cross-linked at the same conditions

Table 6 Summary of the direction-dependent crystal strength modeled using the anisotropic intermolecular cross-linking bonds. Arg-Arg, Arg-Lys and Lys-Lys < 10 Å distance, Lys-Lys* 6, 9, and 16 Å because of the linker mix

	Arg–Arg	Arg–Lys	Lys–Lys	Lys–Lys*
Z	271.77	142	145.4	174.68
X	231.63	137.81	107.04	133.07
$Y + 60°$	230.52	140.04	105.47	132.11
$Y - 60°$	230.49	141.84	107.97	135.75

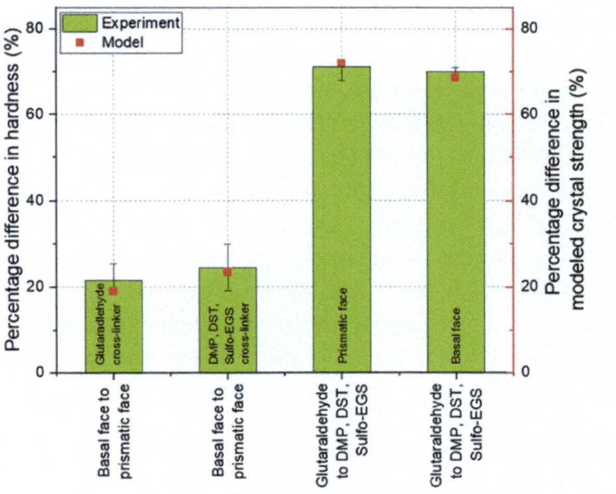

Fig. 34 Comparison of the experimental and the modeled results of differently cross-linked wild-type HheG crystals. The bars represent the percentage differences between the mean hardness of the respective crystal faces considered, as described in the diagram. The square dots represent the modeling results, which are also represented as the percentage difference between the corresponding crystal faces. Figure taken from [16], licensed under the Creative Commons Attribution 4.0 International license (CC BY 4.0)

(GA or DMP, DST, Sulfo-EGS-linker mix, referred to as Lys–Lys* in Table 6). These percentage differences are shown as a bar (experiment) in Fig. 34. Analogously, the percentage difference in the modeling results, e.g., between the anisotropic crystal faces, is shown as dots (model) [16].

The results showed that the crystal strength resulting from the Lys–Lys and Arg–Arg bonds was about 20% higher in the Z-direction (for the basal face), which is consistent with the experimental results. Looking at the cross-linking using the linker mix, it also turns out that the experiment agreed well with the model. The experimentally determined anisotropy between the basal and prismatic faces was about 24%. This corresponded to the difference in crystal strength due to the Lys–Lys bridges from the three defined linker lengths (see Table 6, Lys–Lys*, ratio of bond strength of the Z-direction (174.68) and an average of X, Y + 60°, and Y − 60° (ca. 133.64)). Previously, it was found that the Lys–Lys and Arg–Arg bridges lead to the anisotropic behavior of the HheG crystals. Assuming that there are bonds of all the considered amino acid residue pairs in the crystal, the Arg–Lys bond here did not contribute to the anisotropic behavior of the wild-type HheG crystals, and the bonds in the defined directions were summed up and the total strength of the respective areas for differently cross-linked crystals was proportionally adjusted. The mechanical investigation showed that, due to the use of a linker mix instead of glutaraldehyde, the average hardness of the prismatic and basal crystal faces decreased by about 70%. The results corresponded to the modeling with some slight deviation.

The knowledge that a crystal structure can be used to predict the mechanical behavior of the surfaces saves an enormous amount of time through the targeted selection of the linker and the subsequent model prediction of the crystal strength based on the formed cross-link bridges. The model can be used, for example, to determine which binding-site distances occur most frequently and therewith how many bonds of a cross-linker molecule with a certain length can be formed. In addition, it can be determined whether other linkers, e.g., carboxyl or sulfhydryl instead of amine linkers, will provide better mechanical performance. By selectively combining different variants–linker length, type, with binding sites, e.g., far away from the catalytic site–the desired mechanical strength can be modeled in advance [16].

In the introduction, it was written that protein engineering is a common method for improving the performance of enzymes. Unfortunately, improvements in the properties of solubilized proteins are not always accompanied by improvements in the mechanical stability of the crystals.

Figure 35 summarizes the experimental and modeled results of the CLECs variants. Analogous to the wild-type crystals, the difference in hardness from the prismatic and basal crystal faces is shown with the results of the model for mutant T123W. Since the activity of mutant T123W was not affected by the soaking method, it is assumed that only the Arg–Lys and Lys–Lys bridges are formed here. This assumption is confirmed by the modeling result and thus well describes the crystal anisotropy for this mutant. The question arises whether it is possible, based on the results of the wild-type crystals, to transfer the correlations to other crystal structures without the need to perform the time-consuming mechanical measurements. To test this, the respective areas of the mutant model proteins T123W and E85K were set in relation to the wild-type. The percentage difference in the hardness of prismatic and basal faces of the crystals, respectively, as well as the model can also be seen in Fig. 35. The results of the modeling show that for mutant E85K, predominantly the Arg–Arg and the Lys–Lys bridges are formed. For the different crystal structures, there is a slight deviation between the model and the experiment, but the trends are accurately represented. The hardness of the end face of mutant E85K was not investigated mechanically.

So far, the results of the hardness measured with the AFM at a penetration depth of less than about 30 nm (wild-type crystals) were compared with the model. In this section, the model is used for correlation with other mechanical parameters, such as the fraction of elastic energy at a penetration depth of 900 nm. Thus, it is tested whether modeling from a small crystal section is sufficient for the global description of the mechanical behavior.

In Fig. 36, the results obtained with the nanoindenter are presented, where the anisotropy of the wild-type (WT) crystals was again related to the model based on the differences in the fraction of elastic energy. The fraction of elastic energy of the basal face of the WT CLECs was, on average, about 14% higher than for the prismatic face. Comparing the percentage difference with the summed bond strengths from the bond pairs considered, i.e., Lys–Lys, Lys–Arg, and Arg–Arg, the modeling results also showed a difference of about 14.6% in the Z-direction compared to the X, Y + 60°, and Y − 60° directions ($\sum_{\text{Z-direction}} = 559.17$ and mean of $\sum_{\text{XY-direction}} =$

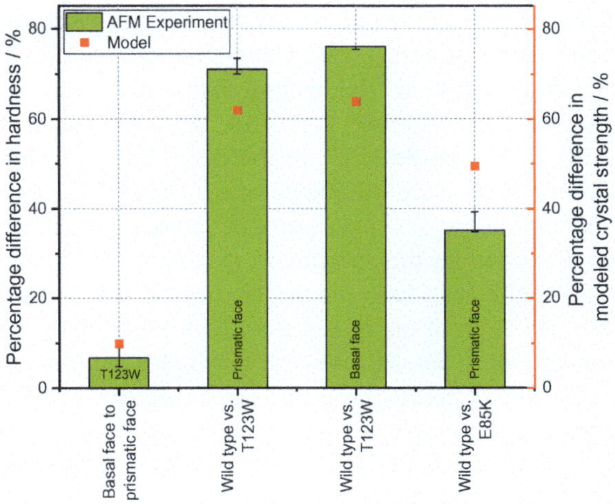

Fig. 35 Comparison of the experimental and modeled results of crystals of the mutant model proteins T123W and E85K

477.59, additional Table not shown). Thus, based on the modeling results, it can be concluded that all bond pairs contributed to the crystal mechanics in the deep crystal regions. This would also explain why the anisotropy between different crystal faces decreased with increasing penetration depth.

The crystal faces of D114C mutant cross-linked by GA show about a 6% difference in the elastic energy fraction. From Table 6, it can be seen that the sum of Lys–Lys and Arg–Arg bonds (365.03 in the Z-direction and 343.88 in the X, Y + 60°, and Y − 60° directions) also showed about a 6% difference. A similar result concerned the cross-linking using GA and BMOE. Comparable to the cross-linking with GA, the difference between the model and the experimental results was negligible here. The deviation between the model and the result was the highest for cross-linking with BMOE and amounted to 30% when mean values were considered. The reason for this was that the cross-linking of cysteines alone was not sufficient to provide

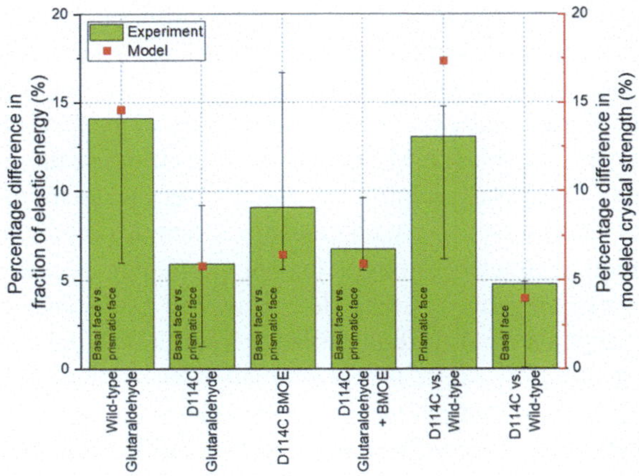

Fig. 36 Comparison of the experimental (by nanoindentation) and the modeled results of wild-type HheG crystals and the mutant D114C. Figure taken from [16], licensed under the Creative Commons Attribution 4.0 International license (CC BY 4.0)

stable support to the crystal lattice such that increased dislocations were observed during the measurement, especially in deeper crystal regions, leading to the increased distribution width of the results. This distribution width could also be seen in the error bars. For example, a positive error bar meant that the percentage difference of the third quartile (75% of the distribution) of ca. 17% was about 90% higher than the percentage difference of the mean value (9%, Fig. 36). Nevertheless, the model result was within the error or distribution width, which demonstrated the good representation by the experiment. A comparison of the respective crystal faces (basal or prismatic face) between the two crystal structures showed that there was about a 16% deviation of the model from both the third quartile for the prismatic face and from the mean value of the distribution for the basal face [16].

The most important causes, which can have a limiting influence on the modeling of the crystal behavior, can occur at three different stages of experimental execution: X-ray structure analysis, sample preparation, and mechanical measurements. For the model, the X-ray structure analysis was one of the most critical stages. Regarding crystal quality, parameters such as protein production batch or crystallization conditions and growth time may have a great influence. For the mutant D114C, the best-achieved resolution was unfortunately only 3.9 Å, which allowed for the determination of the positions of the atoms with a certain tolerance. During data processing, for example, the symmetry of the respective chains was calculated based on the crystallographic space group. Moreover, the co- ordinate origin was manually changed. In this process, the atoms were manually shifted to the same origin. However, the coordinates of the atoms were influenced by the B-factors and thus were not rigid but movable in the crystal and did not represent absolute positions but, rather, were relative to each other. The exact displacement measurement of the atoms was limited by the poor resolution, which is why deviations could occur. Due to the relativity of position within a crystal, this was not a critical problem when experiments and modeling were performed using the same crystal structure. However, the error may be significant if different crystal structures are compared to each other. Nevertheless, taking into account the fact that the accuracy of the crystal structures and the developed model were in the angstrom range, the agreement between the model and experiment was surprisingly high for different model proteins. For the mechanical measurements, influencing factors such as the number of measurement points for a statistically validated estimation of the distribution of mechanical properties and the cantilever quality (geometry or durability of the measurement tip) are to quantify the tilt or by regular SEM images of the cantilevers, these influences can be controlled and, if necessary, minimized. Another serious factor that influences the modeling of the structure–property relationship is sample preparation. Depending on the formulation parameters, different crystal morphologies can be formed, resulting in differences in mechanical behavior. However, it happened that the mechanical behavior of crystals from different protein batches differed despite the crystal morphology remaining the same. This phenomenon was difficult to explain and was probably due to small differences in ions in the solution after, e.g., the purification or desalting step. Since these influencing factors are difficult to control or to avoid despite using established and consistent preparation methods, sample preparation is considered a highly influential factor comparable to that of X-ray structure analysis. Nevertheless, the model reproduced the trends very well. Despite certain limitations, such as the poor resolution of the crystal structure, the model allowed for a representation of the mechanical properties, both the hardness and the elastic part of the deformation energy, within the whole crystal. The accuracy of this model could be refined by the addition of further information, e.g., regarding the bond strength as a function of bond length. Additionally, X-ray crystallographic studies of cross-linked crystals could also be performed, allowing for the resolution of systematic cross-linking bonds. Subsequently, the relevant distances/positions could be selected and modeled. It would also be possible to introduce a position-dependent strength of interactions instead of cross-linking bridges and, thus, additionally ensure the modeling of the mechanical behavior of non-cross-linked crystals. However, a prerequisite is that accurate quantifiable data on the force fields between individual molecules within the crystal are available, e.g., via molecular dynamics simulation. By combining the data from native and cross-linked crystals, the highest accuracy in terms of predicting the mechanical behavior should be achieved. However, a very high computational capacity is recommended to perform such calculations. The excerpt of crystal structures in this work was limited to 100 Å in space. Despite the relatively small crystal excerpt, the number of rows searched for targeted information in MATLAB was almost 30 million. In order to calculate the data more efficiently, the .pdb file had to be directly converted with suitable functions and limited to relevant information. In case a sensitivity analysis should be performed using the model, all data must be available. With the help of a high-performance computer, all distances to each other could be calculated and clustered. On a qualitative basis, the amino acid positions could be located, which should be exchanged by the rational protein design to incorporate quite effective cross-linking sites. Assuming that the crystal structure will not be subject to major changes, it should therefore be possible to perform targeted mutations with this tool, which should contribute to enhancing cross-linking and mechanical properties.

Acknowledgements The authors gratefully acknowledge financial support by the German Research Foundation (DFG) in the Priority Programme 1934 DiSPBiotech–Dispersity, structural and phase modifications of proteins and biological agglomerates in biotechnological processes (SPP 1934 DiSPBiotech–315462951).

References

1. Rehm FBH, Chen S, Rehm BHA (2016) Enzyme engineering for *in situ* immobilization. Molecules 21(10):1370. https://doi.org/10.3390/molecules21101370
2. Brady D, Jordaan J (2009) Advances in enzyme immobilisation. Biotechnol Lett 31(11):1639–1650. https://doi.org/10.1007/s10529-009-0076-4
3. Chauhan V, Kaushal D, Dhiman VK, Kanwar SS, Singh D, Dhiman VK, Pandey H (2022) An insight in developing carrier-free immobilized enzymes. Front Bioeng Biotechnol 10:794411. https://doi.org/10.3389/fbioe.2022.794411

4. Roessl U, Nahalka J, Nidetzky B (2010) Carrier-free immobilized enzymes for biocatalysis. Biotechnol Lett 32(3):341–350. https://doi.org/10.1007/s10529-009-0173-4
5. Quiocho FA, Richards FM (1964) Intermolecular cross linking of a protein in the crystalline state: carboxypeptidase-A. Proc Natl Acad Sci USA 52(3):833–839. https://doi.org/10.1073/pnas.52.3.833
6. Hekmat D (2015) Large-scale crystallization of proteins for purification and formulation. Bioprocess Biosyst Eng 38(7):1209–1231. https://doi.org/10.1007/s00449-015-1374-y
7. Drenth J, Haas C (1992) Protein crystals and their stability. J Cryst Growth 122(1–4):107–109. https://doi.org/10.1016/0022-0248(92)90233-9
8. St Clair N, Shenoy B, Jacob LD, Margolin AL (1999) Cross-linked protein crystals for vaccine delivery. Proc Natl Acad Sci USA 96(17):9469–9474. https://doi.org/10.1073/pnas.96.17.9469
9. Fernandez-Penas R, Verdugo-Escamilla C, Martinez-Rodriguez S, Gavira JA (2021) Production of cross-linked lipase crystals at a preparative scale. Cryst Growth Des 21(3):1698–1707. https://doi.org/10.1021/acs.cgd.0c01608
10. St Clair NL, Navia MA (1992) Cross-linked enzyme crystals as robust biocatalysts. J Am Chem Soc 114(18):7314–7316. https://doi.org/10.1021/ja00044a064
11. Jegan Roy J, Emilia Abraham T (2006) Continuous biotransformation of pyrogallol to purpurogallin using cross-linked enzyme crystals of laccase as catalyst in a packed-bed reactor. J Chem Technol Biotechnol 81:1836–1839. https://doi.org/10.1002/jctb.1612
12. Margolin AL, Navia MA (2001) Protein crystals as novel catalytic materials. Angew Chem Int Ed Engl 40(12):2204–2222. https://doi.org/10.1002/1521-3773(20010618)40:12%3C2204::AID-ANIE2204%3E3.0.CO;2-J
13. Jegan Roy J, Emilia Abraham T (2006) Preparation and characterization of cross-linked enzyme crystals of laccase. J Mol Catal B Enzym 38(1):31–36. https://doi.org/10.1016/j.molcatb.2005.11.001
14. Buch M, Wine Y, Dror Y et al (2014) Protein products obtained by site-preferred partial crosslinking in protein crystals and "liberated" by redissolution. Biotechnol Bioeng 111(7):1296–1303. https://doi.org/10.1002/bit.25186
15. Wine Y, Cohen-Hadar N, Freeman A, Frolow F (2007) Elucidation of the mechanism and end products of glutaraldehyde crosslinking reaction by X-ray structure analysis. Biotechnol Bioeng 98(3):711–718. https://doi.org/10.1002/bit.21459
16. Kubiak M, Kampen I, Schilde C (2022) Structure-based modeling of the mechanical behavior of cross-linked enzyme crystals. Crystals 12(4):441. https://doi.org/10.3390/cryst12040441
17. Cobos-Puc L, Rodriguez-Herrera R, Cano-Cabrera JC, Aguayo-Morales H, Silva-Belmares SY, Gallegos ACF, Hernandez JLM (2020) Classical and new pharmaceutical uses of bacterial penicillin G acylase. Curr Pharm Biotechnol 21(4):287–297. https://doi.org/10.2174/1389201020666191111151642
18. Wang Y, Hu X, Long Z et al (2022) Proteomic analysis of Penicillin G acylases and resulting residues in semi-synthetic beta-lactam antibiotics using liquid chromatography-tandem mass spectrometry. J Chromatogr A 1678:463365. https://doi.org/10.1016/j.chroma.2022.463365
19. Marešová H, Plackova M, Grulich M, Kyslik P (2014) Current state and perspectives of penicillin G acylase-based biocatalyses. Appl Microbiol Biotechnol 98(7):2867–2879. https://doi.org/10.1007/s00253-013-5492-7
20. Srirangan K, Orr V, Akawi L, Westbrook A, Moo-Young M, Chou CP (2013) Biotechnological advances on penicillin G acylase: pharmaceutical implications, unique expression mechanism and production strategies. Biotechnol Adv 31(8):1319–1332. https://doi.org/10.1016/j.biotechadv.2013.05.006
21. Schallmey A, Schallmey M (2016) Recent advances on halohydrin dehalogenases-from enzyme identification to novel biocatalytic applications. Appl Microbiol Biotechnol 100(18):7827–7839. https://doi.org/10.1007/s00253-016-7750-y
22. Hasnaoui-Dijoux G, Majeric Elenkov M, Lutje Spelberg JH, Hauer B, Janssen DB (2008) Catalytic promiscuity of halohydrin dehalogenase and its application in enantioselective epoxide ring opening. ChemBioChem 9(7):1048–1051. https://doi.org/10.1002/cbic.200700734

23. Bornscheuer UT, Hesseler M (2010) Enzymatic removal of 3-monochloro-1,2-propanediol (3-MCPD) and its esters from oils. Eur J Lipid Sci Technol 112:552–556. https://doi.org/10.1002/ejlt.200900245
24. You ZY, Liu ZQ, Zheng YG (2014) Chemical and enzymatic approaches to the synthesis of optically pure ethyl (R)-4-cyano-3-hydroxybutanoate. Appl Microbiol Biotechnol 98(1):11–21. https://doi.org/10.1007/s00253-013-5357-0
25. Calderini E, Wessel J, Süss P, Schrepfer P, Wardenga R, Schallmey A (2019) Selective ring-opening of di-substituted epoxides catalysed by halohydrin dehalogenases. ChemCatChem 11:2099. https://doi.org/10.1002/cctc.201900103
26. Koopmeiners J, Diederich C, Solarczek J, Voß H, Mayer J, Blankenfeldt W, Schallmey A (2017) HheG, a Halohydrin Dehalogenase with Activity on Cyclic Epoxides. ACS Catal 7(10):6877–6886. https://doi.org/10.1021/acscatal.7b01854
27. Solarczek J, Kaspar F, Bauer P, Schallmey A (2022) G-type halohydrin dehalogenases catalyze ring opening reactions of cyclic epoxides with diverse anionic nucleophiles. Chemistry 28(72):e202202343. https://doi.org/10.1002/chem.202202343
28. Staar M, Henke S, Blankenfeldt W, Schallmey A (2022) Biocatalytically active and stable cross-linked enzyme crystals of halohydrin dehalogenase HheG by protein engineering. ChemCatChem 14(9):e202200145. https://doi.org/10.1002/cctc.202200145
29. Staar M, Staar S, Schallmey A (2022) Crystal contact engineering for enhanced cross-linking efficiency of HheG crystal. Crystals 12(12):1553. https://doi.org/10.3390/catal12121553
30. Wichmann J, Mayer J, Hintmann M, Lukat P, Blankenfeldt W, Biedendieck R (2023) Multi-step engineering of a penicillin G acylase for systematic improvement of crystallization efficiency. Cryst Growth Des 23:3230–3243. https://doi.org/10.1021/acs.cgd.2c01408
31. Mayer J, Pippel J, Gunther G, Muller C, Lauermann A, Knuuti T, Blankenfeldt W, Jahn D, Biedendieck R (2019) Crystal structures and protein engineering of three different penicillin G acylases from Gram-positive bacteria with different thermostability. Appl Microbiol Biotechnol 103(18):7537–7552. https://doi.org/10.1007/s00253-019-09977-8
32. Kubiak M, Mayer J, Kampen I, Schilde C, Biedendieck R (2021) Structure-properties correlation of cross-linked penicillin G acylase crystals. Crystals 11(4):451. https://doi.org/10.3390/cryst11040451
33. Kubiak M, Solarczek J, Kampen I, Schallmey A, Kwade A, Schilde C (2018) Micromechanics of anisotropic cross-linked enzyme crystals. Cryst Growth Des 18(10):5885–5895. https://doi.org/10.1021/acs.cgd.8b00637
34. Kubiak M, Staar M, Kampen I, Schallmey, Schilde C (2021) The depth-dependent mechanical behavior of anisotropic native and cross-linked HheG enzyme crystals. Crystals 11(7):718. https://doi.org/10.3390/cryst11070718.
35. Kubiak M, Storm K, Kampen I, Schilde C (2019) Relationship between cross-linking reaction time and anisotropic mechanical behavior of enzyme crystals. Cryst Growth Des 19(8):4453–4464. https://doi.org/10.1021/acs.cgd.9b00232
36. Rojviriya C, Pratumrat T, Saper MA, Yuvaniyama J (2011) Improved X-ray diffraction from *Bacillus megaterium* penicillin G acylase crystals through long cryosoaking dehydration. Acta Crystallogr Sect F Struct Biol Cryst Commun 67:1570–1574. https://doi.org/10.1107/S1744309111040462
37. Wittchen KD, Meinhardt F (1995) Inactivation of the major extracellular protease from *Bacillus megaterium* DSM319 by gene replacement. Appl Microbiol Biotechnol 42(6):871–877. https://doi.org/10.1007/BF00191184
38. Jurrus E, Engel D, Star K et al (2018) Improvements to the APBS biomolecular solvation software suite. Protein Sci 27(1):112–128. https://doi.org/10.1002/pro.3280
39. McDonough MA, Klei HE, Kelly JA (1999) Crystal structure of penicillin G acylase from the Bro1 mutant strain of *Providencia rettgeri*. Protein Sci 8(10):1971–1981. https://doi.org/10.1110/ps.8.10.1971
40. McVey CE, Walsh MA, Dodson GG, Wilson KS, Brannigan JA (2001) Crystal structures of penicillin acylase enzyme-substrate complexes: structural insights into the catalytic mechanism. J Mol Biol 313(1):139–150. https://doi.org/10.1006/jmbi.2001.5043

41. Yang Y, Biedendieck R, Wang W, Gamer M, Malten M, Jahn D, Deckwer WD (2006) High yield recombinant penicillin G amidase production and export into the growth medium using *Bacillus megaterium*. Microb Cell Fact 5:36. https://doi.org/10.1186/1475-2859-5-36
42. Biedendieck R, Borgmeier C, Bunk B, Stammen S, Scherling C, Meinhardt F, Wittmann C, Jahn D (2011) Systems biology of recombinant protein production using *Bacillus megaterium*. Methods Enzymol 500:165–195. https://doi.org/10.1016/B978-0-12-385118-5.00010-4
43. Freudl R (2018) Signal peptides for recombinant protein secretion in bacterial expression systems. Microb Cell Fact 17(1):52. https://doi.org/10.1186/s12934-018-0901-3
44. Terpe K (2006) Overview of bacterial expression systems for heterologous protein production: from molecular and biochemical fundamentals to commercial systems. Appl Microbiol Biotechnol 72(2):211–222. https://doi.org/10.1007/s00253-006-0465-8
45. Yamane K, Bunai K, Kakeshita H (2004) Protein traffic for secretion and related machinery of *Bacillus subtilis*. Biosci Biotechnol Biochem 68(10):2007–2023. https://doi.org/10.1271/bbb.68.2007
46. Mayer J, Knuuti T, Baumgarten L, Menke E, Bischoff L, Bunk B, Biedendieck R (2022) Construction and application of a plasmid-based signal peptide library for improved secretion of recombinant proteins with *Priestia megaterium*. Microorganisms 10(4):777. https://doi.org/10.3390/microorganisms10040777
47. Eppinger M, Bunk B, Johns MA et al (2011) Genome sequences of the biotechnologically important *Bacillus megaterium* strains QM B1551 and DSM319. J Bacteriol 193(16):4199–4213. https://doi.org/10.1128/JB.00449-11
48. Goldschmidt L, Cooper DR, Derewenda ZS, Eisenberg D (2007) Toward rational protein crystallization: a web server for the design of crystallizable protein variants. Protein Sci 16(8):1569–1576. https://doi.org/10.1110/ps.072914007
49. Longenecker KL, Garrard SM, Sheffield PJ, Derewenda ZS (2001) Protein crystallization by rational mutagenesis of surface residues: Lys to Ala mutations promote crystallization of RhoGDI. Acta Crystallogr D Biol Crystallogr 57(Pt 5):679–688. https://doi.org/10.1107/s0907444901003122
50. Cooper DR, Boczek T, Grelewska K, Pinkowska M, Sikorska M, Zawadzki M, Derewenda Z (2007) Protein crystallization by surface entropy reduction: optimization of the SER strategy. Acta Crystallogr D Biol Crystallogr 63(Pt 5):636–645. https://doi.org/10.1107/S0907444907010931
51. Derewenda ZS, Godzik A (2017) The “Sticky Patch” model of crystallization and modification of proteins for enhanced crystallizability. Methods Mol Biol 1607:77–115. https://doi.org/10.1007/978-1-4939-7000-1_4
52. Gao X, Bain K, Bonanno JB et al (2005) High-throughput limited proteolysis/mass spectrometry for protein domain elucidation. J Struct Funct Genomics 6(2–3):129–134. https://doi.org/10.1007/s10969-005-1918-5
53. Elez K, Bonvin A, Vangone A (2018) Distinguishing crystallographic from biological interfaces in protein complexes: role of intermolecular contacts and energetics for classification. BMC Bioinform 19(Suppl 15):438. https://doi.org/10.1186/s12859-018-2414-9
54. Luo J, Liu Z, Guo Y, Li M (2015) A structural dissection of large protein-protein crystal packing contacts. Sci Rep 5:14214. https://doi.org/10.1038/srep14214
55. Malawski GA, Hillig RC, Monteclaro F et al (2006) Identifying protein construct variants with increased crystallization propensity–a case study. Protein Sci 15(12):2718–2728. https://doi.org/10.1110/ps.062491906
56. Sheldon RA (2011) Characteristic features and biotechnological applications of cross-linked enzyme aggregates (CLEAs). Appl Microbiol Biotechnol 92(3):467–477. https://doi.org/10.1007/s00253-011-3554-2
57. Solarczek J, Klunemann T, Brandt F, Schrepfer P, Wolter M, Jacob CR, Blankenfeldt W, Schallmey A (2019) Position 123 of halohydrin dehalogenase HheG plays an important role in stability, activity, and enantioselectivity. Sci Rep 9(1):5106. https://doi.org/10.1038/s41598-019-41498-2

58. Migneault I, Dartiguenave C, Bertrand MJ, Waldron KC (2004) Glutaraldehyde: behavior in aqueous solution, reaction with proteins, and application to enzyme crosslinking. Biotechniques 37(5):790–796, 798–802. https://doi.org/10.2144/04375RV01
59. Jiménez-García B, Elez K, Koukos PI, Bonvin AM, Vangone A (2019) PRODIGY-crystal: a web-tool for classification of biological interfaces in protein complexes. Bioinformatics 35(22):4821–4823. https://doi.org/10.1093/bioinformatics/btz437
60. Abdul Wahab MKH, El-Enshasy HA, Bakar FDA, Murad AMA, Jahim JM, Illias RM (2019) Improvement of cross-linking and stability on cross-linked enzyme aggregate (CLEA)-xylanase by protein surface engineering. Process Biochem 86:40–49. https://doi.org/10.1016/j.procbio.2019.07.017
61. Guex N, Peitsch MC (1997) SWISS-MODEL and the Swiss-PdbViewer: an environment for comparative protein modeling. Electrophoresis 18(15):2714–2723. https://doi.org/10.1002/elps.1150181505
62. Krissinel E, Henrick K (2007) Inference of macromolecular assemblies from crystalline state. J Mol Biol 372(3):774–797. https://doi.org/10.1016/j.jmb.2007.05.022
63. Chayen NE (1998) Comparative studies of protein crystallization by vapour-diffusion and microbatch techniques. Acta Crystallogr D Biol Crystallogr 54(Pt 1):8–15. https://doi.org/10.1107/s0907444997005374
64. Avrami M (1939) Kinetics of phase change. I general theory. J Chem Phys 7(12):1103–1112. https://doi.org/10.1063/1.1750380
65. Khayati GR, Nourafkan E, Karimi G, Moradgholi J (2013) Synthesis of cuprous oxide nanoparticles by mechanochemical oxidation of copper in high planetary energy ball mill. Adv Powder Technol 24(1):301–305. https://doi.org/10.1016/j.apt.2012.07.006
66. Campos R, Narine SS, Marangoni AG (2002) Effect of cooling rate on the structure and mechanical properties of milk fat and lard. Food Res Int 35(10):971–981. https://doi.org/10.1016/S0963-9969(02)00159-X
67. Auclair JR, Boggio KJ, Petsko GA, Ringe D, Agar JN (2010) Strategies for stabilizing superoxide dismutase (SOD1), the protein destabilized in the most common form of familial amyotrophic lateral sclerosis. Proc Natl Acad Sci USA 107(50):21394–21399. https://doi.org/10.1073/pnas.1015463107
68. Lomant AJ, Fairbanks G (1976) Chemical probes of extended biological structures: synthesis and properties of the cleavable protein cross-linking reagent [35S]dithiobis(succinimidyl propionate). J Mol Biol 104(1):243–261. https://doi.org/10.1016/0022-2836(76)90011-5
69. Mattson G, Conklin E, Desai S, Nielander G, Savage MD, Morgensen S (1993) A practical approach to crosslinking. Mol Biol Rep 17(3):167–183. https://doi.org/10.1007/BF00986726
70. Sinz A (2003) Chemical cross-linking and mass spectrometry for mapping three-dimensional structures of proteins and protein complexes. J Mass Spectrom 38(12):1225–1237. https://doi.org/10.1002/jms.559
71. Staros JV (1982) N-hydroxysulfosuccinimide active esters: bis(N-hydroxysulfosuccinimide) esters of two dicarboxylic acids are hydrophilic, membrane-impermeant, protein cross-linkers. Biochemistry 21(17):3950–3955. https://doi.org/10.1021/bi00260a008
72. Kasvinsky PJ, Madsen NB (1976) Activity of glycogen phosphorylase in the crystalline state. J Biol Chem 251(21):6852–6859. https://doi.org/10.1016/S0021-9258(17)33022-3
73. Milčić N, Stepanić V, Crnolatac I, Findrik Blažević Z, Brkljača Z, Majerić Elenkov M (2022) Inhibitory effect of DMSO on halohydrin dehalogenase: experimental and computational insights into the influence of an organic co-solvent on the structural and catalytic properties of a biocatalyst. Chemistry 28(56):e202201923. https://doi.org/10.1002/chem.202201923
74. Eyring H, Stearn AE (1939) The application of the theory of absolute reaction rates to proteins. Chem Rev 24(2):253–270. https://doi.org/10.1021/cr60078a005
75. Sobolov SB, Leonida MD, Bartoszko-Malik A, Voivodov KI, McKinney F, Kim J, Fry AJ (1996) Cross-linked LDH crystals for lactate synthesis coupled to electroenzymatic regeneration of NADH. J Org Chem 61(6):2125–2128. https://doi.org/10.1021/jo951408v

76. Liao Q, Du X, Jiang W, Tong Y, Zhao Z, Fang R, Feng J, Tang L (2018) Cross-linked enzyme aggregates (CLEAs) of halohydrin dehalogenase from *Agrobacterium radiobacter* AD1: preparation, characterization and application as a biocatalyst. J Biotechnol 272–273:48–55. https://doi.org/10.1016/j.jbiotec.2017.12.014
77. Staar M, Schallmey A (2023) Performance of cross-linked enzyme crystals of engineered halohydrin dehalogenase HheG in different chemical reactor systems. Biotechnol Bioeng 120(11):3210–3223. https://doi.org/10.1002/bit.28528
78. D'Arcy A, Bergfors T, Cowan-Jacob SW, Marsh M (2014) Microseed matrix screening for optimization in protein crystallization: what have we learned? Acta Crystallogr F Struct Biol Commun. 70(Pt 9):1117–1126. https://doi.org/10.1107/S2053230X14015507
79. Xu Z, Cen YK, Zou SP, Xue YP, Zheng YG (2020) Recent advances in the improvement of enzyme thermostability by structure modification. Crit Rev Biotechnol 40(1):83–98. https://doi.org/10.1080/07388551.2019.1682963
80. Yan EK, Cao HL, Zhang CY, Lu QQ, Ye YJ, He J, Huang LJ, Yin DC (2015) Cross-linked protein crystals by glutaraldehyde and their applications. RSC Adv 5:26163–26174. https://doi.org/10.1039/C5RA01722J
81. Chruszcz M, Potrzebowski W, Zimmerman MD, Grabowski M, Zheng H, Lasota P, Minor W (2008) Analysis of solvent content and oligomeric states in protein crystals–does symmetry matter? Protein Sci 17(4):623–632. https://doi.org/10.1110/ps.073360508
82. Elgersma AV, Ataka M, Katsura T (1992) Kinetic studies on the growth of three crystal forms of lysozyme based on the measurement of protein and Cl-concentration changes. J Cryst Growth 122:31–40. https://doi.org/10.1016/0022-0248(92)90223-6
83. Diamond R (1974) Real-space refinement of the structure of hen egg-white lysozyme. J Mol Biol 82(3):371–391. https://doi.org/10.1016/0022-2836(74)90598-1
84. Koopmeiners J, Diederich C, Solarczek J, Voß H, Mayer J, Blankenfeldt W, Schallmey A (2017) HheG, a halohydrin dehalogenase with activity on cyclic epoxides. ACS Catal 7:6877–6886. https://doi.org/10.1021/acscatal.7b01854
85. Rupp B (2010) Biomolecular crystallography: principles, practice, and application to structural biology. Garland Science, New York, NY, p 809. https://doi.org/10.1201/9780429258756
86. DeSantis G, Jones JB (1999) Chemical modification of enzymes for enhanced functionality. Curr Opin Biotechnol 10(4):324–330. https://doi.org/10.1016/S0958-1669(99)80059-7
87. Jegan Roy J, Emilia Abraham T (2004) Strategies in making cross-linked enzyme crystals. Chem Rev 104(9):3705–3722. https://doi.org/10.1021/cr0204707
88. Seidel WW, Hahn F (2014) Werkstofftechnik: Werkstoffe-Eigenschaften-Prüfung-Anwendung; mit zahlr. Tabellen, Beispielen, Übungen und Testaufgaben. 10., aktualisierte Aufl. ed. Lernbücher der Technik. Hanser, München, p 473
89. Schilde C, Burmeister CF, Kwade A (2014) Measurement and simulation of micromechanical properties of nanostructured aggregates via nanoindentation and DEM-simulation. Powder Technol 259:1–13. https://doi.org/10.1016/j.powtec.2014.03.042
90. Whipple EB, Ruta M (1974) Structure of aqueous glutaraldehyde. J Org Chem 39(12):1666–1668. https://doi.org/10.1021/jo00925a015
91. Yan E-K, Lu Q-Q, Zhang C-Y et al (2016) Preparation of cross-linked hen-egg white lysozyme crystals free of cracks. Sci Rep 6:34770. https://doi.org/10.1038/srep34770
92. Chui WK, Wan LS (1997) Prolonged retention of cross-linked trypsin in calcium alginate microspheres. J Microencapsul 14(1):51–61. https://doi.org/10.3109/02652049709056467
93. Margolin AL (1996) Novel crystalline catalysts. Trends Biotechnol 14(7):223–230. https://doi.org/10.1016/0167-7799(96)10031-7
94. Salem M, Mauguen Y, Prangé T (2010) Revisiting glutaraldehyde cross-linking: the case of the Arg-Lys intermolecular doublet. Acta Crystallogr Sect F Struct Biol Cryst Commun 66(Pt 3):225–228. https://doi.org/10.1107/S1744309109054037
95. Yonath A, Sielecki A, Moult J, Podjarny A., Traub W (1977) Crystallographic studies of protein denaturation and renaturation. 1. Effects of denaturants on volume and X-ray pattern of cross-linked triclinic lysozyme crystals. Biochemistry 16(7):1413–1417. https://doi.org/10.1021/bi00626a027

96. Bonneté F, Ferté N, Astier JP, Veesler S (2004) Protein crystallization: Contribution of small angle X-ray scattering (SAXS). J Phys, IV-Proc 118:3–13. https://doi.org/10.1051/jp4:2004118001
97. Rumpf HCH (1970) Zur Theorie der Zugfestigkeit von Agglomeraten bei Kraftübertragung an Kontaktpunkten. Chem Ing Tech 42(8):538–540. https://doi.org/10.1002/cite.330420806
98. Schilde C, Kwade A (2012) Measurement of the micromechanical properties of nanostructured aggregates via nanoindentation. J Mater Res 27(4):672–684. https://doi.org/10.1557/jmr.2011.440
99. Schilde C, Westphal B, Kwade A (2012) Effect of the primary particle morphology on the micromechanical properties of nanostructured alumina agglomerates. J Nanopart Res 14(3):2344. https://doi.org/10.1007/s11051-012-0745-4

Multiscale Model-Based Investigation of Functional Macromolecular Agglomerates for Biotechnological Applications

Philipp Nicolas Depta, Maksym Dosta, and Stefan Heinrich

Abstract The formation of macromolecular agglomerates and ordered assemblies has diverse applications in both nature and technology. Especially biotechnology often relies heavily on (semi-)ordered macromolecular assemblies to enable specific functions. Examples are enzymatic biocatalysis enhanced through metabolic channeling as a result of structural organization of enzymes and virus-like particles (VLPs) as self-assembled transport containers for genetic materials or pharmaceuticals. In order to better understand such systems, we developed a novel multiscale method using data-driven parameterization approaches to simulate formation and properties of large macromolecular assemblies on scales of micro-meters and milli-seconds. Thus, enabling the investigation well beyond traditional methods such as molecular dynamics (MD). The method is termed the molecular discrete element method (MDEM). As model systems the calcium mediated gelation of the biopolymer alginate, self-assembly of the hepatitis B core antigen (HBcAg) into VLPs, and semi-structured agglomerate formation of the pyruvate dehydrogenase complex (PDC) were investigated at various process conditions. Experimental validation was performed using both available literature data and experimental results of collaborators showing good agreement with model predictions.

P. N. Depta · M. Dosta · S. Heinrich (✉)
Institute of Solids Process Engineering and Particle Technology, Hamburg University of Technology, Hamburg, Germany
e-mail: stefan.heinrich@tuhh.de

P. N. Depta
e-mail: nicolas.depta@tuhh.de

M. Dosta
e-mail: maksym.dosta@boehringer-ingelheim.com

M. Dosta
Boehringer Ingelheim Pharma GmbH & Co Kg., Biberach an der Riss, Germany

A. Kwade and I. Kampen (eds.), *Dispersity, Structure and Phase Changes of Proteins and Bio Agglomerates in Biotechnological Processes*,
https://doi.org/10.1007/978-3-031-63164-1_2

1 Introduction

Nature provides a wealth of systems reliant on macromolecular structures to enable specific functions from material science to biology. One example is the enhancement of cascade bioreactions through multi-enzymatic clusters and subsequent metabolic channeling [1]. However, gaining the detailed understanding inherently necessary for technological usage and adaptation is tremendously challenging. Reason for this is that especially biological applications often rely on semi-structured assemblies, where possibly multiple molecular species (proteins) form dynamic agglomerates dependent on their chemical composition, three dimensional structure, non-covalent (self-)interaction, and process environment. As a result, such systems are inherently multiscale in length and time with atomistic phenomena governing structural formation and thus function on the micro-meter and milli-second scale and longer [2, 3].

This multiscale nature in combination with partially disordered structural components are the main challenges for both experimental and numerical investigation. While some experimental methods, such as small angle scattering (SAXS) [4] and biomolecular mass spectrometry [5], are capable of resolving fast dynamics, they are insufficient concerning inherently disordered assemblies in dynamic equilibrium (particularly present in enzymatic biocatalysis). Similarly, higher time-scale methods are limited resolving kinetics and thus rather allow steady-state insights, such as crystallography studies, electron microscopy, dynamic light scattering (DLS), and biolayer interferometry (BLI) [6].

With respect to numerical investigations, methods on the atomistic scale, such as all-atom molecular dynamics (AA-MD), are limited to small scales in length and time due to their computational complexity and thus unable to address meso-scale phenomena such as macromolecular assemblies. In order to alleviate the scale limitations, a variety of coarse-graining (CG) methods (combine multiple atoms into one bead) [7] have been proposed around Langevin dynamics (LD) [8, 9], Brownian dynamics (BD) [10], and dissipative particle dynamics (DPD) [11] – including various solvent representations [12] and force-fields (e.g. Martini [13] and SIRAH [14]). However, reaching the necessary scales of micro-meters and milli-seconds with sufficient detail and in an easily transferable bottom-up parametrization approach remains challenging.

In order to advance numeric capabilities and investigate formation and properties of macromolecular assemblies, this project developed a novel multiscale modeling methodology using a hybrid parametrization from MD and optional empirical data. Each macromolecule is abstracted as an anisotropic object with a position and orientation. Data-driven models are employed to describe interaction between abstracted molecules and with the solvent environment. These models are derived from MD using a generic supervised learning approach and maintain the necessarily detail for describing molecular interaction in their six dimensional structure. At the same time, the chosen abstraction enables investigation of the micro meter and milli second scale thus well beyond traditional coarse graining methods. Additionally, the methodology allows variation of many process conditions without re parameterization such as temperature, viscosity, and fluid flow.

The methodology is tested on three model systems with increasing complexity: The first model system is the calcium mediated gelation of the biopolymer alginate [15]. During gelation, polymer fibers are cross linked with an intermediate calcium ion leading to mechanical stability of the hydrogel. For this system, the generic model is specialized and parametrized based on available literature data as an initial testing step during method development. The second model system is the self-assembly of the hepatitis B core antigen (HBcAg) into virus-like particles (VLP) [16] using the fully generic modeling approach. VLPs act as transport containers for the genetic material of the virus and are therefore highly interesting to be understood or re used as transport containers for pharmaceutical purposes. The third model system is the pyruvate dehydrogenase complex (PDC) linking the anaerobic glycolytic energy pathway to the aerobic tricarboxylic acid cycle in mammals [17]. PDC is a multi enzymatic complex of four enzymes (E1, E2, E3, E3BP) relying on structural formation to enable its catalytic activity through phenomena such as metabolic channeling and is therefore highly interesting for biotechnological applications. Additionally, we are currently investigating the effects of structural formation on function (i.e. catalytic activity) and will present some preliminary results. Overall, the three model systems cover a wide range of phenomena and challenges in the field of macromolecular structural formation and are hence ideal for testing purposes. Validation of results is performed with experimental data from available literature and collaborators.

This report will present some of the essential developments and findings from previous publications [18–22], as well as introduce some novel results with regard to on-going research of the PDC system.

2 Methods

2.1 Modeling Framework

The proposed modeling framework relies conceptually on transferring the complexity of many atomistic interactions into data-driven models for macromolecules to maintain sufficient detail, while enabling a high abstraction level. The overall framework is displayed in Fig. 1 and subsequently summarized. Detailed information on methodology can be found in Refs. [18, 21] and high-performance computing (HPC) implementation aspects in Refs. [20, 22].

At the foundation of macromolecular assemblies are the individual molecules themselves. For most biological molecules, their composition and structural information is available from the protein data bank (PDB) or experimental procedures are well established. This detailed information is necessary for parametrization of macromolecular behavior in the context of assembly modeling, but subsequently of little interest for the assembly process itself. As a result, the proposed framework abstracts each macromolecule as an anisotropic object with a position, orientation, and inertia properties. The complex interaction between these abstracted molecules and with the solvent environment is subsequently described by up to three model components

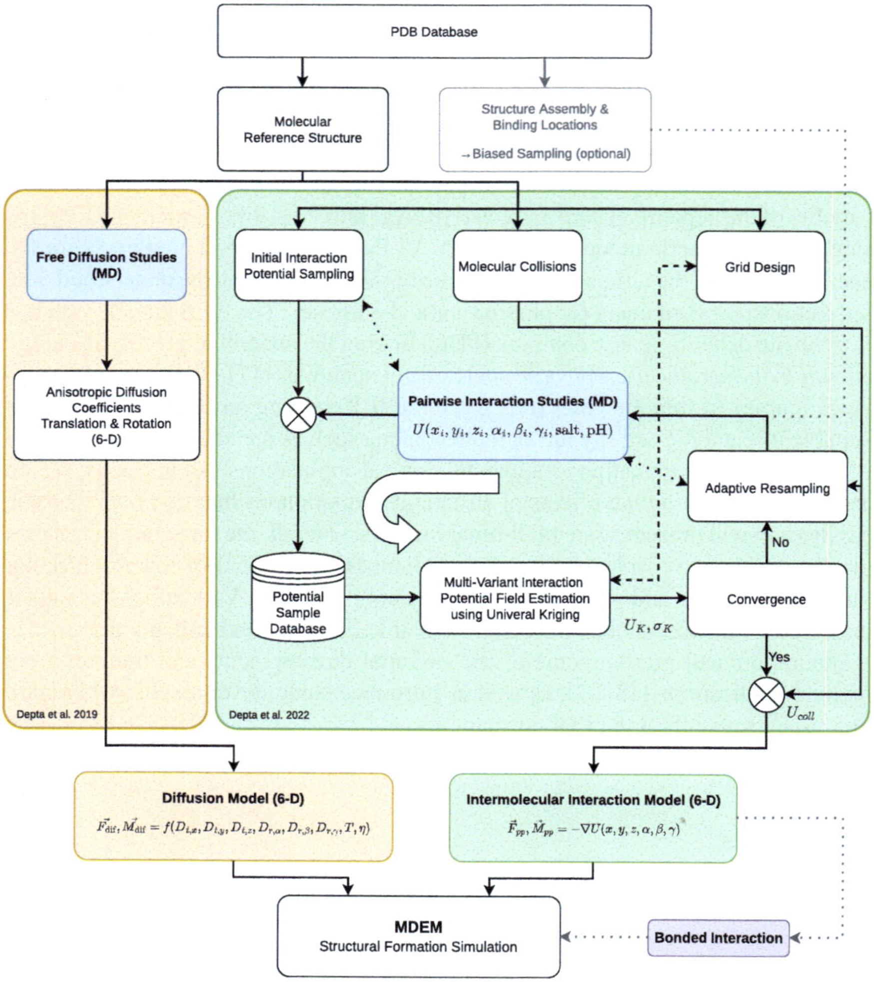

Fig. 1 Overview of MDEM modeling framework (adapted from Ref. [21] under CC-BY 4.0 license). Blue marks MD parametrization

parameterized through molecular dynamics (MD) using the aforementioned PDB reference structure.

The first component is the diffusion model (marked orange in Fig. 1), which describes interaction of macromolecules with the solvent environment. Its objective is a reproduction of the same anisotropic translatory and rotatory behavior as the atomistic molecule, as well as an enforcement of correct thermodynamics, i.e., a Maxwell-Boltzmann distribution of velocities for constant temperature. The developed model [18] achieves this through a Langevin dynamics (LD) based approach and MD parametrization. Temperature and solvent viscosity can be modified without re-parametrization as long as no significant conformational changes occur.

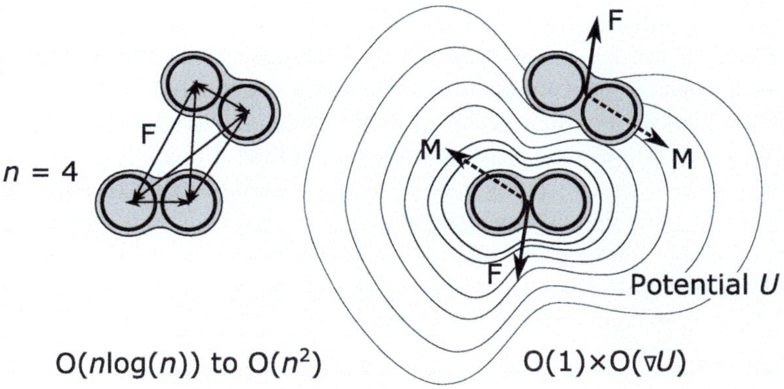

Fig. 2 Effect of abstraction as anisotropic bead on computational complexity (reprinted from Ref. [21] under CC-BY 4.0 license). $n \approx 10^5$ for one pairwise HBcAg interaction including solvent

The second component is the intermolecular interaction model (marked green in Fig. 1), which describes pairwise interaction between macromolecules and is therefore most crucial for modeling molecular assembly. The model provides forces and torques acting between molecules as a function of relative position and orientation. For this, the interaction potentials U between molecule types are determined using a Universal Kriging approach with MD sampling in combination with optional empirical data for hybrid parametrization strategies [21]. The interaction potentials are iteratively improved using a supervised learning strategy and subsequently saved in 6D data fields. This approach allows capturing high levels of detail (e.g. flexibility in molecular structure during binding) beyond functional descriptors, as well as constant runtime in comparison to e.g. a neural network approach. As shown in Fig. 2, this approach reduces computational complexity drastically by moving informational content from many 1D atom based contributions to one gradient operation per pairwise interaction.

The third component is the bonded interaction model (marked purple in Fig. 1), which can *optionally* describe strong interaction between abstracted molecules, i.e. interactions that do not break over the course of a simulation – both non-covalent as well as covalent. As such, it can decouple these interactions from the intermolecular interaction model in a numerically favorable environment concerning stability. This component was only employed for modeling the stretching and bending of alginate polymer fibers using a linear elastic connection. Details for this specialized model will be provided in Sect. 2.2.

Functional Modeling

Subsequent to the modeling of structural formation, functional modeling is of great interest. Concerning the alginate and HBcAg model systems, function mostly relates to mechanical stability and meso-scale properties of structures dependent on process conditions. While the latter was directly studied using post-processing, mechanical stability can be investigated through a modification of system boundary conditions and as such is highly dependent on the research question (e.g. shear (flow), compression). Thus, such investigations were left to subsequent works. Instead, focus was placed on the PDC model system for which function relates to the catalytic activity of the reaction kinetics. For this, we are currently modeling the multi-enzymatic reaction at the level of individual steps as a function of time. However, as this research is still ongoing, we are only going to be present preliminary results on the proximity of reactants as a function of structure and time. This information is ultimately what governs the reaction kinetics and directly correlates to the relationship between structure and function. Subsequently, time development and state changes of reactants are modeled during the remainder of the project.

2.2 Model System Abstractions

2.2.1 Calcium Mediated Alginate Gelation

For investigation of the calcium mediated alginate gelation a specialized version of the model framework was employed with dimer combinations of alginic acid as the unit structures. The polymer chains were represented by bonded connections of the dimer units as shown in Fig. 3. A statistical ion binding and unbinding model was formulated for pairwise interactions of the units. All models were parametrized purely using literature data and theoretical considerations. Studies were performed at 300 K with various ion concentrations, polymer concentrations, compositions, and molecular weights. More details can be found in Ref. [19].

2.2.2 Hepatitis B Virus-Like Particles

For investigation of the hepatitis B virus-like particles the full generic model framework was employed using HBcAg dimers as the unit structure. Studies were

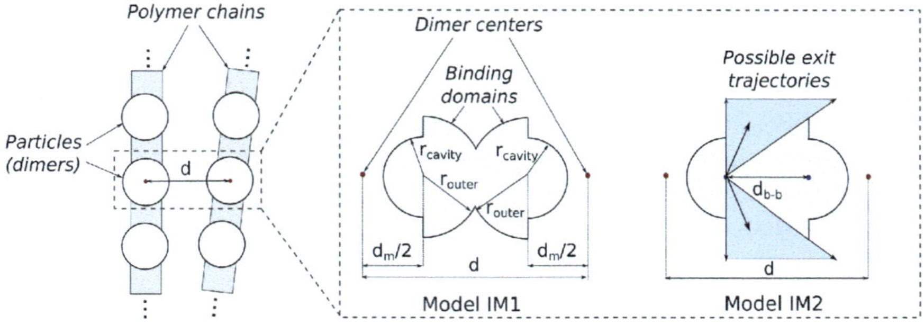

Fig. 3 Schematic of ion models for calcium mediated gelation of alginate (reprinted from Ref. [19] under CC-BY 4.0 license)

performed at 293 K and 150 mM sodium chloride with various protein concentrations. More details can be found in Ref. [21].

2.2.3 Pyruvate Dehydrogenase Complex

For investigation of the pyruvate dehydrogenase complex the full generic model framework was employed. However, as the computational budget of 19.3 million CPU core-h was found to be insufficient to parametrize all interaction permutations in detail, resources were focused on crucial pairwise interactions as discussed in Ref. [22]: E2-E2, E2-E1, and E3BP-E3 were fully parametrized; E3BP-E3BP and E3BP-E2 were assumed equivalent to E2-E2 due to their similar structure; E2-E3 was considered as a weaker form of E2-E1; and all other permutations were assumed repulsive only. Studies were performed at 300 K with various concentrations and stoichiometry of the four enzymes. No additional salts were used. More details can be found in Ref. [22].

2.3 Implementation

The modeling framework was implemented in custom code using C++ and CUDA v11.2 [23] for CPU and GPU platforms building fully upon open-source software: The macromolecular assembly and functional components were implemented as modules for the open-source DEM code MUSEN [24] and specifically tailored for GPU usage [21]. The intermolecular model parameterization using Universal Kriging was implemented custom for a high-performance computing setting at the High-Performance Computing Center Stuttgart (HLRS) using hybrid MPI/OpenMP parallelization [20, 22]. For the underlying MD model the open-source code Gromacs version 2020.1 [25, 26] was used. Details are provided in the respective publications [18–22].

3 Results

3.1 Calcium Mediated Alginate Gelation

As a first model system, the polymer cross-linking of calcium mediated alginate gelation was investigated using the specialized model outlined in Sect. 2.2, which is based purely on literature and theory for parametrization. As it can be seen in the visualization of Fig. 4, the model permits detailed investigation of polymer structure and subsequent functional properties (e.g. pore size, specific surface area, mechanical stability) resulting from gel cross-linking. These properties are expressed in a variety of numerical measures of the gel including: normalized global interaction

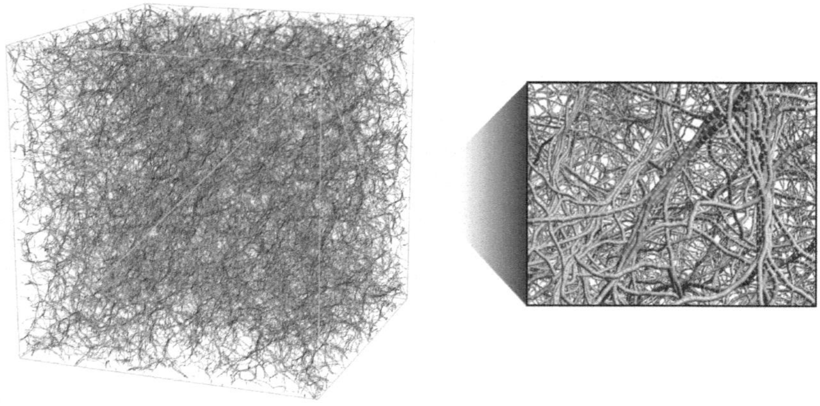

Fig. 4 Exemplary visualization of alginate system after gelation with calcium using developed methodology (0.5 wt. %, 200 kDa, f = 0.5, high G composition, IM2 in $0.125 \mu m^3$ box (500nm edge); reprinted from Ref. [19] under CC-BY 4.0 license)

potential per unit particle/dimer ($U_{glob.int.}/N_{part}$); uptake of calcium ions per monomer of alginic acid (f_{used}); coordination number of unit dimer (average $\overline{k}_{0.9nm}$ and histogram fraction ϕ); normalized contact number of guluronic acid (G) dimers (GG), i.e. GG-GG, interactions essential for cross-linking ($\xi_{GG-GG-attract}$, referred to 'egg-box' cross-linking in literature [27]); volume fractions of particle concentrations indicating pores and fiber bundles (ϕ_V); pore size distributions (PSD) using chord length; as well as bundle thickness – both of the latter can be used for direct comparison with experimental data. In the following, the effects of polymer concentration (c), polymer composition (high G content (H) with 63 % or low G content (L) with 33 %), molecular weight (M_w), and ion concentration (f) will be summarized. For more detail including experimental validation using e.g. solid-state nuclear magnetic resonance (NMR), NMR cryoporometry, small-angle neutron scattering (SANS), and N_2 adsorption-desorption analysis please see Ref. [19].

As it can be observed in Fig. 5, all gel properties are strongly dependent on ion concentration f and reach their asymptotic values dependent on polymer concentration, composition, and molecular weight. While a 0.5 wt. % and 200 kDa polymer solution with high G content (H), further considered the *base case* (blue), requires $f \approx 0.35$ ions for saturation (supported by uptake $f_{used} = 0.34 - 0.38$), a low G content (L, green) reduces this to $f \approx 0.15 - 0.2$, low molecular weight of 100 kDa (red) increases this to $f \approx 0.5$, and high concentration of 1.0 wt. % with intermediate molecular weight of 155 kDa (purple) increases this slightly as well. Furthermore, the ion saturation concentration strongly correlates with the ion uptake (Fig. 5b) and as such directly influences the cross-linking degree.

In order to quantify this cross-linking degree, the coordination number (Fig. 5c, e, g) and contact number (Fig. 5d) are the primary measures. As it can be seen in Fig. 5c, a low G content (L) leads to a significantly reduced coordination number (both average and with regard to distribution in Fig. 5e and g), while the normalized contact

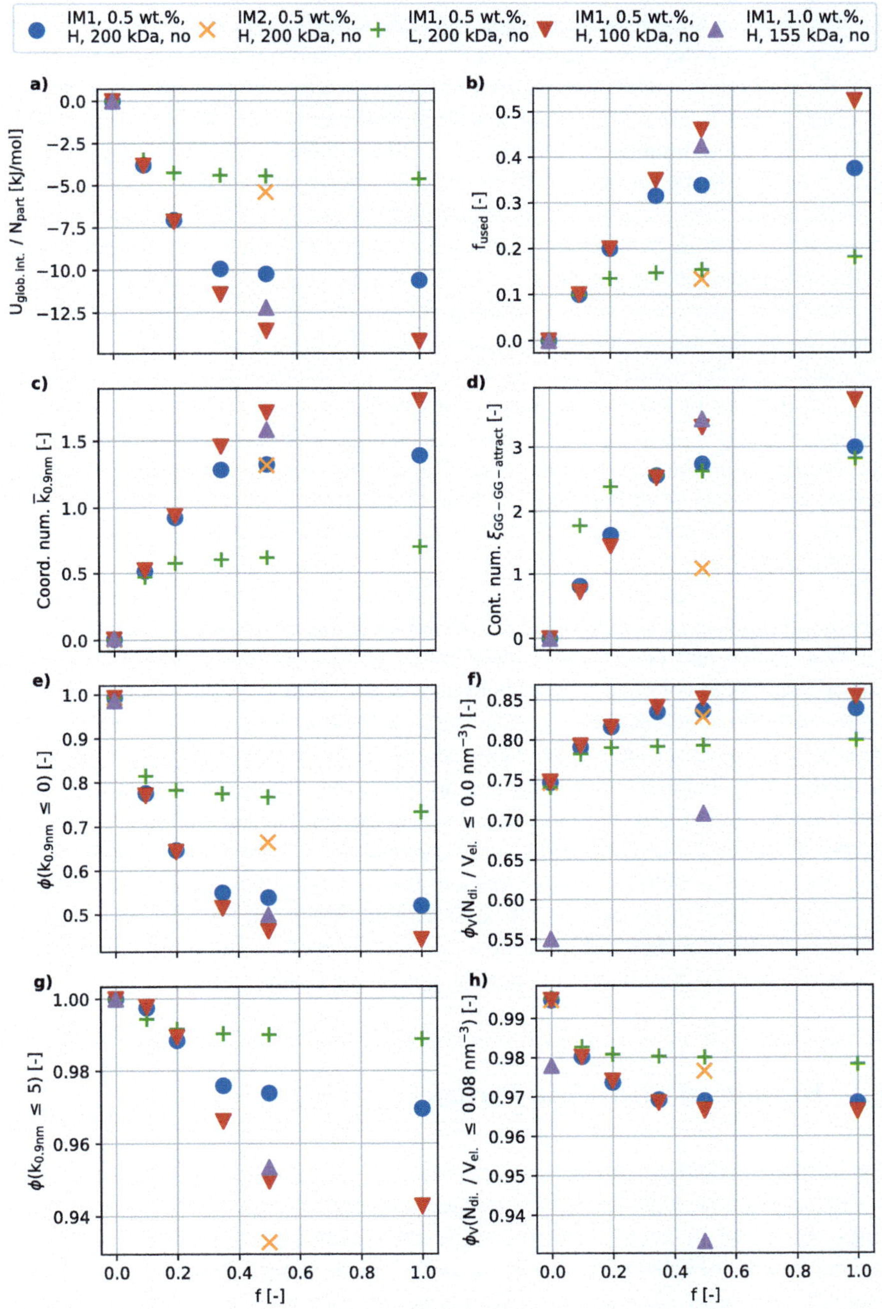

Fig. 5 Overview of simulation results for various parameters of gelation process (reprinted from Ref. [19] under CC-BY 4.0 license)

number between GG-GG units remains asymptotically largely unchanged (Fig. 5d). Consequently, the 'egg-box' formation [27] *per* GG unit remains unchanged, while only the total number decreases with lower G composition and thus having an impact on overall cross-linking. In contrast, lower molecular weight of 100 kDa leads to both an increase in coordination number (Fig. 5c) and normalized GG-GG contact number (Fig. 5d). This increase in normalized cross-linking is attributed to the increase in mobility of the fibers resulting from shorter polymer length. The high concentration of 1.0 wt. % with intermediate molecular weight of 155 kDa has similar properties as the low molecular weight case.

Similarly to the degree of cross-linking, the pore size and volume have significant influence on the macroscopic properties of the hydrogel. With regard to the pore volume, it can be seen that the low G content (L) leads to a lower pore volume (Fig. 5f) in favor of volume fractions of intermediate density (Fig. 5h). In contrast, lower molecular weight of 100 kDa shows a slightly increased pore volume (Fig. 5f) and decreased intermediate density (Fig. 5h), thus favoring both pore size and high-density regions for fiber bundles. The high concentration of 1.0 wt. % shows a yet further decreased pore volume (Fig. 5f) and decreased intermediate concentration volume (Fig. 5h).

Overall, modeling data has shown that a reduced G content causes smaller pore volume, thinner fiber bundles, and decreased cross-linking degree including lower calcium uptake and coordination number. In contrast, reduced molecular weight caused larger pore volume, thicker fiber bundles, and slightly increased cross-linking degree including higher calcium uptake and coordination number. Lastly, an increase in polymer concentration primarily caused a decrease in pore volume and increase in high-density bundle volume, but can at this point not be fully separated from molecular weight effects. For more detail including simulated annealing procedures, effects of ion binding and unbinding, and experimental validation please see Ref. [19]. Note that these modeled systems represent the ideally homogeneous case without any defects caused by impurities or process anomalies (e.g. shear). Such effects could additionally be modeled and might lead to further levels of the pore network (i.e. meso-pores) and consequently impact e.g. mechanical stability.

3.2 Hepatitis B Virus-Like Particles

For the modeling of HBcAg self-assembly into virus-like particles, the diffusion coefficients and intermolecular interaction potential were determined using the methodology described in Sect. 2 and the respective publications [18, 20, 21]. While parametrization of anisotropic diffusion coefficients is computationally inexpensive, determination of the 6D intermolecular interaction potential was found to be computationally challenging - even in the HPC environment at HLRS. In total, 375'000 MD simulations (5.85 million CPU core-h) were performed for iterative sampling and subsequently enhanced through empirical data from crystallography. For details see Ref. [20]. The final 6D intermolecular interaction potential between HBcAg dimer

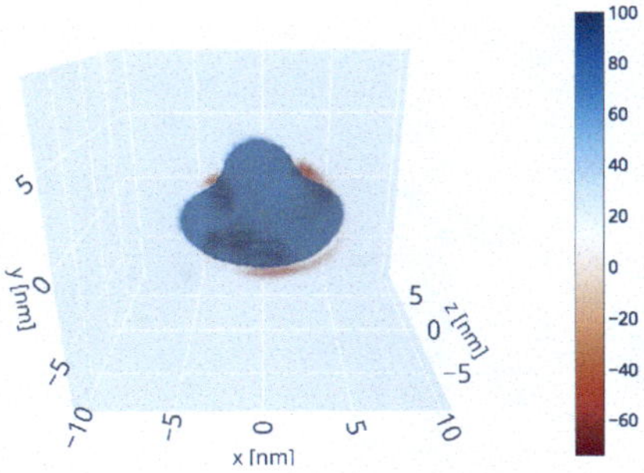

(a) Mean projection into cartesian space.

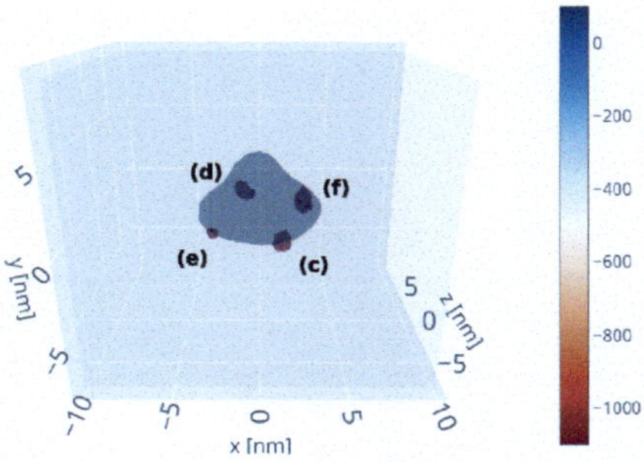

(b) Minimum projection into cartesian space.

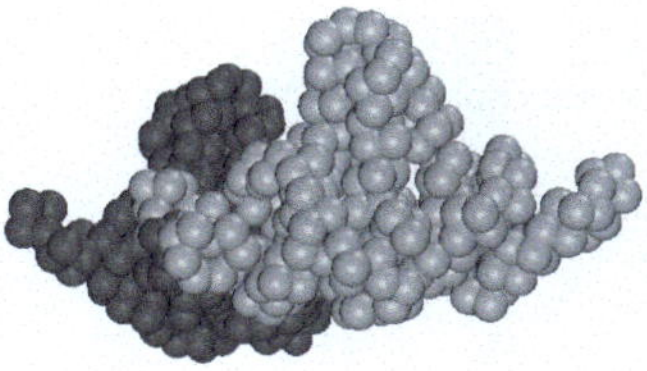

(c) HBcAg–HBcAg interaction (c) in (b).

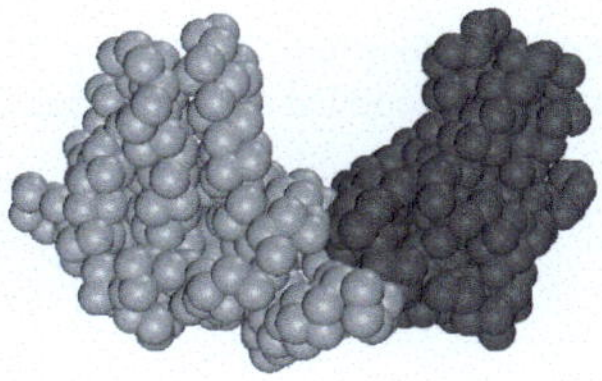

(d) HBcAg–HBcAg interaction (d) in (b).

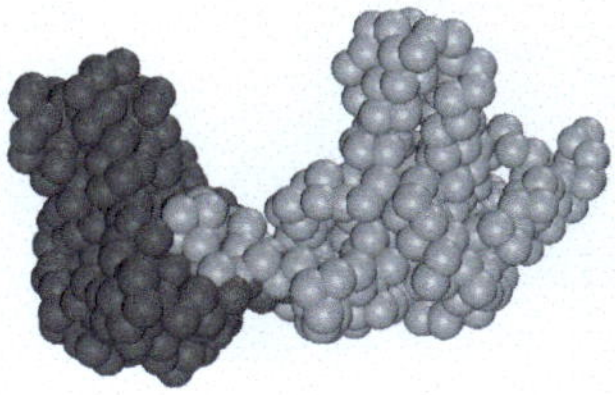

(e) HBcAg–HBcAg interaction (e) in (b).

(f) HBcAg–HBcAg interaction (f) in (b).

Fig. 6 Visualizations of six dimensional HBcAg–HBcAg interaction potential by projections into cartesian 3D space (**a**, **b**, units kJ/mol) and back-bone position of potential minima **c**–**f**. Reprinted from Ref. [21] under CC-BY 4.0 license

units is visualized in Fig. 6. As it can be seen (Fig. 6a and b), the potential has a complex shape with four distinct potential minima (binding sites) corresponding to the respective binding modes (Fig. 6c–f) of known VLP assemblies. Additionally, a slight potential barrier of 0–5 kJ/mol exists at a minimum distance of approximately 1.5 nm, which is attributed to hindering over-growing of capsids.

Based on the parameterized model, self-assembly was investigated for dimer concentrations of 5, 10, 50, and 100 µM [20, 21]. In the following, a summary of the results will be provided with focus on the concentration of 10 µM, which is visualized in Fig. 7. As it can be seen, HBcAg dimers self-assemble from their random original state into capsids of approximately 90–120 dimers in size. The observed capsids are highly structured and agree well with expectations of icosahedral symmetry:

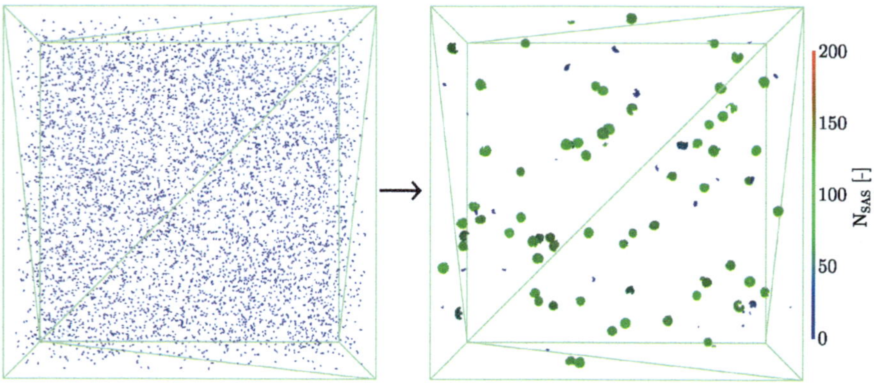

(a) Visualization of before (left) and after (right) self-assembly using back-bone carbon structures. Color indicates self-assembled structure sizes by number of dimer particles (N_{SAS}).

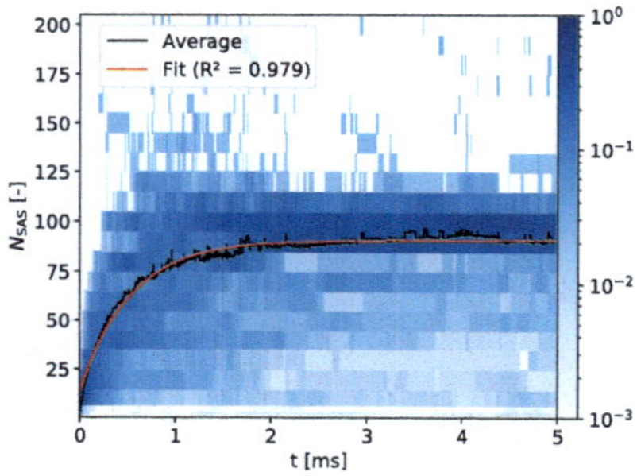

(b) Histogram of self-assembled structures by numbered size over time.

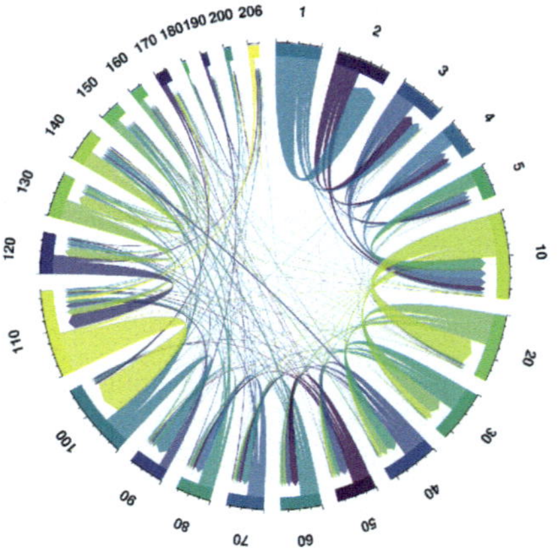

(c) Assembly pathway by net transitions between size classes.

Fig. 7 VLP self-assembly of HBcAg dimers at 10 μM concentration during 5 ms in 1 μm^3 box (1 μm edge) (reprinted from Ref. [21] under CC-BY 4.0 license)

HBcAg is known to form two types of capsids with icosahedral symmetries of T = 3 (90 copies) and T = 4 (120 copies, larger fraction) [28–31]. This agreement is also quantitatively observed with on average 3.5 structured contacts per dimer, which is near the theoretical value of 4.0 for the ideal T = 4 capsid.

With regard to the distribution of capsid size over time (Fig. 7b), it can be observed that assembly occurs following a trend with exponential decay ($R^2 = 0.979$) and forming a primary distribution in the region of 90–120 dimers. After initial formation of this population, kinetics significantly slow down and are highly concentration dependent (exponential time constants reducing from 2.9 ms for 5 μM to 0.2 ms for 100 μM). When looking at the final distribution by numbered size (Fig. 7b at 5 ms) and additionally using the radius of gyration as a criteria [21], it can be found that depending on concentration between 11.5–24.0 % are T = 3 capsids, while the remainder are pre-stages to T = 4 capsids as expected [28–31]. These pre-stages are missing individual dimers or small segments for completion of capsids, which is attributed to the low availability of such small structure at this stage of the assembly process and expected to be alleviated with longer simulation times. Furthermore, availability of small structures after 5 ms was found to be concentration dependent with higher fractions for low concentrations of 5 and 10 μM, which is attributed to higher mean distances between dimers (Fig. 7a right).

This pathway of structural assembly is visualized in more detail using a chord diagram in Fig. 7c. The visualization shows normalized net transitions between all sizes classes and indicates a hierarchical stage-wise buildup: Initially, individual dimers form small structures (<10) with decreasing fraction. In the following sizes up to 50–90 dimers, assembly takes place through addition of similarly sized structures. Subsequent, the range of 90–120 dimers is characterized by frequent transitions on the pathway to equilibrium. The aforementioned pre-stages of the T = 4 capsid are in the process of finalization and hole closure is a question of probability with regard to availability of correctly sized components and time. Equilibration pathways of this semi-stable region are strongly dependent on dimer concentration with low concentrations (5 and 10 μM) rarely exceeding sizes of 140 dimers, while temporary formation of such sizes is very common for high concentrations (50 and 100 μM). However, it should be noted that many of such sizes are also mere contact of otherwise well-structured capsids. This is further supported by the distribution of structure lifetimes (i.e. before class transition) [21], which peaks with tenth of milli-seconds in the range of 70–120 dimers. For more details including concentrations of 5, 50, and 100 μM see Refs. [20, 21].

3.3 Pyruvate Dehydrogenase Complex

For modeling the self-assembly of PDC components (E1, E2, E3, E3BP) into hierarchically structured complexes, the diffusion coefficients and intermolecular interaction potential were determined using the methodology described in Sect. 2 and the respective publications [18, 21, 22]. Due to the computational complexity of determining the 6D intermolecular interaction potentials, the system was simplified as described in Sect. 2.2.3. In total, 350'000 MD simulations (13.25 million CPU core-h) were performed for iterative sampling and subsequently enhanced through empirical data. The resulting 6D intermolecular interaction potential for the pairwise interaction of E2 and E2 is visualized as an example in Fig. 8. As it can be seen (Fig. 8a and b), the potential has a complex shape with two main regions of potential minima (binding sites) corresponding to the respective binding modes (Fig. 8c and d). The binding mode of catalytic domain interaction (Fig. 8d) is responsible for assembly of the 60-mer core structure known to be crucial for PDC function. Note the significant size of the interaction region due to the varying relative configurations present inside the 60-mer core. In addition, the interaction of linker arms (Fig. 8c) enables possibly

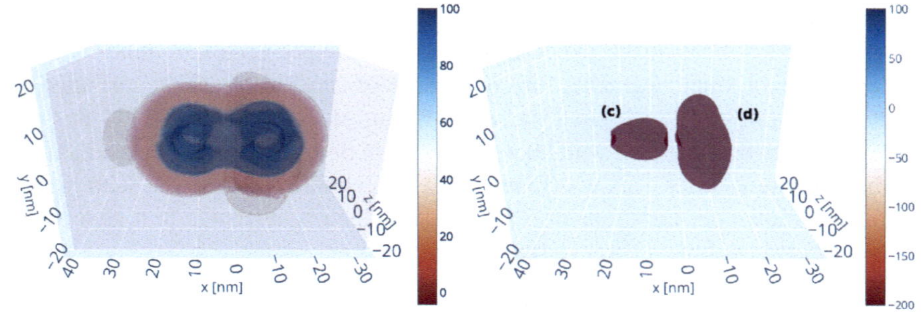

(a) Mean projection into cartesian space. (b) Minimum projection into cartesian space.

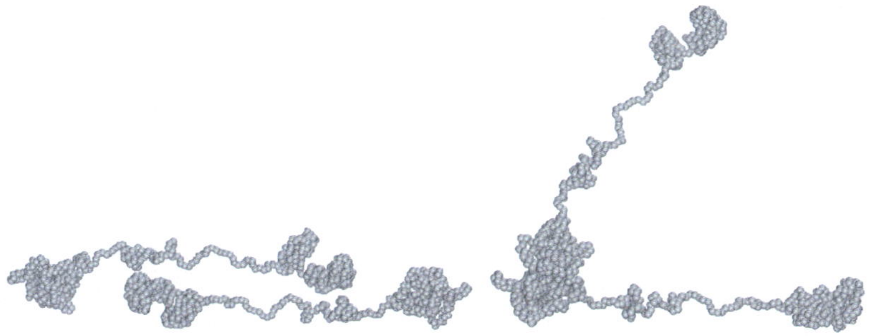

(c) E2-E2 arm interaction (marking c in b, exemplary position). (d) E2-E2 catalytic domain interaction (marking d in b, exemplary position).

Fig. 8 Visualizations of six dimensional E2-E2 interaction potential by projections into cartesian 3D space (**a**, **b**, units kJ/mol) and back-bone position of potential minima (**c**, **d**). Adapted with permission from Springer Nature [22]

assembly on hierarchical scales beyond the 60-mer core, which has been observed using dynamic light scattering (DLS) [32]. Similarly, in Fig. 9 the interaction potential of E1 with E2 is visualized including rotationally symmetric binding of E1. For details on PDC interaction potentials and their sampling see Ref. [22].

Based on the parameterized model, self-assembly was investigated for varying concentrations and stoichiometries. Note that the PDC system is still under investigation and results are hence preliminary.[1] In the following, results of the full PDC system with a concentration of 1 $mgmL^{-1}$ and stoichiometry of $40 \times$ E2 + $20 \times$ E3BP + $30 \times$ E1 + $10 \times$ E3 will be provided, which are visualized in Fig. 10. As it can be seen, the PDC components self-assembly from their random original state

[1] Note added in proofs: Regarding the preliminary results of the PDC system shown in this report, subsequent investigations are discussed in "Physics-Based and Data-Driven Multiscale Modeling of the Structural Formation in Macromolecular Systems" by P.N. Depta. These updated results show a good agreement with experimental data, while finding an additional structural formation beyond the 60-mer through linker arm interaction to be unlikely.

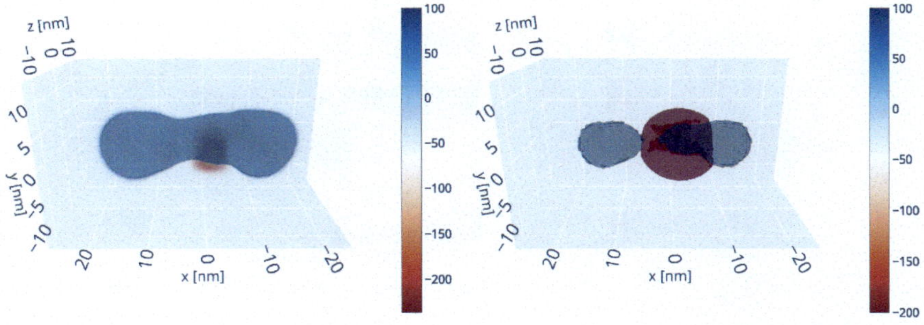

(a) Mean projection into cartesian space. (b) Minimum projection into cartesian space.

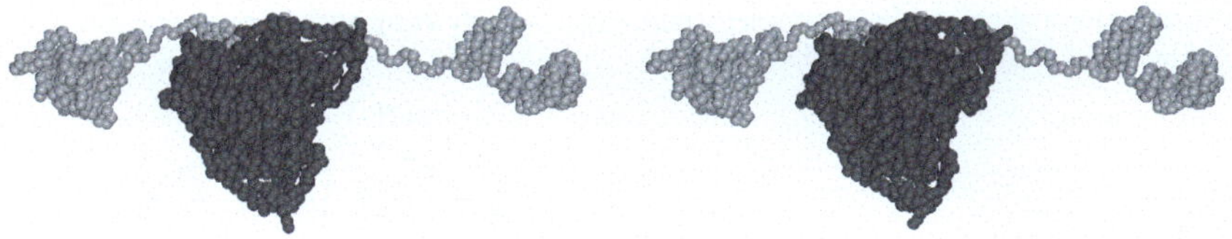

(c) E2-E1 interaction (exemplary position 1). (d) E2-E1 interaction (exemplary position 2).

Fig. 9 Visualizations of six dimensional E2-E1 interaction potential by projections into cartesian 3-D space (**a**, **b**, units kJ/mol) and back-bone position of potential minima (**c**, **d**). Note the rotational symmetry of E1 binding

into semi-ordered structures of predominantly 75–175 enzymes in size (125 average, monomers excluded) with individual structures up to 310 enzymes. This self-assembly process follows a trend with exponential decay (Fig. 10b) and after initial formation of agglomerates kinetics significantly slow down towards the end of the simulation. Thus, possibly leaving further assembly processes on longer time scales. The formed structures posses a spatial extent of back-bone atoms averaged over all orientations (162 direction discretizations) between 60 and 90 nm with maximum extents up to 140 nm, which is in reasonable agreement with DLS data predicting a radius of hydration (i.e. half the spatial extent) between 40.8 nm [33] and 75.2 nm [32] for the most similar PDC system of E2 and E3BP. Note that typically a water layer associated with the protein structure is included in the radius of hydration.

Additionally, it can be observed that the formed assemblies are highly structured (see close-up in Fig. 11) with smaller structures closely resembling the expected 60-mer structure formed by E2 and E3BP (particularly structure in bottom center of Fig. 11). For this structuredness, E2 and E3BP are the crucial components to which E1 and E3 bind in (dynamic) equilibrium. In Fig. 10d, it can be observed that relative to the 60-mer stoichiometry approximately 7 $\times$ E1 and 4 $\times$ E3 bind, which is unexpectedly low. However, it is known that for the function of PDC the reaction of E1 with pyruvate is the limiting step [34]. Consequently, a dynamic

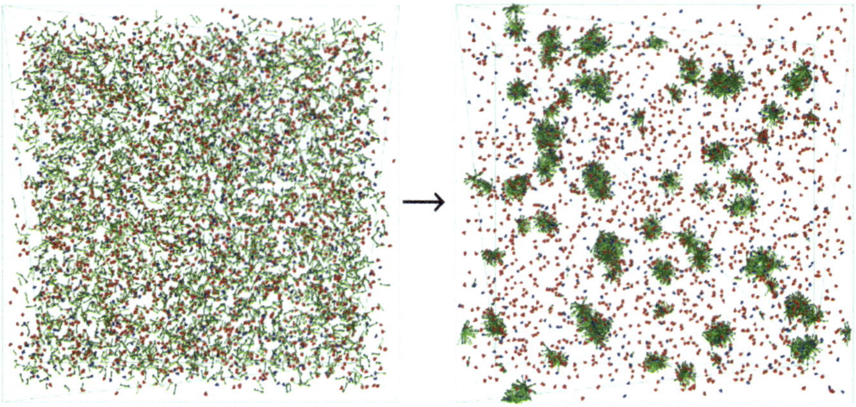

(a) Visualization of before (left) and after (right) self-assembly using back-bone carbon structures. Color indicates enzyme type: E1, E3, E2, E3BP.

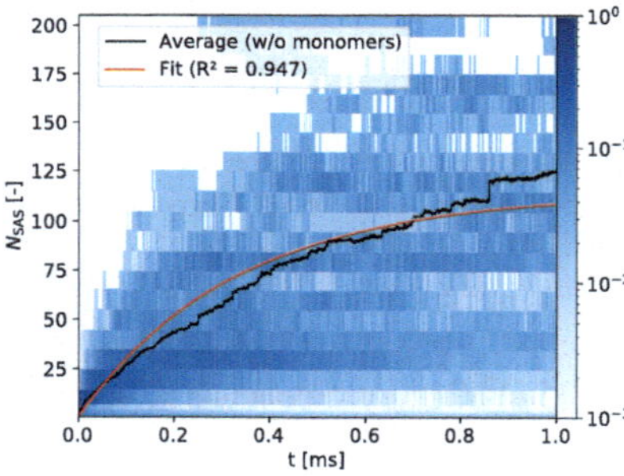

(b) Histogram of self-assembled structures by numbered size over time.

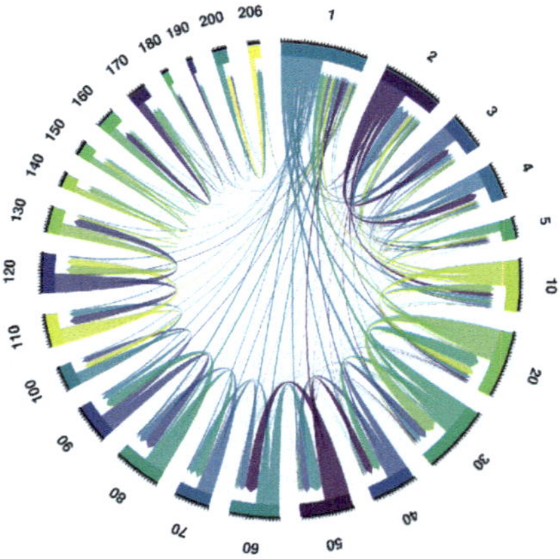

(c) Assembly pathway by total transitions between size classes.

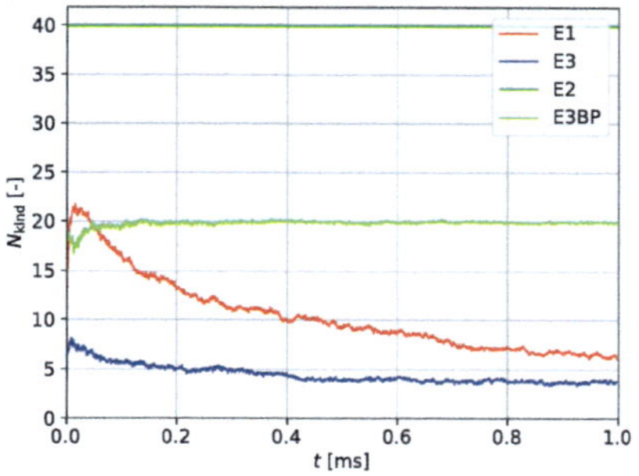

(d) Stoichiometry of structures with at least 5 enzymes relative to E2 content in 60-mer.

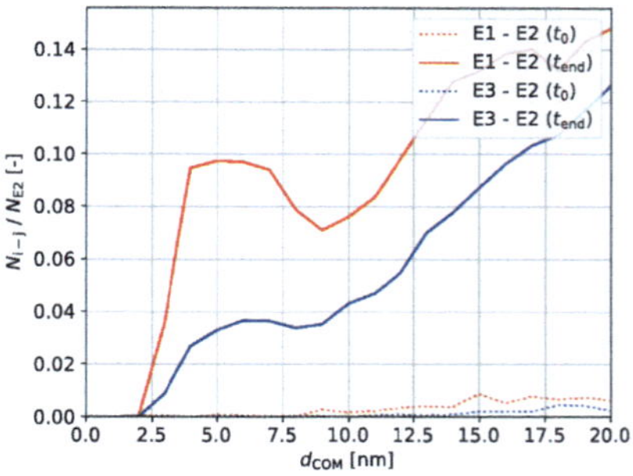

(e) Reactant distance distribution at beginning and end of self-assembly.

Fig. 10 PDC self-assembly for a concentration of 1 $mgmL^{-1}$ with stoichiometry of $40 \times$ E2 + $20 \times$ E3BP + $30 \times$ E1 + $10 \times$ E3 during 1 ms in $1 \mu m^3$ box (1 μm edge)

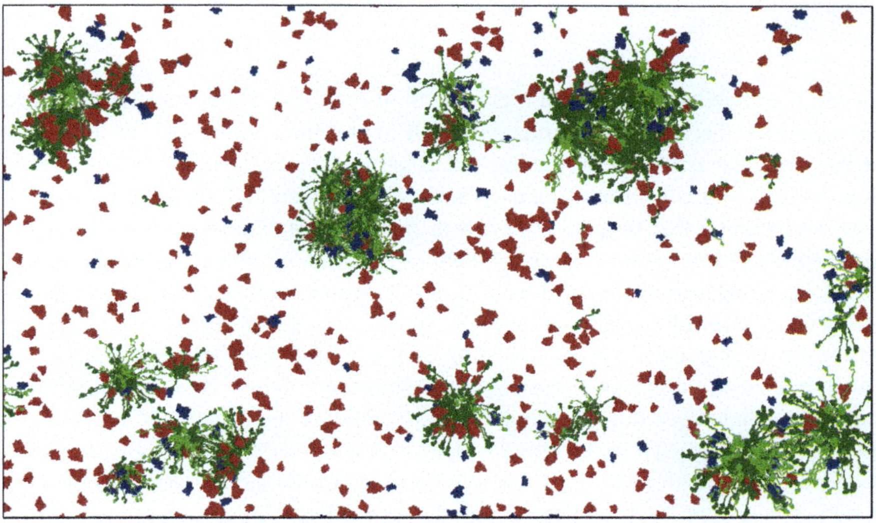

Fig. 11 Close-up of PDC self-assembly shown in Fig. 10

binding and unbinding of E1 is crucial, which occurs with more than 2 % of all E1 per micro-second.[2]

With regard to the biocatalytic function of PDC converting pyruvate into acetyl-CoA, pyruvate is initially bound to E1 with thiamine pyrophosphate (TPP) as a cofactor, followed by the E1 complex binding to E2 and reacting with lipoic acid bound at a lipoyl-domain (LD) of E2, which subsequently reacts with coenzyme A (CoA) forming acetyl-CoA and is regenerated by E3 and nicotinamide adenine dinucleotide (NAD+) with flavin adenine dinucleotide (FAD) as a cofactor (see details e.g. in Ref. [34]). Consequently, for efficient biocatalytical function of the multi-enzymatic system, in the structural assembly the proximity of E1 to E2 and E3 to E2 are of critical importance. As it can be seen in Fig. 10e, the resulting structures significantly enhance proximity of these reactants in comparison to the randomly diluted system. Additionally, the previously mentioned binding and unbinding of E1 enables the initial step of the reaction. Based on this information and ability to vary process parameters, for the remainder of the project we are currently working on explicitly modeling the chemical reaction sequence between all reactants using a probabilistic state change model during the structural formation.

---

[2] Note that due to the accelerated nature of the coarse-grained simulation these time scales are not real world time scales, see discussion in Ref. [21].

4 Conclusion

In summary, we have developed a generic multiscale modeling framework termed the molecular discrete element method (MDEM) for the investigation of functional macromolecular assemblies in natural and technological applications. The methodology abstracts entire macromolecules as anisotropic objects and maintains high detail from atomistic scales through data-driven models parameterized by a supervised learning algorithm from molecular dynamics (MD) and optional empirical data. It enables detailed insight into the dynamic formation and stability, as well as functional properties of macromolecular assemblies - thus, laying the groundwork to adapt and design novel molecular systems. Testing and validation has been performed with collaborators on three model systems including the polymer cross-linking during calcium mediated gelation of alginate, self-assembly of the hepatitis B core antigen into virus-like particles, and hierarchical structural formation of the pyruvate dehydrogenase complex for enhancing catalytic activity. Consequently, application of the generic framework has been demonstrated over vastly different research areas from material science to biotechnology. Current works for the remainder of the project are enhancing the functional modeling of biocatalytic reactions based on structural information.

Acknowledgements The authors thank the German Research Foundation (DFG) for funding within the focus program SPP 1934 (HE 4526/19-2) and the High-Performance Computing Center Stuttgart for providing the computational resources (HLRS, Acid 44178). Furthermore, all collaborators and their contributions are gratefully acknowledged as specified in the related publications [18–22]. Lastly, the authors thank Dr. Uwe Jandt and Cornelius Jacobi for initial collaboration in the context of molecular dynamics and PDC.

References

1. Castellana M, Wilson MZ, Xu Y, Joshi P, Cristea IM, Rabinowitz JD, Gitai Z, Wingreen NS (2014) Enzyme clustering accelerates processing of intermediates through metabolic channeling. Nat Biotechnol 32(10):1011–1018
2. Henzler-Wildman KA, Lei M, Thai V, Kerns SJ, Karplus M, Kern D (2007) A hierarchy of timescales in protein dynamics is linked to enzyme catalysis. Nature, 450(7171):913–916
3. Ruscio JZ, Kohn JE, Aurelia Ball K, Head-Gordon T (2009) The influence of protein dynamics on the success of computational enzyme design. J Am Chem Soc 131(39):14111–14115
4. Franke D, Petoukhov MV, Konarev PV, Panjkovich A, Tuukkanen A, Mertens HDT, Kikhney AG, Hajizadeh NR, Franklin JM, Jeffries CM, Svergun DI (2017) ATSAS 2.8: a comprehensive data analysis suite for small-angle scattering from macromolecular solutions. J Appl Crystallogr 50(4):1212–1225
5. Uetrecht C, Heck AJR (2011) Modern biomolecular mass spectrometry and its role in studying virus structure, dynamics, and assembly. Angew Chem Int Ed 50(36):8248–8262
6. Abdiche Y, Malashock D, Pinkerton A, Pons J (2008) Determining kinetics and affinities of protein interactions using a parallel real-time label-free biosensor, the Octet. Anal Biochem 377(2):209–217

7. Saunders MG, Voth GA (2013) Coarse-graining methods for computational biology. Annu Rev Biophys 42(1):73–93
8. Coffey WT, Kalmykov YP (2004) The Langevin equation: with applications to stochastic problems in physics, chemistry and electrical engineering. World Scientific
9. Paquet E, Viktor HL (2015) Molecular dynamics, monte carlo simulations, and langevin dynamics: a computational review. BioMed Res Int 1–18:2015
10. van Gunsteren WF, Berendsen HJC (1982) Algorithms for Brownian dynamics. Mol Phys 45(3):637–647
11. Groot RD, Warren PB (1997) Dissipative particle dynamics: bridging the gap between atomistic and mesoscopic simulation. J Chem Phys 107(11):4423–4435
12. Onufriev AV, Izadi S (2018) Water models for biomolecular simulations. WIREs Comput Mol Sci 8(2)
13. Marrink SJ, Risselada HJ, Yefimov S, Peter Tieleman D, De Vries AH (2007) The MARTINI force field: coarse grained model for biomolecular simulations. J Phys Chem B 111(27):7812–7824
14. Darré L, Machado MR, Brandner AF, González HC, Ferreira S, Pantano S (2015) SIRAH: a structurally unbiased coarse-grained force field for proteins with aqueous solvation and long-range electrostatics. J Chem Theory Comput 11(2):723–739
15. Donati I, Paoletti S (2009) Material properties of alginates. In Rehm BHA (ed) Alginates: biology and applications, vol 13. Springer, Berlin, Heidelberg, pp 1–53
16. Wynne SA, Crowther RA, Leslie AGW (1999) The crystal structure of the human hepatitis B virus capsid. Mol Cell 3(6):771–780
17. Patel MS, Nemeria NS, Furey W, Jordan F (2014) The pyruvate dehydrogenase complexes: structure-based function and regulation. J Biol Chem 289(24):16615–16623
18. Depta PN, Jandt U, Dosta M, Zeng A-P, Heinrich S (2019) Toward multiscale modeling of proteins and bioagglomerates: an orientation-sensitive diffusion model for the integration of molecular dynamics and the discrete element method. J Chem Inf Model 59(1):386–398
19. Depta PN, Gurikov P, Schroeter B, Forgács A, Kalmár J, Paul G, Marchese L, Heinrich S, Dosta M (2022) DEM-based approach for the modeling of gelation and its application to alginate. J Chem Inf Model 62(1):49–70
20. Depta PN, Dosta M, Heinrich S (2023) Data-driven multiscale modeling of self-assembly and hierarchical structural formation in biological macro-molecular systems. In: Nagel Wolfgang E, Kröner Dietmar H, Resch Michael M (eds) High performance computing in science and engineering'21. Springer International Publishing, Cham
21. Depta PN, Dosta M, Wenzel W, Kozlowska M, Heinrich S (2022) Hierarchical coarse-grained strategy for macromolecular self-assembly: application to hepatitis B virus-like particles. Int J Mol Sci 23(23):14699
22. Depta PN, Dosta M, Heinrich S (2023) Data-driven multiscale modeling of self-assembly and hierarchical structural formation in biological macro-molecular systems: pyruvate dehydrogenase complex. In: Nagel Wolfgang E, Kröner Dietmar H, Resch Michael M (eds) High performance computing in science and engineering'22. Springer International Publishing, Cham. (in print)
23. NVIDIA Corporation (2021) CUDA Toolkit V11.2 programming guide. NVIDIA Corporation
24. Dosta M, Skorych V (2020) MUSEN: an open-source framework for GPU-accelerated DEM simulations. SoftwareX 12:100618
25. Abraham MJ, Murtola T, Schulz R, Páll S, Smith JC, Hess B, Lindahl E (2015) GROMACS: high performance molecular simulations through multi-level parallelism from laptops to supercomputers. SoftwareX 1–2:19–25
26. Berendsen HJC, van der Spoel D, van Drunen R (1995) GROMACS: a message-passing parallel molecular dynamics implementation. Comput Phys Commun 91(1):43–56
27. Grant GT, Morris ER, Rees DA, Smith PJC, Thom D (1973) Biological interactions between polysaccharides and divalent cations: the egg-box model. FEBS Lett 32(1):195–198
28. Böttcher B, Nassal M (2018) Structure of mutant hepatitis B core protein capsids with premature secretion phenotype. J Mol Biol 430(24):4941–4954

29. Dryden KA, Wieland SF, Whitten-Bauer C, Gerin JL, Chisari FV, Yeager M (2006) Native hepatitis B virions and capsids visualized by electron cryomicroscopy. Mol Cell 22(6):843–850
30. Roseman AM, Berriman JA, Wynne SA, Jonathan P, Butler G, Anthony Crowther R (2005) A structural model for maturation of the hepatitis B virus core. Proc Natl Acad Sci USA 102(44):15821
31. Seitz S, Urban S, Antoni C, Böttcher B (2007) Cryo-electron microscopy of hepatitis B virions reveals variability in envelope capsid interactions. EMBO J 26(18):4160–4167
32. Guo J, Hezaveh S, Tatur J, Zeng A-P, Jandt U (2017) Reengineering of the human pyruvate dehydrogenase complex: from disintegration to highly active agglomerates. Biochem J 474(5):865–875
33. Ilhan S (2021) Novel strategies for automated engineering of enzymatic systems: structural and functional insights to human pyruvate dehydrogenase complex. Doctoral Thesis, Technische Universität Hamburg, 2021
34. DeBrosse SD, Kerr DS (2016) Pyruvate dehydrogenase complex deficiency. In: Mitochondrial case studies. Elsevier, pp 93–101

Modeling-Based Monitoring and Control of Protein Crystallization in Bioprocesses

Daniel Bischoff, Brigitte Walla, Christina H. Wegner, Jürgen Hubbuch, and Dirk Weuster-Botz

Abstract Protein crystallization has evolved from its use for structure determination of biomolecules via X-ray diffraction in the 1960s to industrial applications in the last decades. Generalizing crystallization processes across a wide range of target proteins remains challenging due to narrow and specific windows of viable process parameters. Here, progress on rational methods to monitor and control protein crystallization at multiple scales is presented. While microscopic and macroscopic scales are studied in terms of the phase behavior and process analytical technologies, molecular scales are approached through molecular dynamics simulation and protein engineering.

Dedicated to Dr.-Ing. Dariusch Hekmat, one of the principal investigators of this research, who passed away unexpectedly during the project.

D. Bischoff · B. Walla · D. Weuster-Botz (✉)
Chair of Biochemical Engineering, Department of Energy and Process Engineering, School of Engineering and Design, Technical University of Munich, Boltzmannstr. 15, 85748 Garching, Germany
e-mail: dirk.weuster-botz@tum.de

D. Bischoff
e-mail: daniel.bischoff@tum.de

B. Walla
e-mail: brigitte.walla@tum.de

C. H. Wegner · J. Hubbuch
Institute of Process Engineering and Life Sciences, Section IV: Biomolecular Separation Engineering, Karlsruhe Institute of Technology KIT, Fritz-Haber-Weg 2, 76131 Karlsruhe, Germany
e-mail: christina.wegner@kit.edu

J. Hubbuch
e-mail: juergen.hubbuch@kit.edu

A. Kwade and I. Kampen (eds.), *Dispersity, Structure and Phase Changes of Proteins and Bio Agglomerates in Biotechnological Processes*,
https://doi.org/10.1007/978-3-031-63164-1_3

1 Introduction

Protein crystallization has evolved from its use for structure determination of biomolecules via X-ray diffraction in the 1960s [31] to industrial applications in the last decades with its most prominent example being recombinant insulin [7, 22]. Due to the increasing industrial bioproduction of proteins, effective downstream processing remains an important challenge. Classical protein purification by chromatography requires high amounts of cost-intensive consumables, e.g. absorbent material. In comparison, crystallization as a purification step is a fast and cost-effective alternative to chromatography due to its characteristic to assemble identical molecules into crystals from polydisperse solutions (reviewed in [23]). Protein crystallization processes can be performed continuously (e.g. [24]) and scale-up criteria for stirred protein crystallizers are known [54].

To ensure constant product quality and purity, it is essential to monitor and control the protein crystal growth processes. Monitoring protein crystallization remains a challenge. The identification and analysis of suspended crystals through microscopic photography using deep learning has been a subject of increased attention for various crystallization processes [39, 60]. Advancing the detection of suspended protein crystals using deep learning is limited: robust models require large-scale and high-quality data sets usually obtained in an error-prone manual labeling process. This approach is highly impractical for dense systems of transparent protein crystals. In this work recent trends for supervised learning are transferred involving the use of synthetic data sets. Here, virtual protein crystals in suspension (PCS) are placed into a 3D scene, and photorealistic images are generated through the use of ray tracing algorithms. The resulting large-scale and high-quality data set is used to train state-of-the-art object detection models and perform validation tasks on real photomicrographs (see Chap. 4).

The objective to monitor and control (bio)-pharmaceutical processes is in line with the regulatory expectations of the U.S. Food and Drug Administration (FDA) to implement process analytical technology (PAT) [19]. Common PAT analytics in downstream processes are e.g. different spectroscopic methods (UV–Vis, IR, turbidity measurements, focused beam reflectance measurements (FBRM), Raman, light scattering) and are often combined with chemometric techniques. The latter aims to reduce the measurement data, and simplify and extract underlying, valuable information from a multivariate data set when the information of interest is covered by measurement noise or overlapping signals. Principal component regression (PCA), partial least squares (PLS) regression, and other statistics based methods have been used together with spectroscopic analytics to calculate e.g. different protein concentrations in mixtures [9], distinguish between crystal morphologies [52], or to monitor reaction mechanisms [3]. Using spectroscopic measurements and chemometric models for protein crystallization is not widespread and was therefore investigated in Chap. 6. The focus of this study lies in the development of a fast, low-volume analytical tool to accurately determine quantitative process parameters in high-throughput crystallization studies. Combining UV–Vis spectroscopy and PLS models can aid in determining the protein-specific crystal yield and crystal purity of investigated screening conditions.

Despite advances in crystallization monitoring, the major bottleneck for technical purposes is the preliminary step, the determination of appropriate crystallization conditions. As a result, few proteins can be sufficiently crystallized for preparative downstream processes. To improve the crystallization process of recombinant proteins, a better understanding and modeling-based control of the crystallization processes is mandatory. Starting at the molecular level, molecular dynamics (MD) simulations were performed to take a deeper look at the crystal contacts. In MD simulations, the large molecular crystal system can be effectively approximated through periodic boundary conditions and the crystallographic unit cell. Additionally, a common and well-tested application of molecular dynamics simulations is drug discovery, where intermolecular interactions between two molecules are often optimized to increase binding affinities [12]. To this end, free energy perturbation techniques are employed. Here, MD simulations are not limited to small ligands. Through slow perturbation of amino acids at interaction sites, e.g. crystal contacts, information on an associated change in the free energy can be efficiently calculated. In this chapter, free energy simulations are presented as a useful tool to analyze crystal contacts for stability, and furthermore, introduce a computational approach to protein engineering at crystal contacts. Protein engineering as a tool for crystallography was first summarized by Price and Nagai [47]. Relevant here are the exposed amino acids of the protein and thus the different surface properties caused by the side chain residues of the amino acids of the protein [37]. Individual amino acid exchanges on the protein surface were found to have an extreme effect on crystallization properties. Further work on the application of molecular engineering combined with recombinant methods led to the introduction of the terms ‘rational engineering’ and ‘rational protein crystallization’ [14]. The protein engineering presented in Chaps. 2 and 4 aimed to specifically engineer crystal contacts by rationally substituting specific amino acids to enhance the crystal contact interactions on a molecular level.

Modeling-based monitoring and control of protein crystallization in bioprocesses covers rational control of protein crystallization processes at different scales, aiming for a holistic understanding of engineered protein crystallization. While the microscopic and macroscopic scale is studied in terms of the phase behavior and process analytical technologies, the molecular scale is approached through MD simulation and protein engineering.

2 Implementation of Protein Crystallization and Protein Engineering

2.1 Introduction

For about three decades, protein engineering has been conducted to obtain high-quality crystals for X-ray crystallography, beginning with engineering Ferritin [35]. Subsequently, studies focused on generally enabling crystallization and engineering crystallizability by mutating on the protein surface [28, 40], resulting in a review of protein engineering as a new tool for crystallography [47]. The group of Derewenda [14] established a rational engineering approach to enhance crystallizability by reducing surface entropy. This "SER strategy" was proposed to generate "low-entropy" surface patches leading to improved crystal quality for X-ray diffraction analysis by exchanging amino acids with long side chains (Lys/Arg/Glu) on the protein surface with alanine or glycine [11, 14, 15, 37].

In the following, a proof of concept is described that addresses the empirical high throughput screening step for crystallization conditions by rationally engineering the protein itself to enhance crystallizability. It will be shown that both, SER [15], and the introduction of charged amino acids at the crystal contact can improve the crystallizability of a protein. Rational protein engineering will be exemplarily applied to the *Lactobacillus brevis* alcohol dehydrogenase (*Lb*ADH) to improve its applicability in preparative crystallization [46]. *Lb*ADH is an industrially relevant oxidoreductase for the enantioselective reduction of a wide spectrum of ketones showing high regio- and enantioselectivity [8, 13, 36].

2.2 Materials and Methods

All used chemicals for *Lb*ADH variant mutagenesis, production, and further processing are purchased from Carl Roth GmbH & Co. KG (Karlsruhe, Germany) and enzymes from New England Biolabs (Ipswich, USA) unless otherwise indicated.

2.2.1 Mutagenesis, Production, and Further Processing of *Lb*ADH Variants

Site-directed mutagenesis of *Lb*ADH variants was conducted by QuikChange PCR according to Nowotny et al. [46]. After the transformation of chemically competent *E.coli* BL21(DE3) cells with a sequenced DNA plasmid encoding the *Lb*ADH variant, recombinant protein production in *E. coli* BL21 (DE3) of the *Lb*ADH variants was implemented in 0.5 L shaking flasks (TB medium with 35 μg mL^{-1} kanamycin) by inoculation with preculture (6 mL TB medium with 35 mg L^{-1} kanamycin, 16 h, 30 °C, 180 rpm). When the *E. coli* cells reached an optical density (OD_{600}) of 0.8, gene overexpression of the *Lb*ADH variants was induced by the addition of 1 mM isopropyl β-D-thiogalactopyranoside (IPTG). The cells were harvested after 14 h (20 °C, 230 rpm) of protein production. After cell disruption via sonication (2 $\times$ 3 min, 90% intensity, 50% pulse) and removal of the cell debris by centrifugation (12,000 $\times$ g, 4 °C, 20 min), protein purification via immobilized metal affinity chromatography (IMAC) was conducted by a 2-step elution with imidazole as described in Nowotny et al. [46].

2.2.2 Microliter Batch Protein Crystallization

The purified protein was dialyzed against protein buffer (20 mM HEPES NaOH pH 7.0, 1 mM $MgCl_2$) and concentrated as described in Nowotny et al. [46]. Static microbatch crystallization of *Lb*ADH variants was performed in 96-well microbatch plates according to Hermann et al. [26]. The crystallization buffer (0.1 M Tris HCl at pH 7.0, 50 mM $MgCl_2$) was completed with 100–150 g L^{-1} polyethylene glycol (PEG) 550 monomethyl ether (MME) from Sigma-Aldrich (Deisenhofen, Germany).

2.2.3 X-ray Analysis of *Lb*ADH Crystals

Protein crystals (>60 μm) were selected and flash-frozen with reservoir solution (supplemented with 30% (v/v) ethylene glycol) for cryoprotection and stored in liquid nitrogen. X-ray diffraction data was collected at the ID30-A beamline at the Synchrotron Radiation Facility (ESRF, Grenoble, France), and the data refinement was conducted as described in Sect. 3.2.

2.2.4 Protein Analytics

*Lb*ADH concentration was measured with a spectrophotometer (A_{280}, ε = 19,940 M^{-1} cm^{-1}) according to Nowotny et al. [46]. Supernatant protein concentration during stirred milliliter batch experiments was determined by bicinchoninic acid BCA protein assay. Protein purity was validated via sodium dodecyl sulfate polyacrylamide gel electrophoresis (SDS-PAGE), and the enzymatic activity of *Lb*ADH variants was determined photometrically by detection of the NADH oxidation during the reduction of the substrate acetophenone to 1-phenylethanol in microtiter plates at 30 °C according to Nowotny et al. [46].

2.3 Results and Discussion

2.3.1 Implementation of μL Batch Protein Crystallization

The industrially relevant *Lb*ADH represents a suitable protein for the implementation of protein crystallization due to its known primary structure, reproducible recombinant protein production in *E. coli*, high protein stability and solubility, and the availability of a simple spectrophotometric activity assay. Homotetrameric *Lb*ADH catalyzes the enantioselective reduction of prochiral ketones to the corresponding secondary alcohols [27]. It is stable and enzymatically active even at elevated temperatures and has a broad substrate spectrum [38, 43, 44, 51]. For efficient preparative purification via IMAC, a His_6-tag was cloned to the N-terminus of *Lb*ADH. The crystallization properties were not affected by the His_6-tag, confirmed by X-ray crystal structure comparison to an untagged *Lb*ADH [26, 43].

Microbatch crystallization was chosen at lab-scale, as this method has high reproducibility and enables high sample throughput compared to vapor diffusion methods. Published crystallization conditions of *Lb*ADH [44, 51] were adapted for the modified His_6-*Lb*ADH [46]. After successful crystallization and evaluation of X-ray measurements, the crystal structure of His_6-*Lb*ADH was available for further studies [26].

2.3.2 Protein Engineering at Crystal Contacts

To improve preparative protein crystallization, this study aims to enhance crystal contact interactions and thus increase crystallization by a rational crystal contact engineering approach. Besides using the established SER strategy [14], the research project focused on enhancing electrostatic interactions at crystal contacts. The rational protein engineering resulted among others in the promising *Lb*ADH mutants T102E, Q126H, and Q126K [20, 46]. To ensure an inference of the single amino acid exchange to the difference in crystallizability, a robust methodology needed to be implemented for the whole lab-scale production, purification, and crystallization process. The performance of protein production and purification of these mutants was comparable to the *Lb*ADH wild type, confirming no negative effect of the single amino acid exchange on cell integrity. Evaluation of protein purity, monodispersity, and enzyme activity showed no significant differences compared to the *Lb*ADH wild type.

Batch crystallization of purified protein on the µL scale resulted in a shift of the nucleation window to reduced protein and precipitation concentrations for the *Lb*ADH mutants compared to the wild type (Fig. 1).

A shift of the nucleation window to reduced protein and precipitation concentrations was defined as a parameter for "crystallizability", thus *Lb*ADH mutants' crystallizability was enhanced compared to the wild type. To verify the hypothesis, that these experimental results rely on single amino acid exchanges on crystal contacts, the crystal structure was discussed further on a molecular level (Fig. 2). The positively charged imidazole ring of *Lb*ADH mutant Q126H shows a reduction in distance to the hydroxyl group of S40 and negatively charged carboxyl group of E44 (Fig. 2a), indicating a formed electrostatic interaction compared to the wild type, and thus an enhanced interaction. Comparing the crystal structure of the *Lb*ADH mutant Q126K and T102E to the wild type revealed a significant decrease in the distance between the intermolecular neighboring amino acids (Fig. 2b), which are E44/S40 (Q126), and K48 (T102). Furthermore, the corresponding charged amino acids glutamic acid and lysine (Q126K-E44; T102E-K48) are facing each other at the crystal contacts of the mutants compared to the wild type, demonstrating that the incorporation of charged amino acids can enhance crystallizability.

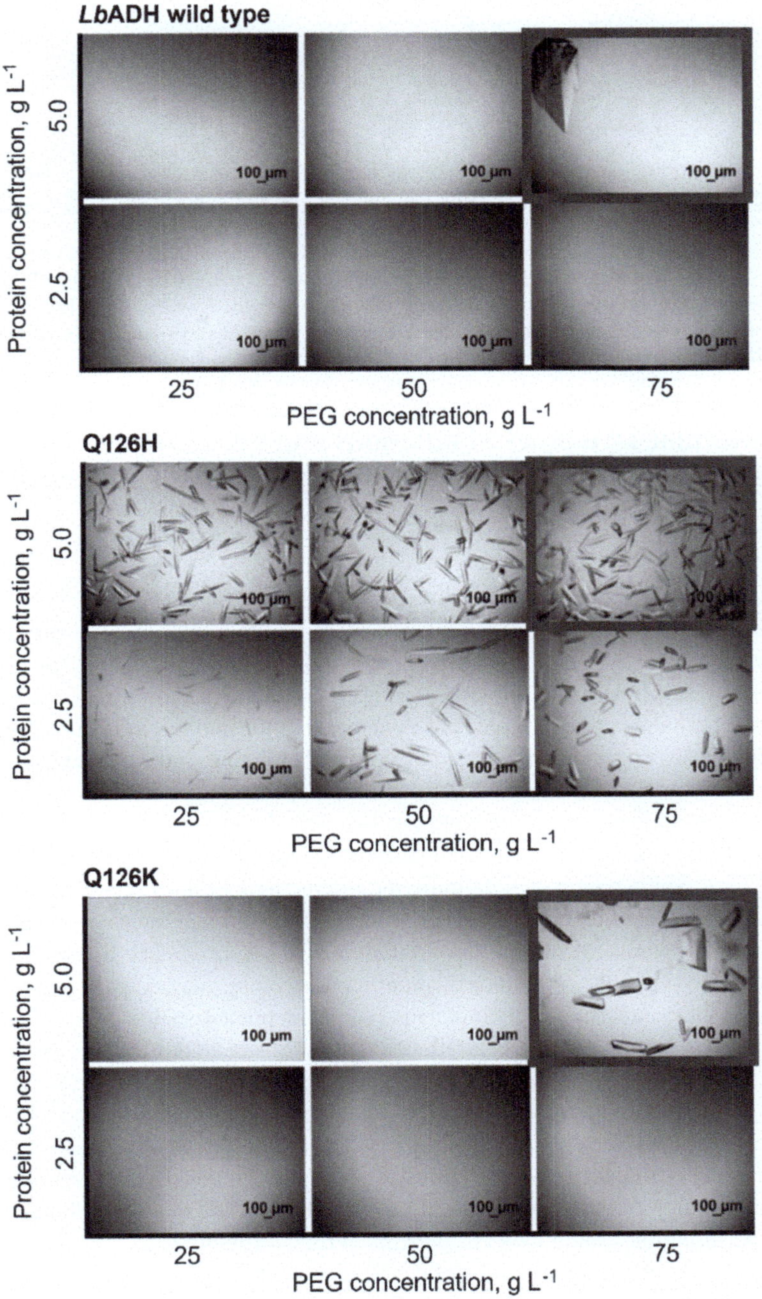

Fig. 1 Microphotographs of *Lb*ADH wild type and mutants Q126H and Q126K after 48 h of crystallization in static μL batch (0.1 M Tris HCl pH 7.0, 50 mM $MgCl_2$, and 50–150 g L^{-1} PEG MME 550). Adapted with permission from Nowotny et al. [46]. Copyright 2019 American Chemical Society

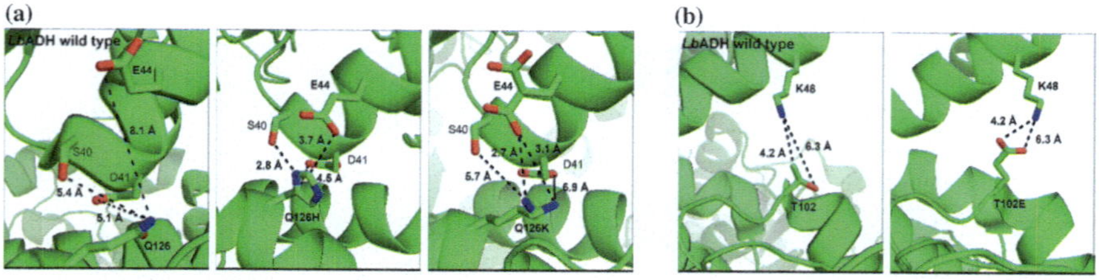

Fig. 2 Illustration of the *Lb*ADH crystal contact of (**a**) wild type (PDB ID: 6H07) and mutant Q126H (PDB ID: 6Y10) and Q126K (PDB ID: 6Y0Z), as well as (**b**) T102E (PDB ID: 6Y0S). The illustration was generated and distances were calculated using PyMOL (v.2.1; Schrödinger). Figure adapted from Grob et al. [19], licensed under the Creative Commons Attribution 4.0 International license (CC BY 4.0)

3 Molecular Dynamics Simulations and Free Energy Perturbation Applied to Protein Crystallization Processes

3.1 Introduction

Experimentally determined 3D structures of biological macromolecules are largely obtained through crystallization experiments and subsequent X-ray diffraction. In this setting, multiple macromolecules are ordered and tightly packed together, exhibiting periodic patterns. Molecular dynamics (MD) simulations are usually initiated from these experimentally determined crystal structures, surrounded by solvent and simulated under periodic boundary conditions. Using specific protocols for the setup, the simulation of macromolecules in their actual crystalline environment might therefore be considered a natural application of MD simulations. However, the accuracy of force-fields parameters defining the interactions in the simulated system is only rarely evaluated by comparing the integrity of the crystalline environment of a particular macromolecule during the simulation [10, 34], instead being optimized to reproduce the secondary structure of proteins. Therefore, the accuracy with which protein crystal contacts can be simulated remains an open question.

At the same time, protein crystallization has gained attention from biological downstream processing applications. Here, the crystallizability of proteins is of central importance to increase yield and product stability, motivations similar to drug discovery in pharmaceutical applications [12], which often involve computational methods such as free energy calculations and MD simulations. Recent work focuses on the calculation of free energy differences of wild type and mutant proteins in free energy perturbation simulations via alchemical transformations [25]. This is used to infer which crystallization process (wild type or mutant) is preferable, by calculating a quantity analogous to relative binding free energies for crystallization processes. Hermann et al. [25] showed that free energy calculations can be used to explain the crystallization behavior of different *Lb*ADH variants, introducing a promising alternative to random mutagenesis and experimental screenings for crystallization conditions in future work.

3.2 *Materials and Methods*

3.2.1 X-ray Data Collection and Structure Solution

Diffraction data was collected at the Swiss Light Source on beam lines PXI X06SA and SLS PXIII X06DA. Crystals were mounted on nylon loops, flash-cooled to 100 K in liquid nitrogen, and cryo-protected with 25–30% (v/v) ethylene glycol. The data was indexed and integrated with XDS and scaled with XSCALE [30]. The software suite CCP4 was used for model building and refinement. More specifically, the PDB entry 6h07 of the wild type (WT) *Lb*ADH structure (PDB-ID 6h07) was used as a search model for molecular replacement with PHASER [40], and refinement and manual model building were performed with REFMAC [41] and COOT [16]. This way, multiple mutants of the *Lb*ADH were deposited in the Protein Data Bank under accession codes 6y1c (D54F), 6y10 (Q126H), 6y0z (Q126K), 7a2b (Q207D).

3.2.2 System Construction

The calculation of relative changes in free energy requires the setup of a particular protein mutant in two ways. First, the protein is surrounded by the solvent only to calculate the change in free energy upon an amino acid mutation in the solvated state. Second, the protein is placed in its crystalline environment. Since the targeted amino acid exchanges are at protein crystal contacts, one can simulate the neighboring chains around that contact and surround it with an additional solvent shell. For amino acid mutations involving a change of the total charge in the system, this has the additional advantage of reducing errors due to self-interaction of changing electric charges within a system under periodic boundary conditions. Counter ions charging is used to mitigate these artifacts, which benefits from counter ions having large regions of accessible solvent. For *Lb*ADH protein variants for which a crystal structure existed (Q126H, Q126K, K32A, Q207D, T102E, D54F) simulations for both wild type (WT) and mutant were constructed so that the mutation could be investigated in both crystal systems. Simulations in a crystalline environment use the reduced setup with only neighboring chains around the crystal contact, except for mutant Q126H, where the simulation of the periodic boundary was performed directly using the crystallographic unit cell of the protein crystal.

For the worse crystallizing variants, H39A, K45A, and D54A, no diffraction data could be obtained and the simulation setup could thus only be performed in the WT crystal system. For the unbound state (protein surrounded by solvent), the number of simulated monomers depends on the position of the target amino acid and the estimated range and influence of a particular amino acid mutation. Mutations

Q126H and Q207D were simulated including a whole tetramer of the *Lb*ADH, while other mutants were simulated using one or two neighboring monomers. GROMACS was used for the simulation setup. Here, proteins are protonated at pH 7.0 and parametrized using the CHARMM36m force field. The simulation boxes were then solvated with a TIP3P water model and neutralized using 25.5 mM $MgCl_2$, according to the experimental setup.

3.2.3 System Equilibration

The experimentally determined X-ray structures prepared as described in the previous section, might not be automatically in an energetically favorable configuration using a specific force field, solvation procedure, and water model. Again, GROMACS is used to perform energy minimization and temperature annealing in the NVT ensemble. For energy minimization, the steepest descent method is used, with a convergence criterion of 8500 steps or 100 kJ mol^{-1} nm^{-1}. This is followed by 0.9 ns of annealing with a final temperature of 293 K and 1 ns of constant volume simulation (NVT). Here, velocity rescaling was used to control the system temperature, while a leap-frog integrator was used for trajectory propagation with a time step of 1 fs. During the simulation, the protein backbone is restrained using a force constant of 1000 kJ mol^{-1} nm^{-2}. Electrostatic interactions were calculated as usual with the Particle mesh Ewald (PME) method, with a Fourier spacing of 0.12 nm. Van der Waals interactions were cut-off at a distance of 1.2 nm using a smooth potential switch starting from 1.0 nm. For the equilibration run, the time step was increased to 2 fs and the positional restraints were decreased to atoms with a force constant of 750 kJ mol^{-1} nm^{-2}. The equilibration simulation lasted for 200 ns and it was validated, that the backbones RMSD of the simulated proteins remained constant on average.

3.2.4 Free Energy Calculations

For the generation of hybrid structures for free energy perturbation (FEP) simulations, the GROMACS-compatible program PMX was used. For mutant Q126H with the bound state represented by the crystallographic unit cell during the simulations, 60 equally spaced alchemical transition windows were used. For the rest of the mutants (reduced simulation setup) 30 transition windows were found to be sufficient. For each of the transition windows, an initial structure was picked from the stable RMSD region at the end of the equilibration simulation. After preparing the hybrid topology, these initial structures then were energy minimized, temperature annealed (0.3 ns), and simulated with another equilibration run (5 ns). For nonbonded interactions, softcore potentials were used with softcore parameter alpha of 0.3, a power of 6 for the radial term, and a soft core sigma for particles of 0.25. Hamiltonian replica exchange was performed with an exchange rate between alchemical transition windows every 1000 steps resulting in transition probabilities of 25–40%. The production run lasted

20 ns with the same parameters as the equilibration run and free energy differences of every window to its two neighboring windows calculated every 0.2 ps. The free energy difference was estimated via the Bennett Acceptance ratio (BAR) implemented in GROMACS and divided by the number of mutations in the respective set-up. Error estimate was calculated with GROMACS BAR. The counter ion approach was used, which corrects the dominant error for free energy calculations involving a change in the total system charge under periodic boundary conditions. The net charge was kept constant by charging or de-charging a divalent magnesium ion (Mg^{2+}) or a chloride ion (Cl^{-}) during the alchemical transformation while keeping their Van-der-Waals parameters. By performing counter ion charging for the solvated state and the crystalline state of the protein, the contribution of the ion charging cancels once the relative free energy differences are calculated.

3.3 Results and Discussion

X-ray structures of crystallizable *Lb*ADH variants (WT, D54F, T102E, Q126H, Q126K, K32A, Q207D) resulted in corresponding crystallographic unit cells (UCs). For all mutants, the resulting crystal system was orthorhombic. Small differences can be observed in the rectangular base of the unit cell between different mutants. The mutations did not influence the overall arrangement but may have altered the interactions at crystal contacts. Additionally, all mutants have a root mean square deviation smaller than 0.19 Å after the structural alignment of C_α atoms. Furthermore, preserved activity indicates the correct formation of the tetrameric biological assembly for the mutants that did not crystallize. Based on the described protocol for calculating relative changes in free energy of crystallization, an order of crystallizability (as described previously) was experimentally determined. For *Lb*ADH mutants with increased crystallizability: T102E $\approx$ D54F > Q126H > Q126K > K32A > WT, and for *Lb*ADH mutants with reduced crystallizability: wild type (WT) > Q207D > H39A, K45A, D54A [20]. Here, the latter three mutants did not crystallize under the experimental conditions within 24 h.

This ordering was compared with the computed relative changes of the free energy upon crystallization $\Delta\Delta G$ using a thermodynamic cycle (see Fig. 3a). Here, the alchemical transformation in the crystallized state was performed in the mutant crystal system, as well as the wild type crystal system for comparison (due to the small changes in the crystallographic unit cell). The theoretical calculations of the relative changes of free energy upon crystallization yielded negative values (indicating increased crystallizability) for all *Lb*ADH mutants that were experimentally determined to crystallize better than the wild type, except for mutants Q126K and K32A in wild type crystal packing. For these mutants, the calculated error allowed no determination of the sign of $\Delta\Delta G$. For all mutants crystallizing worse than the wild type in the experiment, the relative changes of the binding free energy upon crystallization were calculated to be positive, indicating worse crystallizability (see Fig. 3b).

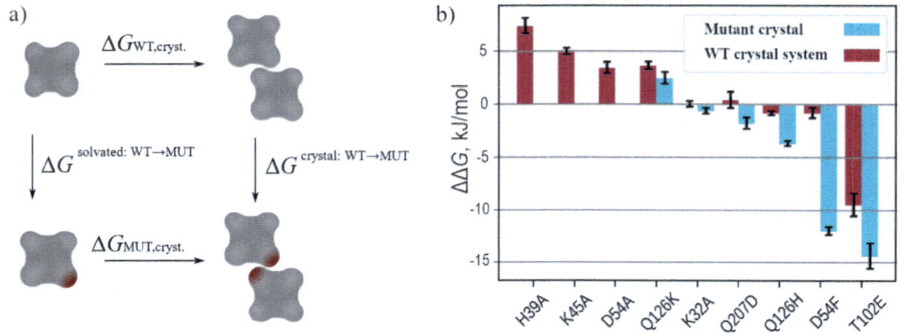

Fig. 3 Calculation of relative binding free energies for a crystallization process **a** Displayed is the thermodynamic cycle, which describes the calculation of $\Delta\Delta G = \Delta G_{\text{MUT,crystallization}} - \Delta G_{\text{WT,crystallization}}$. However, the horizontal free energy differences are not accessible in normal MD simulations. Instead, the same value can be calculated via the vertical arrows: $\Delta\Delta G = \Delta G^{\text{crystal: WT}\rightarrow\text{MUT}} - \Delta G^{\text{solvated: WT}\rightarrow\text{MUT}}$. Ordering of the calculated values of relative changes in the binding free energy from Hermann et al. [25] for all investigated *Lb*ADH mutants, licensed under the Creative Commons Attribution 4.0 International license (CC BY 4.0)

Combining MD simulations and X-ray structure analysis allows reasoning and interpretation of the changes in crystallizability for specific mutants. An interesting example is the *Lb*ADH mutant K32A. Here, the mutation seems to influence the adjacent crystal contact D54. For the wild type crystal structure, the contact is composed of two negatively charged carboxylate groups that interact with each other via dynamic water-mediated hydrogen bonds. However, for mutant K32A, the measured X-ray crystallography structure shows that the interaction is not only mediated by water but also by a divalent magnesium ion. MD simulation allowed alchemically inserting an Mg^{2+} ion at the binding position and weighting the result according to the occupancy measured in the X-ray crystallography experiments. The reason for the improved crystallizability might be the removal of a screening effect due to the electric charge of the nearby lysine for K32A.

On the other hand, mutation D54F leads to better crystallizability by increasing the crystal contact area. Here, the newly formed interface (threonine, proline, and phenylalanine) appears to be hydrophobic. Indeed, MD simulation showed that water molecules only rarely participate in the mediation of the interactions of this crystal contact (1.3 ± 1.1 water molecules).

Better crystallization can also be mediated through hydrogen bonding. In MD simulations, an H-bond between the hydroxy hydrogen of serine (S40) and the amide oxygen of glutamine (Q126) existed on average $8.8 \pm 4.1\%$ of the simulated time frames, assuming a maximum hydrogen bond distance of 2.5 Å. This bond was stabilized for the mutant Q126K. Here, an H-bond/salt bridge between the ammonium group of lysine (K126) and the carboxylate group of glutamate (E44) was detected in $66.0 \pm 7.7\%$ of the simulated time frames.

Similarly, worse crystallization behavior can be interpreted using MD simulations. For the wild type, H39 and K45 mediated attractive interactions. Relative free energy calculations upon crystallization for mutants H39A and K45A resulted in positive $\Delta\Delta G$ values (indicating worse crystallization behavior), with no corresponding interactions. Similarly, mutant Q207D destroyed a symmetry-related contact of Q207 in the wild type. Here, MD simulations showed that for $48.2 \pm 8.8\%$ of the simulated time frames a hydrogen bond is formed, while for mutant Q207D, no such interaction exists.

Lastly, for the wild type, the distance and chemical composition between T102 and K48 does not allow for any interactions. The mutant T102E improves upon this by introducing a salt bridge. MD simulations showed, that this interaction is formed for more than 90% of the simulated time frames. Again, the calculation of relative changes in free energy upon crystallization agrees with the experimental observation, resulting in one of the best crystallizing *Lb*ADH mutants.

4 Transfer of Crystal Contact Engineering Strategy to a Homologous Enzyme

4.1 Introduction

Previous studies [20, 46] focused on enhancing electrostatic interactions specifically at crystal contacts of *Lb*ADH to increase protein crystallizability for preparative purposes (Chap. 2). Especially *Lb*ADH mutants resulting from applying SER strategy (K32A), as well as introducing aromatic (D54F) and electrostatic interactions (Q126H/T102E) showed an increase in crystallizability [20, 46].

In the following, the transfer of a protein engineering strategy to the homologous *Lactobacillus kefir* alcohol dehydrogenase (*Lk*ADH) is shown [57]. High similarity in crystal packing is a prerequisite for a rational crystal contact engineering transfer to ensure changes in crystallizability of *Lk*ADH can be traced back to the (transferred) amino acid exchange [46]. Thus, the *Lk*ADH crystal structure should be aligned to *Lb*ADH (PDB ID: 6h07), selected mutations should be transferred to *Lk*ADH and the resulting differences in crystallizability should be compared to *Lb*ADH.

4.2 Materials and Methods

All used chemicals for *Lk*ADH variant mutagenesis, production, and further processing are purchased from Carl Roth GmbH & Co. KG (Karlsruhe, Germany) and enzymes from New England Biolabs (Ipswich, USA), unless otherwise indicated.

4.2.1 Mutagenesis, Production, and Further Processing of *Lk*ADH Variants

Site-directed mutagenesis of *Lk*ADH variants was conducted by QuikChange PCR according to Walla et al. [57]. After the transformation of the chemical competent *E. coli* BL21(DE3) cells with a sequenced DNA plasmid encoding the *Lk*ADH variant, recombinant protein production in *E. coli* BL21 (DE3) of the *Lk*ADH variants was implemented in 0.5 L shaking flasks (TB medium with 35 mg L^{-1} kanamycin) by inoculation with preculture (6 mL LB medium with 35 mg L^{-1} kanamycin, 16 h, 30 °C, 180 rpm). When the *E. coli* cells reached the exponential growth phase (OD_{600} of 0.4–0.6), gene overexpression of the *Lk*ADH variants was induced by the addition of 200 μM IPTG. The cells were harvested after 20 h (20 °C, 230 rpm) of protein production. After cell disruption via sonication (3 × 3 min, 90% intensity, 50% pulse) and removal of the cell debris by centrifugation (12,000 × g, 4 °C, 1 h), protein purification via IMAC was conducted by a 2-step elution with imidazole as described in Walla et al. [57].

4.2.2 Batch Protein Crystallization (μL and mL Scale)

To obtain a crystal structure of the *Lk*ADH variants, the purified protein was dialyzed against protein buffer (20 mM HEPES NaOH pH 7.0, 1 mM $MgCl_2$) and concentrated as described in Nowotny et al. [46]. Static microbatch crystallization of *Lb*ADH/*Lk*ADH variants was performed in 96-well microbatch plates according to Hermann et al. [26]. The crystallization buffer (0.1 M Tris HCl at pH 7.0, 50 mM $MgCl_2$) was completed with 100–150 g L^{-1} PEG 550 MME or PEG 3000.

Parallel protein crystallization on stirred mL scale with PEG MME 550 as a crystallization agent was conducted for non-purified *Lb*ADH/*Lk*ADH variants according to Grob et al. [20] and Walla et al. [57]. The harvested cells were processed according to Grob et al. [20]. Crystallization experiments with clarified (phosphate buffered saline, pH 7.4) and/or dialyzed (20 mM HEPES NaOH pH 7.0, 1 mM $MgCl_2$) cell lysate were conducted in up to eight parallel 5 mL stirred tank reactors (20 °C, 150 rpm) as described in Smejkal et al. [55]. Protein crystallization was initiated by the addition of 2.5 mL crystallization buffer to 2.5 mL of clarified (and dialyzed) *E. coli* cell lysate ($A_{280} = 20$, 20 °C).

4.2.3 X-ray Analysis of *Lk*ADH Crystals

Protein crystals (length > 60 μm) were selected and flash-frozen with reservoir solution (supplemented with 25–30% (v/v) ethylene glycol) for cryoprotection and stored in liquid nitrogen. X-ray diffraction data was collected and data refinement was conducted as described in Sect. 3.2.

4.2.4 Protein Analytics and Analysis of Crystal Photomicrographs

*Lb*ADH protein analytics are described in Sect. 2.2. *Lk*ADH concentration was measured spectrophotometrically (A_{280}, $\varepsilon = 23{,}950\ M^{-1}\ cm^{-1}$) according to Nowotny et al. [46]. Supernatant protein concentration during stirred milliliter batch experiments was determined by BCA assay. Protein purity was validated via discontinuous sodium dodecyl sulfate polyacrylamide gel electrophoresis (SDS-PAGE), and the enzymatic activity of *Lk*ADH variants was determined photometrically by detection of the NADH oxidation during the reduction of the substrate acetophenone to 1-phenylethanol in microtiter plates (40 °C, 10 min) according to Walla et al. [57].

For the analysis of crystal photomicrographs, samples were taken after 48 h of protein crystallization, diluted tenfold, and monitored microscopically on a crystallization plate inside of an incubator (Volume = 10 μL, 20 °C) as described in Walla et al. [57].

4.3 Results and Discussion

4.3.1 Transfer of Mutations to Homologous Enzyme

Subsequently, the established *Lb*ADH mutants, which enhanced the crystallizability of the protein [20, 46], should be transferred to a homologous enzyme to support the general validity of this approach. The *Lk*ADH, which shares an amino acid sequence homology of 88.5% with the *Lb*ADH as well as a high structural similarity (RMSD = 0.22 Å; [57]), was chosen for this transfer. *Lk*ADH is also active as a homotetramer and is characterized among others by a broad substrate spectrum, which differs from *Lb*ADH [58]. Initially, a purification tag and linker sequence was added to the N-terminus of *Lk*ADH, aligning with the *Lb*ADH sequence. Although crystallization conditions of *Lk*ADH variants are already existing (PDB ID: 4rf4; [45]), the space groups of these variants (monoclinic crystal system) differ to *Lb*ADH (orthorhombic crystal system). Thus, crystallization of His_6-GSG-*Lk*ADH under the same conditions as His_6-GSG-*Lb*ADH (PEG MME 550, magnesium chloride, Tris/HCl, pH 7.0) resulted in His_6-GSG-*Lk*ADH crystallizing in space group I222 (PDB ID: 7P36, *Lk*ADH wild type) with similar crystal contacts compared to His_6-GSG-*Lb*ADH ($P2_122_1$, PDB ID: 6H07, *Lb*ADH wild type). A sequence alignment of these two crystal structures with the crystal contacts in the lattice is depicted in Fig. 4 ($RMSD_{backbone} = 0.183$ Å and $RMSD_{side\ chain} = 0.212$ Å).

Based on the crystal structure alignment, the crystal contacts of homologous *Lb*ADH and *Lk*ADH wild type were compared. Since MD simulations revealed *Lb*ADH mutant Q126H interacting with S40 [25], but *Lk*ADH has an alanine encoded at position 40 instead, Q126H was not considered for mutant transfer to *Lk*ADH. The mutants T102E and Q126K were rationally selected as transferable mutations for *Lk*ADH, as they have comparable crystal contact areas (see Figs. 5a and 6a) with

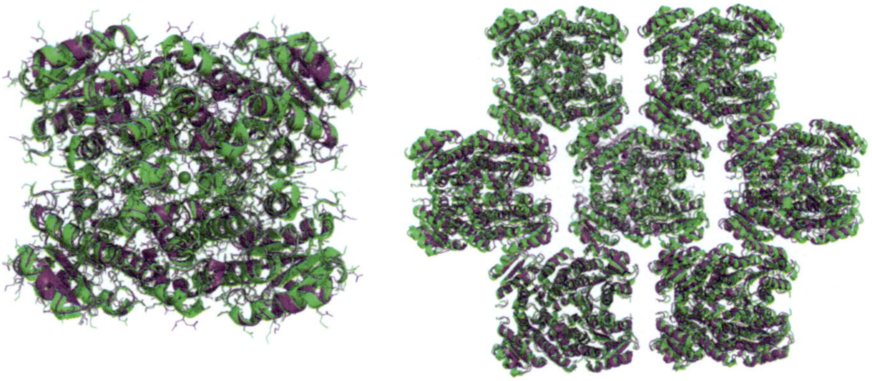

Fig. 4 Structural alignment of *Lk*ADH (magenta; PDB ID: 7P36; $R_{work} = 0.9$; $R_{free} = 0.11$) and *Lb*ADH (green, PDB ID: 6H07) wild type **a** tetramers with $RMSD_{backbone} = 0.183$ Å and $RMSD_{side\ chain} = 0.212$ Å and **b** with selected symmetry-related molecules in the crystal structure. Figures were generated with PyMOL (v.2.3; Schroedinger). The homotetramers are complexing two Mg^{2+} ions (dark green). Illustration is adapted from Walla et al. [57], licensed under the Creative Commons Attribution 4.0 International license (CC BY 4.0)

similar distances to the neighboring amino acid side chains of K48, and E44, respectively. *Lk*ADH mutants T102E and Q126K were generated, overexpressed in *E. coli*, and purified according to [57]. Purified and enzymatically active protein solutions were crystallized in μL batch and crystals were analyzed via X-ray diffraction as described in Walla et al. [57]. The refined *Lk*ADH crystal structures were compared with the corresponding *Lb*ADH variants, as depicted below (Fig. 5).

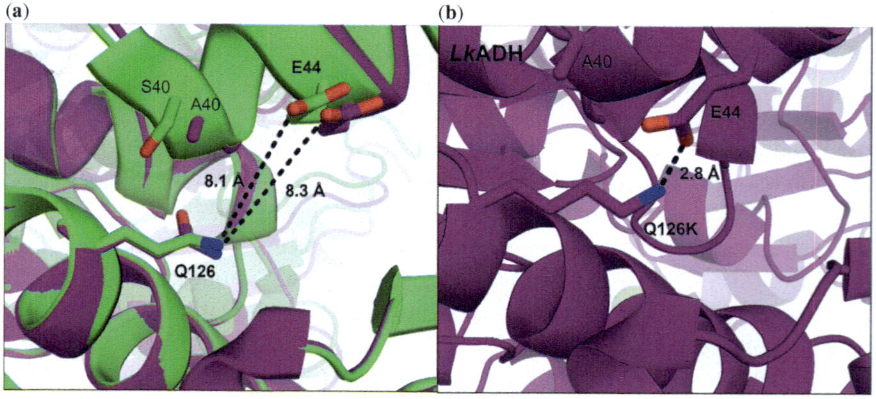

Fig. 5 **a** Illustration of *Lk*ADH (magenta, PDB ID: 7P36) and *Lb*ADH (green, PDB ID: 6H07) wild types' intermolecular crystal contact at position Q126 and **b** *Lk*ADH Q126K (PDB ID: 7P7Y) with neighboring amino acids E44, S/A40, and D41. Figures were generated and distances were calculated using PyMOL (v.2.3; Schroedinger). The illustration is adapted from Walla et al. [57], licensed under the Creative Commons Attribution 4.0 International license (CC BY 4.0)

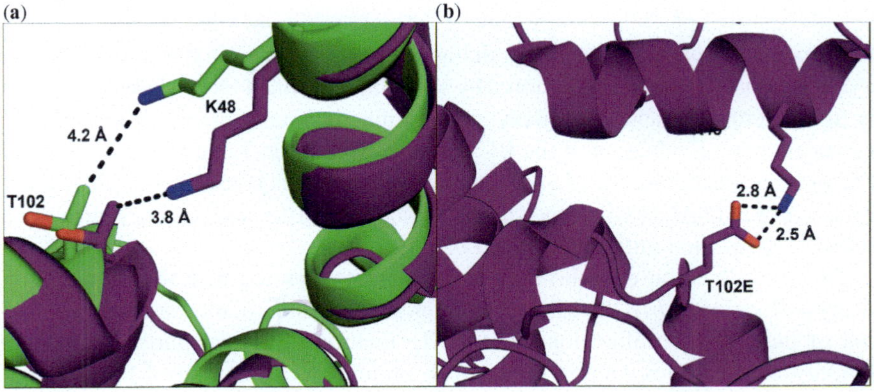

Fig. 6 **a** Illustration of *Lk*ADH (magenta, PDB ID: 7P36) and *Lb*ADH (green, PDB ID: 6H07) wild types' intermolecular crystal contact at position T102 and **b** *Lk*ADH T102E with neighboring amino acids K48. Figures were generated and distances were calculated using PyMOL (v.2.3; Schroedinger)

Q126K is showing for both *Lb*ADH and *Lk*ADH a strong reduction of distance to the neighboring E44 at the crystal contact from >8 Å to 3.9 Å (*Lb*ADH, Fig. 2b) and 2.8 Å (*Lk*ADH, Fig. 5b), respectively. Furthermore, the interacting positively charged ammonium group of lysine and the negatively charged carboxyl group of glutamic acid are facing each other in contrast to the *Lk*ADH wild type. Evaluating the electron density of the interacting functional groups of glutamic acid and lysine is also demonstrating an enhanced interaction [57]. Notably, the *Lb*ADH crystal structure reveals two possible conformations for E44 (Fig. 2a), whereas in *Lk*ADH there is one conformation (Fig. 5b). Combined with the decreased Q126K-E44 distance, a stronger interaction for *Lk*ADH Q126K can be assumed than for the *Lb*ADH mutant.

*Lb*ADH/*Lk*ADH mutant T102E shows a reduction in the distance between the carboxyl group of T102E and the ammonium group of K48 of about 1 Å (Fig. 6) compared to the wild type. Similar to Q126K, the functional groups of glutamic acid and lysine are facing each other in the respective mutant, indicating an established interaction at this crystal contact.

4.3.2 Stirred Crystallization of Clarified *E. Coli* Cell Lysate

In the next step, the crystallization was scaled up from the static μL batch to stirred 5 mL batch crystallizer. For comparability with the static crystallization results, all production process steps were conducted as described before [20, 57]. However, to reduce the process steps for preparative crystallization, the scale-up was performed with non-purified, dialyzed, and clarified *E. coli* cell lysate.

At first, Grob et al. [20] demonstrated a robust and reproducible method for clarified *E. coli* cell lysate crystallization of *Lb*ADH wild type and selected mutants Q126H and T102E, among others. For crystallization initiation, crystallization buffer

(0.1 M Tris–HCl pH 7.0, 50 mM $MgCl_2$, 200 g L^{-1} PEG MME 550) was added to the clarified *E. coli* cell lysate with defined concentration (initial $A_{280} = 20$). The crystallization process was monitored by sampling in defined intervals and evaluated by light microscopy and BCA assay to determine the total protein concentration in the supernatent. Evaluation of the BCA assay results (Fig. 7a) shows a decrease in total protein concentration of *Lb*ADH mutants T102E and Q126H within 6 h of about 50%, whereas *Lb*ADH wild type started crystallizing after about 24 h, indicating a faster crystallization initiation for the *Lb*ADH mutants.

The crystallization equilibrium for the mutants was reached after about 20 h of crystallization (*Lb*ADH wild type: after >44 h), thus demonstrating a reduction of the total crystallization duration. Evaluation of crystal purity resulted in a more than tenfold reduction of host cell protein (HCP) compared to the initial

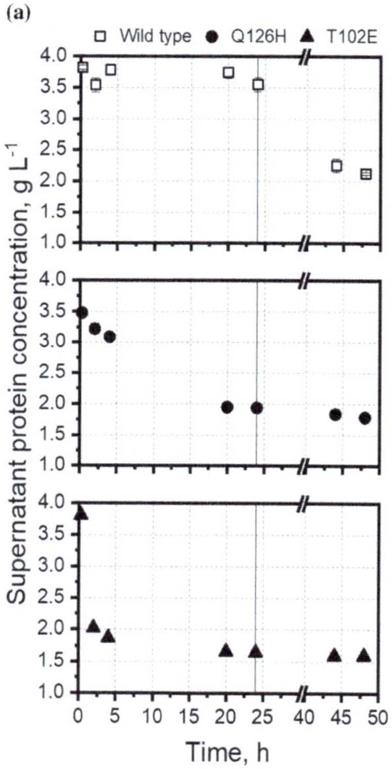

Fig. 7 **a** Protein crystallization of dialyzed and clarified *E. coli* cell lysate of *Lb*ADH wild type, Q126H, and T102E in 5 mL stirred crystallizer, monitored by protein concentration reduction of the supernatant via BCA. **b** Photomicrographs were taken after 24 h of crystallization ($n_{Stirrer}$ = 150 rpm, 20 °C, 200 g L^{-1} PEG MME 550) with tenfold diluted samples. The illustration is adapted from Grob et al. [20], licensed under the Creative Commons Attribution 4.0 International license (CC BY 4.0), licensed under the Creative Commons Attribution 4.0 International license (CC BY 4.0)

dialyzed *E. coli* cell lysate, with a further tenfold HCP reduction after recrystallization [20], considering the crystallization as a successful purification step.

After crystallization equilibrium was reached, the crystal size of the *Lb*ADH variants was analyzed. A reduction of crystal size was observed for the *Lb*ADH mutants along with an increase in crystal number, as depicted below (Fig. 7b). Therefore, the term "increased crystallizability" was extended to the following attributes: decreased crystallization induction time, decreased duration until crystallization equilibrium, crystallization at reduced protein and crystallization agent (PEG MME 550) concentration, and observation of a higher number of crystals with a concomitant decrease of crystal size [20]. Evaluation of these attributes and MD simulations resulted in an "order of crystallizability" as follows: T102E > Q126H (> Q126K) > *Lb*ADH wild type [20, 25]. Additionally, this order matches the results of the static μL batch crystallization, demonstrating successful scalability of the crystallization approach.

To verify the transferability of stirred crystallization to *Lk*ADH, selected *Lb*ADH mutations were transferred to *Lk*ADH and the *Lk*ADH variants were tested on a mL scale [57]. Therefore, the production and crystallization process for the stirred crystallizer was applied for *Lk*ADH wild type and mutants T102E and Q126K. The difference in reducing total protein concentration is not as significant as for *Lb*ADH variants, but a faster decrease in protein concentration is observed for *Lk*ADH T102E than for the wild type (Fig. 8). Comparing the *Lk*ADH variant concentrations densitometrically (evaluation of SDS-PAGE bands) after 48 h, a decrease in target protein concentration of 80% (*Lk*ADH wild type), 85% (Q126K) and 95% (T102E) was revealed.

Since the *Lk*ADH wild type already crystallized fast under the same conditions as the *Lb*ADH wild type, a further step of the sample preparation before crystallization–dialysis–was omitted. Now the *Lk*ADH wild type did not crystallize anymore, but the mutants T102E and Q126K did, as depicted below (Fig. 9).

To compare the apparent difference in crystal size and quantity, the photomicrographs of crystallized *Lk*ADH wild type and mutants (Fig. 8) were evaluated. The crystal length and width were analyzed and plotted against the frequency density (Fig. 10). The log-normal distributed frequency density of *Lk*ADH crystal mean length (97.1 ± 29.5 μm; mean width: 11.4 ± 4.5 μm) is shifted to a smaller mean crystal length by 40% for Q126K (mean length/width: 58.9 ± 9.8 μm/10.2 ± 2.9 μm) and by more than 70% for T102E (mean length/width: 27.5 ± 6.4 μm / 5.9 ± 1.6 μm).

Because the decrease in protein crystal size of the *Lk*ADH mutants is concomitant with an increase in crystal quantity, Walla et al. [57] presumed a higher number of nucleation events for the mutants compared with the *Lk*ADH wild type, indicating increased molecular interactions in T102E and Q126K. The evaluation of mL-scale protein crystallization revealed an order of crystallizability of T102E > Q126K > *Lk*ADH wild type, which corresponds to the order of crystallizability for *Lb*ADH variants. Thereby, for the first time, a successful transfer of a crystal contact engineering strategy was demonstrated between homologous enzymes.

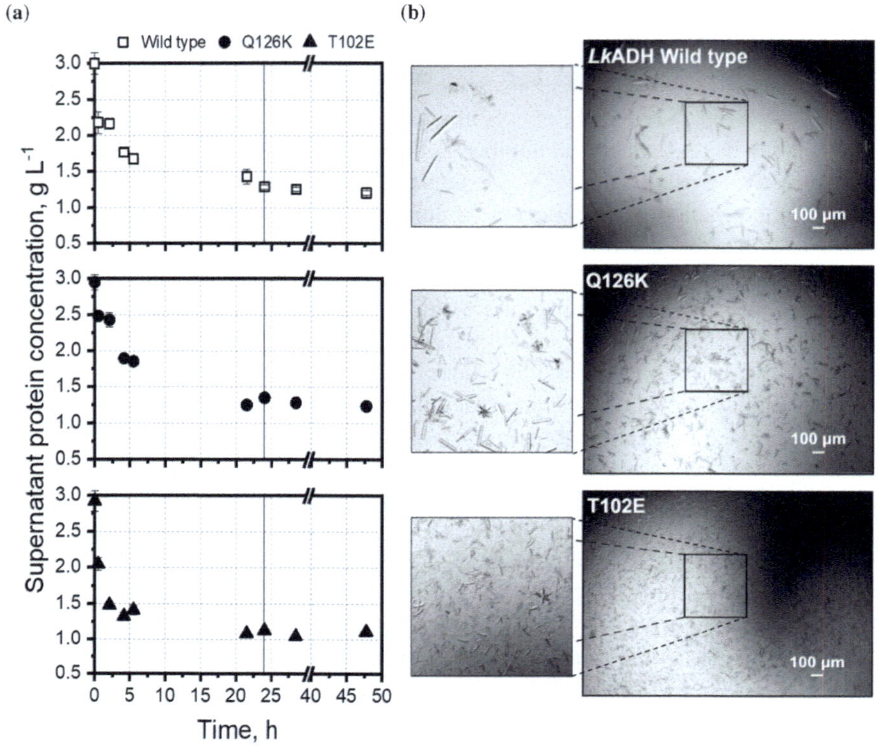

Fig. 8 Protein crystallization of dialyzed and clarified *E. coli* cell lysate of *Lk*ADH wild type, Q126K and T102E in 5 mL stirred crystallizer, monitored by protein concentration reduction of supernatant via BCA. Photomicrographs were taken after 48 h of crystallization ($n_{Stirrer} = 150$ rpm, 20 °C, 200 g L^{-1} PEG MME 550) with tenfold diluted samples. The illustration is adapted from Walla et al. [57], licensed under the Creative Commons Attribution 4.0 International license (CC BY 4.0)

5 Machine Learning-Based Image Analysis

5.1 Introduction

In industrial and academic settings, monitoring protein crystallization processes non-invasively by microscopic photography and automated image evaluation remains a challenging problem while being essential for rapid development cycles and precise process control. Crystals may appear predominantly transparent, surrounded fully or partially by visible edges due to the system's refractive properties and lighting conditions. Additionally, visual monitoring of general in-situ crystallization processes may be subject to varying sources of noise and varying object sizes, resulting in (non-trivial) challenges for image analysis techniques that are further strained by the requirement for fast evaluation times to enable statistically informed feedback for process controls. For the first time, recent trends involving the use of synthetic data

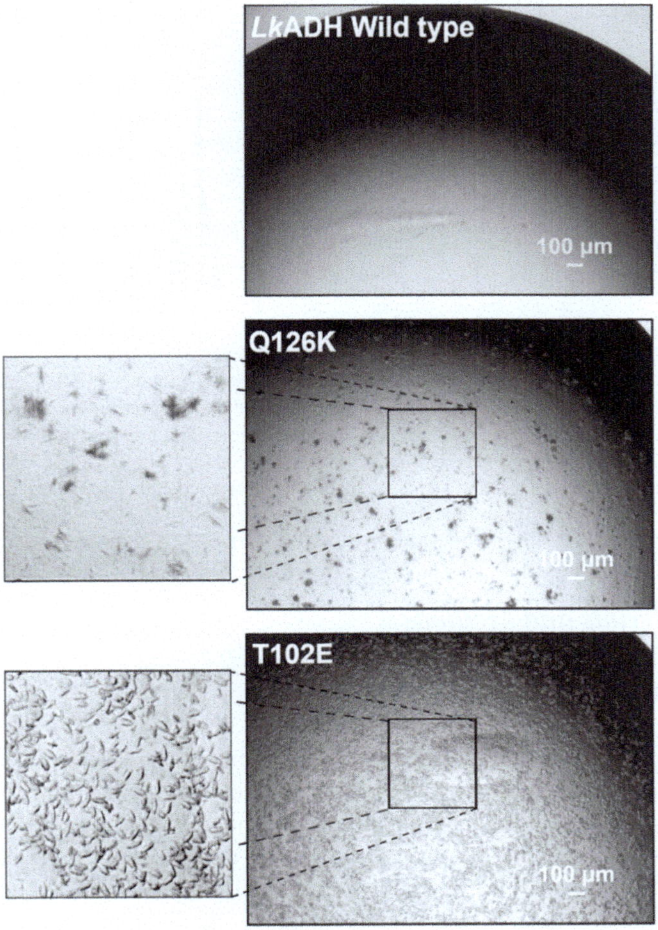

Fig. 9 Protein crystallization of clarified *E. coli* cell lysate of *Lk*ADH wild type, Q126K and T102E in 5 mL stirred crystallizer. Photomicrographs were taken after 48 h of crystallization ($c_0 = 5$ g L^{-1}, $n_{Stirrer} = 150$ rpm, 20 °C, 150 g L^{-1} PEG MME 550) with tenfold diluted samples. The illustration is adapted from Walla et al. [57], licensed under the Creative Commons Attribution 4.0 International license (CC BY 4.0)

sets for supervised learning are transferred, thus generating photorealistic images of virtual protein crystals in suspension (PCS) through the use of ray tracing algorithms, accompanied by specialized data augmentations modeling experimental noise. It is demonstrated that state-of-the-art models trained with the large-scale synthetic PCS data set outperform similar fine-tuned models based on the average precision metric on a validation data set, followed by experimental validation using high-resolution photomicrographs from stirred tank protein crystallization processes [5].

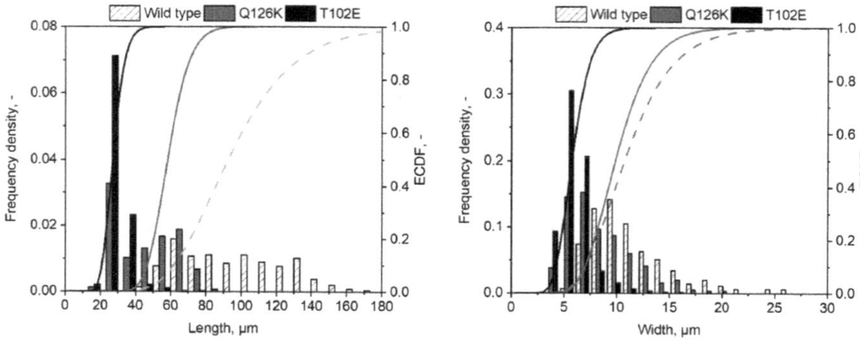

Fig. 10 Protein crystal size distribution of *Lk*ADH wild type and mutants T102E and Q126K (Bar: Frequency density distribution; Line: Empirical cumulative distribution function (ECDF) of wild type (dashed line), Q126K (grey line), and T102E (black line)). Photomicrographs were taken after 48 h of crystallization ($c_0 = 3$ g L^{-1}, $n_{Stirrer} = 150$ rpm, 20 °C, 200 g L^{-1}PEG MME 550) with tenfold diluted samples. The length and width of the protein crystals ($n = 335$) was evaluated with OriginPro 2020. The illustration is adapted from Walla et al. [57], licensed under the Creative Commons Attribution 4.0 International license (CC BY 4.0)

5.2 Materials and Methods

5.2.1 Data Set Generation

For the monitoring of crystallization processes by image analysis methods and deep learning, the first step was the generation of a synthetic data set, designed for modern object detection algorithms that leverage deep learning. Here, a statistical model was defined, that guides the generation of an unbiased data set. The observations of the statistical model are defined to be the crystal ratio, its area, and its angle in the photomicrography. To obtain an unbiased data set, the distributions underlying the outcomes of the observation were chosen as broad as possible. This means uniform distributions for the ratio and the angle of crystal objects. For the area, a different approach is required, since a uniform distribution of areas would lead to the majority of the synthetic image being covered by larger crystals. Instead, a reciprocal distribution leads to an equal contribution of crystal area in the image from all possible crystal size classes. Once a particular crystal object is drawn from the underlying distributions, the remaining task is its placement into a virtual 3D scene, such that the observables exactly match the sample from the statistical model. This is achieved by the formulation of a projection matrix that transforms 3D coordinates onto the 2D image plane. Under this transformation, the set of 3D points can be mapped to a 2D rotated rectangle parametrized by its area, ratio, and angle by calculating the convex hull of the mapped 2D coordinates followed by the rotating calipers method [53, 56]. The calculated parameters of the rotated rectangle are then used to calculate the difference between the sample of the statistical model, and the current projected 3D crystal model. Adjustment of the 3D models' volume and orientation during a minimization procedure leads to acceptance once the differences between the ratio is smaller than 0.2, square root of the area smaller than 2, and difference in angle smaller than 2 deg. Once a sample has been accepted, the corresponding model is assigned a glass material with varying, uniformly distributed index of refraction $n \sim U[1.1,1.8]$ with the 3D modeling software Blender (v2.93; Blender Foundation; Amsterdam; Netherlands). Here, limiting the data set to grayscale images enabled a reduction of the complexity of the modeling process while retaining the required information for the crystal detection task. Subsequent rendering through ray tracing results in photo-realistic base images that are used as inputs for the designed data augmentation pipeline. The final PCS data set consists of 322,558 images and 16,801,555 crystal instances. An additional PCS validation data set was prepared for validation purposes and consists of 10,000 images and 513,337 crystal instances [5].

5.2.2 Data Augmentation

Synthetic noise is modeled such that detectors trained on the PCS data sets can perform under a wide range of experimental conditions. Commonly used augmentations are used, such as brightness or contrast augmentations. Additionally, more realistic noise that one might perceive during an experiment is introduced through the use of Perlin augmentations (precipitation and/or aggregation), as well as additional interference patterns (thin-film interference) modeled by the superposition of multiple sine waves. The scale of these noise layers can be individually adjusted and is set to a few pixels to the diagonal spanning the input image dimensions. The final blending of these noise layers was performed using a weighted average, with a default weight of 1 for the input image and varying, randomized weights for the noise layer depending on the brightness value at a particular pixel. By observing that both types of noise primarily result in dark patterns propagating over the image, noise layer weights are additionally suppressed for high-intensity values down to a factor of 1/2 and reinforced up to a factor of 2 for low-intensity values. Therefore, the weights for the noise layer of the weighted average are calculated as

$$w_l = 5\sigma_b l_\alpha 2^{-2l+1}$$

In addition to local or spatially correlated noise layers, object-level augmentations alter the intensity values of crystal objects relative to their environment, or faintly outline their borders. Furthermore, a small chance for random splines exists, which are drawn over the image with varying intensity values and might represent scratches or other foreign residues. Similarly, it is possible to round corners to simulate images taken from round reactor volumes, usually characterized by darkened pixel values. Finally, the edges of crystal objects can be distorted by displacing entire rows or columns of pixels [5].

5.2.3 Machine-Learning

Multiple models based on the S2A-Net oriented-object detection model [21] were trained in either fine-tuning or trained from random initialization experiments. The used model has a ResNet-50 backbone and uses a feature pyramid network to detect objects at different scales. The trained models were altered in the following way:. First, two different anchor box configurations were tested and compared. Second, crystal detection relies on crucial information on the scale of a few pixels to detect faint edges in the photomicrographs. Therefore, the usual max-pooling step (resolution reduction) in the initial model is skipped, at the cost of additional computational demands. Model parameters were initialized from pre-trained ImageNet backbones for fine-tuning experiments with the first stage of the model frozen during the training procedure. Fine-tuning experiments furthermore employ common fine-tuning protocols for this model. Therefore, eight passes over the PCS training data set were conducted with learning rate (LR) 0.01, followed by three passes with LR of 0.001, and a final pass with LR 0.0001.

Training from random initialization requires longer training schedules. For these experiments, 80 epochs with LR 0.01 are used, then 15 epochs with LR 0.001, and finally 15 epochs with LR 0.0001. All machine learning experiments were conducted on four GPUs (Tesla V100) with two images per GPU, utilizing group normalization to stabilize training from random initialization. The model performance, as well as the mentioned model modifications, were evaluated on the dedicated PCS validation data set. Here, the common average precision metric is used to measure detector performance. For validation and the calculation of the AP metric, code from the Facebook AI Research Detectron2 project was used [61]. Here, the AP metric can be additionally categorized into APs for crystals with their oriented bounding box smaller than $a < 8\hat{}^2$, AP m for $8\hat{}^2 < a < 24\hat{}^2$ and APl for $24\hat{}^2 < a$ crystals [5].

5.2.4 Crystallization and Imaging

To evaluate the trained model experimentally, crystals of the *Lactobacillus kefir* alcohol dehydrogenase (*Lk*ADH) were used. *Lk*ADH wild type was mutated by QuikChange site-directed mutagenesis with partial overlapping primers according to Walla et al. [57] to generate the mutants T102E, and Q126K, respectively. *E. coli* BL21(DE3) was transformed with the selected *Lk*ADH variants and the proteins were produced in shaking flasks according to Grob et al. [20]. Ultrasound and centrifugation were used for cell lysis of the harvested *E. coli* cell pellets (12000 g, 4 °C, 1 h). The resulting supernatant of the *E. coli* cell lysate was dialyzed (20 mM HEPES/NaOH pH 7.0, 1 mM $MgCl_2$) and filtrated (0.2 μm polypropylene syringe filter). Stirred batch crystallization ($V = 5$ mL, $n_{stirrer} = 150$ rpm) was initiated with the addition of equal amounts of precipitation agent (0.1 M Tris/HCl pH 7.0, 50 mM $MgCl_2$, 200 g L^{-1} PEG MME 550) to clarified and dialyzed *E. coli* cell lysate. The stirred crystallizer was located in a temperature controlled water bath (20 °C) according to Walla et al. [57]. Samples were taken at regular intervals within 32 h for protein analysis and image generation. Further crystallization was prevented by diluting the samples tenfold with buffer (20 mM HEPES/NaOH pH 7.0, 1 mM $MgCl_2$). Photomicrographs were taken automatically by a digital camera (DS-Fi3, 0.84 μm per pixel at 2880 × 2048 pixel) attached to the microscope.

5.3 Results and Discussion

5.3.1 Validation of Trained Crystal Detection Models

Using the described procedure, in total 332,558 synthetic images were created with a resolution of 384 × 384 pixels each. Additionally, a specialized data augmentations pipeline was created that produces highly realistic scenes of crystallization experiments. Individual crystal instances were generated using a statistical model, therefore guaranteeing an unbiased data set. Different variations of the S2A-Net oriented-object detection (OOD) model were trained and compared. Therefore, model modifications were determined, which further increased the crystal detection performance, as measured by the average precision metric (AP). First, decreasing the size of the smallest anchor from 16 to 8 pixels increased the AP metric for small objects. Second, a max-pooling layer after the first convolutional layer is normally used to downsample the input image and increase the overall inference and training time of the model. Skipping this resolution reduction step further boosted crystal detection performance for smaller objects. This might be due to the very fine details of the crystal edges in the input image being essential for the detection task. Lastly, the large-scale synthetic PCS data set enabled longer training procedures initiated from random initialization, which are further shown to increase the model performance [5].

5.3.2 Experimental Validation

Multiple crystallization experiments were conducted to test whether the models trained on synthetic data can also perform on real photomicrographs. Here, three technically replicated protein crystallization processes were conducted. During the experiment, high-resolution photomicrographs with 2272 × 1632 resolution were taken to capture the crystal growth process of the *Lk*ADH (Fig. 11).

Diluted samples result in clear images, whereas undiluted samples contain noise due to the unprocessed sample having high amounts of host cell protein. Notably, for the diluted samples, seemingly all crystals are detected by the model. However, Fig. 11a shows that the modified S2A-Net OOD model detects fewer crystals than in the undiluted case, although Fig. 11b shows, that the extracted crystal size distributions match between the diluted and undiluted samples. Additionally, the model was evaluated for varying crystal sizes and aspect ratios. Therefore, crystallization of four *Lk*ADH variants (wild type, T102E, Q126K, and Q126H) with different crystallization kinetics was studied. The resulting crystal shape was compared with previous

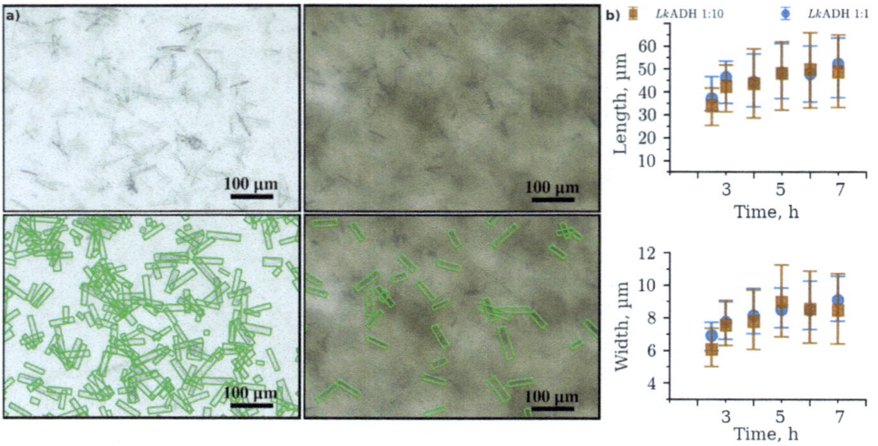

Fig. 11 Diluted 10 µL droplet samples from a *Lk*ADH crystallization process (**a**). In photomicrographs of diluted samples, almost all crystals are detected (left), while fewer crystals are detected in the presence of noise (middle). However, the resulting crystal size distributions are shown to be equal (**b**). Illustration is adapted from Bischoff et al. [5], licensed under the Creative Commons Attribution 4.0 International license (CC BY 4.0)

work on the homologous *Lactobacillus brevis* ADH (*Lb*ADH) that covered manually evaluated crystal shape distributions. Experiments with the same *Lk*ADH mutants yield matching order of crystal sizes: WT > Q126K > T102E. However, while the *Lb*ADH wild type previously had the highest aspect ratio among these variants, the current work shows higher ratios and smaller crystals for *Lk*ADH mutant Q126K compared to the *Lk*ADH wild type. Here, a possible explanation might be that the model detects more of the smaller crystals compared to human labeling. [5]

6 Analyzing Multi-Component High-Throughput Crystallization Screenings Combining PLS Regression Modeling and UV/Vis Spectroscopy

6.1 Introduction

Early stage process development of selective protein crystallization processes requires fast and reliable analytics. With this project, we aimed to develop a low-volume, quantitative, analytical tool to quickly assess crystallization processes regarding crystal yield, purity and selectivity. The basis of the analytical tool was the combination of ultraviolet visible light (UV/Vis) spectroscopy and partial least squares modeling. In order to selectively quantify protein concentrations in heterogeneous supernatants, a ternary model protein system was chosen for the model calibration and subsequent crystallization screenings using the same model protein system. The screening parameters pH, precipitant concentration, and target molecule concentration were varied to test the analytical tool in different screening conditions. With a 4 times shorter analysis time and a minimum sample uptake of 4 μL the developed tool quantified the three different model proteins with high precision. This said, the methodology could be easily adapted to higher throughput scales, i.e., 384 well or 1536 well plates.

6.2 Materials and Methods

The experimental and analytical setup is described in detail by Wegner et al. [59]. All chemicals were purchased from Merck KGaA (Darmstadt, DE), unless stated otherwise. The protein stock solutions were prepared in a multi-component buffer (MCB: 21 mM N-1,1-dimethyl-2-hydroxyethyl-3-amino 2-hydroxypropanesulfonic acid (AMPSO), 17 mM 3-N-morpholino propansulfonic acid (MOPS, Carl Roth GmbH + Co. KG, Karlsruhe, DE), 15 mM succinate acid (AppliChem GmbH, Darmstadt, DE)) at pH 7 or 9 with the three model proteins Lys from chicken egg white (Hampton Research, Aliso Viejo, CA US), Ribonuclease A (RibA) from bovine pancreas and Cytochrome C (CytC) from equine heart. The protein and buffer stock solutions were filtered with 0.2 μm (Pall Corporation, Port Washington, NY US) and 0.02 μm syringe filters (Cytiva, Marlborough, MA US), respectively.

The calibration ranges were chosen to be 0–1.5 mg/mL for Lys and 0.042–0.273 mg/mL for the two impurity species RibA and CytC. The concentration ranges of the validation solutions were 0.317–2.049 mg/mL and 0.042–0.273 mg/mL for Lys and the impurities, respectively. The calibration and validation stock solutions were mixed manually.

The screening solutions were prepared in duplicates and were pipetted by a liquid handling station (Tecan Freedom Evo 100, Tecan, Männedorf, CH). Three protein stock solutions were mixed with the MCB and 3.5 M ammonium sulfate (AMS, AppliChem GmbH, Darmstadt, DE) precipitant concentration at target pH according to Fig. 12 to screen 96 different conditions per screening. The different conditions were incubated for 13 days at 8 °C in a cooled system (RI 54, FORMULATRIX, Bedford, MA US; T 1000 mytron Bio- und Solartechnik GmbH, Heiligenstadt, DE) and automatically imaged. After incubation, the supernatants were 50 times diluted with MCB, and filtered in 0.2 μm filter plates (Pall Corporation) prior analysis.

The calibration and validation solutions, and the diluted supernatants were analyzed using a Dionex Ultimate 3000 RS UHPLC system (Thermo Fisher Scientific Inc.), equipped with a diode array detector. The spectral analysis was performed with a pre-column filter cartridge (0.5 μm OPTI SOLV EXP, Supelco, Bellefonte, PA US) in the mobile phase (20 mM Tris, 100 mM NaCl, pH 8.0) but no chromatography column was installed. The UV/Vis absorbance spectra between 240 and 450 nm were recorded. As a reference, the same samples were analyzed with cation exchange chromatography (CEX) using a ProSwift SCX 1S 4.6 × 50 mm column

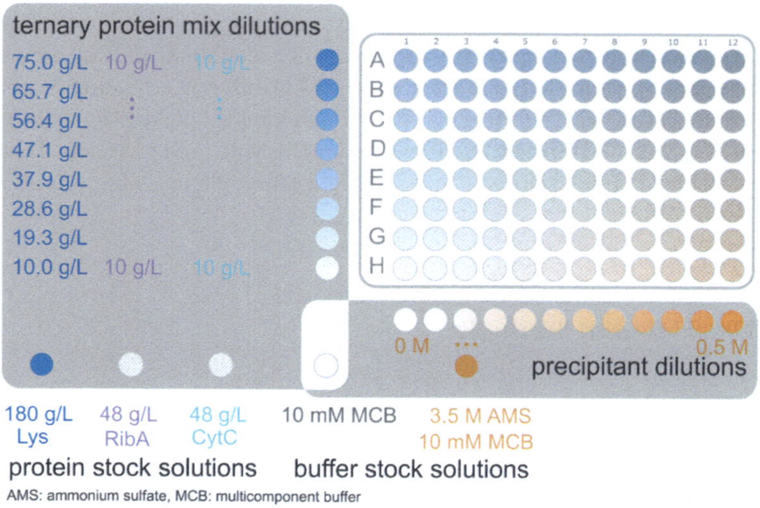

Fig. 12 Eight ternary protein dilutions were prepared from three protein stock solutions and combined with MCB at the target pH. To prepare the crystallization screening solutions, the protein solutions were mixed with 12 different precipitant dilutions prepared from MCB and AMS stock solutions. 96 conditions in 24 μL batches were investigated at two different pH levels. Subsequently, the crystallization screening plates were stored at 8 °C for 13 days. The illustration is adapted by permission from ACS Publications from Wegner et al. [59]

(Thermo Fisher Scientific Inc.) and the same UHPLC system (low salt buffer: 20 mM Tris, pH 8.0; high salt buffer: 20 mM Tris, 1000 mM NaCl, pH 8) with a flow rate of 1.5 mL/min.

Preprocessing of the spectral data consisted of background subtraction, summation along the time axis, selection of the essential wavelength range, and application of the Savitzky-Golay filter [50]. Subsequently, the sought-after protein concentrations derived from the reference CEX data were regressed on the preprocessed spectral data to calculate one PLS model per protein species. All data preprocessing, model calculation, and model application were performed in MATLAB, R2019b (The MathWorks, Inc., Natick, MA US).

For faster, quantitative assessment of crystallization screening conditions and optimal process time, the PLS models were applied to diluted supernatants of phase diagrams and from a kinetic study enabling elaborate crystallization yield and purity estimation.

6.3 Results and Discussion

In order to develop crystallization processes fast HT screenings and reliable analytics are crucial. The developed analytical tool may not be affected by temperature, different aqueous conditions, e.g. pH or ionic strength of the solution, and crystallization kinetics. Thus, these parameters were tested in a HT screening to show the flexibility of this tool. First, the accuracy of the models to predict the specific protein concentrations is discussed. Secondly, the application of the screenings to evaluate phase diagrams is shown. Last, the potential of this analytical tool on different crystallization processes is assessed.

6.3.1 Data Analysis and Model Accuracy

Contrasting the absorbance values integrated over time at a wavelength of 280 nm of the spectral and the reference CEX analysis, increased absorbance areas were observed for the reference analysis, in particular for the outer boundary wells of the 96 well plate. Sample evaporation caused the observed difference when samples were stored prior to analysis. Assuming similar relative specific concentration changes due to evaporation, the time-wise integrated absorbance areas were used to remove the absorbance area contribution caused by sample evaporation and correct the protein concentrations calculated from the reference CEX analysis. By applying this preprocessing method the different sample storage times could be neglected when an increased number of samples were analyzed during the HT screening Furthermore, the model prediction could be improved as proven by Table 1. The model calibration was evaluated with the parameters root mean squared error of cross-validation (RMSECV) and the coefficient of determination R2; the model validation with the root mean squared error of prediction (RMSEP) and predictive relevance Q2.

Each PLS model showed a different wavelength range as the protein absorption spectrum was different for each protein. Larger wavelength ranges increase the risk of overfitting [2] but showed the highest model accuracy, regarding the RMSECV and Q2, in the case of RibA (data not shown).

Spectral derivation can enhance spectral differences [6, 32, 49]. The Savitzky-Golay filter can smooth and derive spectra [48] to reduce the RMSEP and to improve PLS models and, thus, the model accuracy when protein species demonstrate similar absorbance spectra [9]. As the spectrum of RibA overlayed in the chosen wavelength range with the other species spectra, the 2nd spectral derivative was required to calculate a robust PLS model. Compared to one latent variable for CytC, the models

Table 1 Parameters for preprocessing the spectral data and calibrating the PLS models. The table is adapted from Wegner et al. [59]

	Lys	RibA	CytC
Wavelength range/nm	240–300	250–430	385–425
Latent variables	4	4	1
R2	1.0000	0.9949	0.9972
RMSECV/mg/mL	0.0019	0.0045	0.0038
Q2	0.9999	0.9955	0.9956
RMSEP/mg/mL	0.0048	0.0037	0.0040

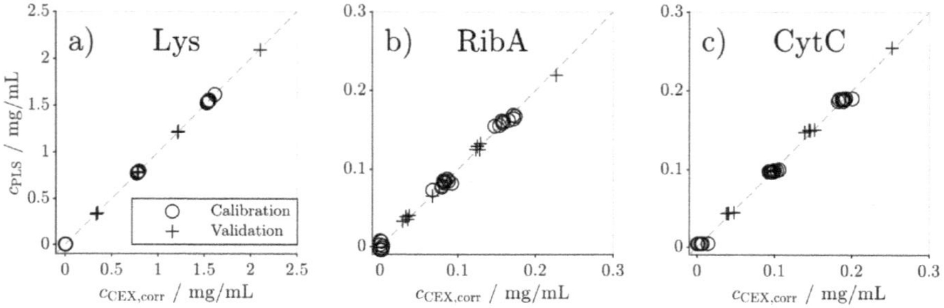

Fig. 13 The PLS model predicted c_{PLS} and CEX measured protein concentration $c_{\mathrm{CEX,corr}}$ are shown for the proteins Lys (**a**), RibA (**b**), and CytC (**c**). The reference data were corrected according to Wegner et al. [59]. The circles and crosses illustrate the calibration and validation data, respectively. The ideal relationship between the predicted and reference measurement is illustrated by the dashed lines. The illustration is adapted by permission from ACS Publications from Wegner et al. [59]

for Lys and RibA needed 4 latent variables as the absorbance spectra were affected by all species present in the solution. More latent variables in each PLS model include more spectral information and, thus, can help calculate robust models, but bear the risk of overfitting. When the coefficient of determination R2 and predictive relevance Q2 of the calibration and validation samples are similar and high, the validation data are reasonable and the models can predict the desired data well. RMSEP for the Lys model was twice as high compared to the RMSECV. It still lies in an acceptable range as the validation solutions contained Lys concentration values outside the calibration range.

Figure 13 shows the predicted concentrations of the calibrated PLS models in relation to concentrations calculated from the CEX reference analytics. The corresponding RMSEP values are listed in Table 1. The largest discrepancies occurred in the RibA concentration model and is presumably caused by the overlapping absorption area of the three model proteins.

Analogous to Fig. 13, the prediction accuracy of the models applied to diluted supernatants in the HT screening is shown in Fig. 14. The model application for the species CytC led to similar RMSEP values compared to the validation solutions. The RMSEP values for the remaining protein species were 1.8 to 4.9 times higher. Minor differences in the protein spectrum induced by the HT screening conditions and its effect on the ternary protein structure may cause these discrepancies. As opposed to the incubation time over 13 days with the precipitant AMS [4] at lower temperatures [17] during the screening, the calibration solutions were prepared and directly analyzed. The highest RMSEP values are calculated for the Lys model and can be explained by the higher average Lys concentration during the screening in contrast to RibA and CytC.

In a nutshell, the PLS models could be applied to the HT screenings with varying pH, target protein concentration, and precipitant concentration. The dilution of the sample supernatant quenched the crystallization process and allowed the analysis

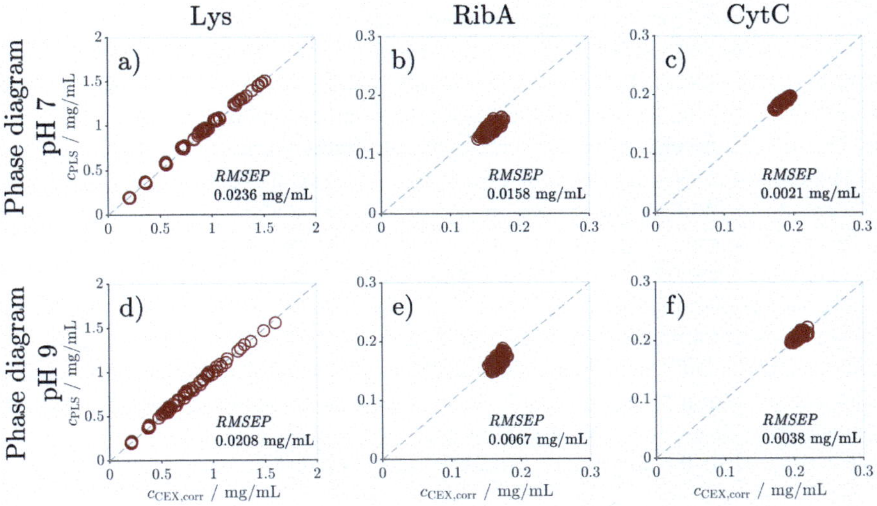

Fig. 14 The PLS predicted and reference data of the two-phase diagrams at pH 7 and 9 are shown in subfigures (**a**)–(**c**) and (**d**)–(**f**), respectively. The PLS model predicted protein concentration c_{PLS} is displayed over the reference protein concentration calculated from the CEX measurement $c_{CEX,corr}$. The reference data were corrected according to Wegner et al. [59]. The ideal relationship between the predicted and reference measurement is illustrated by the dashed lines. The illustration is adapted by permission from ACS Publications from Wegner et al. [59]

of timed process snapshots as the dilution step lowered the protein concentration below the saturation concentration of each protein species. The model accuracy of the analyzed HT screening samples was lower compared to the calibration and validation solutions as the incubation conditions may induce subtle conformational changes.

6.3.2 Crystallization Process Parameters

After all sample supernatants were analyzed, the specific, predicted concentrations were visualized in phase diagrams to investigate the phase behavior of multi-component solutions.

Figure 15 illustrates the protein specific phase behavior in phase diagrams at two different pH levels. The background color of each condition indicates the final supernatant concentration and is the saturation concentration assuming that the conditions reached the equilibrium after the incubation time. The missing drop in protein concentration is caused by the supersaturation and thus leads to crystallization. The circle areas visualize the crystal yield which is caused by supersaturation and is derived for Fig. 15 from the specific protein concentration of the supernatant and its estimated initial concentration of each screening condition. By this, the conditions could be distinguished between stable conditions and crystallizing conditions where phase transition occurred out of the supersaturated liquid. The screening conditions were monitored with microscopic images taken regularly. In the images mostly crystallization was observed. Due to low image resolution, the images of 4 conditions could be not evaluated and 9 conditions at the crystallization window boundary could not be classified as the micro-crystals and precipitate could not be clearly distinguished.

The crystallization windows of Lys in Fig. 15a and d are visible at lower salt concentrations. With increasing precipitant concentration, the crystal yield decreased as the supersaturation concentration increased. The highest Lys crystal yield values per screening were 64.6% at pH 9 and 36.6% at pH 7.

The two contaminant species RibA (Fig. 15b and e) and Cyt C (Fig. 15c and f) did not show a continuous area of crystallization and the smaller circles are assumed to be artifacts of dilution inaccuracy, analytics or model uncertainty. The higher RibA concentration drops lay outside the crystallization window of Lys. The highest concentration drops for CytC and Riba were 4.1 and 11.12%, respectively. Note, that compared to the target molecule Lys, the contaminant species were present

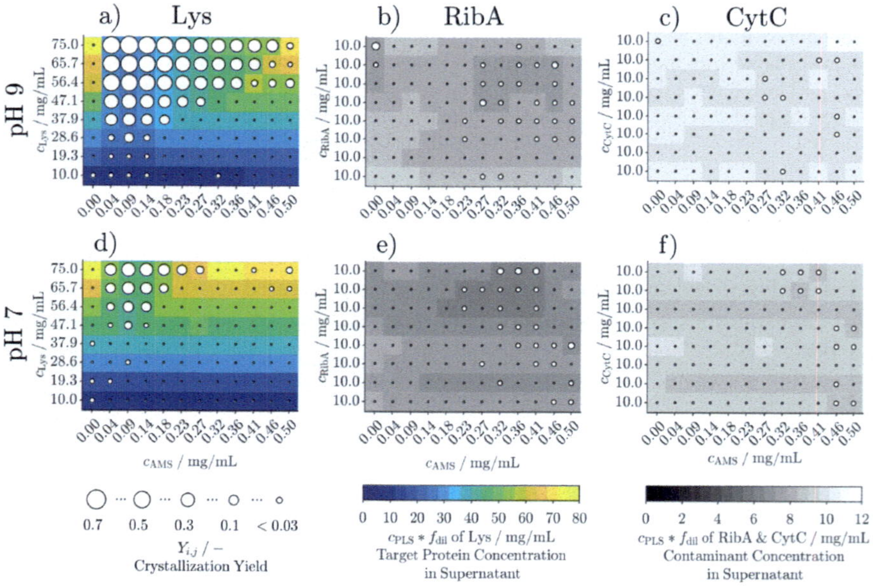

Fig. 15 The subfigures displays the examined plates at pH 7 (**a**)–(**c**) and at pH 9 (**d**)–(**f**) and, thus, the protein specific phase behavior, e.g., crystallization windows. The screening conditions are arranged according to Fig. 12. Each circle represents the examined screening condition and its position in the phase diagram. The circle area visualizes the protein specific crystal yield $Y_{i,j}$. The variable i and j stand for the protein and the screened well. The protein specific final concentration in the supernatant, calculated from the PLS models and the dilution factor, is illustrated by the circle background color. The different color scale represent the different concentration ranges for the target and contaminant proteins. The illustration is adapted by permission from ACS Publications from Wegner et al. [59]

at lower protein concentration ranges and, thus, pipetting errors and measurement inaccuracies have a larger effect on the yield calculation.

To further assess the purity of the generated crystals, the concentration drop of all three species was compared against the sum of the protein concentration drop. This led to theoretical Lys crystal purity values between 99.68 and 109.16% at pH 7, and 81.62 and 142.91% at pH 9. Purity values above 100% are caused by the purity calculation method as it is prone to inaccuracies in the experiment. Still, the model-derived purity shall be rather seen as a guideline to quickly evaluate crystallization screening conditions than an exact purity determination.

The pH strongly influences the Lys crystallization area as can be seen in Fig. 15. As the pI of Lys lies nearer to pH 9, the increased crystallization window at pH 9 can be caused by the fact that the pI of Lys lies nearer to pH 9. This is further supported by Lys crystallization studies of pure Lys screenings [33] bearing the longer incubation time and higher storage temperature in mind. The phase behavior of the contaminant species CytC and RibA was not affected by the pH during the screening. The Lys concentration strongly influenced the crystallization process as its effect on the supersaturation led the supersaturated solution to form crystals. As expected, the higher initial Lys protein concentration led to a higher yield and is caused by the different degree of supersaturation.

The presence of contaminants can influence crystal growth or crystal nucleus formation [1, 29]. Comparing a pure Lys crystallization screening at pH 9 with the multi-component screening performed for this study, the presence of the contaminants did not show an effect on conditions with high supersaturation but showed small deviations at the boundaries of the crystallization window when the supersaturation was low. The crystallization window of the pure Lys solutions was larger by 9 screened conditions and confirms that structurally unrelated contaminants influence Lys crystallization at low supersaturation levels [29].

To sum it up, screening conditions could be classified reliably in crystallizing and stable conditions with the developed analytical tool. Furthermore, this method allowed crystal yield and purity calculation and can be used to evaluate crystallizing conditions objectively regarding the screening environment, e.g. pH, protein or precipitant concentration, and impurities.

6.3.3 Potential of PLS UV/Vis Spectroscopy for Crystallization

The developed analytical workflow requires only small sample amounts and may be transferred to other protein systems or other protein solubility based processes, e.g. flocculation or precipitation as long as the species of interest and the impurities have an absorption spectrum and show a phase transition in a screening. As the analytical tool is HT compatible, it can quickly assess a large amount of screening conditions regarding process performance, purity and selectivity. The application of the developed method for crystallization kinetics is introduced by Wegner et al. [59]. When material resources are scare in early stage process development, the calibration and validation solutions can even be a subset of the actual screening if the proteins of interest undergo phase transition. The visualization of more dimensional data sets in one phase diagram per species can provide a broad overview over different screening conditions in a complex mixture and aid rating different screening conditions with quantitative values, e.g. crystal yield and purity. The workflow can be adjusted to the initial starting material to ensure a minimal sample uptake, quenching of the supernatant solution, and the linear absorbance range for the UV/Vis spectroscopic measurement. This makes it suitable for plate-based UV/Vis spectrophotometer measurements. Compared to standard image based analytics, the combination of UV/Vis spectroscopy and PLS modeling can provide easily quantitative information and distinguish between salt and protein crystals.

7 Conclusion and Outlook

On a molecular scale, studies were performed to rationally engineer the technically relevant *Lb*ADH to increase its applicability for preparative protein crystallization purposes, due to decreased crystallization process duration and thus increased space–time yields. For protein engineering, the SER strategy [14], aromatic interactions as well as electrostatic interactions at specific crystal contacts were investigated. This rational protein engineering at crystal contact areas resulted in enzymatically active *Lb*ADH mutants with enhanced crystallizability in comparison to *Lb*ADH wild type [46]. Using the definition of crystallizability from Grob et al. [20], the order of crystallizability could be determined as follows: T102E > Q126H > Q126K > *Lb*ADH wild type. Subsequent studies [57] demonstrated, for the first time, that rational protein engineering at crystal contacts leading to enhanced crystallizability can be transferred to homologous proteins shown on the example of *Lk*ADH with the order of crystallizability for mutants T102E and Q126K maintained.

Molecular dynamics (MD) free energy simulations enabled an in silico determination of relative free energy differences between crystal contacts of different protein variants. This was shown to correlate with the experimentally determined crystallizability [25]. Analysis of the MD simulation at an atomistic level suggests that there are no simple, universal rules for accurate control of the crystallization behavior, instead, the success of the protein engineering can be evaluated in silico by free energy change calculations. In future work, further optimizations and high-throughput calculations may be used for in silico screenings to narrow down the experimental design space.

Here, a novel approach to monitoring of crystallization processes on a microscopic scale was introduced, based on a large-scale and high-quality synthetic data set, as well as a state-of-the-art machine learning model for oriented object detection. The trained model was shown to be robust against varying source of noise and was able to detect crystals at different scales in experimental photomicrographs [5]. Future work in this area might apply this model with online microscopy probes in technical crystallization processes that require non-invasive quantitative analysis.

An additional non-invasive measurement for the monitoring of protein crystallization processes is UV/Vis spectroscopy. Coupled with chemometrics, it can selectively quantify different protein species in a HT multi-component screening and has versatile applications. This developed analytical tool evaluated the protein phase behavior of a ternary protein mixture in a HT screening [59]. The process variables parameters pH, precipitant concentration, and target protein concentration were varied and did not affect the model accuracy. With the calculated PLS models, the quantitative process state variablesparameters, e.g. crystal yield and purity, could be determined requiring only a minimal sample uptake of 4 μL per screening condition. The performed UV/Vis spectroscopy measurement was 4 times faster than the reference analytics. At small scale, the combination of UV/Vis spectroscopy and chemometric models can be further applied to complex solutions, and, next to crystallization, to other protein-solubility based processes, e.g. precipitation. At larger scales, the provided calibration and data handling workflow can help setting up real-time monitoring for selective crystallization processes in complex mixtures.

Modeling-based monitoring and control of protein crystallization in bioprocesses enables a holistic engineering understanding of fundamental principles of technical protein crystallization from the molecular scale (protein engineering of crystal contacts and MD simulation) to the microscopic scale (phase behavior monitored by machine learning-based image analysis) and the macroscopic scale (combining PLS regression modeling and UV/Vis spectroscopy in a PAT approach). Ongoing work expands the transfer of crystal contact engineering strategies to non-homologous and so far non-crystallized enzymes applying protein engineering of crystal contacts and MD simulation. Further work should address the scale up of stirred batch crystallization from cell lysate to further improve the applicability for industrial preparative purposes. In addition, further PAT strategies may be implemented to enable modeling-based online monitoring and control of protein crystallization in bioprocesses integrating microscopic and macroscopic scales.

Acknowledgements Funding by the German Research Foundation (DFG) is gratefully acknowledged: research projects WE 2715/14-1, WE 2715/14-2, and HU 985/3-2, respectively, within the framework of the priority program SPP 1934. The authors thank the X-ray Crystallography Platform at the Helmholtz Center Munich, Germany. Support of Brigitte Walla and Daniel Bischoff from the TUM Graduate School is gratefully acknowledged as well.

References

1. Abergel C, Nesa MP, Fontecilla-Camps JC (1991) The effect of protein contaminants on the crystallization of turkey egg white lysozyme. J Cryst Growth 110(1–2):11–19
2. Andersen CM, Bro R (2010) Variable selection in regression-a tutorial. J Chemom 24(11–12):728–737
3. Andris S, Rüdt M, Rogalla J, Wendeler M, Hubbuch J (2018) Monitoring of antibody-drug conjugation reactions with UV/Vis spectroscopy. J Biotechnol 288:15–22. https://doi.org/10.1016/j.jbiotec.2018.10.003

4. Baumgartner K, Galm L, Nötzold J, Sigloch H, Morgenstern J, Schleining K, Suhm S, Oelmeier S, Hubbuch J (2015) Determination of protein phase diagrams by microbatch experiments: Exploring the influence of precipitants and pH. Int J Pharm 479(1):28–40
5. Bischoff D, Walle B, Weuster-Botz D (2022) Machine learning-based protein crystal detection for monitoring of crystallization processes enabled with large-scale synthetic data sets of photorealistic images. Anal Bioanal Chem 414:6379–6391
6. Bosch Ojeda C, Sanchez Rojas F (2004) Recent developments in derivative ultraviolet/visible absorption spectrophotometry. Anal Chim Acta 518(1–2):1–24
7. Brange J (1987) Galenics of insulin. The physico-chemical and pharmaceutical aspects of insulin and insulin preparations, 1st edn. Springer, Berlin, Heidelberg
8. Bräutigam S, Dennewald D, Schürmann M, Lutje-Spelberg J, Pitner W-R, Weuster-Botz D (2009) Whole-cell biocatalysis: Evaluation of new hydrophobic ionic liquids for efficient asymmetric reduction of prochiral ketones. Enzyme Microb Technol 45(4):310–316. https://doi.org/10.1016/j.enzmictec.2009.06.015
9. Brestrich N, Rüdt M, Büchler D, Hubbuch J (2018) Selective protein quantification for preparative chromatography using variable pathlength UV/Vis spectroscopy and partial least squares regression. Chem Eng Sci 176(6):157–164
10. Cerutti DS, Freddolino PL, Duke RE Jr, Case DA (2010) Simulations of a protein crystal with a high resolution X-ray structure: evaluation of force fields and water models. J Phys Chem B 114(40):12811–12824. https://doi.org/10.1021/jp105813j
11. Cooper DR, Boczek T, Grelewska K, Pinkowska M, Sikorska M, Zawadzki M, Derewenda Z (2007) Protein crystallization by surface entropy reduction: optimization of the SER strategy. Acta Crystallogr D Biol Crystallogr 63(Pt 5):636–645. https://doi.org/10.1107/s0907444907010931
12. Cournia Z, Allen B, Sherman W (2017) Relative binding free energy calculations in drug discovery: recent advances and practical considerations. J Chem Inf Model 57(12):911–2937. https://doi.org/10.1021/acs.jcim.7b00564
13. Dennewald D, Pitner W-R, Weuster-Botz D (2011) Recycling of the ionic liquid phase in process integrated biphasic whole-cell biocatalysis. Process Biochem 46(5):1132–1137. https://doi.org/10.1016/j.procbio.2011.01.032
14. Derewenda ZS (2004) Rational protein crystallization by mutational surface engineering. Structure 12(4):529–535. https://doi.org/10.1016/j.str.2004.03.008
15. Derewenda ZS, Vekilov PG (2006) Entropy and surface engineering in protein crystallization. Acta Crystallogr D Biol Crystallogr 62(Pt 1):116–124. https://doi.org/10.1107/s0907444905035237
16. Emsley P, Lohkamp B, Scott WG, Cowtan K (2010) Features and development of Coot. Acta Crystallogr Sect D Biol Crystallogr 66:486–501
17. Esfandiary R, Hunjan JS, Lushington GH, Joshi SB, Middaugh CR (2009) Temperature dependent 2nd derivative absorbance spectroscopy of aromatic amino acids as a probe of protein dynamics. Protein Sci 18(12):2603–2614
19. Food and Drug Administration (2004) Guidance for industry, PAT-A framework for innovative pharmaceutical development, manufacturing and quality assurance. Accessed from http://www.fda.gov/downloads/Drugs/GuidanceComplianceRegulatoryInformation/Guidances/ucm070305.pdf
20. Grob P, Huber M, Walla B, Hermann J, Janowski R, Niessing D, Hekmat D, Weuster-Botz D (2020) Crystal contact engineering enables efficient capture and purification of an oxidoreductase by technical crystallization. Biotechnol J e2000010. https://doi.org/10.1002/biot.202000010
21. Han J, Ding J, Li J, Xia G-S (2022) Align deep features for oriented object detection. IEEE Trans Geosci Remote Sens 60:1–11. https://doi.org/10.1109/TGRS.2021.3062048
22. Harrison RG, Todd PW, Rudge SR, Petrides DP (2003) Bioseparations science and engineering, 2nd edn. Oxford University Press, New York. ISBN: 9780195391817
23. Hekmat D (2015) Large-scale crystallization of proteins for purification and formulation. Bioprocess Biosyst Eng 38(7):1209–1231. https://doi.org/10.1007/s00449-015-1374-y

24. Hekmat D, Huber M, Lohse C, von den Eichen N, Weuster-Botz D (2017) Continuous crystallization of proteins in a stirred classified product removal tank with a tubular reactor in bypass. Cryst Growth Des 17(8):4162–4169. https://doi.org/10.1021/acs.cgd.7b00436
25. Hermann J, Bischoff D, Grob P, Janowski R, Hekmat D, Niessing D, Weuster-Botz D (2021) Controlling protein crystallization by free energy guided design of interactions at crystal contacts. Crystals 11(6):588. Accessed from https://www.mdpi.com/2073-4352/11/6/588
26. Hermann J, Nowotny P, Schrader TE, Biggel P, Hekmat D, Weuster-Botz D (2018) Neutron and X-ray crystal structures of Lactobacillus brevis alcohol dehydrogenase reveal new insights into hydrogen-bonding pathways. Acta Crystallogr Sect F 74(12):754–764. https://doi.org/10.1107/S2053230X18015273
27. Hummel W (1997) New alcohol dehydrogenases for the synthesis of chiral compounds. Adv Biochem Eng Biotechnol 58:145–184. https://doi.org/10.1007/BFb0103304
28. Jenkins TM, Hickman AB, Dyda F, Ghirlando R, Davies DR, Craigie R (1995) Catalytic domain of human immunodeficiency virus type 1 integrase: identification of a soluble mutant by systematic replacement of hydrophobic residues. Proc Natl Acad Sci U S A 92(13):6057–6061. https://doi.org/10.1073/pnas.92.13.6057
29. Judge RA, Forsythe EL, Pusey ML (1998) The effect of protein impurities on lysozyme crystal growth. Biotechnol Bioeng 59(6):776–785. Accessed from https://doi.org/10.1002/(SICI)1097-0290(19980920)59:6<776::AID-BIT14>3.0.CO;2-3
30. Kabsch W (2010) XDS. Acta Crystallogr Sect D Biol Crystallog 66(2):125–132. https://doi.org/10.1107/S0907444909047337
31. Kendrew JC, Bodo G, Dintzis HM, Parrish RG, Wyckoff H, Phillips DC (1958) A three-dimensional model of the myoglobin molecule obtained by x-ray analysis. Nature 181(4610):662–666. https://doi.org/10.1038/181662a0
32. Kessler RW, Kessler W, Zikulnig-Rusch E (2016) A critical summary of spectroscopic techniques and their robustness in industrial PAT applications. Chemie-Ingenieur-Technik, 88(6):710–721. Accessed from https://doi.org/10.1002/cite.201500147
33. Klijn ME, Hubbuch J (2018) Application of empirical phase diagrams for multidimensional data visualization of high-throughput microbatch crystallization experiments. J Pharm Sci 107(8):2063–2069
34. Krieger E, Darden T, Nabuurs SB, Finkelstein A, Vriend G (2004) Making optimal use of empirical energy functions: force-field parameterization in crystal space. Proteins: Struct Funct Bioinform 57(4):678–683. https://doi.org/10.1002/prot.20251
35. Lawson DM, Artymiuk PJ, Yewdall SJ, Smith JM, Livingstone JC, Treffry A et al (1991) Solving the structure of human H ferritin by genetically engineering intermolecular crystal contacts. Nature 349(6309):541–544. https://doi.org/10.1038/349541a0
36. Leuchs S, Greiner L (2011) Alcohol dehydrogenase from Lactobacillus brevis: a versatile robust catalyst for enantioselective transformations. Chem Biochem Eng Quart 25:267–281
37. Longenecker KL, Garrard SM, Sheffield PJ, Derewenda ZS (2001) Protein crystallization by rational mutagenesis of surface residues: Lys to Ala mutations promote crystallization of RhoGDI. Acta Crystallogr D Biol Crystallogr 57(Pt 5):679–688
38. Machielsen R, Looger LL, Raedts J, Dijkhuizen S, Hummel W, Hennemann H-G, Daussmann T, van der Oost J (2009) Cofactor engineering of Lactobacillus brevis alcohol dehydrogenase by computational design. Eng Life Sci 9(1):38–44. https://doi.org/10.1002/elsc.200800046
39. Manee V, Zhu W, Romagnoli JA (2019) A deep learning image-based sensor for real-time crystal size distribution characterization. Ind Eng Chem Res 58(51):23175–23186. https://doi.org/10.1021/acs.iecr.9b02450
40. McCoy AJ, Grosse-Kunstleve RW, Adams PD, Winn MD, Storoni LC, Read RJ (2007) Phaser crystallographic software. J Appl Crystallogr 40:658–674
41. Murshudov GN, Skubák P, Lebedev AA, Pannu NS, Steiner RA, Nicholls RA, Winn MD, Long F, Vagin AA (2011) REFMAC5 for the refinement of macromolecular crystal structures. Acta Crystallogr Sect D Biol Crystallogr 67(4):355–367
42. McElroy HE, Sisson GW, Schoettlin WE, Aust RM, Villafranca JE (1992) Studies on engineering crystallizability by mutation of surface residues of human thymidylate synthase. J Cryst Growth 122(1):265–272. https://doi.org/10.1016/0022-0248(92)90255-H

43. Niefind K, Muller J, Riebel B, Hummel W, Schomburg D (2003) The crystal structure of R-specific alcohol dehydrogenase from Lactobacillus brevis suggests the structural basis of its metal dependency. J Mol Biol 327(2):317–328. https://doi.org/10.1016/s0022-2836(03)00081-0
44. Niefind K, Riebel B, Müller J, Hummel W, Schomburg D (2000) Crystallization and preliminary characterization of crystals of R-alcohol dehydrogenase from Lactobacillus brevis. Acta Crystallogr D Biol Crystallogr 56(Pt 12):1696–1698. https://doi.org/10.1107/s0907444900011860
45. Noey EL, Tibrewal N, Jiménez-Osés G, Osuna S, Park J, Bond CM, Cascio D, Liang J, Zhang X, Huisman GW, Tang Y (2015) Origins of stereoselectivity in evolved ketoreductases. Proc Natl Acad Sci 112(51):E7065-E7072. https://doi.org/10.1073/pnas.1507910112
46. Nowotny P, Hermann J, Li J, Krautenbacher A, Klöpfer K, Hekmat D, Weuster-Botz D (2019) Rational crystal contact engineering of lactobacillus brevis alcohol dehydrogenase to promote technical protein crystallization. Cryst Growth Des 19(4):2380–2387. https://doi.org/10.1021/acs.cgd.9b00067
47. Price SR, Nagai K (1995) Protein engineering as a tool for crystallography. Curr Opin Biotechnol 6(4):425–430. https://doi.org/10.1016/0958-1669(95)80072-7
48. Rolinger L, Rüdt M, Hubbuch J (2020) A critical review of recent trends, and a future perspective of optical spectroscopy as PAT in biopharmaceutical downstream processing. Anal Bioanal Chem 412(9):2047–2064
49. Rüdt M, Andris S, Schiemer R, Hubbuch J (2019) Factorization of preparative protein chromatograms with hard-constraint multivariate curve resolution and second-derivative pretreatment. J Chromatogr A 1585:152–160. Accessed from https://doi.org/10.1016/j.chroma.2018.11.065
50. Savitzky A, Golay MJE (1964) Smoothing and differentiation of data by simplified least squares procedures. Anal Chem 36(8):1627–1639. Accessed from https://doi.org/10.1021/ac60214a047
51. Schlieben NH, Niefind K, Muller J, Riebel B, Hummel W, Schomburg D (2005) Atomic resolution structures of R-specific alcohol dehydrogenase from Lactobacillus brevis provide the structural bases of its substrate and cosubstrate specificity. J Mol Biol 349(4):801–813. https://doi.org/10.1016/j.jmb.2005.04.029
52. Simone E, Saleemi AN, Nagy ZK (2015) In situ monitoring of polymorphic transformations using a composite sensor array of Raman, NIR, and ATR-UV/vis spectroscopy, FBRM, and PVM for an intelligent decision support system. Org Process Res Dev 19:167–177. https://doi.org/10.1021/op5000122
53. Sklansky J (1982) Finding the convex hull of a simple polygon. Pattern Recogn Lett 1(2):79–83. https://doi.org/10.1016/0167-8655(82)90016-2
54. Smejkal B, Helk B, Rondeau JM, Anton S, Wilke A, Scheyerer P, Fries J, Hekmat D, Weuster-Botz D (2013) Protein crystallization in stirred systems–scale-up via the maximum local energy dissipation. Biotechnol Bioeng 110(7):1956–1963. https://doi.org/10.1002/bit.24845
55. Smejkal B, Agrawal NJ, Helk B, Schulz H, Giffard M, Mechelke M, Ortner F, MHeckmeier F, Trout BL, Hekmat D (2013) Fast and scalable purification of a therapeutic full-length antibody based on process crystallization. Biotechnol Bioeng 110(9):2452–2461. https://doi.org/10.1002/bit.24908
56. Toussaint GT (1983) Solving geometric problems with the rotating calipers. In: Proceedings of IEEE melecon, vol 83, p A10
57. Walla B, Bischoff D, Janowski R, von den Eichen N, Niessing D, Weuster-Botz D (2021) Transfer of a rational crystal contact engineering strategy between diverse alcohol dehydrogenases. Crystals 11(8):975. Accessed from https://www.mdpi.com/2073-4352/11/8/975
58. Weckbecker A, Hummel W (2006) Cloning, expression, and characterization of an (R)-specific alcohol dehydrogenase from Lactobacillus kefir. Biocatal Biotransform 24(5):380–389. https://doi.org/10.1080/10242420600893827
59. Wegner CH, Zimmermann I, Hubbuch J (2022) Rapid analysis for multicomponent high-throughput crystallization screening: combination of UV-Vis spectroscopy and chemometrics. Crystal Growth Des 22(2):1054–1065. Accessed from https://doi.org/10.1021/acs.cgd.1c00907

60. Wu Y, Gao Z, Rohani S (2021) Deep learning-based oriented object detection for in situ image monitoring and analysis: a process analytical technology (PAT) application for taurine crystallization. Chem Eng Res Des 170:444–455. https://doi.org/10.1016/j.cherd.2021.04.013
61. Wu Y, Kirillov A, Massa F, Lo W-Y, Girshick R (2019) Detectron2. Accessed 22 Feb 2022, from https://github.com/facebookresearch/detectron2

Changes in Structure, Dispersity and Phase Behavior of Proteins: The Virus-Like Particles in the Presence of Nucleic Acids

Srdjan Pusara, Angela Valentic, Wolfgang Wenzel, Mariana Kozlowska, and Jürgen Hubbuch

1 Introduction

Proteins are multifunctional bioactive macromolecules made of a wide variety of amino acid (aa) combinations that deviate their molecular properties as a function of subtle changes of the solution, as also the three-dimensional structure of the folded protein and the local environment made by the neighboring residues. In most cases, these subtle changes are driven by the network of interactions of both attractive and repulsive character, therefore, the understanding of dispersity and phase behavior of proteins is extremely cumbersome and requires the combination of various experimental and theoretical methods. Properties of proteins such as solubility, as well as aggregation, precipitation and crystallization are mainly driven by protein–protein

S. Pusura, A. Valentic: These authors equally contributed.

S. Pusara · W. Wenzel · M. Kozlowska
Institute of Nanotechnology, Karlsruhe Institute of Technology KIT, Kaiserstraße 12, 76131 Karlsruhe, Germany
e-mail: srdjan.pusara@hotmail.com

W. Wenzel
e-mail: wolfgang.wenzel@kit.edu

M. Kozlowska
e-mail: mariana.kozlowska@kit.edu

A. Valentic · J. Hubbuch (✉)
Institute of Process Engineering and Life Sciences, Section Biomolecular Separation Engineering 4, Karlsruhe Institute of Technology KIT, Kaiserstraße 12, 76131 Karlsruhe, Germany
e-mail: juergen.hubbuch@kit.edu

A. Valentic
e-mail: angela.valentic@kit.edu

A. Kwade and I. Kampen (eds.), *Dispersity, Structure and Phase Changes of Proteins and Bio Agglomerates in Biotechnological Processes*,
https://doi.org/10.1007/978-3-031-63164-1_4

interactions (PPIs), which are modulated by the nature of proteins, but also properties of solution, e.g. its ionic strength, pH, temperature etc. Understanding PPIs is important from a technological point of view, because for many biotechnological processes it is important, for example, to prepare stable protein formulations and prevent unwanted aggregation or to fabricate desired self-assembly constructs, reducing uncontrolled aggregation [18].

In the present research, we aim to understand processing conditions, dispersity and phase behavior of HBcAg proteins that self-assemble with the formation of hepatitis B virus (HBV) virus-like particles (VLPs), see Fig. 1a. These proteins are also known as core proteins (Cp), because they compose the core of the viral capsid [69, 98] and, upon the existence of the C-terminal (C-ter) nucleic acid binding chain, tend to encapsulate the genome information carried by the virus [59]. Typically, wild-type HBcAg core protein consists of a rigid assembly domain (1–140 aa) and a flexible C-terminal (150–183 aa) [69], see Fig. 1a. The C-terminal (C-ter) domain is rich in positively charged arginine residues, therefore, often they are responsible for the binding of negatively charged nucleic acids. In contrast to the assembly domain, which has a stable tertiary structure and is responsible for the surface charge of the capsid, the C-ter has no defined tertiary structure, i.e. it is extremely flexible [97]. During the HBV life cycle it is mostly localized in the capsid interior, but sometimes it can also get exposed at the capsid outer surface through the pores between the pentamer oligomers formed from the homodimers of the HBcAg [23]. On contrary to HBV made out of Cp183, stable VLPs can be self-assembled also from Cp with 149 aa, which was shown to be the main domain of the protein taking part in the capsid self-assembly [98]. In such a case, the formed construct is either empty or encapsulates a significantly lower amount of nucleic acids [35, 92], because the total charge of Cp149 is -7e and it does not possess RNA/DNA binding chain. This opens a variety of possible applications of the VLP particles either as new vaccines, drug delivery systems up to the nucleic acid-based therapeutics [9, 27, 33, 42, 73].

Lately, nucleic acid-based therapeutics are investigated for selective and efficient therapies [11, 16, 62, 75, 78, 95]. Outstandingly, first vaccines against the coronavirus disease 2019 were two mRNA-based vaccines [2], demonstrating their enormous potential for rapid development to large-scale manufacturing [2, 94]. For packaging and delivering the nucleic acids lipid nanoparticles, that have been studied intensively in the last years [3, 8, 46], were used [37]. However, due to remaining challenges such as efficient targeting in tissues and lipid toxicity [3], promising alternatives for targeted delivery of small molecules, proteins and nucleic acids are investigated [73]. Therefore, in this work, we investigate VLPs as promising nanocarriers of the nucleic acids. With advantages like the possibility of surface modification on genetic level or post-translational modification and uniform structures of VLPs with large loading capacities [73], several VLP types have been already demonstrated for RNA packing [49, 65, 67], protection [26] and delivery of nucleic acids [17].

A typical production process for VLP, being developed for nucleic acid delivery, is shown in Fig. 1b. After fermentation and harvesting, it consists of cell lysis, precipitation or ultracentrifugation, disassembly, loading with therapeutic nucleic acids and reassembly, polishing and formulation [35, 55, 92, 93]. However, effective VLP downstream processing is challenging and traditional unit operations, such as size

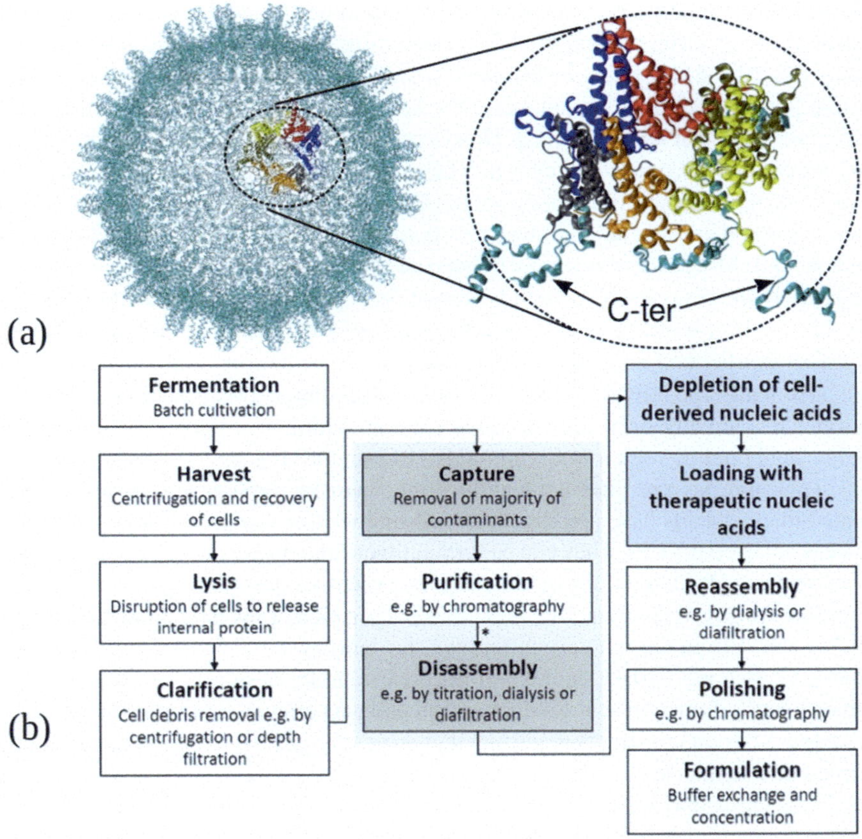

Fig. 1 **a** All-atom structure of the HBV capsid (PDB 6htx). For the sake of illustration, the trimer of dimers is highlighted within the capsid structure. The structure of the dimer of core proteins (Cp_2) is magnified on the right side. **b** Downstream purification process for HBcAg VLP. The steps necessary for use as a nucleic acid delivery system are highlighted in blue and central purification steps investigated in this study in gray. Figure taken from [85], licensed under the Creative Commons Attribution 4.0 International license (CC BY 4.0)

exclusion chromatography (SEC) and ultracentrifugation [91], get replaced by more scalable ultrafiltration-based unit operations [13, 34, 35, 60, 74]. Factors affecting the purification process, such as encapsulated nucleic acids or general VLP properties, are critical to the implementation of these new developments for effective large-scale production of VLPs for nucleic acid delivery. For that, protein–protein and protein-nucleic acid interactions, modulating the disassembly and assembly processes in the preparation of the VLP-based nanocarriers have to be better understood.

In general, HBcAg VLPs are already known as promising candidates for gene therapeutic nucleic acid delivery. They are usually mixed with nucleic acids while disassembled and subsequently reassembled to encapsulate the nucleic acids [65, 67, 92]. Different versions of these HBcAg VLPs, for example, the wild-type HBcAg protein Cp183 with the naturally occurring full-length nucleic acid binding region

[67, 92] (see Fig. 1a), variants with few replaced amino acids [67, 81], but also constructs with different lengths of this naturally occurring nucleic acid binding region [50, 61, 65] are being investigated for an effective loading. Here, the binding region, located inside the capsid, plays the key role for the encapsulation of a sufficient amount of nucleic acids [92]. However, the consideration of different lengths of C-ter could overcome challenges in VLP purification after formation in host cells, such as the reported solubility issues of the Cp183 construct [67]. Due to undesirable association and encapsulation of host cell-derived nucleic acids during production in the host cell. Moreover, this seems to have an influence on the purification process, particularly the disassembly which is crucial for separating encapsulated impurities [35, 67, 81, 92], improving structural integrity [93] and enabling effective loading with therapeutic nucleic acids. A clear dependence of the disassembly reaction of truncated wild-type HBcAg protein (Cp149) and a construct with a chimeric surface epitopes has already been demonstrated [35]. In addition, different lengths of the nucleic acid binding region were reported to impact the HBcAg VLP capsid stability [61]. This suggests that different lengths of nucleic acid binding regions and encapsulated nucleic acids have significant effects on the phase behavior and purification process of HBcAg VLPs. They can be further investigated after deeper understanding of intermolecular interactions modulating the processes described.

In this work, we combine theoretical and experimental investigations of HBcAg VLPs formed from Cp with six different C-ter lengths, i.e. Cp149, Cp154, Cp157, Cp164, Cp167 and Cp183, capable of different binding of nucleic acids as a function of length of nucleic acid binding region and processing conditions. We characterize the different constructs regarding their size, charge, and loading with host-cell derived nucleic acids and study the dependence of the precipitation, re-dissolution, and disassembly purification steps on the various constructs and bound nucleic acids. To investigate an interplay of protein–protein and protein-nucleic acid interactions as a function of Cp length and ionic strength of the solution, we perform molecular dynamic simulations for Cp dimers and trimers of Cp dimers with and without nucleic acids (RNA and DNA). The umbrella sampling simulations are used to calculate the free energy of binding in a complex. To reduce the experimental efforts to screen PPIs at a wide range of solution conditions, we develop a theoretical model, which permits the calculation of osmotic second virial coefficient (B_{22}) in a fast and accurate way.

The chapter is organized as follows: Firstly, the newly developed method, i.e. xDLVO-CG, is explained and used to calculate nonspecific interactions and protein assembly conditions by means of the change of the osmotic second virial coefficient. Next, molecular dynamics simulations are utilized to get more insights about specific interactions between beta lactoglobulin proteins and to calculate the free energy of HBcAg protein binding in the VLP in the presence and absence of nucleic acid. Later, six different HBcAg-VLP constructs with different lengths of the nucleic acid binding region are characterized with size, charge and loading with nucleic acids determined by dynamic light scattering, electrophoretic mobility and UV–Vis spectroscopy. Finally, the influence of the VLP variants and the bound nucleic acids on the downstream purification steps, especially on the disassembly of the capsids are explained.

2 Material and Methods

2.1 Computational Details

2.1.1 Calculation of Osmotic Second Virial Coefficient

Colloidal PPIs can be estimated using the osmotic second virial coefficient, B_{22}, which is defined by solution deviation from ideal behavior, quantifying how much osmotic pressure differs from that of the ideal solution, as expressed by Eq. 1:

$$\Pi = RTc_p\left(\frac{1}{M_W} + B_{22}c_p + \cdots\right) \tag{1}$$

where Π is the osmotic pressure, c_p is the protein concentration (in mass units), R is the gas constant, T is the absolute temperature, and M_w is the molecular weight of the protein. If repulsion forces between proteins surpass their attraction, the osmotic pressure is higher than ideal-gas pressure, increasing the B_{22} coefficient. Upon stronger protein–protein attraction, the osmotic pressure of the solution, Π, and virial coefficient decrease, therefore, it can be the direct measure of the state of the macromolecule in the solution, i.e. the preference of the solvated or aggregated (assembled) state. Using statistical thermodynamic concepts, the osmotic second virial coefficient can be related to the effective interaction potential between two molecules in solution as follows [56]:

$$B_{22} = \frac{1}{2}\frac{N_A}{M^2}\int\limits_{\Omega_1}\int\limits_{\Omega_2}\int\limits_0^{\infty}\left(1 - e^{-\frac{W(r,\Omega_1,d\Omega_2)}{k_B T}}\right)4\pi r^2 dr d\Omega_1 d\Omega_2 \tag{2}$$

where $W(r, \Omega_1, \Omega_2)$ represents the effective interaction potential (potential of mean force, PMF) as a function of the intermolecular center-of-mass distance r and relative orientation (Ω_1, Ω_2) between two proteins, and k_B is the Boltzmann constant. Thus, in addition to the experimental measurement of the osmotic second virial coefficient (e.g. with membrane osmometry [25], static light scattering [52] or self-interaction chromatography[72]), it can be also calculated. For that, the PMF between proteins should be known. With the estimated PMF, it can be integrated over all distances and relative orientations of two molecules in a liquid phase, with each configuration weighted by a Boltzmann factor, therefore B_{22} can be obtained using, e.g., Eq. 2.

So far, B_{22} was calculated by using different levels of theory and protein structural details. The common approach is based on the Derjaguin–Landau–Verwey–Overbeek (DLVO) [87], where protein is represented as an ideal sphere. According to the DLVO theory, the PMF consists of the repulsive electric double-layer forces and attractive dispersion interactions:

$$W(r) = W_{disp}(r) + W_{el}(r) \tag{3}$$

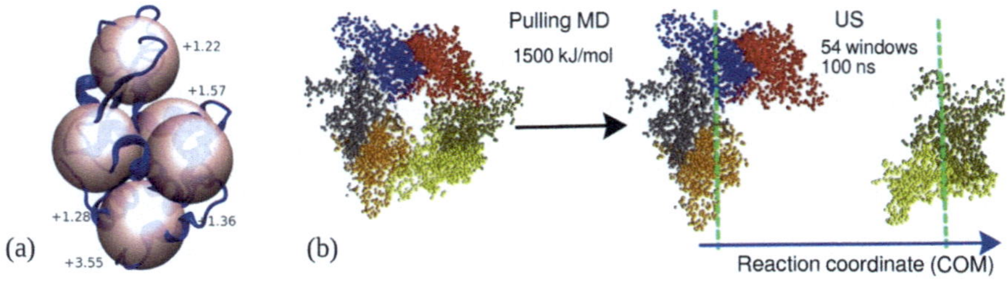

Fig. 2 **a** Shape-based coarse-grained structure of lysozyme (LYZ) protein, used in xDLVO-CG, superimposed on an all-atom structure (represented by New Cartoon style). **b** Computational scheme of umbrella sampling simulations performed on trimer of HBcAg dimers by using SIRAH force field in the present work

This simple model can fit the dependence of B_{22} on salt concentrations, but the derived model parameters are often physically unrealistic and hardly transferable from protein to protein. The most important limitations of this model is that proteins are represented by a simplified spherical shape. Moreover, proteins also encounter other types of interactions beyond electrostatic repulsion and van der Waals (vdW) interactions, which contribute to unphysical fitted parameters.

Here, we calculate B_{22} coefficients by using the newly developed xDLVO-CG model [70]. By this approach, proteins are represented by a shape-based coarse-grained (CG) model, which allows to consider protein shape in a more realistic way and also account for charge anisotropy. An example of CG protein structure for lysozyme protein is shown on Fig. 2a.

The interaction potential, *W(r)*, is determined as the sum of electrostatic, dispersion, osmotic attraction and ion–protein interactions:

$$W(r) = W_{el}(r) + W_{disp}(r) + W_{osm}(r) + W_{i-pr}^{disp}(r) \tag{4}$$

Electrostatic interactions are calculated by using Debye–Hückel theory, [41] which takes into account the screening of electrostatic potential in the presence of electrolytes:

$$W_{el}(r) = \sum_{i=1}^{N_1}\sum_{j=1}^{N_2} \frac{Z_i Z_j e^2 \exp(\kappa(d_{ij} - r_{ij}))}{4\pi\varepsilon_0\varepsilon_r r(1 + \frac{\kappa d_{ij}}{4})^2},\ r_{ij} > d_{ij} + 2\sigma \tag{5}$$

Here, r is the center-of-mass (COM) distance between two proteins, $N1$ and $N2$ are the total numbers of CG beads of each protein, Z_i and Z_j are charges of each bead, σ is the thickness of water layer on a protein surface (e.g. 0.1 nm), d_{ij} is the initial distance between bead i of the first protein and bead j of the second protein (according to the crystal structure), and r_{ij} is the current (variable) distance between beads during pulling of proteins. ε_r is the relative permittivity, κ is the inverse Debye length given by:

$$\kappa = \sqrt{\frac{2N_A e^2 I}{\varepsilon_0 \varepsilon_r k_B T}} \tag{6}$$

and I is the ionic strength. The dispersion potential, $W_{\text{disp}}(r)$ describing the attractive forces between macroscopic colloidal particles, is calculated by using Hamaker potential derived by the summation of van der Waals interactions over infinite volumes of two ideal spheres was implemented according to [31]

$$W_{disp}(r) = -\sum_{i=1}^{N_1}\sum_{j=1}^{N_2} \frac{A_H}{12}\frac{1}{N_1 N_2}\left[\frac{d_{ij}}{r_{ij}^2 - d_{ij}^2} + \frac{d_{ij}^2}{r_{ij}^2} + 2\ln(1 - \frac{d_{ij}^2}{r_{ij}^2})\right], r_{ij} > d_{ij} + 2\sigma \tag{7}$$

where A_H is the Hamaker constant, usually determined by fitting from experimentally measured B_{22} values and sometimes by theoretical models [20, 48]. The osmotic attraction potential $W_{osm}(r)$ takes into account additional attraction, which arises due to salt ion exclusion from the protein interspace at short separations. This causes local osmotic pressure imbalance and causes additional attractive interaction between proteins, and is calculated according to:

$$W_{osm}(r) = -\sum_{i=1}^{N_1}\sum_{j=1}^{N_2} \frac{1}{N_1 N_2}\frac{4\pi k_B}{3} T D_{ij}^3 \rho_3 \left(1 - \frac{3r_{ij}}{4D_{ij}} + \frac{r_{ij}^3}{16 D_{ij}^3}\right), d_{ij} + 2\sigma \le r_{ij} \le 2D_{ij} \tag{8}$$

The last term in the PMF, i.e. the ion–protein potential, describes the total dispersion interaction between protein and all ions in its surrounding. Here, the protein is represented as an ideal sphere with the charge Z, while ions are placed non-uniformly around the protein sphere according to the Boltzmann distribution. The total potential is calculated by integrating the contributions of each ion according to:

$$W_{i-pr}^{disp}(r) = -4\pi B_a \int_{R_P+\sigma}^{R_P+D} \frac{c_b \exp(-\frac{z_a \phi(r)}{k_B T})}{r} dr - 4\pi B_c \int_{R_P+\sigma}^{R_P+d} \frac{c_b \exp(-\frac{z_c \phi(r)}{k_B T})}{r} dr \tag{9}$$

where B_{anion} and B_{cation} are the ion-macroion dispersion coefficients taken from literature, [10, 58] Rp is the protein radius, d is the thickness of the spherical shell around a protein, z_c and z_a are the charges of the cation/anion, c_{bulk} is the salt concentration and $\varphi(r)$ is the electrostatic potential around a protein sphere with the charge Z.

Apart from reducing experimental efforts, theoretical determination of osmotic second virial coefficients also permits theoretical models to give insight about origin of the PPIs and not only the average effective interaction. Therefore, we have used the xDLVO-CG method explained to calculate B_{22} coefficients in the present work.

2.1.2 Coarse-Grained Molecular Calculations

Several different proteins were considered for the calculations performed in this study. Among them are: lysozyme (LYZ), human immunoglobulin type I (IgG1), beta-lactoglobulin (BLG), and HBcAg core protein of the HBV (Cp). All-atom protein structures were obtained from the Protein Data Bank (PDB) under the codes 4nhi, 3ph5, 1mco, and 6htx for proteins mentioned, respectively. The missing residues and atoms, often present in the structures deposited, were modeled by the Swiss Model program [88]. The reconstructed structures were protonated at desired pH by using PROPKA method (version 3.3) [82] and PDB2PQR online web-server [24]. This resulted in the charge of the LYZ of +11e and +9e at pH 4.5 and pH 7, respectively, of the IgG1 of +56e, +48e and +27e at pH 5, pH 5.75 and pH 5,75, respectively, and of the BLG of +18e and +8e at pH 3 and pH 7, respectively. Upon that, the protonated all-atom structures were mapped into the coarse-grained (CG) representation (with approximately 500 atoms per bead), by using shape-based CG model, implemented in VMD program (version 1.9.3) [40]. Center of each bead was placed in the center-of-mass (COM) of the corresponding set of atoms, the bead radius was assigned to the radius of gyration and charge as a sum of partial charges. PMF (see Eq. 4) was calculated by summing interactions between the corresponding bead pairs from each protein pair. The osmotic second virial coefficient was determined by numerical integration of the PMF over different protein–protein orientations (Eq. 2). Conformational sampling of starting protein–protein configurations was done by performing 16 different rotations at each of the 83 starting radial positions, resulting in 1328 starting configurations. Each of them was used for a separate PMF calculation based on translating the first protein along a vector connecting the COM of the second protein pair. The PMF up to a COM distance of 30 nm was calculated. More description of the method can be found in Ref. [70].

The calculated B_{22} values were compared with the experimental values taken from literature. In addition, two other theoretical methods, i.e. spherical extended DLVO (xDLVO) model [32] and all-atom FMAPB2 (FFT-based Modeling of Atomistic Protein–protein interactions) [71], were used for comparison. xDLVO was implemented according to the Ref. [32], while the FMAPB2 model, available on the web-server (http://pipe.rcc.fsu.edu/fmapb2), was used.

2.1.3 All-Atom Molecular Dynamics Simulations

B_{22} coefficient is in most cases the measure of nonspecific PPIs. Specific PPIs are directional and they are governing processes such as dimerization, gelation, enzyme substrate binding or macromolecular assembly [15]. These processes are often steered by specific residue-residue interactions, hydrogen bonds or salt bridges. In order to model these processes, it is necessary to model proteins at an atomic level, as well as to take into account dynamical effects due to macromolecular movements. For this, molecular dynamics (MD) is a useful tool, as it permits to simulate macromolecules in a dynamical environment [43]. We have performed MD simulations for the HBV Cp dimers and trimer of dimers. MD of the dimers was treated on the all-atom level (AA), i.e. using CHARMM36m [38] force field and SPC [6] water model. Simulations of the trimer of dimers were performed by using SIRAH CG force field [21]. All simulations were performed in the GROMACS package (version 2019.2) [1].

All-atom structure of Cp protein was taken from PDB 6htx crystal structure, and missing residues and atoms were modeled by Swiss modeler. Using the 6htx crystal structure of the Cp, the protein structure was modeled till 183 aa. Since various lengths of the C-terminal chain were aimed to be investigated, i.e. Cp167, Cp164, Cp157, Cp154 and Cp149 (see Table 1), the truncated proteins were obtained by cutting off the sequence of the protein at desired length. The structure of the dimer of dimers (dimer of Cp_2) was taken from the HBV capsid structure, which was created by applying symmetry transformations given in the PDB file. AA-MD calculations were performed using periodic boundary conditions with a rectangular box of the size of $20.0 \times 20.0 \times 20.0$ nm^3. Na^+ and Cl^- ions were added to fulfill the salt concentrations of 37.5 mM used in the experiment. All systems were initially minimized by the steepest descent algorithm during 30.000 steps with position restraints applied to proteins heavy atoms. Equilibration under canonical (NVT) and later under constant–pressure (NPT) ensembles at 300 K and for 1 ns each were performed using a Berendsen thermostat. The Berendsen weak coupling method was also used to isotropically maintain pressure at 1.0 bar. All simulations were performed with the timestep of 2 fs. Short-range nonbonded interactions were cut off at 1.2 nm. The full electrostatic interactions beyond 1.2 nm were evaluated by the particle mesh Ewald (PME algorithm). After equilibration, MD production runs were performed under NPT conditions for 100 ns. Apart from the simulations of pure protein systems, complexes of Cp_2 dimers with RNA were simulated. For this, RNA structure was taken from PDB under the code 1q9a. To generate initial complexes of RNA and $HBcAg_2$ dimers, the docking calculations were performed. For that, HDOCK web-server (available at http://hdock.phys.hust.edu.cn/) was used [90]. Upon that, the docked complex with the highest score (calculated by a hybrid strategy based on template based and free docking) was simulated by MD the same as described earlier. All simulation trajectories were post-analyzed using Molecular Mechanics Poisson-Boltzmann method (MM-PBSA) [36], using g_mmpbsa analysis tool [47]. The binding energy between Cp dimer-dimer or between Cp dimer and RNA was determined by a single trajectory approach, i.e. by subtracting the total molecular mechanics and solvation energy of the complex with the total energy of single components.

2.1.4 Coarse-Grained Molecular Dynamics Simulations

Coarse-grained MD permits to simulate biomolecular systems at larger timescales [79]. In the present work, we use SIRAH CG force field [21], which models proteins by high structural resolution, i.e. positions of backbone C and N atoms are preserved, while amino acid side chains are modeled by one or two additional beads depending

Table 1 Amino acid sequence in one letter code of nucleic acid binding region located C-terminally at core domain with 149 amino acids [19] in HBcAg virus-like particle constructs [85]

Construct	Amino acid sequence
Cp154	RRRGR
Cp157	RRRGRSPR
Cp164	RRRGRSPRRRTPSPR
Cp167	RRRGRSPRRRTPSPRRRR
Cp183	RRRGRSPRRRTPSPRRRRSQSPRRRRSQSRESQC

on group size. This allows simulating proteins without introducing additional bias to fix the secondary structure. Here, we have used CG-MD in combination with the accelerated MD technique, i.e. umbrella sampling (US) [57]. It allows us to observe rare events, which are not accessible to conventional MD. In the US method a series of configurations is generated along a reaction coordinate, for example the COM distance between proteins, with its position maintained by a biasing potential. This allows to generate an ensemble of structures along the reaction coordinate, and explore rarely visited parts of phase space, and then obtain unbiased probability by reweighting the snapshots according to statistical significance (see Fig. 2b). US and CG-MD were used here for the simulation of the six trimers of dimers (trimer of Cp_2) to estimate PPIs and an impact of interactions with nucleic acid on the PPIs in the trimer disassembly.

The $HBcAg_2$ trimer was taken from capsid structure (6htx pdb code) and was protonated at pH 7 using the PROPKA method (version 3.3) and PDB2PQR online web server, using Amber force field parameters and atom names. The protonated all-atom trimer of dimers was mapped into CG structure by using SIRAH tools [54]. Trimers of dimers with various C-ter lengths were placed in a rectangular box of 26 $\times$ 22 $\times$ 22 nm^3, and solvated by the WT4 water model used in the SIRAH force field. Here, one WT4 molecule replaces 11 all-atom water molecules [21]. Sodium and chloride ions were added to fulfill a concentration of 37.5 mM. The system was minimized by using a steepest descent algorithm for 5000 steps, while harmonically restraining protein heavy atoms, and by another 5000 steps without restraints. The equilibration under NVT ensemble at 300 K by using V-rescale thermostat [12] was performed. For the first 5 ns, all protein atoms were restrained by harmonic potential of 1000 kJ/mol nm. For the next 25 ns, only backbone beads were weakly restrained with 50 kJ/mol nm, while protein side chains and solvent were allowed to move and relax. Finally, unrestrained simulations were performed for 50 ns under NPT conditions at 300 K and 1 bar, using V-rescale thermostat and Parrinello-Rahman barostat. Timestep of 20 fs was used. Nonbonded interactions were cut off at 1.2 nm, beyond that, electrostatic interactions were calculated by the PME method.

The last snapshot of a production run was used for pulling simulations, where dimer (Cp_2) was pulled away from the trimer of dimer along the vector connecting COM distances of those two groups (see Fig. 2b). Pulling force of 500 kJ/mol nm was used with a pulling rate of 0.001 nm/ps. The snapshots from the collected pulling trajectories were used to generate starting configurations for the respective US windows necessary for the PMF [64]. An asymmetric distribution of sampling windows was used: the window spacing of 0.0625 nm was used to generate the first 35 snapshots (used later as a starting point of CG-MD simulations), while 0.125 nm was used for the remaining US windows. In total, 54 US windows were generated. In each window, the system was equilibrated using NPT ensemble at 300 K and 1 bar for 3.5 ns with the following 100 ns MD run using NPT ensemble with V-rescale thermostat and Parrinello–Rahman barostat. Analysis of all US simulations was performed with the weighted histogram analysis method (WHAM) [39].

To understand an impact of the nucleic acid on the PPIs in the trimer, trimers of $HBcAg_2$ were simulated with a DNA molecule (PDB code 1hw2) attached. SIRAH force field and the same computational setup as described above were applied. Note that such calculations using RNA molecules are not yet possible due to the lack of the force field parameters in SIRAH. The same as above, DNA was docked on the trimer of dimers using HDOCK web-server and 100 best docking structures (i.e. with the lowest energy representing the best binding affinity in the system) were simulated using MD for 50 ns. The binding energy between the DNA and the $HBcAg_2$ trimers was determined by analyzing MD trajectories with MM-GBSA method. The best three structures were used for the US, where the HBcAg dimer was separated from the trimer in the way described above. The DNA molecule was attached in the vicinity to determine the impact of the bonded DNA on the trimer disassembly. The DNA molecule was not part of any pulling group and was not restrained during pulling and US.

2.2 *Experimental Setup*

2.2.1 Materials and Buffers

All chemicals were purchased from Merck (Darmstadt, Germany), unless otherwise stated and ultrapure water (PURELAB Ultra, ELGA LabWater) was used to prepare solutions and buffers. HCl with a concentration of 32% was used to adjust the pH of the buffers. Buffers were filtered through a 0.2 μm pore-size cellulose acetate filter (Pall Corporation, Port Washington, NY, United States) before use. As wash buffer, lysis buffer, which consisted of 50 mM Tris, 100 mM NaCl, 1 mM EDTA (AppliChem, GmbH, Darmstadt, Germany) at pH 8, adjusted to 1 M $(NH_4)_2SO_4$ (AppliChem, Darmstadt, Germany) and 0.25% (v/v) polysorbate 20 (AppliChem, Darmstadt, Germany) was used. The re-dissolution buffer consisted of 50 mM Tris, 150 mM NaCl at pH 7.2. Protein solutions, dissolved in this re-dissolution buffer, were diluted with a 50 mM Tris, 10 M urea stock solution for disassembly screening experiments.

2.2.2 Cloning of Cp154, Cp157, Cp164, Cp167 and Cp183 Plasmids

Based on a Cp149 expression vector coding for the C-terminally truncated wild-type HBcAg with 149 amino acids, kindly provided by Prof. Adam Zlotnick (Indiana University Bloomington, United States) plasmids coding for VLP constructs with different lengths of the C-terminal nucleic acid binding region were cloned. In Table 1 the amino acid sequences introduced as C-terminal nucleic acid binding regions can be found. The pET11c plasmid coding Cp149 was used to insert bp sequences coding for the C-terminal binding regions. Thereby plasmids coding for Cp154, Cp157, Cp164 and Cp167 were made. Overlapping oligonucleotides were used to introduce site-directed mutagenesis by the polymerase chain reaction (PCR) to amplify the Cp149 encoding regions on the pET11-based vector. In the same way a Cp183 plasmid was produced by amplifying the Cp167 plasmid.

PCRBio HiFi Polymerase (Nippon Genetics Europe GmbH, Düren, Germany) was used to perform the amplification. Digestion of the PCR template and product was done using DpnI (New England BioLabs, Ipswitch, MA, United States), Further, the product was purified by native gel electrophoresis, extracted using a Wizard SV gel and PCR Clean-Up kit (Promega, Madison, WI, United States) and the Gibson Assembly Master Mix (New England BioLabs, Ipswitch, MA, United States) was used for ligation. Subsequently, the new vectors were transformed into BL21[DE3] cells (New England Biolabs, Ipswitch, MA, United States).

2.2.3 Intra-Cellular Production and Purification of VLPs

After overexpression in *E.coli* with a TB-based auto-induction medium VLPs were released by cell lysis as described previously [34]. VLPs were precipitated by dilution of the lysate solution with $(NH_4)_2SO_4$ and polysorbate 20 stock solutions. The final precipitation conditions were 1 M $(NH_4)_2SO_4$ and 0.25% (v/v) polysorbate 20. After stirring for 2 h at 4 °C the solution was centrifuged at 17,000 rcf for 30 min using a centrifuge 5810 R (Eppendorf, Hamburg, Germany). The centrifugation pellet was mixed with a wash buffer and incubated at 10 rpm in an overhead shaker LD-79 (Labinco, Breda, Netherlands) for 10 min and centrifuged with the identical settings. VLPs were re-dissolved from the pellet by mixing with re-dissolution buffer and stirring at 150 rpm overnight at 4 °C. After centrifugation at 1700 rcf for 30 min and filtration with 0.2 μm cellulose acetate syringe filters the solution was dialysed step with 3.5 kDa molecular weight cut-off SnakeSkin Dialysis Tubing (Thermo Fisher Scientific, Waltham, MA, United States). Further purification of the material was conducted with a CaptoCore 400 column (Cytiva, Marlborough, MA, United States), where the VLP capsid containing fractions in the flow-through were freshly used or stored in aliquots at −30 °C. Frozen material was thawed and filtered through a 0.2 μm syringe filter before use.

2.2.4 Characterization of VLP Constructs

Western blot analysis confirmed the expression of each HBcAg VLP construct. After analysis of re-dissolved VLPs by SDS-PAGE the material was transferred onto a nitrocellulose membrane. As primary antibody an anti-HBcAg antibody (abcam, Cambridge, United Kingdom) was used, followed by an anti-mouse antibody (Merck Millipore, Darmstadt, Germany). NuPage 4–12% BisTris Protein Gels, LDS sample buffer and MES running buffer were used for SDS-PAGE. The gels were run on a PowerEase 500 Power Supply (all Invitrogen, Waltham, MA, United States) with 50 mM DTT in the sample solution and a Coomassie blue solution was used for subsequent protein staining. Dimers of Cp, capsids and aggregates were quantified by SEC coupled with a diode array detector to evaluate the phase behavior of the VLPs. Furthermore, this was used to assess the loading of the VLPs with nucleic acids by calculation of the ratio of HBcAg capsid peaks at 260 and 280 nm. SEC was performed at a Vanquish UHPLC system (Thermo Fisher Scientific, Waltham, MA, United States) with an Agilent Bio SEC-5, 5 μm, 1000 Å, 4.6 × 300 mm column (Agilent, Santa Clara, CA, United States). With a DynaPro Plate Reader III (Wyatt Technology, Santa Barbara, CA, United States) the hydrodynamic radius and size distribution profiles of the VLPs were determined by dynamic light scattering. Samples were measured six times, with 25 runs of 5 s each at 25 °C for each measurement, using a sample volume of 30 μl in a Corning 384 well plate (Corning, NY, United States). Disposable folded capillary cells were used on a Zetasizer Nano ZS instrument (both Malvern Instruments Ltd., Malvern, United Kingdom) to measure electrophoretic mobility at 40 mV and 25°, with a 120 s equilibration and five runs with 15 sub runs for each measurement. With the obtained electrophoretic mobility, and a material refractive index of 1.45, absorption of 0.001, a viscosity of 0.8872 mPas, a dielectric constant of 78.54, and a Smoluchowski approximation of 1.5 [80], zeta potential was calculated by Zetasizer Software (Version 7.12, Malvern Instruments Ltd., Malvern, United Kingdom).

2.2.5 Characterization of the Purification Process

As precipitation conditions $(NH_4)_2SO_4$ concentrations of 0.75–1.15 M and 0.25% (v/v) polysorbate 20 respectively were tested. The filtered lysate was diluted and incubated for 3 h at 4 °C, spun down at 17,000 rcf for 30 min in a centrifuge 5810 R (Eppendorf, Hamburg, Germany). The supernatant was examined by SDS-PAGE and the pellet resuspended with re-dissolution buffer. After stirring at 150 rpm overnight at 4 °C, the solution was centrifuged, filtered and the supernatant was examined by SDS-PAGE. ImageJ was used to analyze the VLP band intensity on the gels. To investigate the disassembly reaction, different liquid phase conditions were screened. VLPs were diluted to 3 M, 3.5 M or 4 M Urea with a 50 mM Tris, 10 mM Urea stock solution and titrated to pH 7, 7.5 or 8.0. After stirring 18 h at 4 °C the phase behavior, in particular the peak areas of aggregates, capsids and dimers, of the VLPs were examined by SEC in triplicates. To assess the amount of bigger aggregates control runs at the high-performance liquid chromatography system without a prefilter and column were performed. With the corrected peak areas at 280 nm dimer yields for every condition and constructs were calculated by the ratio of HBcAg dimer peak area to total peak area at 280 nm.

3 Results and Discussions

3.1 Calculation of Protein Aggregation Conditions

Changing ionic strength of the solution is one of the most common ways to modulate PPIs and drive protein solutions either towards stable solutions or aggregation. Before the calculation of solution conditions resulting in the aggregation or assembly of VLP core proteins, we have validated the xDLVO-CG method on three well-studied proteins possessing different shape and complexity. For that lysozyme (LYZ), human immunoglobulin type I (IgG1) and beta lactoglobulin (BLG) were considered. LYZ is a small protein of globular shape which was extensively used in many scientific investigations and has also found its practical use as an antibacterial agent [5]. In Fig. 3a and b the calculated B_{22} coefficients for LYZ as a function of ionic strength at two different pH values of 4.5 and 7 are shown.

B_{22} values are positive at lower ionic strengths, which means that under these conditions LYZ-LYZ interactions are repulsive and, hence, give stable solutions. However, with an increase of ionic strength, the B_{22} values decrease towards negative values crossing the zero point at ca. 0.23 M NaCl at pH 4.5, indicating a switch from repulsive to attractive attractions, i.e. aggregation. Increasing salt concentration permits opposite charge screening, reducing strong electrostatic repulsion between equally charged protein species (see Fig. 3c, W_{el} is significantly decreasing with an increase of NaCl concentration). This allows the attractive dispersion interactions to be more pronounced (see W_{disp} in Fig. 3c and the final impact on the PMF in Fig. 3d), inducing molecule self-assembly. B_{22} values for LYZ are slightly more negative at pH 7 compared to pH 4.5 with the crossing point at around 0.18 M NaCl instead of 0.23 M NaCl, i.e. LYZ at pH assembles faster. It is connected to the fact that the protein charge at pH = 7 is reduced (from +11e to +9e), therefore, less counterions are necessary to screen electrostatic repulsion between LYZ protein pairs.

If attraction between proteins becomes sufficiently strong, they might precipitate or crystallize from solutions, which usually occurs if $B_{22} < 0.0001$ (as determined by Wilson [28]). Molecular mechanism of salt screening consists of oppositely charged ions binding to the charged protein residues accessible to the solvent, and xDLVO-CG model takes that effect into account by using a continuum Debye-Hückel theory. Similarly, the effect of pH is also accounted for, because protein charge is different at each pH as determined by the Propka method. The calculated values show good correlation of xDLVO-CG with experimental data, which show somehow deviated values due to slight differences in the experimental methodology to obtain B_{22}. Due to the

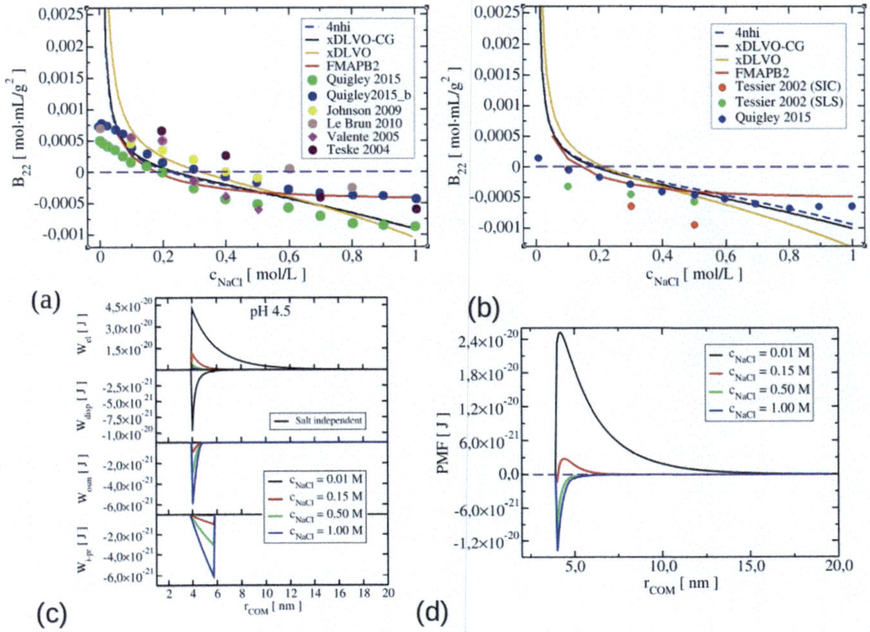

Fig. 3 **a** B_{22} coefficients for LYZ at pH 4.5 and **b** at pH 7 as a function of NaCl concentration calculated using xDLVO-CG, xDLVO (protein as a sphere) and FMAPB2 (all-atom representation). Data marked with dashed blue curve denotes xDLVO-CG calculation for only one relative orientation (in crystal structure). Otherwise, B_{22} of the sampled configurations (see Materials and Methods) were calculated. Experimental data is marked with colored circles. **c** The change of each of the potentials (Eqs. (5)–(9)) and **d** the total PMF in LYZ PPIs at pH 4.5 as a function of NaCl concentration. Reprinted from [70] with permission from Royal Society of Chemistry

spherical shape of LYZ, B_{22} values calculated by xDLVO model show similar correlation with experiments, but they are getting overestimated at higher ionic strengths. The FMAPB2 model, which uses all-atom protein representation, is giving a similar trend to the xDLVO-CG model, but with higher computational cost [70].

The complexity of the protein directly impacts PPIs and the predictive power of the approximative computational approaches at various conditions. This was observed in the case of more challenging proteins, e.g. IgG1 (see Fig. 4a), which is a large protein belonging to a group of monoclonal antibodies with characteristic T-shape. Presently, the preparation of the stable formulations of monoclonal antibodies is still an open issue for the pharmaceutical industry, therefore, the calculation of B_{22} is helpful. Calculated B_{22} values for IgG1 as a function of NaCl concentration and at three pH, i.e. pH 5, 5.75 and 6.5, are depicted in Fig. 4.

IgG1 is a highly charged protein with charge +56e, +48e and +27e at pH 5, 5.75 and 6.5, respectively, therefore values of B_{22} are more positive compared to LYZ. In general, B_{22} values of IgG1, calculated with the developed xDLVO-CG, can reproduce main trends of experimental measurements, despite the fact that this

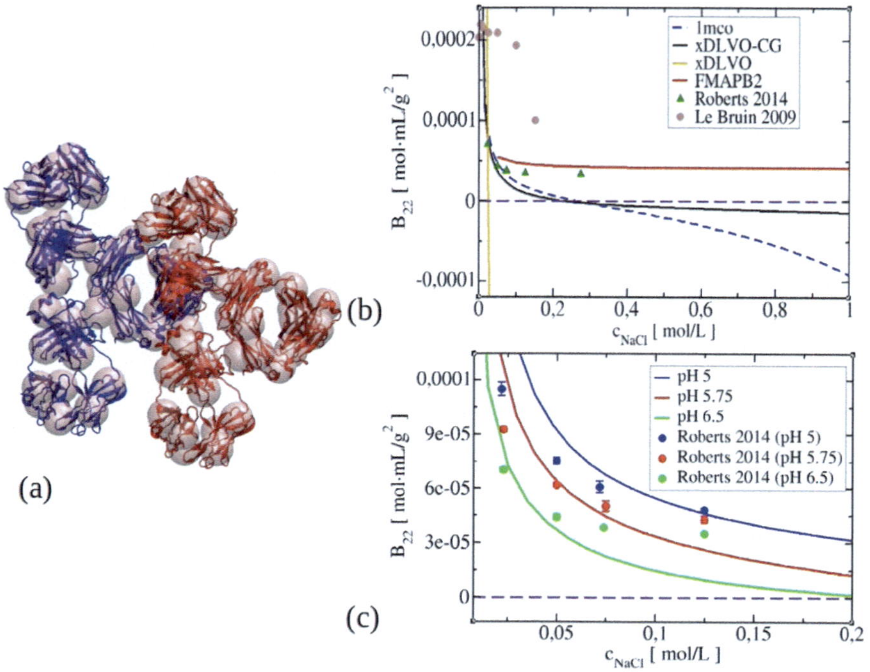

Fig. 4 **a** CG structure of IgG1 dimer. Osmotic second virial coefficients for: **b** IgG1 at pH 6.5 as a function of NaCl concentration calculated using xDLVO-CG, xDLVO and FMAPB2 model, **c** IgG1 at pH 5, 5.75 and 6.5 calculated using xDLVO-CG. Experimental data is marked with colored circles. Reprinted from Ref. [70] with permission from Royal Society of Chemistry

large protein is represented only by 40 CG beads, which mimic its basic shape and charge anisotropy (see Fig. 4a). We have found that CG representation of IgG1 is an important factor impacting calculated B_{22} values, because a simpler spherical-based model, such as xDLVO, fails to reproduce experimental trends of non-globular proteins without using parameters fitted from experiments (see curve in yellow in Fig. 4b). All-atom FMAPB2 model gives similar B_{22} values like xDLVO-CG at lower ionic strengths (marked in red in Fig. 4b), but it deviates towards more positive values at higher ionic strengths. In present, no B_{22} experiments were reported at higher ionic strengths, which makes it more challenging to compare models. Finally, we see that the effect of pH change can be semi-quantitatively modeled with xDLVO-CG (Fig. 4c), which opens more possibilities for further improvements that are currently under development.

3.2 Interplay of Specific and Nonspecific Interactions

As mentioned in Introduction, due to the existence of diverse amino acids in the sequence of proteins, both nonspecific (i.e. electrostatic and dispersion) and specific (e.g. hydrogen bonding) interactions drive the self-assembly and the system dispersity. Beta lactoglobulin (BLG, see Fig. 5a), for example, which is a small globular protein found in milk, exists in the form of a monomer or a dimer in diluted solutions as a function of pH and ionic strength [30]. BLG monomers are dominantly present at pH ≤ 3 and at pH > 8, otherwise, dimers and higher oligomers occur. It is known that here mostly electrostatic interactions play the key role for PPIs, as many amino acids of BLG have pH-dependent charge states. However, the dimer binding site of BLG is stabilized by several H-bonds [45, 76]. Amino acids at the binding interface are getting positively charged at low pH, which prevents dimer formation unless higher salt concentration is added which can neutralize those charges. Therefore, at intermediate pH an interplay of PPIs of various characters occurs and BLG exists in solution as a mixture of monomers and dimers. The exact ratio of molecules in both states cannot be properly obtained.

In order to determine B_{22} coefficients of such dynamically interacting systems, we have combined the xDLVO-CG calculations of B_{22} with the umbrella sampling MD simulations to estimate the binding energy of BLG in its dimer state as a function of pH and ionic strength. Moreover, we have compared the calculated values with the experimental data obtained in the group of Wolfgang Peukert (Institute of Particle Technology, Friedrich-Alexander-Universität Erlangen-Nürnberg) and Anna Wagemans (Institute of Food Technology and Food Chemistry, Technical University Berlin) [84]. In the modified experimental protocol, firstly, the monomer–dimer equilibrium constant was determined by analytical ultracentrifugation (AUC), which allowed us to determine the exact ratio of monomers vs dimers. This was used to interpret membrane osmometry experiments, which gave the averaged B_{22} of solution. In Fig. 5b, the calculated B_{22} coefficients for BLG at pH 3 and pH 7 and at NaCl concentrations of 10 mM and 100 mM using xDLVO-CG are compared to membrane osmometry measurements. B_{22} values were calculated to estimate nonspecific interactions between BLG monomers, dimers and monomer-dimers mixture. Due to the total charge of BLG monomers of $+18$ and -8 at pH 3 and pH 7, respectively, they clearly show the highest preference of electrostatic repulsion interactions, characterized with more positive B_{22} values. As a consequence, the pH dependence of B_{22}, estimated from experiments and the calculated ones, is noticeable. Moreover, higher tendency of self-assembly, e.g. towards dimeric state, with the increase of ionic strength exists. However, B_{22} changes with ionic strength slightly differently for monomer–monomer, monomer–dimer and dimer–dimer case, and a difference is more pronounced at pH 3. Still, the B_{22} values at the measured salt concentrations do not differ significantly, indicating an equilibrium state.

Finally, US calculations were performed between BLG protein pairs in order to calculate energy of dimer dissociation as a function of pH and ionic strength, as shown on Fig. 5c. This allows us to quantify specific interactions as a function of solution

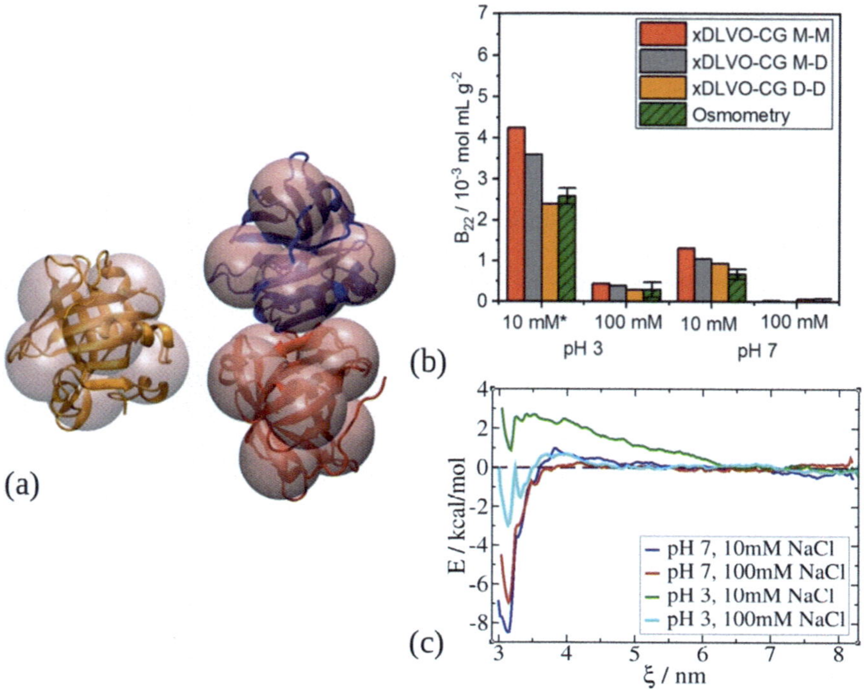

Fig. 5 **a** Coarse-grained representation of BLG monomer (on the left) and dimer (on the right) used to calculate B_{22} coefficient with xDLVO-CG. **b** The osmotic second virial coefficient, B_{22}, at different pH values and NaCl concentrations. The CG-xDLVO calculations provide values for monomer–monomer (M–M), monomer–dimer (M–D) and dimer–dimer (D–D) interactions. Experimental values are retrieved from a combination of the results from AUC and membrane osmometry (green bars). **c** Free energy of BLG dimerization at pH 3 and pH 7 and for 10 and 100 mM NaCl obtained from umbrella sampling simulations at 300 K and atmospheric pressure. Reprinted from [84] with permission from Royal Society of Chemistry

conditions. BLG dimers are more stable at pH 7 giving binding energy of −8.3 kcal/mol and −6.9 kcal/mol for NaCl concentrations of 10 and 100 mM, respectively. These values correlate with the experimental data reported in Ref. [84] In contrast, BLG dimers are less stable at pH 3, giving binding energy of −4.2 kcal/mol at 100 mM NaCl, however, dimer monomer equilibrium is shifted more to monomers compared to pH 7. At pH 3 and 10 mM NaCl, the energy of binding is repulsive (+1.2 kcal/mol), which means that dimers are thermodynamically unstable at these conditions.

3.3 PPIs Between Core Proteins of Virus-Like Particles and Impact of Nucleic Acids

The formation of HBV VLPs is also modulated by the interplay of diverse interactions. Here, often electrostatic forces are the most dominant [44]. To investigate PPIs between core protein homodimers (Cp_2) [77] of VLPs of different sequences and explain their self-assembly versus interactions with nucleic acids, several computational methods were applied. Firstly, xDLVO-CG calculations of six Cp_2, differing with the length of the C-ter (see Table 1 and Fig. 6a) were performed. In Fig. 6c, the calculated B_{22} coefficients for Cp_2 dimers are shown. Due to the low charge state of $Cp_2$164, $Cp_2$157, $Cp_2$154 (see charges of Cp monomers in Fig. 6b), the B_{22} coefficients are fastly decreasing with the increase of ionic strength. This means that the PPIs between these proteins are of the strongest attractive character. However, locally, i.e. in the flexible C-ter, charged amino acids are present, therefore, interactions with nucleic acids are still possible. The assembly of $Cp_2$149 (without flexible C-ter chain) is known to be highly modulated by ionic strength [98]. The formation of VLPs is possible only in specific salt conditions. From the calculated B_{22} values, we see a preference of attractive PPIs for solutions with the concentration higher than 0.15 M NaCl. Similar observations were also reported previously for the capsid formation from the Cp149 assembly domain [22, 98]. The strongest repulsion interactions, i.e. the need of high ionic strengths solutions, occur for $Cp_2$183. Here, the C-ter of a single Cp is +15e charged, therefore, these molecules will strongly bind oppositely charged species, i.e. nucleic acids, to permit self-assembly and further capsid or VLP formation. Note that charges given in Fig. 6b are calculated for the whole core protein, i.e. including the contributions from the assembly domain and C-ter.

All-atom MD simulations (see Materials and Methods) were performed in order to investigate dynamical properties and energetics of the VLP HBV core proteins, as a function of a core protein length. Calculated root mean square deviation (RMSD) shows that the assembly domain of Cp_2 (at 1–149aa) maintains its tertiary structure during simulation of 100 ns (see Fig. 7a). Here, RMSD fluctuates around 0.2 nm in comparison to the crystal structure, however, it remains stable. The C-ter, e.g. of the $Cp_2$183 frequently changes its conformation, which is visible in the large fluctuations of RMSD of the whole structure (see blue curve in Fig. 7a). This effect is getting more pronounced as the length of the C-ter domain is increasing. The flexibility of the C-ter is also known from the literature [97].

In order to get insight into energetics of the dimer-dimer binding and contributions of individual residues, MD trajectories were analyzed by Molecular Mechanics Poisson Boltzmann method (MMPBSA) [36]. In Fig. 5b, individual contributions of all residues of $Cp_2$149 in the protein–protein binding occurring during capsid formation are depicted. It can be seen that the majority of residues have rather small contributions, however some residues have stronger contributions (20–40 kJ/mol per one residue) either of favorable or unfavorable character. For the sake of comparison, residues with high contribution are visualized in Fig. 7c, where residues with

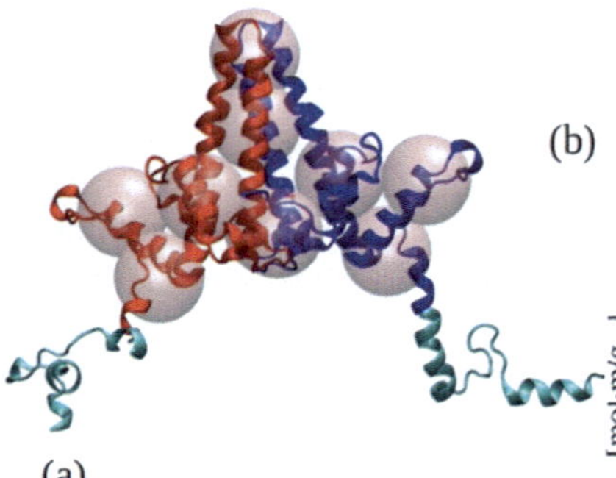

(a)

(b)

Cp length, aa	Charge
149	-7
154	-3
157	-2
164	1
167	4
183	8

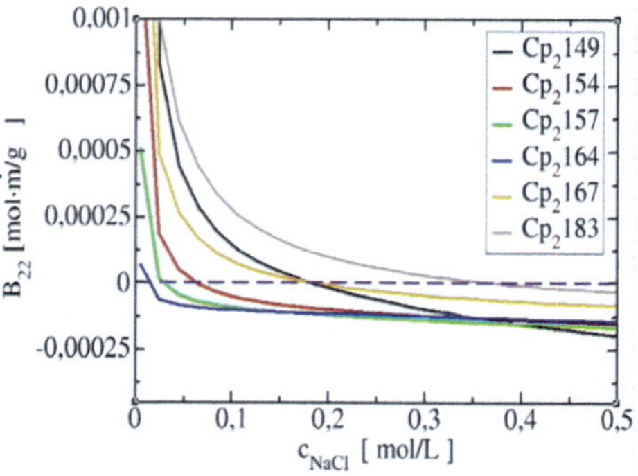

(c)

Fig. 6 **a** Shape-based CG model of Cp_2149 superimposed on its all-atom structure (red and blue chains). For the sake of illustration, the flexible C-terminal (residues 150–183aa) of the full length Cp_2183 are shown in cyan. **b** Charges of core proteins (per one Cp monomer) with truncated C-ter (see Table 1). **c** Calculated B_{22} coefficients for the full-length HBV core protein dimers (Cp_2183) and truncated dimers (Cp_2149, Cp_2154, Cp_2157, Cp_2164, Cp_2167)

favorable and unfavorable contribution are colored in red and blue, respectively. The majority of the residues with attractive contributions are located near the dimer-dimer interface or spike region (e.g. residues LEU95, THR128 and TYR132). This observation explains the aggregation processes that do not lead to correct capsids, and therefore, VLPs, known in the literature [51, 96]. Residues with both attractive and repulsive character are located at the dimer-dimer interface and their subtle interactions drive Cp_2 assembly, i.e. the capsid formation [14, 22, 51]. The hotspot residues (THR128, TYR132, ARG112, PRO134), responsible for HBcAg PPIs are depicted in Fig. 6d. These residues are mostly hydrophobic and uncharged. It is in accordance with previous studies, which have concluded that capsid assembly is mostly driven by weak hydrophobic interactions [44, 96].

Adding the residues at the C-ter domain from Cp149 to Cp183 (mentioned in Table 1) causes the change of the total protein charge (see Fig. 6b), e.g. the Cp_2149 dimer has a charge of $-14e$, while the Cp_2183 dimer has a charge of $+16e$. The shift from the negative to positive charge is mostly due to the addition of positively charged arginine residues, which are abundantly present in the flexible 150–183aa C-ter located inside the capsid (see Fig. 1). The addition of positively charged residues makes it possible for dimers to bind nucleic acids. Therefore, MD simulations of Cp dimer-RNA complexes were performed to investigate the affinity of Cp to bind nucleic acids as a function of core protein length. For this, RNA was firstly docked

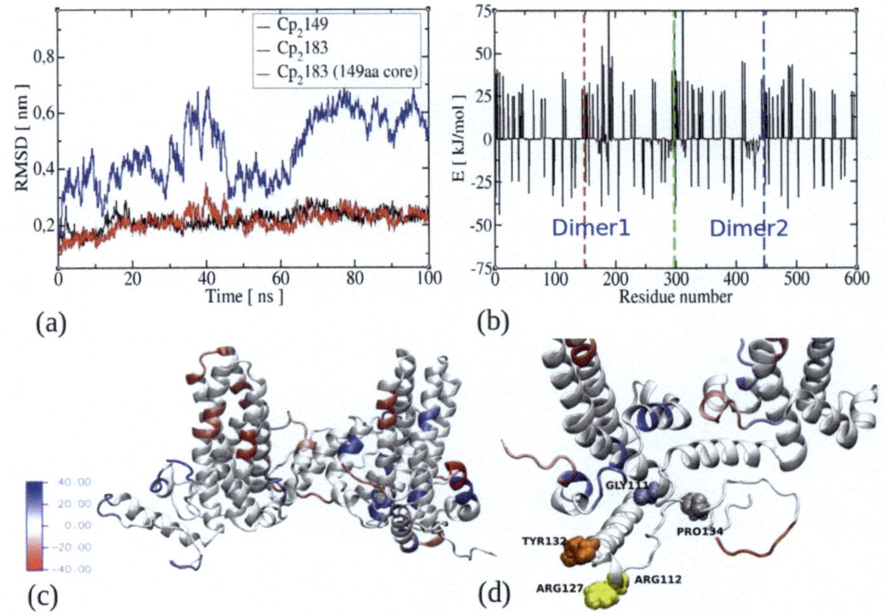

Fig. 7 All-atom MD simulations of Cp dimers (i.e. Cp_2) with full length (Cp_2183) and truncated proteins (Cp_2149): **a** The change of RMSD during 100 ns MD runs. **b** The contribution of each residue of Dimer1 (residues 1–298) and Dimer2 (residues 299–596) to the total free energy of the protein–protein binding. **c** Visualization of Cp_2149 dimer with the colored residues that contribute to the PPIs. Blue and red color denote residues which have unfavorable and favorable contributions, respectively **d** Dimer-dimer interface (Cp_2149) with hotspot residues that have the strongest contribution to the binding free energy. The free energy was determined by MMPBSA

to the dimer and then the docked complex (shown in Fig. 8c) was used as a starting structure for 100 ns MD simulations. This allowed the dimer-RNA structure to relax at finite temperature and the RNA to find a binding spot towards interaction with proteins. MD trajectories were used to calculate Cp dimer-RNA binding energy and its dependence for different C-ter lengths and salts conditions.

In Fig. 8a, the contribution of electrostatic interactions (mainly of an attractive character) in the total binding energy between Cp dimers, with different C-ter lengths, and RNA is demonstrated. It is clearly seen that Cp_2183 binds RNA (27 base pairs) most strongly: The binding energy is ca. -18 kJ/mol. The affinity for RNA binding is decreasing with the truncation at the arginine rich C-ter domain. Cp_2167 and Cp_2164 still have rather high favorable free energy of binding with RNA, however binding energy in these cases is of ca. -14 kJ/mol and -12 kJ/mol, respectively. Dimers of the core proteins with the shortest C-ter length (Cp154 and Cp157) bind RNA in the smallest degree, i.e. the binding energy is of ca. -2.2 kJ/mol and -1.2 kJ/mol, respectively. This indicates that these dimers have low affinity for binding with nucleic acids, thus should be easier disassembled in the VLP processing.

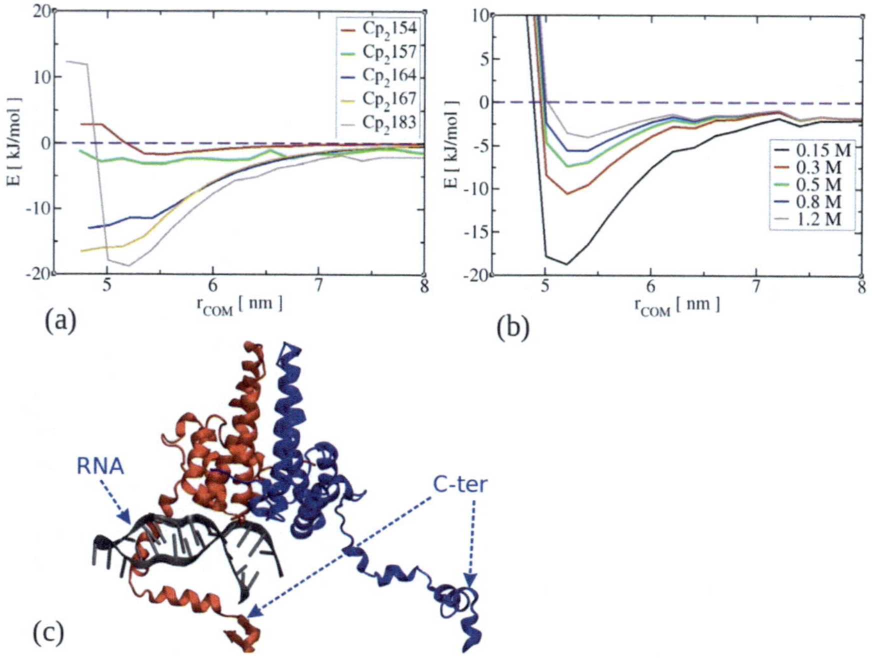

Fig. 8 **a** Electrostatic interactions in Cp dimer-RNA complex as a function of C-ter length in the presence of 0.15 M NaCl. **b** The change of Cp_2-RNA binding energy for Cp183-RNA complex at different NaCl concentration. **c** All-atom structure of RNA docked to $Cp_2$183, used as a starting structure for MD simulations

Furthermore, the nucleic acid binding affinity was investigated also as a function of ionic strength of the solution. In Fig. 8b, the electrostatic contribution to the free energy of binding of the Cp183-RNA complex at various NaCl concentrations is shown. The binding energy is stronger at lower ionic strengths and it gradually decreases with an increase of NaCl concentration. It significantly weakens at ionic strengths higher than 1 M NaCl. This indicates that an increase of ionic strength, on the one hand, reduces the electrostatic repulsion between Cp dimers towards self-assembly (as was also shown in Fig. 6a) and, on the other hand, reduces the attractive interactions with nucleic acids for their encapsulation. However, the electrostatic attraction interactions in the $Cp_2$183-RNA complex are still around −4 kJ/mol at 1.2 M NaCl, therefore, extreme solution conditions have to be applied to disassemble the complex during the VLP purification process [35, 85]. Note that electrostatic contribution to the free energy of Cp dimer-RNA binding is showing a similar trend (as for Cp183 in Fig. 8b) of the energy decrease with the addition of salt for other dimers too.

Since the trimer of core protein dimers (the trimer of Cp_2) is known to be an important intermediate formed during capsid assembly or disassembly, identified by several mechanistic studies [4, 63], MD simulations of trimers of dimers with various

C-ter were performed. The molecular system in this case is very large, therefore, CG-MD simulations using SIRAH force field were carried on. In addition, the free energy of protein–protein and impact of protein-nucleic acid interactions on PPIs was calculated with US technique. Firstly, steered MD was used to dissociate dimer (i.e. Cp_2) from the trimer of dimers (see Fig. 2b). For this, the external force was applied and the rare event of dimer separation was simulated. Snapshots from the pulling simulations were used to generate the starting structure for umbrella sampling simulations, which consists of a series of constrained MD simulations along the reaction coordinate. The same approach was also performed for the Cp_2 trimer-nucleic acid complex (see Fig. 9a). Here, due to the absence of the RNA force field parameters, DNA was used to mimic the nucleic acid. In Fig. 9b, the calculated free energy minima for the Cp_2 dissociation from its trimer with and without presence of DNA are depicted. Cp_2 with different lengths (see Table 1) were considered.

The free energy of Cp_2 dissociation from the trimer (blue bars), steered by various PPIs, confirms the dependency on the length of the core protein, as was also suggested with B_{22} calculations with xDLVO-CG (see Fig. 6). However, the effect of the C-ter length is not uniform. Dimer-dimer interactions in $Cp_2$183 and $Cp_2$154 trimers are lower compared to the shortest sequence of $Cp_2$149: the free energy of dissociation is of −4.2 kcal/mol and −5.5 kcal/mol, respectively. In contrast to that, the other three types of core proteins interact stronger within the trimer formed, stabilizing it, therefore, increasing the final trimer dissociation energy (in the absolute value). This effect is the strongest for the trimer formed from $Cp_2$164 proteins, which have

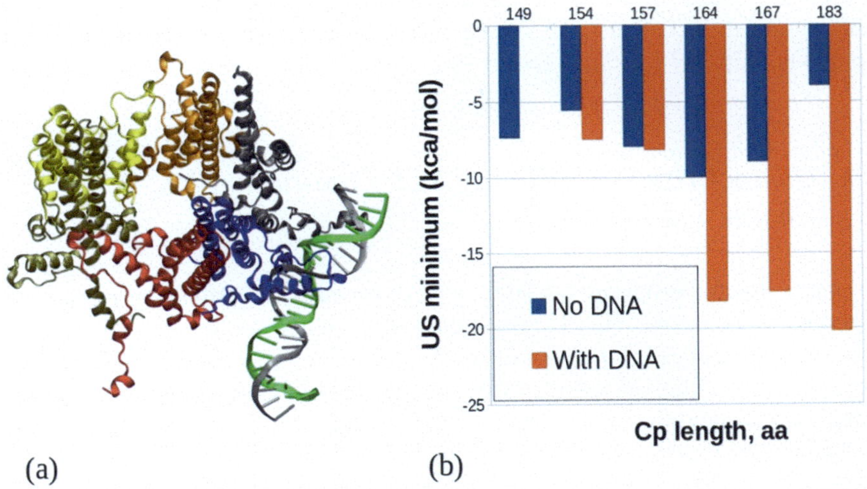

Fig. 9 Coarse-grained MD of trimer of dimers with explicit water and 37.5 mM NaCl: **a** Example of the Cp_2 trimer-DNA complex used for MD simulations. **b** The calculated free energy of separating Cp_2 from Cp_2 trimer by using umbrella sampling technique. Simulations were performed with (in blue) and without (in red) bonded DNA to the trimer. Results for Cp_2 trimer-DNA complex are average of three different US simulations, in which DNA was attached to the trimer at different positions

the lowest charge within all simulated proteins, i.e. a charge of +2 per Cp_2 dimer (see Fig. 6b). It again suggests that PPIs during capsid formation are dominated by electrostatic interactions. Here, it should be noted that despite the relatively high charge of $Cp_2$149 (of −14e), the PPIs have modest attractive character. Probably this is a reason for efficient VLP formation from these proteins upon delicate screening of local charge repulsions by rather low ionic strengths [96].

When a DNA molecule interacts with the trimer (see Fig. 9a), the free energy of dimer disassembly from the trimer is significantly changed. The dimer-dimer interactions in $Cp_2$154 and $Cp_2$157 trimers are only slightly impacted, i.e. the binding free energy increases by 1.9 kcal/mol for $Cp_2$154 and by 0.2 kcal/mol for $Cp_2$157, giving the average free energy of dissociation similar to $Cp_2$149 (i.e. when DNA is not present). In contrast, $Cp_2$164, $Cp_2$167 and $Cp_2$183 trimers are clearly stabilized in the presence of the DNA, giving substantially more attractive free energy of dissociation, i.e. making it harder to disassemble a capsid made out of these core proteins. $Cp_2$164 and $Cp_2$167 trimers are stabilized by nearly two-fold, while the full length $Cp_2$183 in the trimer has a five times attractive energy increase, when DNA is attached. Obviously, the $Cp_2$149 lacks nucleic acid binding region (i.e. region with positively charged amino acids, C-ter) to establish a sufficiently strong interaction with nucleic acids, which can stabilize the system, therefore, it was not considered in the simulations with DNA.

Finally, the efficiency of VLP (dis)assembly will depend on the interplay of interactions between core proteins and core proteins with the nucleic acid. Considering both high PPIs in the $Cp_2$183 trimer and its strong affinity to interact with the negatively charged DNA or RNA, the capsid formed is very stable, therefore, processing of these species is cumbersome. Similar effects should be also present in the case of $Cp_2$164 and $Cp_2$167 (see Figs. 8a and 9b).

3.4 Characterization of HBcAg VLP Constructs: Size, Charge and Loading

To enable effective production of HBcAg VLPs for nucleic acid delivery, it is critical to better understand the impact of the nucleic acid binding region and bound nucleic acids on the purification process of VLPs. Therefore, six different HBcAg-VLP constructs with different lengths of the nucleic acid binding region (see Fig. 6a, Table 1) were produced and characterized regarding their size, zeta potential and loading with host cell-derived nucleic acids.

SDS-PAGE and western blot analysis were used to verify the successful intracellular formation of all protein variants and show the different molecular weights of the Cp149, Cp154, Cp157, Cp164, Cp167 and Cp183 monomers. On the SDS-PAGE Cp149 displayed a length close to the 14.4 kDa marker, the intermediate constructs higher molecular weights with their ascending lengths and with Cp183 close to the 21.5 kDa marker (Fig. 10). These obtained monomer sizes of the constructs are

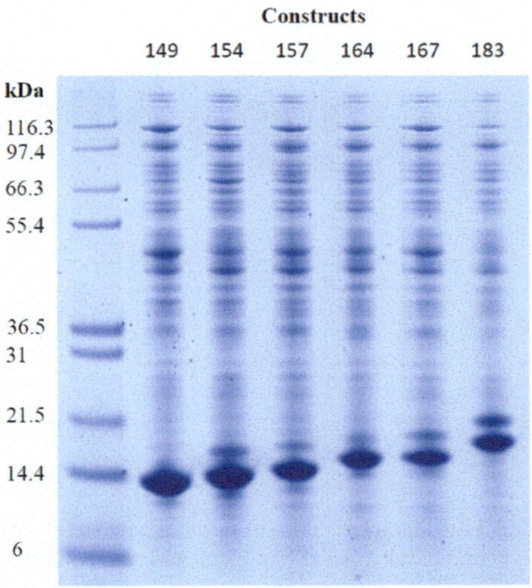

Fig. 10 SDS-PAGE of HBcAg virus-like particle monomers after re-dissolution in reducing conditions. Invitrogen Mark 12 Unstained Standard in lane 1 and different VLP constructs in lanes 2–7. Protein staining was performed by Coomassie blue and molecular weights of the samples are displayed on the left. Figure taken from [85], licensed under the Creative Commons Attribution 4.0 International license (CC BY 4.0)

consistent with the size values given in literature for Cp183 of 21 kDa [7, 89] and the shorter migration time of the truncated construct [83].

Dynamic light scattering was used to investigate the hydrodynamic diameters of the VLP capsids. Despite the different monomer sizes, similar values were obtained for all capsids assembled from different constructs, with an average hydrodynamic diameter of 36.9 $\pm$ 3.2 nm (Fig. 11), consistent with previous reports of 36 nm for Cp183 VLP [67]. These results confirm the former suggested independence of the capsid size from HBcAg VLP constructs, shown by evaluation of transmission electron microscope pictures and measuring electrophoretic mobilities resulting in capsid sizes of about 38 nm for all investigated HBcAg genotypes in literature [65]. Despite the different lengths of the nucleic acid binding regions (see Table 1), electrophoretic mobility measurements of the HBcAg VLP capsids resulted in fairly constant zeta potentials of -11.4 $\pm$ 1.5 mV (Fig. 11) for all constructs. It is likely that the positive charges of the binding regions do not affect the negative surface charge due to its position inside the capsid [86] (see Fig. 1) and the presence of charge compensation by encapsulated nucleic acids [66].

The A260/A280 ratio is a commonly used parameter to examine protein or nucleic acid purity, due to distinct absorption profiles of the two molecule types [29]. A260/A280 coefficients of the capsid peak areas on a SEC were evaluated for the different HBcAg VLP constructs to investigate their loading with host cell-derived nucleic acids (Fig. 12). This method allows the relative amount of encapsulated nucleic acids to be determined in relation to proteins without the need for separate protein and nucleic acid specific analysis. Cp149, lacking the nucleic acid binding region, displayed an A260/A280 coefficient of 0.60, which is similar to previously reported values of 0.61 [35] and 0.67 [92] before further downstream purification steps. For pure protein, A260/A280 ratios of 0.6 are assumed [29], so only a minimum or no nucleic acids seem to be encapsulated in Cp149. Accordingly, an A260/A280 ratio of

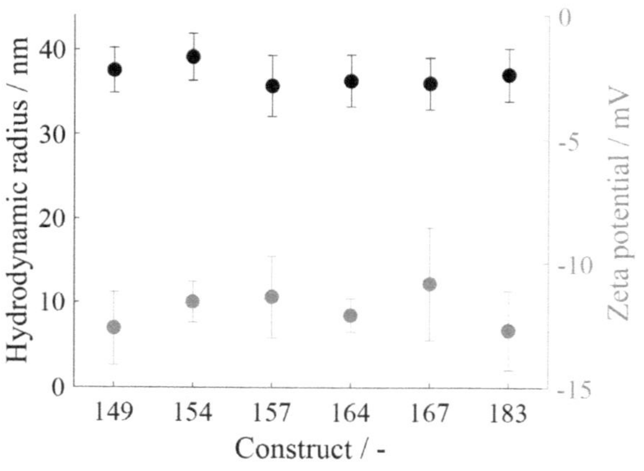

Fig. 11 Characterization of virus-like particle construct capsids formed from different Cp: Hydrodynamic radius is shown in black and zeta potential is in gray. They were determined by measuring dynamic light scattering in six replicates and electrophoretic mobility in five replicates, respectively. Measurements directly after CaptoCore 400 chromatography purification step. Figure taken from [85], licensed under the Creative Commons Attribution 4.0 International license (CC BY 4.0)

approximately 0.6 was reported for purified HBcAg capsid protein after performing the proposed scatter correction [68]. Deviating from this, A260/A280 ratios of 0.55 [35] and 0.57 [92] were reported after disassembly followed by cross-flow filtration or affinity chromatography, indicating that some nucleic acids are still encapsulated in Cp149 capsids, which are then separated by the subsequent purification procedure. The scatter corrected A260/A280 coefficient of 0.60 obtained here shows that nucleic acids are present in the Cp149 capsids even without a binding region. However, the amount of encapsulated nucleic acids in Cp149 is low compared to the Cp183 with the full-length wild-type nucleic acid binding region with an A260/A280 coefficient of 1.65, consistent with previous reports based on cryogenic electron microscopy analysis of Cp149 and Cp183 [89, 97]. Considering all investigated constructs, A260/A280 ratios rise with longer nucleic acid binding regions, displaying an increase in the amount of encapsulated nucleic acids (see Fig. 12). This has also been previously demonstrated by cryogenic electron microscopy [50]. Interestingly, the correlation between the encapsulation of host cell-derived nucleic acids after intra-cellular formation and the length of or the amount of positive charges within the nucleic acids binding region is not linear (similarly as PPIs between core proteins upon RNA binding simulated by CG-MD, see Fig. 9b). A two-sector behavior is distinguishable with one sector (marked red in Fig. 12), including the constructs with no or short binding regions, where an increased length of the binding region displayed a substantial effect on the amount of encapsulated nucleic acids. For the constructs Cp149, Cp154 and Cp157 within this sector A260/A280 ratios show a steep linear increase with an A260/A280 of 0.95 for Cp154 and 1.31 for Cp157. Whereas for the other sector (marked blue in Fig. 12), with the constructs with longer binding regions up to full-length wild-type the longer binding region, shows less impact on the A260/

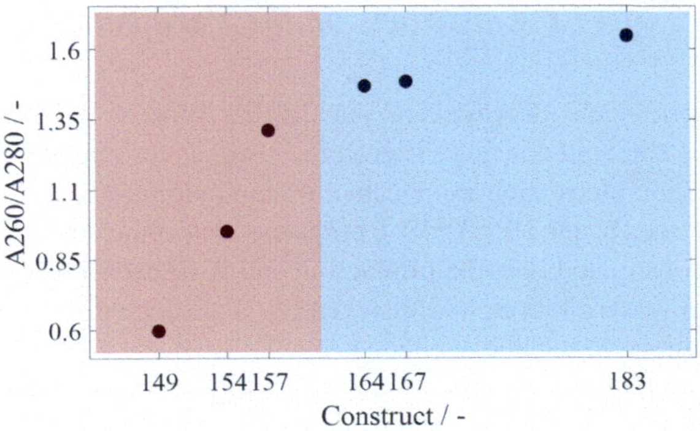

Fig. 12 A260/A280 coefficients of the virus-like particle capsids with host cell-derived nucleic acids for all constructs. Size exclusion chromatography was used to determine the A260/A280 coefficient after scatter correction [68] and to evaluate the loading of the purified VLP capsids. Size exclusion chromatography runs were conducted in triplicates. Error bars are negligible and therefore omitted. Adapted from [85], licensed under the Creative Commons Attribution 4.0 International license (CC BY 4.0)

A280, and thus the amount of encapsulated nucleic acids, displaying an increase of A260/A280 from 1.47 for Cp164, to 1.48 for Cp167 and to 1.65 for Cp183 within this sector.

These two sectors and the respective amount of encapsulated nucleic acids influence charge ratios between the positive charges of the binding region and the negative charges of the nucleic acids in the capsids. Previously, it was reported that the stability of the HBcAg-VLP capsid depends on this charge ratio, where it was hypothesized that a balanced charge density inside the capsid lead to stable capsid structures [66]. Furthermore, the inner shell of nucleic acids bound to the C-ter binding regions, previously described [50, 97], seemingly stabilizes the HBcAg VLP capsid's structure. Moreover, the loading of the capsids with host-cell derived nucleic acids at the beginning of the downstream purification process might indicate the ability of the respective constructs to encapsulate target therapeutic nucleic acids for medical use. For efficient encapsulation of nucleic acids, a nucleic acid binding region seems to be of great importance. However, the results displayed in Fig. 12 indicate that an intermediate length of this binding region could bind and encapsulate a sufficient amount of target nucleic acids (as also suggested by MD simulations, Figs. 8a and 9b). This could circumvent the reported stability problems after host cell-derived nucleic acid depletion for Cp183 [67], as well as for Cp164 and Cp167 [66], especially, because Cp154 showed no such stability problems in the mentioned study [66].

3.5 Influence of VLP Constructs on the Purification Process

Effective production and downstream purification process of HBcAg VLPs is inevitable for nucleic acid delivery. Therefore, it was investigated how the production and purification steps such as selective precipitation, redissolution and disassembly are influenced by the HBcAg VLP properties determined in the chapter above. Therefore, we investigated how the production and purification steps such as selective precipitation, redissolution, and disassembly are affected by the HBcAg-VLP properties determined in the above chapter.

To examine the selective precipitation of the VLP constructs, ammonium sulfate concentrations from 0.75 M up to 1.15 M were studied. Both the remaining VLPs in the supernatant and the precipitated and re-dissolved material were analyzed by SDS-PAGE and subsequent image analysis. The intensities obtained by image analysis of precipitation supernatant samples after normalization to the highest derived intensity for the respective construct are shown in Fig. 13a. For all constructs, the amount of remaining VLPs was reduced with an increase in ammonium sulfate concentration in the precipitation step and a concentration of 1.05 M fully precipitated the VLPs from the clarified lysate. Even 0.95 M ammonium sulfate was sufficient to completely precipitate Cp167 and Cp183. However, for Cp167 and Cp183, with rising precipitant concentrations up to 0.90 M the intensities decline and stay relatively constant when further increasing the concentration. The same behavior is shown by Cp149 and Cp154 with fairly stable intensities for 1.10 and 1.15 M ammonium sulfate. The corresponding gel images evidence for the conditions with fairly stable intensities at high ammonium sulfate concentrations, that VLPs have been fully precipitated and other protein impurities are causing the residual intensity signal. For the two other constructs: Cp157 and Cp164, the normalized intensities diminish over the entire investigated precipitant concentration range, whereas the gel images show no remaining VLPs for ammonium sulfate concentrations equal and higher than 1.05 M. Under these conditions, other protein impurities could be co-precipitated and cause a further decrease in the displayed intensities.

Overall, the reported differences in ammonium sulfate concentrations for the precipitation of the investigated constructs with different lengths of the nucleic acid binding regions is minor. Whereas there is a big gap to the former reported precipitant concentration of merely 0.15 M for a chimeric version of Cp149 [34]. A possible reason for this could be that the nucleic acid binding regions and bound nucleic acids of the constructs studied here are present as an inner shell within the VLP capsids [50, 97], and changes on the surface of the HBcAg capsids appear to have a greater impact on the susceptibility to ammonium sulfate as a precipitant. Another deviation in the findings reported for the chimeric Cp149 VLPs and the constructs studied here could possibly also be explained by this. Precipitation of the constructs in the present report occurred within a broad range of at least 4 increments of 50, 200 mM ammonium sulfate, with only a portion of the VLPs precipitating. In contrast, for the chimeric VLP, precipitation behaved less dynamically within a maximum of two increments [34].

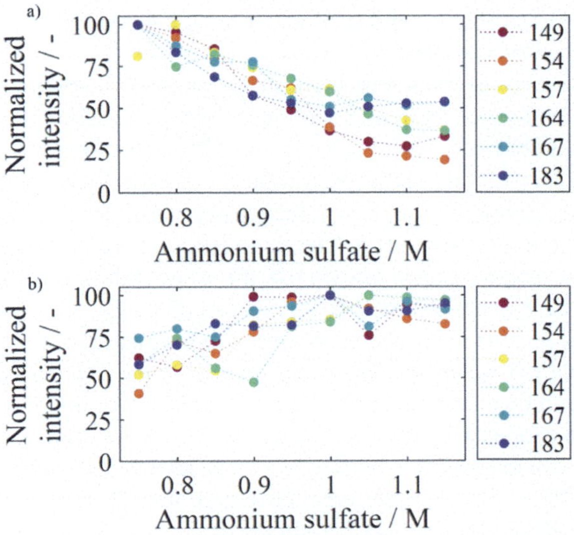

Fig. 13 Normalized intensities for virus-like particle constructs depending on precipitant concentration. After precipitation the supernatant (**a**) and re-dissolution solution (**b**) were analyzed by SDS-PAGE and evaluation of the gel scans by the intensity of the virus-like particle band on the gel lane with respect to the highest determined intensity resulted in displayed remaining virus-like particle intensities Figure taken from [85], licensed under the Creative Commons Attribution 4.0 International license (CC BY 4.0)

After re-dissolution of the precipitated material, high normalized intensities were observed for all constructs for ammonium sulfate concentrations equal and higher than 0.95 M (Fig. 13b), with the overall highest intensities at 1 M precipitant when all constructs are considered. Because of the relatively minor differences in the precipitation behavior of the different constructs, the same precipitation condition with 1 M ammonium sulfate was selected for further processing of the various HBcAg VLP constructs. A suitable re-dissolution buffer was determined in preliminary studies, where Tris buffer with different sodium chloride molarities were investigated. 150 mM NaCl in Tris resulted in the highest amount of VLP capsids and was chosen as uniform buffer condition for all constructs. A more detailed investigation of the re-dissolution behavior was not feasible because the precise determination of dimer content was difficult at this stage of the purification process due to remaining impurities with overlapping chromatography peaks in the SEC.

After precipitation and re-dissolution, the HBcAg VLPs studied were further purified by dialysis and chromatography with CaptoCore 400 to separate smaller contaminants. Subsequently, capsid disassembly is crucial for the separation of encapsulated impurities [35], the separation of bound nucleic acids [67, 81, 93] and improvement of structural integrity [93]. The impact of encapsulated host cell-derived nucleic acids on capsid stability and the phase behavior of the different constructs was investigated in a comprehensive disassembly screening. Therefore, the disassembly of the purified VLP constructs was conducted under reducing conditions with a variation of urea concentrations and pH.

In the disassembly experiments, high dimer yields were obtained for high pH values and high urea concentrations, resulting in the expected correlation between the liquid phase conditions and dimer yields. In Fig. 14a and b, dimer yields for all investigated constructs at various pH values and urea concentration of 4 M (the highest tested urea concentration) are displayed. For Cp149, Cp dimer yields increased from 53.4% at pH 7 up to 64.9% at pH 8. The same behavior was observed for all other constructs, but to a lesser extent. For Cp164, the yield of dimers increased by 2.8% and for Cp154 by 8.3% when the pH was increased from 7 to 8. In Fig. 14c and d, dimer yields for all investigated urea concentrations and constructs at pH 8 (i.e. the highest tested pH value) are displayed. Increasing the urea concentration from 3 to 4 M resulted in an increase in the dimer yield of 4.9% for Cp149 to 15.5% for Cp154 for the constructs without or with short nucleic acid binding regions (Fig. 14c). The constructs with longer C-ter binding regions demonstrated fairly constant dimer yields within ± 1% when comparing urea concentrations of 3 and 4 M. Equal behavior could be observed for pH 7 and pH 7.5 and for urea concentrations of 3 and 3.5 M.

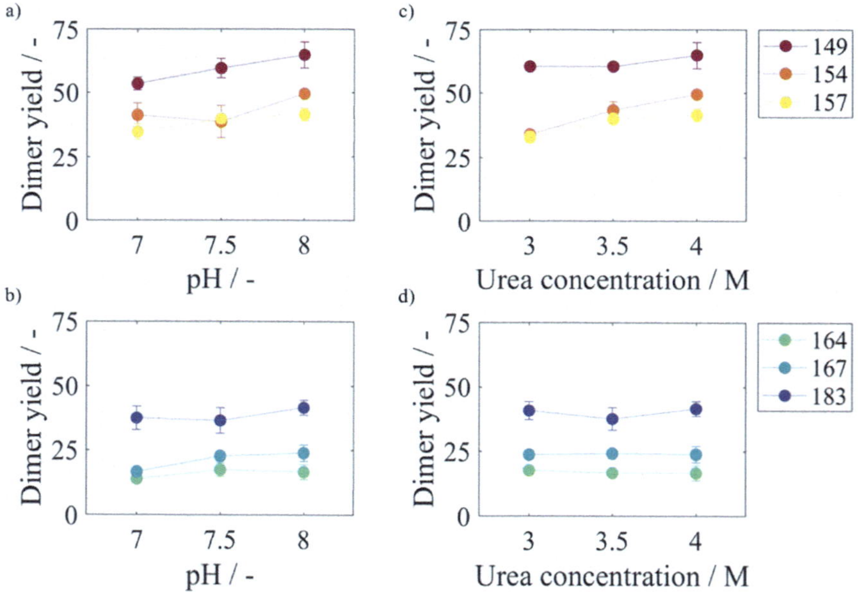

Fig. 14 HBcAg dimer yields after disassembly reaction were determined by size exclusion chromatography for the different virus-particles and for different liquid phase conditions. Results for different pH values and a urea concentration of 4 M (**a**, **b**) and for different urea concentrations and a pH of 8 (**c**, **d**) are displayed. Subfigures a and c show results for Cp149, Cp154 and Cp157 and subfigures b and d for Cp164, Cp167 and Cp183, according to the two sectors, distinguished in Fig. 12. Figure taken from [85], licensed under the Creative Commons Attribution 4.0 International license (CC BY 4.0)

Reducing agents, such as urea or guanidine hydrochloride, low ionic strength and high pH are frequently reported in the literature to induce the disassembly of HBcAg VLPs [7, 35, 81, 92] and other types of VLPs [53, 55]. Consistent with the results presented here, an optimal urea concentration of 4 M was suggested for both a chimeric Cp149 [35] and wild-type Cp149 and Cp183 [67, 92]. Although urea concentrations above 4 M result in protein aggregation and even denaturation [92], under the conditions tested here (with urea concentration between 3 and 4 M at different pH values) only partial disassembly and a maximum dimer yield of 64.9% for Cp149 with concomitant aggregation and residual capsids were observed. Similar results have been reported for Cp149 and a chimeric Cp149 [35]. The thereby obtained little higher yields of 71% for Cp149 and 69% for the chimeric Cp149 possibly arise out of differences in the tested pH range, in the dilution strategies and observed reaction times. In the mentioned study, synergistic effects of pH and urea for the disassembly of HBcAg VLPs were described for Cp149 [35]. This effect is also visible in the study presented here. Considering all conditions tested and averaged over all constructs and all urea concentrations, the average increase in the dimer yield was 7.6%, when pH was increased from 7 to 8. Similarly, the average increase in the dimer yield, when the molar concentration of urea was increased from 3 to 4 M, was 4.9% averaged over all pH values. However, the simultaneous increase in pH and urea concentration resulted in an increase in the dimer yield of about 13.3%.

Interestingly, in addition to the influence of liquid phase conditions on the disassembly of the HBcAg VLPs just discussed, significant differences in the disassembly behavior of the HBcAg VLP constructs investigated have emerged. As for the loading of the VLP capsids a two-sector behavior can be observed (see Fig. 12), for the dimer yields obtained in the disassembly screening experiments (Fig. 14) the different sectors are also noticeable. The highest dimer yields were observed for Cp149 without nucleic acid binding region with values up to 64.9% at a urea concentration of 4 M and pH 8. For the constructs without or with a short nucleic acid binding region (red sector in Fig. 12), the yield of dimers steadily decreases with increased length of binding region and the associated higher amount of bound nucleic acids, shown in Fig. 14a and c. This implies an increased capsid stability due to the nucleic acid binding region and the thus elicited binding of host cell-derived nucleic acids forming an inner shell within the capsid [50, 97]. It can be also explained in terms of protein–protein interactions between core proteins, modulated by the presence of the nucleic acid, suggested by MD simulations shown in Fig. 9. For the constructs Cp164, Cp167 and Cp183 (blue sector in Fig. 12), the disassembly behavior changes differently: With an increased length of the nucleic acid binding region (i.e. C-ter chain), the yield of Cp dimers increases steadily, as opposed to the correlation for the constructs with no or short binding regions. The lowest dimer yields were observed for Cp164 with intermediate length of the nucleic acid binding region, as shown for varying pH values in Fig. 14b and varying urea concentrations in Fig. 14d. It may be connected to additional PPIs between Cp164 dimers driven by decreased repulsive interactions of Cp164 due to the lowest protein charge (Fig. 6b). According to the change of the osmotic second virial coefficient (i.e. B_{22}), Cp_2164 self-assembles to the highest extent (see Fig. 6c). For the constructs with even further increased C-ter lengths up to the full-length wild-type Cp183, dimer yields increase, resulting in similar dimer yields for Cp154, Cp157 and Cp183 at urea concentrations higher than 3.5 M. Considering the results presented in Figs. 12 and 14, the increase of the length of the nucleic acid binding region in Cp does not necessarily increase the nucleic acid loading, but lead to more efficient disassembly. Moreover, the binding region with positively charged arginines seems to affect capsid stability [66] studied here. According to the previously reported charge balance hypothesis [66], the small amount of more bound nucleic acids and their negative charges may not be able to compensate for the additional positively charged arginines in the binding region. Therefore, the VLP capsids are destabilized and disassembly is promoted by the repulsive forces within the capsid caused by the excess of positive charges [61], as seen here in the rising dimer yields for the Cp164, Cp167 and Cp183 constructs. The disassembly behavior displayed in Fig. 14 also applies to the results for urea concentrations of 3 and 3.5 M and to pH 7 and pH 7.5. The investigation of the disassembly behavior of the various HBcAg VLP constructs showed both capsid-stabilizing effect of the bound nucleic acids (i.e. protein-nucleic acid interactions) and capsid-destabilizing effect of the positive charges within the nucleic acid binding region (i.e. protein–protein interactions) and their substantial impact on the VLP dispersity, i.e. dimer yields of the disassembly reaction, and phase behavior as a crucial step for HBcAg VLPs purification and loading.

4 Conclusions

HBcAg VLP constructs with variable lengths of the natural nucleic acid binding region, i.e. the wild-type Cp183, were investigated to understand the impact of the changes in the protein structure and the presence of nucleic acids on the dispersity and phase behavior of the virus-like particles. The truncated wild-type Cp149, the wild-type Cp183 and four intermediates (Cp154, Cp157, Cp164, Cp167) of core proteins used for the VLP formation were characterized using the combination of several experimental techniques and theoretical methods. The influence of different constructs and bound nucleic acids on early downstream purification steps was investigated to promote an effective production of HBcAg VLPs for nucleic acid delivery. No significant differences were observed in terms of capsid size and zeta potential, and both capturing by precipitation and re-dissolution were hardly affected by the various constructs and their properties. In contrast, the amount of host cell-derived bound nucleic acids was strongly influenced by the length of the nucleic acid binding region. This revealed a two-sector behavior of nucleic acid loading with a steep increase in the amount of bound molecules for constructs in the sector with short nucleic acid binding regions and less effect on the increase of nucleic acid encapsulation for constructs with longer C-ter binding region.

The comprehensive screening of the disassembly reaction for all constructs and different liquid phase conditions showed that it is highly dependent on the construct type and its loading with nucleic acids. As expected, the highest dimer yields were observed for high urea concentrations and for high pH. However, capsid stability has shown also a significant effect on the disassembly and achievable dimer yields. The complex interplay of capsid-stabilizing effects of the encapsulated nucleic acids and capsid-destabilizing effects, resulting from repulsive forces generated by the positive charges of the C-terminal of core proteins, following the two-sector behavior of the VLP loading and stability, was observed.

Interplay of various protein–protein and protein-nucleic acid interactions, modulating assembly and disassembly processes towards VLP formation, were investigated using all-atom and coarse-grained molecular dynamics simulations together with accelerated MD technique, i.e. umbrella sampling, and the calculation of the osmotic second virial coefficients using the specially developed xDLVO-CG method. Six types of core proteins, as studied experimentally, were considered and their binding energies with RNA and DNA and impact on the Cp_2 disassembly were systematically analyzed. Clear differences in the character of the PPIs and their strength with an increase of C-ter binding site in the presence and absence of nucleic acids were detected. Strong influence of electrostatic interactions, of attractive and repulsive character, both between core proteins and their assemblies with nucleic acids could be elucidated. Similar trends as observed in experiments were obtained and the molecular bases of the dispersity and phase behavior effects were explained.

Acknowledgements The authors gratefully acknowledge the Deutsche Forschungsgemeinschaft (DFG, German Research Foundation) for financial support (WE 1863/30-2 and HU 985/1-2) within the framework of the priority program SPP 1934. M.K. acknowledges, additionally, funding by the Ministry of Science, Research and Art of Baden-Württemberg (Germany) under the Brigitte-Schlieben-Lange-Programm.

The authors thank Prof. Adam Zlotnick for providing VLP production plasmids.

References

1. Abraham MJ, Murtola T, Schulz R, Páll S, Smith JC, Hess B, Lindahl E (2015) GROMACS: High performance molecular simulations through multi-level parallelism from laptops to supercomputers. SoftwareX 1–2:19–25. https://doi.org/10.1016/j.softx.2015.06.001
2. Albert-Ludwigs-Universität Freiburg, Germany , Fortner A, Schumacher D (2021) First COVID-19 Vaccines Receiving the US FDA and EMA Emergency Use Authorization. Discoveries 9:e122. https://doi.org/10.15190/d.2021.1. Department of Anesthesiology, University Hospital RWTH Aachen, Germany
3. Aldosari BN, Alfagih IM, Almurshedi AS (2021) Lipid nanoparticles as delivery systems for RNA-based vaccines. Pharmaceutics 13:206. https://doi.org/10.3390/pharmaceutics13020206
4. Asor R, Schlicksup CJ, Zhao Z, Zlotnick A, Raviv U (2020) Rapidly forming early intermediate structures dictate the pathway of capsid assembly. J Am Chem Soc 142:7868–7882. https://doi.org/10.1021/jacs.0c01092
5. Barbiroli A, Bonomi F, Capretti G, Iametti S, Manzoni M, Piergiovanni L, Rollini M (2012) Antimicrobial activity of lysozyme and lactoferrin incorporated in cellulose-based food packaging. Food Control 26:387–392. https://doi.org/10.1016/j.foodcont.2012.01.046
6. Berendsen HJC, Postma JPM, van Gunsteren WF, Hermans J (1981) Interaction models for water in relation to protein hydration. In: Pullman B (ed) Intermolecular forces, the jerusalem

symposia on quantum chemistry and biochemistry. Springer Netherlands, Dordrecht, pp 331–342. https://doi.org/10.1007/978-94-015-7658-1_21
7. Bin Mohamed Suffian IF, Garcia-Maya M, Brown P, Bui T, Nishimura Y, Palermo ARBMJ, Ogino C, Kondo A, Al-Jamal KT (2017) Yield optimisation of hepatitis B virus core particles in E. coli expression system for drug delivery applications. Sci Rep 7:43160. https://doi.org/10.1038/srep43160
8. Blasi P, Giovagnoli S, Schoubben A, Ricci M, Rossi C (2007) Solid lipid nanoparticles for targeted brain drug delivery☆. Adv Drug Deliv Rev 59:454–477. https://doi.org/10.1016/j.addr.2007.04.011
9. Boisgérault F, Morón G, Leclerc C (2002) Virus-like particles: a new family of delivery systems. Expert Rev Vaccines 1:101–109. https://doi.org/10.1586/14760584.1.1.101
10. Boström M, Tavares FW, Ninham BW, Prausnitz JM (2006) Effect of salt identity on the phase diagram for a globular protein in aqueous electrolyte solution. J Phys Chem B 110:24757–24760. https://doi.org/10.1021/jp061191g
11. Burnett JC, Rossi JJ, Tiemann K (2011) Current progress of siRNA/shRNA therapeutics in clinical trials. Biotechnol J 6:1130–1146. https://doi.org/10.1002/biot.201100054
12. Bussi G, Donadio D, Parrinello M (2007) Canonical sampling through velocity rescaling. J Chem Phys 126:014101. https://doi.org/10.1063/1.2408420
13. Carvalho SB, Silva RJS, Moleirinho MG, Cunha B, Moreira AS, Xenopoulos A, Alves PM, Carrondo MJT, Peixoto C (2019) Membrane-based approach for the downstream processing of influenza virus-like particles. Biotechnol J 14:1800570. https://doi.org/10.1002/biot.201800570
14. Ceres P, Stray SJ, Zlotnick A (2004) Hepatitis B virus capsid assembly is enhanced by naturally occurring mutation F97L. J Virol 78:9538–9543. https://doi.org/10.1128/JVI.78.17.9538-9543.2004
15. Chaturvedi SK, Parupudi A, Juul-Madsen K, Nguyen A, Vorup-Jensen T, Dragulin-Otto S, Zhao H, Esfandiary R, Schuck P (2020) Measuring aggregates, self-association, and weak interactions in concentrated therapeutic antibody solutions. mAbs 12:1810488. https://doi.org/10.1080/19420862.2020.1810488
16. Chen S-H, Zhaori G (2011) Potential clinical applications of siRNA technique: benefits and limitations: clinical application of SIRNA Technique. Eur J Clin Invest 41:221–232. https://doi.org/10.1111/j.1365-2362.2010.02400.x
17. Choi K, Kim K, Kwon IC, Kim I-S, Ahn HJ (2013) Systemic delivery of siRNA by chimeric capsid protein: tumor targeting and RNAi activity in vivo. Mol Pharm 10:18–25. https://doi.org/10.1021/mp300211a
18. Cromwell MEM, Hilario E, Jacobson F (2006) Protein aggregation and bioprocessing. AAPS J 8:E572–E579. https://doi.org/10.1208/aapsj080366
19. Crowther R (1994) Three-dimensional structure of hepatitis B virus core particles determined by electron cryomicroscopy. Cell 77:943–950. https://doi.org/10.1016/0092-8674(94)90142-2
20. Damodaran S (2013) Electrodynamic pressure modulation of protein stability in cosolvents. Biochemistry 52:8363–8373. https://doi.org/10.1021/bi400656a
21. Darré L, Machado MR, Brandner AF, González HC, Ferreira S, Pantano S (2015) SIRAH: a structurally unbiased coarse-grained force field for proteins with aqueous solvation and long-range electrostatics. J Chem Theory Comput 11:723–739. https://doi.org/10.1021/ct5007746
22. Depta PN, Dosta M, Wenzel W, Kozlowska M, Heinrich S (2022) Hierarchical coarse-grained strategy for macromolecular self-assembly: application to hepatitis B virus-like particles. Int J Mol Sci 23:14699. https://doi.org/10.3390/ijms232314699
23. Diab A, Foca A, Zoulim F, Durantel D, Andrisani O (2018) The diverse functions of the hepatitis B core/capsid protein (HBc) in the viral life cycle: Implications for the development of HBc-targeting antivirals. Antivir Res 149:211–220. https://doi.org/10.1016/j.antiviral.2017.11.015
24. Dolinsky TJ, Czodrowski P, Li H, Nielsen JE, Jensen JH, Klebe G, Baker NA (2007) PDB2PQR: expanding and upgrading automated preparation of biomolecular structures for molecular simulations. Nucleic Acids Res 35:W522–W525. https://doi.org/10.1093/nar/gkm276

25. Ersch C, Meijvogel LLC, van der Linden E, Martin A, Venema P (2016) Interactions in protein mixtures. Part I: second virial coefficients from osmometry. Food Hydrocoll 52:982–990. https://doi.org/10.1016/j.foodhyd.2015.07.020
26. Fang P-Y, Bowman JC, Gómez Ramos LM, Hsiao C, Williams LD (2018) RNA: packaged and protected by VLPs. RSC Adv 8:21399–21406. https://doi.org/10.1039/C8RA02084A
27. Garcea RL, Gissmann L (2004) Virus-like particles as vaccines and vessels for the delivery of small molecules. Curr Opin Biotechnol 15:513–517. https://doi.org/10.1016/j.copbio.2004.10.002
28. George A, Wilson WW (1994) Predicting protein crystallization from a dilute solution property. Acta Crystallogr D Biol Crystallogr 50:361–365. https://doi.org/10.1107/S0907444994001216
29. Goldfarb AR, Saidel LeoJ, Mosovich E (1951) The ultraviolet absorption spectra of proteins. J Biol Chem 193:397–404. https://doi.org/10.1016/S0021-9258(19)52465-6
30. Gottschalk M, Nilsson H, Roos H, Halle B (2009) Protein self-association in solution: the bovine β -lactoglobulin dimer and octamer. Protein Sci 12:2404–2411. https://doi.org/10.1110/ps.0305903
31. Hamaker HC (1937) The London—van der Waals attraction between spherical particles. Physica 4:1058–1072. https://doi.org/10.1016/S0031-8914(37)80203-7
32. Herhut M, Brandenbusch C, Sadowski G (2016) Modeling and prediction of protein solubility using the second osmotic virial coefficient. Fluid Phase Equilib 422:32–42. https://doi.org/10.1016/j.fluid.2016.01.020
33. Hill BD, Zak A, Khera E, Wen F (2017) Engineering virus-like particles for antigen and drug delivery. Curr Protein Pept Sci 19. https://doi.org/10.2174/1389203718666161122113041
34. Hillebrandt N, Vormittag P, Bluthardt N, Dietrich A, Hubbuch J (2020) Integrated process for capture and purification of virus-like particles: enhancing process performance by cross-flow filtration. Front Bioeng Biotechnol 8:489. https://doi.org/10.3389/fbioe.2020.00489
35. Hillebrandt N, Vormittag P, Dietrich A, Wegner CH, Hubbuch J (2021) Process development for cross-flow diafiltration-based VLP disassembly: A novel high-throughput screening approach. Biotechnol Bioeng 118:3926–3940. https://doi.org/10.1002/bit.27868
36. Homeyer N, Gohlke H (2012) Free energy calculations by the molecular mechanics poisson−boltzmann surface area method. Mol. Inform. 31:114–122. https://doi.org/10.1002/minf.201100135
37. Hou X, Zaks T, Langer R, Dong Y (2021) Lipid nanoparticles for mRNA delivery. Nat Rev Mater 6:1078–1094. https://doi.org/10.1038/s41578-021-00358-0
38. Huang J, Rauscher S, Nawrocki G, Ran T, Feig M, de Groot BL, Grubmüller H, MacKerell AD (2017) CHARMM36m: an improved force field for folded and intrinsically disordered proteins. Nat Methods 14:71–73. https://doi.org/10.1038/nmeth.4067
39. Hub JS, de Groot BL, van der Spoel D (2010) G_wham—a free weighted histogram analysis implementation including robust error and autocorrelation estimates. J Chem Theory Comput 6:3713–3720. https://doi.org/10.1021/ct100494z
40. Humphrey W, Dalke A, Schulten K (1996) VMD: visual molecular dynamics. J Mol Graph 14:33–38. https://doi.org/10.1016/0263-7855(96)00018-5
41. Hückel E (1924) Zur Theorie der Elektrolyte. In: Ergebnisse der exakten naturwissenschaften, Ergebnisse der Exakten Naturwissenschaften. Springer, Berlin, Heidelberg, pp 199–276. https://doi.org/10.1007/BFb0111753
42. Kaczmarczyk SJ, Sitaraman K, Young HA, Hughes SH, Chatterjee DK (2011) Protein delivery using engineered virus-like particles. Proc Natl Acad Sci 108:16998–17003. https://doi.org/10.1073/pnas.1101874108
43. Karplus M, McCammon JA (2002) Molecular dynamics simulations of biomolecules. Nat Struct Biol 9:646–652. https://doi.org/10.1038/nsb0902-646
44. Kegel WK, van der Schoot P (2004) Competing hydrophobic and screened-coulomb interactions in hepatitis B virus capsid assembly. Biophys J 86:3905–3913. https://doi.org/10.1529/biophysj.104.040055
45. Khan S, Ipsen R, Almdal K, Svensson B, Harris P (2018) Revealing the dimeric crystal and solution structure of β-lactoglobulin at pH 4 and its pH and salt dependent monomer-dimer equilibrium. Biomacromol 19:2905–2912. https://doi.org/10.1021/acs.biomac.8b00471

46. Kulkarni JA, Cullis PR, van der Meel R (2018) Lipid nanoparticles enabling gene therapies: from concepts to clinical utility. Nucleic Acid Ther 28:146–157. https://doi.org/10.1089/nat.2018.0721
47. Kumari R, Kumar R, Open Source Drug Discovery Consortium, Lynn A (2014) g_mmpbsa—a GROMACS tool for high-throughput MM-PBSA calculations. J Chem Inf Model 54:1951–1962. https://doi.org/10.1021/ci500020m
48. Laven J, Vissers JPC (1999) The Hamaker and the Lifshitz approaches for the Van der Waals interaction between particles of composite materials dispersed in a medium. Colloids Surf Physicochem Eng Asp 152:345–355. https://doi.org/10.1016/S0927-7757(98)00824-3
49. Li J, Sun Y, Jia T, Zhang R, Zhang K, Wang L (2014) Messenger RNA vaccine based on recombinant MS2 virus-like particles against prostate cancer: the packaged RNA vaccine for prostate cancer. Int J Cancer 134:1683–1694. https://doi.org/10.1002/ijc.28482
50. Liu S, He J, Shih C, Li K, Dai A, Hong Zhou Z, Zhang J (2010) Structural comparisons of hepatitis B core antigen particles with different C-terminal lengths. Virus Res 149:241–244. https://doi.org/10.1016/j.virusres.2010.01.020
51. Lutomski CA, Lyktey NA, Zhao Z, Pierson EE, Zlotnick A, Jarrold MF (2017) Hepatitis B virus capsid completion occurs through error correction. J Am Chem Soc 139:16932–16938. https://doi.org/10.1021/jacs.7b09932
52. Ma Y, Acosta DM, Whitney JR, Podgornik R, Steinmetz NF, French RH, Parsegian VA (2015) Determination of the second virial coefficient of bovine serum albumin under varying pH and ionic strength by composition-gradient multi-angle static light scattering. J Biol Phys 41:85–97. https://doi.org/10.1007/s10867-014-9367-7
53. Mach H, Volkin DB, Troutman RD, Wang BEI, Luo Z, Jansen KU, Shi LI (2006) Disassembly and reassembly of yeast-derived recombinant human papillomavirus virus-like particles (HPV VLPs). J Pharm Sci 95:2195–2206. https://doi.org/10.1002/jps.20696
54. Machado MR, Pantano S (2016) SIRAH tools: mapping, backmapping and visualization of coarse-grained models. Bioinformatics 32:1568–1570. https://doi.org/10.1093/bioinformatics/btw020
55. McCarthy MP, White WI, Palmer-Hill F, Koenig S, Suzich JA (1998) Quantitative disassembly and reassembly of human papillomavirus type 11 viruslike particles in vitro. J Virol 72:32–41. https://doi.org/10.1128/JVI.72.1.32-41.1998
56. McMillan WG, Mayer JE (1945) The statistical thermodynamics of multicomponent systems. J Chem Phys 13:276–305. https://doi.org/10.1063/1.1724036
57. Miao Y, McCammon JA (2016) Unconstrained enhanced sampling for free energy calculations of biomolecules: a review. Mol Simul 42:1046–1055. https://doi.org/10.1080/08927022.2015.1121541
58. Moreira LA, Boström M, Ninham BW, Biscaia EC, Tavares FW (2007) Effect of the ion-protein dispersion interactions on the protein-surface and protein-protein interactions. J Braz Chem Soc 18:223–230. https://doi.org/10.1590/S0103-50532007000100026
59. Nassal M (1992) The arginine-rich domain of the hepatitis B virus core protein is required for pregenome encapsidation and productive viral positive-strand DNA synthesis but not for virus assembly. J Virol 66:4107–4116. https://doi.org/10.1128/jvi.66.7.4107-4116.1992
60. Negrete A, Pai A, Shiloach J (2014) Use of hollow fiber tangential flow filtration for the recovery and concentration of HIV virus-like particles produced in insect cells. J Virol Methods 195:240–246. https://doi.org/10.1016/j.jviromet.2013.10.017
61. Newman M, Chua PK, Tang F-M, Su P-Y, Shih C (2009) Testing an electrostatic interaction hypothesis of hepatitis B virus capsid stability by using an in vitro capsid disassembly/reassembly system. J Virol 83:10616–10626. https://doi.org/10.1128/JVI.00749-09
62. Nikam RR, Gore KR (2018) Journey of siRNA: clinical developments and targeted delivery. Nucleic Acid Ther 28:209–224. https://doi.org/10.1089/nat.2017.0715
63. Oliver RC, Potrzebowski W, Najibi SM, Pedersen MN, Arleth L, Mahmoudi N, André I (2020) Assembly of capsids from hepatitis B virus core protein progresses through highly populated intermediates in the presence and absence of RNA. ACS Nano 14:10226–10238. https://doi.org/10.1021/acsnano.0c03569

64. Pellegrini M, Gro/nbech-Jensen N, Doniach S (1996) Potentials of mean force for biomolecular simulations: theory and test on alanine dipeptide. J Chem Phys 104:8639–8648. https://doi.org/10.1063/1.471552
65. Petrovskis I, Lieknina I, Dislers A, Jansons J, Bogans J, Akopjana I, Zakova J, Sominskaya I (2021) Production of the HBc protein from different HBV genotypes in E. coli. Use of reassociated HBc VLPs for packaging of ss- and dsRNA. Microorganisms 9:283. https://doi.org/10.3390/microorganisms9020283
66. Le Pogam S, Chua PK, Newman M, Shih C (2005) Exposure of RNA templates and encapsidation of spliced viral RNA are influenced by the arginine-rich domain of human hepatitis B virus core antigen (HBcAg 165–173). J Virol 79:1871–1887. https://doi.org/10.1128/JVI.79.3.1871-1887.2005
67. Porterfield JZ, Dhason MS, Loeb DD, Nassal M, Stray SJ, Zlotnick A (2010) Full-length hepatitis B virus core protein packages viral and heterologous RNA with similarly high levels of cooperativity. J Virol 84:7174–7184. https://doi.org/10.1128/JVI.00586-10
68. Porterfield JZ, Zlotnick A (2010) A simple and general method for determining the protein and nucleic acid content of viruses by UV absorbance. Virology 407:281–288. https://doi.org/10.1016/j.virol.2010.08.015
69. Pumpens P, Grens E (2016) The true story and advantages of the famous hepatitis B virus core particles: outlook 2016. Mol Biol 50:489–509. https://doi.org/10.1134/S0026893316040099
70. Pusara S, Yamin P, Wenzel W, Krstić M, Kozlowska M (2021) A coarse-grained xDLVO model for colloidal protein–protein interactions. Phys Chem Chem Phys 23:12780–12794. https://doi.org/10.1039/D1CP01573G
71. Qin S, Zhou H-X (2019) Calculation of second virial coefficients of atomistic proteins using fast fourier transform. J Phys Chem B 123:8203–8215. https://doi.org/10.1021/acs.jpcb.9b06808
72. Quigley A, Williams DR (2015) The second virial coefficient as a predictor of protein aggregation propensity: a self-interaction chromatography study. Eur J Pharm Biopharm 96:282–290. https://doi.org/10.1016/j.ejpb.2015.07.025
73. Rohovie MJ, Nagasawa M, Swartz JR (2017) Virus-like particles: next-generation nanoparticles for targeted therapeutic delivery. Bioeng Transl Med 2:43–57. https://doi.org/10.1002/btm2.10049
74. Rüdt M, Vormittag P, Hillebrandt N, Hubbuch J (2019) Process monitoring of virus-like particle reassembly by diafiltration with UV/Vis spectroscopy and light scattering. Biotechnol Bioeng 116:1366–1379. https://doi.org/10.1002/bit.26935
75. Sahin U, Karikó K, Türeci Ö (2014) MRNA-based therapeutics—developing a new class of drugs. Nat Rev Drug Discov 13:759–780. https://doi.org/10.1038/nrd4278
76. Sakurai K, Oobatake M, Goto Y (2008) Salt-dependent monomer-dimer equilibrium of bovine β-lactoglobulin at pH 3. Protein Sci 10:2325–2335. https://doi.org/10.1110/ps.17001
77. Selzer L, Katen SP, Zlotnick A (2014) The hepatitis B virus core protein intradimer interface modulates capsid assembly and stability. Biochemistry 53:5496–5504. https://doi.org/10.1021/bi500732b
78. Sharma VK, Sharma RK, Singh SK (2014) Antisense oligonucleotides: modifications and clinical trials. Med Chem Commun 5:1454–1471. https://doi.org/10.1039/C4MD00184B
79. Singh N, Li W (2019) Recent advances in coarse-grained models for biomolecules and their applications. Int J Mol Sci 20:3774. https://doi.org/10.3390/ijms20153774
80. Smoluchowski MV (1921) Handbuch der Electrizität und des Magnetismus (Graetz) vol II. Leipzig, Germany, Barth pp 366–427
81. Strods A, Ose V, Bogans J, Cielens I, Kalnins G, Radovica I, Kazaks A, Pumpens P, Renhofa R (2015) Preparation by alkaline treatment and detailed characterisation of empty hepatitis B virus core particles for vaccine and gene therapy applications. Sci Rep 5:11639. https://doi.org/10.1038/srep11639
82. Søndergaard CR, Olsson MHM, Rostkowski M, Jensen JH (2011) Improved treatment of ligands and coupling effects in empirical calculation and rationalization of p K_a values. J Chem Theory Comput 7:2284–2295. https://doi.org/10.1021/ct200133y

83. Tan WS, Dyson MR, Murray K (2003) Hepatitis B virus core antigen: enhancement of its production in Escherichia coli, and interaction of the core particles with the viral surface antigen. Biol Chem 384. https://doi.org/10.1515/BC.2003.042
84. Uttinger MJ, Hundschell CS, Lautenbach V, Pusara S, Bäther S, Heyn TR, Keppler JK, Wenzel W, Walter J, Kozlowska M, Wagemans AM, Peukert W (2022) Determination of specific and non-specific protein–protein interactions for beta-lactoglobulin by analytical ultracentrifugation and membrane osmometry experiments. Soft Matter 18:6739–6756. https://doi.org/10.1039/D2SM00908K
85. Valentic A, Müller J, Hubbuch J (2022) Effects of different lengths of a nucleic acid binding region and bound nucleic acids on the phase behavior and purification process of HBcAg virus-like particles. Front Bioeng Biotechnol 10:929243. https://doi.org/10.3389/fbioe.2022.929243
86. Venkatakrishnan B, Zlotnick A (2016) The structural biology of hepatitis B virus: form and function. Annu Rev Virol 3:429–451. https://doi.org/10.1146/annurev-virology-110615-042238
87. Verwey EJW (1948) Theory of the stability of lyophobic colloids. J Colloid Sci 6
88. Waterhouse A, Bertoni M, Bienert S, Studer G, Tauriello G, Gumienny R, Heer FT, de Beer TAP, Rempfer C, Bordoli L, Lepore R, Schwede T (2018) SWISS-MODEL: homology modelling of protein structures and complexes. Nucleic Acids Res 46:W296–W303. https://doi.org/10.1093/nar/gky427
89. Wynne SA, Crowther RA, Leslie AGW (1999) The crystal structure of the human hepatitis B virus capsid. Mol Cell 3:771–780. https://doi.org/10.1016/S1097-2765(01)80009-5
90. Yan Y, Zhang D, Zhou P, Li B, Huang S-Y (2017) HDOCK: a web server for protein–protein and protein–DNA/RNA docking based on a hybrid strategy. Nucleic Acids Res 45:W365–W373. https://doi.org/10.1093/nar/gkx407
91. Yoon KY, Tan WS, Tey BT, Lee KW, Ho KL (2013) Native agarose gel electrophoresis and electroelution: a fast and cost-effective method to separate the small and large hepatitis B capsids: General. Electrophoresis 34:244–253. https://doi.org/10.1002/elps.201200257
92. Zhang Y, Liu Y, Zhang B, Yin S, Li X, Zhao D, Wang W, Bi J, Su Z (2021) In vitro preparation of uniform and nucleic acid free hepatitis B core particles through an optimized disassembly-purification-reassembly process. Protein Expr Purif 178:105747. https://doi.org/10.1016/j.pep.2020.105747
93. Zhao Q, Allen MJ, Wang Y, Wang B, Wang N, Shi L, Sitrin RD (2012) Disassembly and reassembly improves morphology and thermal stability of human papillomavirus type 16 virus-like particles. Nanomedicine Nanotechnol Biol Med 8:1182–1189. https://doi.org/10.1016/j.nano.2012.01.007
94. Zhong Z, Mc Cafferty S, Combes F, Huysmans H, De Temmerman J, Gitsels A, Vanrompay D, Portela Catani J, Sanders NN (2018) MRNA therapeutics deliver a hopeful message. Nano Today 23:16–39. https://doi.org/10.1016/j.nantod.2018.10.005
95. Zhou L-Y, Qin Z, Zhu Y-H, He Z-Y, Xu T (2019) Current RNA-based therapeutics in clinical trials. Curr Gene Ther 19:172–196. https://doi.org/10.2174/1566523219666190719100526
96. Zlotnick A (2003) Are weak protein–protein interactions the general rule in capsid assembly? Virology 315:269–274. https://doi.org/10.1016/S0042-6822(03)00586-5
97. Zlotnick A, Cheng N, Stahl SJ, Conway JF, Steven AC, Wingfield PT (1997) Localization of the C terminus of the assembly domain of hepatitis B virus capsid protein: Implications for morphogenesis and organization of encapsidated RNA. Proc Natl Acad Sci 94:9556–9561. https://doi.org/10.1073/pnas.94.18.9556
98. Zlotnick A, Venkatakrishnan B, Tan Z, Lewellyn E, Turner W, Francis S (2015) Core protein: a pleiotropic keystone in the HBV lifecycle. Antiviral Res 121:82–93. https://doi.org/10.1016/j.antiviral.2015.06.020

Innovative Pharmaceutical Stabilization and Formulation Processes for Protein Drugs and Their Impact on Sequence and Structure. Part: Analytics

Yomnah Y. Elsayed, Toni Kühl, and Diana Imhof

Abstract Peptide and protein design for biopharmaceutical and food applications gained considerable attention in recent years. Numerous peptides and proteins are approved every year, however, their quality and long-term stability as e.g., a pharmaceutical product is still a matter of concern. Innovative methods in drug formulation were developed, including hot-melt extrusion and spray freeze drying, which have been tested to study their potential for protein stabilisation. Since the formulation procedures could also influence the features of the carried entity, it is also required to provide appropriate analysis methods to test stability, structural integrity, and activity of the peptide or protein cargo after formulation. In addition, many proteins carry post-translational modifications, such as disulfide bridges, which are crucial for the protein's fold and specificity and should thus be sustained throughout the formulation process. To address both issues, the development of a bioanalytical workflow for the analysis of hot-melt lysozyme extrudates as well as an Edman sequencing protocol for determining the disulfide connectivity in peptides and proteins were investigated and reported in this chapter. The latter study included the establishment of suitable cysteine standards for sequencing and their examination in the context of suitable model peptides before investigating relevant functionally active representatives.

Abbreviations

ED	Edman degradation
GFD	Glass fiber disk

Y. Y. Elsayed · T. Kühl · D. Imhof (✉)
Pharmaceutical Biochemistry and Bioanalytics, Pharmaceutical Institute, University of Bonn, Bonn, Germany
e-mail: dimhof@uni-bonn.de

Y. Y. Elsayed
e-mail: y.elsayed@uni-bonn.de

T. Kühl
e-mail: tkuhl@uni-bonn.de

A. Kwade and I. Kampen (eds.), *Dispersity, Structure and Phase Changes of Proteins and Bio Agglomerates in Biotechnological Processes*,
https://doi.org/10.1007/978-3-031-63164-1_5

HPLC	High performance liquid chromatography
HME	Hot-melt extrusion
SDS-PAGE	Sodium dodecyl sulfate–polyacrylamide gel electrophoresis
MALDI-TOF	Matrix-assisted laser desorption ionization—time-of-flight
PTH-AA	Phenyl thiohydantoin amino acid derivative
PVDF	Polyvinylidene fluoride membrane
BLG	β-Lactoglobulin

1 Introduction

Over the last decade, proteins and peptides gained higher contribution in the pharmaceutical market (about 25%), which is also reflected by an increasing number of approved biological products every year for the treatment of various disorders [1–3]. However, the formulation of such entities is still challenging due to the well-known physical and chemical features, including instability, of proteins. A high tendency for aggregation or denaturation could affect the quality of the final product and activity on top [4]. Consequently, many formulation techniques were developed and studied in order to provide higher stability and protection of the protein or peptide cargo, especially before large-scale production of such valuable products is set up.

Hot-melt extrusion (HME) emerged as an interesting formulation technology that was applied in the pharmaceutical industry as a convenient and robust manufacturing tool for the preparation of solid dosage forms [5]. It provides advantage of taste masking, higher bioavailability, and the possibility of drug targeting [6], and thus was used as an innovative stabilization technique for proteins. Through HME, the target is mixed with selected polymers and plasticizers, and heated in a stepwise manner then extruded into a final homogeneous solid rod. Cossé et al. [7] confirmed the applicability of the hot-melt extrusion process to prepare homogeneous solid dispersions of different proteins in a PLGA-matrix.

The first part of our studies thus focussed on the development of a general procedure for analysing proteins from HM-extrudates (Chap. 11). Thereby, lysozyme (N-acetylmuramide glycanhydrolase or muramidase) was selected as a model protein because it was already used in earlier studies employing techniques such as freeze drying [8, 9], hot-melt extrusion [10, 11], or other polymer-based formulation strategies [12–14]. A model HME formulation of lysozyme with polyethylene glycol (PEG20.000) for further analysis was provided by project part process (Chap. 11) and examined thoroughly for its structural and functional properties before and after formulation following the workflow shown in Fig. 1a [15]. Evaluation of the model HME formulation was performed following the international guidelines for identity, purity, and activity of the International Council of Harmonization ICH Topic Q1A (R2) [16]. The guidelines for biotechnological products stating the parameters to be tested were followed to examine the stability and activity of lysozyme after exposure of lysozyme to hot-melt extrusion and long-term storage of the formulation. As a result, the analysis data of the hot-melt extruded lysozyme should allow for judging the extent how the additives or applied procedures, in particular mechanical stress, and temperature changes, could impair with the proteins' properties during processing and the work-up procedure. The presence of such stability evaluation protocols is urgently needed to support the development of formulation strategies for protein drugs.

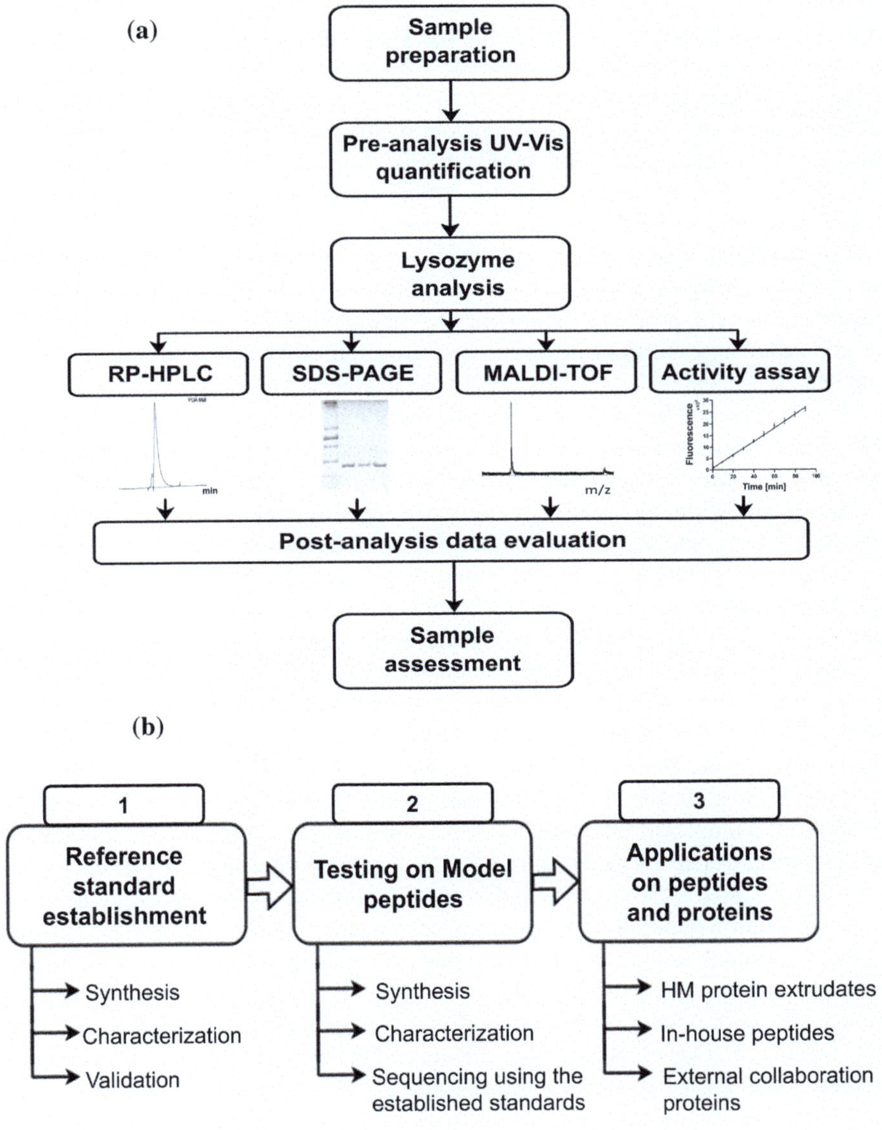

Fig. 1 **a** Schematic representation of the workflow for the analysis of the lysozyme hot-melt extrudates (modified from [15], with permissions related to the material excerpted from ACS publications), **b** Workflow for establishing a strategy for the elucidation of the disulfide bond connectivity using automated Edman sequencing

Continuing this line, the development of proper analysis workflows and protocols for protein drugs needs to be established as well, however, may require even more customized strategies due to the diversity and individual complexity of the protein of interest (Fig. 1b). Among others, proteins are subjected to post-translational modifications that are significant for maintaining their structural integrity and functional activity [17]. The formation of disulfide bonds often leads to a well-defined 3-dimensional structure responsible for highly specific interactions and thus certain tasks in the context of the cell, tissue or even the entire organism [18]. In nature, there is a plethora of disulfide-rich peptides and proteins occurring, which fulfil very different tasks and functions, ranging from regulation of metabolic reactions (e.g., insulin) [19] to management of chronic diseases such as plecanatide and linaclotide (2- and 3-disulfide bonded peptides for chronic idiopathic constipation and irritable bowel syndrome, respectively) [20, 21] or diagnostics such as ^{68}Ga-dotatoc [22]. The correct disulfide bond connectivity is highly essential for proper folding. Different analytical techniques were developed to determine the disulfide connectivity of a peptide or protein, primarily employing enzymatic digestion and applying reducing or non-reducing conditions including derivatization of the released cysteine of the digest [23].

This is followed by an intense analysis using mass spectrometric (MS) and MS/MS methods to investigate the molar masses obtained for the digested fragments and perform sequencing of these peptides [24]. Apart from MS, also automated Edman degradation was used, however, fails to adequately determine cysteine or disulfide bonds. So far, both techniques are not approved for full and unequivocal recognition of a complex modified protein sequence [25], proving the necessity of combining Edman degradation and MS methods for complete sequence confirmation and elucidation of the disulfide bond connectivity in peptides and proteins.

Therefore, in the second part of our studies, we focussed on the assessment of the earlier established MS methods and protocols for disulfide bond analysis [26, 27] as well as the revival of the method of Edman degradation to check its applicability for the same scientific question. Edman degradation (ED) is an N-terminal sequencing technique introduced in the 1950ies that is based on the stepwise release of the N-terminal amino acid after modification with phenyl isothiocyanate, subsequent cyclization and cleavage of the amino acid as phenyl thiohydantoin (PTH) derivative [28]. The released PTH-amino acid derivative is eluted and compared with the characteristic elution pattern of a PTH-amino acid standard. After cleavage and release of the N-terminal amino acid, the remaining protein is recycled, and the next amino acid can undergo the same procedure. One major drawback of ED is the incomplete analysis and detection of cysteines due to incomplete conversion or cleavage of cysteine [29]. One of the attempts investigated in the present study was to develop suitable standards of modified cysteines that could be applied for cysteine detection in protein samples to minimize or bypass digestion with a maximum of accuracy concerning connectivity preservation and detection. The bioanalytical method development according to ICH M10 [30] and Q14 [31] was applied on the PTH cysteine standards synthesized herein in order to test their applicability as reference standard. Additionally, incorporation of the developed standards into model peptides and establishment of protocols for partial reduction followed by an alkylation step were undertaken to facilitate future application on peptides and proteins. In parallel, the earlier introduced MS protocols were applied to a selection of β-lactoglobulin (BLG) proteins from natural sources or recombinant expression [32–34]. Since BLG is a suitable example of a 2-disulfide bonded miniprotein, which has a special feature, namely an additional free cysteine. Although intensive studies have been carried out to recombinantly produce and modify BLG and thereby change its physicochemical properties, continuous analysis and monitoring of changes in disulfide connectivities have not been primarily focussed on yet [32, 34]. This becomes even more interesting when modifications, such as mutations, are introduced or specific treatments, such as chemical reduction or UV light irradiation, are applied in which the structural features, i.e. the disulfide bond patterns of the resulting BLG versions, are changed.

2 Material and Methods

2.1 Chemicals

Freeze dried chicken egg white lysozyme hydrochloride, Tris(2-carboxyethyl) phosphine (TCEP), *N,N′*-methylene-bisacrylamide, Ammonium persulphate (APS) and Coomassie brilliant blue G-250 dye were purchased from AppliChem GmbH, Darmstadt, Germany. Polyethylene glycol 20,000 were obtained from Carl Roth, Karlsruhe, Germany, while acrylamide, tris base/HCl, sodium dihydrogen phosphate, disodium hydrogen phosphate, glycine, EDTA, sodium citrate, citric acid, trifluoroacetic acid (spectroscopic grade) were purchased from Merck KGaA (Darmstadt, Germany), and *N,N,N′,N′*-tetraacetylethylenediamine (TEMED), HPLC grade acetonitrile and reagents for solid phase peptide synthesis (piperidine, N,N-dimethylformamide, and N-methylmorpholine) from VWR (Germany). Sodium dodecyl sulphate (SDS), 4-methyl-umbelliferyl *β*-D-*N,N′,N″*-triacetylchitotrioside, 4-methyl-umbelliferone, 2,5 dihydroxybenzoic acid (2,5-DHB), 2,5-dihydroxyacetophenone (2,5-DHAP) and α-cyano-4-hydroxycinnamic acid (HCCA), phenyl isothiocyanate (8.36 M, 99%), 2-iodoacetamide, Sequa-brene, S-methyl cysteine hydrochloride and Fmoc-S-carbamidocysteine were obtained from Sigma-Aldrich, GmbH, Germany, sodium trifluoroacetate (98%) and 2-iodoacetamide were obtained from Thermo Fisher Scientific (Germany).

Reagents for the Edman protein sequencer (phenyl isothiocyanate (5% in n-heptane), acetonitrile (37% (v/v)), ethyl acetate, n-chlorobutane, trifluoroacetic acid, triethylamine, TFA treated glass fiber disks (GFD), polyvinylidene fluoride (PVDF) membranes, standard phenyl thiohydantoin (PTH) amino acid derivative mixture and PTH-amino acids mobile phase for isocratic elution were purchased from FUJIFILM, Wako Pure Chemical Corporation, Japan. The 2-disulfide bonded Tridegin was produced in-house as earlier described [18, 35].

2.2 Instrumentation

Shimadzu LC-8A chromatograph (Duisberg, Germany) equipped with a Knauer 300–5 reversed phase C18 column (particle size 5 μm, 300 Å pore size, 32 × 250 mm) at 25 °C was used for preparative separation and purifications. Shimadzu LC-10AT chromatograph equipped with a Vydac 218TP54 reversed phase C18 column (particle size 5 μm, 300 Å pore size, 4.6 × 250 mm) was used for purity confirmation of synthesized PTH-Cys derivatives and model peptides. Shimadzu LC-20AD chromatograph equipped with a Vydac 218TP54 reversed phase C18 column (particle size 5 μm, 300 Å pore size, 4.6 × 250 mm) at 25 °C was used for quantitative analysis of lysozyme and fraction collection of Tridegin samples after partial reduction and alkylation. Spectrofluorimeter FP-8300 (Jasco, Germany) was used for fluorescence measurements. Mass spectrometers: UltrafleXtreme TOF/TOF mass spectrometer (Bruker Daltonics GmbH, Bremen, Germany) was used for MALDI-TOF analysis and for LC–ESI–MS measurement, micrOTOF-Q III spectrometer (Bruker Daltonics GmbH, Bremen, Germany) was used. The NMR experiments for the ^{1}H and ^{13}C chemical shift assignments were performed on an Avance III HD 700 MHz NMR spectrometer (Bruker Daltonics GmbH, Bremen, Germany). Protein sequencing was executed using Shimadzu PPSQ-53A isocratic protein sequencer (Japan) equipped with Wakopak® Wakosil PTH-II column 4.6 × 250 mm (S-PSQ).

2.3 Hot-Melt Extruded Lysozyme Formulation and Analysis

Effect of hot melt extrusion on lysozyme was studied by mixing lysozyme with PEG20,000 and as described earlier in Chap. 11 [15]. 20% Lysozyme/PEG20,000 hot melt extrudates were prepared (referred as H-Lyso/PEG20) and a physical mixture of 20% lysozyme with PEG20,000 (referred as P-Lyso/PEG20) was prepared and tested to evaluate the effect of ram extrusion and heating on lysozyme characteristics. Both H- and P-Lyso/PEG20 as well as the untreated lysozyme (referred as P-Lyso) were stored at 4, 25 and 40 °C at 65% relative humidity (RH) ± 5% RH and their stability was evaluated according to ICH guidelines [16] at first day of formulation and after 2 weeks, 4 weeks, 3 months and 6 months. Four techniques were presented in our study to evaluate the quality of lysozyme in the accelerated storage samples [15]. Initially, extraction procedures of lysozyme from formulation were optimized to provide quantitative analysis of lysozyme, that was confirmed with RP-HPLC analysis of extracted samples using elution system of A: 0.1% TFA in water and eluent B:0.1% TFA in acetonitrile in gradient of 20–60% eluent B in 40 min and detection was at 220 nm. 50 μl samples were injected and peak area of lysozyme peak observed at 24.7 min was recorded. Concentration of lysozyme was determined using established calibration curve and regression equation over 50–500 μg/ml. Activity of lysozyme in samples was measured fluorometrically using 4-methyl-umbelliferyl β-D-N,N',N''-triacetylchitotrioside as substrate releasing 4-methylumbelliferone that was quantitatively correlated to lysozyme content. SDS-PAGE was performed using 15% acrylamide gels for lysozyme bands and indication of degradation products. MALDI-TOF mass spectrometric analysis was executed using 2, 5-DHB matrix in presence of sodium trifluoroacetic acid to confirm sample identity and stability as well. The established workflow was applied on lysozyme and the tested formulations at the selected temperatures and specified time intervals. Quantitative and qualitative data evaluation for samples was implemented for assessment of their stability.

2.4 Edman Degradation of Cysteine-Containing Sequences

PTH amino acids were synthesized according to an earlier reported protocol [36–38], purified and chemically analyzed using TLC, HPLC, ESI–MS, ^{1}H– and ^{13}C–NMR as described for PTH-derivatives [39–41]. The synthesized PTH-derivatives were validated according to ICH M10 for chromatographic methods. Model pentapeptides containing cysteine or modified cysteine in different positions and amino acid environment were automatically synthesized by solid-phase peptide synthesis following a standard Fmoc-protocol, purified and characterized using amino acid analysis, TLC, RP–HPLC and MALDI–TOF–MS using HCCA as the matrix [42, 43]. Subsequently, the model peptides were sequenced, and differential chromatograms were established for data evaluation by means of the developed standards. Edman sequencing was performed on an automated Edman protein sequencer PPSQ-53A using PTH-amino acid mobile phase as provided by FUJIFILM, Wako Pure Chemical Corporation, Japan at flow rate of 1 ml/min at 40 °C. Analysis was done by a PDA detector at 269 nm. For validation of PTH-Cys derivatives, they were directly eluted over the chromatographic system without prior chemical treatment by the sequencing unit. For sequencing of model peptides, peptide solutions were applied on GFD (200 pmol) and Edman degradation was executed. Each released N-terminal PTH-derivative was detected by the aforementioned isocratic HPLC system and identified based on a comparison to standard mixture of PTH derivatives of all proteinogenic amino acids (excluding cysteine, including dehydrothreonine).

As an example for an active peptide containing disulfide bridges, Tridegin was selected and subjected to full reduction and alkylation using 2-iodoacetamide in order to modify the reduced cysteines to carbamidomethyl cysteine. Separation of oxidized and fully reduced fractions was performed using HPLC. Fractions were subjected to Edman degradation by applying GFD membranes, the modified cysteines will be released as PTH-Cys(Cam) and detected during the ED analysis.

3 Results and Discussions

3.1 *Qualitative and Quantitative Characterization of Hot-Melt Extruded Lysozyme by Applying Methods for Protein Bioanalysis*

Analytical methods reported for hen egg white lysozyme characterization were re-evaluated to develop a proper analysis protocol for formulated lysozyme over a storage time of 6 months. The established analysis procedure based on RP–HPLC, SDS–PAGE, MALDI–TOF MS, and a fluorescence activity assay were tested initially on pure lysozyme (control) and then applied on lysozyme hot-melt extrudates.

For chromatographic analysis by RP-HPLC, the establishment of a calibration curve for untreated lysozyme was required. Serial dilutions of lysozyme (in 0.1 M sodium phosphate buffer pH 5.20, 50–500 μg/ml) were prepared, injected and their peak area (at $t_R = 24.7 \pm 0.2$ min) was correlated to the corresponding concentration to establish the calibration curve that was used later for lysozyme quantification. Accuracy samples of unextracted lysozyme samples were prepared and tested. The recovery rate of the found lysozyme concentrations was $99.7 \pm 1.56\%$ ($n = 6$), that indicated applicability of the procedure for lysozyme quantification. In addition, identification of lysozyme in the pre-formulated state was performed using reducing and non-reducing conditions for SDS–PAGE on a 15% acrylamide resolving gel [44], and molar mass was confirmed with MALDI–TOF mass spectrometric analysis of the protein samples. All protocols were proven useful for lysozyme identification and quantification and later used for the hot-melt extrudates (see below).

In order to test the lysozyme hydrolytic activity, the reported turbidimetric and fluorogenic assays were used and evaluated. The conventional turbidimetric assay [45] is based on incubation of lysozyme with bacterial culture. Changes in the culture's optical density are monitored and specific activity of lysozyme is then calculated as normalized density reduction to culture biomass. This method suffers from a high degree of deviation, low robustness and reproducibility (RSD = 18.2%, n = 12), thus another assay was evaluated for quantitative activity assessment using a fluorogenic substrate, namely 4-methyl-umbelliferyl β-D-N,N',N''-triacetylchitotrioside (in 0.1 M sodium phosphate buffer pH 5.2) [46]. The substrate was hydrolyzed by lysozyme releasing 4-methyl-umbelliferone (4-MU), a fluorescent product measured at an excitation wavelength of 355 nm and an emission wavelength of 460 nm. The release kinetics parameters of 4-MU were studied by the measurement of the released 4-MU using different substrate concentration and a fixed lysozyme concentration. The K_M value was found to be 20.2 μM that matched the reported value of 15.9 μM [46]. The activity of lysozyme (control) was thus determined and calculated using the fluorogenic method since it provided more accurate and reproducible results (RSD = 1.9%, n = 6). This fluorogenic assay was adapted and applied for the analysis of formulated lysozyme (Fig. 2) [15].

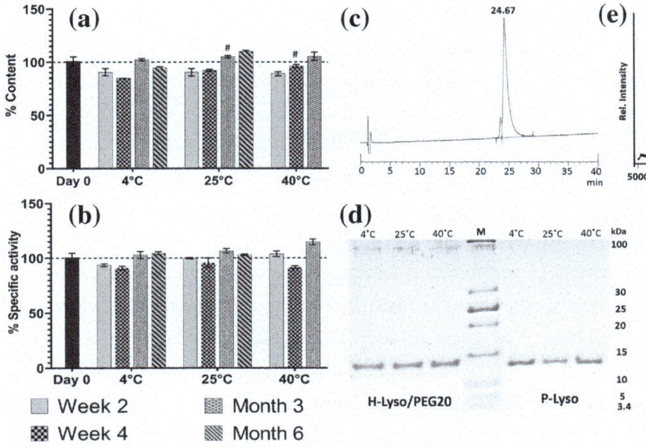

Fig. 2 **a** Protein content analysis and **b** specific activities results of extracted samples from 6-months stored H-Lyso/PEG20 at 4, 25, and 40 °C (n = 3, # n = 2). **c** HPLC chromatogram of extracted H-Lyso/PEG20 (stored for 6 months at 25 °C, in phosphate buffer, pH 5.2), **d** Reducing SDS-PAGE gel for extracted P-lyso and H-Lyso/PEG20 (stored for 6 months at 4, 25, and 40 °C), where M is the protein ladder, **e** MALDI-TOF mass spectrum for extracted H-Lyso/PEG20 (stored for 6 months at 25 °C) using 2,5-DHB as matrix (modified from [15], with permissions related to the material excerpted from ACS publications)

Initially, the effect of PEG20,000 on the aforementioned selection of methods for lysozyme analysis needed to be evaluated as the formulation of lysozyme by hot-melt extrusion was carried out using PEG20,000 as the polymer. It turned out that presence of PEG20,000 interfered with analysis of lysozyme, i.e., mass spectrometry analysis was impossible. Therefore, lysozyme extraction was required for analysis of formulated extrudates. Solubility of lysozyme and PEG20,000 were tested using different commonly used solvents such as acetonitrile, ethanol, methanol, acetone, diethylether and ethylacetate. Differential solubility was observed with acetonitrile, that solubilizes only PEG20,000 and for maximizing the extraction procedure. Hot-melt extruded lysozyme rods were ground to fine particles, so higher surface area exposure as well as complete precipitation was achieved. The precipitated lysozyme was collected by centrifugation and supernatants containing PEG20,000 were discarded, while lysozyme pellets were dried for further analysis. A suitable diluent was selected for each method; 0.1 M sodium phosphate, pH 5.2 was used for activity and chromatographic assays, while pellets were dissolved in double distilled water for SDS-PAGE runs followed by freeze drying and redissolution in 30% acetonitrile/0.1%TFA (v/v) for MALDI-TOF analysis [15].

The established analytical methods and extraction protocols were applied on hot-melt extrudates (H-Lyso/PEG20), physically mixed lysozyme with PEG20,000 (P-Lyso/PEG20) and pure, untreated lysozyme (P-Lyso). P-Lyso samples were considered as controls to be tested along with each time point's sample set. Analysis of freshly formulated extrudates (Day 0) matched with the analysis of pure lysozyme and showed a good recovery rate of 99.9 ± 3.6%, n = 3 of the claimed content (Fig. 2a). The obtained specific activities for the extrudates (100.3 ± 4.1%, n = 3) emphasized that lysozyme maintained its activity upon undergoing the process of hot-melt extrusion (Fig. 2b) that correlated with HPLC (Fig. 2c), SDS–PAGE (Fig. 2d) and MALDI–TOF–MS (Fig. 2e) findings verifying mass purity, compound identity and absence of aggregates or degradation products [15].

The suggested workflow was subsequently applied to test the stability and lysozyme properties of extruded lysozyme after storage over different conditions as specified by ICH guidelines. Contents and specific activities were correlated to Day 0 results (Fig. 2a, b). Over a period of 6 month obtained contents for lysozyme were 84.6–110.0% ± 4.2% (n = 3 at each tested temperature and time point) and specific activities were 97.7–104.9% ± 4.7% (n = 3) indicating no significant activity changes upon storage. All SDS–PAGE gels and MALDI–TOF mass spectra also showed no change in lysozyme characteristics. For P-Lyso/PEG20 and P-Lyso full activity with 98.0−105.1% (RSD ± 3.1%, $n = 3$) and 90.2−110.6% (RSD ± 4.3%, $n = 3$), respectively, was observed [15].

3.2 Establishment of a Strategy to Analyze Free Cysteines and Disulfide Bonds in Peptides by Automated Edman Degradation

Analysis of cysteines and elucidation of the correct disulfide bond connectivity in a peptide or protein is essential for confirming their correct folding and active or functional state. This could, however, be affected during different processing procedures, such as formulation or handling of large quantities. Cysteine detection during ED is possible after modification of the thiol moiety and applying suitable PTH-cysteine standards for unequivocal identification. So far, the latter are missing, in particular if considering quantification of the sequencing steps. We herein established an in-depth analysis of two PTH-Cys derivatives to examine their applicability for proper detection and quantification [47]. Both, the PTH-derivative of methylated cysteine, i.e. PTH-Cys(Me), and the PTH-derivative of carbamidomethylated cysteine, PTH-Cys(Cam), were investigated for their use as PTH-Cys standards in automated ED [47]. These derivatives were synthesized by the reaction of cysteine with phenyl isothiocyanate (PITC) in triethylamine and extraction of the mixture with ethyl acetate. After purification, the PTH-cysteine standards were purified and chemically characterized using TLC, HPLC, LC–ESI–MS, ^{1}H and ^{13}C–NMR spectroscopy to confirm purity and identity before use for Edman sequencing [47].

Validation of the PTH-Cys derivatives as ED standards was performed to investigate their analytical properties according to the ICH guidelines ICH M10 and Q14. Therefore, stock solutions of both derivatives were prepared, diluted to obtain test samples and stored for different time periods before analysis on an automated protein sequencer accordingly. The elution of the respective peaks was monitored and the peak areas were recorded to evaluate the following parameters: [30, 31] (i) selectivity of analysis in 37% acetonitrile, (ii) specificity compared to other PTH-amino acid derivatives, (iii) linearity range for the peak area related to the concentration of the PTH-amino acid derivative, (iv) accuracy and precision, (v) presence of carry over effects during the sequencing cycles, (vi) integrity of the response after dilution, and (vii) stability. The stability was studied concerning the following criteria: short-term storage, room temperature, freezing–thawing, and long-term storage at –20 °C [48]. Both PTH-Cys standards displayed adequate characteristics as bioanalytical standard, however, PTH-Cys(Me) showed an elution time close to PTH-DTh (PTH-dehydrothreonine) that affected quantification accuracy to a certain extent (data not shown) [47].

Next, a series of model peptides (**1–12**) of different length containing cysteine, threonine and modified cysteines, respectively, were synthesized, chemically characterized and analyzed by ED (Table 1). For the selection of the peptide sequences we were considering the sequence environment in proteins as well as the contamination from adjacent amino acids eluting close to the PTH-Cys standards: While PTH-Cys(Cam) eluted next to threonine, quantification of PTH-Cys(Me) could be influenced by its close proximity to DTh, which is a product that could be formed upon sequencing of threonine. Therefore, model peptides containing threonine in relevant positions (1–2 previous cycles) to the modified cysteines were selected (Table 1). The peptide length varied between 5 and 9 amino acids with the residue of interest, i.e., threonine, cysteine or modified cysteine derivatives in the central position. Longer model peptides (9 amino acids) were sequenced to test the accessibility and yield of the modified PTH-Cys derivatives after sequencing of PTH amino acid derivatives with close by retention times such as Ser, Thr, His or DTh [47].

Upon sequencing of the model peptides, each cycle was observed for characteristic peaks. Differential chromatograms were generated by subtraction of the previous cycle's chromatogram, which minimized background noise and provided

Table 1 Sequences of selected model peptides for analysis in automated Edman degradation. All peptides were synthesized as C-terminal amides and as resin-bound compounds

No	Peptide sequence
1	AA**C**AA
2	AA**T**AA
3	AA**C(Me)**AA
4	**TAC(Me)**AA
5	A**TC(Me)**AA
6	AYR**C(Me)**MVAA
7	AYR**TC(Me)**MVAA
8	AA**C(Cam)**AA
9	**TAC(Cam)**AA
10	A**TC(Cam)**AA
11	NQS**C(Cam)**HGAA
12	NQS**TC(Cam)**HGAA

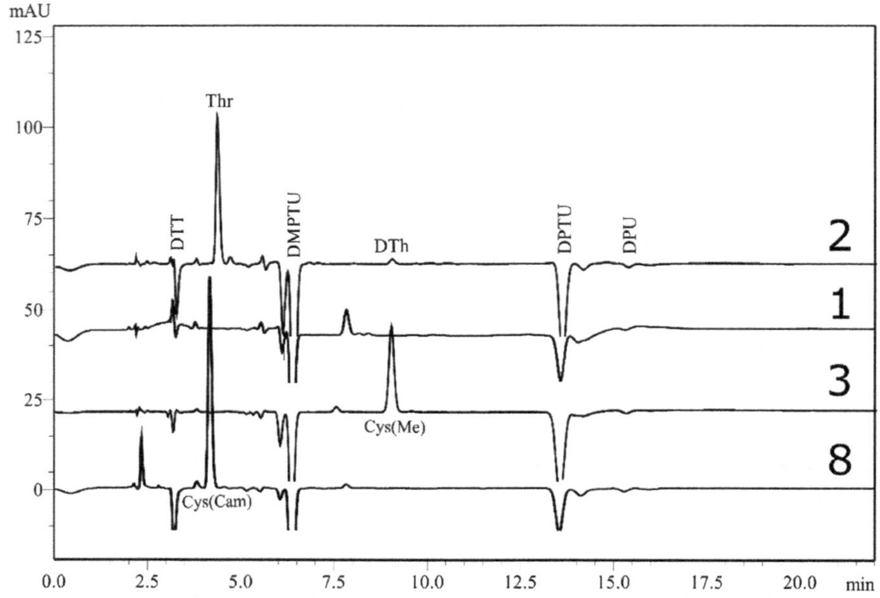

Fig. 3 N-terminal Edman degradation sequence analysis: Differential chromatograms of cycle 3 for peptides **1** (AACAA), **2** (AATAA), **3** (AAC(Me)AA), and **8** (AAC(Cam)AA) displaying the peaks of PTH-Cys(Me) and PTH-Cys(Cam) (modified from [47] with permission from ACS publications)

an increased resolution of the obtained results. For model peptide **1**, peaks for PTH-threonine and dehydrothreonine derivatives could be observed, while detection of the unmodified cysteine in peptide **2** (Fig. 3) was not possible, which is in agreement with the literature [49]. Thr and DTh in peptide **1** were detected at 4.30 and 9.15 min, respectively.

The PTH-Cys derivatives PTH-Cys(Me) and PTH-Cys(Cam) could also be clearly observed with an appropriate resolution to still distinguish Cys sufficiently from surrounding byproduct peaks as can be seen from peptides **3** and **8** (Fig. 3). However, a high degree of overlay was observed between PTH-DTh and PTH-Cys(Me) (9.10 min) (Fig. 3) [47].

Analysis of peptides **4** and **5** with threonine at positions –1 or –2 relative to the modified cysteine revealed that DTh was produced in a minor amount in all cycles of peptides that contained Thr and accumulated in the next cycle as a broad, though, small peak. The latter was also confirmed by sequencing of longer peptides **6**, **7**, **11**, and **12** (8–9 amino acids, Fig. 4). The maximum effect for the accumulation was found in peptides containing threonine in the immediate prior cycle to a cycle, in which a cysteine derivative was supposed to occur. However, detection of cysteine was still feasible and identification of DTh from Cys(Me) could be achieved due to the significant intensity differences between PTH-DTh and PTH-Cys(Me) in peptides containing threonine [47].

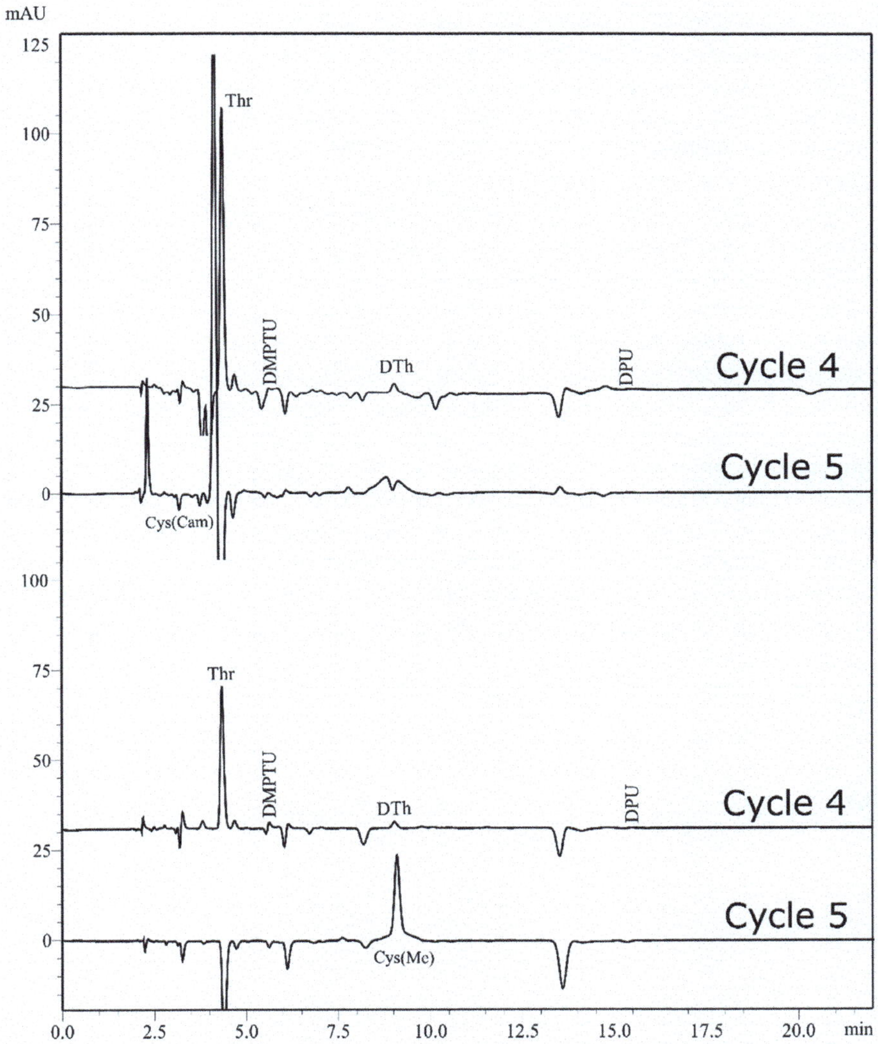

Fig. 4 N-terminal Edman degradation sequence analysis chromatograms of cycles covering Thr4 and Cys5 for peptides **12** NQS**TC(Cam)**HGAA (top) and **7** AYR**TC(Me)**MVAA (bottom) displaying the peaks of PTH-Cys(Cam) (cycle 5, top) and PTH-Cys(Me) (cycle 5, bottom), [47] with permission from ACS Publications

The proximity of the peaks for Thr (4.30 min) and Cys(Cam) (4.18 min) was evaluated with the model peptides **8–12**. Here, identification and quantification of the Cys(Cam) derivative peaks were achieved without interference from Thr (Fig. 4 top) [47].

Taken together, although there is some interference between PTH-DTh and PTH-Cys(Me), which will slightly affect the quantification of the PTH-Cys(Me) derivative, in standard proteins DTh appears very rarely and will, thus, not significantly affect standard analyses procedures in which modified cysteines such as PTH-Cys(Me) or PTH-Cys(Cam) are produced and analyzed. The information about the herein described PTH-Cys derivatives obtained from the sequencing of the model peptides thus confirms applicability of the derivatives for cysteine detection (for both) and quantification (for Cys(Cam) with higher accuracy) [47].

In the next step, natural disulfide-containing peptides should be examined to check the applicability of the established PTH-Cys standards. An in-house produced analog of the leech-derived compound tridegin was selected for the determination of the disulfide connectivity. This tridegin derivative is a 66mer 2-disulfide-bonded peptide inhibitor of coagulation factor XIIIa that was synthesized according to an orthogonal protecting group strategy to selectively form disulfide bridges between C5 and C37 as well as C17 and C31 (Fig. 5a) [41]. Edman sequencing was implemented as a tool to confirm formation of disulfide bridges in tridegin compared to the completely reduced compound. In order to perform the analysis, the oxidized and the fully reduced peptide obtained after treatment with an excess of TCEP have been applied to excess concentrations of iodoacetamide for derivatization of available free thiol groups [47]. As expected, the analysis revealed that complete reduction and cleavage of disulfide bridges followed by quantitative conversion of free cysteines into carbamidomethyl cysteine (Cys(Cam)) occurred at all available positions 5, 17, 31, and 37 (Fig. 5b–e).

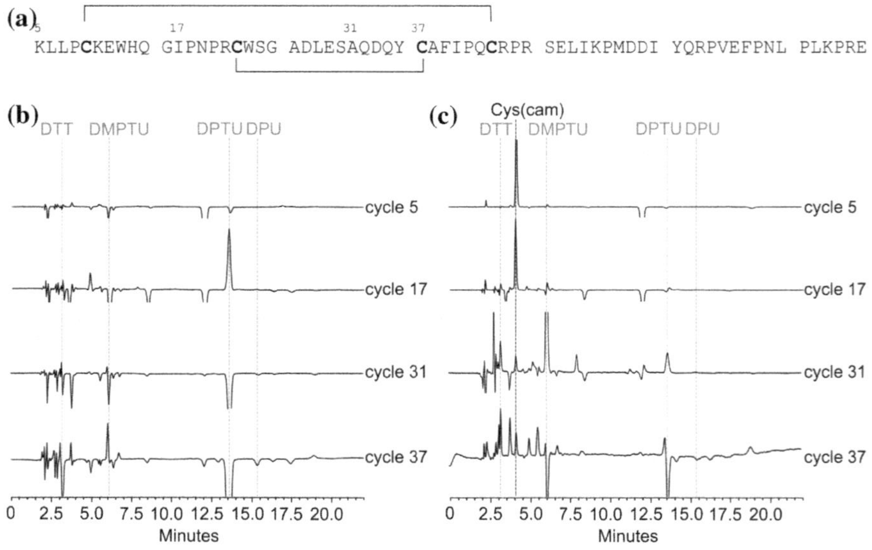

Fig. 5 **a** Sequence of the 2-disulfide bonded tridegin isomer used in this study, **b**, **c** N-terminal sequencing chromatograms of cycles Cys5, Cys17, Cys31, and Cys37 in oxidized tridegin (**b**) and completely reduced, carbamidomethylated tridegin (**c**), where all cysteines are modified as Cys(Cam). The background noise increases at higher cycle numbers due to a higher degree of magnification of the y-axis as a result of lower yields of PTH-amino acid derivatives

3.3 *Analysis of Technologically Treated Proteins by Tandem Mass Spectrometry and Edman Degradation*

Optimization, production, and treatment of proteins for industrial applications require varying parameters and procedures the protein is exposed to. For cysteine-containing proteins, this represents a special challenge as highlighted in the aforementioned section. Another molecule of interest in our studies, which had been performed in close collaboration with Rebekka Biedendieck and Rainer Krull (Chap. 6) and Julia Keppler and Karin Schwarz (Chap. 7), represented the milk whey protein β-lactoglobulin (BLG), which is a significant component in many foods and important for the texture of several of them [32, 50]. This protein consists of 162 amino acids. Two variants, A and B, that differ in two positions (64, 118) only are well described in the literature (Fig. 6a) [33, 51]. Apart from this, BLG contains 5 cysteines at positions 66, 106, 119, 121, and 160, which were found to form two intramolecular disulfide bonds (Cys66-Cys160 and Cys106-Cys119) according to previously described structures derived from X-ray crystallography and NMR spectroscopy [52–55]. However, the correct disulfide connectivity and the question, which of the cysteines remains in its reduced state, was a matter of debate for several decades [56, 57]. In a recent approach, the role of the reduced Cys121 in the folding process was discussed to essentially contribute to the formation of properly formed disulfide bonds [58].

The determination of the disulfide connectivity in multiply bridged peptides and proteins is difficult per se, however, is particularly complicated by potential disulfide scrambling and the occurrence of disulfide isomers in the sample [18, 59]. Especially when reduced cysteines are present in the sequence there is a high chance of disulfide

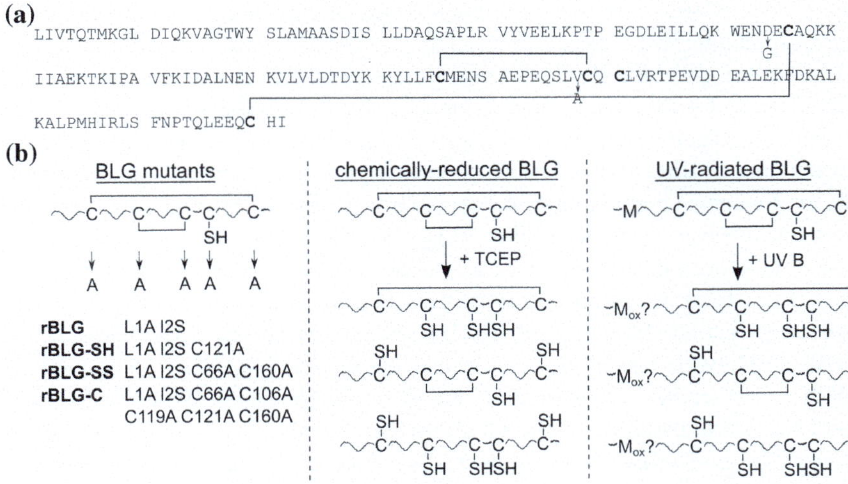

Fig. 6 **a** Sequence of BLG AB. Variants A and B differ in positions 64 and 118 (indicated by arrows). Cysteines are shown in bold and disulfide bridges are represented by connecting lines. **b** Schematic overview of analyzed BLGs and their different treatments

shuffling in the process of protein folding. The risk for this phenomenon to occur increases when proteins are expressed in non-native environment and organisms or when mutations are introduced. In case of BLG, the successful expression of the variant B in *E. coli* Origami B along with its crystal structure was an important step to proof the formation of correctly folded protein in non-native settings [32–34]. In addition, proper formation of disulfide bridges in BLG may significantly affect its physicochemical properties as well as its aggregation behavior as outlined by others in this edition (Chaps. 6 and 7). Thus, the analysis of the disulfide bonds and the changes in the sequence of the BLG are essential for understanding the changes associated with mutations or modifications by physical or chemical treatment. Herein, a combined approach applying tandem mass spectrometry and N-terminal Edman sequencing was used to evaluate the different BLGs provided by Brune and coworkers and Fitzner and coworkers (Fig. 6b) [60, 61]. In addition to the recombinantly produced BLG variant B (rBLG), different mutants thereof were produced with alanine mutations at either position Cys121 (rBLG-SH), Cys66 and Cys160 (rBLG-SS), or at all cysteine positions at the same time (rBLG-C). Initially, the number of reduced cysteines was investigated by mass spectrometry after carbamidomethylation of the protein [60]. These data revealed different numbers of disulfide bridges depending on the cysteine that was mutated. For bovine BLG (BLG AB, wild-type) and rBLG a single carbamidomethylation could be observed consistent with the presence of the single reduced Cys121. Accordingly, mutation of Cys121 in rBLG-SH abolished the occurrence of reduced cysteines completely, while mutations in rBLG-SS maintained the occurrence of one reduced cysteine. No modifications could be found for the cysteine-deficient mutant rBLG-C.

Subsequently, analysis of BLG wild-type (BLG AB) after tryptic digest was performed [60]. The resulting peptide fragments were separated by HPLC and analyzed by LC–ESI–MS/MS or Edman degradation. Successful annotation of the corresponding HPLC peaks to specific peptide fragments yielded a complete sequence coverage of the BLG wild type protein (Fig. 7a). A comparative analysis finally allowed for the precise annotation of a disulfide bond between Cys66 and Cys160 (outer disulfide bridge). Comparison of the chromatograms of the different BLG mutants to BLG AB or rBLG revealed significant changes in the retention times for fragments containing cysteine, i.e. loss of disulfide bridges due to mutation of the involved cysteine residues (Fig. 7a). The latter could be annotated to the corresponding mutated sequence as can be seen from their MS- and MS/MS-spectra (Fig. 7b) as well as their chromatograms from Edman sequencing (Fig. 7c). Specifically, disulfide bridge Cys66-Cys160 could be preserved in BLG AB, rBLG, and rBLG-SH, while loss of the outer bridge was detected for rBLG-SS and rBLG-C. Within their studies, Creamer et al. described the absence of oxidized fragment 102–124 covering disulfide bridge Cys106-Cys119 (inner disulfide bridge) which only could be observed when the protein was reduced [62]. The same was observed in our studies, i.e. no peak for the desired oxidized fragment only occurred in the chromatogram upon separation for BLG AB, rBLG, rBLG-SS. Only upon reduction with TCEP or when Cys121 was mutated as in fragment 102–124 of rBLG-SH, a major peak in the chromatogram appeared and could be assigned accordingly. Consequently, formation of the inner disulfide bridge as in BLG AB was derived for rBLG, rBLG-SS and rBLG-SH.

Analysis of the BLG mutants represented an ideal start to continue investigations on chemically and physically treated BLGs. For chemical reduction, BLG variant B (BLG B) was reduced by TCEP and analyzed by LC–ESI–MS/MS [60]. Using the same methodology as for the analysis of BLG mutants, two major observations were made compared to the investigated mutant proteins: 1. a peak for fragment 102–124 could be clearly observed after complete reduction of the protein, and 2. reduction of the outer disulfide bridge occurred before reduction of the inner disulfide bridge (Fig. 7a). The latter could be detected by analysis of partially reduced samples of BLG B. Only minor amounts of the reduced fragment 102–124 were observed for partially reduced BLG B, while the fraction containing the disulfide-connected fragments 61–70 and 149–162 disappeared nearly completely and two new peaks for single fragments 61–70 and 149–162 emerged.

In addition to controlled reduction approaches, physical treatment of BLG AB by UV B light was more challenging. Radiation with UV B light may lead to a breaking of the disulfide bonds but can also cause further oxidative damage of the protein. To analyze the effect of UV B light, BLG AB was treated accordingly and the induced damages and/or structural changes were investigated by the aforementioned approach for sequence analysis (Fig. 7a) [61]. Interestingly, only two major changes could be observed: 1. oxidation of Met7 indicated by a shift of the corresponding HPLC peak and a mass increase in the MS/MS spectrum at the corresponding position and 2. reduction of the signal intensity for the HPLC peak representing the disulfide-linked fragments 61–70 and 149–162 and a concomitant increase of a peak for a modified fragment 149–162. Thus, radiation with UV B light induced opening of the outer bridge as well as oxidation of Met7 in the protein, which both may lead to significant structural and physicochemical changes if the radiation process lasts longer.

The analysis of disulfide-linked BLG from different production and treatment approaches using mild chemical as well as harsh physical conditions was successfully performed and elucidation of disulfide connectivities in the corresponding samples yielded further impetus on how structural changes may influence protein properties as well as aggregation behavior. However, it became evident that in some cases LC–MS/MS analysis is not sufficient for a proper analysis of peptide sequences, especially when intensities of the relevant compounds had been very low as was found for fragment 102–124. In these cases supporting analysis by N-terminal Edman sequencing was a valuable additional tool to complete the analysis of model protein BLG.

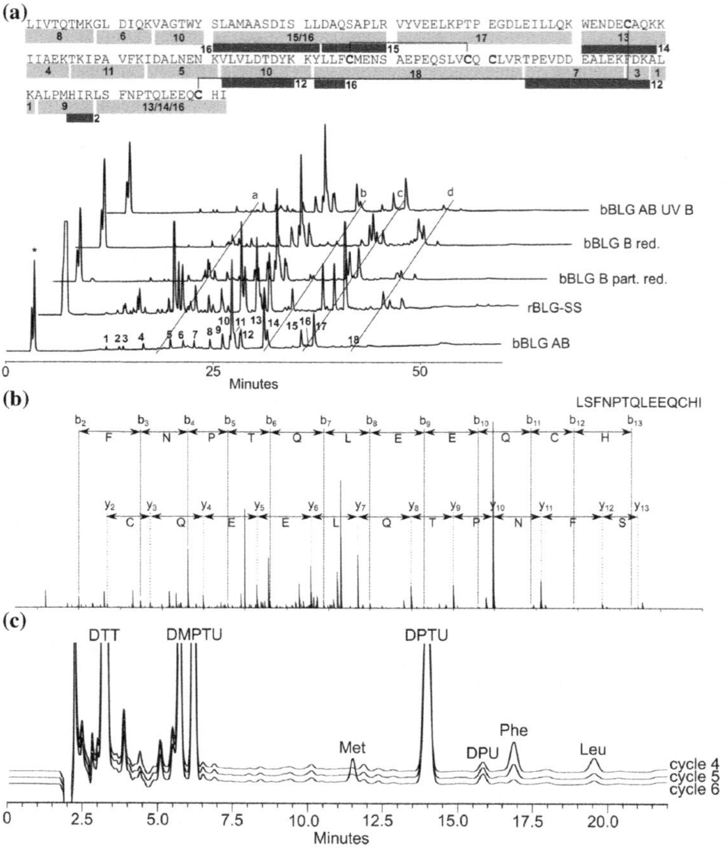

Fig. 7 Analysis of the tryptic digest of BLG AB. **a** RP-HPLC profiles of bBLG AB, rBLG-SS, bBLG B partially and completely chemically reduced and UV B-treated bBLG AB after tryptic digest. Corresponding peaks that were collected and selected for further analysis by MS and MS/MS are labeled with numbers for bBLG AB and annotated to the sequence as depicted by the light and dark grey bars below the BLG sequence. Diagonal lines indicate positions of peaks containing fragments with C66 (a, fragment 61–70), C160 (c, fragment 149–162), the disulfide-linked fragment C66-C160 (b, fragment 61–70 linked to fragment 149–162) and the fragment covering cysteines 106, 119, and 121 (d, fragment 102–124). Asterisk marks the injection peak. **b** LC-ESI-QTOF-MS/MS spectrum of m/z 829.9028 (2+) from the MS spectrum of peak No. 16 of reduced bBLG B representing the reduced fragment 149–162. **c** Cycle runs of N-terminal sequencing covering amino acids at positions 4–6 of peak 18 displaying an increase in the intensity of the corresponding PTH-derivatives that are representative for the occurrence of the amino acids phenylalanine, cysteine (no peak increase) and methionine, respectively. Occurrence of residual PTH-leucine from cycle 3 that was not cleaved in that cycle can be still detected in cycle 4. The total run for this peptide was performed for 10 cycles confirming the existence of peptide sequence YLLFCMENSA that allowed for further determination in combination with the observed mass detected by MS and MS/MS analysis. **c** is taken from [60], licensed under the Creative Commons Attribution 4.0 International license (CC BY 4.0)

4 Conclusions

The studies presented in this chapter first explored analytical methods applied for lysozyme analysis in innovative formulations, e.g. in hot-melt extrudates prepared with PEG20,000. The extrudates were analyzed for their lysozyme content using an optimal RP-HPLC method. Activity of lysozyme in the extrudate was examined using a fluorogenic method and showed preferable accuracy and reproducibility compared to a turbidimetric method that was reported earlier. SDS–PAGE and MALDI–TOF–MS methods were established for mass analysis. Extrudates were analyzed following the workflow described and compared to pure lysozyme and a physical mixture of the protein with PEG20,000 as controls including stability studies of the extrudates following ICH Topic Q1A (R2) guidelines. Extrudates succeeded to retain lysozyme characteristics over 6 months, indicating suitability of hot-melt extrusion as formulation technique for stabilization of the applied protein.

Additionally, conservation of essential post-translational modifications in a protein or peptide was examined and Edman degradation chemistry was adapted to study disulfide connectivity. Suitable PTH-standards were established and validated for the detection of cysteine and its quantification. Evaluation of the standards was performed by sequencing selected model peptides containing modified cysteines, that revealed applicability of the standards. Elucidation of disulfide bridge formation in a model miniprotein containing two disulfide bonds that was derived from the FXIIIa inhibitor tridegin was performed employing the examined PTH-Cys derivatives, PTH-Cys(Me) and PTH-Cys(Cam). Application of our standards for the qualitative determination of modified cysteines and, thus, changes in the formed disulfide bridges before and after complete reduction was possible and allowed for a distinct differentiation between reduced and oxidized cysteines. The established PTH-Cys derivatives are also in agreement with the ICH Q14 guidelines as bioanalytical standard compounds and can be involved for free cysteines and disulfide bridges assessment.

Moreover, a combined technology applying mass spectrometry and N-terminal Edman sequencing turned out as a useful approach to completely elucidate and identify disulfide connectivies in a tryptic digest of differently produced and treated BLGs. This allowed for complete annotation and identification of disulfide bridges and also their changes in connectivities in BLGs which represents valuable information to clearly identify structural changes in the future without a need of sophisticated crystallographic or NMR analyses.

Acknowledgements The authors would like to thank The Deutsche Forschungsgemeinschaft (DFG) within priority program SPP 1934 under project number 425781873 for their financial support. We thank M. Engeser (University of Bonn) for access to the MALDI-TOF/TOF mass spectrometer, D. Menche (University of Bonn) for access to the 700 MHz NMR spectrometer, and T. Schmitz and M. Beck (University of Bonn) for technical support and Solvay GmbH for generous provision of TFA used for solid-phase peptide synthesis.

References

1. Leader B, Baca QJ, Golan DE (2008) Protein therapeutics: a summary and pharmacological classification. Nat Rev Drug Discov 7:21–39. https://doi.org/10.1038/nrd2399
2. Al MO, Al SD, Albericio F, de la Torre BG (2021) 2020 FDA TIDES (Peptides and Oligonucleotides) Harvest. Pharmaceuticals 14:1–14. https://doi.org/10.3390/ph14020145
3. Mullard A (2021) 2020 FDA drug approvals. Nat Rev Drug Discov 20:85–90. https://doi.org/10.1038/D41573-021-00002-0
4. Manning MC, Matsuura JE, Kendrick BS et al (1995) Approaches for increasing the solution stability of proteins. Biotechnol Bioeng 48:506–512. https://doi.org/10.1002/BIT.260480513
5. Breitenbach J (2002) Melt extrusion: from process to drug delivery technology. Eur J Pharm Biopharm 54:107–117. https://doi.org/10.1016/S0939-6411(02)00061-9
6. Patil H, Tiwari RV, Repka MA (2016) Hot-melt extrusion: from theory to application in pharmaceutical formulation. AAPS PharmSciTech 17:20. https://doi.org/10.1208/S12249-015-0360-7
7. Cossé A, König C, Lamprecht A, Wagner KG (2017) Hot melt extrusion for sustained protein release: matrix erosion and in vitro release of PLGA-based implants. AAPS PharmSciTech 18:15–26. https://doi.org/10.1208/S12249-016-0548-5
8. Liao YH, Brown MB, Martin GP (2004) Investigation of the stabilisation of freeze-dried lysozyme and the physical properties of the formulations. Eur J Pharm Biopharm 58:15–24. https://doi.org/10.1016/j.ejpb.2004.03.020
9. Ferrati S, Wu T, Fuentes O et al (2018) Influence of formulation factors on the aerosol performance and stability of lysozyme powders: a systematic approach. AAPS PharmSciTech 19:2755–2766. https://doi.org/10.1208/S12249-018-0980-9
10. Ghalanbor Z, Körber M, Bodmeier R (2010) Improved lysozyme stability and release properties of Poly(lactide-co- glycolide) implants prepared by hot-melt extrusion. Pharm Res 27:371–379. https://doi.org/10.1007/S11095-009-0033-X
11. Farinha S, Moura C, Afonso MD, Henriques J (2020) Production of lysozyme-PLGA-loaded microparticles for controlled release using hot-melt extrusion. AAPS PharmSciTech 21. https://doi.org/10.1208/s12249-020-01816-8
12. Srinivasan C, Katare YK, Muthukumaran T, Panda AK (2008) Effect of additives on encapsulation efficiency, stability and bioactivity of entrapped lysozyme from biodegradable polymer particles. J Microencapsul 22:127–138. https://doi.org/10.1080/02652040400026400
13. Gaudana R, Gokulgandhi M, Khurana V et al (2013) Design and evaluation of a novel nanoparticulate-based formulation encapsulating a HIP complex of lysozyme. Pharm Dev Technol 18:752–759. https://doi.org/10.3109/10837450.2012.737806
14. Jiang G, Woo BH, Kang F et al (2002) Assessment of protein release kinetics, stability and protein polymer interaction of lysozyme encapsulated poly(D, L-lactide-co-glycolide) microspheres. J Control Release 79:137–145. https://doi.org/10.1016/S0168-3659(01)00533-8
15. Elsayed YY, Kühl T, Dauer K et al (2022) Bioanalytical workflow for qualitative and quantitative assessment of hot-melt extruded lysozyme formulations. ACS Omega 7:40836–40843. https://doi.org/10.1021/acsomega.2c03559
16. (2003) ICH topic Q 1 A (R2) Stability testing of new drug substances and products step 5 note for guidance on stability testing: Stability testing on new dug substances and products. Eur Med Agency
17. Manteca A, Alonso-Caballero Á, Fertin M et al (2017) The influence of disulfide bonds on the mechanical stability of proteins is context dependent. J Biol Chem 292:13374–13380. https://doi.org/10.1074/JBC.M117.784934
18. Bäuml CA, Schmitz T, Paul George AA et al (2019) Coagulation factor XIIIa inhibitor tridegin: on the role of disulfide bonds for folding, stability, and function. J Med Chem 62:3513–3523. https://doi.org/10.1021/acs.jmedchem.8b01982
19. Wang L, Wang N, Zhang W et al (2022) Therapeutic peptides: current applications and future directions. Signal Transduct Target Ther 7:1–27. https://doi.org/10.1038/s41392-022-00904-4

20. Miner PB (2020) Plecanatide for the treatment of constipation-predominant irritable bowel syndrome. Expert Rev Gastroenterol Hepatol 14:71–84. https://doi.org/10.1080/17474124.2020.1722101
21. Corsetti M, Tack J (2013) Linaclotide: A new drug for the treatment of chronic constipation and irritable bowel syndrome with constipation. UEG J 1:7–20. https://doi.org/10.1177/2050640612474446
22. Hennrich U, Benešová M (2020) [68Ga]Ga-DOTA-TOC: The first FDA-approved 68Ga-radiopharmaceutical for PET imaging. Pharmaceuticals 13. https://doi.org/10.3390/ph13030038
23. Yen TY, Yan H, Macher BA (2002) Characterizing closely spaced, complex disulfide bond patterns in peptides and proteins by liquid chromatography/electrospray ionization tandem mass spectrometry. J Mass Spectrom 37:15–30. https://doi.org/10.1002/jms.241
24. Tsai PL, Chen SF, Huang SY (2013) Mass spectrometry-based strategies for protein disulfide bond identification. Rev Anal Chem 32:257–268. https://doi.org/10.1515/revac-2013-0011
25. Lakbub JC, Shipman JT, Desaire H (2018) Recent mass spectrometry-based techniques and considerations for disulfide bond characterization in proteins. Anal Bioanal Chem 410:2467. https://doi.org/10.1007/S00216-017-0772-1
26. Böhm M, Bäuml CA, Hardes K et al (2014) Novel insights into structure and function of factor XIIIa-inhibitor tridegin. J Med Chem 57:10355–10365. https://doi.org/10.1021/jm501058g
27. Heimer P, Schmitz T, Bäuml CA, Imhof D (2018) Synthesis and structure determination of μ-conotoxin PIIIA isomers with different disulfide connectivities. J Vis Exp 140:58368. https://doi.org/10.3791/58368
28. Grant GA, Crankshaw MW, Gorka J (1997) Edman sequencing as tool for characterization of synthetic peptides. Methods Enzymol 289:395–419. https://doi.org/10.1016/S0076-6879(97)89056-8
29. Brune DC (1992) Alkylation of cysteine with acrylamide for protein sequence analysis. Anal Biochem 207:285–290. https://doi.org/10.1016/0003-2697(92)90013-W
30. (2022) ICH guideline M10 on bioanalytical method validation and study sample analysis Step5. Eur Med Agency
31. (2022) ICH guideline Q14 on analytical procedure development. Eur Med Agency
32. Keppler JK, Heyse A, Scheidler E et al (2021) Towards recombinantly produced milk proteins: Physicochemical and emulsifying properties of engineered whey protein beta-lactoglobulin variants. Food Hydrocoll 110:106132. https://doi.org/10.1016/j.foodhyd.2020.106132
33. Ponniah K, Loo TS, Edwards PJB et al (2010) The production of soluble and correctly folded recombinant bovine beta-lactoglobulin variants A and B in Escherichia coli for NMR studies. Protein Expr Purif 70:283–289. https://doi.org/10.1016/j.pep.2009.12.006
34. Loch JI, Bonarek P, Tworzydło M et al (2016) Engineered β-lactoglobulin produced in E. coli: purification. Biophysical and Structural Characterisation. Mol Biotechnol 58:605–618. https://doi.org/10.1007/s12033-016-9960-Z
35. Bäuml CA, Paul George AA, Schmitz T et al (2020) Distinct 3-disulfide-bonded isomers of tridegin differentially inhibit coagulation factor XIIIa: The influence of structural stability on bioactivity. Eur J Med Chem 201:112474. https://doi.org/10.1016/j.ejmech.2020.112474
36. Jangale AD, Wagh YB, Tayade YA, Dalal DS (2015) Simple and efficient synthesis of 5-substituted-3-phenyl-2-thioxoimidazolidin-4-one derivatives from s -amino acids and phenylisothiocyanate in et 3N/DMF-H_2O. Synth Commun 45:1876–1886. https://doi.org/10.1080/00397911.2015.1052506
37. Beloglazkina EK, Majouga AG, Moiseeva AA, et al (2007) 5-[2-(Methylthio)ethyl]-3-phenyl-2-thioxoimidazolidin-4-one and its complexes with transition metals (CoII, NiII, and CuII). Synthesis and electrochemical investigation. Russ Chem Bull 562 56:351–355. https://doi.org/10.1007/S11172-007-0057-5
38. Khatik GL, Kaur J, Kumar V et al (2011) Aldol derivatives of Thioxoimidazolidinones as potential anti-prostate cancer agents. Eur J Med Chem 46:3291–3301. https://doi.org/10.1016/j.ejmech.2011.04.050

39. Bhushan R, Reddy GP (1987) TLC of phenylthiohydantoins of amino acids: a review. J Liq Chromatogr 10:3497–3528. https://doi.org/10.1080/01483918708077810
40. Barton JS, Tang CF, Reed SS (2000) Dipeptide sequence determination: analyzing phenylthiohydantoin amino acids by HPLC. J Chem Educ 77:268–269. https://doi.org/10.1021/ed077P268
41. Schmitz T, Paul George AA, Nubbemeyer B et al (2021) NMR-based structural characterization of a two-disulfide-bonded analogue of the FXIIIa inhibitor Tridegin: New insights into structure–activity relationships. Int J Mol Sci 22:880. https://doi.org/10.3390/ijms22020880
42. Kühl T, Wißbrock A, Goradia N et al (2013) Analysis of Fe(III) heme binding to cysteine-containing heme-regulatory motifs in proteins. ACS Chem Biol 8:1785–1793. https://doi.org/10.1021/cb400317x
43. Kühl T, Sahoo N, Nikolajski M et al (2011) Determination of hemin-binding characteristics of proteins by a combinatorial peptide library approach. ChemBioChem 12:2846–2855. https://doi.org/10.1002/cbic.201100556
44. Laemmli UK (1970) Cleavage of structural proteins during the assembly of the head of bacteriophage T4. Nature 227:680–685. https://doi.org/10.1038/227680a0
45. Shugar D (1952) The measurement of lysozyme activity and the ultra-violet inactivation of lysozyme. Biochim Biophys Acta 8:302–309. https://doi.org/10.1016/0006-3002(52)90045-0
46. Yang Y, Hamaguchi K (1980) Hydrolysis of 4-methylumbelliferyl N-acetyl-chitotetraoside catalyzed by hen lysozyme. J Biochem 88:829–836. https://doi.org/10.1093/oxfordjournals.jbchem.a133037
47. Elsayed YY, Kühl T, Imhof D (2024) Edman degradation reveals unequivocal analysis of the disulfide connectivity in peptides and proteins. Anal Chem 96:4057–4066. https://doi.org/10.1021/acs.analchem.3c04229
48. Wieling J, Hendriks G, Tamminga WJ et al (1996) Rational experimental design for bioanalytical methods validation illustration using an assay method for total captopril in plasma. J Chromatogr A 730:381–394. https://doi.org/10.1016/0021-9673(96)00006-4
49. Gray WR (1993) Disulfide structures of highly bridged peptides: a new strategy for analysis. Protein Sci 2:1732–1748. https://doi.org/10.1002/PRO.5560021017
50. van den Akker CC, Schleeger M, Bonn M, Koenderink GH (2014) Structural Basis for the polymorphism of β-lactoglobulin amyloid-like fibrils. Bio-nanoimaging Protein Misfolding Aggreg:333–343. https://doi.org/10.1016/B978-0-12-394431-3.00031-6
51. Imre T, Zsila F, Szabó PT (2003) Electrospray mass spectrometric investigation of the binding of cis-parinaric acid to bovine beta-lactoglobulin and study of the ligand-binding site of the protein using limited proteolysis. Rapid Commun Mass Spectrom 17:2464–2470. https://doi.org/10.1002/rcm.1217
52. Kuwata K, Era S, Hoshino M et al (1999) Solution structure and dynamics of bovine beta-lactoglobulin A. Protein Sci 8:2541. https://doi.org/10.1110/PS.8.11.2541
53. Oliveira KMG, Valente-Mesquita VL, Botelho MM et al (2001) Crystal structures of bovine beta-lactoglobulin in the orthorhombic space group C222(1). Structural differences between genetic variants A and B and features of the Tanford transition. Eur J Biochem 268:477–484. https://doi.org/10.1046/j.1432-1033.2001.01918.x
54. Qin BY, Bewley MC, Creamer LK et al (1998) Structural basis of the Tanford transition of bovine beta-lactoglobulin. Biochemistry 37:14014–14023. https://doi.org/10.1021/BI981016T
55. Uhrínová S, Smith MH, Jameson GB et al (2000) Structural changes accompanying pH-induced dissociation of the beta-lactoglobulin dimer. Biochemistry 39:3565–3574. https://doi.org/10.1021/BI992629O
56. McKenzie HA, Ralston GB, Shaw DC (1972) Location of sulfhydryl and disulfide groups in bovine β-lactoglobulins and effects of ureat. Biochemistry 11:4539–4547. https://doi.org/10.1021/bi00774a017
57. Phelan P, Malthouse JPG (1994) 13C-n.m.r. of the cyanylated beta-lactoglobulins: evidence that Cys-121 provides the thiol group of beta-lactoglobulins A and B. Biochem J 302:511–516. https://doi.org/10.1042/BJ3020511

58. Iwaoka M, Mitsuji T, Shinozaki R (2019) Oxidative folding pathways of bovine milk β-lactoglobulin with odd cysteine residues. FEBS Open Bio 9:1379–1391. https://doi.org/10.1002/2211-5463.12656
59. Heimer P, Tietze AA, Bäuml CA et al (2018) Conformational μ-Conotoxin PIIIA isomers revisited: impact of cysteine pairing on disulfide-bond assignment and structure elucidation. Anal Chem 90:3321–3327. https://doi.org/10.1021/acs.analchem.7b04854
60. Brune SE, Hoppenreijs LJG, Kühl T et al (2023) Precision fermentation as a route to modify β-lactoglobulin structure through substitution of specific cysteine residues. Int Dairy J 147:105772. https://doi.org/10.1016/j.idairyj.2023.105772
61. Fitzner L, Kühl T, Hasler M et al (2023) Modification and oxidative degradation of β-lactoglobulin by UVB irradiation. Food Chem 428:136698. https://doi.org/10.1016/j.foodchem.2023.136698
62. Creamer LK, Nilsson HC, Paulsson MA et al (2004) Effect of genetic variation on the tryptic hydrolysis of bovine beta-lactoglobulin A, B, and C. J Dairy Sci 87:4023–4032. https://doi.org/10.3168/JDS.S0022-0302(04)73543-2

Production, Purification, and Characterization of Recombinant β-Lactoglobulin Variants

Sarah E. Brune, Dieter Jahn, Rainer Krull, and Rebekka Biedendieck

1 Introduction

Due to the growing world population and an increasing demand for proteins, a rethinking of animal-based foods must take place. Recently, "cellular agriculture" is increasingly becoming the focus of current research. A distinction is made between cellular products, which include for example cultured meat, and acellular products, such as food proteins or fats, which are produced by different microorganisms (bacteria, fungi, yeasts) [1]. In particular, the demand for so called "animal-free milk" is increasing as a basis for further processing into yogurt, cottage cheese, cheese, and other food products. Plant-based milk alternatives made from soya, oat, and others do not mimic the functionalities of cow's milk, such as nutrients and texture, which are mainly caused by the milk proteins caseins, β-lactoglobulin, and α-lactalbumin [2]. In this respect, the recombinant production of the whey protein

S. E. Brune
Institute of Microbiology, Institute of Biochemical Engineering, Braunschweig Integrated Centre of Systems Biology (BRICS), Center of Pharmaceutical Engineering (PVZ), Technische Universität Braunschweig, Braunschweig, Germany
e-mail: s.brune@tu-braunschweig.de

R. Krull
Institute of Biochemical Engineering, Braunschweig Integrated Centre of Systems Biology (BRICS), Center of Pharmaceutical Engineering (PVZ), Technische Universität Braunschweig, Braunschweig, Germany
e-mail: r.krull@tu-braunschweig.de

D. Jahn · R. Biedendieck (✉)
Institute of Microbiology and Braunschweig Integrated Centre of Systems Biology (BRICS), Technische Universität Braunschweig, Braunschweig, Germany
e-mail: r.biedendieck@tu-braunschweig.de

D. Jahn
e-mail: d.jahn@tu-braunschweig.de

A. Kwade and I. Kampen (eds.), *Dispersity, Structure and Phase Changes of Proteins and Bio Agglomerates in Biotechnological Processes*,
https://doi.org/10.1007/978-3-031-63164-1_6

β-lactoglobulin (BLG) is particularly interesting. Making up about 50% of whey protein (with approximately 3.2 g L^{-1}) it is the most abundant whey protein in cow's milk and as such an important ingredient in many foods [3].

BLG is a member of the lipocalin family and possesses the ability to selectively bind hydrophobic ligands such as fatty acids or retinol. This ligand binding primarily occurs within the hydrophobic binding pocket of the protein, also known as the calyx or barrel, which is constructed by nine antiparallel β-sheets (Fig. 1). Additionally, weaker binding sites located on the surface of the protein have been reported in previous studies [4]. BLG exists mainly as a dimer at neutral pH and as a monomer in acidic environments [5]. The bovine BLG-monomer has a molecular weight of around 18,300 Da and, after processing, consists of 162 amino acid residues without any post-translational modifications (PTMs) [6]. Bovine BLG exists in at least 11 different isoforms, with isoform A (BLG A) and B (BLG B) being the most common ones [7]. Bovine BLG A and BLG B differ in only two amino acid residues at positions 118 and 64. Moreover, in all described bovine BLG variants five conserved cysteine residues are present. The cysteine residues at positions 66 and 160 form a disulfide bond which is located on the surface of the natively folded protein (will be called "outer disulfide bond" within this chapter), while the disulfide bond which is formed between cysteine residues 106 and 119 is located inside the correctly folded BLG (will be called "inner disulfide bond" within this chapter). The fifth cysteine residue at position 121 is embedded inside the folded protein and is not described to be involved in disulfide formation [8] (Fig. 1).

Its functional properties and abundance make BLG an important compound in many foods, e.g., as foaming agent, gelling agent, or emulsifier. Furthermore, it is one of the most important model proteins for structure–function studies in food science [10]. Bovine BLG can be extracted from bovine milk in highly pure form in

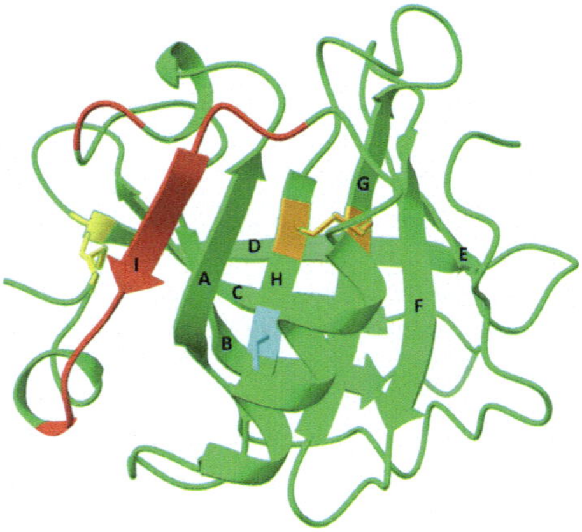

Fig. 1 3D structure of a monomer of native bovine BLG B (PDB ID: 1BSO). The free thiol (Cys121) is labeled in blue, the inner disulfide bond (Cys106–Cys119) in orange, and the outer disulfide bond (Cys66–Cys160) in yellow. The dimerization motifs are highlighted in red. Letters A–H denote β-strands forming the antiparallel β-sheet calyx. The 3D structure was created with ChimeraX [9]

different isoforms [11, 12]. However, its recombinant production in microorganisms provides the opportunity to analyze this protein more comprehensively as variants of the protein can be produced much more quickly and easily, bypassing recombinant animals. Position-directed mutagenesis can be used to specifically alter certain amino acid residues and generate different BLG variants. In this way, the influence of specific regions of the protein on its functionality can be analyzed. For example, systematic mutagenesis allowed Bonarek et al. [13] to investigate the influence of point mutations in the hydrophobic binding pocket of BLG [13].

Due to its status as a model protein, the well understood folding mechanisms, and the potential applications in the food and pharmaceutical industries, there have been constant research efforts to optimize the recombinant production of bovine BLG in different microorganisms over the last decades. Eukaryotic microorganisms such as yeasts and fungi have emerged as promising hosts for the production and secretion of recombinant BLG [14, 15]. Nevertheless, despite achieving high final product titers (>1 g L^{-1}) of secreted BLG in the yeast strain *Pichia pastoris* (*Komagataella phaffi*), a non-native N-terminus due to incomplete cleavage of the signal peptide has been reported for the produced BLG [15]. This deviation could potentially alter the physicochemical properties, differentiating it from the native BLG. However, just recently Aro et al. [16] were able to produce BLG in the fungus *Trichoderma reseei* with a final BLG concentration of up to 1 g L^{-1} which was found to be N-methylated but showed the same secondary structure and the same emulsifying properties as the native bovine BLG [16].

The introduction of PTMs such as glycosylation or methylation can be overcome by using prokaryotes as production strains. In 1990, for the first time, bovine BLG A was recombinantly produced intracellularly in *Escherichia coli* in the form of insoluble inclusion bodies and with additional N-terminal methionine. By using the production system described by Bessette et al. [17] in which a *trxB*/*gor* mutant of *E. coli* BL21 (*E. coli* Origami B) was combined with intracellular coproduction of the disulfide bond isomerase DsbC [18], Ponniah et al. [19] were able to recombinantly produce soluble and properly folded BLG B. Due to the *trxB*/*gor* mutation, the cytosol provides an oxidizing environment in which disulfide bonds can be formed. On the other hand, DsbC can help to correctly link disulfide bonds due to its chaperone and isomerase activity [20, 21]. However, BLG B recombinantly produced in this manner also carries an additional N-terminal methionine residue as its starting amino acid residue during protein synthesis, which negatively affected the physicochemical properties upon crystallization and ligand binding. By exchanging the native first N-terminal amino acids present in the bovine BLG B (L1A/I2S) by Loch et al. [22], successful cleavage of the start methionine was achieved. This system provided the starting point for further biochemical, biophysical, and structural analyses.

Table 1 Overview of different BLG variants used in this study

Abbreviations	Description	Modification	Reference
bBLG AB	Bovine BLG AB		[23]
bBLG B	Bovine BLG B		[23]
bBLG R1	Bovine BLG AB partly reduced	Chemically reduced at pH 3.5	[23]
bBLG R2	Bovine BLG AB fully reduced	Chemically reduced at pH < 3.5	[23]
rBLGwt	Recombinant wild-type BLG B		[22, 23, 25]
rBLG	Recombinant BLG B	L1A/I2A	[22, 25]
rBLG–SH	Recombinant BLG B without free thiol group	L1A/I2A + C121A	[23]
rBLG–SS	Recombinant BLG B without outer disulfide bond	L1A/I2A + C66A + C160A	[23]
rBLG–C	Recombinant BLG B without cysteine residues	L1A/I2A + C121A + C66A + C160A + C106A + C119A	[23]

2 Materials and Methods

2.1 Recombinant Plasmids and Production Strains

For all cloning experiments to construct the recombinant plasmids for the production the BLG variants, *E. coli* DH10B (Life Technologies, Carlsbad, USA) was used. While for the recombinant production of BLG variants rBLG, rBLG-SH and rBLG-SS *E. coli* Origami B (DE3) (Novagen, Darmstadt, Germany) was utilized, for rBLG-C *E. coli* BL21 (DE3) (ThermoFisher Scientific, Waltham, USA) was chosen (Table 1). For plasmid construction see Loch et al. [22] and Brune et al. [23].

2.2 Structural Characterization of BLG

For measuring changes in the secondary structure of the recombinant variants, Circular dichroism (CD) spectroscopy was performed with a Jasco J-810 spectropolarimeter (Jasco, Tokyo, Japan). Using a non-linear Levenberg–Marquardt fitting routine, corresponding secondary structure elements (i.e., α-helix, β-sheet, random coil, β-turn) were quantified applying the software package CD Fit (V1.1, Wageningen Centre for Food Sciences: Structure and Functionality, The Netherlands), and an independent t-test SPSS statistics (V28.0.1.1, IBM, New York, USA) was used to display differences with bBLG B.

Intrinsic tryptophan fluorescence, measured by using a Jasco Spectrofluorometer FP-8500 (Jasco, Tokyo, Japan), was used to determine the tertiary structure of BLG with an excitation at 290 nm and an emission between 300 and 400 nm to follow wavelength and intensity shifts in the fluorescence.

Also native polyacrylamide gel electrophoresis (PAGE) was performed for tertiary structural analyses. It was performed based on a modified protocol published by Righetti et al. [24] using 15% gels with omitting the use of SDS. After electrophoretic separation (constantly at 150 V) the gels were stained with Coomassie Brilliant Blue.

2.3 *Retinol Binding Assay*

Protein solutions of 50 μM (corresponds to 1 mg mL^{-1}) were prepared in 10 mM sodium phosphate buffer at pH 7. Fluorescence of retinol was measured using a Spark multimode microplate reader (Tecan, Männedorf, Switzerland) in 96 well plates with sample triplicates of 100 μL. Retinol solution of 1 mM (in ethanol) was added stepwise until a final concentration of 75 μM per well was reached. The excitation wavelength was set to 370 nm and emission was measured at 480 nm.

2.4 *8-Anilinonaphthalene-1-Sulfonic Acid (ANS) Fluorescence Spectroscopy*

The surface hydrophobicity of BLG variants was analyzed using ANS fluorescence. Proteins were dissolved in a 10 mM sodium phosphate buffer at pH 7, reaching a concentration of 100 μg mL^{-1}. 20 μL of an 8 mM ANS solution (in dimethylformamide) was added to 2 mL of protein solution. After 1 min of incubation, duplicate samples were analyzed using a Jasco Spectrofluorometer FP-8500 (Jasco, Tokyo, Japan). Excitation was set to 390 nm, while emission was measured at 470 nm.

3 Results and Discussions

BLG is the most abundant whey protein in bovine milk. A distinguishing feature of the protein is the presence of two highly conserved disulfide bonds (C66–C160 and C106–C119), along with an unpaired cysteine residue (C121). First, to get an idea about the similarities and differences of recombinantly produced and purified BLG B compared to the native bovine protein, the recombinant [22] and native BLG B were analyzed in detail with respect to structure, physicochemical behavior, and emulsifying properties [25]. Further, to get a better understanding of the influence of the cysteine residues on the functionality of BLG, different BLG variants with modifications to their cysteine residues were recombinantly produced, purified, and structurally compared with chemically modified bovine BLG [23].

3.1 Recombinant Production and Purification of BLG Variants

Loch et al. developed a recombinant system using *E. coli* for the IPTG-inducible production of bovine BLG. The authors engineered two variants: one with an additional methionine residue at the N-terminus and another lacking the methionine due to an N-terminal modification (L1A/I2A), which facilitates the cleavage of the methionine [22]. The plasmid system utilized for the recombinant production of these BLG variants was also used to produce cysteine-modified BLG variants. Thus, on plasmid basis, the corresponding codons encoding cysteine residues were substituted accordingly against codons encoding alanine residues (Table 1). All new expression plasmids are based on the plasmid responsible for the recombinant production of BLG with a modified N-terminus (L1A/I2A) [22]. *E. coli* production strains were individually transformed with the corresponding new plasmids [22, 23, 25].

The production and following isolation of the different BLG variants were carried out using two different purification processes (Fig. 2). Variants rBLGwt (recombinant wild-type bovine BLG B, carrying an additional N-terminal methionine lacking in the native bovine BLG), rBLG (recombinant wild-type bovine BLG B with modified N-terminus (L1A/I2A) for successful removal of start-methionine), rBLG–SH (rBLG with C121A), and rBLG–SS (rBLG with C66A–C160A) were co-produced in *E. coli* Origami B (DE3) cells along with the plasmid-encoded disulfide isomerase DsbC. The activity of DsbC in combination with the *trxB*/*gor* mutation within the genome of the *E. coli* host Origami (DE3) lead to a more oxidative cytoplasm to ensure proper formation of disulfide bonds [19]. This prevented misfolding and aggregation of the corresponding BLG variants and ensured their recombinant production in soluble form. After ultrasonic treatment for cell disruption and centrifugation, the individual BLG variants were present in soluble form in the supernatant. In a first purification step, the coproduced DsbC could be removed by anion exchange chromatography. The remaining contaminating proteins could be removed by precipitation with ammonium sulfate while the corresponding BLG target protein remained in solution after centrifugation. Finally, size exclusion chromatography was used to remove the salt from the BLG protein before freeze-drying (Fig. 1, Process A) [25]. Up to 120 mg target protein L^{-1} culture broth could be purified in high purity for variants rBLGwt, rBLG, and rBLG-SH cultivated in a 2 L shake flask filled with 1 L (shaking frequency 150 min^{-1}, amplitude 50 mm) [23]. The yield for variant rBLG–SS was about 60 mg L^{-1}, which can be explained by destabilization of the protein due to removal of the outer disulfide bond. All variants showed purity >90 wt% when analyzed by Fourier-transform infrared (FTIR) spectroscopy [23].

Substitution of all cysteine against alanine residues (rBLG–C) leads to the formation of the BLG variant as non-classical inclusion bodies (ncIBs) when recombinantly produced in *E. coli* BL21 (DE3). This type of inclusion bodies has the unique feature that a certain part is still properly folded [26]. Here, it was proven that the inclusion bodies are already soluble at higher pH values (pH > 9) or lower detergent concentrations (data not shown). As rBLG-C accumulated as insoluble IBs after recombinant

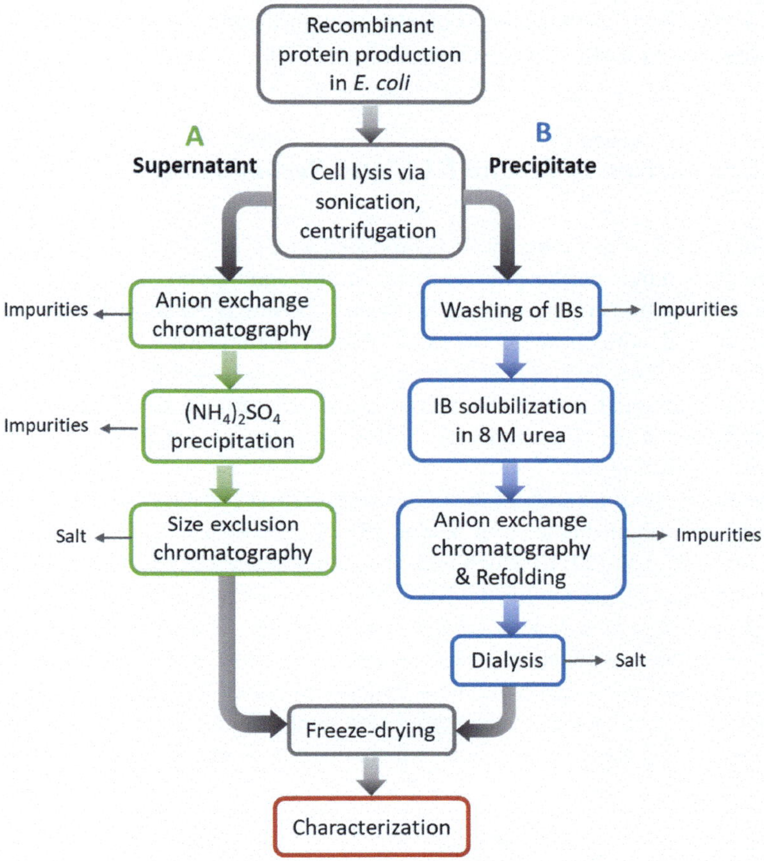

Fig. 2 Purification process of recombinant BLG according to Keppler et al. [25] and Brune et al. [23]. After recombinant production in *E. coli*, cell lysis, and centrifugation, process A (green) was used for the purification of variants rBLGwt, rBLG, rBLG–SH, and rBLG–SS (Table 1) and process B (blue) for variant rBLG–C (Table 1). After removing the salt, all variants were freeze-dried individually and characterized. IB = Inclusion body

prodution it was part of the precipitate after ultrasonic cell disruption and centrifugation. A washing step of the precipitate with 2% (v/v) triton was required to remove impurities e.g. cell debris. To ensure a complete solubilization of the target protein, the corresponding precipitate was afterwards treated with 8 M urea. During gravity flow anion exchange chromatography, the refolding of rBLG–C occurred on the purification column while simultaneously removing urea. The new properly folded rBLG–C was capable of binding to the column material while contaminating as well as aggregated proteins were washed off. With a high salt concentration (300 mM NaCl) the correctly folded rBLG–C could be eluted. After the removal of salt via dialysis, the process was finished by freeze-drying the pure rBLG–C (Fig. 1, process

B). With this newly established production and purification process, up to 400 mg L^{-1} of pure refolded rBLG–C could be achieved [23].

3.2 Comparison of Bovine BLG and Recombinant BLG

The potential use of recombinant BLG produced with *E. coli* for applications in food technology requires similar physicochemical and emulsifying properties. For this reason, the properties of commercially obtained bovine BLG B (bBLG B, Table 1) were compared with these of the two different recombinant wild-type variants (rBLG and rBLGwt, Table 1) using multidisciplinary analyses [25]. While rBLGwt exhibits an additional methionine residue at the N-terminus, rBLG has two amino acid exchanges at the N-terminus (L1A/I2S) that allow cleavage of the N-terminal methionine [22] (Fig. 3).

Structurally, both recombinant BLG variants (rBLG and rBLGwt) were found to be almost indistinguishable from bovine BLG [25] which was also shown by the results of Loch et al. [22] in which both secondary and tertiary structure showed no significant differences from bovine BLG. Moreover, the two BLG variants recombinantly produced in *E. coli* do not differ from bovine BLG in terms of their nativity, post-translational modifications, and oxidative state. At the quaternary level, all three analyzed variants show similar monomer/dimer behavior. Molecular dynamics (MD) simulations detected a slightly decreased structural stability of the rBLGwt variant, which could be induced by the additional methionine at the N-terminus [25], while

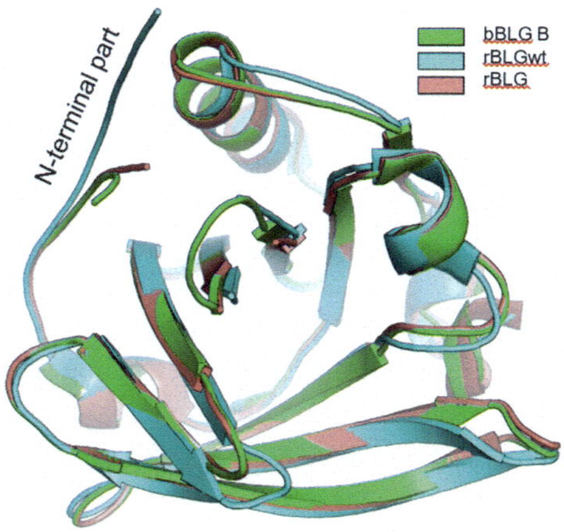

Fig. 3 Superposition (Cα) of BLG crystal structures: bBLG B (PDB ID: 3NPO), rBLGwt (PDB ID: 5K06), and rBLG (PDB ID: 6QI6). The overall structure of all compared variants is almost the same but the N-terminal part of the variant with additional N-terminal methionine (cyan) has a different conformation than observed in the natural protein isoform B (green) or the variant with N-terminal modification L1A/I2S (pink). Figure taken and modified from [25], licensed under the creative commons attribution 4.0 international license

the isoelectric point (IEP) of both rBLG variants is in the range of the literature (IEP 4.9–5.4) [27]. Incomplete resolubilization after freeze-drying was observed for rBLGwt. Furthermore, slight deviations within the denaturation temperature of the variants could be found, which can be mainly attributed to different protein purities and analysis methods. This could also explain the lower thermal stability of rBLG compared to bBLG B [25]. Minor differences were also observed in the hydrophobicity of the variants. Here, mainly the amino acid exchanges at the N-terminus lead to slightly changed residence times in RP-HPLC.

For use as an ingredient in foods, the emulsifying properties of the recombinant variants are particularly crucial. Keppler et al. [25] showed that surface properties, such as lag-time, adsorption rate, and equilibrium interfacial tension for bBLG B as well as for both recombinant wild-type variants were within the range of literature data. In addition, similar oil droplet size distribution and creaming stability were observed for all measuring variants. Only the viscoelastic properties of rBLGwt seem to differ slightly, which may be due to less densely packed interface or a lower mechanical stability [25].

In summary, the measurements and simulations showed that both recombinant variants (rBLG and rBLGwt) behave very similar to bovine BLG B in terms of structure and emulsifying properties. However, rBLG with cleaved N-terminal methionine seems to exhibit higher equivalence to native bovine BLG, as the variant rBLGwt with extended N-terminus shows small deviations on radius and structural stability (Fig. 3). Therefore, from a scientific point of view, these recombinant variants can be used as substitutes for native bovine BLG and thus act as a starting point for further analyses with regard to potential applications in food technology [25].

3.3 Characterization of Secondary and Tertiary Structure of Native and Recombinant BLG Variants

Different recombinant BLG variants with substituted cysteine residues (Table 1) could be successfully produced and purified as described above. After showing that both recombinant wild-type BLGs can substitute the native bovine BLG B, the different recombinant cysteine variants, for which a correct disulfide bond connectivity was confirmed [23], were compared with chemically reduced native bovine BLG variants with respect to their secondary and tertiary structure. For this, bovine BLG AB was either partly or completely chemically reduced with Tris(2-carboxyethyl)phosphin (TCEP) according to Brune et al. [23] (Table 1).

3.3.1 Secondary Structure

Circular dichroism (CD) spectroscopy is a technique used to study the secondary structure of molecules. It involves the use of circularly polarized light to measure the differences in absorbance between left- and right-handed circularly polarized light. The resulting CD spectrum can provide information about conformation and dynamic properties [28]. CD analysis tools (e.g., CD Fit) can be used to analyze the distribution of secondary structure elements in proteins.

The CD spectra in the far UV region (190–260 nm) were very comparable for native bovine BLG B, BLG AB, and rBLG, indicating a similar β-sheet based secondary structure (Fig. 4a + b). Substitution of the free thiol by a methyl group (C121A) resulted in a minor decrease in random coil structures (–8%) and minor increase in helical content (+6%) (Fig. 4d). Accordingly, the secondary structure remains almost unaffected, which has also been shown by Jayat et al. [29]. The removal of the outer disulfide bond by substituting two cysteine with alanine residues (C66A, C160A) but keeping the free thiol group leads to a loss of β-sheet content (–12%), while resulting in an increase in random coil structures (+10%) and α-helical content (+5%) (Fig. 4d). Similar results show the partly reduced variant (bBLG R1) with a decrease in β-sheet structure (–8%) and an increase in random coil structure (+12%) [23]. Therefore, the substitution of the outer disulfide bond (rBLG–SS) seems to have a minor impact on the secondary structure as it hinders the folding of only a small part of the β-sheet structure, while the subsequent chemical removal has much less influence on the main β-sheet structure.

Upon additional removal of the inner disulfide bond and the free thiol group in rBLG–C, the protein fails to fold the intramolecular β-sheet structure typical for BLG (–39%) and forms random coil structures (+31%) instead (Fig. 4d). It can be assumed that the inner disulfide bond is crucial for the formation of the main barrel structure, as it connects β-strands G and H which are forming the core β-sheet during early stages of folding (compare Fig. 1). Furthermore, an increase in α-helical structures was observed in the cysteine-free variant rBLG–C. According to Hamada et al. the sequence of BLG is supposed to form α-helices, which convert to β-sheets during folding [30]. A possible reason for these transitions are non-local interactions in the tertiary structure, which might be hindered by removal of the inner bond. Interestingly, the β-sheet structure remains mostly intact after chemical reduction of both disulfide bonds with a decrease in β-sheets of 8% and an increase in random coil structures of only 16% (Fig. 4c). Therefore, subsequent chemical reduction seems to lead to a destabilization but not to a complete unfolding of the protein as the inner disulfide bond is crucial for the folding but not to maintain the structure.

3.3.2 Tertiary Structure

Tryptophan residues show an emission at a specific excitation wavelength due to their aromatic structure. The wavelength and intensity of emission can change depending on the environment of the tryptophan residues. Therefore, measuring the emission of the tryptophan residues is a widely used tool to analyze tertiary structure and structural dynamics of a protein [31].

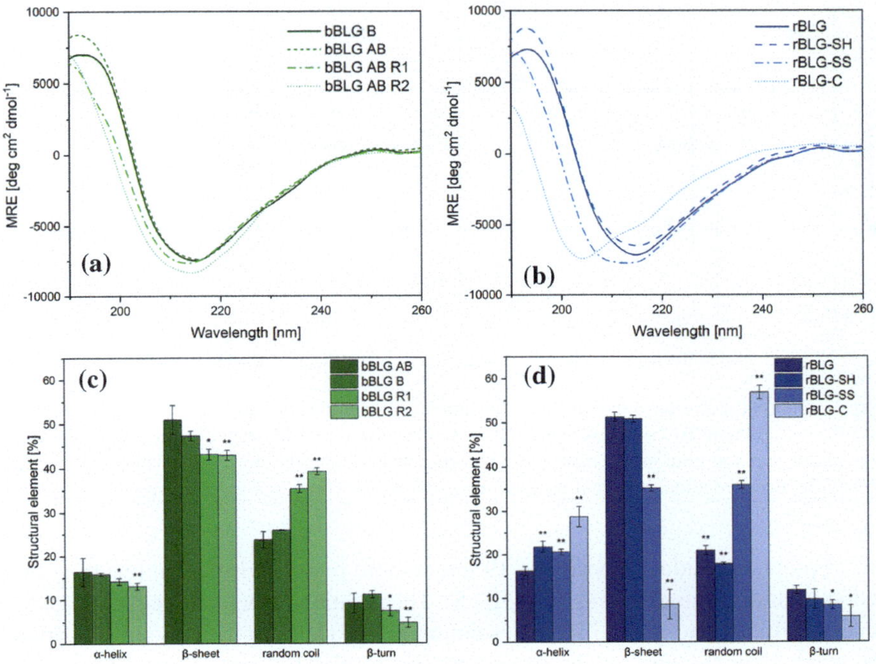

Fig. 4 Far UV CD spectra (190–260 nm) of **(a)** bovine and **(b)** recombinant BLG variants at pH 7.0. Secondary structure element distribution of **(c)** bovine and **(d)** recombinant BLG variants, derived from circular dichroism (CD) analyses. Values significantly different from bBLG B at $p \leq 0.05$ (*) or $p \leq 0.01$ (**) are labelled accordingly. Figure taken from Brune et al. [23], licensed under the creative commons attribution 4.0 international license

The intrinsic fluorescence spectra of bovine and recombinant BLG variants are shown in Fig. 5a+b. For native bBLG B and AB as well as rBLG maxima at wavelength 331 nm were detected, indicating that these proteins show the same globular structure. The BLG variant rBLG–SH shows a comparable spectrum to those of native bBLG B, bBLG AB, and rBLG, indicating that the substitution of C121 by an alanine residue in rBLG has no effect on the tertiary structure (Fig. 5), as also shown by Jayat et al. for C121 substitution by a serine residue [29].

Tryptophan fluorescence emission in BLG is mainly caused by Trp19 which is buried in a hydrophobic pocket. In contrast Trp61 is present at the surface of the protein, solvent exposed, and its fluorescence strongly quenched by the disulfide bond C66–C160 [32]. For this reason, the removal of the outer disulfide bond by chemical reduction in native bBLG AB or substitution in rBLG-SS and rBLG-C leads to a significant higher fluorescence intensity (Fig. 5). Additionally, all these variants show a significantly higher emission wavelength at 336–337 nm, as compared to native bBLG B and AB as well as rBLG and rBLG-SH. The outer disulfide bond C66–C160 connects the C-terminus of BLG to the D strand (Fig. 1). By removing this bond the protein partly unfolds. This conformational change seems to bring

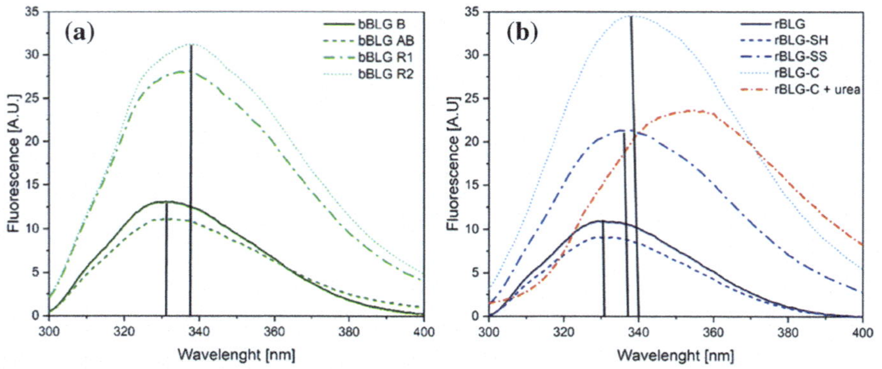

Fig. 5 Fluorescence emission spectra of (**a**) native and chemically reduced bovine BLG variants and (**b**) recombinant BLG variants with rBLG–C solubilized in 8 M urea. Fluorescence excitation was at 290 nm and emission spectra were recorded from 300–400 nm. Parts of the Figure taken from [23], licensed under the creative commons attribution 4.0 international license

the tryptophan residue at positions 19 into a more polar environment resulting in a red-shift of fluorescence emission (Fig. 5). The additional chemical reduction of the inner disulfide bond (bBLG R2) does not seem to have any effect on the polarity of the tryptophan residues, as the secondary structure of the core mainly stays intact (Fig. 5a).

The cysteine-free variant rBLG–C shows a higher fluorescence intensity and a slightly higher emission wavelength of 338 nm compared to all of the other variants (Fig. 5b). Due to the cysteine substitutions and the loss of most of the native secondary structure (see above) the fluorescence of the tryptophan residues is no longer quenched by disulfide bonds or other amino acids. The higher emission wavelength seems to indicate a stronger unfolding due to the lack of the inner disulfide bond compared to rBLG-SS. Nevertheless, it can be assumed that a globular-like structure still exists, since complete denaturation of the protein in 8 M urea leads to a fluorescence intensity of >350 nm (Fig. 5b).

Another method to analyze the tertiary structure of BLG monomers is native polyacrylamide gel electrophoresis (PAGE), where proteins are separated by their native size and their charge. Due to a high pH of 8.8 within the gel, the dimer-monomer equilibrium is towards the dimer dissociation in native PAGE [33]. Native bBLG AB showed two close bands on the native PAGE, which is due to surface charge differences of the two BLG isoforms A and B within this protein mixture [34]. rBLG and rBLG-SH migrated with a similar velocity as bBLG B indicating a similar tertiary structure (Fig. 6), which confirms the results of the tryptophan fluorescence already discussed above.

However, recombinant BLG migrated slower when the outer disulfide bond was substituted (rBLG-SS) which confirms an opening of the globular structure. Similarly, removal of both disulfide bonds and the free cysteine residue (rBLG–C) led to further unfolding and thereby an even slower migration (Fig. 6). These results are

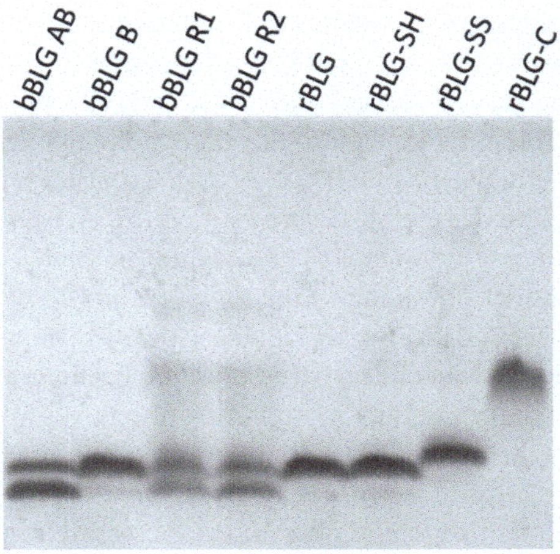

Fig. 6 Native PAGE analysis of native and reduced bovine BLG and recombinant BLG variants. 5 μg of freeze-dried protein were applied per lane. The lanes are labeled with the abbreviations found in Table 1. Figure taken from Brune et al. [23], licensed under the creative commons attribution 4.0 international license

also consistent with the results from the intrinsic fluorescence measurements (Fig. 5). Both reduced bBLG AB variants bBLG R1 and R2 showed a comparable pattern to native bBLG AB, which confirms that these variants were correctly folded initially and were not unfolded upon reduction and removal of the disulfide bonds. Additionally, the analysis of the reduced native bBLG variants shows higher molecular weight oligomer bands that might indicate the formation of non-native intermolecular disulfide bonds. Native PAGE is performed at high pH of 8.8 under which thiols are mostly deprotonated and thereby highly reactive [35].

3.3.3 Thermal Denaturation of Recombinant BLG Variants

To analyze the influence of heat treatment on the secondary and tertiary structure of the different BLG variants, CD spectroscopy and native PAGE were used to monitor aggregation behavior and changes in secondary structure at temperatures ranging from 20 to 90 °C. These data can provide information on the stability and conformational changes that occur upon heating [20].

At temperatures between 20 and 70 °C, a strong monomer band and a much weaker upper dimer band could be detected on a native gel for rBLG (Fig. 7). The low dimer content can be explained by the pH of 8.8 in the gel, which shifts the monomer–dimer equilibrium toward dimer dissociation as already shown above [33]. At 80 and 90 °C a very weak third band appears above the dimer band possibly caused by the initial aggregation of the protein through the formation of intermolecular disulfide bonds. Exposure of the free cysteine (Cys121) leads to non-native dimers, which can be recognized by a slight shift in the dimer band [36]. In addition, larger aggregates form when monomers and dimers oligomerize. These observations are consistent with the results of de Wit when bovine BLG is heated [37]. At 90 °C, the monomer band is only weakly present and the intensity of the upper bands increase. Progressive unfolding might lead to increasing exposure of the free thiol and thus to increasing aggregation, which could explain the decrease of the monomer band at 90 °C. At the secondary level, rBLG is dominated by β-sheet structures (Figs. 4d and 7), which steadily decrease above 50 °C (Fig. 7). Between 50 and 60 °C, the transition of the protein into the more flexible "molten globule" state occurs, as described by de Wit for native bovine BLG [37]. This transition is reflected by a decrease of β-sheets and an increase of disordered structures. At a temperature of 70 °C, disordered structures already dominate (48%), while the β-sheet fraction decreases to 31%. At 90 °C, further unfolding of the protein could be detected by the increase of the disordered structures to 54% (Fig. 7). However, more than 40% of the protein still exhibits defined secondary structures, suggesting that BLG has some basic stability due to certain structural motifs. According to de Wit and Klarenbeek, a complete transition to the unfolded state does not occur until a temperature of 130 °C is reached [38].

Substitution of the free cysteine (Cys121) by alanine leads to constant monomer/dimer bands as temperature is increased (Fig. 7), confirming the results of Jayat et al., who showed that the free thiol group is responsible for the irreversible aggregation of BLG [29]. Within the secondary structures, only small deviations from the rBLG can be observed at a temperature of 30 °C. As for rBLG, a further increase in temperature leads to an increase in random coil structures with a concomitant decrease in β-sheet structures (Fig. 7).

Removal of the outer disulfide bond in rBLG–SS leads to a significant increase in disordered structures (36%) with a simultaneous decrease of other structural elements (Fig. 4d). However, a β-sheet proportion of 35% indicates, that the main β-barrel structure was preserved, which was already described by Yamada et al. for equine BLG [39]. Besides, a slight opening of the globular structure was detected using intrinsic fluorescence and native PAGE analysis (Figs. 5b and 6). This opening of the tertiary structure leads to a reduced stability of the protein, which explains the unfolding at lower temperatures. Here, a distinct reduction of the secondary structure elements can already be observed at 50 °C, which was only observed at 70 °C for rBLG. At 70 °C the rBLG–SS variant was completely unfolded (Fig. 7). As a consequence of destabilization, the free cysteine residue seems to be exposed even at the relatively low temperature of 60 °C, facilitating the formation of intermolecular disulfide bonds. This process leads to the aggregation of proteins, which is observable as high molecular weight bands on the native gel (Fig. 7).

Upon removal of the inner disulfide bond and the free thiol group, a very diffused band was observed on the native gel (Figs. 6 and 7), indicating a lack of stability due to the absence of secondary structure elements in this cysteine-free variant. As this variant does not contain any cysteine residue, the formation of new intermolecular disulfide bonds within and between proteins is not possible. However, an increasing temperature leads to the progressive formation of insoluble aggregates, visible as high molecular weight traces in the gel pockets indicated by a decrease in the intensity of the monomer band (Fig. 7). rBLG–C predominantly showed random coil structures (54–60%) over the entire temperature range between 30 and 90 °C. Upon heating, only minor fluctuations of the secondary structure were detected (Fig. 7). The increase in β-sheets at 50 °C (+12%) might be attributed to the initial formation of intermolecular β-sheets. The removal of intramolecular disulfide bonds as in

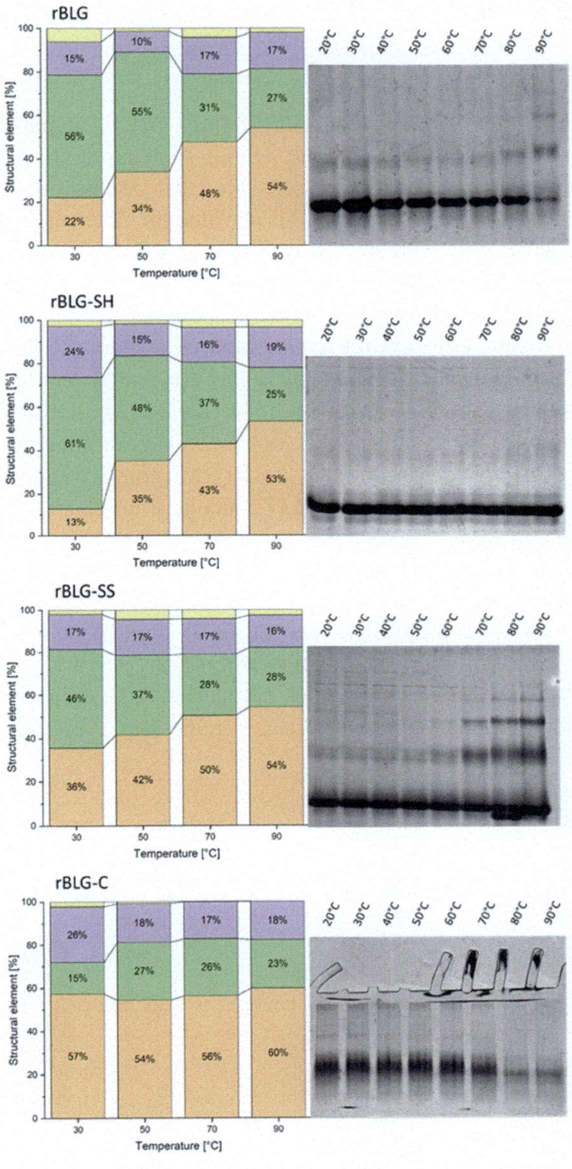

Fig. 7 **Diagrams**: Secondary structure elements (in %) (α-helix (violet), β-sheets (green), random coil (orange), β-loops (yellow)) of recombinant BLG variants rBLG, rBLG-SH, rBLG-SS, and rBLG-C at different temperatures; derived from CD analyses. **Gel images**: Native PAGE of recombinant BLG variants rBLG, rBLG–SH, rBLG–SS, and rBLG–C at temperatures between 20 and 90 °C. 1 mg mL^{-1} of protein was heated for 5 min at the indicated temperatures values

rBLG–C leads to an increased structural flexibility and accelerates the α-helix → β-sheet transition, as described by Hoppenreijs et al. [40].

3.4 *Influence of Cysteine-Residues on BLG Variants During Ligand Binding*

As a member of the lipocalin family, BLG is capable of binding and transporting small, mainly hydrophobic ligands such as retinol [4]. Due to its structure, retinol has an extended system of conjugated double bonds [41], which can be excited to fluorescence at a specific wavelength. The intensity of fluorescence is thereby significantly increased by the binding of retinol to proteins [42]. This property helps to analyze the impact of cysteine residues present in the different recombinant BLG variants on the retinol binding property.

For bBLG B and the recombinant equivalent rBLG an initial continuous increase in fluorescence with increasing retinol concentration (0–75 μM) was detected (Fig. 8a), indicating a binding of the molecule within the barrel-like binding pocket of the protein. Within this pocket, one molecule of retinol per BLG molecule can be bound [4]. Above a concentration of 30 μM retinol, a saturation of fluorescence intensity occurred (Fig. 8a). The BLG concentration of 1 mg mL^{-1} corresponds to approximately 50 μM, which is consistent with the retinol concentration at which the linear slope begins, indicating that above this concentration, all binding pockets should be occupied.

Compared to rBLG and bBLG B, slightly increased fluorescence was observed for rBLG-SH, although this variant is structurally almost identical to the wild-type variants apart from a slight destabilization (see above). While Yagi et al. could not detect any differences in the retinol binding behavior of Cys121 mutants

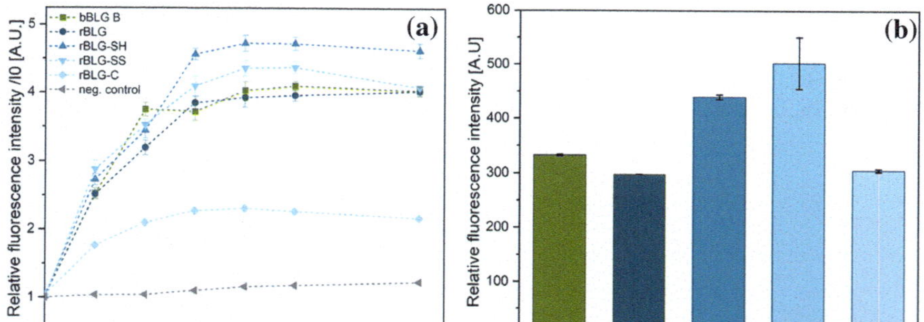

Fig. 8 **(a)** Increase in retinol fluorescence. Retinol was added stepwise to 1 mg mL^{-1} of recombinant BLG variants and bBLG solubilized in sodium phosphate buffer (10 mM, pH 7). Samples were excited at 480 nm and emission was measured at 330 nm. Negative (neg.) control: reaction buffer. **(b)** ANS (1-anilino-8-naphthalenesulfonate) fluorescence. Protein solutions of recombinant BLG and bBLG (100 μg mL^{-1}) supplemented with 80 μM ANS were excited at 390 nm and emission was measured at 370 nm

(C121A, C121S, C121V) [43], Jayat et al. demonstrated enhanced ANS (1-anilino-8-naphthalenesulfonate) binding to a C121S variant due to increased surface hydrophobicity [29]. This increase of surface hydrophobicity as shown in Fig. 8b was confirmed by Brune et al. [23]. It was suggested that removal of the free thiol group leads to a more flexible structure in which hydrophobic sites are more accessible to ANS. With a more mobile structure, potential binding sites of retinol in the calyx or on the protein surface could be more accessible which could explain the minor increase in retinol fluorescence [23].

Removal of the outer disulfide bond (rBLG–SS) also leads to a slightly increased fluorescence due to an opening of the globular structure and resulting in an increase of the accessibility of the calyx. This result is in agreement with the ANS fluorescence study where a slightly higher surface hydrophobicity for rBLG–SS was detected (Fig. 8b). For other lipocalins, including human tear lipocalin, similar results have been shown. Removal of the highly conserved disulfide bond (Cys61–Cys153) by recombinant methods resulted in the proper formation of the calyx, leading to increased structural flexibility and better binding of ligands. Studies have shown that this disulfide bond plays a critical role in regulating the accessibility and structural flexibility of the calyx, which in turn controls receptor-ligand interactions in human tear lipocalins [44].

Finally, the removal of all cysteine residues (rBLG–C) resulted in only half the retinol fluorescence enhancement compared to all other recombinant variants. The inner disulfide bond thus seems to be essential for the proper formation of the hydrophobic calyx. Nevertheless, retinol still seems to bind to the protein as an increase in fluorescence is detectable compared to the negative control where no BLG was added (Fig. 8a).

4 Conclusions

The production and purification strategies for different recombinant BLG variants were successfully established. Depending on the variant, up to 400 mg of pure BLG from one liter culture broth could be purified. Recombinant bovine BLG B (rBLG) produced with *E. coli* was found to be structurally and in its physicochemical and emulsifying properties very similar to native bovine BLG B (bBLG B). Thus, from a scientific point of view, recombinant BLG produced in *E. coli* can be used as a substitute for bovine BLG for further structure–function analysis.

Based on the already established production system, BLG variants with different cysteine substitutions were produced and purified to analyze the impact of disulfide bonds and of the free thiol group on the protein structure. Removal of the free cysteine (rBLG–SH) demonstrated no effect on secondary or tertiary structure of the protein, while substitution of the outer disulfide bond (rBLG–SS) resulted in a minor decrease in β-sheet structures, leading to an opening of the globular structure. Thermal denaturation experiments also revealed a destabilization of rBLG–SS, which was accompanied by an earlier unfolding and formation of intermolecular disulfide bonds at lower temperatures. Removal of all cysteine residues (rBLG–C) strongly disrupts the major secondary structure, resulting in further unfolding and aggregation of the protein in *E. coli* cells as ncIBs. While rBLG–C was successfully refolded and showed mostly disordered structures, despite some of the secondary and tertiary elements still being present, chemically reduced BLG variants (bBLG R1 and R2) showed only minor structural differences compared to bovine BLG.

In summary, the characterization of the BLG variants presented here [22, 23, 25] provides a robust foundation for subsequent investigations aimed at identifying protein behavior that yields improved functionality. For instance, the unfolded and flexible rBLG–C can generate functional fibrillar structures upon heating at pH 7.0, which is unattainable with bovine BLG due to its constrained conformation [40, 45].

Acknowledgements The authors gratefully acknowledge financial support by the German Research Foundation (DFG) in the Priority Programme 1934 *DiSPBiotech—Dispersity, structural and phase modifications of proteins and biological agglomerates in biotechnological processes* (SPP 1934 DiSPBiotech—315460011).

References

1. Rischer H, Szilvay GR, Oksman-Caldentey K-M (2020) Cellular agriculture - industrial biotechnology for food and materials. Curr Opin Biotechnol 61:128–134. https://doi.org/10.1016/j.copbio.2019.12.003
2. Hettinga K, Bijl E (2022) Can recombinant milk proteins replace those produced by animals? Curr Opin Biotechnol 75:102690. https://doi.org/10.1016/j.copbio.2022.102690
3. de Wit JN (1997) Nutritional and functional characteristics of whey proteins in food products. J Dairy Sci 81(3):597–608. https://doi.org/10.3168/jds.S0022-0302(98)75613-9
4. Kontopidis G, Holt C, Sawyer L (2004) Invited review: beta-lactoglobulin: binding properties, structure, and function. J Dairy Sci 87(4):785–796 https://doi.org/10.3168/jds.S0022-0302(04)73222-1
5. Sawyer L (2013) β-Lactoglobulin. In: McSweeney PH, Fox PF (eds) Advanced Dairy Chemistry. Springer, US, Boston, MA, pp 211–259
6. Broersen K (2020) Milk processing affects structure, bioavailability and immunogenicity of β-lactoglobulin. Foods 9(7). https://doi.org/10.3390/foods9070874
7. Farrell Jr H, Jimenez-Flores R, Bleck GT, Brown EM, Butler JE, Creamer LK, Hicks CL, Hollar CM, Ng-Kwai-Hang KF, Swaisgood HE (2004) Nomenclature of the proteins of cows' milk-sixth revision. J Dairy Sci 87(6):1641–1674. https://doi.org/10.3168/jds.S0022-0302(04)73319-6
8. Loveday SM, Sawyer L (2016) β-Lactoglobulin: nomenclature, structure, and function. https://doi.org/10.1016/B978-0-08-100596-5.00940-9
9. Pettersen EF, Goddard TD, Huang CC, Meng EC, Couch GS, Croll TI, Morris JH, Ferrin TE (2021) UCSF ChimeraX: Structure visualization for researchers, educators, and developers. Protein Sci 30(1):70–82. https://doi.org/10.1002/pro.3943
10. Hoppenreijs L, Fitzner L, Ruhmlieb T, Heyn TR, Schild K, van der Goot A-J, Boom RM, Steffen-Heins A, Schwarz K, Keppler JK (2022) Engineering amyloid and amyloid-like morphologies of β-lactoglobulin. Food Hydrocoll 124:107301. https://doi.org/10.1016/j.foodhyd.2021.107301
11. Stojadinovic M, Burazer L, Ercili-Cura D, Sancho A, Buchert J, Cirkovic Velickovic T, Stanic-Vucinic D (2012) One-step method for isolation and purification of native β-lactoglobulin from bovine whey. J Sci Food Agric 92(7):1432–1440. https://doi.org/10.1002/jsfa.4722

12. Kristiansen KR, Otte J, Impsen R, Qvist KB (1998) Large-scale preparation of beta-Lactoglobulin A and B by ultrafiltration and ion-exchange Chromatography. Int Dairy J 8(2):113–118. https://doi.org/10.1016/S0958-6946(98)00028-4
13. Bonarek P, Loch JI, Tworzydło M, Cooper DR, Milto K, Wróbel P, Kurpiewska K, Lewiński K (2020) Structure-based design approach to rational site-directed mutagenesis of β-lactoglobulin. J Struct Biol 210(2):107493. https://doi.org/10.1016/j.jsb.2020.107493
14. Totsuka M, Katakura Y, Shimizu M, Kumagai I, Miura K, Kaminogawa S (1990) Expression and secretion of bovine beta-lactoglobulin in *Saccharomyces cerevisiae.* Agric Biol Chem 54(12):3111–3116
15. Kim TR, Goto Y, Hirota N, Kuwata K, Denton H, Wu S-Y, Sawyer L, Batt CA (1997) High-level expression of bovine beta-lactoglobulin in *Pichia pastoris* and characterization of physical properties. Protein Eng. 10(11):1339–1345. https://doi.org/10.1093/protein/10.11.1339
16. Aro N, Ercili-Cura D, Andberg M, Silventoinen P, Lille M, Hosia W, Nordlund E, Landowski CP (2023) Production of bovine beta-lactoglobulin and hen egg ovalbumin by *Trichoderma reesei* using precision fermentation technology and testing of their techno-functional properties. Food Res Int 163:112131. https://doi.org/10.1016/j.foodres.2022.112131
17. Batt CA, Rabson LD, Wong WS, Kinsella JE (1990) Expression of recombinant bovine beta-lactoglobulin in *Escherichia coli.* Agric Biol Chem 54(4):949–955. https://doi.org/10.1080/00021369.1990.10870080
18. Bessette PH, Aslund F, Beckwith J, Georgiou G (1999) Efficient folding of proteins with multiple disulfide bonds in the *Escherichia coli* cytoplasm. Proc Natl Acad Sci U S A 96(24):13703–13708. https://doi.org/10.1073/pnas.96.24.13703
19. Ponniah K, Loo TS, Edwards PJB, Pascal SM, Jameson GB, Norris GE (2010) The production of soluble and correctly folded recombinant bovine beta-lactoglobulin variants A and B in *Escherichia coli* for NMR studies. Protein Expr Purif 70(2):283–289. https://doi.org/10.1016/j.pep.2009.12.006
20. Chen J, Song JL, Zhang S, Wang Y, Cui DF, Wang CC (1999) Chaperone activity of DsbC. J Biol Chem 274(28):19601–19605. https://doi.org/10.1074/jbc.274.28.19601
21. Kadokura H, Beckwith J (2010) Mechanisms of oxidative protein folding in the bacterial cell envelope. Antioxid Redox Signal 13(8):1231–1246. https://doi.org/10.1089/ars.2010.3187
22. Loch JI, Bonarek P, Tworzydło M, Polit A, Hawro B, Łach A, Ludwin E, Lewiński K (2016) Engineered β-lactoglobulin produced in *E. coli*: purification, biophysical and structural Characterisation. Mol Biotechnol 58(10):605–618. https://doi.org/10.1007/s12033-016-9960-z
23. Brune SE, Hoppenreijs LJ, Kühl T, Lautenbach V, Walter J, Peukert W, Schwarz K, Imhof D, Boom RM, Krull R, Keppler JK, Biedendieck R (2023) Precision fermentation as a route to modify β-lactoglobulin structure through substitution of specific cysteine residues. Int Dairy J, 105772. https://doi.org/10.1016/j.idairyj.2023.105772
24. Righetti P, Gianazza E, Gelfi C, Chairi M (ed) (1990) Gel electrophoresis of proteins: A practical approach, 2nd ed. Oxford University Press, Oxford
25. Keppler JK, Heyse A, Scheidler E, Uttinger MJ, Fitzner L, Jandt U, Heyn TR, Lautenbach V, Loch JI, Lohr J, Kieserling H, Günther G, Kempf E, Grosch J-H, Lewiński K, Jahn D, Lübbert C, Peukert W, Kulozik U, Drusch S, Krull R, Schwarz K, Biedendieck R (2021) Towards recombinantly produced milk proteins: Physicochemical and emulsifying properties of engineered whey protein beta-lactoglobulin variants. Food Hydrocoll 110:106132. https://doi.org/10.1016/j.foodhyd.2020.106132
26. Trinh NTM, Thuoc TL, Thao DTP (2021) Production of recombinant human G-CSF from non-classical inclusion bodies in *Escherichia coli.* Braz J Microbiol 52(2):541–546. https://doi.org/10.1007/s42770-020-00413-y
27. Yan Y, Seeman D, Zheng B, Kizilay E, Xu Y, Dubin PL (2013) PH-Dependent aggregation and disaggregation of native β-lactoglobulin in low salt. Langmuir 29(14):4584–4593. https://doi.org/10.1021/la400258r
28. Miles AJ, Janes RW, Wallace BA (2021) Tools and methods for circular dichroism spectroscopy of proteins: a tutorial review. Chem Soc Rev 50(15):8400–8413. https://doi.org/10.1039/d0cs00558d

29. Jayat D, Gaudin J-C, Chobert J-M, Burova TV, Holt C, McNae I, Sawyer L, Haertlé T (2004) A recombinant C121S mutant of bovine beta-lactoglobulin is more susceptible to peptic digestion and to denaturation by reducing agents and heating. Biochemistry 43(20):6312–6321. https://doi.org/10.1021/bi0362469
30. Hamada D, Segawa S, Goto Y (1996) Non-native alpha-helical intermediate in the refolding of beta-lactoglobulin, a predominantly beta-sheet protein. Nat Struc Biol. 10:868–873. https://doi.org/10.1038/nsb1096-868
31. Vivian JT, Callis PR (2001) Mechanisms of tryptophan fluorescence shifts in proteins. Biophys J 80(5):2093–2109. https://doi.org/10.1016/S0006-3495(01)76183-8
32. Albani JR, Vogelaer J, Bretesche L, Kmiecik D (2014) Tryptophan 19 residue is the origin of bovine β-lactoglobulin fluorescence. J Pharm Biomed Anal 91:144–150. https://doi.org/10.1016/j.jpba.2013.12.015
33. Mercadante D, Melton LD, Norris GE, Loo TS, Williams MAK, Dobson RCJ, Jameson GB (2012) Bovine β-lactoglobulin is dimeric under imitative physiological conditions: dissociation equilibrium and rate constants over the pH range of 2.5–7.5. Biophys J 103(2):303–312. https://doi.org/10.1016/j.bpj.2012.05.041
34. Chen WL, Hwang MT, Liau CY, Ho JC, Hong KC, Mao SJT (2005) Beta-Lactoglobulin is a thermal marker in processed milk as studied by electrophoresis and circular dichroic spectra. J Dairy Sci 88(5):1618–1630. https://doi.org/10.3168/jds.S0022-0302(05)72833-2
35. Surroca Y, Haverkamp J, Heck A (2002) Towards the understanding of molecular mechanisms in the early stages of heat-induced aggregation of beta-lactoglobulin AB. J Chrom 970(1–2):275–285. https://doi.org/10.1016/s0021-9673(02)00884-1
36. Peram MR, Loveday SM, Ye A, Singh H (2013) *In vitro* gastric digestion of heat-induced aggregates of β-lactoglobulin. J Dairy Sci 96(1):63–74. https://doi.org/10.3168/jds.2012-5896
37. Wit JN de (2009) Thermal behaviour of bovine β-lactoglobulin at temperatures up to 150°C. a review. Trends Food Sci Technol 20(1):27–34. https://doi.org/10.1016/j.tifs.2008.09.012
38. de Wit JN, Klarenbeek G (1980) A differential scanning calorimetric study of the thermal behaviour of bovine β-lactoglobulin at temperatures up to 160 °C. Biochim Biophys Acta 624(1):40–50. https://doi.org/10.1016/0005-2795(80)90223-8
39. Yamada Y, Nakagawa K, Yajima T, Saito K, Tokushima A, Fujiwara K, Ikeguchi M (2006) Structural and thermodynamic consequences of removal of a conserved disulfide bond from equine beta-lactoglobulin. Proteins 63(3):595–602. https://doi.org/10.1002/prot.20905
40. Hoppenreijs LJG, Overbeck A, Brune SE, Biedendieck R, Kwade A, Krull R, Boom RM, Keppler JK (2023) Amyloid-like aggregation of recombinant β-lactoglobulin at pH 3.5 and 7.0: Is disulfide bond removal the key to fibrillation? Int J Biol Macromol 242(Pt 2):124855. https://doi.org/10.1016/j.ijbiomac.2023.124855
41. Mensi A, Choiset Y, Rabesona H, Haertlé T, Borel P, Chobert J-M (2013) Interactions of β-lactoglobulin variants A and B with vitamin A. Competitive binding of retinoids and carotenoids. J Agric Food Chem 61(17) 4:114–4119. https://doi.org/10.1021/jf400711d
42. Futterman S, Heller J (1972) The enhancement of fluorescence and the decreased susceptibility to enzymatic oxidation of retinol Complexed with Bovine Serum Albumin, β-Lactoglobulin, and the Retinol-binding Protein of Human Plasma. J Biol Chem 247(16):5168–5172. https://doi.org/10.1016/S0021-9258(19)44953-3
43. Yagi M, Sakurai K, Kalidas C, Batt CA, Goto Y (2003) Reversible unfolding of bovine beta-lactoglobulin mutants without a free thiol group. J Biol Chem 278(47):47009–47015. https://doi.org/10.1074/jbc.M308592200
44. Gasymov OK, Abduragimov AR, Glasgow BJ (2008) Ligand binding site of tear lipocalin: contribution of a trigonal cluster of charged residues probed by 8-anilino-1-naphthalenesulfonic acid. Biochemistry 47(5):1414–1424. https://doi.org/10.1021/bi701955e
45. Hoppenreijs LJ, Brune SE, Biedendieck R, Krull R, Boom RM, Keppler JK (2023) Fibrillation of β-lactoglobulin at pH 2.0: Impact of cysteine substitution and disulfide bond reduction. Food Hydrocoll 141:108727. https://doi.org/10.1016/j.foodhyd.2023.108727

(Amyloid) Protein Aggregates from β-Lactoglobulin and Their Behavior Along the Process Chain

Julia K. Keppler, Timon R. Heyn, Jacqueline Lux, Therese Ruhmlieb, Laura Meissner, Loes J. G. Hoppenreijs, Anja Steffen-Heins, and Karin Schwarz

Abstract Amyloid aggregates are anisotropic protein aggregates that comprise a recurrent alpha-helix or beta-sheet motif. These aggregates can vary in shapes or morphologies ranging from rigid, semi-flexible fibrils with a high aspect ratio to short, worm-like or curly and flexible structures. Their specific properties lead to enhanced functionalities suitable for food and non-food industries. However, exploration of their processing stability is currently insufficient. We hypothesize, that the (food) processing chain leads to an overlapping of assembly and disassembly mechanisms of amyloid structures within a process, which can have positive and negative effects on the resulting aggregate structure, its functionality as well as its formation kinetics. To produce protein aggregates with particular morphologies and target functionality we investigated the assembly process of proteins from the nanoscale (protein structures)

J. K. Keppler (✉) · L. J. G. Hoppenreijs
Laboratoryof Food Process Engineering, Wageningen University and Research, WG Wageningen 6708, Netherlands
e-mail: julia.keppler@wur.nl

L. J. G. Hoppenreijs
e-mail: loes.hoppenreijs@gmail.com

J. K. Keppler · T. R. Heyn · J. Lux · T. Ruhmlieb · L. Meissner · A. Steffen-Heins · K. Schwarz
Division of Food Technology, Kiel University, Kiel, Germany
e-mail: theyn@foodtech.uni-kiel.de

J. Lux
e-mail: jacqueline.lux@ki.tng.de

T. Ruhmlieb
e-mail: truhmlieb@foodtech.uni-kiel.de

L. Meissner
e-mail: lfitzner@foodtech.uni-kiel.de

A. Steffen-Heins
e-mail: aheins@foodtech.uni-kiel.de

K. Schwarz
e-mail: kschwarz-2@foodtech.uni-kiel.de

A. Kwade and I. Kampen (eds.), *Dispersity, Structure and Phase Changes of Proteins and Bio Agglomerates in Biotechnological Processes*,
https://doi.org/10.1007/978-3-031-63164-1_7

to the mesoscale (aggregate morphologies) to the macroscale (network formation) across the processing chain of the production of foams, emulsions and films. These results will allow us to better control the formation and process behavior of these functional aggregates leading to targeted applications.

Keywords Fibrils · Amyloid aggregates · Beta-Lactoglobulin · Protein modification · Process conditions

1 Introduction

Amyloid structures usually have a recurrent beta-sheet motif and are primarily anisotropic, although some alpha helix fibrils have been reported as well. They range from semi-flexible, micro-metre-long fibrils to smaller and less densely packed curly- or worm-like aggregates, also known as amyloid-like because of their higher flexibility and (thus far) not well characterized underlying structure. Amyloid protein structures can be formed through directed interactions under denaturing conditions by almost any protein, including the whey protein beta-lactoglobulin (BLG). BLG was selected as a model protein for this study as it is an abundantly available food protein and well characterized, especially with respect to amyloid aggregation.

The formation of amyloid structures for BLG is mainly dependent on the protein concentration and a partial or complete unfolding of the protein at low ionic strength and low pH value (< 3.5). This unfolding can be attributed to the process itself, e.g. the influence of high temperatures [1] microwave ultrasound [2], denaturing substances such as urea [3], combinations of stirring and high temperatures [4] and interactions at an interface [5–7]. In many biotechnological processes, such as homogenization and spray drying, protein unfolding occurs during the process due to high shear rates and temperatures, as well as potentially due to contacts with different interfaces (liquid/liquid). However, the extent to which various process-related parameters influence amyloid aggregation has not been addressed in much detail and requires further clarification. On the one hand, uncontrolled amyloid aggregation during processing can lead to undesirable protein losses and process hindrances; on the other hand, the targeted production of amyloid protein aggregates leads to useful functional ingredients. In fact, the properties of amyloid protein aggregates have captured growing attention in biotechnology and food process engineering over recent years [8]. BLG fibrils exhibit strong film- and structure-forming properties and attach to interfaces in a stabilizing manner in disperse systems [9]. Furthermore, compared to the non-fibrillar forms [9, 10], they exhibit increased viscosity and gel-forming ability. Therefore, utilizing in creating innovative carrier materials for pharmaceuticals and nutraceuticals offers benefits [9, 11]. However, their formulation processes stability is limited. High shear rates (300 s^{-1}) [6], high pressure treatment (>50 MPa) [12] and temperatures (>120 °C for 8 h) [1] break intact fibrils into shorter fragments with potentially different functionality. Furthermore, fragmentation of fibrils can result in so-called 'seeds' which can re-initiate amyloid aggregation [6].

The food processing chain can lead to a superposition of assembly and disassembly mechanisms of amyloid structures within a single process, with both positive or negative consequences for the resulting functionality. Since there is not much known about these factors, studies on amyloid aggregate formation from BLG and its properties in model systems are presented along the process chain with the aim of deepening our understanding of aggregate formation at various pH values by systematically investigating the formation and fragmentation. The formation of amyloid aggregates and their analysis are described in Sect. 2.1. For the analysis of aggregates and their building blocks we established a wide array of methods, including new analytical approaches (Sect. 2.2). Methods based on various principles of spectroscopy, imaging techniques and chemical analyses enabled us to explore aggregate systems from the molecular level (Sect. 3) up to the meso-, (Sect. 4) and macroscale (Sect. 5), i.e. from individual building blocks (proteins, peptides), and aggregates (amyloid and amyloid-like) up to entire model systems (foams, emulsions and films). Using independent analytical approaches in cooperation with research groups of the priority program (SPP1934) provides the opportunity to verify the findings of the presented study.

2 Formation and Analysis of Amyloid Aggregates

The formation and main characteristics of amyloid aggregates are summarized in Sect. 2.1. The analysis and challenges in characterizing amyloid aggregates are further described in Sect. 2.2. A novel method for analysing engineered amyloid structures based on EPR spin labelling, inspired by approaches used for natural amyloids, is introduced in Sect. 2.3.

2.1 Amyloid Aggregates

The usual conditions for engineered fibrils formation includes heating (>70 °C) a BLG solution of ~1–4 wt% at pH 2 for several hours at low ionic strength [13, 14]. Thereby, intact proteins are hydrolysed and peptides are released (Fig. 1). Subsequently, specific (amyloid-prone) peptides assemble through intermolecular β-sheets, forming straight and semi-flexible amyloid fibrils [15]. Other conditions such as enzymatic hydrolysis can also be applied to release these amyloid-prone peptide sequences and form fibrils at lower temperature and acidic pH value [15]. Amyloid-prone sequences are hydrophobic and low in charge, and thereby mainly N-terminal peptide sequences 1–33 and 1–53 of BLG are included in amyloid aggregation [13, 15]. Amyloid aggregates are formed through a nucleation and growth mechanism, which is kinetically analogous to crystallisation [16]. Thereby, the addition of amyloid-prone peptides can act as nucleation sites and accelerate amyloid formation [17]. Since amyloids are long and semi-flexible, they tend to align and

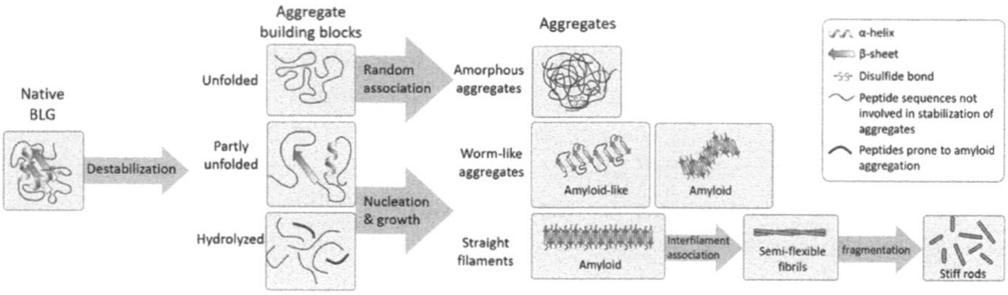

Fig. 1 Different thermally induced amyloid morphologies ('fibrils' or 'worm-like') and their presumed association according to Hoppenreijs et al. [20]. Figure reused with permission from Elsevier

subsequently associate into bundles. The number of filaments in the mature fibril depends on the environmental conditions, as well as the pitch of the twisted filaments [18]. In contrast to fibrils, worm-like aggregates are built from intact or less hydrolysed, partially unfolded BLG under lower electrostatic repulsion. These conditions often include higher pH values (e.g. pH 3.5), high protein concentrations (e.g. ~7.5wt%; [13]), or higher ionic strength (e.g. >50 mM sodium chloride [19]). Due to the lowered repulsion, β-sheet stacking is less specific and the aggregates are less dense and more flexible. Also, they are shorter due to the enhanced aggregation kinetics. Their lower aspect ratio could potentially lead to distinct functionalities compared to fibrils, but this remains to be examined.

2.2 *Methods to Analyse Amyloids and Problem of Polydispersity*

Functional amyloid aggregates rarely have a complete conversion from native protein to aggregates. Often final conversion rates only reach 35% for fibrils at pH 2 and 80% for worm-like aggregates at pH 3.5 [21] after 5 h of incubation. However, this ratio of native, denatured, hydrolysed and partially or fully aggregated proteins changes constantly during the whole duration of the incubation period. Here, we aimed to understand if and how the polydisperse nature of these sample impacts their analysis and furthermore, we also wanted to know the duriation time required to form initial aggregates during the process. To investigate this, a range of typical methods for amyloid aggregation was established and tested with the polydisperse samples as well as isolated amyloid and non-amyloid fractions generated by separating these fractions via filtration at various time points (Fig. 4).

Usually, the first analysis method to examine the morphology of the aggregates is through the use microscopy. Besides transmission electron microscopy (TEM) [1], atomic force microscopy (AFM) is a common method to detect and describe the morphology of fibrils [18]. The measurement principle of AFM is based on the use of a cantilever to measure amyloid aggregates deposited and dried on a mica surface [9] by scanning the surface. This technique was used to study the formation of specific fibril dimensions and to detect the presence or absence of aggregates in the amyloid and non-amyloid fractions isolated after ultrafiltration. AFM studies showed that straight or worm-like aggregates formed within 5 h in a pH-dependent manner and that these could be successfully separated from the non-aggregated fraction (filtrate) by ultrafiltration (Fig. 4).

However, the protein particles in the filtrate were too small to detect with TEM or AFM to provide a concrete estimate of the particle size [21]. At the same time, it was expected that the drying process on the mica surface would lead to a change in the aggregate dimension compared to the initial morphology of the aggregates present in solution. Therefore, the samples were additionally analysed before and after isolation by Small Angle X-ray Scattering (SAXS) in a liquid sample to define and investigate further properties regarding the morphology (Contour Length and Kuhn Length) of the isolated amyloid and non-amyloid fractions [22]. For this purpose, the analysis of the scattering angle and intensity of X-rays provided information about the structural dimensions of the aggregates. The analysis by SAXS confirmed the protein dimensions of isolated fibrils as measured by AFM and concretised them for the non-amyloid fractions (Table 1), but also showed that it is necessary to isolate the heated protein solutions to prepare a monodisperse sample. Thus, the polydisperse nature of the sample greatly affected the interpretation of the SAXS data.

AFM and SAXS do not provide reliable data on the amount of amyloid structures formed. To answer the question about the formation kinetics of amyloid structures during the 5 h incubation, another analysis method was applied. The commonly used ThT assay according to Loveday et al. [1] was used to quantify protein aggregation in the solution during the incubation time. ThT dye binds to the specific beta-sheet structures of the aggregates, increasing the fluorescence of the ThT molecule. The amount of amyloid aggregates correlates with the amount of bound ThT molecules and therefore the measured fluorescence intensity [23]. It was observed that the number of amyloid aggregates in a 2.5% BLG solution increased sigmoidally over a period of 5 h at 90 °C and pH 2 when not stirred, but had a shorter lag phase when stirred at 350 min^{-1} (Fig. 2a). Subsequent isolation of the amyloid aggregates by ultra-filtration (300 kDa) and measurements of the protein concentration in the

Table 1 SAXS model parameters determined at different time intervals of amyloid aggregation and fractions after ultrafiltration with a 300 kDa membrane according to Heyn et al. [21]. <L> : The mean length of amyloid aggregate is given in nm; Elliptical cross-section (A) and axis ratio (y). Table reused with permission from Elsevier, with slight modifications

Incubation time [h] and fractions	Mean number of monomers per fibril	<L>	Small semiaxis A [nm] and axis ratio y [nm]
0	1.3	4.0 ± 0.2	1.7 ± 0.1, 1
4	130	550 ± 50	1.45 ± 0.1, 1
Retentate	> 2000	> 2000	1.4 ± 0.1, 1.2 ± 0.1
Filtrate	2.1	4.4 + –0.4	1.7 ± 0.1, 1

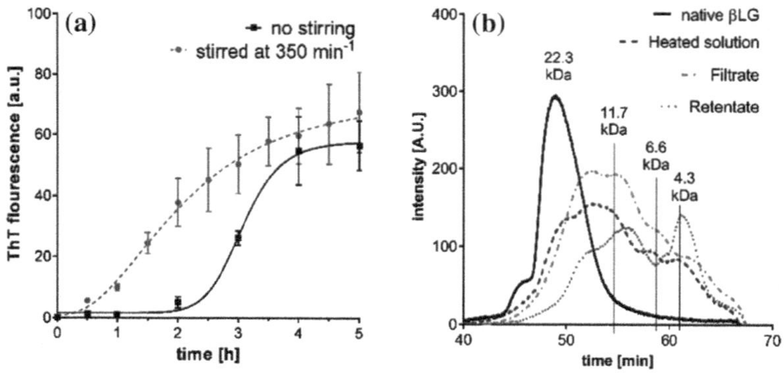

Fig. 2 **a** THT fluorescence of beta-lactoglobulin during incubation at 90 °C, pH 2 and 0 and 350 min^{-1} according to Heyn et al. [21]. **b** Size exclusion chromatogram of in TRIS–HCL eluent dissociated beta-lactoglobulin before (black line) and after incubation at 90 °C and pH 2 as well as ultrafiltration. Figure reused with permission from Elsevier, with slight modifications: Ref. [21]

retentate [9, 21] confirmed the correlation of the amount of aggregates in the solution with the ThT fluorescence, thus this method was not affected by the polydisperse nature of the protein solutions.

To investigate the onset of fibril formation, its different building blocks, and to receive information about the differences of other aggregate building blocks or non-aggregate material in the solutions, SEC was performed under dissociating conditions according to the method of Akkerman et al. [15]. The studies of the heated and unheated protein solutions showed well that acid hydrolysis of BLG occurred during fibril synthesis (Fig. 2b). The analysis of samples in which amyloid aggregates were separated from the other peptide components by ultrafiltration showed that mainly hydrolysed material is incorporated into the fibrils (retentate). Here, specific peptide sizes (e.g. 4.3 kDa) appeared to be more prevalent in the fibrils than in non-aggregated material (filtrate) (Fig. 2b). The differences in the composition of amyloid aggregates and non-aggregated material was also confirmed using AUC-SV experiments [21, 24].

These results led to the hypothesis that the different protein fractions also had different internal structure. The differences in the structure of amyloid aggregated and non-aggregated material were effectively demonstrated by infrared (ATR-FTIR) and fluorescence (intr. Trp-Fluorescence) spectroscopy after separation by ultrafiltration. The FTIR reveals information about the secondary structure elements of the proteins via the specific observation of the amide I band [14]. The results showed that during fibril formation, native structures such as alpha-helix (1655 cm^{-1}) and intramolecular beta sheets (1632 cm^{-1}) decreased and amyloid aggregates absorbed strongly, especially in the region of 1622 cm^{-1} [21]. This result indicated a high density of intermolecular beta-sheets in the aggregated material, which were weakly expressed in the non-THT-reactive material.

Whether these differences in secondary structure also affect protein conformation was investigated using tryptophan fluorescence. The intrinsic tryptophan fluorescence dependents on the microenvironment around the tryptophan residues located in the protein [25]. The measurements of the protein solutions were performed with a fluorescence spectrophotometry at an excitation of 294 nm and a measurement of the emission between 300 and 500 nm. The observed shift of the absorption maximum indicates the superposition of different effects (e.g. unfolding, acid hydrolysis, amyloid aggregation and altered proportions of non-amyloid material). For the measurements of the isolated aggregate fractions a strong blue shift of the absorption maximum occurred, which indicates the presence of solvent inaccessible apolaric side in the vicinity of Trp as a side-effect of the compact amyloid structure [21].

All these results showed when heating of BLG at an acidic pH a heterogeneous mixture of aggregates is formed, which differ in morphology, size, composition, structure and conformation, sometimes even with completely opposite characteristics (e.g. Trp fluorescence). To determine the true properties of the amyloid and non-amyloid fractions, it is therefore necessary to separate the fractions from each other to avoid superposition effects and to discover the true properties of the individual fractions.

2.3 *Structure and Dynamics of Aggregates*

2.3.1 Spin Labeling of Naturally Occurring Cysteines

Electron paramagnetic resonance spectroscopy (EPR) is used in protein analysis to answer relevant structural and dynamic questions related to biological systems [26]. Because of lower number of spins in proteins, nitroxide radicals are frequently used as spin labels, the most common ones being e.g. methanothiolsulfonate spin label (MTSSL) and iodoacetamido proxyl spin label (IPSL). These can bind covalently to specific cysteine-thiol residues and thus generate a stable EPR-active spin label side chain [27], which is called site-directed spin labelling (SDSL). EPR spectra encode the interactions of the nitroxides towards the near environment, such as micro viscosity, micro polarity, nitroxide side-chain mobility, accessibility to e.g. water, oxygen, and reducing agents. Line-shape analyses gives information on structure and dynamics of the protein backbone, so that complexation, aggregation, adhesion, and fibrillation of proteins can be studied [28].

In contrast to the investigation of natural (pathogenic) fibrils, the application of SDSL-EPR for engineered amyloids is completely new, and also has not been applied so far in food chemistry and food biophysics to evaluate the dynamics of non-covalent protein aggregation. For that reason, we aimed to develop in this project a new approach to investigate structure and dynamics of engineered amyloid and non-amyloid material. Unlike the conventional studies for pathogenic fibrils using SDSL, we aimed to investigate the possibilities and obstacles of labelling the naturally occurring cysteines of BLG without performing mutagenesis of primary structure of BLG.

We demonstrated that the spin labels MTSSL and IPSL bind to the free cysteine groups of BLG [29, 30]. Since a pH value of 7.5 for MTSSL and 8.5 for IPSL for efficient labeling, a shuffle effect of the labels or a higher binding propensity of IPSL to Cys66 based on C-S binding may occur, i.e. labels bind to all cysteine residues. However, MTSSL and IPSL preferred different but reproducible binding positions, resulting in MTSSL and IPSL being 80 and 60% reliably bound to peptide 102–124, respectively, which carries the formally free Cys121. The combination of mass spectrometric and EPR analyses made it possible to assign the two spectral components (C(b), of the labels bound to BLG. One labeled side chain, indicated by a non-polar environment, was located outside of the β-barrel of β-lg (Cys106, Cys119, and Cys121) and thus describes the microenvironment of the β-sheets ($C(b_2)$). The other component was the superposition of labeled Cys66 embedded in strand D and Cys160 contained in the random coil ($C(b_1)$). Both binding positions were in the DMSO/water environment and were in close proximity to the tertiary protein structure (Fig. 3) [30].

Under given labeling conditions, the behavior of labeled and non-labeled BLG was the same before and after the heat incubation, independent of the pH values. We also proved in detail that spin labelling with either MTSSL or IPSL does not alter the amyloid aggregation at pH 2 and pH 3.5 as amyloid aggregation, potential changes in hydrophobicity, building blocks, and aggregate morphology was the same for labeled and unlabeled BLG [29].

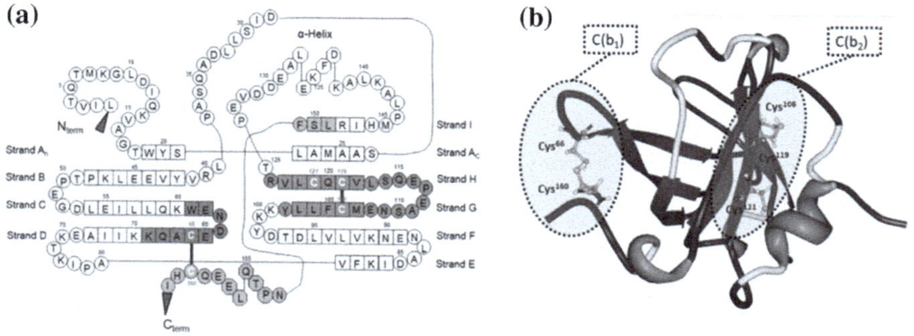

Fig. 3 **a** Schematic representation of BLG A with grey-coded cysteine-containing peptides resulting from the tryptic digestion. Dark grey, peptide 61–69 contains Cys66; middle grey peptide 102–124 contains Cys106, Cys119, and Cys121; and light grey, peptide 149–162. White letters represent the cysteine residues. Symbols represent the secondary structure: squares, β-strands; hexagon, the α-helix; and circle, the random coil. Dark grey lines represent the intramolecular disulphide bridges (modified according to [31, 32]). **b** The tertiary structure of β-lg A indicates the plausibility of two spectral components of the spin labels: $C(b_2)$ describes the spectral properties of peptide 102–124, which is part of the β-sheets. The main labeled position, Cys121, is located between the β-barrel and the α-helix. $C(b_1)$ describes the spectral properties of the combination of peptides 61–69 and 149–162, located in the outer loop. Figure reused with permission from Elsevier, with slight modifications: Ref. [30]

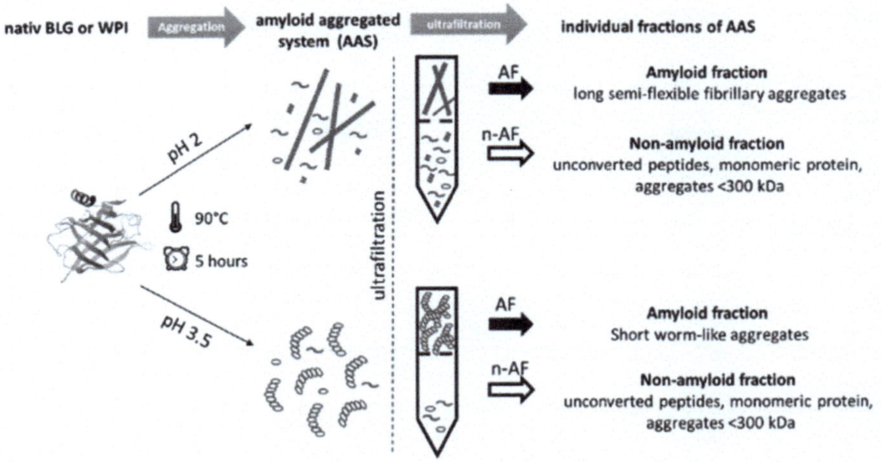

Fig. 4 Isolation of amyloid fraction (AF) and non-amyloid fraction of the amyloid aggregated system (AAS) at pH 2 and 3.5 by ultrafiltration method (MWCO = 300 kDa)

2.3.2 Evaluation of Aggregation Mechanism of BLG

We used side directed spin labeling (SDSL) for the first time in food science to gain deep insights into the aggregation mechanism of BLG under food processing conditions. We hypothesised that it is possible to follow aggregation mechanisms by evaluating different labelling affinities of two spin labels MTSSL and IPSL to these cysteine residues. This technique can contribute to the elucidation of individual aggregation mechanisms depending on different BLG building blocks, such as specific peptides or the entire protein. For this purpose, amyloid aggregation of spin-labelled BLG was performed at pH 2 (Sect. 2.3.3) and pH 3.5 (Sect. 2.3.4) and then the amyloid aggregated system (AAS) was fractionated by ultrafiltration to separate the amyloid (AF) and non-amyloid fraction (n-AF) (Fig. 4). Based on these fractions, the recorded EPR spectra were simulated with the EasySpin tool to obtain information about the individual components of the superimposed EPR spectra [30].

2.3.3 Fibrils and Non-Amyloid Protein at pH 2.0

The approach of SDSL to characterise fibril formation in the entire amyloid aggregated system at pH 2 remained limited, since the acid hydrolysis that occurs during heat incubation at this pH value led to different peptides than those obtained by tryptic digestion, i.e. the monitoring of cysteine-containing peptides was difficult [30]. Since MTSSL was mainly bound in the hydrophobic interior of BLG ($C(b_2)$), it may be deduced that these respective peptides were not involved in amyloid aggregation. In contrast, IPSL was able to bind to Cys160 and Cys66 to a higher degree than MTSSL, which are part of the outer loops originating from the random coil ($C(b_1)$).

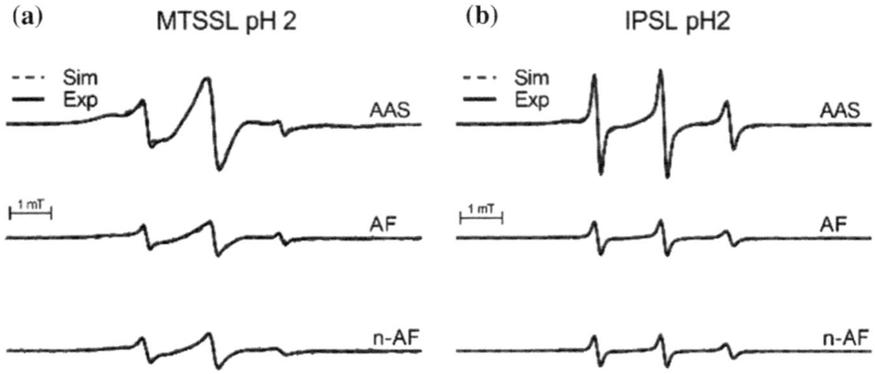

Fig. 5 EPR spectra of BLG spin labeled with MTSSL (**a**) and IPSL (**b**) after fibrilisation (at pH 2, 90 °C for 5 h) and ultrafiltration: amyloid aggregate system (AAS), amyloid (AF) and a non-amyloid fraction (n-AF). Experimental (Exp, black line) and simulated (Sim, grey dotted line) EPR spectra. All spectra were normalised to the protein concentration. Figure reused with permission from Elsevier, with slight modifications: Ref. [30]

This side chain was more hydrophobic and restricted in its rotation after amyloid aggregation, so that it may be assumed that these labeled peptides are more likely to be entrapped in the formed aggregates of hydrolysed peptides.

The similarity of the spectra (Fig. 5) and the comparable results of AF and n-AF simulation for each spin label suggested the same microenvironment of spin labels in both fractions. Approximately the same protein content of ~50% was measured in both fractions after fractionation at 300 kDa. This indicates that the spin-labeled peptides were not part of the amyloid aggregates, which were larger than 300 kDa, as expected for fibril structures. While only an aggregation of the IPSL-labeled side chain could be observed, which was also present in both fractions, neither intermolecular β-sheets in n-AF could be verified nor could ThT fluorescence be measured [29]. Since none of the individual spectral components can be clearly assigned to the amyloid aggregates, it can be assumed that the spin labels reflected other aggregates and smaller peptides that were present in both fractions.

2.3.4 Amyloid-Like Aggregates at pH 3.5

SDSL was successful for intact protein during thermal and acidic treatment at pH value of 3.5, as the two spectral components of the labels bound to native BLG, the non-polar environment outside of the β-barrel of BLG (Cys106, Cys119, and Cys121) and thus describes the microenvironment of the β-sheets $C(b_2)$. $C(b_1)$ was the superposition of labeled Cys66 embedded in strand D and Cys160 contained in the random coil. Both binding positions were located in the DMSO/water environment and were in close proximity to the tertiary protein structure. During amyloid-like aggregation at pH 3.5 three different binding positions of the spin labels can be more clearly distinguished after possible rearrangement when compared to the natural BLG monomer. Therefore, it is likely that the C-terminus C(b_3) will fold outward, as evidenced by high accessibility to the aqueous phase and fast rotational motion (Fig. 6a and b (II)). This rearrangement allows the α-helix to be shifted out of the autonomous folding domain during partial unfolding, which subsequently leads to greater accessibility of the β-sheets to the aqueous environment (Fig. 6c (IIa)). Meanwhile, the β-sheets are most likely compressed by the accumulation of intermolecular β-sheets, which subsequently support the assumption that amyloid-like structures can be formed from the building blocks of the entire primary BLG structure (Fig. 6c (II)). Furthermore, pulsed EPR technique confirmed the spatial distribution of the labels within the amyloid-like fraction in a one-dimensional arrangement of the entire protein aggregates, which was similar to a string of pearls (Fig. 6c (II)). These findings were also in accordance with SAXS measurements (see also Table 1 and Sect. 2.1). Therefore, SDSL proteins containing several cysteine residues can be successfully used to gain deep insights into the aggregation mechanism of engineered amyloids if the binding positions are known.

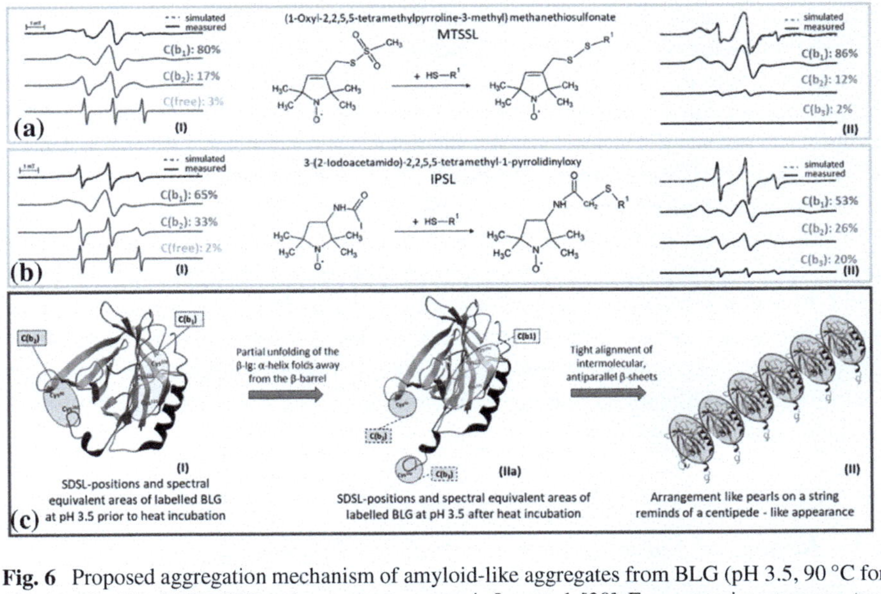

Fig. 6 Proposed aggregation mechanism of amyloid-like aggregates from BLG (pH 3.5, 90 °C for 5 h) described by SDSL-EPR and mass spectrometry in Lux et al. [30]. For comparison purpose, two spin labels MTSSL (**a**) and IPSL (**b**) were used. Measured and simulated EPR spectra of MTSSL (**a (I)**) and IPSL (**b (I)**) bound to BLG before heat treatment result in two bound compounds (C(b1) and C(b2) and one free compound C(free). Measured and simulated EPR spectra of MTSSL (**a (II)**) and IPSL (**b (II)**) bound to BLG after thermal aggregation showed three individual bound compounds (**c (IIa)**). **c** Schematic representation of the formation of centipede-like aggregates (**c (II)**). The structural model of BLG (PDB 3NPO) was created using the ICM browser from molsoft L.L.C. (La Jolla, California). Figure reused with permission from Springer, with slight modifications: Ref. [28]

3 Nano/Microscale: Influence of Protein Modifications on Amyloid Aggregation

Established conditions for engineered amyloid aggregation include protein modifications, in particular protein hydrolysis and release of specific amyloid-prone peptides by heating at pH 2.0 [13, 14] or enzymatic treatment [15]. These amyloid-prone sequences of BLG (mainly the N-terminal peptide sequences 1–33 and 1–53) are hydrophobic and low in charge [13, 15]. At different aggregation conditions, intact BLG can assemble into similar but shorter and more flexible ('worm-like' or 'curly') aggregates, often referred to as 'amyloid-like'. These aggregates can, for example, form upon heating at pH values where hydrolysis is significantly slower than at pH 2 (e.g. pH 3.5) [14]. In Hoppenreijs et al. [33], we have reviewed the conditions under which straight amyloid and worm-like aggregates are formed to identify common traits that help to modify the morphology of the aggregates.

We found that for both, the formation of amyloid and amyloid-like aggregates, destabilisation through hydrolysis and/or denaturation is crucial and is often induced through heating at ≥80 °C. However, as an alternative to heating, specific chemical environmental conditions are reported to also induce amyloid-like aggregation, for example, by incubation of BLG in urea at pH 7.0 and 37 °C [3].

In the following section, we further elucidate on the impact of protein destabilisation on the amyloid (-like) aggregation mechanism either during production process (temperature treatment) or as preceding treatment. Destabilisation was induced by the addition of solvents [34] (Sect. 3.1), by disulphide bond removal [20] (Sect. 3.2.1), or by ultraviolet (UV) radiation [35] (Sect. 3.2.2). Furthermore, the role of protein oxidation is on amyloid aggregation is being addressed (Sect. 3.2.3).

It was hypothesised that these conditions or modifications manipulate especially intramolecular interactions, stimulating protein unfolding. The enhanced unfolding could drive conformational changes, accelerate acid hydrolysis kinetics, or lead to further exposure of amyloid-prone sequences. In turn, this could alter the aggregation kinetics or even aggregate morphology, as will be discussed in this section. Revealing conditions or modifications that drive amyloid aggregation is useful for the future production of functional protein aggregates.

3.1 Impact of Solvents and Alcohols on Protein Aggregation

Alcohols and solvents have versatile effects on proteins, as they reduce the dielectric constant of the aqueous mixture as compared to water and influence the electrostatic and hydrophobic interactions of BLG (both intra– and intermolecular) by changing the hydration layer of the protein [34]. We hypothesised, that the production of amyloid-like aggregates can be affected by changing the solvent concentration and temperature. Therefore, the aggregation behavior of BLG induced by alcohols or solvents was studied at moderate temperature between 30 and 50 °C in the presence of different concentrations of solvents. Ethanol, methanol and dimethyl sulfoxide

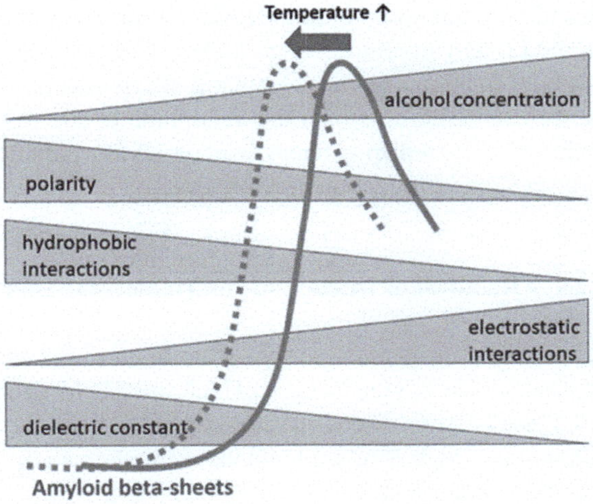

Fig. 7 Schematics of the forces that affect amyloid aggregation and how an increase in temperature affects the amyloid-aggregation kinetics (solid lines for higher temperature and dotted line for lower temperature) by changing the equilibrium between the hydrophobic and electrostatic interactions. The amyloid aggregation reaches a maximum, which is temperature and alcohol concentration dependent

(DMSO) where chosen based on their polarity. Similar methods typically used to study functional fibrils, such as beta-sheet stacking by Thioflavin-T test (ThT), conformation by FTIR, amyloid building blocks by SEC, and morphology by TEM, are used in the experimental setup (Sect. 2.2). We could show that a concentration of up to 40% ethanol favors the amyloid-like aggregation of BLG, whereby this is already possible at temperatures around 40 °C compared to purely aqueous solutions, which require a temperature of 80–90 °C for similar aggregation effects. The resulting protein aggregate structures were spherical to worm-like and formed from non-hydrolysed BLG. A further increase in ethanol content or temperature reduced the aggregation propensity again, i.e. there is an optimum for self-association of β-sheets via hydrogen bonds and hydrophobic interactions which differs for each alcohol or solvent molecule based on the polarity, as depicted in (Fig. 7).

Thus, alcohols and solvents can be used to modify the aggregation of proteins towards specific morphologies, but the resulting functionalities for food applications for example still needs to be tested.

3.2 *Impact of Disulphide Bonds on Protein Aggregation*

Besides peptide bonds, disulphide bonds are the only covalent interactions that stabilise the protein structure. BLG contains one disulphide bond that is mostly exposed at its surface (Cys66-Cys160; 'outer SS bond'), while the other is located in its hydrophobic core (Cys106-Cys119; 'inner SS bond'). Upon heating, its free thiol (SH; Cys121) can form intermolecular SS bonds through SH-SS exchange reactions, which mainly occurs at alkaline pH due to deprotonation of SH (pKa~8.3). We hypothesised that disulphide bonds maintain partial folding of the protein, even after denaturation, and thereby can keep amyloidogenic sequences partially inaccessible.

In addition, inter- and intramolecular disulphide bonds put a conformation restraint on the protein structure, which probably hinder conformational changes necessary for the formation of intermolecular β-sheets (Sect. 2.1).

We used BLG in which the disulphide bonds were (partially) cleaved after the addition of the reducing agent or upon UVB irradiation, or recombinantly removed to investigate its effect on amyloid or amyloid-like aggregation.

3.2.1 Removal of Intra- and Intermolecular Disulphide Bonds

We obtained insights in the role of both intra- and intermolecular disulphide bonds in amyloid-like aggregation (at pH 3.5 and 7.0; [36]) and amyloid aggregation (at pH 2.0; [20]) of BLG. This was achieved by studying the aggregation behavior of BLG variants in which disulphide bonds were chemically cleaved (i.e., reduced) and/or recombinantly removed, as shown in Table 2.

These variants were produced and structurally characterised at pH 7.0 (prior to heating), as shown in [37]. Production of BLG by *Escherichia coli* allowed for the selective removal of the exposed disulphide bond (Cys66–Cys160) or free SH (Cys121), while largely preserving the globular fold. When targeting all cysteines, a structurally disordered BLG variant was obtained. In contrast, chemical reduction of both disulphide bonds mostly maintained the globular structure, but the release of reactive SH limits it application to pH 2.0, as disulphide bond formation was limited at this acidity [20].

In short, Hoppenreijs et al. [20, 36] showed that intramolecular disulphide bonds can be targeted to modulate the aggregate morphology. Partial structural destabilisation induced by disulphide removal stimulated acidic hydrolysis (pH 2.0) and/or

Table 2 Cysteine-modified variants that were used to study either amyloid or amyloid-like aggregation. Based on [20, 36, 37]

Variant	Modification	Amyloid-like aggregation (pH 3.5 + 7.0; [36])	Amyloid aggregation (pH 2.0; [20])	Tertiary structure ([37])
rBLG-SH	SH (Cys121) removed	✓	✘[1]	Globular
rBLG-SS	Exposed disulphide bond (Cys66-160) removed	✓	✓	Globular
rBLG-C	All cysteines removed	✓	✓	Disordered
bBLG R2	All disulphide bonds reduced into SH	✘[2]	✓	Globular

[1]At this acidic condition, SH is not reactive towards thiol/disulphide bond exchange reactions anyways

[2]Especially at pH 7.0 but also 3.5, released SH are highly reactive towards disulphide bond formation

non-covalent association (pH 3.5 and 7.0). Generally, this led to longer aggregates. Surprisingly, complete destabilisation even led to the formation of fibril-like aggregates at pH 7.0, while an alternative aggregation pathway hindered fibrillation at pH 2.0 and instead led to aggregates with worm-like morphology. A detailed explanation of the role of (specific) disulphide bonds in amyloid-like (pH 3.5 and 7.0) and amyloid (pH 2.0) aggregation will be given in the following.

Amyloid-Like Aggregation (pH 3.5 and 7.0)

When heating BLG at pH 3.5 and 7.0, BLG denatures, changes its conformation and assembles into short linear aggregates with worm-like morphology. In Hoppenreijs et al. [36], relative changes in secondary structure (α-helix and β-sheets) and *inter*molecular β-sheet formation (Fig. 8) were followed by FTIR spectroscopy. Upon heating at pH 3.5 at 80 °C, SS bond formation was found to disrupt *intra*molecular β-sheet structure. In addition, non-covalent association occurred through the transformation of destabilised α-helices into *inter*molecular β-sheets. The latter was accelerated upon removal of the *intra*molecular SS bond (i.e. rBLG-SS). Removal of both bonds instead of only the outer SS bond (i.e., rBLG-C), resulted in an even further increase in *inter*molecular β-sheets (Fig. 8a), while the decrease in α-helical structure was similar. Thereby, removal of the inner disulphide bond was expected to expose more amyloid-prone regions. For example, through exposure of the peptide sequence KYLLFC (position 101–106). This sequence was earlier identified as amyloidogenic [38], but the bound Cys106 in native BLG might suppress the aggregation propensity of this region.

Upon heating at pH 7.0 at 80 °C, the formation of intermolecular β -sheets was also enhanced upon removal of the intramolecular disulphide bonds (Fig. 8b), although to a lesser extent. The non-covalent assembly is lower at neutral pH, as β-sheets are expected to be less stable and α-helices more stable, while the opposite is expected at acidic pH [39]. The α-helical structure was preserved further by

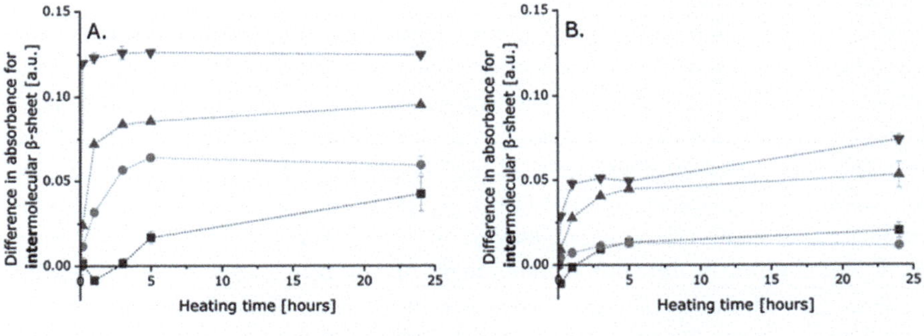

Fig. 8 Relative *inter*molecular beta-sheet formation upon heating at 80 °C and **a** pH 3.5 and **b** pH 7.0 of BLG variants: (■) rBLG, (●) rBLG-SH, (▲) rBLG-SS, (▼) rBLG-C. Figure reused with permission from Elsevier: Ref. [36]

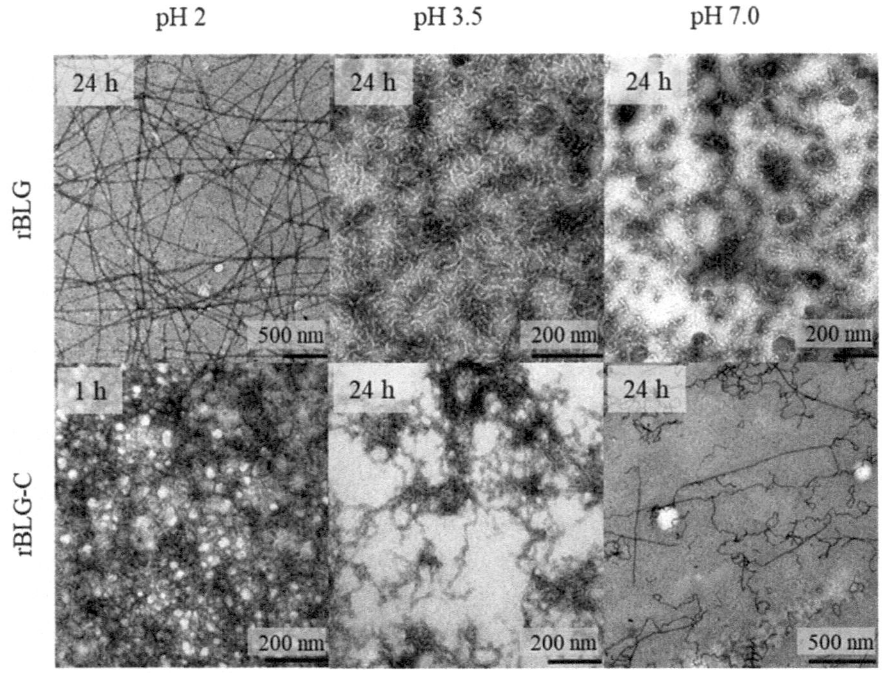

Fig. 9 TEM images of recombinant wild-type (rBLG) and recombinant variant without cysteine (rBLG-C) heating at pH 2.0, pH 3.5 or pH 7.0. Figure reused with permission from Elsevier, with modifications: Ref. [36]

*intra*molecular disulphide bonds, while it was disrupted by conformational changes forced through *inter*molecular disulphide bonds. Similarly, *intra*molecular β-sheets were also disrupted by *inter*molecular disulphide bonds.

Both at pH 3.5 and 7.0, removal of *inter*- and/or *intra*molecular dilsufide bonds led to an enhanced β-sheet formation. Visualisation with TEM showed that this also resulted in longer linear aggregates (Fig. 9).

This is in line with Li et al. [40], who reported that small BLG aggregates were covalently linked at pH 7.0, while longer aggregates were mostly stabilised by non-covalent interactions. Possibly, covalent interactions can terminate growth of β-sheets through their position [41]. Interestingly, a morphological shift from worm-like to fibril-like was observed between 5–24 h of heating of BLG at pH 7.0, which did not contain any cysteines (i.e., rBLG-C). We propose that under these conditions, the increased length of the aggregates allow for alignment, attraction and subsequent association. In turn this entanglement of worm-like aggregates can reduce the flexibility of the aggregates, resulting in the straight fibril-like structures. This morphological shift did not occur at pH 3.5, indicating that this association is influenced by electrostatic interactions. Additional analysis with AFM supported this hypothesis, as worm-like aggregates were thinner as compared to the fibrillar structures in these samples. These results highlight the need to select specific proteins for engineered fibrillation, in which amyloidogenic sequences are not constraint by disulphide bonds.

Fibrillization (pH 2.0)

When heating BLG at pH 2.0, electrostatic repulsion is too high to allow for intact BLG to assemble. Instead, acid catalyses the partial hydrolysis of the peptide chain (specifically next to aspartic acid). Released peptides with enhanced propensity to form β-sheets can then assemble, being mostly the N-terminal peptides [15]. Protonation at this acidic pH hinders thiol/disulphide bond exchange and thereby the formation of *inter*molecular disulphide bonds in BLG limited at pH 2.0, even up to heating for 5 h [20]. In Hoppenreijs et al. [20], we showed that removal of *intra*molecular disulphide bonds has an indirect impact on hydrolysis rate and *inter*molecular assembly at pH 2.0, through destabilisation of the protein structure. The findings will be discussed in the following.

Removal of *intra*molecular disulphide bonds through chemical reduction (i.e., bBLG R2), as well as recombinant removal of the outer disulphide bond (ie., rBLG-SS), resulted in variants with an opened globular fold. It was shown for the first time, that this destabilisation accelerated the acid hydrolysis rate. Removal of the outer disulphide bond (i.e., rBLG-SS) did not impact the formation rate of *inter*molecular β-sheets (as measured with ThT), indicating a particular increased release of peptides that are not involved in the fibrillation. In contrast, additional removal of the *intra*molecular disulphide bond (i.e., BLG R2) accelerated both the acid hydrolysis rate and the formation of *inter*molecular β-sheets. Thereby it was concluded that cleavage of the inner disulphide bond (Cys106-Cys119) resulted in enhanced release of fibril building blocks. This is possibly achieved by release of peptides that show enhanced aggregation propensity but are bound through the inner disulphide bond in native BLG. Lastly, the kinetic impact had no obvious impact on the fibrillar morphology, although image analysis was not performed to reveal minor differences.

Surprisingly, the disordered recombinant variant lacking all cysteine residues (rBLG-C) was shown to assemble through intermolecular β-sheets (shown by both FTIR and ThT fluorescence) prior to heating. The fold of native BLG is stabilised at acidic pH [42]. Exposes the regions that are usually involved in these *intra*molecular bonds could thereby lead to *inter*molecular bonds instead in the disordered variant rBLG-C. This yielded small aggregates (10–50 nm) that aggregated further into aggregates with worm-like morphology during the first hour of heating at 80 °C (Fig. 9). As acid hydrolysis only occurred after 1–3 h of heating, it was concluded that these aggregates were built from intact (unfolded) BLG. Similarly, aggregation of disordered BLG at pH 2.0, without the need for acid hydrolysis, is also reported upon the addition of solvents [34] or glycine [43]. During extended heating, these worm-like aggregates from rBLG-C did not transform into fibrillar structures, indicating that this particular aggregation pathway competes with fibrillization.

3.2.2 Impact of UVB Radiation on Aggregation

The denaturing effect of UVB irradiation on whey proteins has been reported in the literature before [44–47] but little information is available on the effect of UVB irradiation on the underlying mechanism.

We hypothesised that UVB radiation specifically targets disulphide bonds of BLG via the excitation of tryptophan, resulting in loosening of compact folding and cleavage of disulphide bonds. This could promote the formation of intermolecular interactions, which in turn could facilitate the aggregation of BLG. Therefore, we investigated the potential of UVB irradiation as non-thermal and non-chemical treatment for conformational destabilisation of BLG taking into account oxidative degradation [48]. Furthermore, we investigated the effect of the preceding conformational destabilisation on the temperature-induced formation of functional aggregates of BLG [35].

BLG was exposed to UVB irradiation (UVB-BLG) with a maximal radiation intensity of 1.9 ± 0.1 mW cm^{-2}. By varying the exposure time between 0.5 h and 8 h different radiation doses between 3 J cm^{-2} and 55 J cm^{-2} were applied. Afterward, aggregates were prepared upon heating 2.5% BLG solutions (with or without 2-h UVB pre-treatment) at pH 2 or pH 3.5 at 80 °C for up to 5 h.

Information about disulphide bonds and thiol groups were gained by mass spectrometry and spectroscopically quantification after reaction with dithionitrobenzoic acid (Ellman's assay). The Ellman's assay was conducted under denaturing (with urea) and non-denaturing conditions (without urea) to examine the total thiol group content (SH) and accessible thiol group content (aSH). Thereby the release of thiol groups and unfolding processes of BLG can be revealed. Unfolding was also followed by measuring the surface hydrophobicity using the fluorescent probe 8-anilinonaphthalene-1-sulfonic acid that interacts with surface exposed hydrophobic areas of the protein and thereby changing its fluorescence. The thermal stability investigated by differential scanning calorimetry. The occurrence of oxidation reactions of BLG during UVB treatment as well as during temperature exposure was spectroscopically followed by the formation of protein-bound carbonyls after derivatisation with 2,4-dinitrophenylhydrazine. Aggregation kinetics of BLG at both pH values were followed by ThT fluorescence. Furthermore, gel electrophoresis was performed to obtain more information about aggregate formation at pH 3.5.

UVB irradiation of BLG induces conformational modifications, while keeping the oxidative alterations based on protein-bound carbonyls rather low (Fig. 10a), max 7 nmol mg^{-1}). Both the conformational modifications as well as oxidation reactions followed a dose-dependent kinetics (Fig. 10a and b).

The Ellman's assay revealed an increased amount of free thiol groups (about 17 $\pm$ 4%) within the first 0.5–2 h (Fig. 10b), which indicates disulphide cleavage.

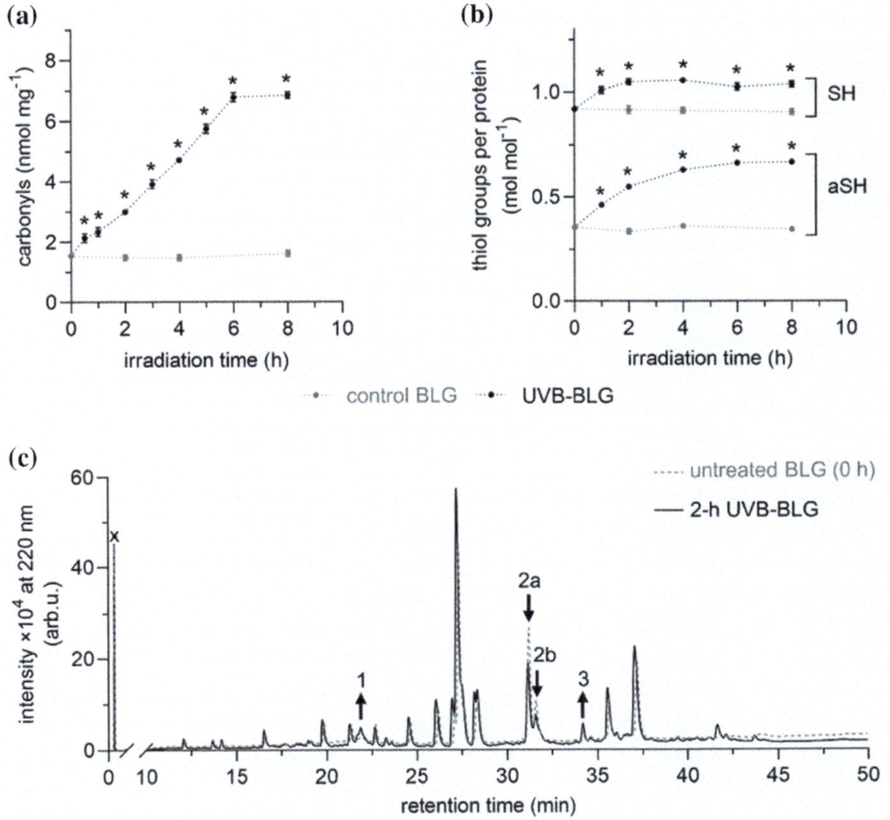

Fig. 10 Protein oxidation and thiol group content in BLG during UVB exposure. **a** Formation of protein-bound carbonyls and **b** total thiol group content (SH, with urea) and accessible thiol group content (aSH, without urea) as a function of UVB exposure time. **c** Peptide profile of BLG and 2-h UVB-BLG after enzymatic hydrolysis. Peptides were identified using LC–ESI–MS/MS. For (a and b), values are given as mean ± standard error of the mean (n ≥ 3), * statistically significant differences (one tailed) between UVB-BLG and untreated BLG (0 h), $p < 0.05$. The untreated BLG (0 h) and control BLG did not differ significantly (two-tailed), $p < 0.05$. Figure reused with permission from Elsevier, with slight modifications: Ref. [48]

Concomitant also the accessible thiol group content increased, but to a higher extent than expected by cleavage of 2 disulphide bridges, indicating additionally the exposure of the previously buried thiol group at Cys121 (Fig. 10b) as a result of protein unfolding. Unfolding processes were also reflected by an increased surface hydrophobicity between 9 ± 0% and 44 ± 5% after 0.5 h to 8 h UVB exposure.

For the 2-h UVB-irradiated BLG, the mass spectrometric analysis of peptides after enzymatic hydrolysis revealed for the first time the cleavage of the 'outer' disulphide bond Cys66–Cys160 by the decrease in signal intensity for the disulphide bond connected peptides Trp61–Lys70 & Leu149–Ile162 (Fig. 10c, peak 2a and 2b).

Furthermore, a new signal occurred that is attributable to the single peptide Leu149–Ile162 with a free thiol group at Cys160, while the peptide Trp61–Lys70 with Cys66 was not detectable (Fig. 10c, peak 3). That only one of the released thiol groups was measurable in UVB-induced cleaved disulphide bonds was also reported for UVB-irradiated α-lactalbumin, where one of the released thiol groups interacts with nearby amino acid side chains such as lysine or tryptophan [46]. Thus, the release of 17 ± 4% thiol groups in UVB-BLG indicates that about 26–42% of UVB-treated BLG have a cleaved 'outer' disulphide bond, which fits very well with the decrease in signal intensity of the peptides associated with the disulphide bond by about 30% in mass spectrometric analysis.

In conclusion, UVB irradiation can be applied to induce conformational rearrangement of BLG resulting in less rigid protein folding. Furthermore, the cleaved 'outer' disulphide bond contributes to an increased flexibility and together with the partial unfolding the reactivity increased due to the release and exposure of reactive side chains.

The 2-h UVB-BLG was then subjected to temperature at 80 °C at pH 2 and pH 3.5 and the aggregation behavior was examined using ThT fluorescence and gel electrophoresis. Based on ThT fluorescence, the 2-h UVB-BLG showed accelerated aggregation tendency at pH 3.5, while the aggregation at pH 2 was slightly impaired (Fig. 11a). The preceding UVB treatment had no clear effect on fibrillar or worm-like aggregate morphology as measured using atomic force microscopy.

When heating BLG, first the protein denatures and unfolds leading to aggregation. The unfolding has been partly initiated in the preceding UV treatment. Consequently, 2-h UVB-BLG exhibited faster denaturation as demonstrated by faster decrease of remaining enthalpy at pH 3.5 during thermal treatment (Fig. 11b). Hence the assembly of intact proteins at pH 3.5 was speeded-up. The gel electrophoresis demonstrates, that especially large aggregates were formed earlier in 2-h UVB-BLG (0.5 h instead of 2 h, Fig. 11c) and were stabilised through both non-covalent but also covalent bonds, from which the latter are disulphide bonds, since these aggregates disappeared after the addition of a reducing agent (Fig. 11d). We propose that on the one hand the unfolding exposes hydrophobic side chains, which facilitate the formation of non-covalent interactions. On the other hand, the higher flexibility of UVB-BLG and the release of thiol groups upon disulphide bond cleavage provides reactive side chains, leading to the formation of new intermolecular disulphide bonds or increased chance for thiol-/disulphide bond exchange reactions.

At pH 2, aggregation requires not only unfolding but also acid hydrolysis. It is likely that at pH 2, the thermal denaturation proceeds similarly to that at pH 3.5. Therefore, it is assumed that faster unfolding of UVB-BLG facilitates the acid-catalysed hydrolysis, resulting in accelerated provision of aggregate building blocks (peptides). However, it is not only the provision of peptides that is necessary. The peptides must be hydrophobic and low in charge [15]. We assume that oxidation reactions in 2-h UVB-BLG deteriorate the aggregation propensity of i.e., the fibril-prone N-terminal peptide sequence (e.g. oxidation at Met7, Fig. 10c, peak 1). In conclusion, it is possible that both promoting and hindering factors play a role in the aggregation process at pH 2, although at the end only slight changes were visible for 2-h UVB-BLG.

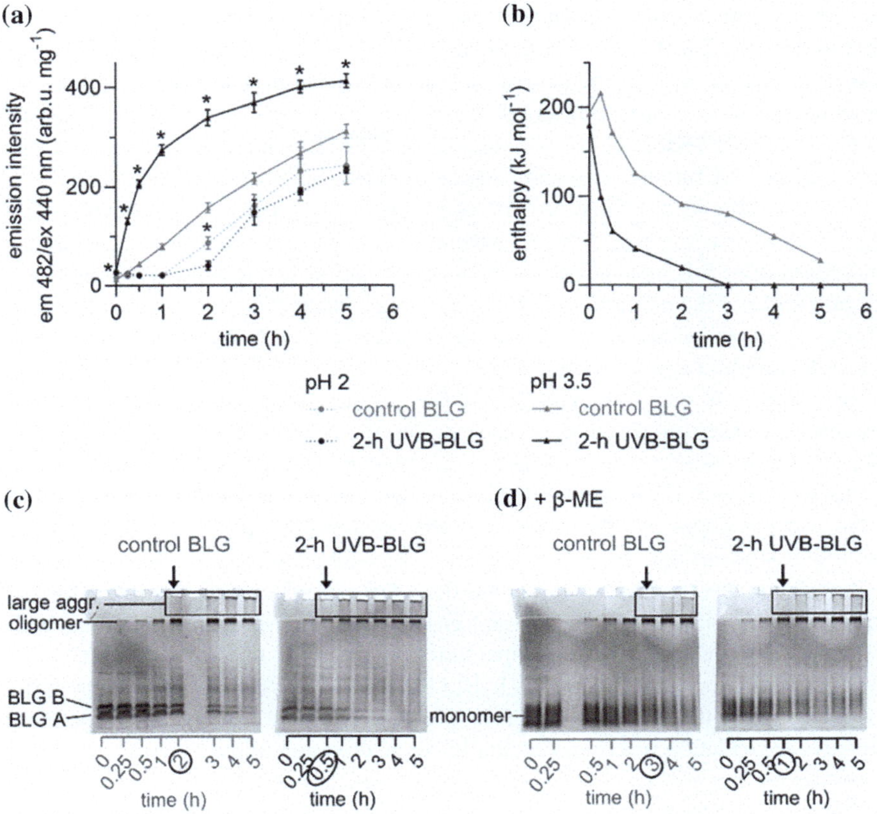

Fig. 11 Aggregation and thermal stability of control and 2-h UVB-BLG during temperature treatment at pH 2 and pH 3.5. **a** ThT fluorescence and **b** remaining enthalpy as a function of incubation time at 80 °C. **c** and **d** Native page with or without the addition of the reducing agent β-mercaptoethanol during temperature treatment. For **a**, results are given as mean ± standard deviation (n = 3), * statistically significant differences (two-tailed) between control-BLG and 2-h UVB-BLG for each pH value separately, $p < 0.05$. For (b) n = 1, therefore no statistical analysis was conducted. Figure reused with permission from Elsevier, with slight modifications: Ref. [35]

3.2.3 Oxidation During Amyloid Aggregation

There is little information to date on the oxidative damage that can occur when BLG is incubated for more than 5 h at a temperature of 80 °C to produce amyloid fibrils. Furthermore, it is unclear how the preceding conformational destabilisation through the UVB treatment might alter the accessibility of oxidation-prone regions. For other proteins, such as kappa-casein, selective oxidation of the methionine residue is known to enhance its amyloid aggregation [49]. A link between oxidative protein damage, amyloid aggregation and diseases such as Alzheimer's has also been described [50]. We therefore hypothesised, that significant protein oxidation reactions occur during temperature induced amyloid aggregation of BLG which could potentially be a part of the amyloid aggregation process [51]. To investigate this, we studied the occurrence of typical oxidation markers NFK (N-formyl kynurenine), dityrosine and protein carbonyls in the protein solution. More specifically we analysed the separated amyloid and non-amyloid fractions to examine if one of those fractions shows a specific accumulation of oxidation markers. The amyloid aggregation was followed by ThT fluorescence and FTIR, as described previously (Sect. 2).

We were able to show that protein carbonyls in BLG increase significantly only after 5 h of incubation at >80 °C, whereas tyrosine oxidation already reaches a maximum after 5 h of incubation. After fractionation of the protein solution into the amyloid and the non-amyloid fraction, no significant differences were observed. Thus, we concluded that the temperature-induced protein oxidation was unlikely to affect the amyloid aggregation of BLG.

In contrast to the untreated BLG, the 2-h UVB-irradiated BLG exhibited pronounced oxidation reactions already during the first 5 h of incubation at 80 °C and a sulfurous odor was noticed. The accelerated oxidation was mainly attributed to the loose folding of UVB-treated BLG, resulting in higher accessibility of oxidation-prone regions. Especially the exposure of the free thiol group (C^{121}) and the release of thiol groups through the disulphide bond cleavage upon the preceding UVB irradiation facilitates the formation of hydrogen sulfide, which is responsible for the sulfurous odor [40]. Furthermore, thiol groups are highly prone for reactions with radicals and oxygen during the temperature treatment. As described in Sect. 3.2.2, the aggregation behavior of 2-h UVB-BLG at pH 3.5 was accelerated, while at pH 2 it was slightly decelerated. Therefore, the aggregation at pH 2 seems to be more sensitive to oxidation reactions especially when alteration in the N-terminal protein region occurs.

We were able to demonstrate that while temperature-induced oxidation does not lead to a significant accumulation of protein-oxidation products or induce amyloid aggregation, the UVB-treated BLG in combination with temperature treatment does produce oxidation products and off-flavors. In case of a targeted protein modification, it is advisable to keep the protein irradiation to a minimum in order to invoke functionalisation but to avoid accumulation of oxidation products.

4 Mesoscale: Formation and Disruption of Amyloid Structures in the Process

After the formation of amyloid or amyloid-like aggregates, the aggregate morphology can be altered by physical processing. Physical processing mainly induces the fragmentation of amyloid aggregates by applied stresses, while flexible amyloid-like aggregates mostly remain unchanged. Examples of applied stresses are shearing (e.g. stirring), the application of high pressure (e.g. during emulsification), exposure to surface tension (i.e. at the interface of emulsions or foams), or capillary forces (i.e. during drying). Fragmentation results in the formation of stiff and rod-like fragments which could have their own functionality, but they could also lose their specific functional application range. Next to fragmentation, physical processes can also induce protein aggregation, and we assume that there is even a superposition of aggregate degrading and aggregate forming processes at the same time. Thus, the aim of the following section is to study the formation and disruption of protein aggregate morphologies during simplified process conditions such as mechanical forces (stirring, glass bead collisions) and the presence of interfaces and surfaces.

4.1 The Onset of Aggregation as a Function of Temperature and Stirring Rate

It has long been known that rather extreme conditions such as heating strongly acidic 2.5% BLG solutions to above 80 °C within 5 h leads to the formation of nanofibrils when the solution is stirred at 350 min^{-1} [9, 52]. However, it has not been clear until now where the exact threshold range lies for the critical parameters such as protein concentration, pH, shear stress, incubation time and temperature, i.e. at which parameter combination amyloid aggregation starts or is possible.

We hypothesised that amyloid aggregates can also be synthesised under 'milder' conditions and that interactions between parameters such as stirring speed and temperature are important. To answer this hypothesis, a fractional 2-level 5-factorial experimental design with three center points and one replicate was created [14]. As output parameters, the ThT fluorescence, the aggregate yield and the structural changes of the protein were investigated by FTIR. In order to better understand the influence of shear stress on the kinetics of amyloid aggregation, the shear stress and

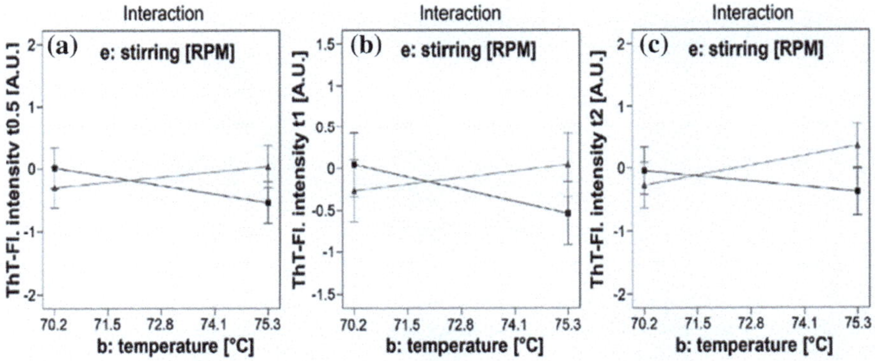

Fig. 12 Interaction of stirring*temperature as an influencing factor on Thioflavin-T Fluorescence t0.5 (**a**), t1 (**b**) and t2 (**c**). Stirring at 300 RPM (Δ) and 200 RPM (□) (Other parameters are pH 2.75, 1.25 wt% protein concentration and 5 mMol NaCl). Figure reused with permission from Elsevier, with slight modifications: Ref. [14]

shear rate determined by the shear flows in the incubation vessel were calculated by a computational fluid dynamics (CFD) simulation (Fig. 13). It was found that temperature is the dominant factor in fibril synthesis, which could be due to manifold effects that superpose each other. However, temperatures ≥70 °C seemed to be sufficient to trigger amyloid aggregation of BLG. During the first 3 h (lag-phase), interaction effects between stirring speed and temperature occurred (Fig. 12). This seems to be due to the increased stress associated with higher viscosities at higher temperatures. After four hours of incubation (growth phase), interaction effects between pH and temperature were found. DSC results showed that BLG is more unstable to temperature-induced denaturation at lower pH values (pH 2.5 versus pH 3.0). Although the parameter combinations are not able to push the threshold of amyloid aggregation below 70 °C, the study was able to provide a more detailed view of the most important factors that cause amyloid aggregation and at which time range they come into effect.

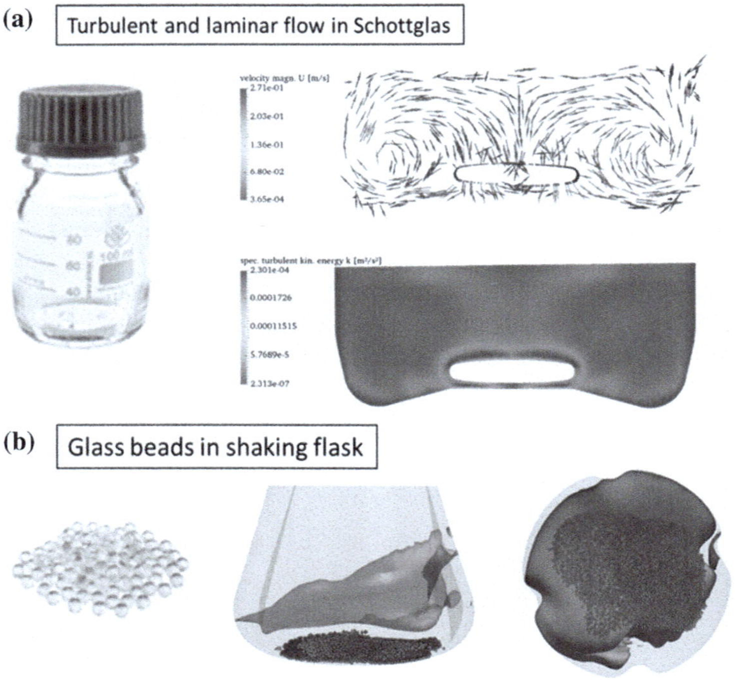

Fig. 13 Design set-up of producing protein aggregates in a magnetic stirrer device or with a shaking flask with glass beads **a** magnetic stirrer design and resulting velocities and turbulent kinetic energy and **b** glass beads in a shaking flask. Figure reused with permission from Elsevier, with modifications: Ref. [14]

4.2 Comminution at Liquid Interfaces

The fibrillar or worm-like aggregations that BLG forms at pH 2 and pH 3.5 can be comminuted in the course of shear stress, which can occur, for example, during processing [24]. Fibrils and worm-like aggregates differ not only in size and shape but also in their flexibility. The fibrils are described as semi-flexible, the worm-like aggregations rather as flexible. Due to the different shape, size and flexibility it was assumed in the study of Heyn et al. [22] that both morphologies are differently sensitive to shear stress. Different processes used for the production of an emulsion differ in their stress energy input per volume (Ev), but also in their stress intensity. In addition, the oil–water interface in an oil-in-water system can have an influence. To investigate these effects, fibrils (pH 2) or worm-like aggregate solutions (pH 3.5) with and without 18% oil were processed using a colloid mill, ultrasound or dynamic high-pressure homogenisation, at different energy inputs (0.1–100 J ml^{-1}) and subsequently tested for their degree of aggregate disintegration (AFM, AUC, DLC), structural change (FTIR) and nucleus destruction (ThT Fluorescence) (Sect. 1). It was found that the degree of disintegration depends on the energy input, the particular process, the presence of the oil phase and the aggregate morphology. Thus, it can be concluded that the highest fragmentation of amyloid fibrils and amyloid-like aggregates was achieved by sonication, followed by high-pressure homogenisation, while the processing with rotor–stator-dispersion resulted in the lowest size reduction (Table 3) [22].

The addition of oil to the process further enhanced the effect of fibril size reduction. The more flexible pH 3.5 aggregates were more stable at lower energy inputs and are only disrupted when the shear stress locally exceeds a specific value. In general, no effects on the intermolecular beta-sheets and the measured ThT fluorescence (amyloid nucleus) were observed.

We concluded that any further food processing in the presence of liquid interfaces, such as preparing emulsions for example, will reduce the fibril size during the preparation process and strongly affect the resulting functionality.

4.3 Glass Beads and Surfaces

Our previous studies have shown that shear can be used to increase the formation kinetics of fibrils [14]. Effects such as secondary nucleation or increased particle motion may play a major role. We also showed in the previous section that the exposure to interfaces can have a destructive effect on the fibrils. We therefore assumed that the inclusion of additional surface area in the incubation vessel can further increase fibril formation. To clearly define the surface area, solid glass particles were used for this experiment [53]. These generate a specific solid surface area [cm^2/gram glass particle] in the incubation system, but also additional shear stress due to particle friction and collision (Fig. 13). In order to distinguish the surface effects from the

Table 3 Minimal and maximal median aggregate length and respective 1st and 3rd quartiles received after high-pressure homogenisation (HPH), rotor–stator dispersion and ultra sonication with and without added oil. Table reused with permission from Elsevier, with slight modifications: Ref. [22]

		Production process					
		Rotor stator dispersion		Ultra-sonication		HPH	
		Without oil	With oil	Without oil	With oil	Straight	Orifice
pH2 fibrils	Min. Median (1st qt; 3rd qt) [nm]	187 (104;313)	192 (131;284)	84 (54;129)	88 (58;143)	140 (109;182)	111 (80;150)
	Max-median (1st qt; 3rd qt) [nm]	6213 (4826;7490)	680 (368;997)	136 (87;200)	150 (100;210)	283 (212;435)	252 (182;344)
pH 3.5 aggregates	Min. median (1st qt; 3rd qt) [nm]	57 (45;83)	71 (41;105)	46 (31;61)	58 (40;78)	n.r	n.r
	Max-median (1st qt; 3rd qt) [nm]	70 (53;84)	97 (68;136)	98 (80;121)	81 (59;108)	n.r	n.r

Qt: quartile; n.r.:no results

kinetic particle effects, different particle sizes (20–1000 μm), rotation frequencies (0–280 min^{-1}) and surface properties (hydrophobicity and surface roughness) were applied. In addition, the shear stresses of certain system settings were modulated and quantified using CFD-DEM simulation. The influence on fibril formation was investigated using ThT fluorescence, infrared spectroscopy, and AFM. It was found that the addition of glass beads accelerates fibril formation (pH 2).

The formation of worm-like aggregates (pH.3.5) was not affected. AFM confirmed that glass bead assisted fibril formation systems may lead to smaller fibril length and thus a reduction of the high aspect ratio, an important feature of fibrils (Fig. 14a). The CFD-DEM confirmed that the fibril formation had a parabolic relation to the stress power input (Fig. 14b), which was also described by Akkermans et al. [15] that shear stress can have a constructive but also destructive effect on the fibril formation. The smallest glass beads used offered a sufficient surface area to support the nucleation of fibrils during the lag phase by interfacial effects. The covalent bonding of stearic acid at the surface of the glass beads led to a strong reduction of the supportive effects on fibril formation.

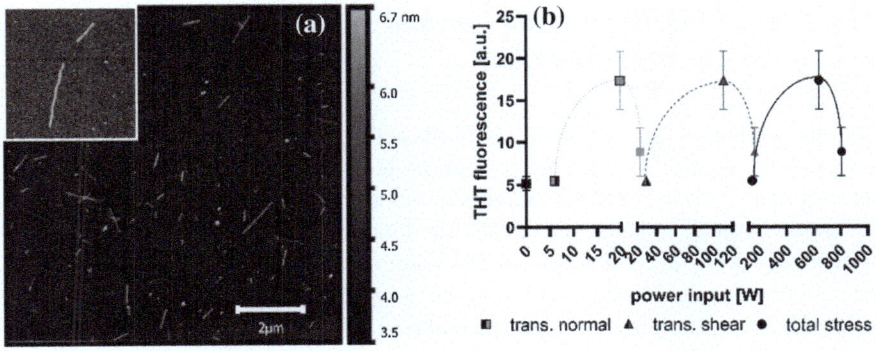

Fig. 14 **a** AFM images of samples with 0.01wt% BLG after 5 h at ~70 °C. Samples were agitated at 180 min^{-1} with glass beads with a mean diameter of 170 μm at pH 2 reused from Heyn et al. (2023) [53]. **b** Increase of ThT-Fl. in pH 2 solutions after 5 h depending on the bead power input in the presence of unmodified glass beads with diameter of ~870 μm at different shaking frequencies (60 min^{-1} [right-side-filled], 120 min^{-1} [left-side-filled], 180 min^{-1} [not-filled], 280 min^{-1} [filled]). Stress power input was calculated out of CFD-DEM simulations and is the product of stress frequency and mean stress energy. Figure reused with permission from Elsevier: Ref. [53]

To conclude, the use of glass beads in a shaken system can accelerate fibril formation compared to a stirred system, potentially saving heating energy, but may also lead to altered aggregate morphologies.

5 Meso-Macroscale: Amyloid BLG Aggregates and Their Interaction with Interfaces and Films

In the following application studies, we applied fibrillated BLG or whey protein isolate solutions in disperse systems and investigated the resulting functional and technological properties. Previous studies showed that BLG aggregates accumulate at interfaces. Since the stabilisation of interfaces is a key function for physical stability of disperse systems, BLG aggregates are assumed to have beneficial functional potential due to their higher surface activity compared to the native protein. For this reason, BLG aggregate solutions are interesting food grade materials with i.e., gelling, emulsifying, foaming and film formation capacity [33, 54, 55]. Apart from the behavior of different aggregate morphologies in such complex matrices, we were aiming at understanding the surface activity and interactions of engineered fibrils and the non-amyloid structures at liquid interfaces and in liquid condensed material. We therefore focused our studies on their functions at air–water interfaces as in foams, oil–water interfaces as in emulsions and liquid condensed material as in in 2D-films or as film formers, which are formed from a dispersion via gel formation by drying.

5.1 Foams: Fibrils and Non-amyloid Structures at the Air-Water Interface

It has been reported that the mixture of fibrils and non-amyloid material can improve foam stability regardless of the origin of the protein [54, 55], but experimental data on the interaction of fibrils and non-amyloid protein materials at the air–water interface are scarce. Aiming to better understand the function of individual protein material of fibrillated whey protein isolate (WPI) solutions, we investigated the behavior of the complete amyloid system (AAS) at the air–water interface in comparison to its isolated fractions, i.e. amyloid aggregated fibrils (AF) and the remaining non-amyloid material (n-AF) (Fig. 4) [56]. We anticipated that there is an optimal ratio of amyloid to non-amyloid structures for improved foaming at a low protein concentration, where non-amyloid fractions play an important role and that the optimal ratio of both fractions can be achieved by varying the initial WPI concentrations prior to fibrillation.

Fibrillation was carried out at different WPI concentrations at pH 2 and 90 °C for 5 h with which we obtained AAS with varying conversion rates (Fig. 15a) and fractionated into AF and n-AF. Unheated (UH) and a short time heated WPI sample (STH) was used as controls at pH 2. STH corresponds to a partly denatured but non-hydrolysed sample. Adsorption kinetics of protein material on the air–water interphase was studied by pendant drop tensiometry (Fig. 15b) and film formation by Langmuir trough (Fig. 15c and d). The foamability and foam stability of the amyloid system were evaluated by performing foam fractionation experiments in a foam fractionation column (Fig. 15e).

The initial WPI concentrations governed the ratio of amyloid and non-amyloid fraction, where a ratio with a higher content of non-amyloid material and few fibrils resulted in an optimal interaction at the air–water interface. Acid hydrolysis at pH 2.0 is rate-limiting. With a higher initial WPI concentration, more peptides are formed in general, but also significantly more longer peptides in which N– and C–terminal segments are still connected [13]. These peptide segments favor fibril formation and consequently increase the conversion rate.

We demonstrated that low-concentrate fibrillated WPI systems at pH 2 without separation of the non-amyloid material greatly improved interfacial activity, foamability and foam stability. Faster migration and twice as high adsorption rate of the smaller non-amyloid material (Fig. 15b) could be the reason for the higher foamability, e.g. in presence of small air bubbles (Fig. 16e). Apart from the adsorption rate, the less energy required to assemble in an LE phase in a monolayer, i.e., they compress stronger in the monolayer than the fibrils (Fig. 15c and d), also improves the ability to rapidly stabilise air bubbles. The diffusion of fibrils was possibly slowed by the high aspect ratio and charge anisotropy but compensated by the high surface hydrophobicity. The higher energy needed of fibril orientation and assembly in a monolayer in the presence of small protein material in AAS than in its absence suggested an interplay between amyloid and non-amyloid proteins at the air–water interface (Fig. 15c and d). Non-amyloid fractions can be incorporated into a fibrillar

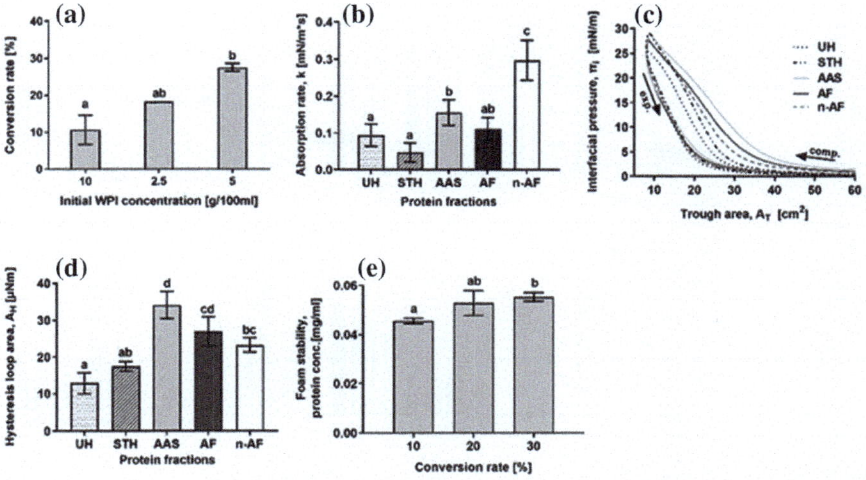

Fig. 15 Interplay of amyloid and non-amyloid WPI structures at different ratios at the air–water interface adapted and modified from Lux and colleagues [56]. **a** Conversion rates of WPI to amyloid aggregates in AAS as a function of initial protein concentration. **b** Adsorption rate at the air–water interface of whey protein isolate solution at pH 2 heated at 90 °C for 5 h (AAS) and for individual protein fractions and controls of unheated (UH) and short-term (30 min) heated (STH) WPI. The amyloid fraction (AF) and non-amyloid fraction (n-AF) were isolated from AAS with a conversion rate of approximately 20%. Measurements were performed by pendant drop. **c** Pressure-area isotherms (Π_i–A_T) of spread WPI monolayers with different thermal treatment and fractionation. **d** Energy consumption for protein orientation at the air–water interface calculated from the area of the hysteresis loop from the Π_i–A_T isotherms. Measurements were performed using Langmuir trough. **e** Foam stability expressed as protein concentration of the retentate of the WPI solution after foaming. Before measurements, all samples were adjusted to a final protein concentration of 0.01 g/100 ml. Different letters indicate statistically significant differences ($n = 3$, $p < 0.05$, Tukey). Figure reused with permission from Elsevier, with modifications: Ref. [56]

film, which in turn could increase the density and stiffness of the network and thus the interfacial stability of the film. For foams, we suggest that WPI should preferably be used as a fibrillated protein system with low protein concentration, as it synergistically occupies the air–water interface earlier and more densely than pure fibrils or non-structured protein material at the same protein concentration. Avoiding the isolation of fibrils will lower the production expense and makes the application of fibrillated systems based on low protein concentration in food more attractive.

5.2 *Emulsion: Fibrils and Non-amyloid Structures of pH 2 at the Oil–Water Interface*

An improved emulsifying capacity of fibrillated BLG compared to the native form has been described in several studies [9, 55]. However, the behavior of fibrils and

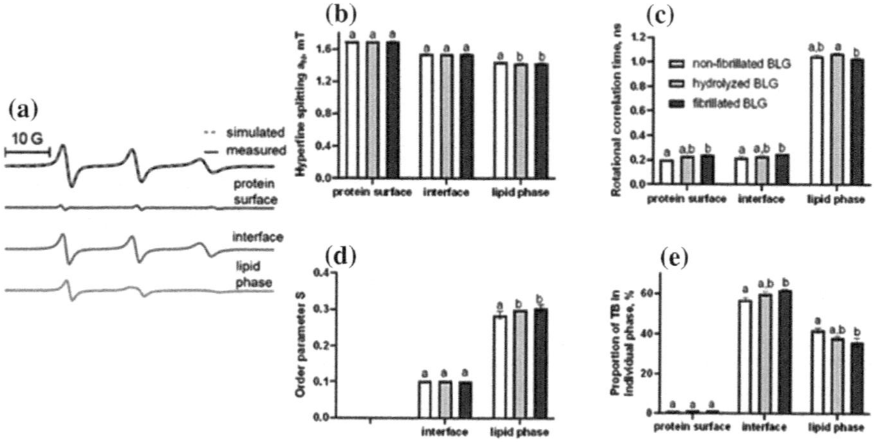

Fig. 16 Spin probe tempolbenzoate (TB) in emulsions with non-fibrillated (white), hydrolysed but not yet fibrillated (light gray) and fibrillated BLG (dark grey) as sole emulsifier. **a** Line Shape analysis of EPR spectra using EPRSim-C to analyse properties of spin probe populations in the interface, lipid and water phase. **b** Micropolarity determined from hyperfine splitting (in mTesla). **c** Microvisvosity determined from rotational correlation time (in nanoseconds). **d** Order parameter S (dimensionless, S = 0 strongly structured environment, S = 1 weakly structured environment). **e** Proportion of spin probe TB solubilised in different phases

non-amyloid aggregates and their possible interaction at the oil–water interface has not been studied so far. The aim of this study was to investigate whether the interface's properties are altered owing to the structural modifications in proteins or protein arrangements when with untreated or fibrillated BLG as the sole emulsifier. We hypothesised that adsorption of proteins at the oil–water interface could lead to different protein mobilities and interfacial environments in both spherical and amyloid and non-amyloid aggregated proteins and peptides as a function of their surface activity. Similar to the air–water interface (Sect. 5.1), the emulsifying ability of amyloid BLG solutions should come from the non-amyloid peptides, which migrate very fast to the interface and stabilise the disperse system.

Based on electron paramagnetic resonance spectroscopy (EPR) measurements, two different approaches were carried out to obtain detailed information on protein mobility and the micro environment of proteins after emulsification. First, the lipophilic nitroxide spin probe TEMPOL benzoate (TB) was used in the emulsion. The spin probe is not bound and, analogous to several previous studies [57–60] freely partitioned with approximately 1.5 ± 0.2% in the aqueous phase, 57 ± 3% in the interface, and 41 ± 2% in the oil phase of an emulsion (Fig. 16e) which can be calculated by simulation of measured EPR spectrum (Fig. 16a). The oil phase consisted of medium-chain triglyceride oil (MCT) to prevent loss of nitroxide function by lipid radicals or antioxidants.

The micropolarity remained unchanged for the interface and water phase regardless of the BLG structures, but the lipid phase became somewhat more hydrophobic in the presence of the hydrolysed and amyloid protein structures (Fig. 16b). Also, small but significant differences in the microviscosity were observed, increasing in the lipid phase when BLG is fibrillated and slightly decreasing in the interface and water phases (Fig. 16c). Fibrillated BLG formed a more appropriate environment or increased volume of the interface in which more spin probe was solubilised, while the probe content in the lipid phase was significantly lower than in non-fibrillated BLG (Fig. 16c). However, in all phases, the hydrolysed but not yet fibrillated BLG behaved intermediately, indicating the importance of the peptides, which increases when they are also thermally aggregated. Results thus indicate that the fibrillated BLG may form a more structured and more hydrophobic part at the transition from interface to lipid phase than the native BLG. However, the extent to which the influence of non-amyloid and amyloid structures can be distinguished was not tested with this approach.

In a second EPR approach, BLG was spin-labeled with MTSSL (Sect. 2.2.3) and subsequently fibrillated. Either the whole protein solution or its individual amyloid and non-amyloid fractions were used as emulsifier. Non-fibrillated BLG solution at pH 2 again served as a control. Because EPR spectra are a superposition of different magnetically equivalent binding sites (Sect. 2.2.3), the emulsion was mechanically separated again into three individual immiscible phases by centrifugation (2000 g for 15 min). The EPR spectra were then recorded and interpreted separately for interface, the water and oil phases. EPR spectra were then recorded and interpreted separately for the interface, the water and oil phases. However, it was not possible to recover completely pure phases after emulsification. Macroscopically, it was evident that emulsions with non-amyloid fractions had the largest interfacial volume compared to all other emulsions, which could indicate that the non-amyloid peptides have a higher surface activity, i.e. form a more stable interface that requires greater energy to separate. During EPR-monitoring the distribution of the spin-labeled protein, it was found that in all emulsions peptides preferrentially accumulated from the aqueous phase towards the interface. Additionally it was observed that only a small amount of free MTSSL was present in the lipid phase without any bound peptides. With respect to the protein environment, the EPR spectra showed a marked difference in the conformation of the BLG control solution at each phase of the emulsion, from a more free tumbling spin label to an environment with lower polarity and higher microviscosity. A stronger restriction of spin label tumbling indicated that the structure of the non-fibrillated protein aggregates at the interface. In contrast, all emulsions with fibrillated proteins displayed near-identical EPR spectra across the different phases. However, it should be noted that MTSSL is only bound to peptides in non-amyloid aggregates (Sect. 2.2.4) and therefore only the behavior of the non-amyloid fractions at the interface can be followed.

The results may indicate that the spin-labeled non-amyloid, BLG aggregates have similar protein mobility at the interface and in the water phase. This could imply that there is no further aggregation or rearrangement of these aggregates at the interface. Moreover, in the absence of amyloid fibrils, the amount of non-amyloid fraction that formed the interface appears to be higher than in the presence of fibrils.

5.3 Films: Fibrils and Non-amyloid Structures in 2D-Film Formation

Proteins, such as BLG and WPI, and their aggregates have the ability to form transparent, colorless, and odorless films upon drying (see e.g. [61, 62]). Nevertheless, these films have limited quality due to their brittle nature. Currently, there is insufficient evidence regarding the impact of aggregate morphologies, aggregate mixtures, and structural arrangement of protein aggregates in films that determine certain properties such as stickiness, free-standing structures, and mechanical peak load. For this reason, this study was to investigate the impact of individual mesostructures of WPI aggregates (Fig. 1) on 2D macrostructures in films during the drying process. A mechanistic understanding of film formation from structured proteins will allow films to be made with specific properties, minimal use of plasticizers or other additives, and improved functional attributes of protein films and coatings.

We therefore hypothesise that the morphology of the aggregates determines their orientation and molecular flows during the drying process as well as the force distribution within the dried films. This, in turn, has a significant effect on the appearance of the films. The long, flexible fibrils (pH 2) are more suitable for forming films than the short, worm-like, amyloid-like aggregates (pH 3.5) due to their ability to align with the network facilitated by their morphology. However, not only the presence of amyloid fraction but also the non-amyloid fraction may be necessary for the condensed protein film's stability and film formation.

Using convection drying at 40 °C, we were able to form 2D-films from differently structured WPI dispersions through gelation and drying on various substrates, including polyethylene foil, polystyrene dishes, Teflon and glass. The base WPI dispersions consisted of aggregate mixtures (AAS), which were fibrillated previously at high temperatures at pH 2, pH 3.5, or at ambient temperatures in the presence of ethanol or solvents (Sect. 3.1), (Fig. 17e). As in other application studies, we prepared fractions consisting solely of either non-amyloid (n-AF) or amyloid (AF) material to assess different structural effects (Fig. 4). We utilised untreated WPI at respective pH values and those heated for 30 min as controls.

Films prepared from native WPI exhibited high brittleness, with angular and spiral crack patterns that were influenced by substrate hardness. On the other hand, films made from all individual AAS used resulted in films with reduced fragility and morphology-adapted straight or curvy aligned crack patterns. We have demonstrated, for the first time, that the appearances of the films were significantly impacted by individual fractions. The fraction containing the amyloid aggregates (AF) resulted in transparent films upon drying, whereas the non-amyloid (like) fraction (n-AF) forms milky opaque films. These films differed also in terms of their peelability from the substrate, with fibrils tended to delaminate and the non-amyloid (like) fraction led to strongly adhered films (Fig. 17a).

A phenomenon known as the coffee ring effect [63, 64] was observed during the drying of the differently structured WPI dispersions. In general, distinct drying zones were observed in the dried protein droplets and films. Although no typical coffee-ring

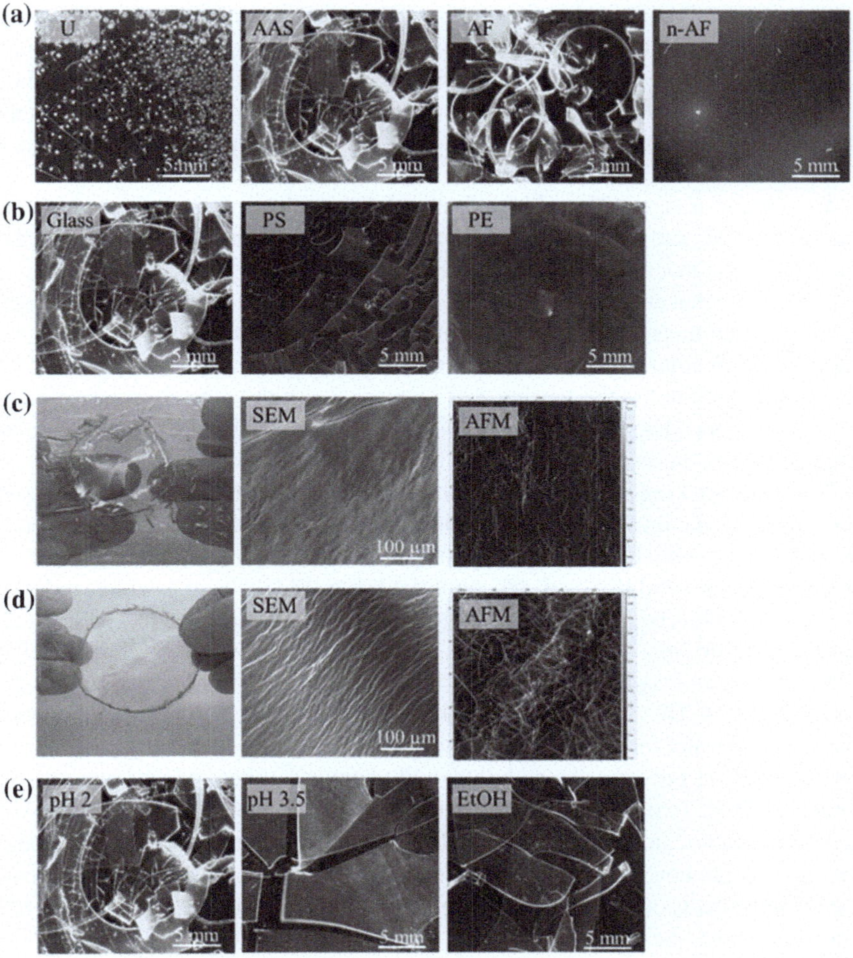

Fig. 17 Images of protein films from various structured WPI dispersions at pH 2. **a** Comparison of films prepared from unheated WPI (U), amyloid aggregated system (AAS), amyloid fraction (AF) and non-amyloid fraction (n–AF) on glass as the substrate. **b** Comparison of AAS films on different substrates: glass dish (glass), polystyrene dish (PS), polyethylene foil (PE). **c** Different resolutions (picture on polyethylene foil, SEM, AFM) of AAS film without additives. **d** Different resolutions (picture free-standing, SEM, AFM) of AAS film with additives. **e** Comparison of AAS films with different solvent-treatments (pH 2, pH 3.5, ethanol) on glass as the substrate

occurred in the non-amyloid fraction, an optical gradient with a particularly opaque center and outer edge was apparent. Structure dependent varying deposition patterns could be explained by different dominant flow dynamics due to particle shape, size, interaction, pH value, charge, solvents, viscosity, surface activity, interfaces, contact line pinning and substrates.

A comprehensive characterisation of liquid dispersions with different WPI structures was performed to derive physical interactions between protein and substrate during drying. We demonstrated that comparatively hard substrates such as glass caused changes in adhesion and greater tension in the protein film. On the other hand, softer substrates, such as polystyrene dishes or polyethylene foils produced a more homogeneous film with altered crack patterns (Fig. 17b).

Microscopic images were obtained at various resolutions, including photomicrographs, SEM, and AFM, to examine the impact of the mesoscale (i.e., fibrillar structure) on the macroscopic appearance (i.e., crack pattern) (Fig. 17c).

It was found that cracks are a useful visual indicator for studying the morphology of protein structures at the meso level. We assume that the structures present in the protein solution align and crosslink during the drying process, potentially aided by the flows. Additionally, the tension that increases with the solvent evaporation is morphology-dependent. This is analogous to a lightning conductor that conducts energy along the mesostructure.

To prevent cracking and achieve uniform films, various plasticizers were assessed with emphasis on minimising the usage of plasticizers and sugars. Only food-grade substances with hydroxyl groups (polyols) and low molecular weight, such as trehalose, glycerol and PEG [65, 66] were used (Fig. 17d). The plasticizers reduced cracking in all films by partially replacing a natural plasticizer found in water, which is lost during drying. This replacement process likely allowed for more spacing within the protein networks, resulting in increased flexibility and decreased brittleness of the films. This effect can be attributed to the competition between polymer molecules and water for protein binding, which in turn reduced intermolecular forces between adjacent polymer chains or amyloid aggregates via a reduction in protein interactions [65]. Two theories are debated regarding the stabilisation mechanism of sugar: Vitrification and water exchange. In order to prevent possible Maillard reactions, it is recommended to avoid reducing sugars as additives. Different additives have varying effects on the flexibility of protein films. Peydayesh and colleagues [62] demonstrated that the fibrillation process is possible even when plasticizers such as glycerol are present. Self-assembly is achieved despite the additives to be tested, but the contour length distribution is reduced [62]. Additives may be added after fibril production and before the drying step.

Using factorial statistically designed experiments (Design of Experiments), we were able to figure out optimal combinations of amyloid aggregates with various edible additives to produce application-oriented films/coatings. Inspired by the trend of "clean labeling", this allowed us to decrease the number of additives which is necessary for stable, flexible, free-standing films. In this process, we took advantage of the synergistic effect of several ingredients, such as plasticizers. We evaluated those films of amyloid aggregates for their crack-free peelability, high UV-light shielding (resulting in low transmission in the UV range), high optical transparency (resulting in high transmission in visible light), and mechanical strength (demonstrating high peak load). Differential scanning calorimetry (DSC), FTIR, and water vapor transmission tests were conducted for further analysis of the films. With these properties, the developed films reinforced their suitability as alternative packaging materials with function.

Currently, we are studying the drying kinetics of individual fractions containing plasticizers to provide a more comprehensive explanation of particle and film formation with amyloid (like) aggregates. To investigate the complex and mainly unexplored process of drying amyloid aggregate solutions, including the changing physical and chemical conditions over time, we have chosen single drop drying and thin film drying as our methods.

The data indicate that the films prepared with amyloid aggregates had higher resistance to peak loads during fracture experiments than those made with unheated protein in terms of mechanical strength. The extra effort involved in preparing amyloid aggregates may yield benefits in the production of protein structures that occupy space advantageously for film formation and crosslinking compared to native globular proteins. The promotion of crosslinking by the fibrils, due to the internal film structure, can reduce potential additive levels in comparison to the use of native proteins. The additives utilised to enhance film quality will specify the usage (such as edibility) and biodegradability of WPI films. Sugar additives meet both food grade and biodegradable requirements.

6 Conclusion

We demonstrated the effect of process relevant conditions on the (amyloid) aggregation of BLG from the nanoscale to the meso–, and macroscale. Conditions that accelerate or decelerate the aggregate formation kinetics were investigated, and processing parameters and conditions that led to different types and shapes of protein aggregates were identified. We found that the kinetics of aggregate formation, and fragmentation varied with the applied process conditions, allowing a more controlled approach of forming different protein aggregate structures. Overall, the reported results show that the model protein BLG can be modified into various aggregate shapes and structures with altered functionalities as compared to its native form. This enhanced functionality of WPI is highly promising for employment in both, food and non-food applications. It should be noted that the interplay between amyloid and non-amyloid structures at different interfaces is responsible for the improved properties. The non-amyloid material in foams and emulsions rapidly adsorbs at the interface and is subsequently stabilised by fibrils when forming a common network. However, arranging the amyloid fibrils beforehand is essential for forming a film that is both transparent and less brittle once the non-amyloid structures have been incorporated. With respect to general functionality, it also seems that amyloid fibrils have a higher functionality range as compared to worm-like aggregates, likely due to the high aspect ratio of the fibrils as well as polydisperse nature. Further research is required to substantiate this for various applications.

It would also be of interest to identify specific molecular properties (nanoscale), such as low temperature stability (no or reduced cysteine bonds), in combination with a high load of aggregation prone regions that are indicative for high amyloid aggregation propensity. It remains to be seen how applicable that knowledge is to other proteins. This information can be used to select existing proteins or to design favorable structures via precision fermentation. Furthermore, the knowledge will also contribute towards a more targeted design of processes that lead to aggregated structures and mixtures of these structures, suited for specific applications.

Acknowledgements The authors would like to thank the German Research Foundation (DFG, Germany) for financial support of this work within the SPP 1934 priority program DiSPBiotech (project number 315456892). The authors are also grateful for the excellent collaboration with various universities within the SPP, e.g. University of Braunschweig, TU Berlin, TU Munich, Friedrich-Alexander University Erlangen-Nuremberg, Hamburg University of Technology, University of Bonn and DESY. These collaborations led (among other publications mentioned in Chapter 7) to the production and joint characterisation of recombinant BLG variants, published in Keppler et al [67]. Towards recombinantly produced milk proteins: Physicochemical and emulsifying properties of engineered whey protein beta-lactoglobulin variants. Food Hydrocolloids 110, 106132. https://doi.org/10.1016/j.foodhyd.2020.106132.

References

1. Loveday SM, Wang XL, Rao MA et al (2012) β-Lactoglobulin nanofibrils: Effect of temperature on fibril formation kinetics, fibril morphology and the rheological properties of fibril dispersions. Food Hydrocolloids 27:242–249. https://doi.org/10.1016/j.foodhyd.2011.07.001
2. So M, Yagi H, Sakurai K et al (2011) Ultrasonication-dependent acceleration of amyloid fibril formation. J Mol Biol 412:568–577. https://doi.org/10.1016/j.jmb.2011.07.069
3. Hamada D, Dobson CM (2002) A kinetic study of beta-lactoglobulin amyloid fibril formation promoted by urea. Protein Sci 11:2417–2426. https://doi.org/10.1110/ps.0217702
4. Hill EK, Krebs B, Goodall DG et al (2006) Shear flow induces amyloid fibril formation. Biomacromol 7:10–13. https://doi.org/10.1021/bm0505078
5. Hamley IW (2010) Amyloid formation: Interface influence. Nat Chem 2:707–708. https://doi.org/10.1038/nchem.816
6. Schladitz C, Vieira EP, Hermel H et al (1999) Amyloid-beta-sheet formation at the air-water interface. Biophys J 77:3305–3310. https://doi.org/10.1016/S0006-3495(99)77161-4
7. Thomas CR, Geer D (2011) Effects of shear on proteins in solution. Biotechnol Lett 33:443–456. https://doi.org/10.1007/s10529-010-0469-4
8. Gilbert J, Campanella O, Jones OG (2014) Electrostatic stabilization of β-lactoglobulin fibrils at increased pH with cationic polymers. Biomacromol 15:3119–3127. https://doi.org/10.1021/bm500762u
9. Serfert Y, Lamprecht C, Tan C-P et al (2014) Characterisation and use of β-lactoglobulin fibrils for microencapsulation of lipophilic ingredients and oxidative stability thereof. J Food Eng 143:53–61. https://doi.org/10.1016/j.jfoodeng.2014.06.026
10. Gosal WS, Clark AH, Ross-Murphy SB (2004) Fibrillar beta-lactoglobulin gels: Part 1. Fibril formation and structure. Biomacromolecules 5:2408–2419. https://doi.org/10.1021/bm049659d
11. Nicolai T, Durand D (2013) Controlled food protein aggregation for new functionality. Curr Opin Colloid Interface Sci 18:249–256. https://doi.org/10.1016/j.cocis.2013.03.001

12. Oboroceanu D, Wang L, Kroes-Nijboer A et al (2011) The effect of high pressure microfluidization on the structure and length distribution of whey protein fibrils. Int Dairy J 21:823–830. https://doi.org/10.1016/j.idairyj.2011.03.015
13. Ye X, Hedenqvist MS, Langton M et al (2018) On the role of peptide hydrolysis for fibrillation kinetics and amyloid fibril morphology. RSC Adv 8:6915–6924. https://doi.org/10.1039/c7ra10981d
14. Heyn TR, Mayer J, Neumann HR et al (2020) The threshold of amyloid aggregation of beta-lactoglobulin: Relevant factor combinations. J Food Eng 283:110005. https://doi.org/10.1016/j.jfoodeng.2020.110005
15. Akkermans C, Venema P, van der Goot AJ et al (2008) Peptides are building blocks of heat-induced fibrillar protein aggregates of beta-lactoglobulin formed at pH 2. Biomacromol 9:1474–1479. https://doi.org/10.1021/bm7014224
16. Chatani E, Lee Y-H, Yagi H et al (2009) Ultrasonication-dependent production and breakdown lead to minimum-sized amyloid fibrils. Proc Natl Acad Sci U S A 106:11119–11124. https://doi.org/10.1073/pnas.0901422106
17. Bolder SG, Vasbinder AJ, Sagis LM et al (2007) Heat-induced whey protein isolate fibrils: Conversion, hydrolysis, and disulphide bond formation. Int Dairy J 17:846–853. https://doi.org/10.1016/j.idairyj.2006.10.002
18. Adamcik J, Mezzenga R (2018) Amyloid polymorphism in the protein folding and aggregation energy landscape. Angew Chem Int Ed Engl 57:8370–8382. https://doi.org/10.1002/anie.201713416
19. Arnaudov LN, de Vries R, Ippel H et al (2003) Multiple steps during the formation of beta-lactoglobulin fibrils. Biomacromol 4:1614–1622. https://doi.org/10.1021/bm034096b
20. Hoppenreijs LJ, Brune SE, Biedendieck R et al. (2023) Fibrillation of β-lactoglobulin at pH 2.0: Impact of cysteine substitution and disulphide bond reduction. Food Hydrocoll 141:108727. https://doi.org/10.1016/j.foodhyd.2023.108727
21. Heyn TR, Garamus VM, Neumann HR et al. (2019) Influence of the polydispersity of pH 2 and pH 3.5 beta-lactoglobulin amyloid fibril solutions on analytical methods. Eur Polym J 120:109211. https://doi.org/10.1016/j.eurpolymj.2019.08.038
22. Heyn TR, Uttinger MJ, Kwade A et al (2021) Whey protein (amyloid)-aggregates in oil-water systems: The process-related comminution effect. J Food Eng 311:110730. https://doi.org/10.1016/j.jfoodeng.2021.110730
23. Krebs MRH, Bromley EHC, Donald AM (2005) The binding of thioflavin-T to amyloid fibrils: localisation and implications. J Struct Biol 149:30–37. https://doi.org/10.1016/j.jsb.2004.08.002
24. Uttinger MJ, Heyn TR, Jandt U et al (2020) Measurement of length distribution of beta-lactoglobulin fibrils by multiwavelength analytical ultracentrifugation. Eur Biophys J 49:745–760. https://doi.org/10.1007/s00249-020-01421-4
25. Stănciuc N, Aprodu I, Râpeanu G et al (2012) Fluorescence spectroscopy and molecular modeling investigations on the thermally induced structural changes of bovine β-lactoglobulin. Innov Food Sci Emerg Technol 15:50–56. https://doi.org/10.1016/j.ifset.2012.03.001
26. Klare JP (2013) Site-directed spin labeling EPR spectroscopy in protein research. Biol Chem 394:1281–1300. https://doi.org/10.1515/hsz-2013-0155
27. Berliner LJ (1976) Spin labeling: theory and applications. Academic Press, Nwe York
28. Heyn TR, Garamus VM, Steffen-Heins A et al (2022) Analysis of natural and engineered amyloid aggregates by spectroscopic and scattering techniques. In: Bulavin L, Lebovka N (eds) Soft matter systems for biomedical applications, vol 266. Springer International Publishing, Cham, pp 295–314
29. Lux J, Heyn TR, Kampen I et al. (2021) Amyloid aggregation of spin-labeled β-lactoglobulin. Part I: Influence of spin labeling on amyloid aggregation. Food Hydrocoll 112:106178. https://doi.org/10.1016/j.foodhyd.2020.106178
30. Lux J, Azarkh M, Fitzner L et al. (2021) Amyloid aggregation of spin-labeled β-lactoglobulin. Part II: Identification of spin-labeled protein and peptide sequences after amyloid aggregation. Food Hydrocoll 112:106174. https://doi.org/10.1016/j.foodhyd.2020.106174

31. Keppler JK, Koudelka T, Palani K et al (2014) Characterization of the covalent binding of allyl isothiocyanate to β-lactoglobulin by fluorescence quenching, equilibrium measurement, and mass spectrometry. J Biomol Struct Dyn 32:1103–1117. https://doi.org/10.1080/07391102.2013.809605
32. Fogliano V, Monti SM, Visconti A et al (1998) Identification of a beta-lactoglobulin lactosylation site. Biochim Biophys Acta 1388:295–304. https://doi.org/10.1016/S0167-4838(98)00177-0
33. Hoppenreijs L, Fitzner L, Ruhmlieb T et al (2022) Engineering amyloid and amyloid-like morphologies of β-lactoglobulin. Food Hydrocolloids 124:107301. https://doi.org/10.1016/j.foodhyd.2021.107301
34. Kayser JJ, Arnold P, Steffen-Heins A et al (2020) Functional ethanol-induced fibrils: Influence of solvents and temperature on amyloid-like aggregation of beta-lactoglobulin. J Food Eng 270:109764. https://doi.org/10.1016/j.jfoodeng.2019.109764
35. Fitzner L, Hasler M, Heyn TR et al (2023) UVB pretreatment of β-lactoglobulin affects the temperature-induced formation of functional amyloid-like aggregates and promotes oxidative degradation. Food Chem 429:136898. https://doi.org/10.1016/j.foodchem.2023.136898
36. Hoppenreijs LJG, Overbeck A, Brune SE et al. (2023) Amyloid-like aggregation of recombinant β-lactoglobulin at pH 3.5 and 7.0: Is disulphide bond removal the key to fibrillation? Int J Biol Macromol 242:124855. https://doi.org/10.1016/j.ijbiomac.2023.124855
37. Brune SE, Hoppenreijs LJ, Kühl T et al (2023) Precision fermentation as a route to modify β-lactoglobulin structure through substitution of specific cysteine residues. Int Dairy J 147:105772. https://doi.org/10.1016/j.idairyj.2023.105772
38. Euston SR, Ur-Rehman S, Costello G (2007) Denaturation and aggregation of β-lactoglobulin—a preliminary molecular dynamics study. Food Hydrocolloids 21:1081–1091. https://doi.org/10.1016/j.foodhyd.2006.07.018
39. Xing R, Yuan C, Li S et al (2018) Charge-induced secondary structure transformation of amyloid-derived dipeptide assemblies from β-sheet to α-Helix. Angew Chem Int Ed Engl 57:1537–1542. https://doi.org/10.1002/anie.201710642
40. Li C, Paulsen PA, Akıllıoğlu HG et al (2022) Cysteine residues are responsible for the sulfurous off-flavor formed in heated whey protein solutions. Food Chem (Oxf) 5:100120. https://doi.org/10.1016/j.fochms.2022.100120
41. Yamaguchi T, Yagi H, Goto Y et al (2010) A disulphide-linked amyloid-beta peptide dimer forms a protofibril-like oligomer through a distinct pathway from amyloid fibril formation. Biochemistry 49:7100–7107. https://doi.org/10.1021/bi100583x
42. Kella NK, Kinsella JE (1988) Enhanced thermodynamic stability of beta-lactoglobulin at low pH. A possible mechanism. Biochem J 255:113–118. https://doi.org/10.1042/bj2550113
43. Jaklin M, Hritz J, Hribar-Lee B (2022) A new fibrillization mechanism of β-lactoglobulin in glycine solutions. Int J Biol Macromol 216:414–425. https://doi.org/10.1016/j.ijbiomac.2022.06.182
44. Kehoe JJ, Remondetto GE, Subirade M et al (2008) Tryptophan-mediated denaturation of beta-lactoglobulin A by UV irradiation. J Agric Food Chem 56:4720–4725. https://doi.org/10.1021/jf0733158
45. Permyakov EA, Permyakov SE, Deikus GY et al (2003) Ultraviolet illumination-induced reduction of alpha-lactalbumin disulphide bridges. Proteins 51:498–503. https://doi.org/10.1002/prot.10371
46. Vanhooren A, Devreese B, Vanhee K et al (2002) Photoexcitation of tryptophan groups induces reduction of two disulphide bonds in goat alpha-lactalbumin. Biochemistry 41:11035–11043. https://doi.org/10.1021/bi0258851
47. Zhao Z, Engholm-Keller K, Poojary MM et al (2020) Generation of aggregates of α-Lactalbumin by UV-B light exposure. J Agric Food Chem 68:6701–6714. https://doi.org/10.1021/acs.jafc.0c00757
48. Fitzner L, Kühl T, Hasler M et al (2023) Modification and oxidative degradation of β-lactoglobulin by UVB irradiation. Food Chem 428:136698. https://doi.org/10.1016/j.foodchem.2023.136698

49. Koudelka T, Dehle FC, Musgrave IF et al (2012) Methionine oxidation enhances κ-casein amyloid fibril formation. J Agric Food Chem 60:4144–4155. https://doi.org/10.1021/jf205168t
50. Cheignon C, Tomas M, Bonnefont-Rousselot D et al (2018) Oxidative stress and the amyloid beta peptide in Alzheimer's disease. Redox Biol 14:450–464. https://doi.org/10.1016/j.redox.2017.10.014
51. Keppler JK, Heyn TR, Meissner PM et al (2019) Protein oxidation during temperature-induced amyloid aggregation of beta-lactoglobulin. Food Chem 289:223–231. https://doi.org/10.1016/j.foodchem.2019.02.114
52. Akkermans C, Venema P, Rogers SS et al (2006) Shear pulses nucleate fibril aggregation. Food Biophys 1:144–150. https://doi.org/10.1007/s11483-006-9012-5
53. Heyn TR, Schrader M, Kampen I et al (2023) Glass beads increase the formation kinetics of beta-lactoglobulin amyloid fibrils. Food Hydrocoll 139:108511. https://doi.org/10.1016/j.foodhyd.2023.108511
54. Yue J, Yao X, Gou Q et al (2022) Recent advances of interfacial and rheological property based techno-functionality of food protein amyloid fibrils. Food Hydrocoll 132:107827. https://doi.org/10.1016/j.foodhyd.2022.107827
55. Mohammadian M, Madadlou A (2018) Technological functionality and biological properties of food protein nanofibrils formed by heating at acidic condition. Trends Food Sci Technol 75:115–128. https://doi.org/10.1016/j.tifs.2018.03.013
56. Lux J, Kieserling H, Koop J et al (2023) Identification of an optimized ratio of amyloid and non-amyloid fractions in engineered fibril solutions from whey protein isolate for improved foaming. Colloids Surf, A 660:130849. https://doi.org/10.1016/j.colsurfa.2022.130849
57. Keppler JK, Steffen-Heins A, Berton-Carabin CC et al (2018) Functionality of whey proteins covalently modified by allyl isothiocyanate. Part 2: Influence of the protein modification on the surface activity in an O/W system. Food Hydrocoll 81:286–299. https://doi.org/10.1016/j.foodhyd.2018.03.003
58. Milsmann J, Oehlke K, Greiner R et al (2018) Fate of edible solid lipid nanoparticles (SLN) in surfactant stabilized o/w emulsions. Part 2: Release and partitioning behavior of lipophilic probes from SLN into different phases of o/w emulsions. Colloids Surf, A 558:623–631. https://doi.org/10.1016/j.colsurfa.2017.05.050
59. Frenzel M, Steffen-Heins A (2015) Impact of quercetin and fish oil encapsulation on bilayer membrane and oxidation stability of liposomes. Food Chem 185:48–57. https://doi.org/10.1016/j.foodchem.2015.03.121
60. Jores K, Mehnert W, Mäder K (2003) Physicochemical investigations on solid lipid nanoparticles and on oil-loaded solid lipid nanoparticles: a nuclear magnetic resonance and electron spin resonance study. Pharm Res 20:1274–1283. https://doi.org/10.1023/A:1025065418309
61. Murrieta-Martínez CL, Soto-Valdez H, Pacheco-Aguilar R et al (2018) Edible protein films: Sources and behavior. Packag Technol Sci 31:113–122. https://doi.org/10.1002/pts.2360
62. Peydayesh M, Bagnani M, Mezzenga R (2021) Sustainable bioplastics from amyloid fibril-biodegradable polymer blends. ACS Sustain Chem Eng 9:11916–11926. https://doi.org/10.1021/acssuschemeng.1c03937
63. Deegan RD, Bakajin O, Dupont TF et al (1997) Capillary flow as the cause of ring stains from dried liquid drops. Nature 389:827–829. https://doi.org/10.1038/39827
64. Gao M, Huang X, Zhao Y (2018) Formation of wavy-ring crack in drying droplet of protein solutions. Sci China Technol Sci 61:949–958. https://doi.org/10.1007/s11431-017-9202-y
65. Gounga ME, Xu S-Y, Wang Z (2007) Whey protein isolate-based edible films as affected by protein concentration, glycerol ratio and pullulan addition in film formation. J Food Eng 83:521–530. https://doi.org/10.1016/j.jfoodeng.2007.04.008
66. Carpenter JF, Prestrelski SJ, Arakawa T (1993) Separation of freezing- and drying-induced denaturation of lyophilized proteins using stress-specific stabilization. I. Enzyme activity and calorimetric studies. Arch Biochem Biophys 303:456–464. https://doi.org/10.1006/abbi.1993.1309
67. Keppler et al (2021) Towards recombinantly produced milk proteins: physicochemical and emulsifying properties of engineered whey protein beta-lactoglobulin variants. Food Hydrocoll 110:106132

Crystal Breakage of Mechanically Labile Protein Crystals in the Field of Mechanical Solid-Liquid Separation

Benjamin Radel and Hermann Nirschl

Nomenclature

A	Area (mm^2)
C	Relative Centrifugal Acceleration (–)
f	Solids Flux ($m\ s^{-1}$)
f_{bk}	Flux Density Function ($m\ s^{-1}$)
g	Gravitation ($m\ s^{-2}$)
h	Height (mm)
K	Specific Permeability (m^2)
κ	Cake Concentration Number (–)
m	Mass (g)
p	Pressure (Pa)
p_0	Reference Pressure (Pa)
p_s	Compressive Yield Stress (Pa)
Q	Particle Size Distribution (–)
q	Particle Size Density Distribution (μm^{-1})
R	Resistance (m^{-1})
r	Radial Distance (mm)
T	Temperature (°C)
t	Time (s)
V	Volume (m^3)
v	Velocity ($m\ s^{-1}$)

B. Radel · H. Nirschl (✉)
Institute for Mechanical Process Engineering and Mechanics, Karlsruhe Institute of Technology, Karlsruhe, DE, Germany
e-mail: hermann.nirschl@kit.edu

B. Radel
e-mail: benjamin.radel@gmx.de

A. Kwade and I. Kampen (eds.), *Dispersity, Structure and Phase Changes of Proteins and Bio Agglomerates in Biotechnological Processes*,
https://doi.org/10.1007/978-3-031-63164-1_8

x Diameter (μm)
α_m Mass Specific Resistance (m kg^{-1})
α_h Height Specific Resistance (m^{-2})
ϵ Porosity (–)
η Dynamic Viscosity (Pa s)
ω Angular Frequency (s^{-1})
ϕ Solids Volume Fraction (–)
ρ Density (kg m^{-3})
z Centrifugation Factor (–)

Indices

0 Initial Value
c Filter Cake Related
g Gel Point Related
l Liquid Related
m Filter Media Related
s Suspension or Solids Related

1 Introduction

The market demands for proteins to catalyze reactions is increasing worldwide. A common advantage in using proteins to catalyze chemical reactions is their ability to convert substances in an enantiomerically pure form. In 2015, global sales estimated by Novozymes were 3.3 billion Euro [1]. The portal BioSpace published a report by Emergen Research, according to which sales in 2020 were already 10.47 billion US dollar. By 2028, it is predicted to increase to 17.84 Billion US dollar [2]. The demand for proteins for general technical applications in the fields of food technology, industry (detergents, catalysts, ...) [1] and pharmaceuticals is also growing continuously. In the food sector, the demand for alternatives to meat protein is increasing. In technical food production, proteins ensure constant and reliable product quality in bread, cheese and wine production as well as in the fabrication of meat products. Detergents contain proteins that break down starch and fats, thus ensuring highly effective cleaning. In pharmaceuticals, proteins are relevant as therapeutic agents. They are used in vaccines and medicines. Pharmaceutical proteins for cancer therapy, the monoclonal antibodies, are particularly high-priced proteins with increasing demand. Since proteins are complex macromolecules with a three-dimensional structure, their production is only possible in organisms. Simpler proteins can be produced via bacteria, whereas more complex proteins such as antibodies require the use of animal cells. The use of bacteria or animal cells for production always results in protein by-products. These are proteins that the cell needs for its own metabolism. Cost effective purification of the target protein with high purity is therefore a challenge for separation technology.

The purification of proteins requires several methods from the field of solid-liquid separation. These steps are centrifugation, chromatography and filtration [3–5]. Whereas centrifugation uses the density difference between disperse and continuous phase, filtration is based on the particle size. Also, methods from the field of thermal process engineering are used for the purification and formulation of proteins. One such method is the selective crystallization of the target protein [5, 6]. A successful preparative protein crystallization step does not only increase the purity but also serves as a formulation step. This means that the crystallization process produces crystals with a desired size and shape distribution. The major challenge is to find proper crystallization conditions for the target protein [7]. Many native proteins are difficult to crystallize which is why biotechnologists are trying to improve the crystallization of proteins [8]. Conventional crystallization processes using evaporation or cooling are often not directly applicable for proteins because such molecules are much more sensitive to high and low temperatures, pH shifts and ion strength. Small scale crystallization setups typically use precipitation by adding a high salt solution to the crystallization buffer. Such methods, however, do not scale well because high gradients at the inlet lead to inhomogeneous crystal size distributions and to a dilution of the crystallization batch. Alternative methods for large scale protein crystallization are therefore in the focus of current research [9, 10].

The further processing of the protein crystals with mechanical solid-liquid separation requires special care. Protein crystals are built up by large molecules and have a high crystal water content. Therefore, these crystals have less mechanical strength compared to conventional organic or inorganic crystals. For solid-liquid separation, this means that filtration and centrifugation of the crystallizate may lead to crystal breakage. Breakage changes the product properties and, in most cases, worsens the separation efficiency [11, 12].

In the early development stages, the available amount of protein crystal suspension is limited. The reasons are, that protein development, like enhancing the crystallization properties [13], or crystal development [14], like cross linking for improving the mechanical strength, are conducted using only few milliliters. To allow an early characterization of the filtration behavior, a downscaled filtration setup is necessary [15]. Ideally, not only the cake resistance and solids volume fraction, but also the compressibility and flux density function are derived for the protein crystal suspension [16]. With such data from the small scale, a transfer to a larger filtration device is possible [17]. Finally, the breakage behavior of the protein crystals has to be considered carefully. Many devices induce normal and shear stress on the crystal suspension. This is why crystal breakage under combined normal and shear stress has to be investigated and considered [18].

2 Theoretical Background

The flux density function characterizes settling and permeation behavior of a compressible particle system. Therefore, the theory of cake filtration, settling and the flux density function are outlined in the following. Additionally, the basic principles of comminution are discussed.

2.1 *Cake Filtration*

The working principle of cake filtration is the separation of particles on a porous filter medium. The requirements for a efficient cake filtration is a high enough solids content in the suspension and a big enough particle size for a low permeation resistance [19]. The Darcy law, empirically developed by Henry Darcy, describes the permeation of a packed bed. Today it is known, that Darcy's law is a special solution of the Navier-Stokes equations. The Darcy law

$$\dot{V} = \frac{\mathrm{d}V}{\mathrm{d}t} = \frac{\Delta p A}{R\eta} \tag{1}$$

connects the volumetric flow ($\dot{V}$), the cross sectional area (A), the dynamic viscosity (η) and the permeation resistance (R) [19]. The permeation resistance is the combination of the filter media resistance (R_m) and the filter cake resistance (R_c). The filter cake resistance depends on the height of the filter cake (h_c). To make the values of the filter cake resistance more meaningful and comparable, it is usually normalized with the cake height (α_h) or the cake mass (α_m):

$$\alpha_h = \frac{R_c}{h_c} = K^{-1} \tag{2}$$

$$\alpha_m = \frac{\alpha_h}{\phi_c \rho_s} \tag{3}$$

The reciprocal value of the height specific cake resistance is the height specific permeability (K). The symbol ρ_s describes the solids density and ϕ_c is the solids volume fraction in the filter cake. The latter is defined as

$$\phi_c = 1 - \epsilon = \frac{V_s}{V_s + V_p}, \tag{4}$$

with V_s as the solids volume and V_p as the volume of the void. The counterpart to the solids volume fraction is the porosity (ϵ). In the case of incompressible particle systems, the normalized resistance as well as the solids volume fraction are constant for the whole filter cake.

Insertion into the Darcy equation and subsequent integration yields the typical filter equation which follows in its linearized form:

$$\frac{t}{V} = \frac{\alpha_h \kappa \eta}{2 \Delta p A^2} V + \frac{R_m \eta}{\Delta p A} \tag{5}$$

with κ as cake concentration number:

$$\kappa = \frac{V_c}{V_l} = \frac{\phi_s}{1 - \phi_s - \epsilon}. \tag{6}$$

The linearized form in Eq. (5) is commonly used to determine cake and filter media resistances. The decisive advantage of the linearization is the easy evaluation with a simple linear representation. When t/V is plotted versus V, the slope represents the cake resistance and the ordinate the filter media resistance.

For compressible material systems, the filter cake is inhomogeneous. The cake resistance and solids volume fraction are therefore not constant, but depend on the cake height. Typically, the highest cake resistance and solids volume fraction can be found on the bottom of the filter cake. Thus, using Eq. (5) for the evaluation of compressible material systems provides only the height averaged resistance and solids volume fraction.

Typical cake filtration experiments require liters of suspension. Due to the limited amount of available protein crystal suspension a reduction of sample size is necessary. One method to achieve such a reduction is the application of analytical photocentrifugation. Centrifugal filtration uses the centrifugal force to create a pressure gradient. Therefore, the pressure does not remain constant but decreases during the filtration. The reason for this is the reduction in hydrodynamic pressure due to the decreasing fluid height above the filter. Hence, to calculate filter cake resistances a differential approach is necessary.

Figure 1 shows the working principle of centrifugal filtration schematically. During the filtration process three different areas form. Area I contains only clear supernatant due to superimposed sedimentation. Area II contains the suspension and area III the filter cake. Particles are retained on the filter medium at position r_m. The reservoir catching the filtrate is shown on the right-hand side of Fig. 1. Loginov et al. [20] published a method to quantify the filter cake resistance by conducting a permeation experiment with a photocentrifuge. For such an experiment, only areas I and III are present. By measuring the transmission through the filter cell during permeation, the position of the liquid-gas interface r_l and the position of the filter cake top r_c are traceable. The resulting equation to obtain the filter cake resistance according to Loginov et al. [20] follows

$$\underbrace{\ln \frac{r_m + r_c - h_l}{r_m - r_c + h_l} \frac{r_m - r_c + h_{l,0}}{r_m + r_c - h_{l,0}}}_{=Y} = \underbrace{\frac{\rho_l \omega^2 r_m}{\eta(\alpha_h (r_m - r_c) + R_m)}}_{=k} t, \tag{7}$$

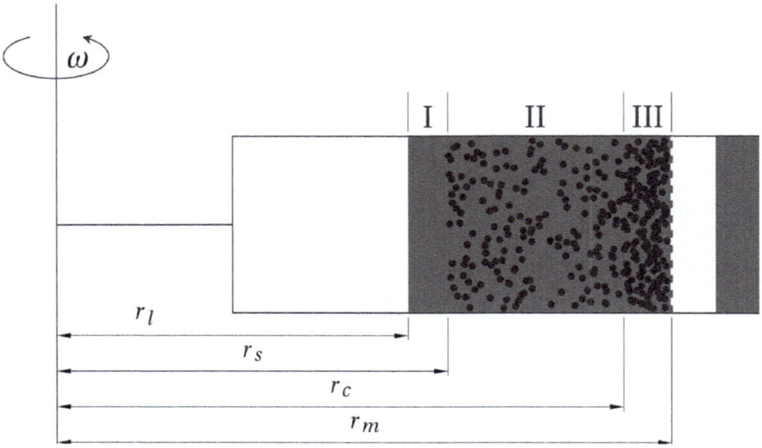

Fig. 1 Schematic working principle of centrifugal dead-end filtration. Reprinted with permission from [32], license: CC-BY-SA 4.0

with h_l as the height of the liquid above the filter cake, $h_{l,0}$ as the initial height of the liquid, ρ_l as liquid density, ω as the angular frequency and t as the time. When the data is plotted in a graph Y versus kt, the resulting curve can be interpreted and evaluated analogous to Eq. (5).

2.2 *Settling*

Settling is the sedimentation of a solid phase dispersed in a continuous fluid phase due to a density difference. For low solids volume fractions, the particles do not influence each other and hence settle independently. For such low concentrations and spherical particles in a creep flow Stokes derived the equation

$$v_{st} = \frac{(\rho_s - \rho_l)x^2 gz}{18\eta} \tag{8}$$

which allows to calculate the settling velocity v_{st} using the particle diameter x and earth gravity g. In the centrifugal field, the centrifugation factor z which transforms the earth gravitational field to the corresponding centrifugal gravitational field is needed. Although Eq. (8) is only valid for monodispersed spherical particles in a creep flow at low volume concentrations, it is typically applied to all sorts of particle systems. At higher concentrations, the particles influence each other which ultimately leads to hindered settling. For these systems Richardson and Zaki [21] proposed the modification

$$v_{rz} = v_{st}(1 - \phi)^N \tag{9}$$

connecting the stokes settling velocity with the solids volume fraction and an additional parameter N. This model still requires the particles to be monodisperse and spherical. In the case of laminar creep flow N equals 4.65. In technical practice the Richardson and Zaki model is also applied to all sorts of particle systems. Stokes and Richardson and Zaki are probably the most common models to describe settling. Several additional models are published in the literature which introduce further parameters to better describe experimental observations.

2.3 Process Functions

Process functions model the material behavior depending on a certain material parameter, like the solids volume fraction. In the case of cake filtration, Tiller and Kwon [22] found the correlation for compressible cake filtration of

$$\alpha = \alpha_0 \left(1 + \frac{p}{p_0}\right)^N, \tag{10}$$

containing the three parameters α_0, p_0 and N. These parameters are fitted to experimental data. For α_0 and p_0 it is common to select the first experimental data point. This correlation allows to calculate the cake resistance of compressible filter cakes depending on the applied pressure.

Analogous to Eq. (10), Tiller and Kwon developed a relation of the solids volume fraction depending on the compressive yield stress p_s

$$\phi = \phi_0 \left(1 + \frac{p_s}{p_0}\right)^\beta. \tag{11}$$

The compressive yield stress is zero for solid volume fractions below the gel point (ϕ_g). The gel point is the solids volume fraction at which the system transitions from a suspension towards a particulate network. For an ideal incompressible material system, the compressive yield stress jumps from zero to infinity at the gel point which is also the maximal solids volume fraction. In the case of compressible material systems, the compressive yield stress is zero at the gel point and then rises towards infinity at the maximal solids volume fraction. The maximal solids volume fraction depends on the material, particle sizes and shapes.

Kynch [23] developed the flux density function to describe the settling of flocculated systems. Over the years this theory was enhanced and amended by [24–27]. By combining the mass balance for fluid and particulate phase, Kynch derived the relation

$$f_{bk} = \phi(1 - \phi)v_{rel} \tag{12}$$

which describes the Kynch batch flux density function with v_{rel} as the relative velocity between solid and fluid phase. For a system with no inlet and outlet the flux density function equals the solids flux f

$$f = f_{bk} = \phi v_s. \tag{13}$$

Tiller and Holdich and Butt [25, 28] added a method to express the flux density function also for particulate systems above the gel point. It follows

$$\phi v_s = f_{bk} = -\frac{K(\phi)\Delta\rho g \phi^2}{\eta}, \tag{14}$$

with K as height specific permeability. With this enhancement, the flux density function describes the settling and permeation behavior with only the solids volume fraction as parameter. For this model, the requirements are zone sedimentation, small particles in relation to the vessel, incompressible particles and liquid, no mass transfer between the phases and that the settling velocity is only dependent on the solids volume fraction. Combining Eqs. (9)–(14) leads to the complete flux density function

$$f_{bk} = \begin{cases} -v_{st}\phi(1-\phi)^N & 0 \leq \phi < \phi_g \\ -\frac{K(\phi)\Delta\rho g\phi^2}{\eta} & \phi_g \leq \phi. \end{cases} \tag{15}$$

2.4 Comminution

Particle breakage occurs when the material strength is exceeded and is divided into the phases cracking, crack initiation and crack propagation. Cracking starts as soon as the material strength is exceeded. Cracks, therefore, form at locations which are already under stress or defective. At sufficient high energy, crack initiation and subsequent crack propagation occurs. Two conditions apply to particle breakage. The force condition states that for the formation of a crack the binding forces must be overcome. The energy condition means that energy consumed at the crack front must be continuously replenished [29]. The mechanical strength of particles depends on the size. The smaller the particles are, the higher is the mechanical strength.

3 Material and Methods

To investigate the filtration properties and breakage of protein crystals on a small scale, several analytics and methods are required. For most of the experiments, we used lysozyme as a model protein. The small-scale filtration experiments were performed with an in house developed filtration cell for an analytical photocentrifuge. To validate the results, we also investigated the filtration for selected crystal shapes with a typical laboratory nutsche filter.

3.1 Protein Crystallization

Lysozyme can crystallize in different shapes depending on the crystallization conditions. With precipitation it is possible to generate isometric and needle shaped crystals. Low temperature evaporative crystallization allows to produce rod-like lysozyme crystals additionally. This method is also suitable for crystallizing the protein alcohol dehydrogenase (ADH) from Lactobacillus brevis, which was kindly provided by Prof. Weuster-Botz.

Precipitation crystallization of lysozyme requires a 50 mM acetic acid buffer at pH 4. From this buffer two stock solutions are prepared. In stock solution one 100 g L^{-1} lysozyme (Ovobest Eiprodukte GmbH, Germany) are dissolved. Stock solution two contains 80 g L^{-1} sodium chloride. Isometric lysozyme crystals are produced by adding 125 mL stock solution two to 125 mL stock solution one at a rate of 1 mL min^{-1} with a membrane pump at 20 °C.

Low temperature evaporative crystallization for proteins is described in detail by Barros-Groß [7, 9]. This method allows to crystallize isometric, needle shaped and rod-like lysozyme crystals as well as rod-like ADH crystals. Whereas isometric and needle shaped lysozyme crystals grow at 20 °C, the rod-like lysozyme crystals need a temperature of 35 °C. In the case of isometric and rod-like lysozyme crystals, a citric acid buffer at pH 4.5 with no additional sodium chloride is used. The crystallizate of such proteins was kindly provided by Prof. Kind.

3.2 Particle System

The four main investigated particle systems are isometric, needle and rod-like shaped lysozyme crystals, as well as rod-like ADH crystals. Microscope images and typical size distributions of these particle systems can be seen in Figs. 2 and 3.

Microscope images were obtained with a Leitz Ortohplan microscope at 100 × magnification and particle size measurements were conducted with laser scattering (Sympatec Helos H0309).

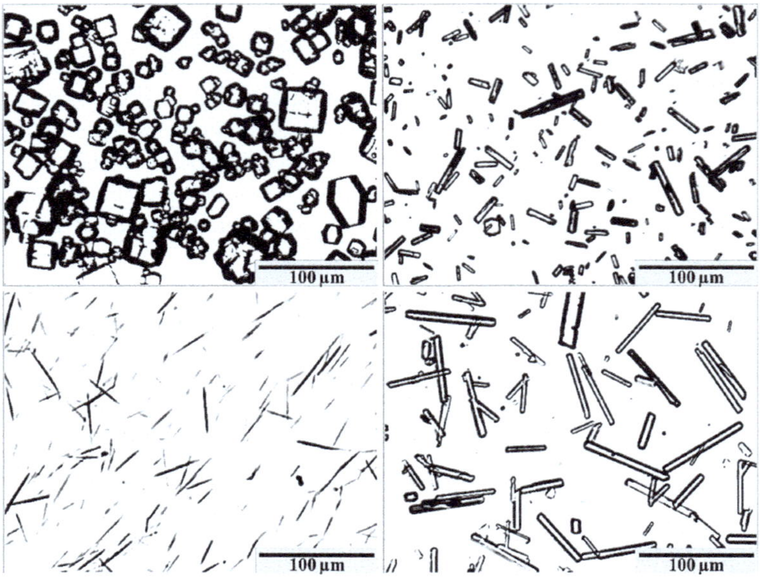

Fig. 2 Microscope images of isometric (top left), rod-like (top right), needle shaped (bottom left) lysozyme crystals and rod-like ADH crystals (bottom right). Reprinted with permission from [32], license: CC-BY-SA 4.0

Further material properties are summarized in Table 1. The solid densities were obtained with a Micromeritics gas pycnometer MP 1305 and liquid densities as well as dynamic viscosity were measured with a Mettler Toledo Densito 30PX according to the oscillating tube method.

4 Filtration

Filtration experiments in a filter nutsche were performed according to VDI 2762 [30]. To ensure that all particles are retained, a Pall Ultipor membrane with a nominal pore size of 0.2 μm was used as a filter medium. After determining the membrane resistance and building the filter cake, permeation experiments of the prebuilt filter cake at constant pressure were performed. The resistance was determined using the time resolved filtrate mass.

For the small-scale characterization, the analytical photocentrifuge LUMiSizer (LUM GmbH) was used. A special translucent filter cell contains the suspension, a filter medium and a reservoir which catches the filtrate. To ensure that all particles are retained, the Pall Ultipore membrane with 0.2 μm pore size was used as well. In a first step the filter cake is prebuilt. Afterwards, about 300 μL clear supernatant are added carefully on the top of the filter cake. The permeation of the filter cake occurs in the range of 500–4000 revolutions per minute.

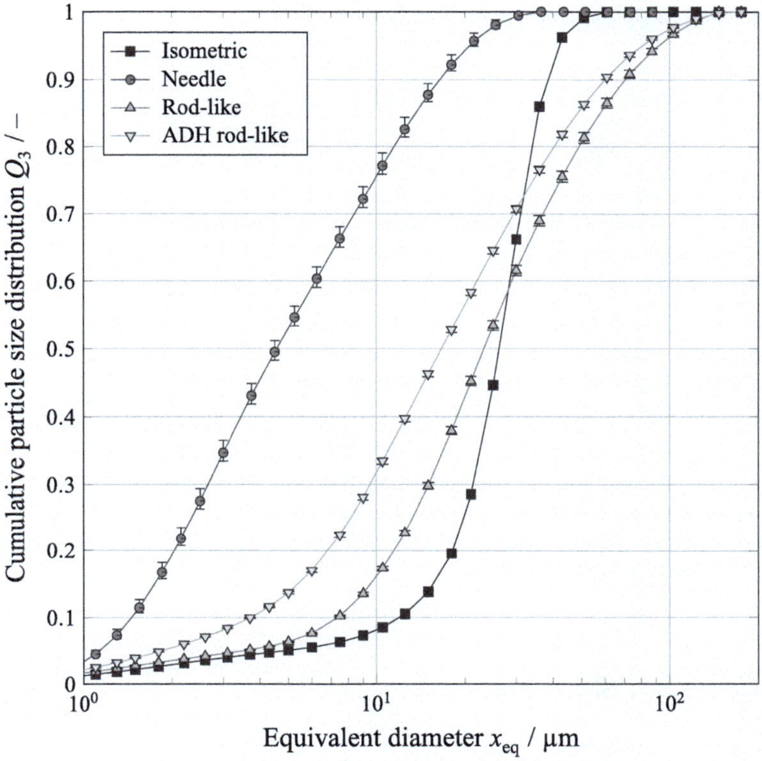

Fig. 3 Particle size distributions from laser scattering for isometric, needle and rod-like shaped lysozyme and rod-like alcohol dehydrogenase (ADH) crystals. Reprinted with permission from [16], license: CC-BY 4.0

Table 1 Material properties of the different crystal systems

Particle system	Solid density ρ_s/kg m^3	Liquid density ρ_l/ kg m^3	Dynamic viscosity η/ mPa s
Lysozyme isometric	1236	1023	1.19
Lysozyme rod-like	1236	1023	1.19
Lysozyme needle	1236	1023	1.19
Alcohol dehydrogenase	1250	1026	1.92

4.1 Computed Micro Tomography

The sample preparation for Computed Micro Tomography (μCT) involves several steps to ensure a sufficient contrast in the scans. Firstly, a small sample is cut out of a protein crystal filter cake or sediment with a polyimide tube with an inner diameter of 3.48 mm. Afterwards, the sample is lyophilized and the tube is sealed with resin. The sealed tube is mounted on a sample holder and scanned in the Zeiss XRadia Versa 520 X-Ray microscope at a voxel size of about 0.75 μm. The amount of projections was at least 2100, and the source energy settings were 4 W at 50 kV.

For the analysis of the obtained reconstruction, the noise was reduced by applying the *non local means* image filter. The region of interest is defined by applying a threshold to the filtered image. In the resulting binary image, touching crystals are separated with a three-dimensional watershed transformation using Euclidian distance transformation. Crystals touching an image border are deleted and the volume, equivalent diameter and maximal diameter of the remaining crystals are calculated with either Dragonfly from Object Research Systems or ImageJ. In the case of rod-like lysozyme crystals, the segmentation was performed with an artificial neural network instead of a threshold.

4.2 Combined Shear and Normal Loading

For the combined shear and normal loading of saturated protein crystal sediments, the Schulze Ring Shear Tester RST-01.pc was used with a modified shear cell. The modifications of the shear cell are described in detail by Hammerich et al. [31]. The sediment is loaded by applying a defined weight to the shear cell lid and rotating the bottom of the shear cell at a defined speed. The shear path was held constant for all experiments. To reduce the influence of the typical loading and unloading cycles, usually performed for the characterization of powders, on particle breakage, only one loading cycle was performed per experiment. To analyze the crystal breakage, the particle size distribution of an unloaded sample and a sample after loading were determined with laser scattering. Selected samples were also subjected to μCT analysis.

5 Results and Discussions

5.1 Small Scale Filtration

To achieve a reduction of the required sample amount, a special filter cell for an analytical photocentrifuge is required. Since the permeation progress is detected with light transmission, the filter cell has to be translucent. To achieve high flexibility and to allow an exchangeable filter medium, the filter cell consists of two 3D printed parts. The bottom part catches the filtrate and the top part holds the filter cake and supernatant. In between those parts the support structure and filter medium are located. An image of the filter cell may be seen in Fig. 4. On the top of Fig. 4 a commercial standard cuvette for sedimentation analysis is shown. In the middle of the image, the 3D printed filter cell, a metal support structure and the filter medium are depicted. When the bottom and the top of the filter cell are assembled, the black O-Rings are used to further stabilize the joint. On the bottom of the image a variant of the filter cell with a glass window is shown. The glass window further improves the light transmission through the interesting part of the filter cell and reduces artifacts caused by the structure of the 3D print.

The phase boundaries between air/supernatant and supernatant/filter cake yield a clear signal in the transmission profile. As the permeation progresses, the air/supernatant boundary moves towards the filter cake. At the same time, the air/filtrate interface at the bottom of the filter cell moves towards the left as the amount of filtrate

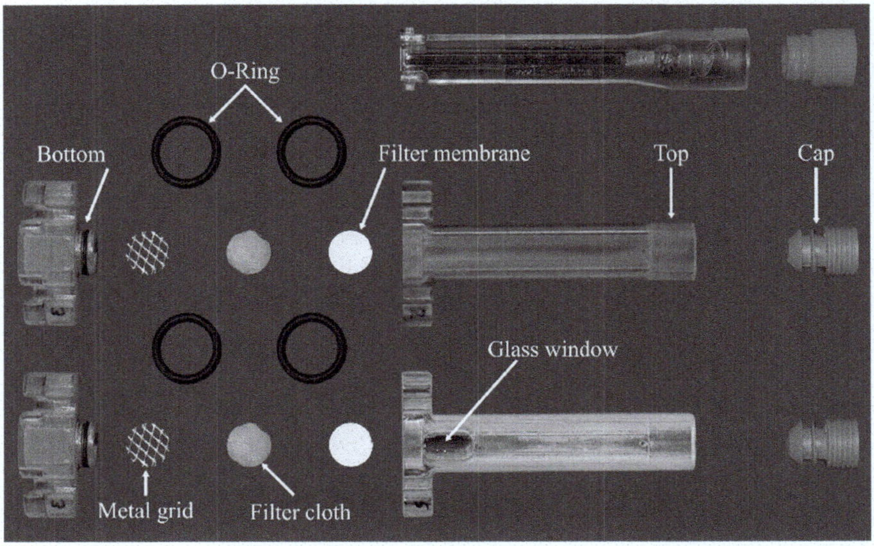

Fig. 4 Top: Commercial cuvette for sedimentation analysis. Middle: 3D printed filter cell without glass window. Bottom: 3D printed filter cell with glass window. Reprinted with permission from [32], license: CC-BY-SA 4.0

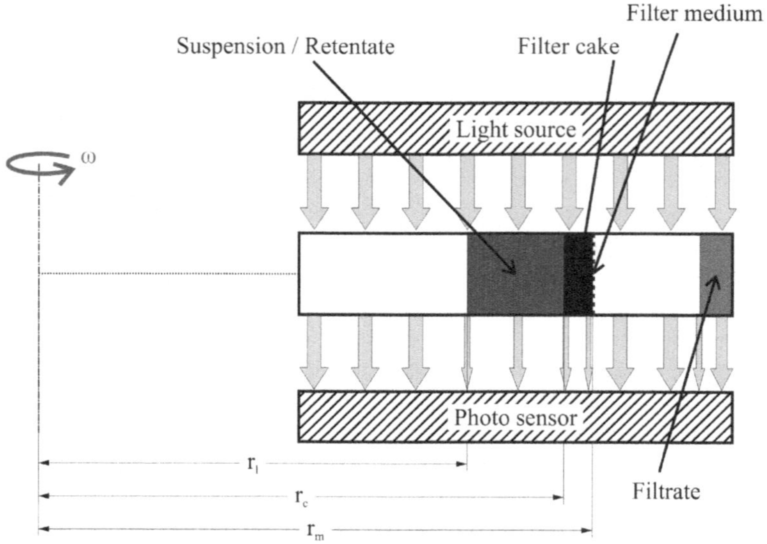

Fig. 5 Schematic drawing of the filter cell's working principle. The phase boundaries are visible in the detected light transmission [15]. Reprinted with permission from Elsevier Ltd.

increases. The basic working principle and the observable phase interfaces can be seen in Fig. 5.

Figure 6 shows the transmission profiles for a permeation experiment. Each line in the transmission profile represents one timestep (black: first profile, light gray: most recent profile). The minima indicate the air/supernatant interface r_l. At the upper side of the filter cake, r_c, the transmission drops to zero, because no light is able to pass through the filter cake. The membrane position r_m is obtained from the transmission profile of an empty filter cell.

The data depicted in Fig. 6, allows to calculate the filter cake resistance according to Eq. (7). Also, the filter cake height can be directly calculated with $h_c = r_m - r_c$. The solids volume fraction of the filter cake follows from the cake height with the assumption that all particles are retained. Hence, this setup allows to measure and calculate all relevant properties and requires only about 200–500 μL sample volume per experiment.

5.2 *Filtration Properties of Different Protein Crystals*

With the 3D printed filter cell, we investigated the height specific cake resistances of the four particle systems: Isometric, rod-like and needle shaped lysozyme crystals and rod-like ADH crystals. The resulting mass specific filter cake resistances for different permeation pressures can be seen in Fig. 7. The solid lines are fits calculated according to Eq. (10). For the isometric lysozyme crystals, the obtained filter cake resistances

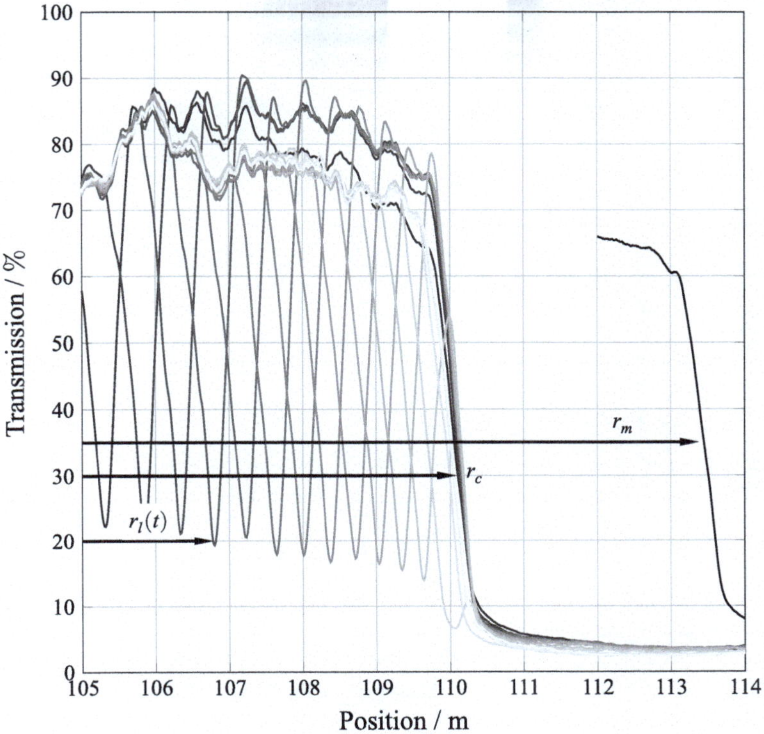

Fig. 6 Exemplary transmission profile of a permeation experiment. The position r_l moves to the right with increasing permeation time (black: first, light gray: most recent profile). The transmission drops to zero at the upper side of the filter cake. The membrane position (black line at the right) is obtained from the measurement of an empty filter cell. Reprinted with permission from [32], license: CC-BY-SA 4.0

were compared with the results from a conventional nutsche filter (diamonds), which yielded similar results.

In the case of isometric and rod-like lysozyme crystals, Fig. 7 shows that the filter cake resistance remains almost constant with increasing permeation pressure. For the rod-like ADH crystals, however, a significant increase of the cake resistance at higher pressures is observed. This increase is even more evident for the needle shaped lysozyme crystals. Here, the filter cake resistance increases in the order of almost three magnitudes. An increase of the filter cake resistance at higher pressures is an indicator for compressible material behavior. The very steep increase over several magnitudes in the case of needle shaped lysozyme crystals is a hint for crystal breakage.

To further evaluate the compressibility and filtration efficiency, the compressible yield stress in dependency of the solids volume fraction has to be evaluated additionally. Fortunately, this data can be calculated from the transmission profile data, which means that no additional experiments are required. However, with the filter

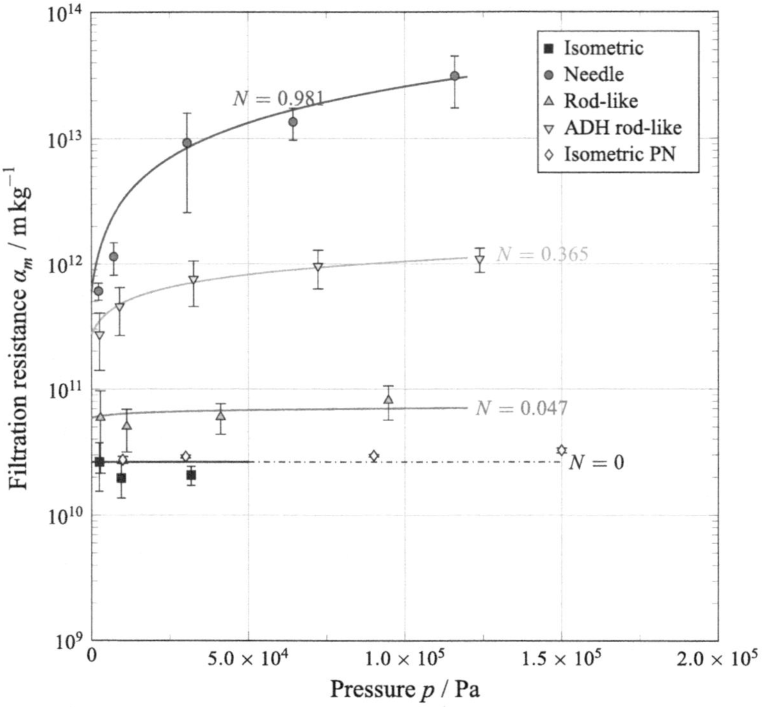

Fig. 7 Mass specific filter cake resistance of the four protein crystal systems: Isometric, needle and rod-like lysozyme crystals and rod-like ADH crystals. The curves are fits calculated according to Eq. (10). The exponent N of the fit is shown next to the corresponding curve. The diamonds show the results from a laboratory pressure nutsche for isometric lysozyme crystals for comparison. Reprinted with permission from [16], license: CC-BY 4.0

cell and the photocentrifuge the maximum pressure is limited to about 2×10^5 Pa. To apply a higher pressure, a different apparatus is necessary.

Figure 8 shows the compressible yield stress versus the solids volume fraction. The solid lines are model fits according to Eq. (11). In the case of the isometric lysozyme crystals a steep increase of the compressive yield stress at a gel point of about 0.26 can be observed. For the rod-like lysozyme crystals, the increase is also quite steep, but the gel point is lower at a solids volume fraction of about 0.21. Since both of those particle systems show a steep increase of the compressive yield stress, this further strengthens the assumption of incompressible or weakly compressible material behavior. For the other two systems (needle and ADH), the results are quite different. The gel point is much lower and lies below 0.1. The increase of the compressible yield stress is less steep and the maximum solids volume fraction reached with the small-scale setup is about 0.25. Hence, those particle systems show a much more pronounced compressible material behavior.

The filtration properties are influenced by the particle size distribution, mechanical stability and shape. For an efficient filtration, incompressible material behavior, a

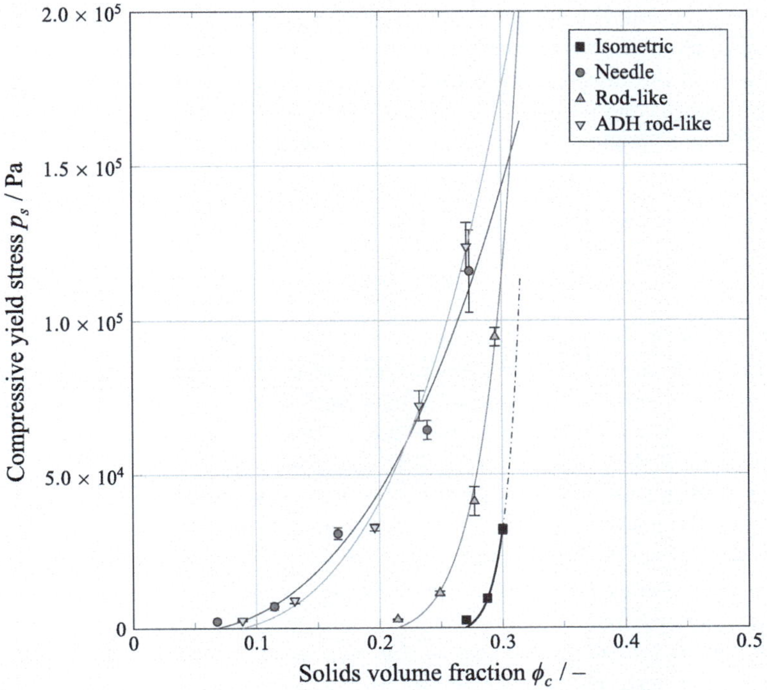

Fig. 8 Compressible yield stress versus the solids volume fraction for the four different crystal systems. The values are calculated from the small scale permeation experiments. Reprinted with permission from [16], license: CC-BY 4.0

high solids volume fraction (less moisture in the filter cake) and a low filter cake resistance (fast filtration) are desirable. Therefore, the isometric and rod-like lysozyme crystals have better filtration properties than the needle shaped lysozyme crystals or the rod-like ADH crystals. Especially the needle shaped lysozyme crystals are highly compressible and probably break at higher pressures. The filter cake resistance rises by several magnitudes with increasing filtration pressure. From the investigated particle systems, the needle shaped lysozyme crystals, therefore, have the worst filterability.

5.3 *Flux Density Function*

The flux density function describes settling and permeation of a compressible particle system. It depends only on the solids volume fraction of the suspension or of the sediment/filter cake. Therefore, it is especially useful for processes in which both effects play a significant role. By measuring the settling velocity for different solids volume fractions, the flux density function is characterized for solids volume fractions below the gel point. For solids volume fractions above the gel point, the data from the small-scale permeation experiments is used. However, one requirement is, that all particles in the sample settle with almost the same velocity. This is called zone sedimentation. Using the model equations of Richardson and Zaki [21], Tiller and Kwon [22] and Eq. (15), the flux density function can be formulated as

$$f_{bk} = \begin{cases} -v_{st}\phi(1-\phi)^N & 0 \le \phi < \phi_g \\ -\frac{\Delta\rho g \phi^2}{\alpha_0 \left(\frac{\phi}{\phi_0}\right)^{\frac{N}{\beta}} \eta} & \phi_g \le \phi. \end{cases} \tag{16}$$

The complete flux density function is only available for the needle shaped lysozyme crystals, because only this particle system provides zone settling at low solids volume fractions. The corresponding complete flux density function is shown in Fig. 9 left. The dashed line corresponds to the modified Richardson and Zaki model and the dash dotted line to the Tiller and Kwon model.

The right-hand side of Fig. 9 shows the flux density function for both needle shaped lysozyme and rod-like ADH crystals for the permeation branch. The values of the flux density function for the ADH crystals are much lower than for the lysozyme needles. This can be explained by the higher permeability of the ADH crystals compared to the lysozyme crystals.

Using an analytical photocentrifuge for measuring the settling velocity and the filter cake resistance with an 3D printed filter cell, enables the characterization of the flux density function on a small scale. One settling experiment requires about 800 μL and one permeation experiment up to 300 μL sample. The small-scale setup allows to reduce the required number of experiments and to accurately estimate the settling and permeation properties in an early stage of product development, when only small amounts of crystal suspension are available. This reduces the time to market and enables a more efficient process design. Furthermore, the flux density

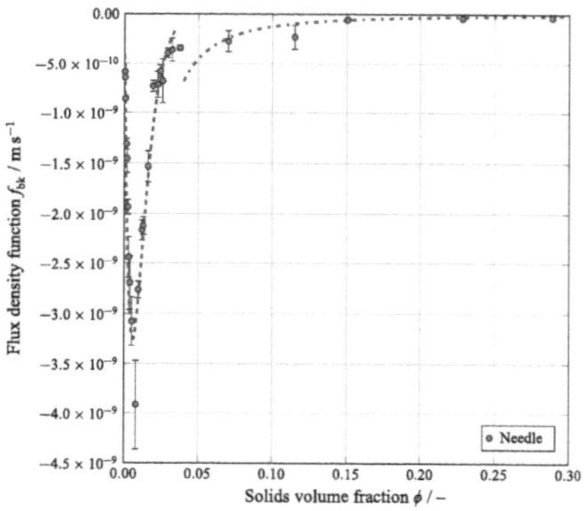

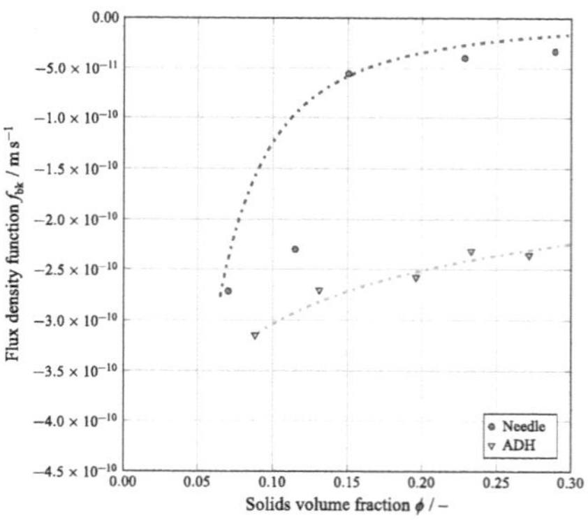

Fig. 9 Flux density function for needle shaped lysozyme crystals and rod-like ADH crystals. Left: Complete flux density function for needle shaped lysozyme, Right: Flux density function of both particle systems from permeation data. Reprinted with permission from [16], license: CC-BY 4.0

function can be used to carry out computer simulations of separation processes which further decreases development time and costs.

5.4 *Process Transfer to an Integrated Apparatus*

The process functions and data gathered on a small scale are now used to estimate the filtration on a larger and integrated apparatus, the belt crystallizer. This device integrates the unit operations low temperature evaporative crystallization, aging and filtration on one apparatus. The filtration step occurs with vacuum filtration and a filter cloth. For a detailed description of the device and apparatus design please refer to [17, 33]. A newly developed "process chamber" allows the evacuation of the crystallization batch and thus enables the low-temperature evaporation. For the generation of rod-like lysozyme crystals, for example, a pressure of 5500 Pa and a temperature of 35 °C are necessary. One challenge is to achieve a homogenous filter cake because the crystallization and aging take place directly on the filter medium.

In a first step, the filter cake resistance data from the small-scale experiments of rod-like lysozyme crystals are validated. Therefore, rod-like lysozyme crystals, generated by the newly developed process chamber, are filtrated on a laboratory nutsche filter to obtain the filtration resistance. The comparison of the results for the small-scale and integrated setup are shown in Fig. 10.

Both methods yield similar filtration resistances. In the case of the filter cell the resistances are in the range of $1.73 \times 10^{13} - 3.14 \times 10^{13}$ m^{-2} for filtration pressures between $0.03 \times 10^5 - 0.95 \times 10^5$ Pa. For the crystals generated in the process chamber and separated on the nutsche filter, the resistances are in the range of $1.93 \times 10^{13} - 2.44 \times 10^{13}$ m^{-2} for the three permeation pressures 0.25×10^5; 0.5×10^5 and 0.75×10^5 Pa. These resistances are very similar to each other. Thus, it can be concluded, that the small-scale setup yields meaningful results and that the crystal fabrication in the new device provides rod-like lysozyme crystals with equal properties than the batch vessel. The latter is also confirmed by comparing the crystal size distributions for the rod-like lysozyme crystals (not shown).

To ensure a homogenous cake structure on the belt crystallizer, we took several samples from the filter cake with a polyimide tube. These samples were subjected to lyophilization and subsequent μCT analytic. The resulting cumulative particle size distributions for different layers in the filter cake are shown in Fig. 11 left.

The cumulative crystal size distributions for the three different layers are very similar to each other. Therefore, the filter cake structure is homogenous with regard to the particle size. The original crystal suspension (black squares) seems to have a broader distribution. However, this curve is obtained from 2D microscope imaging with evaluation of the projection areas. Afterwards, the resulting Q_2 distribution has been recalculated to a Q_3 distribution. Hence, small deviations of the cumulative size distribution are to be expected.

The right-hand side of Fig. 11 shows an exemplary 3D rendering obtained from the μCT analytic. Crystals touching the border of the region of interest have

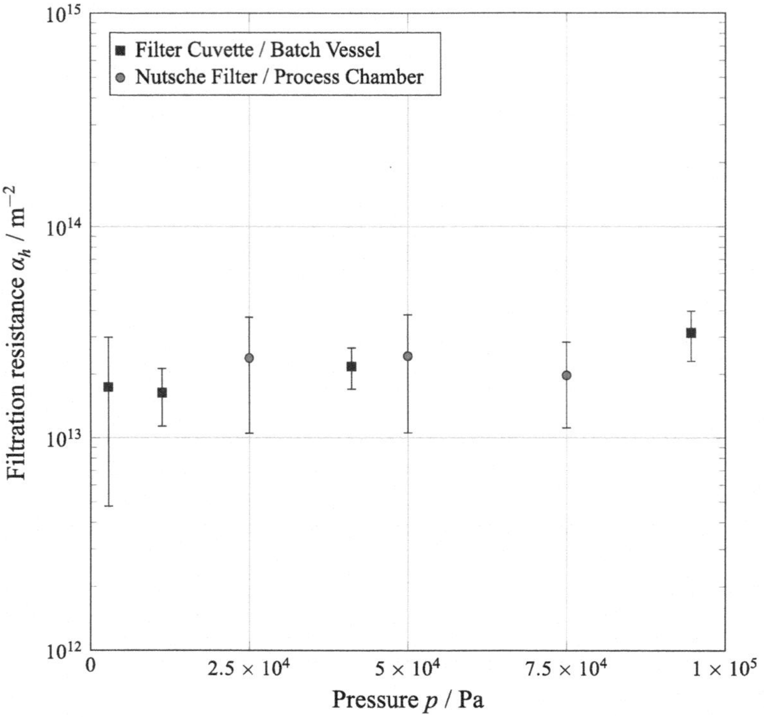

Fig. 10 Comparison of the filtration resistances for rod-like lysozyme crystals. Black squares: Crystals generated in a typical batch crystallizer and filtered with the small-scale filter cell. Gray circles: Crystals generated in the new process chamber and filtered on a standard laboratory nutsche filter. Reprinted with permission from [17], license: CC-BY 4.0

been removed. The rod-like shape of the crystals is clearly visible in the image. Connected crystals are separated into individual crystals with a three-dimensional watershed separation algorithm. For each individual crystal, the crystal's volume and maximal diameter are calculated. Additionally, the integral porosity for each layer was determined.

Because the small-scale experiments yield similar results for the filtration resistance and because the filter cake structure on the belt crystallizer is homogenous, it is possible to calculate the theoretical permeation duration on the new integrated apparatus concept. The resulting times for three individual permeation tests are summarized in Table 2.

The calculated times t_{calc} are obtained from the Darcy equation, using the average height specific filter cake resistance from the small-scale experiments of 2.50×10^{13} m^{-2}, the previously determined filter cloth resistance of 2.60×10^{8} m^{-1} and a filter area of 4.80×10^{-3} m^{2}. Experimental time t_{exp} is the time a defined volume of supernatant V requires to permeate through a pre-built filter cake with the height h_c. This time is measured with a stopwatch. Apparently, both the calculated and

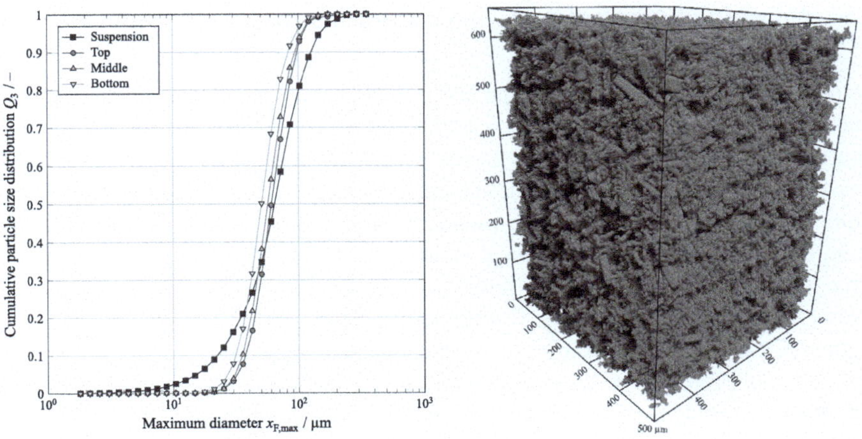

Fig. 11 Left: Cumulative particle size distributions from initial crystal suspension and three different layers of the filter cake. Right: 3D rendering of the filter cake structure from one layer. Reprinted with permission from [17] (right) [32] (left), licenses: CC-BY-SA 4.0 (left), CC-BY 4.0 (right)

Table 2 Calculated and theoretical permeation times [17]

No./–	1	2	3
V/m^3	50.8×10^{-6}	49.9×10^{-6}	48.9×10^{-6}
h_c/m	7.40×10^{-3}	6.40×10^{-3}	7.33×10^{-3}
t_{calc}/s	46.7	39.7	44.5
t_{exp}/s	44.2	37.8	42.3

the experimental determined permeation times differ only marginally. The slight differences between the times can be explained by the fact that the cake height is not constant over the whole filter cake surface.

With the newly developed process chamber, it is possible to implement the low temperature evaporation crystallization for protein crystals or other temperature sensitive compounds on the belt crystallizer. The resulting filter cake resistances for rod-like lysozyme crystals on the larger and integrated apparatus are very similar to the ones obtained from small-scale experiments with the filter cell. Also, the crystal size distribution is the same. Therefore, the small-scale setup is suitable to determine the filtration properties. With these properties the filtration can be transferred to a different apparatus. In the case of the larger belt crystallizer, the filtration time and the permeation duration can be predicted with values from the small-experiments using only about 300 µL of sample per experiment.

5.5 Combined Normal and Shear Loading and Comminution

Compared to conventional crystals, protein crystals have a lower mechanical strength. One cause is the high crystal water content which makes the crystals soft and elastic. This is also the reason, why the mechanical properties like Young's modulus or hardness differ considerably between moist and dried crystals. Additionally, the crystal lattice is composed of large macro molecules which also leads to less mechanical strength. To consider the mechanical behavior and comminution in a solid-liquid separation step, the experimental setup to examine the mechanical strength has to use moist and saturated crystals. Comminution during filtration is unwanted, since it generally worsens the filtration performance and changes the material properties. In the case of protein crystals, comminution might already occur at low pressures or low shear forces [11, 12]. Combined normal and shear stress, however, could have a much more severe impact on comminution. To assess the effects of combined normal and shear load, the Schulze ring shear tester is a useful device, because it allows to adjust normal and shear load separately. Typically, it is used to measure the flow properties of dry bulk powders. A modified ring shear cell with a drainage system by Hammerich et al. [31] enables the application of this device for saturated particulate networks.

In the ring shear tester, only a fraction of the sample is subjected to shearing. This can be illustrated by μCT images from a miniaturized ring shear device with highly X-ray absorbing tracer particles (green) and bulk particles (gray) shown in Fig. 12.

The left-hand side of Fig. 12 shows the initial, unloaded state. The green tracer particles are located in one place. The grooves on the top and bottom side result from drivers in the shear cell lid and shear cell bottom. They increase the shear load. The right-hand side shows the result after shearing. As can be seen, a large part of the tracer particles is still at the initial position. Also, directly between the drivers, the particles experience no shear load. The maximum shear load is applied to particles

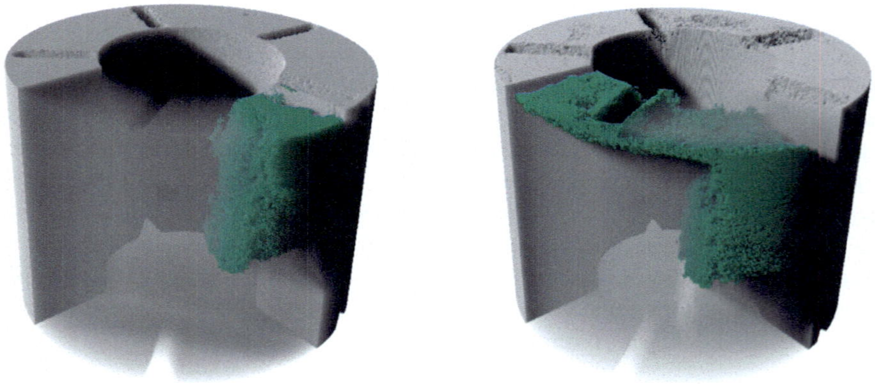

Fig. 12 μCT images of a particulate network (gray) with tracer particles (green) in the unloaded (left) and loaded (right) state. Reprinted with permission from [18], license: CC-BY 4.0

directly below the drivers. These particles are also dragged along the shear path, which is also visible in Fig. 12 right. A combined normal and shear load in the modified ring shear tester, therefore, leads to only partial crystal breakage. Crystals located at the bottom of the device or between the drivers on the lid are not crushed and remain intact.

The crystals were subjected to four different normal loads (20,572; 41,787; 64,288 and 88,396 Pa) and three different shear velocities (0.48; 1.5 and 4.5 mm min^{-1}) to investigate the influence of those parameters on the comminution. Before and after each loading cycle, the particle size distribution was determined with laser scattering. The particle size density distributions of four relevant combinations of normal load and shear velocity are shown in Fig. 13. The top left image shows the particle size density distribution for a normal load of 20,572 Pa and a shear velocity of 1.5 mm min^{-1} compared to the unloaded suspension. A reduction of the particle sizes can be observed. The increase of the distribution's span is about 40%. At the same time, the characteristic diameters x_{10}, x_{50} and x_{90} are shifted towards smaller particle sizes. The x_{10} diameter of the loaded sample corresponds to 30 ± 3 % compared to the x_{10} of the unloaded crystal suspension. Although the number of large particles is reduced, they do not vanish completely. This is because of the aforementioned shear zones in the sample, which lead to a fraction of particles experiencing no shear load. An increase of the shear velocity to 4.5 mm min^{-1} at the same normal load (Fig. 13 top right) does not increase the comminution much. However, an increase of the normal load to 41,787 Pa (not shown) at a shear velocity of 1.5 mm min^{-1}, has a much more pronounced effect on comminution than an increase of the shear velocity. At a normal load of 64,288 Pa at 1.5 mm min^{-1} (bottom left), the observed comminution is much stronger. The span increased 147 % compared to the unloaded sample and the characteristic diameters x_{10}, x_{50} and x_{90} are reduced to only 12, 38 and 68%. A further increase of the normal load does not lead to significantly more comminution (not shown). Also, the same is true for a higher shear velocity of 4.5 mm min^{-1} at the same normal load (bottom right).

The investigated combined normal and shear loads always lead to comminution, even for normal loads at which no comminution would occur without shearing. The applied normal load has a big influence on comminution, whereas the influence of the shear velocity is negligible. The high increase of fines does not only have to be due to comminution, but can also be explained by crystal abrasion. Higher normal loads increase the friction in the particulate network which contributes to crystal abrasion and fracture. Combined normal and shear loading occurs in many technical devices like pumps, centrifuges and cross flow filters. The mechanical stress, therefore, has to be reduced to avoid crystal comminution. Another possibility is to increase the mechanical stability of the crystals, for example with cross-linking.

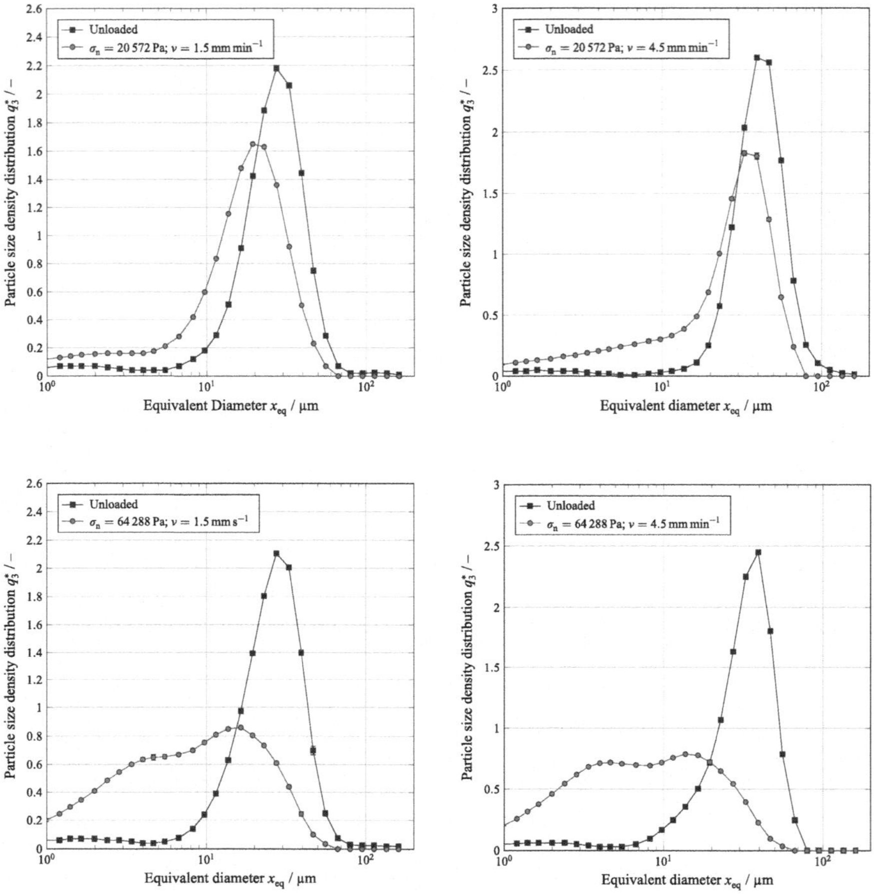

Fig. 13 Influence of different normal loads and shear velocity on the particle size distribution of isometric lysozyme crystals. Reprinted with permission from [18], license: CC-BY 4.0

6 Conclusions

The developed modular, translucent, and 3D-printed filtration cell for an analytical photocentrifuge makes it possible to characterize filtration on a small-scale, in contrast to the commercial state of the art. The required amount of sample is reduced significantly to 200–500 µL, whereas a laboratory nutsche filter requires several hundred milliliters. The photocentrifuge is able to hold multiple filtration cells which allow to highly parallelize an experimental series. The highest achievable pressure is about 2×10^5 Pa at 4000 rpm. The small-scale setup has the advantage to allow feedback concerning the filtration performance early in the product development. This leads to a reduction of development costs, less required resources and a faster time to market.

The application of the small-scale filtration cell to the four protein crystal systems: isometric, rod-like and needle shaped lysozyme and rod-like ADH crystals, yields the filter cake resistance and solids volume fraction as a function of pressure. For the compressible particle systems, the settling and permeation characteristics can be expressed by the flux density function with only the solids volume fraction as parameter. The isometric and rod-like lysozyme crystals are good to filter. They have a comparatively low filter cake resistance and a high solids volume fraction in the filter cake. This enables a fast filtration with a low moisture content in the filter cake. The rod-like ADH crystals show compressible material behavior, which means an increase of the filter cake resistance and solids volume fraction at higher pressures. The highest compression can be observed for the needle shaped lysozyme crystals. This particle system shows a strong increase of the filter cake resistance over several magnitudes. The small-scale results from isometric and rod-like lysozyme crystals were compared to a larger laboratory nutsche filter. The filter cake resistances and solids volume fractions are the same in both filtration setups for the same pressure range. Because of the low sample volumes, the results from the small-scale have a higher standard deviation.

With the process functions determined on a small scale, we designed the solid-liquid separation and calculated the permeation time on a larger integrated device. The experimental permeation time is a little bit shorter than the calculated theoretical permeation time. This can be explained by small variations of the cake height. The newly developed device allows the integrated production of protein crystals using low temperature evaporation crystallization.

Combined normal and shear load of lysozyme crystals leads to crystal breakage. The normal load has a big influence on comminution, whereas the shear velocity's influence is negligible. Increasing the normal load, leads to a much more pronounced comminution and crystal abrasion. At a normal load of approximately 64,288 Pa, an additional increase does not lead to further comminution. Countermeasures to prevent comminution are to avoid combined normal and shear stresses with high normal loads and to increase the mechanical strength of the protein crystals, e.g., by cross-linking.

References

1. Graf P (2016) Enzyme – die Supertalente der Bioindustrie. https://biooekonomie.de/themen/dossiers/enzyme-die-supertalente-der-bioindustrie. Accessed 24 Feb 2022
2. Emergen Research: Enzymes Market Size to Reach USD 17.84 Billion in 2028 | Industry Trend—Rising Demand for High-Quality Pharmaceutical, Food & Beverage and Personal Care Products. BioSpace (2022). https://www.biospace.com/article/enzymes-market-size-to-reach-usd-17-84-billion-in-2028-industry-trend-rising-demand-for-high-quality-pharmaceutical-food-and-beverage-and-personal-care-products/. Accessed 24 February 2022
3. Walsh G (2014) Proteins. Biochemistry and biotechnology, 2nd edn. Wiley, Chichester West Sussex
4. Kayser O (ed) (2005) Pharmaceutical biotechnology. Drug discovery and clinical applications, 1st edn. Wiley-VCH, Weinheim

5. Schmidt S, Havekost D, Kaiser K, Kauling J, Henzler H-J (2005) Crystallization for the downstream processing of proteins. Eng Life Sci. https://doi.org/10.1002/elsc.200500116
6. Hekmat D (2015) Large-scale crystallization of proteins for purification and formulation. Bioprocess Biosyst Eng. https://doi.org/10.1007/s00449-015-1374-y
7. Barros Groß M, Kind M (2018) From microscale phase screening to bulk evaporative crystallization of proteins. J Cryst Growth. https://doi.org/10.1016/j.jcrysgro.2018.06.010
8. Nowotny P, Hermann J, Li J, Krautenbacher A, Klöpfer K, Hekmat D, Weuster-Botz D (2019) Rational crystal contact engineering of lactobacillus brevis alcohol dehydrogenase to promote technical protein crystallization. Cryst Growth Des. https://doi.org/10.1021/acs.cgd.9b00067
9. Groß M, Kind M (2016) Bulk crystallization of proteins by low-pressure water evaporation. Chem Eng Technol. https://doi.org/10.1002/ceat.201500582
10. Chen J, Sarma B, Evans JMB, Myerson AS (2011) Pharmaceutical crystallization. Cryst Growth Des. https://doi.org/10.1021/cg101556s
11. Cornehl B, Overbeck A, Schwab A, Büser J-P, Kwade A, Nirschl H (2014) Breakage of lysozyme crystals due to compressive stresses during cake filtration. Chem Eng Sci. https://doi.org/10.1016/j.ces.2014.02.016
12. Cornehl B, Grünke T, Nirschl H (2013) Mechanical stress on lysozyme crystals during dynamic cross-flow filtration. Chem Eng Technol. https://doi.org/10.1002/ceat.201300178
13. Grob P, Huber M, Walla B, Hermann J, Janowski R, Niessing D, Hekmat D, Weuster-Botz D (2020) Crystal contact engineering enables efficient capture and purification of an oxidoreductase by technical crystallization. Biotechnol J. https://doi.org/10.1002/biot.202000010
14. Kubiak M, Solarczek J, Kampen I, Schallmey A, Kwade A, Schilde C (2018) Micromechanics of anisotropic cross-linked enzyme crystals. Cryst Growth Des. https://doi.org/10.1021/acs.cgd.8b00637
15. Radel B, Funck M, Nguyen TH, Nirschl H (2019) Determination of filtration and consolidation properties of protein crystal suspensions using analytical photocentrifuges with low volume samples. Chem Eng Sci. https://doi.org/10.1016/j.ces.2018.12.019
16. Radel B, Nguyen TH, Nirschl H (2021) Calculation of the flux density function for protein crystals from small scale settling and filtration experiments. AlChE J. https://doi.org/10.1002/aic.17378
17. Dobler T, Radel B, Gleiss M, Nirschl H (2021) Quasi-continuous production and separation of lysozyme crystals on an integrated laboratory plant. Crystals. https://doi.org/10.3390/cryst11060713
18. Radel B, Gleiß M, Nirschl H (2022) Crystal breakage due to combined normal and shear loading. Crystals. https://doi.org/10.3390/cryst12050644
19. Anlauf H (2020) Wet cake filtration. Fundamentals, equipment, strategies. WILEY VCH, Weinheim
20. Loginov M, Samper F, Gésan-Guiziou G, Sobisch T, Lerche D, Vorobiev E (2017) Centrifugal ultrafiltration for determination of filter cake properties of colloids. J Membr Sci. https://doi.org/10.1016/j.memsci.2017.04.064
21. Richardson JF, Zaki WN (1997) Sedimentation and fluidisation: part I. Chem Eng Res Des 75
22. Tiller FM, Kwon JH (1998) Role of porosity in filtration: XIII. Behavior of highly compactible cakes. AlChE J 44:2159–2167
23. Kynch GJ (1952) A theory of sedimentation. Trans Faraday Soc. https://doi.org/10.1039/tf9524800166
24. Concha F, Bustos MC (1991) Settling velocities of particulate systems, 6. Kynch sedimentation processes: batch settling. Int J Miner Process. https://doi.org/10.1016/0301-7516(91)90068-T
25. Tiller FM (1981) Revision of kynch sedimentation theory. AlChE J. https://doi.org/10.1002/aic.690270517
26. Bürger R, Concha F (2001) Settling velocities of particulate systems: 12. Int J Miner Process. https://doi.org/10.1016/S0301-7516(01)00038-2
27. Garrido P, Bürger R, Concha F (2000) Settling velocities of particulate systems. Int J Miner Process. https://doi.org/10.1016/S0301-7516(00)00014-4

28. Holdich RG, Butt G (1997) Experimental and numerical analysis of a sedimentation forming compressible compacts. Sep Sci Technol. https://doi.org/10.1080/01496399708000761
29. Rumpf H (1962) Grundlegende physikalische Probleme bei der Zerkleinerung. Chem Ing Tech. https://doi.org/10.1002/cite.330341104
30. Verein Deutscher Ingenieure: mechanical solid-liquid separation by cake filtration (2010). Beuth-Verlag GmbH 71.040.40(VDI 2762)
31. Hammerich S, Stickland AD, Radel B, Gleiss M, Nirschl H (2020) Modified shear cell for characterization of the rheological behavior of particulate networks under compression. Particuology. https://doi.org/10.1016/j.partic.2019.10.005
32. Radel B (2022) Kleinskalige Charakterisierung der Fest-Flüssig-Trennung und Betrachtung des Scherbruchs von Proteinkristallen. KIT Scientific Publishing, Karlsruhe
33. Dobler T, Buchheiser S, Gleiß M, Nirschl H (2021) Development and commissioning of a small-scale, modular and integrated plant for the quasi-continuous production of crystalline particles. Processes. https://doi.org/10.3390/pr9040663

Impact of the Mechanical Stress and Stress Residence Time on Proteins in Emulsion Formulation Within the Premix Membrane Emulsification

Anja Heyse, Patrick Giefer, Sabrina Bäther, Helena Kieserling, Tobias Wollborn, Anja Maria Wagemans, Udo Fritsching, and Stephan Drusch

Nomenclature

Symbol (Unit)	Property
V (eV)	Electrostatic potential surface
pKa (–)	acid dissociation constant
– (wt%)	weight percent

A. Heyse · H. Kieserling · S. Drusch (✉)
Department of Food Technology and Food Material Science, Institute of Food Technology and Food Chemistry, Technische Universität Berlin, Straße Des 17. Juni 135, 10623 Berlin, Germany
e-mail: stephan.drusch@tu-berlin.de

A. Heyse
e-mail: anja.heyse@campus.tu-berlin.de

H. Kieserling
e-mail: helena.schestkowa@tu-berlin.de

P. Giefer · T. Wollborn · U. Fritsching
Leibniz Institute for Materials Engineering-IWT, Badgasteiner Str. 3, 28359 Bremen, Germany
e-mail: p.giefer@iwt.uni-bremen.de

T. Wollborn
e-mail: wollborn@iwt.uni-bremen.de

U. Fritsching
e-mail: ufri@iwt.uni-bremen.de

S. Bäther · H. Kieserling · A. M. Wagemans
Department of Food Colloids, Institute of Food Technology and Food Chemistry, Technische Universität Berlin, Straße Des 17. Juni 135, 10623 Berlin, Germany
e-mail: sabrina.baether@tu-berlin.de

A. Kwade and I. Kampen (eds.), *Dispersity, Structure and Phase Changes of Proteins and Bio Agglomerates in Biotechnological Processes*,
https://doi.org/10.1007/978-3-031-63164-1_9

Abbreviations

β-lg	β-Lactoglobulin
BLG B	β-Lactoglobulin subvariant B
B_{22}	Osmotic second virial coefficient
CFD	Computational fluid dynamics
CG	Coarse-grained
CLSM	Confocal laser scanning microscopy
DB	Disulfide bonds
D-D	Dimer-dimer
DNA	Deoxyribonucleic acid
E. coli	*Escherischia coli*
LVE	Linear viscoelastic region
LYS8	Lysin at position 8
LYS14	Lysin at position 14
LYS100	Lysin at position 100
NLVE	Non-linear viscoelastic region
MCT	Medium-chain triglycerides
MD	Molecular dynamics
M-D	Monomer-dimer
M-M	Monomer-monomer
Mn	Molecular weight
OWRK	Owens–Wendt–Rabel and Kaelble model
PME	Premix membrane emulsification
SEM	Scanning electron microscopy
SiO_2	Silicon dioxide
xDLVO	Extended Derjaguin, Landau, Verwey, and Overbeek theory

A. M. Wagemans
e-mail: anja_maria.wagemans@tu-dresden.de

H. Kieserling
Department of Food Chemistry and Food Analysis, Institute of Food Technology and Food Chemistry, Technische Universität Berlin, Straße Des 17. Juni 135, 10623 Berlin, Germany

U. Fritsching
Particles and Process Engineering, University of Bremen, Bibliothekstraße 1, 28359 Bremen, Germany

1 Introduction

In biotechnological processes, one major challenge in the production and formulation process chain is the downstream processing. An efficient downstreaming is mandatory for achieving high yields and for being cost-effective and every unit operation step of the downstreaming may exert complex thermal and mechanical stress on the biomaterial. In the case of protein and enzyme processing, this can cause protein denaturation or structural alterations leading to an undesired decrease in the product quality or even a product loss. Thus, the challenge is to design an economic chain of process unit operations that is preserving the biochemical structure and the bioactivity.

Regarding individual unit operations, one step in the downstream process chain is the formulation, where often homogenization of the biomaterial is employed. Commonly used devices for homogenization are e.g., the rotor-stator homogenizer, which exerts high mechanical stress, or high pressure homogenization. A more gentle method to produce emulsions, is premix membrane emulsification (PME) [1]. In PME, a coarsely dispersed premix is forced through a membrane using a pressure gradient which creates a fine emulsion or dispersion due to droplet disruption in the membrane pores. An additional advantage of PME is the production of a more monodisperse droplet size distribution in comparison to common homogenization methods [2–4].

Despite the above-mentioned advantages of PME, the industrial application is still limited due to the lack of knowledge, especially on the control and minimization of the complex mechanical stress on proteins and enzymes. The resulting damage on structural and functional changes due to their high sensitivity when facing stress is still beyond the state-of-the-art. Furthermore, not only the amount of stress but rather the stress residence time in the membrane pores and the impact on the proteins and enzymes need to be considered for mechanistic models to gain a deeper understanding of PME. In this context, this study aims to investigate a fundamental understanding of the stress residence time behavior of proteins and enzymes. Proteins are chosen as a model for biomaterial due to their known shear-sensitivity and the structure-function-relationship.

To fulfil the goal, the stress residence time behavior in PME in the membrane pores needs to be considered in three stages: (1) stress in the pore on the droplet leading to droplet deformation and breakup, (2) stress on the protein at the solid-liquid interface and (3) stress on the protein at the liquid-liquid interface. For each stage, fundamental studies are performed. A schematic drawing of PME and the stresses occurring inside the membrane pore is depicted in Fig. 1.

The breakup mechanism of droplets as a consequence of liquid/liquid and liquid/solid shear stress inside membranes is still widely unknown, even though, the liquid fragmentation plays an essential role in the creation of complex multiphase systems. The fundamentals of the occurring maximum shear stresses with regard to the stress residence time are investigated in **part 1** of the study. The time of exposure to the shear stress, the stress residence time is a critical value for droplet breakup

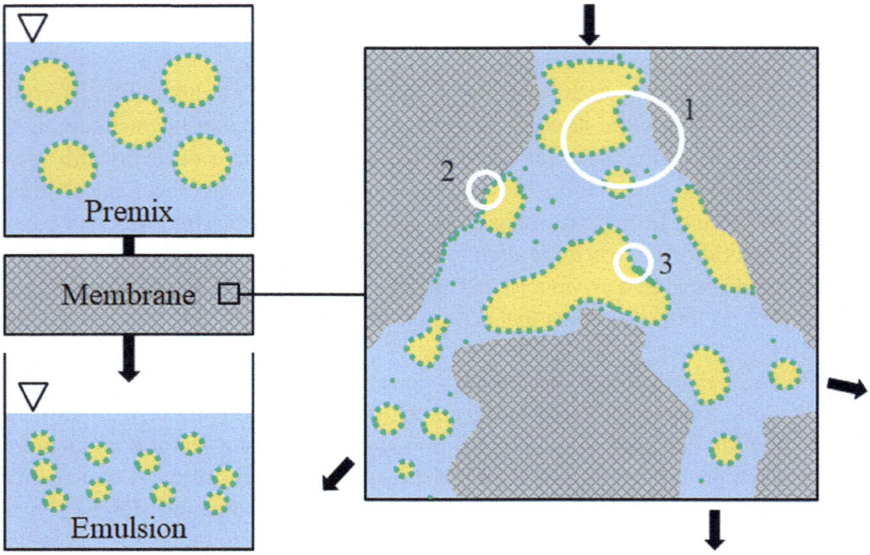

Fig. 1 Schematic drawing of the premix membrane emulsification process (left). Zoom-in (right) depicts the stress (indicated by the white rings) of (1) the stress on the droplet leading to deformation and breakup, (2) the stress on the protein at the solid-liquid interface affecting the protein adsorption and (3) the stress on the protein at the liquid–liquid interface due to droplet deformation resulting in shear and strain against the proteins

and protein degradation. For the stress at the solid/liquid interface, the influence of process parameters like the transmembrane pressure on both stresses has to be clarified as well as membrane properties like wettability and their cumulative effect on the stress-related dispersion mechanism. The liquid/liquid and liquid/solid stress is divided in local stress distributions, to identify the interfacial regions related to the highest occurring stress. Since membranes have a complex pore network where breakup events have different underlaying mechanism, it is challenging to relate governing mechanisms to single droplet breakup events. Therefore, idealized pore structures are a suitable model system to directly relate interfacial shear stress to process parameters and membrane properties. Subsequently, complex pore structures can be used to connect those mechanisms to the application of PME. Computational Fluid Dynamics (CFD) is utilized for the simulation of the multiphase flow (oil/water) through porous networks and idealized pore structures. The evaluation of interfacial shear stress at different operating conditions and varying wettability provides an understanding of critical stress levels for the formation of droplets within PME, which in turn, can be used to evaluate protein degradation under certain process conditions.

The adsorption of proteins to the solid/liquid interface, affects the stress residence time behavior of the proteins, which will be covered in **part 2**. On one hand, the protein adsorption is affecting the wettability of the membrane, which is influencing the stress at the solid/liquid interface and the droplet formation. On the other hand, the adsorption and desorbtion can also influence the characteristics of the protein on a molecular level. Protein sorption is affected by the membrane properties, which can be modified by chemical treatment [5]. Depending on the membrane properties, proteins can adsorb due to the variety of forces involved in the protein-surface interaction (e.g., hydrophobic interactions or electrostatic interactions). Moreover, the conformational state and more precisely the quaternary structure of the proteins, affect the protein sorption and can be described by specific and non-specific interactions. Specific interactions are attractive interactions, such as hydrogen bonds or ionic bonds, between specific protein residues. Non-specific interactions are mainly caused by electrostatic forces and van der Waal forces and act ubiquitous. If adsorption prevails, fouling occurs as a function of time and leads to "inner" pore blocking or the formation of a layer on the surface of the membrane [6]. Most often fouling of the membrane limits continuous operation and upscaling for industrial applications [1, 7]. Furthermore, it can lead to a quantitative loss of biomolecules [8, 9].

A well-described model protein to study the protein adsorption within PME is the whey protein β-lactoglobulin (β-lg). The conformational state of β-lg, in particular the quaternary structure, changes with the solvent properties, such as the pH value and ionic strength. The quaternary structure is dominated by an equilibrium of monomers, dimers, or octamers. The dimer state is stabilized by two intermolecular hydrogen bonds and electrostatic interactions. At pH values close to the isoelectric point p*I* of β-lg (pH 5.2), the octameric form dominates. At low ionic strength and pH values below or above the p*I* where β-lg is negatively (pH > 8) or positively charged (pH < 3), the monomeric form dominates. Although, the stabilization process of β-lg stabilized oil/water interfaces is well described in the literature, the molecular interactions between β-lg molecules and its interactions with the membrane, the change of the membrane properties through β-lg adsorption and the effect on the performance within the PME process have not been systematically analyzed so far. To study the stress caused by adsorption and desorbtion, enzymes are suitable model systems, since both conformational changes and the biocatalyst activity indicate characteristics of the protein. Lipases are of special interest as they are activated only by adsorption to the interface causing structural rearrangements. Characteristically, lipases possess a flexible region on their molecular surface for the detection of hydrophobic substrates. That includes the hydrophilic protein lid, which protects the hydrophobic active site inside the protein [10]. This protein lid is "opened" by adsorption to the interface, thereby exposing the active site. The active site is then orientated toward the hydrophobic phase. The binding process to solid interfaces, such as membrane surfaces, also leads to denaturation and possible loss of activity of enzymes, which are caused by conformational change [11]. Bowen & Gan suggest that, in general, repeated adsorption and desorption at the membrane is a possible explanation for the deactivation of enzymes because of conformational changes in the tertiary structure due to shear and stretching stresses on the molecules as they pass through the pores (stress residence time behavior) [12].

For the investigations in part 2, the specific and non-specific interactions influencing the conformational state and the protein adsorption in dependence on the hydrophilization and the solvent properties (pH, ionic strength, protein concentration) onto silica-based membranes are determined to elucidate the effect on the membrane characteristics, as well as the structural conformation of the protein. Firstly, specific and non-specific interactions of β-lg affecting the conformational state are determined using analytical ultracentrifugation, membrane-osmometry and MD simulation. Secondly, the β-lg adsorption onto a silica-based membrane is characterized using CLSM (conformation of β-lg at membrane), contact angle (membrane wettability), and ellipsometry (β-lg layer thickness). Moreover, the droplet size distribution of β-lg stabilized emulsions within PME is characterized using static light scattering. Thirdly, *Candida rugosa* lipase (only called lipase hereafter) adsorption and microstructure of the membrane surface are monitored by SEM. Molecular dynamic (MD) simulations of both model proteins onto silica/water interfaces is performed to reveal adsorption configurations, binding residues, and the driving mechanism behind protein adsorption onto solid/liquid interfaces. Conformational rearrangements during adsorption with a focus on the exposure of hydrophilic and hydrophobic solvent-accessible surfaces and a change in the radius of gyration are used to estimate the impact of adsorption on the primary and conformational structure, respectively. The structural rearrangements in the protein which result in an exposure of hydrophobic and hydrophilic residues upon protein adsorption can be taken into account to explain changes in the wettability of the membrane after PME. The adsorption of the enzyme lipase onto solid/liquid interfaces is used to determine the relationship between the immobilization and structural rearrangements which might result in a change in enzymatic activity.

In part 3, the impact of the stress on the proteins at the liquid/liquid interface and the droplet's interfacial stabilization will be investigated. In general, the interfacial stabilization by proteins is mainly based on a three-staged process [13, 14]: (i) the migration of the proteins through the bulk phase, (ii) the adsorption at the oil/water-interface accompanied by structural rearrangement, and (iii) the intermolecular film formation. As shown in Fig. 1, the oil/water interface of a droplet in the membrane pores is exposed to shear stress. Thus, the adsorbed proteins at the interface will be pulled in the direction of the applied stress, which in turn, leads to structural rearrangements. Furthermore, during droplet breakage, new interfaces are formed and proteins from the bulk will adsorb at the newly formed interface as well as at the partially occupied interface.

An increasing interfacial preoccupation affects the kinetics of both (i) the protein migration through a bulk phase (i) and the adsorption at the interface (ii) [13, 15, 16]. Stage (iii) depends on the amount of adsorbed proteins and, on the density of the interfacial protein film. Hence, the protein structure and the intra- and intermolecular interactions determine the structure in the bulk water phase and consequently, the film structure during interfacial stabilization and the emulsification process. In part 3, the effect of interfacial preoccupation at the oil/water-interface along with the conformational state and charge of β-lg on the interfacial stabilization (phase (i)–(iii)) are elucidated. Pendant drop analysis of the *lag* time is used for the characterization of the protein migration in stage (i), the analysis of the slope of decreasing interfacial tension for the adsorption rate in stage (ii), and the reduction of interfacial tension after protein addition in stage (iii). To characterize the attractive and repulsive forces in more detail, we use molecular dynamic investigations to simulate force-distance curves and force maps of our samples at various preoccupation levels and pH.

In order to investigate the impact of protein structure on the adsorption behavior, the protein structure can be systematically modified. For that, firstly, high hydrostatic pressure treatment of β-lg and, secondly, recombinantly produced structural variants of β-lg are used. High hydrostatic pressure induces processing-related structural changes in proteins which are mainly non-covalent interactions such as hydrogen bonds, electrostatic, and van der Waals interactions. Therefore, we treated β-lg in the bulk water phase with hydrostatic pressures up to 600 MPa. Recombinant variants of β-lg missing one or several cysteines, which is the amino acid responsible for the formation of intra- and intermolecular disulfide bonds, were recombinantly produced in *E.coli*. The interfacial adsorption behavior (stage (i)–(iii)) as well as the interfacial film analysis for both the pressure-treated and the recombinant β-lgs were analyzed with drop tensiometry, dilatational, and shear rheology, respectively. CFD and MD simulations are used to clarify the adsorption behavior on the molecular level.

2 Material and Methods

2.1 Methods for Part 1

The occurring stresses on the droplet interface that eventually lead to rupture of the interface into smaller droplets were quantified using CFD under varying flow conditions and membrane wettability.

2.1.1 Computational Fluid Dynamics

Computational fluid dynamic simulations are performed with the OpenFoam software package version 6.0 utilizing the Volume of Fluid method. Three dimensional domains of single capillaries with a contraction as well as complex pore structures reconstructed from a CT scan of a membrane are used to simulate a single droplet breakup.

To gain fundamental insight into the flow inside the membrane pores, high-resolution two-dimensional simulations of liquid/liquid displacement in capillaries without a geometric contraction were used. Therefore, advanced advection schemes with geometric interface reconstruction (isoAdvector) [17] were implemented to guarantee the boundedness of the solution with a low advection error. A Navier-slip boundary condition is imposed on the side walls to remove the stress singularity created by an interface with a non-zero velocity propagating over a wall with a noSlip condition [18]. To reduce the occurrence of spurious currents, a sharp surface tension force model is used [19]. The effect of varying membrane wettability is implemented by the Kistler correlation to account for dynamic contact angles [20, 21]. The evaluation of the interfacial shear stress within the CFD-framework is developed.

2.2 Methods for Part 2

The protein–protein interactions in bulk, the protein adsorption at the membrane, its effect on the droplet formation and the resulting structural changes of the protein were characterized in dependence of the environmental conditions (pH and ionic strength). Molecular interactions were analyzed with membrane osmometry, analytical ultracentrifugation and coarse-grained simulations and MD simulations to determine the conformational state in bulk. After performing membrane filtration, the adsorbed mass was determined and the adsorption and desorption onto the silica membrane material was described with confocal microscopy (SEM) and contact angle measurement (wettability).

2.2.1 Membrane Osmometry Experiments

β-lg stock solutions with a protein powder content of 40 g/L were prepared in 10 and 100 mM sodium chloride solvent at pH 3 and pH 7. The stock solutions were stirred for 1 h and stored in a refrigerator overnight. The samples were diafiltrated (ÄKTACrossflowTM, GE Healthcare, Uppsala, Sweden) for precise adjustment of the ionic strength and the pH value. The volume of the stock solution was exchanged 10 times with the corresponding, pH-adjusted sodium chloride-containing solvent. The diafiltration was performed according to Hundschell et al. [22]. To remove any potential protein aggregates, the diafiltrated samples were centrifuged for 30 min at 4 °C and 10,000 g. The residue was discarded and the protein concentration of the supernatant was determined by measuring the absorbance at 278 nm with a Helios Omega UV-vis spectrophotometer (Thermo Fisher Scientific, Waltham Massachusetts, USA) using a specific extinction coefficient of 0.96 L/(g cm). Osmotic pressure measurements were performed at various protein concentrations. The osmotic second virial coefficient was calculated as previously described by Hundschell et al. [22].

2.2.2 Analytical Ultracentrifugation

The β-lg stock solution with a protein concentration of 1000 μM was performed following the protocol by Zhao et al. [23]. The filtered stock solution concentrations were determined via UV-Vis spectrometry using a specific extinction coefficient of 17,679 and prepared according to Uttinger et al. [24]. The sedimentation velocity experiments were performed using a commercial ultracentrifuge, type

OptimaTM AUC from Beckman Coulter (Krefeld, Germany). The samples were measured according to Uttinger et al. [24] at 20 °C and a fixed rotor speed of 40,000 or 50,000 rpm for 10 h until complete sedimentation of all species was observed. Sedimentation velocity experiments were conducted using centerpieces with an optical path length of 12 or 3 mm in order to adjust the optical signal following. Data analysis were performed as described by Kar et al. [25] and Schuck et al. [23, 26, 27]. The sedimentation coefficient distributions and the weight-averaged sedimentation coefficients for varying protein concentrations were determined using the partial specific volume of 0.751 mL/g for β-lg as calculated from Sednterp [28]. The solvent density and viscosity were set to the values of the solution at the respective salt concentrations at 20 °C. Determination of the dimer dissociation constant was subsequently conducted via fitting the isotherms, i.e. weight-averaged sedimentation coefficient for varying protein concentration at a fixed pH value and salt concentration, using the software SEDPHAT and following the protocol of Zhao et al. [23, 27].

2.2.3 Coarse-Grained XDLVO-CG Calculations

The coarse grained xDLVO-CG calculations were performed after an extended DLVO approach. The so-called xDLVO-CG was used to calculate the interaction potential between β-lg monomers and dimers [29]. The β-lg protein and protein oligomers are represented by the shape-based CG model according to Uttinger et al. [24]. The interaction potential is calculated as a sum of electrostatic, dispersion, osmotic, and ion-protein interactions between the β-lg monomers and dimers binary systems.

2.2.4 Molecular Dynamic Simulations

Molecular dynamic simulations were performed with the GROMACS software package [30]. The initial configurations were packed utilizing packmol [31]. The protein structures for inactive lipase (code: 1TRH), active Lipase (code: 1CRL), and β-lg (code:1BSY) have been taken from the protein data bank. Disulfide bonds were assigned manually and protonation states of the titratable groups were adjusted to a pH of 7.1 using H++ web server version 3.2 [32]. The structures were solvated in TIP3P water and Na^+ ions were added to achieve charge neutrality. For all simulations, the AMBER-99SB force field was used. After initialization, the simulation systems were energy minimized and equilibrated in a canonical and isobaric-isothermal environments under periodic boundary conditions. After equilibration microcanonical ensembles were used for production runs. The slab structure of the amorphous silica was derived by Cole et al. [33] with the interaction potentials for silica-water by Butenuth et al. [34].

2.2.5 Membrane Filtration of the Protein Solution

The filtration was performed in accordance with [35, 36]. Briefly, a new silicate membrane (borosilicate membranes type P4, ROBU Glasfilter-Geräte GmbH, Germany) was used for each filtration experiment. The membrane was placed in an in-house made dead-end filtration holder with an adjunctive storage tank on the retentate side of the membrane. After the feed (water as control or protein solution) was filled in the storage tank, the filtration was started by applying specific transmembrane pressure. The permeate was collected in a beaker and used for further analysis.

2.2.6 Measurement of the Adsorbed Mass on the Membrane

The adsorbed mass was measured in accordance with [35]. Before usage, the silicate membrane was weighed with an analytical balance. After the filtration, the membrane was dried overnight in a drying cabinet and subsequently weighed again. The quantitative amount of adsorbed mass on the membrane was calculated from the difference in the mass before and after filtration.

2.2.7 SEM and CLSM of Membrane Surface

Scanning electron microscopy (SEM) at the Center for Electron Microscopy (ZELMI), Technische Universität Berlin, Berlin, Germany was employed to visualize the microstructure of the membrane surface using an S-2700 scanning electron microscope (Hitachi, Tokyo, Japan) in accordance with [35, 36]. For that, the membranes were coated with a 60 μm gold layer for ensuring conductivity before performing the SEM. A magnification of 300x was used for all measurements.

For visualization of the adsorbed proteins, the membranes were scanned from the top- and side-view with a laser scanning confocal microscope (CLSM) as described in [36]. The fluorescent Nile red First was used to dye the used membranes to highlight the hydrophobic parts. Then, the membranes were washed with a water: ethanol mixture to remove any unbound dye and were subsequently dried overnight in a desiccator. The microscopic image data were acquired using a Leica SP8 CLSM (Leica Microsystems CMS GmbH, Wetzlar, Germany) with an HC PL APO 10x/0.40 objective. The image resolution was set to 512 $\times$ 512 pixels. The sample excitation was measured with a 514 nm laser and the detectors were set to the emission spectra of Nile red (600–700 nm).

2.2.8 Wettability of the Membrane

The sessile drop method was used to determine the contact angles (wettability) of the membranes with the drop tensiometer OCA20 (Dataphysics Instruments GmbH,

Germany) using a high-speed camera as described in [35]. A plane silicate wafer was used as a model of the membrane. The treatment of wafers was equal to the membrane filtrations to ensure comparability. For the measurement, a 5 μL water droplet was dosed and gently settled on the wafer. The contact angle of the droplet on the wafer surface was tracked over time and calculated from the average the contact angle on the left and the right side of the water droplet.

2.3 *Methods for Part 3*

Structure variants of β-lg were produced to study the underlying mechanisms and molecular interactions involved in the adsorption behavior. The interfacial adsorption process of β-lg at liquid/liquid interfaces (stages (i) and (iii)) was investigated via drop tensiometry. The protein film response to shear was measured with interfacial rheology (dilatational and shear rheology).

2.3.1 Production of Structure Variants of β-lg

For the pressure-treated samples, protein solutions with 8 wt% β-lg in water were poured into high-pressure stable Nunc CryoflexTM Vials (1.8 ml, Lot 135093, Thermo Fisher Scientific, Roskilde, Denmark) and treated with high hydrostatic pressure from 0 to 600 MPa for 10 min at 20 °C (CH-8124, SITEC-Sieber Engineering AG, Maur/Zurich, Switzerland).

Structural variants of bovine β-lg B (BLG B) were produced and provided by the group of Krull/Biedendieck at the Technische Universität Braunschweig, Germany as described in [37, 38]. The five cysteines were systematically exchanged with alanine to produce structural variants affecting the disulfide bonds (DB) (Table 1).

2.3.2 L/L Adsorption Behavior and Interfacial Rheology

An automated drop tensiometer OCA 20 (Dataphysics GmbH, Filderstadt, Germany) or PAT1M (Sinterface Technologies, Berlin, Germany) with a two-needle system was used for the analysis of the adsorption phases (i–iii). A water droplet was dosed with the outer needle into a quartz-glass cuvette filled with purified MCT-oil. Then, 10% v/v of the water droplet was added (OCA20) or exchanged (PAT1M) with protein solution through the inner needle with an automatic dosing system. The droplet shape was recorded with a highspeed camera and was used to calculate the interfacial tension. The *lag* time was calculated as the time period from the start of inserting the protein solution through the inner needle until the interfacial tension started to change. To characterize the kinetics of the adsorption rate (ii), the slope k was calculated by taking the first derivative of the exponential fit after interfacial tension started to decrease. To estimate the interfacial stabilization state (iii), the

Table 1 Recombinantly produced variants of β-lactoglobulin B (BLG B)

Variant	Genetic modification	Nomenclature in this work	Theoretical impact on secondary structure
BLG B I	L1AI2S[a] (basic mutation for all following variants)	Control	Like bovine BLG B
BLG B II	L1AI2S + C121A	C121A	No intermolecular DB
BLG B IIIa	L1AI2S + C66A	C66A	DB at molecule surface deleted, free C160 could form new intramolecular DB
BLG B IIIb	L1AI2S + C160A	C160A	DB at molecule surface deleted, free C66 could form new intramolecular DB
BLG B IIIc	L1AI2S + C66AC160A	C66AC160A	DB at molecule surface deleted
BLG B IV	L1AI2S + C66AC160A + C106AC119A + C121A	noCys	No intra- or intermolecular DB

[a] L1AI2S = Leucin at position 1 is exchanged by Alanine and Isoleucine at position 2 is exchanged by Serin

difference in interfacial tension before and after the addition of protein solution was calculated. More details can be found in [39, 40]. The droplets were further used for interfacial dilatational rheology. After a specific drop aging, a subsequent amplitude sweep ranging with varying deformation was performed at a frequency of 0.01 Hz ($f_D = 0.01\%$). The change in $\Delta\sigma$ against the relative change in drop area A/A_0 of the droplet was visualized in Lissajous plots. The elastic (storage) modulus E' and viscous (loss) modulus E'' were derived from E^*, which was calculated from the oscillation cycles.

Interfacial shear rheology was performed with a shear rheometer (Physica MCR 301/MCR 102, Anton Paar Germany GmbH, Ostfildern, Germany) using an interfacial rheology system cell (diameter 80 mm) with a biconus as described in [41, 42]. Time and frequency sweeps were carried out at a constant temperature of 22 °C. The time sweep was performed for 10 h at constant frequency $f_S = 1$ Hz and amplitude $\gamma_S = 0.1\%$. Frequency sweeps were carried out after 17 h. and were recorded over a range of $f_S = 0.001 - 1$ Hz at constant amplitude ($\gamma_S = 0.1\%$).

2.3.3 Molecular Dynamics Simulation

Molecular dynamic simulations were performed with the GROMACS software package [30]. The protein structure for β-lg (code:1BSY) has been taken from the protein data bank. Disulfide bonds were assigned manually and protonation states of the titratable groups were adjusted to a pH of 7.1 using H++ web server version 3.2 [32]. The recombinant proteins were mutated starting from native β-lg (code:1BSY) applying mutations in the starting sequence and the deletion of the disulfide bonds.

The oil interface was modeled based on a single chain of medium-chained triglycerides utilizing the Avogadro software version 1.2.0 [43]. Antechamber was used to calculate the force field of the triglycerides [44]. For the simulations, at ambient pressure, the AMBER-99SB force field was used and for the high-pressure simulation, the CHARMM27 force field was used. After initialization, the simulation systems were energy minimized and equilibrated in a canonical and isobaric-isothermal environments under periodic boundary conditions. After equilibration microcanonical ensembles were used for production runs.

2.4 Statistical Analysis of Experimental Data of Part 2 and Part 3

Statistical analysis of experimental data in triplet was performed by two-way ANOVA with significance determined by a Post-hoc Tukey's test ($p < 0.05$) using Statgraphics (Statgraphics Technologies Inc., version 5.1, the Plains, Virginia, USA). A normal distribution was assumed for a large statistical population.

3 Part 1: Droplet Deformation and Shear Stress in the Membrane

The objective of the following chapter is to relate the liquid/liquid shear stress and solid/liquid shear stress to interfacial deformations and consequent droplet breakup. Idealized pore structures as well as complex membranes are utilized within CFD simulations.

3.1 Droplet Deformation and Interfacial Shear Stress in a Single Pore

To evaluate the interfacial stress a simplified pore geometry with a step-like diameter reduction of $d/d_{pore} = 2.25 - 2.75$ is used. This allows the identification of relevant liquid/liquid and liquid/solid shear stresses on the droplet breakup and shear stress distributions in the interfacial region. The simulations were taken out over a range of capillary numbers (Ca) of Ca = 0.091–0.365 with corresponding Reynolds numbers (Re) of $R = 15$–60 and Weber numbers (We) of $We = 0.11$–1.79. Following results are published in Wollborn et al. [45].

As depicted in Fig. 2 the interfacial deformation and the propagation through the pore are different for different capillary numbers. For the lowest simulated capillary numbers (A), the droplet propagates with a slug-like flow, in constant contact with the pore wall. The interfacial shape remains quasi-stationary after full entry into the contraction. The highest shear stresses are located at the interfacial regions in contact with the tube walls. With increasing capillary number (Fig. 2b) the droplet undergoes further deformation during penetration into the pore resulting in partial detachment of the droplet interface from the wall. A region of lower shear stresses in the droplet front occurs, as the droplet front detaches from the wall and shifts into regions with lower shear rates, into the pore center. The area of the detached region increases with time, therefore, reducing the overall shear stress on the droplet interface with time. Further increase in the capillary number (Fig. 2c) results in faster detachment of the droplet from the pore wall and a contraction of the interface in the detached region, indicating higher velocities in these interfacial regions, compared to the interfacial regions in contact with the wall. General stress-dependent trends of droplet deformation and breakup can be derived from these findings: (i) a critical capillary number initializes the breakup by the detachment of the interface from the wall caused by contact line instabilities, (ii) shear and strain histories lead to deformation of the detached droplet interface, (iii) the breakup into droplets at the interfacial constriction occurs at higher capillary numbers.

For further analysis of the interfacial shear stress, the interface is divided into the wall interface region and the liquid interface region as shown in Fig. 3. The cumulative stress distribution, in which the stress is weighted by its surface region (local stress distribution), shows distinct interfacial shear- and strain stress regions, where the lowest shear- and strain stress occur at the droplet front and the highest occurs at the interface in the vicinity of the wall.

In order to weight the influence of the stress conditions of both interface regions, the relative droplet area is depicted in Fig. 4. At low capillary numbers (Fig. 4a)

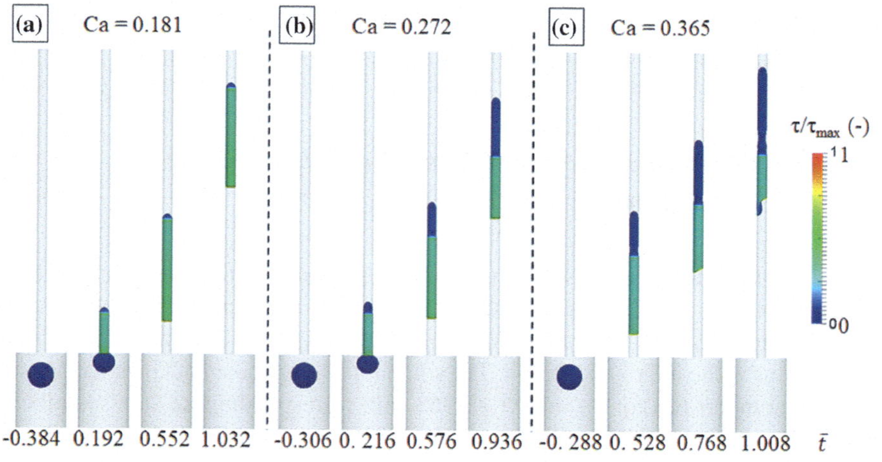

Fig. 2 Simulation of a single droplet ($d/d_{pore} = 2.5$) in a pore with $d_{pore} = 0.2$ mm for different capillary numbers Ca at different time steps $\bar{t}$. The coloring qualitatively indicates the acting interfacial shear stress. Reprint by permission from AIP Publishing: Ref. [45]

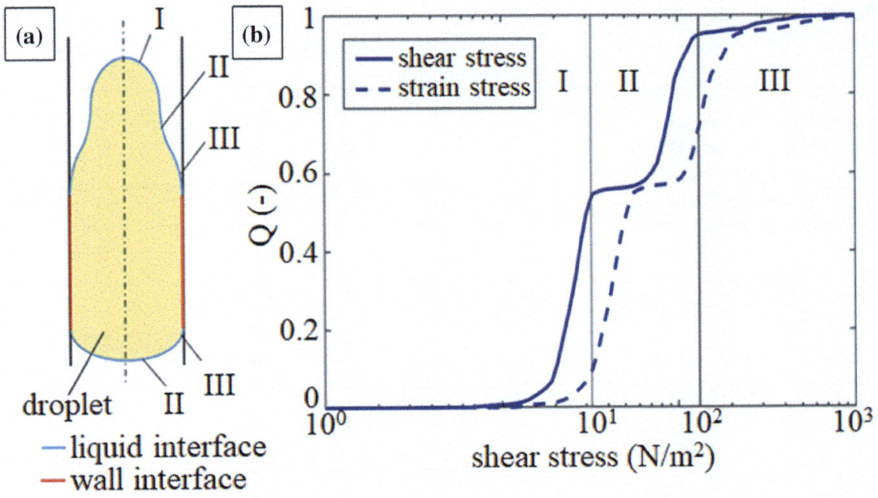

Fig. 3 Cross-sectional illustration of a droplet within the pore channel split into liquid (blue line) and wall (red line) interfaces (**a**) and example of the instantaneous distribution of shear and strain stresses at the liquid interface at a capillary number of Ca = 0.272 (**b**). Numbers I–III indicate characteristic shear stress fractions, which can be related to certain regions of the droplet interface. Reprint by permission from AIP Publishing: Ref. [45]

the droplet attaches to the wall during the entrance in the pore, remaining a quasi-stationary shape afterwards. This leads to an increase in wall-related droplet area and a decrease in the interfacial area with the continuous phase, approaching a constant value. The dominant stress therefore originates from the wall interaction. With increasing capillary number (Fig. 4b) the droplet area in contact with the wall decreases after the droplet has fully emerged and the interfacial region in contact with the continuous phase increases due to the aforementioned wall detachment. For higher capillary numbers (Fig. 4c) the same process is observed. Therefore, it can be concluded that the wall shear stress dominates during the entrance in the pore and for capillary numbers $Ca < 0.272$, while the dominance of the liquid/liquid shear stress increases with capillary numbers of $Ca > 0.181$ and is therefore governing during droplet breakup.

The variation of the droplet size ($d/d_{pore} = 2.25, 2.5, 2.75$) revealed qualitatively similar droplet deformation states for all simulated capillary numbers. The droplet detachment from the wall is independent of the droplet size, hence a function of the capillary number.

In conclusion, the analysis of the local stress distribution and the stress residence time show that the deformation and breakup process is mainly determined by liquid/liquid interface stresses. The highest shear stress occurs in the wall region but only over a short time.

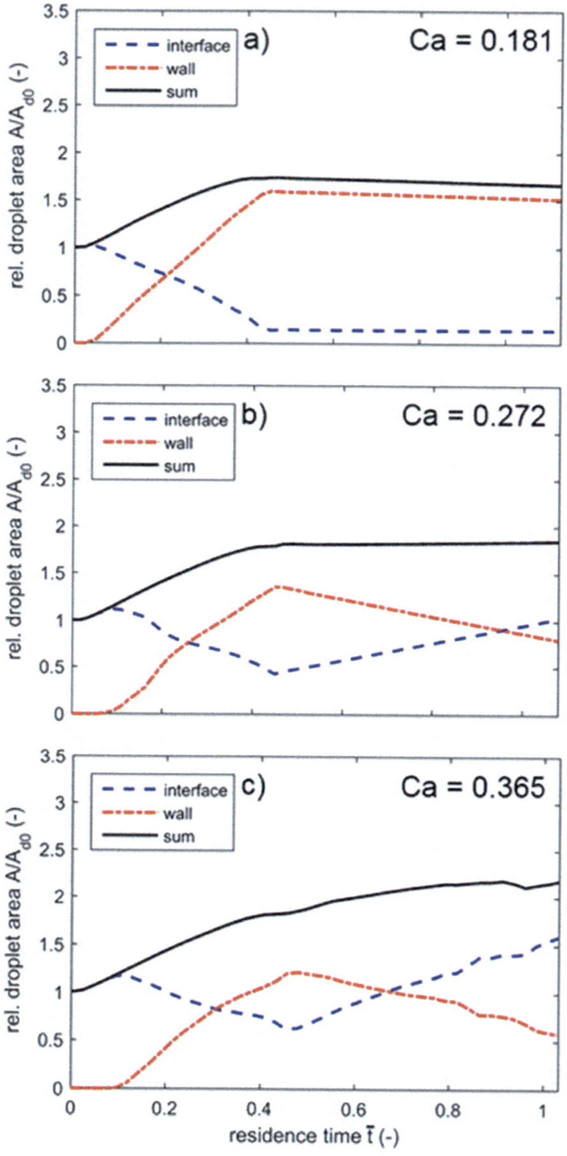

Fig. 4 Relative droplet interface area divided into the contribution of liquid interface stress (blue) and wall interface stress (red) in the pore for different capillary numbers. Reprint by permission from AIP Publishing: Ref. [45]

3.2 *The Influence of Wettability on the Interfacial Deformation*

Due to the interaction of the droplet with the wall and the fact that the initial wall detachment is caused by contact line instabilities a closer look at the dynamics of the liquid/solid interaction is necessary. The displacement of the continuous phase with the disperse phase underlies dynamic changes of the wetting angle in the region of the contact line due to the velocity dependence of the contact angle. The boundaries of the contact line instability have to be identified and the influence of the wetting behavior of the membrane on the interface propagation and detachment has to be evaluated. Following results are published in Giefer et al. [46].

Therefore, high-resolution two-dimensional wedge simulations were performed under varying capillary numbers Ca = 0.04–1 and apparent dynamic contact angles $\theta_a = 0°–170°$. Three scenarios of interfacial propagation were found: I. if the contact line velocity is equal to the mean velocity, and the dynamic pressure is low, a spherical meniscus stably displaces the continuous phase (Ca < 0.1), II. The contact line velocity is equal to the mean velocity but the dynamic pressure contributes significantly to the total pressure exerted on the interface an interface deformation in the shape of a cap is present which propagates with a constant velocity ($0.1 <$ Ca < 0.25). III. If the mean pore velocity exceeds the contact line velocity, the interface detaches from the wall and undergoes dynamic deformations in which a finger in the center of the capillary is formed, penetrating into the continuous phase Ca > 0.25. The dynamic growth of this finger in the flow direction eventually leads to breakup into droplets. The transition between the regimes is related to the balance of viscous forces and interfacial tension forces.

In Fig. 5 the shear rate at the contact line is shown in relation to the capillary number and the contact angle. Overall, the shear rate increases with the capillary number and increased contact angle (dispersed phase wetting), which both can be attributed to higher contact line velocities. The global minimum is located at Ca = 0.04 and $\theta_a = 90°$, for lower contact angles the contact line is dragged behind the meniscus resulting in higher shear rates. For higher contact angles at the same capillary number, the meniscus is dragged behind the contact line resulting in higher shear rates. A local minimum at a capillary number of Ca = 0.26 and contact angles of $\theta_a = 0°, 30°$ is present due to the detachment of the interface from the wall and a formation of the meniscus instability. In this case, the contact line velocity significantly decreases under the mean pore velocity, therefore resulting in lower shear rates at the contact line. At the same capillary number, no contact line instability was present for contact angles $\theta_a > 30°$, hence the occurrence of the instability depends on the wetting properties of the membrane and therefore smears the regime transitions between stable and unstable displacement/droplet formation. For capillary numbers Ca > 0.26, a meniscus instability was present for all applied contact angles.

The evaluation of the secondary flow field allows the identification of underlying mechanisms behind the wetting dependency of the contact line instability. In Fig. 6 the secondary flow field is depicted by a streamline representation by subtracting the mean velocity from the velocity field. It is evident, that a recirculating flow for all applied contact angles is present within the disperse phase and in front of the interface, in the continuous phase. Only for contact angles $\theta_a < 90°$ a secondary vortex, limited by the concave side of the interface is present. This secondary vortex creates two stagnation points at the interface. One is located in the vicinity of the wall, where the recirculating flow inside the disperse phase and the secondary vortex meet. The other one is located on the center axis of the pore at the droplet interface, where

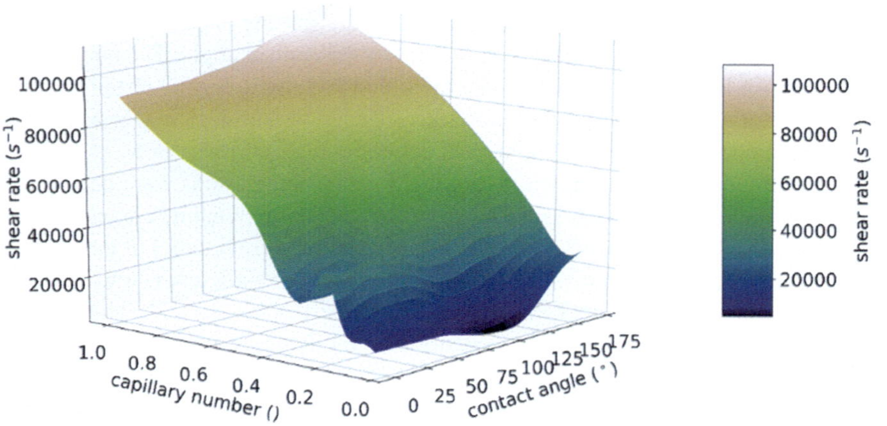

Fig. 5 Shear rate at the contact line over capillary number and apparent contact angle, surface is generated by interpolation of data points. Reprint by permission from AIP Publishing: Ref. [46]

two secondary vortices meet (only one visible due to wedge cut). This stagnation point contributes to the increase in total pressure towards the interface due to viscous displacement. Further, the Laplace pressure contributes to the increased pressure at the droplet front due to the interface curvature, resulting in higher pressures for lower contact angles. The combination of these observations explains the dependence of the wettability on contact line instabilities and the earlier transition to the meniscus instability regime for lower contact angles (continuous phase wetting).

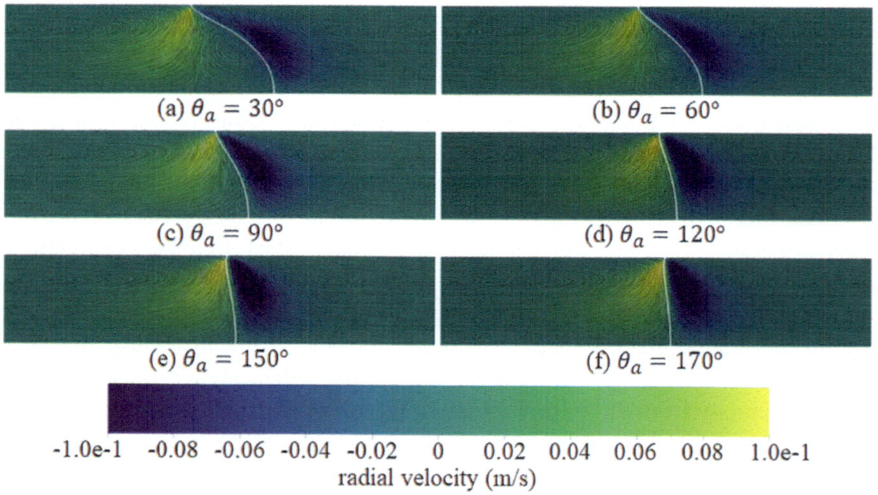

Fig. 6 Radial velocity and streamlines of the secondary flow field in LIC representation for varying contact angles without finger formation, the location of the interface is highlighted in white. Ca = 0.24, We = 0.7, Re = 3. Reprint by permission from AIP Publishing: Ref. [46]

3.3 Droplet Deformation and Interfacial Shear Stress in Membranes

In order to transfer the observations of the interfacial shear stress into the application of PME, simulations with CT scans of complex membrane structures were taken out. Following results are published in Wollborn et al. [47]. The single droplet dispersion process is depicted in Fig. 7a–d. With increasing capillary number an increased branching of the droplet can be observed. A more pronounced dispersion occurs due to higher shear stresses at higher capillary numbers. Moreover, at low capillary numbers (A), the branches of the disperse phase completely fill the pore channels. Therefore, droplet breakup is expected at pore branching. With increasing capillary number, the branches of the droplet detach from the pore walls as ascribed in Sects. 3.1 and 3.2. However, to quantify the effect of wall detachment with an increasing capillary number in complex pore structures, the median shear stress divided into liquid/liquid and liquid/solid shear stress is shown in Fig. 7e, f.

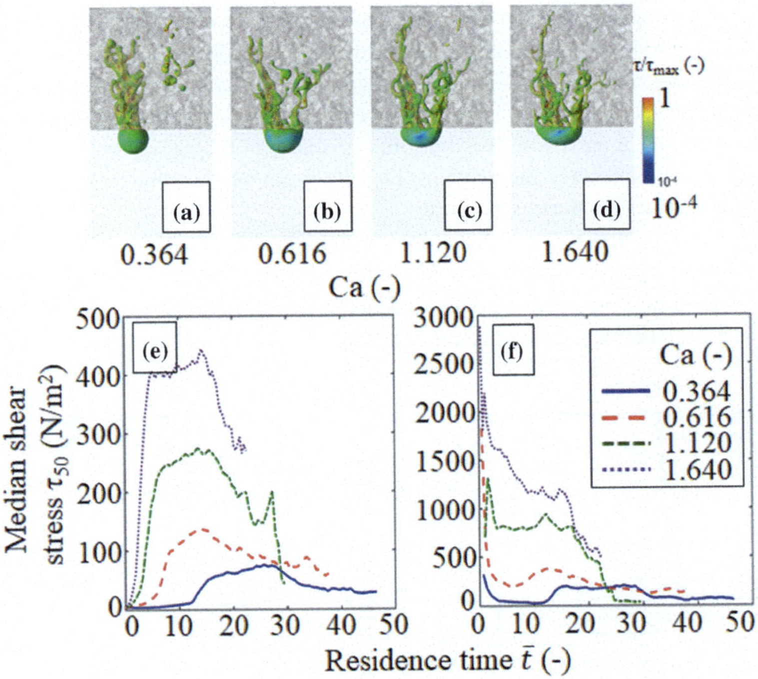

Fig. 7 (Top) **a–d** Visualization of the droplet dispersion at different capillary numbers at a similar dispersion progress. The coloring of the droplet indicates the interfacial shear stress on a logarithmic scale. (Bottom) **e** Median shear stress at the liquid (disperse-continuous) and **f** wall (disperse-solid) interface over dimensionless residence time. Reprint by permission from John Wiley and Sons: Ref. [47]

With increasing capillary numbers, the median shear stress shifts to higher values for the liquid and the wall interface. Due to higher velocity gradients at the wall, the wall shear stress exceeds the liquid shear stress. Further, the liquid median shear stress increases during the entrance into the pore, while the wall shear stress decreases during the entering process. This indicates a short wall contact during the early stage of the entering process with subsequent wall detachment. This effect is more pronounced for higher capillary numbers, as stated earlier. The decrease in the median stress can be explained by the fact that the liquid fragmentation happens at the highest shear stress, consequently, droplets are created, which are less effected by the velocity gradients. Further, dispersed droplets left the membrane and do not contribute to the median shear stress anymore.

In order to evaluate the contribution of the wall shear stress and liquid shear stress further, the stress residence time is shown in Fig. 8. Valuable information about the level of the stress and the exposure time can be extracted, to evaluate the impact of these stresses on shear-sensitive media like proteins. It can be seen, that for all capillary numbers and both fractions of the stress, the shape of the stress residence time distribution is the same, with a shift to higher shear values with an increasing capillary number. After a sharp increase in the normalized residence time of $t_{res}/t_{res,max} = 1$, it remains constant for up to four orders of magnitude. After, it sharply declines back to zero. This indicates, that the maximum shear stresses arise in form of local stress peaks in the membrane. Further, the residence time of the wall shear shows, a narrower distribution, especially at $t_{res}/t_{res,max} = 1$. This can be explained by the entrance mechanism, i.e., the momentary contact of the interface with the wall during entering the pore.

Idealized pore structures with a contraction and without a contraction, as well as complex pore networks were analyzed in order to reveal the droplet breakup mechanism inside membranes within the PME. Localized and cumulative stress levels were quantified in dependence on varying capillary numbers and membrane wettability.

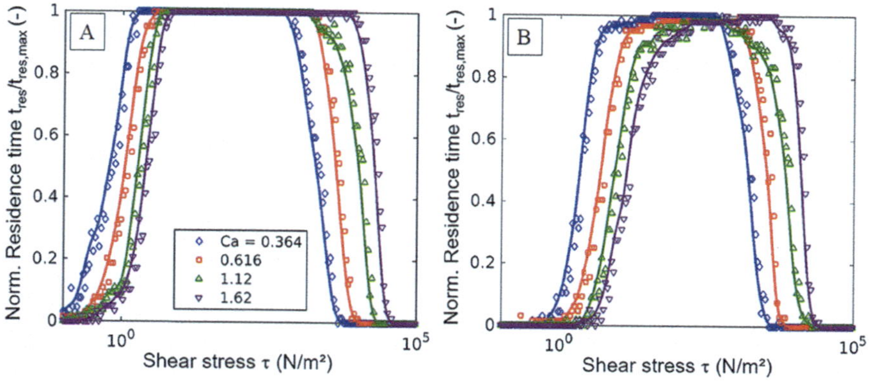

Fig. 8 Normalized stress residence time during droplet dispersion for different capillary numbers (Ca) at **a** liquid and **b** wall interface. The symbols represent the simulation data and the lines a curve fit. Reprint by permission from John Wiley and Sons: Ref. [47]

Idealized pore structures revealed the meniscus instability as the governing mechanism behind the droplet detachment from the wall and consequent droplet breakup. The wettability plays an important role in the formation of this instability and smears the regime transitions that depend on the capillary number and therefore the balance of viscous forces and capillary forces. With increasing capillary number an earlier wall detachment was observed. The secondary flow field revealed the mechanism behind the wettability dependence of the meniscus instability. The single droplet dispersion in CT scans of a micro-porous membrane structure gained further inside into the stress residence time during premix membrane emulsification. It was found that the highest shear stress occurs at the wall, while the stress-related droplet dispersion occurs at the liquid/liquid interface. The relevance of the entrance mechanism with its highest local stress peak has been highlighted.

4 Part 2: Stress Due to Protein Adsorption at the Solid-Liquid Interface

To evaluate the local stress residence time behavior and the resulting structural changes the adsorption onto the wall material were elucidated in the following part. On the hand, the protein adsorption is affecting the wettability of the membrane, which is influencing the stress at the liquid/solid interface. On the other hand, adsorption and desorbtion can also modify the protein on a molecular level.

The protein–protein interactions in bulk, the protein adsorption at the membrane, its effect on the droplet formation and the resulting structural changes of the protein were characterized in dependence of the environmental conditions, like pH and ionic strength. In addition to β-lg, the enzyme lipase was used. Molecular interactions were analyzed with membrane osmometry, analytical ultracentrifugation and MD simulations to determine the conformational state in bulk. The ad- and desorbing process onto the silica membrane material was described with confocal microscopy, contact angle measurement, and ellipsometry. The use of the enzyme lipase made it possible to track the impact of structural changes on the enzymatic activity due to ad- and desorbing.

4.1 Sorption of β-Lactoglobulin Onto the Solid/Liquid Interface

The sorption of β-lg at the solid/liquid interface is determined by the affinity to the wall material, which is affected by the β-lg conformational state and the membrane properties.

The conformational state or quaternary structure of β-lg is determined by the so-called monomer-dimer equilibrium. At pH 7 and low ionic strength, monomers are present (Fig. 9a). With increasing ionic strength more and more dimers are formed, due to the screening of the negative charges reducing the repulsive forces so that short-distance hydrogen bonds can be formed between two monomers. In general, the monomer-dimer equilibrium of β-lg can be described by specific and non-specific interactions. Specific interactions are hydrogen bonds leading to the dimerization of two monomers. Non-specific interactions, such as monomer–monomer, dimer-dimer, and monomer-dimer interactions, encompass electrostatic interactions and van der Waals forces. The monomer-dimer equilibrium is dependent on the protein concentration. The second virial coefficient, as measured by membrane osmometry at varying protein concentrations, is directly affected by the interplay between specific and non-specific interactions that vary with the protein concentration. In order to differentiate between these different types of interactions, the dissociation constant of the dimer has to be known. Through analytical ultracentrifugation, sedimentation profiles can be acquired and the sedimentation and diffusion coefficients can be determined for each individual species of monomers or dimers. Here, the dissociation constant of the β-lg dimer corresponds to the specific interactions. Knowing the dissociation constant of each species, the molecular weight-corrected osmotic pressure ($\Pi_{red} - (2w_D)/M_D$) can be calculated and the second virial coefficient B_{22} can be derived as the slope of the molecular weight-corrected osmotic pressure as a function of the protein concentration. The B_{22} from membrane osmometry is presented in Fig. 9b. Non-specific monomer–monomer, dimer-dimer, and monomer-dimer interactions were calculated using the coarse-grained xDLVO-CG method, with CG representation of the protein [29]. The theoretically calculated and the experimentally measured second virial coefficients correlate well. Evidently, B_{22} significantly depends on the charge state of the system, hence higher B_{22} values are obtained at pH 3 compared to pH 7. The charge screening reduces the extent of the non-specific interactions, which is controlled through the salt concentration. These findings are in line with the literature [48, 49].

The adsorption of β-lg within PME is not only determined by the conformation state, but also by the membrane properties, which can vary in the pore size, the surface roughness, and the hydrophilicity. Silica surfaces are often hydrophilized using the Piranha treatment. Figure 10 shows CLSM micrographs of treated and untreated borosilicate membranes after PME processing with β-lg solutions. Nile red was used as a fluorescence dye since conformational changes of proteins can be detected. Nile red binds through hydrophobic interactions to β-lg and causes an increase in emission intensity and a shift to smaller wavelengths (blue) in a hydrophobic environment [50]. In contrast, if the environment is more hydrophilic, the emission intensity shifts to larger wavelengths (red). For all samples, adsorption was verified by the fluorescence signal in the CLSM micrographs.

For dyed membranes without protein, florescence could not be detected. The average red and blue value of each sample was quantified (Fig. 10, bottom row). In general, β-lg at pH 9 leads to comparable red and blue values regardless of the membrane hydrophilization, however, with slightly higher blue values for the untreated membrane indicating a more hydrophobic environment. At pH 7, β-lg

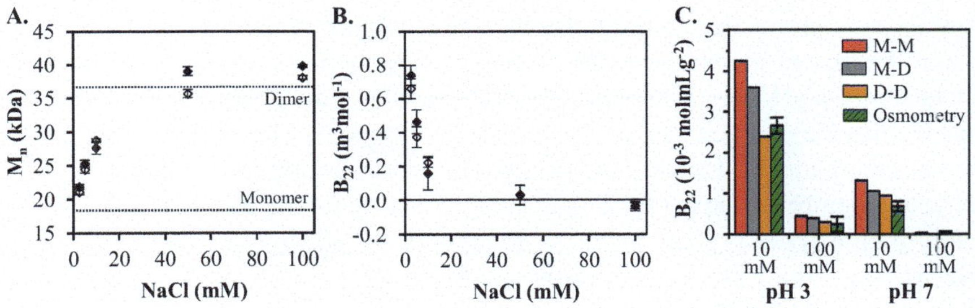

Fig. 9 **a** Molecular weight and second virial coefficient of β-lg as a function of the sodium chloride concentration. Published in [22]. **b** The osmotic second virial coefficient B_{22} for different pH values and sodium chloride concentrations. The coarse-grained xDLVO calculations provide values for monomer-monomer (M-M), monomer-dimer (M-D), and dimer-dimer interactions (D-D). Experimental values are retrieved from a combination of the results from AUC and membrane osmometry (green bars). For a solution pH of 3 and a salt concentration of 10 mM sodium chloride, the dimer dissociation constant for the calculation of the osmotic second virial coefficient was taken from the literature. Reprint by permission from RSC Publication: Ref. [24]

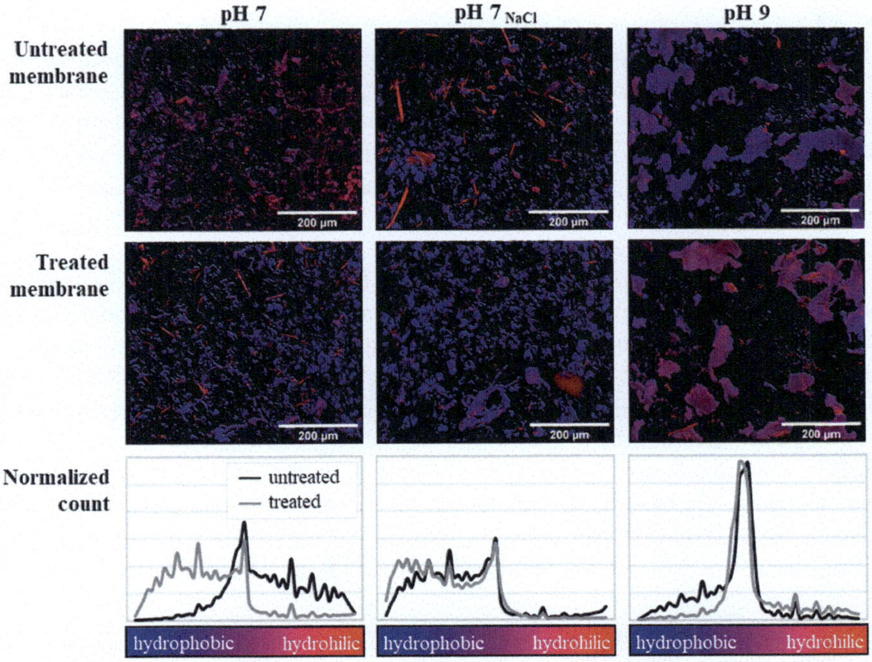

Fig. 10 CLSM images of the untreated and treated borosilicate membranes and the red and blue values of the untreated and treated borosilicate membranes used in PME of β-lg (0.1 wt%) solutions at pH 7, pH 7_{NaCl} (containing 100 mM NaCl) and pH 9. For display purposes, histograms were stretched equally for each channel. Reprint by permission from Elsevier: Ref. [36]

shows a more pronounced red shift for the untreated membrane, whereas the adsorption at the treated membrane leads to greater blue values. For the electrostatically shielded β-lg at pH 7_{NaCl}, (containing 100 mM NaCl) the higher ionic strength screens the electrostatic charges, facilitating a more hydrophobic vicinity regardless of the membrane treatment. Overall, the β-lg conformation is similar for the treated and untreated membrane expect for β-lg at pH 7. A reason for the differences in the β-lg conformation at pH 7, could be the more pronounced adsorption compared to pH 9 and pH 7_{NaCL}, due to less electrostatic repulsion. To further understand the adsorption of β-lg at pH 7, MD simulations were used to identify the molecular interactions between β-lg and silica.

According to the MD results, the adsorption of the protein can be ascribed to electrostatic interactions between the protein and the surface as shown in Fig. 11. Since the silica interface is mainly negatively charged at pH 7, the protein adsorbs with the remaining positively charged residues (blue) to the interface. The negatively charged side of the protein (red) is facing away from the interface into the bulk parallel and normal to the interface. The adsorption occurs after around 150 ns. The positively charged contact residues at pH 7, LYS8, LYS14, and LYS100 are identified. At pH 7, lysin is protonated and, therefore, positively charged, due to a pKa of 10.5 [51]. Furthermore, the non-charged hydrophobic residue TYR99 shows a significant contact with the interface. Due to its neutral charge, the proximity to the surface can be explained by the neighboring binding residue LYS100. After the adsorption to the silica interface at around 100 ns, an exposure of the hydrophobic residues can be noted, as shown in Fig. 11. These results emphasize that the adsorption is driven by electrostatic interactions and then stabilized by hydrophobic interactions being exposed during adsorption.

Ellipsometry was performed to characterize the apparent layer thickness of the adsorbed β-lg. For this purpose, SiO_2-coated wafers were used as a model system for the borosilicate membranes [6]. For all samples, values below 2 nm are found (Fig. 12). As expected, β-lg adsorbing onto untreated surfaces leads to a higher film thickness compared to the treated surface, which is similar for all three solvents (0.3–0.6 nm). Overall, the β-lg adsorption is the highest for pH 7 at an untreated surface, with thicknesses up to 1.7 nm. At pH 9, the film thickness is the lowest with a slightly higher value for the untreated surface. Since β-lg in solution has a diameter of 2–3.6 nm [18, 56], either the formation of a spread β-lg monolayer on the silica surfaces [57] or patches of adsorbed β-lg molecules with uncovered areas in

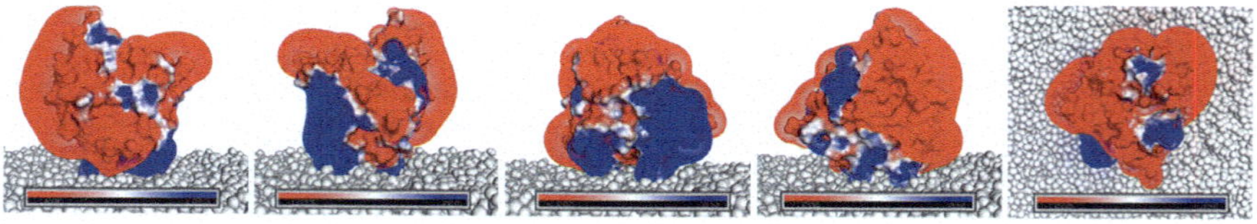

Fig. 11 Electrostatic potential surface from −1 eV (red) to 1 eV (blue) of β-lg at silica-water interface viewed from four different sides and the top, the silica interface is represented in grey. Reprint by permission from Elsevier: Ref. [36]

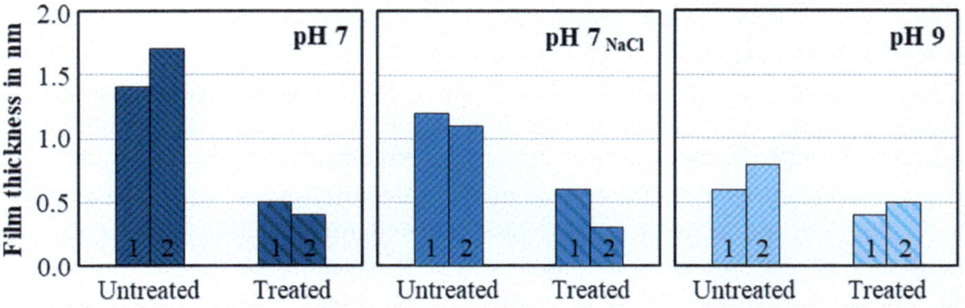

Fig. 12 Film thickness (sample 1 and 2 of duplicate) analyzed by ellipsometry on the treated and untreated SiO_2-coated wafer after 30 min treatment of 0.1 wt% β-lg solution at pH 7, pH 7_{NaCl} (containing 100 mM NaCl) and pH 9. Reprint by permission from Elsevier: Ref. [36]

between can be assumed. The latter might be more reasonable for the untreated silica, as the surface is more heterogenic compared to the treated hydroxylated silica. A higher film thickness for the untreated surface can be explained for two reasons. Firstly, electrostatic repulsion between the protein and the silica surface is less pronounced, since the silica surface is less negatively charged as the silicon atoms are not oxidized in the hydrophilization treatment. Secondly, the hydrophobic attraction between organic contaminants of the membrane and hydrophobic amino acid residues promotes further adsorption. At pH 7, the local electrostatic attraction between the negatively charged silanol groups and the positively charged amino residues of lysine contributes to a greater adsorption compared to pH 9 and pH 7_{NaCl}. For the latter, the electrostatic interactions are screened through the counter ions, increasing the density of the β-lg molecule and, therefore, reducing the film thickness [41].

The sessile drop method was used to determine the surface energy from the contact angle by applying the Owens–Wendt–Rabel and Kaelble model (OWRK), allowing conclusions about the wettability before and after β-lg adsorption [52]. The treated silica surfaces before β-lg adsorption exhibit the highest surface energy (65.6 J/m^2). In contrast, the untreated silica surface shows the lowest surface energy with 43.3 J/m^2. This is in line with our assumptions, as the hydrophilization treatment with piranha removes the organic contaminants and silicon atoms are oxidized, leading to an increase in the overall hydrophilicity. Comparing the surface energy of the treated membrane before and after protein adsorption, it is evident that the surface energy decreases (50.1–51.4 J/m^2), therefore resulting in a more hydrophobic membrane surface. By contrast, the adsorption onto the untreated membrane leads to a more hydrophilic surface (46.6–49.7 J/m^2). Consequently, the surface properties of the treated and untreated membranes after β-lg adsorption equalize. Although the CLSM results indicate differences in the protein conformation of the adsorbed β-lg, the outer layer of the adsorbed protein film might actually show uniform surface properties. The proteins might bind to the membrane (via local electrostatic attraction), and expose predominantly hydrophilic residues since the hydrophobic patches are either hidden in the calyx or bond to the membrane surface regardless of the solvent properties and the surface treatment.

As a last step, we analyzed the effect of the membrane properties and the protein adsorption on the resulting oil droplet distribution for pH 7 and pH 9 using an in-house-made PME device (see Sect. 2.2.4). Since the NaCl screens the electrostatic interactions between the oil droplets, aggregation occurred at pH 7_{NaCl} and, therefore, the results were not taken into consideration. Comparing the cumulative oil droplet distribution of the emulsions prepared with PME (results not shown), only small differences are evident. The PME with the treated membrane at pH 9 produces a slightly smaller mean droplet size (~9 μm) in comparison to pH 7 (~10 μm). This is in line with the results of Kieserling et al. [41], where the same PME set-up was used. Wollborn et al. [53] also produced β-lg stabilized O/W emulsions with oil droplet sizes in a similar range using a comparable PME device and the same (untreated) borosilicate membranes (P4 from ROBU Glasfilter-Geräte GmbH), although different pressures (2 and 8 bar instead of 5 bar) were used.

To sum up, protein adsorption upon silicate membranes within PME of β-lg stabilized oil/water emulsions at pH 7 and 9 is driven by local electrostatic attraction and stabilized through hydrophobic interactions. On a molecular level, the influence of the conformational state (pH 7, pH 7_{NaCl}, pH 9) and the membrane hydrophilicity are evident through differences in ellipsometry and CLSM results. However, the surface energy of the resulting protein layer is almost similar for all samples. As a consequence, the oil droplet size distribution of the oil/water emulsion is merely affected by the membrane treatment on a macroscopic level. Although, the kinetic stability of the β-lg stabilized oil/water emulsions might actually not be influenced by the membrane hydrophilization, the results suggest an effect of the stress-residence time of β-lg on its molecular level, which depends on the conformational state of the protein.

4.2 Sorption of Lipase at Solid/Liquid Interface

To further study the stress-residence time of proteins on a molecular level caused by ad- and desorption onto surfaces, the enzyme lipase was used. The adsorption as a function of the lipase concentration and pH of the feed solution is depicted in Fig. 13a. The absolute adsorbed mass of lipase onto the membrane was increasing with increasing lipase concentration (0.5–4 g/L) (Fig. 13a). Similar to these, the adsorption of the protein bovine serum albumin increased also with increasing protein concentration [54]. For increasing the pH of the feed solution, the adsorbed mass on the membrane did not show an increasing trend (Fig. 13a). However, studies showed that pH can affect adsorption because of decreasing electrostatic repulsion between membrane and enzyme with increasing pH [55, 56]. The principle of decreased electrostatic repulsion is already intensively used in biotechnology for the immobilization of the lipase irreversibly bound to solid materials [57].

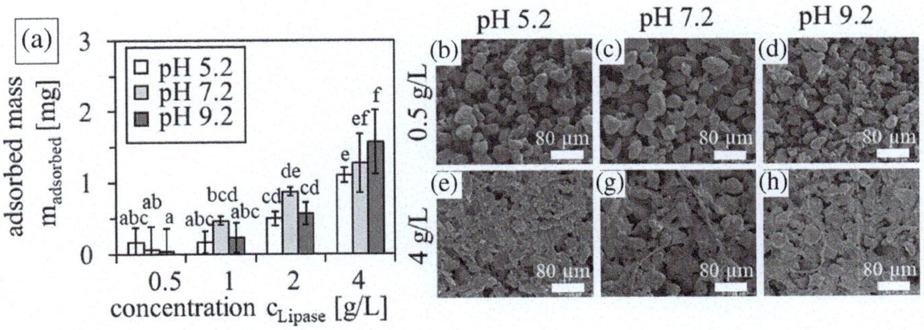

Fig. 13 **a** Absolute adsorbed mass of lipase onto the membrane in dependency on the bulk lipase concentration. Different letters, **a–f** represent statistically significant differences ($n = 3$; $p < 0.05$). **b–g** SEM pictures of silica membrane surface used for a one run filtration of solutions with lowest and highest lipase concentrations at different pH values before ($n = 1$). Reprint by permission from Elsevier: Adapted from Ref. [35]

For detecting fouling and pore blocking, SEM images of the membranes used to filtrate the feed solution with varying lipase concentration and pH were taken (Fig. 13b–g). With increasing lipase concentration, the SEM images showed an increase in the adsorption of lipase on the membrane surface and inside the membrane pores. No deposition at the lowest lipase concentration (0.5 g/L) could be observed on the surface of the membrane, which might indicate a negligible amount of lipase adsorption for this concentration. With the increase of the lipase concentration in the feed (up to 4 g/L), the deposition on the membrane surface was increasing and the fouling layer was covering most of the pores. Militano et al. [58] confirmed the fouling on a membrane surface due to the adsorption of lipase with atomic force microscopy [58]. They also described the formation of a multilayer with increasing lipase concentration due to increased aggregation caused by protein–protein interaction.

The contact angle of a water droplet on the silica surfaces before and after adsorption was measured. A change in the contact angle equals a change in the membrane wettability due to surface modifications. The contact angle of a water drop against air on the surface was decreased with increasing adsorbed lipase on the membrane surface in comparison to the control (Fig. 14a). The contact angle for all tested concentrations felt below a value of 30° regardless of the pH (Fig. 14a). The lower the contact angle in the air/water system, the higher is the hydrophobicity of the membrane surface in a water/oil system. In a study with polyether sulfone membranes, the hydrophilic character of the membrane became more hydrophobic due to lipase adsorption [5]. However, the increase of the lipase concentration caused only a slight but not significant further decrease in contact angle (Fig. 14a). In their study, the change towards a hydrophilic character of the membrane might be due to the hydrophilic globular protein character of the lipase with a hidden hydrophobic binding site covered by a hydrophilic α-helix lid, which only opens during adsorption [59].

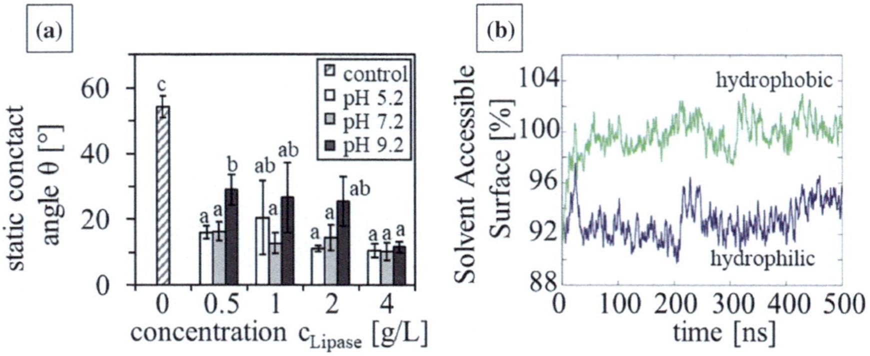

Fig. 14 **a** Static contact angle after 15 s of a water droplet on a membrane surface pre-soaked in solution with different lipase concentrations at different pH values (n = 3). The control sample was taken with a membrane pre-soaked in ultra-pure water (pH ~ 6–7). Different letters, **a**–**f** represent statistically significant differences ($n = 3$; $p < 0.05$). **B** Solvent-accessible surface calculated from molecular dynamic simulations for lipase at SiO_2/water interface for hydrophilic and hydrophobic residues. Reprint by permission from Elsevier: Adapted from Ref. [35]

For further prove of the exposition of hydrophobic parts during the adsorption of lipase, an MD simulation was performed. Therefore, the hydrophilic and hydrophobic solvent accessible surface (SAS) of lipase during adsorption was calculated.

Due to structural rearrangements during adsorption, hydrophobic and hydrophilic residues are exposed. Figure 14b depicts the solvent-accessible surface of the protein, divided into hydrophobic and hydrophilic residues with reference to the initial structure. Absolute values can therefore exceed 100%. For a surface-adsorbed monomer, the solvent-accessible surface increases from 93 to 100%, while the solvent-accessible surface of the hydrophilic residues increases from 92 to 94% (Fig. 14b). The exposure of hydrophobic residues emphasizes the hydrophobic stabilization due to adsorption to the solid/liquid interface. As an indicator for changes in the primary structure the root mean square deviation of the C-alpha atoms of the adsorbed lipase was calculated. Compared to the initial configuration the root mean square deviation increased by about 0.2 nm during adsorption to the silica membrane, indicating a rearrangement of the protein structure (Fig. 15a). To evaluate whether the structural change is associated with an unfolding at the interface, the radius of gyration has been calculated. From Fig. 15b it can be seen, that the radius of gyration increases from about 2.215 to 2.24 nm which can be related to a structural rearrangement rather than an unfolding at the interface.

Therefore, the MD results underline the structural rearrangement during adsorption with hydrophobic stabilization.

In order to evaluate the protein interaction and consequent adsorption in detail, further molecular dynamic simulations were performed with regard to the enzymatic active center. Therefore, active (code 1CRL) and passive (code: 1TRH) lipase were simulated in close proximity to a SiO_2/water interface. Figure 16a, b depict the individual distances between lid residues of active and passive lipase. After around

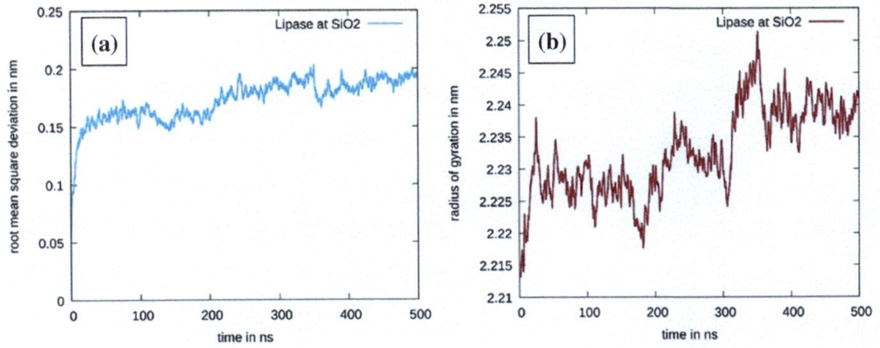

Fig. 15 Root mean square deviation of C-alpha atomic position **a** and radius of gyration **b** from molecular dynamic simulations for Lipase during adsorption at SiO_2/water interface. Reprint by permission from Elsevier: Ref. [35]

200 ns the lid of the active lipase moves significantly, whereas the no lid movement is evident for the passive lipase. In Fig. 17 it can be seen, that the lid moves closer to the active center for 1CRL. Consequently, tunnel calculations were conducted to find possible paths to the active center to guarantee the accessibility of the active center which are shown in Fig. 18. It was found, that the tunnels from the outer surface of the enzyme to the buried active center are induced by the solvent and stabilized by the interface. Furthermore, the active lipase shows a complex tunnel network whereas the tunnel network is less complex for the passive lipase. This can be assessed as an initial state of surface-induced protein activation for 1TRH.

For 1CRL, an accessibility of the active center is still guaranteed, which is confirmed by docking simulations of triglycerides. For this, different triglyceride ligands of different chain lengths from C6 to C14 were docked into the active center.

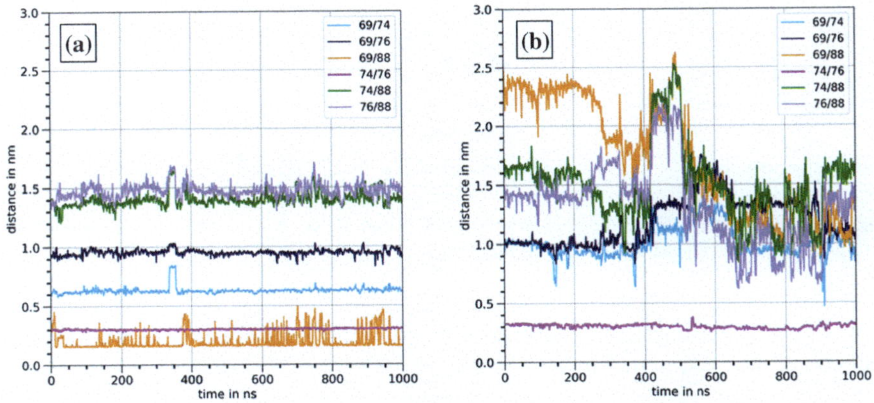

Fig. 16 Distance plot of each lid residue during adsorption to the SiO_2 interface for 1TRH (**a**) and 1CRL, **b** over the time

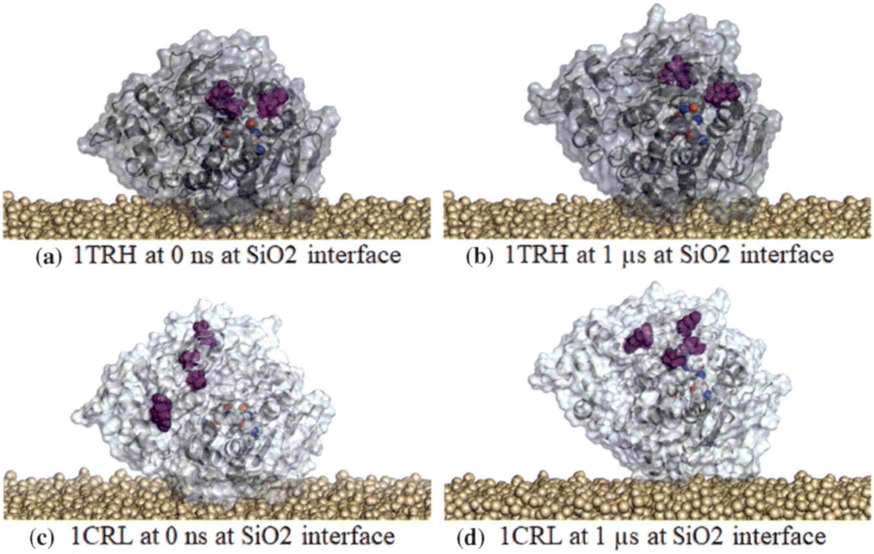

Fig. 17 Starting and adsorption configuration of 1CRL and 1TRH at SiO_2 interface after 1 μs simulation time, lid residues are highlighted in purple ball representation, red/white/blue ball representation indicate the catalytic triad

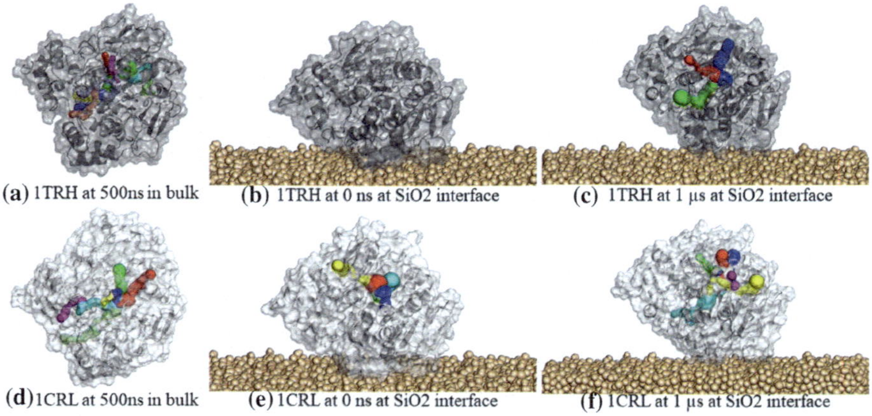

Fig. 18 Surface and cartoon representation of 1TRH and 1CRL at silica surface and in bulk for different times, calculated tunnels are highlighted

It was found that during the adsorption to the solid interface the conformational changes of the lid structure induce selectivity of the enzyme towards ligand binding. Whereas the bulk equilibrated starting structure 1CRL bound a C8 medium chained triglyceride, the surface adsorbed structure did not show any interaction between this ligand and the enzyme. Further, the docking of a C6, C8, C10 triglyceride did not show any interaction with the equilibrated starting structure, while it did show an interaction with the surface-adsorbed active structure. The C6 and C8 chains are stabilized by hydrophobic interactions outside the catalytic triad, while the C10 chain is facing into the binding pocket of the catalytic triad, stabilized by hydrophobic interactions with SER209 and GLY123. Further, Glu88 a lid residue that showed significant movement during adoption forms a hydrogen bond to an oxygen of the C8 side chain.

This indicates, an enzymatic selectivity induced by immobilization onto silica. In addition, the adsorption configuration still enables further interaction with the oil phase, as the lid is facing away from the solid into the liquid phase.

5 Part 3: Stress at the Liquid–Liquid Interface

Due to the amphiphilic character of proteins and enzymes, they are not only able to adsorb to solid interfaces, but also to liquid/liquid interfaces, e.g., in emulsions. The interfacial stabilization of proteins is described by three phases: (i) migration through the bulk phase (*lag* time), (ii) adsorption at the interface, and (iii) structural rearrangements at the interface for interfacial stabilization. However, the interfacial stabilization of proteins within PME at an increasingly occupied oil/water-interface is not fully understood. Thus, the impact of the stress at the liquid/liquid interface on the proteins and on the droplet's interfacial stabilization were investigated in the following part.

The interfacial stabilization of β-lg at liquid/liquid interfaces (stages (i) and (iii)) in dependence on the interfacial preoccupation was investigated via drop tensiometry using native β-lg. To elucidate the mechanism of film formation and stability towards the two stresses—three-dimensional dilatation and two-dimensional shear deformation—the protein film response was measured independently with dilatational rheology and interfacial shear rheology. Afterwards, the protein structure was systematically modified to study the underlying mechanisms and molecular interactions involved in the adsorption behavior. Firstly, hydrostatic pressure was used as it mainly affects noncovalent interactions such as hydrogen bonds, electrostatic, and van der Waals interactions [60]. Secondly, to elucidate the impact of disulfide bonds, β-lg was recombinantly produced in *E. coli* and the cysteines were systematically exchanged with alanine.

5.1 Interfacial Adsorption of Native β-Lactoglobulin

In the first step, pendant drop analysis was performed to investigate the three phases of the interfacial stabilization process of native β-lg, namely the migration through the bulk phase (i), the adsorption at the oil/water-interface (ii), and the interfacial

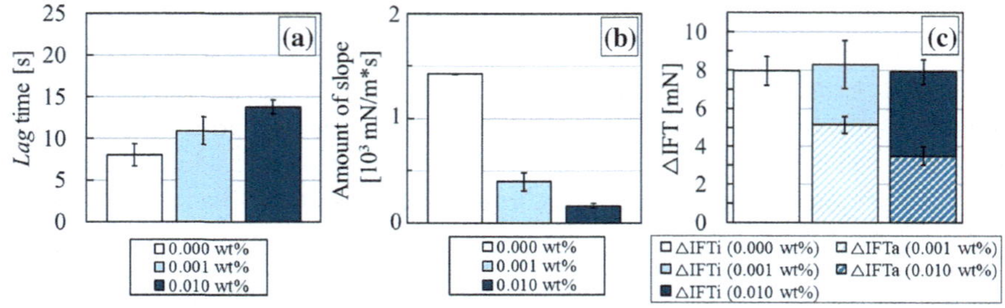

Fig. 19 **a** Stages (I) and (II) of the interfacial stabilization process and estimation of the stabilization state of β-lg at pH 7 at interfacial preoccupation level of 0.000, 0.001, and 0.010 wt% against MCT-oil via two-fluid needle experiments. Analysis of lag time t_a (i), calculated from the point of addition protein solution until the drop of interfacial tension. **b** Evaluation of adsorption rate (ii) via slope k of exponential decrease in interfacial tension after addition of protein solution. **c** Estimation of interfacial stabilization state via delta interfacial tension $\Delta\gamma_u$ and $\Delta\gamma_p$ before and after adding protein solution. Different letters, **a–e** show significant differences between the values ($p < 0.05$). Reprint by permission from Elsevier: Adapted from Ref. [39]

stabilization state of native β-lg (iii). The first phase is a diffusion- and convection-controlled migration of the protein through the bulk phase, estimated via lag time [61]. The migration through the bulk phase of β-lg at pH 7 with interfacial preoccupation level of 0.000, 0.001, and 0.001 wt% is shown in Fig. 19a, b. The *lag* time significantly increased with preoccupation from 0.000 to 0.010 wt% and can be explained by the dependence of the diffusion rate on the protein concentration gradient [62, 63]. A high protein concentration at the oil/water-interface due to a high interfacial preoccupation might have decreased the concentration gradient from bulk towards the oil/water-interface. This might have resulted in a slow diffusion and therefore low *lag* time of β-lg.

In the second step of the interfacial stabilization process, β-lg adsorption is accompanied by the structural rearrangement of the protein at oil/water-interfaces [64, 65]. The adsorption rate was investigated by the slope of the exponentially decreasing interfacial tension (Fig. 19b) [13]. Comparing the interfacial preoccupation levels, the adsorption rate decreases with increasing interfacial preoccupation of the oil/water-interface (Fig. 19b). The slope k of the interfacial tension was reduced from 0.000 to 0.010 wt% β-lg by 89%. A slower adsorption rate might be explained by steric hindrance and electrostatic repulsion between bulk protein and adsorbed protein. The difference in interfacial tension $\Delta\gamma_\nu$ was not affected by the interfacial preoccupation level (Fig. 19c).

To consider individual protein orientations, a force map was obtained indicating the forces at each protein orientation at the minimum of the force-distance curve (Fig. 20). For the unoccupied interface, only attractive forces are observed for all protein orientations. Moreover, regions with high attractive forces of up to −2.7 pN (red color) are visible. At an interfacial preoccupation of 0.0135 nm^{-2}, the main part of the force map indicates repulsive interactions. However, also regions with

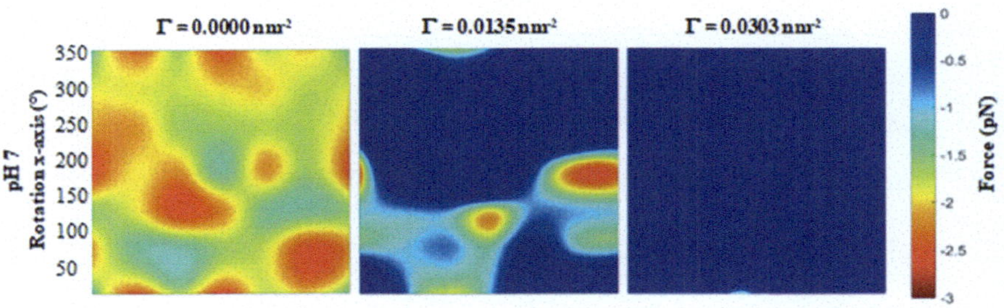

Fig. 20 Force maps, indicating the interaction force at different β-lg orientations to the MCT-oil interface for different interfacial preoccupation levels (0.0000, 0.0135, and 0.0303 nm^{-2}) at pH 7. Reprint by permission from Elsevier: Adapted from Ref. [39]

attractive forces are present. With an interfacial preoccupation of 0.0303 nm^{-2} no attractive forces are observed.

To characterize the sorption processes at the oil/water-interface and to estimate the long-term stability, protein film formation, and the stress-sensitivity of the protein film via interfacial shear rheology and dilatational rheology was performed. The interfacial storage (elastic) modulus G_i' increased within the first 2.5 h, while the loss (viscous) modulus G_i'' remained constant over the investigated time range (Fig. 21a). The protein films were completely formed after 10 h and showed a more elastic than viscous behavior $\left(G_i' > G_i''\right)$. In frequency sweeps, the elastic modulus G_i' slightly increased with increasing frequency and was higher than G_i'' (Fig. 21b). The high elasticity indicates the formation of strong intermolecular interactions between adjacent proteins at the oil/water-interface [66–68]. Attractive van der Waals interactions between all kinds of molecules contribute substantially to the film formation [69, 70]. The unfolding at the interface increases attractive electrostatic interactions between oppositely charged exposed amino acid residues, hydrophobic interactions between nonpolar residues, and the formation of covalent disulfide bonds between adjacent proteins [68, 71, 72]. In addition, the unfolding also favors the formation of intermolecular hydrogen bonds, resulting in a minor increase in intermolecular β-sheets (data not shown). These results are in line with Fang and Dalgleish [73], who highlighted the formation of intermolecular β-sheets after the adsorption of β-lg with time.

Within interfacial dilatational rheology, amplitude sweeps were performed as illustrated in Lissajous plots in Fig. 22. At amplitudes of 0.95%, the sample showed nonlinear scattering. Lissajous plots at an amplitude of 0.95 and 11.35% were symmetric (data not shown, see [41]). From the high elastic and low viscous character of the interfacial film, a lower influence of diffusional exchange compared to structural relaxation was concluded. This was explained by a high structural flexibility and intermolecular interactions in β-lg films at pH 7 leading to a stable interfacial protein film at pH 7 during dilatational deformation.

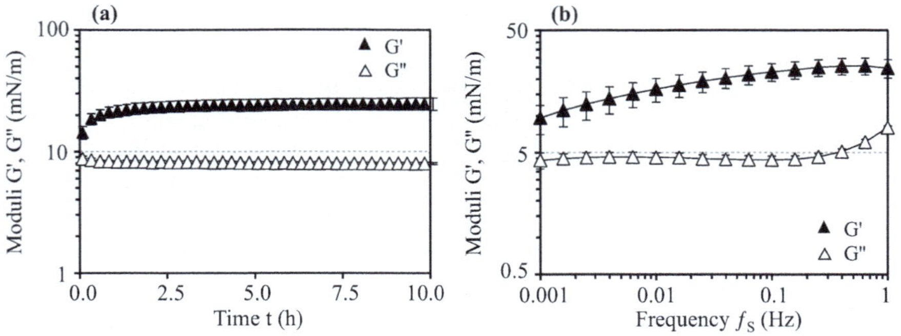

Fig. 21 **a** Interfacial protein film formation with storage modulus G_i' (black symbols) and loss modulus G_i'' (white symbols) against time, analyzed by interfacial shear rheology of 0.1wt% β-lactoglobulin against MCT-oil. Frequency and amplitude were kept constant at $fS = 0.1$ Hz and $f_S = 0.1\%$. **b** Frequency sweeps with storage modulus G_i' (black symbols) and loss modulus G_i'' (white symbols) against frequency ($f_S = 0.001 - 1$ Hz), analyzed by interfacial shear rheology of 0.1wt% β-lactoglobulin against MCT-oil. Amplitude was kept constant at $\gamma_S = 0.1\%$. Reprint by permission from Elsevier: Adapted from Ref. [41]

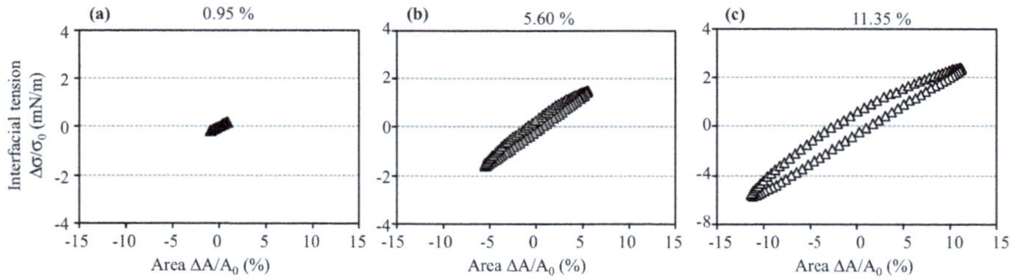

Fig. 22 Lissajous plots with change in interfacial tension Δσ against relative change in drop area $\Delta A/A_0$ analyzed by interfacial dilatational rheology of 0.1wt% β-lactoglobulin at pH 7 (**a**), pH 7_{NaCl} (**b**) and pH 9 (**c**) against MCT-oil at different deformation amplitudes; (1) 0.95%, (2) 5.60% and (3) 11.35%. Frequency was kept constant at $f_D = 0.01\%$. Reprint by permission from Elsevier: Adapted from Ref. [41]

5.2 *Interfacial Stabilization of Structurally Modified β-Lactoglobulin*

For a better understanding of the effect of down-stream-processing on the interfacial stabilization, structurally modified proteins, namely high pressure-treated and recombinant β-lactoglobulin, have been investigated. The interfacial activity of high pressure-treated β-lg was studied using the pendant drop method. A pressure of 600 MPa significantly increases the *lag* time of β-lg, whereas no significant increase was observed for pressures from 100 to 500 MPa (Fig. 23a). We assume that the significant increase in the *lag* time results from the higher molecular weight and higher hydrodynamic diameter of the high pressure-treated β-lg dimers (at 600 MPa) in comparison to monomeric β-lg [74]. The shear rheological monitoring of the storage modulus G' and the loss modulus G'' during interfacial protein film formation is displayed in detail in [40].

The adsorption rate at the oil/water interface in stage (ii), as estimated via the slope of the decreasing interfacial tension, increases significantly at pressures above 300 MPa (Fig. 23b). The increasing pressure results in a higher structural flexibility and thus a fast structural rearrangement of the adsorbed proteins [75, 76]. This increases the affinity to the hydrophobic oil phase and thus accelerates the adsorption at the oil/water interface [74]. In summary, hydrophobic interactions and a high structural flexibility compensate for possible steric interference of the pressure-induced β-lg dimers and favors a fast adsorption at the oil/water interface.

The intra- and intermolecular interactions and the structural flexibility highly affect stage (iii) the interfacial protein film formation. The pressure treatment of β-lg above 400 MPa results in a significantly stronger decrease in the interfacial tension in comparison to native β-lg (Fig. 23c). It was expected that pressure-treated β-lg exhibits a higher interfacial molecule density leading to a high interfacial film thickness. This effect is mainly caused by the pressure-induced hydrophobic intermolecular interactions of adsorbed β-lg [77, 78]. In combination with the partial unfolding of the secondary structure and the loss of the tertiary structure, we assume the formation of a pressure-induced hydrophobic molten globule state [74].

To analyze the stability of the interfacial protein film and thus the intermolecular interactions of the adsorbed proteins, the protein film was exposed to mechanical stress, such as dilatational and shear stress within interfacial rheology. Complex modulus G^* and loss factor tanδ of the pressure-treated β-lg film showed higher values compared to the native β-lg film (Table 2). The protein film stability during shear and dilatational rheology within the linear viscoelastic region (LVE) and nonlinear viscoelastic region (NLVE) were reported in [40]. A low loss factor tanφ and a high complex modulus E^* of pressure-treated β-lg were found in comparison to native β-lg (Table 2). It was assumed that the interactions between adsorbed pressure-treated β-lg appear to form earlier and in a higher quantity and strength compared to those of

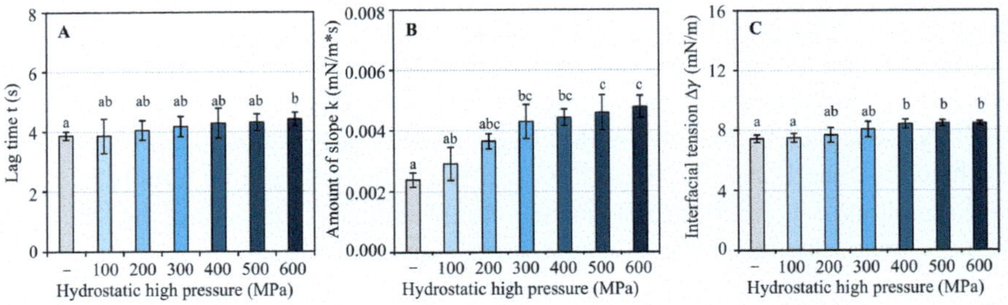

Fig. 23 **a** Lag time, **b** amount of slope of decreasing interfacial tension of the oil/water interface, and **c** difference in interfacial tension of reference (–) and hydrostatic high pressure-treated (100–600 MPa) β-lactoglobulin measured by pendant drop analysis (n = 3, n = 1 illustrated). Different letters **a–c** represent statistically significant differences (n = 3, $p < 0.05$). Reprint by permission from Elsevier: Ref. [40]

Table 2 Interfacial shear and dilatational parameters of pressure-treated (600 MPa) and native β-lg. Different letters indicate significant differences in a row ($n = 3$, $p < 0.05$). From [74]

	Parameter	Pressure-treated β-lg	Native β-lg
Dilatation	$E^*_{1.85\%}$ (mN/m)	41.1 ± 2.8^a	29.1 ± 3.4^b
	$\tan\varphi_{1.85\%}$	0.10 ± 0.01^a	0.15 ± 0.01^b
	γ_D at LVE_{End} (%)	3.34 ± 0.32^a	3.59 ± 0.88^b
Shear	$G^*_{10\,h}$ (mN/m)	14.6 ± 1.0^a	10.4 ± 1.0^b
	$\tan\delta_{10\,h}$	0.25 ± 0.02^a	0.36 ± 0.04^b
	γ_S at LVE_{End} (%)	2.43 ± 0.2^a	2.49 ± 0.04^a

native β-lg. As a consequence, the enhanced unfolding enables the pressure-treated β-lg to a fast formation of more and stronger intermolecular interactions resulting in a higher elasticity compared to native β-lg over time. Following Dickinson and Matsumura [64] and Dumay et al. [79], the formation of covalent disulfide bonds can result from the unfolding and subsequent exposition of free sulfhydryl groups (as Cys^{121}) of β-lg at the oil/water interface. As mentioned by Galazka et al. [80] and later by Considine et al. [81], pressure-induced structural changes could favor such sulfhydryl–disulfide exchange reactions and cause a high strength of intermolecular interactions between adsorbed pressure-treated β-lg.

In summary, the pressure-treated β-lg forms more and stronger intermolecular interactions (e.g. van der Waals interactions, disulfide bonds) between the adjacent proteins at the oil/water interface as well as between the adsorbed proteins and the oil phase (e.g. hydrophobic associations). The latter leads to a high interfacial molecular density. These interactions stabilize the protein film against shear deformation and small dilatational deformation, whereas the high hydrodynamic radius reduces the velocity of diffusional exchange, and thus the stability against high dilatational deformation. These results clarify the influence of high hydrostatic pressure-induced structural changes on the structure and interactions of adsorbed β-lactoglobulin. Our findings expand current literature that either only focuses on the first two stages of the interfacial stabilization process or the interfacial protein film stability without investigating its cause, namely the structure and interactions of the adsorbed proteins.

To study the impact of disulfide bonds, using recombinantly produced β-lg is a powerful technique, where the disulfide bond causing cysteine in the amino acid chain is exchanged by alanine. With this, disulfide bonds cannot be built regardless of chemical conditions. The bacteria *E. coli* is a commonly used microorganism in biotechnology, e.g., for the production of antibodies, since mutations of the DNA can be easily made and the production guarantees high yields. However, the metabolism of *E. coli* requires a methionine at position one of the N-terminal of the amino acid chain, but with that, the correct post-translational folding of β-lg is hindered. Therefore, Loch et al. introduced the mutation L1A/I2S to β-lg to ensure the correct cleavage of the N-terminal methionine [82].

Before introducing mutations of the cysteines, a collaborative study was performed, where the recombinant β-lg B (wildtype) still carrying a methionine at its

N-terminus without any further mutations and the recombinant β-lg with the L1A/I2S modification were produced in *E. coli* and their structure and techno-functionality were compared to bovine β-lg B in [37]. For that study, we contributed to the analysis of the interfacial behavior and emulsifying properties. The *lag* times and adsorption rates of all variants were similar with no statistical differences (data not shown, see [37]). The interfacial tension after 60 min for the β-lg with the L1A/I2S modification (14.9 ± 1.0 mN/m) was lower than for the wildtype (16.7 ± 1.1 mN/m) and the bovine β-lg B (20.2 ± 0.3 mN/m) [37]. Thus, the β-lg with the L1A/I2S modification is able to reduce the interfacial tension even more than the bovine β-lg B. The equilibrium interfacial tension correlates with the long-term effects, like protein packing density at the interface, interface rearrangement, and network formation. Furthermore, in dilatation rheology experiments, the recombinant variants possessed viscoelastic properties, which showed fewer elastic properties in comparison to the bovine β-lg B (data not shown, see [37]). The structural characterization of both recombinant variants revealed that the recombinant β-lg with the L1A/I2S modification has higher similarity to the bovine β-lg B [37], and thus, it was used as the basic mutant for the recombinant exchange of cysteine with alanine. This variant is called "control" hereafter.

As listed in Table 1 (Sect. 2.3.1), five cysteine mutants were recombinantly produced as described in [38]. In C121A, the cysteine at position 121 was exchanged with alanine. The cysteine121 is a "free" cysteine within the β-lg B causing intermolecular disulfide bonds between two β-lg molecules. The cysteine106 and cysteine 119 are located inside the β-lg B molecule and we were not able to produce them due to molecular entrapment in inclusion bodies [38]. The cysteine66 and cystein160 are building an intramolecular disulfide bond at the molecule surface. Here, three variants were produced: C66A and C160A, where either one of the cysteines was exchanged by alanine, and C66AC160A, where both cysteines were exchanged by alanine. For C66A and C160A, the β-lg possesses now two free cysteines (C121 and C60 or C160). Thus, here, either two intermolecular disulfide bonds or one new intramolecular disulfide bond could be possibly built. The fifth cysteine variant (noCys) has all cysteines exchanged by alanine and should not build any disulfide bonds. The variants were used to investigate the interfacial adsorption behavior (stages i to iii) with drop tensiometer and interfacial dilatational rheology.

The *lag* time (stage i), amount of slope (ii), and the delta of the interfacial tension (iii) are depicted in Fig. 24a, b, respectively. The *lag* time and the amount of slope of C121A and C66A were in the same range as the control. In contrast, the C160A, C66AC160A, and noCys showed a shorter *lag* time and a larger amount of slope than the control. Brune et al. characterized the chemical structure of all variants, except C66A and C160A [38]. They found out that the tertiary structure of C121A is identical to the control, and both are present mostly in dimers at pH 7. This could explain their similar behavior in stages (i) and (ii). In contrast, at pH7, C66AC160A were mainly monomers, whereas noCys was a mixture of monomers and dimers. Being only monomeric instead of dimeric could explain the faster migration through the bulk (stage (i)) and the adsorption at the interface (stage (ii)). The different adsorption behavior of monomers and dimers was already explained in [39, 74].

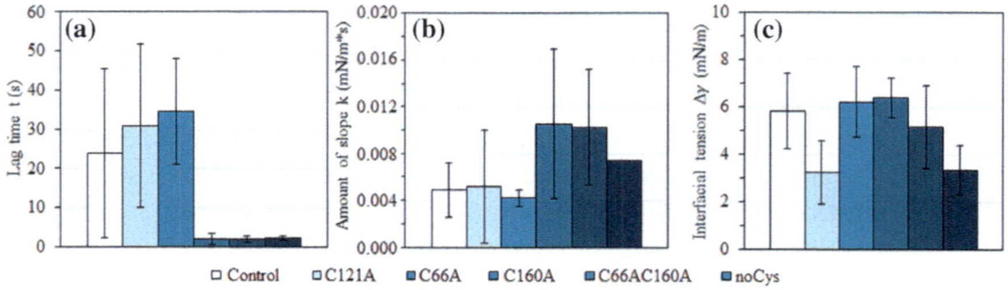

Fig. 24 **a** *Lag* time, **b** amount of slope of decreasing interfacial tension of the oil/water interface, and **c** difference in interfacial tension of β-lactoglobulin variants measured by pendant drop analysis. Partly published in [37]

Interestingly, C66A was behaving more like the control and C121A, than like C160A or C66AC160A (Fig. 24). Thus, the impact of the outer disulfide bond might be only controlled by the C160, which is located nearer to the molecule surface, while C66 is more located inside the molecule. The difference in the interfacial tension between the start of the measurement and after 60 min (stage (iii), Fig. 24c) of C66A, C160A, and C66AC160A is similar to the control, while the difference is larger for C121A and noCys. Brune et al. reported that C66AC160A has a slightly open and noCys a disordered secondary structure [38]. For noCys, most of the β-sheet structure, and thus, the secondary structure, was lost [38]. C121A and noCys are lacking free cysteines for intermolecular interactions. Thus, this might explain the higher difference in the interfacial tension. With free and accessible cysteines, the intermolecular interactions are highly likely accelerating the film formation due to a higher interfacial density (see Sect. 5.1).

The result of the dilatational rheology in form of Lissajous plots for the non-linear viscoelastic region is depicted in Fig. 25. The Lissajous plots show a slightly asymmetric elliptical shape with a higher included area for control, C121A, C66AC160A, and noCys. In contrast, the shape of C66A and C160A is narrower with almost no included area, thus, showing predominantly elastic properties. In comparison, C121A, C66AC160A, and noCys showed slightly more viscoelastic properties, however, the viscoelasticity is not strongly pronounced. The shape of C66AC160A and noCys revealed a more pronounced asymmetry and was more vertical. The asymmetry in C66AC160A and noCys might be due to their monomeric state and a lower structural. To further prove the latter, an MD simulation of the C66A and the noCys being adsorbed to an MCT-Oil/water interface was done (Fig. 26). The alanine residues at positions 66 and 106 of the noCys are highlighted in green showing clearly the higher unfolded conformation in comparison to C66A.

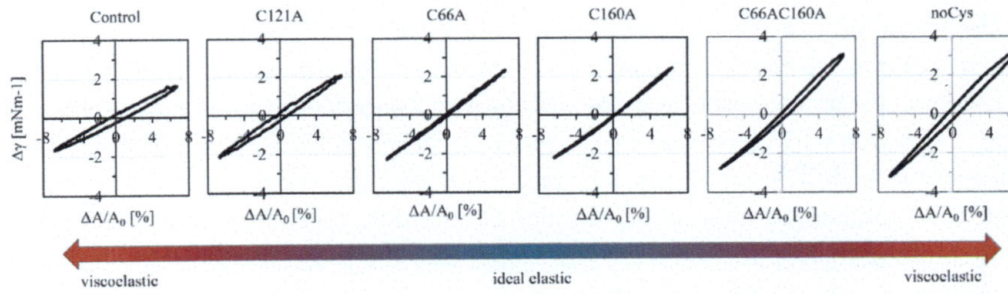

Fig. 25 Lissajous plots with change in interfacial tension $\Delta\sigma$ against relative change in drop area A/ A_0 analyzed by interfacial dilatational rheology of 0.1 wt% β-lactoglobulin at pH 7 against MCT-oil at 7% deformation amplitudes. Frequency was kept constant at 1 Hz

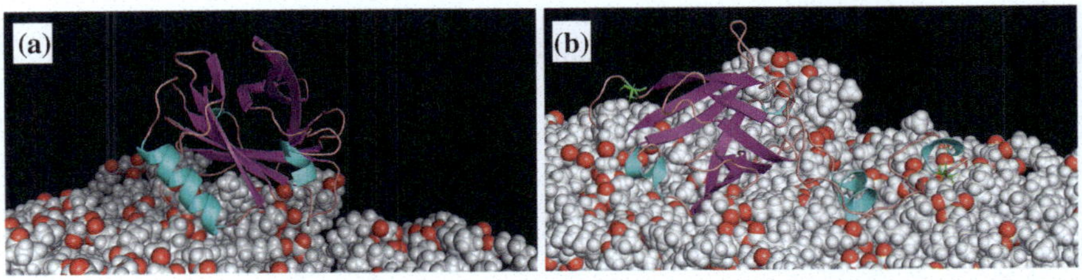

Fig. 26 Molecular dynamics simulation of **a** C66A and **b** noCys variant adsorbed to the MCT-oil/ water interface. The residues at positions 66 and 106 of the noCys are highlighted in green

6 Summary and Conclusion

The study investigated the mechanistic understanding of structural changes of low and high molecular proteins in dependence on their stress residence time behavior during the premix membrane emulsification process. For that, (1) the stress in the pore on the droplet leading to droplet deformation and breakup, (2) stress on the protein at the solid-liquid interface and (3) stress on the protein at the liquid–liquid interface were studied in detail. Experimental data were accompanied by molecular dynamics and computation fluid dynamics simulation.

The analysis of the local stress distribution and the stress residence time behavior in membrane pores show that the deformation and droplet breakup process is mainly driven by liquid/liquid stresses at the interface. The highest shear stress occurs in the wall region over a short time, but the interfacial breakup with subsequent droplet formation is caused by the liquid/liquid shear stress. We found that not only the level of stress, but the duration is critical for droplet breakup and protein conformation (stress residence time behavior). Further, it was shown that the dynamics of the contact line movement is a crucial part of the liquid fragmentation and depends on the wettability of the membrane. The secondary flow field revealed the occurrence of secondary vortices in the vicinity of the interface which contribute to the regime transition of interfacial wall detachment.

The wettability of the membrane was mainly influenced by the protein adsorption. In general, molecular dynamic simulations revealed that the driving mechanism behind protein adsorption to liquid/liquid and liquid/solid interfaces is electrostatic interaction with consecutive exposure of hydrophobic residues and subsequent hydrophobic stabilization at the interface. Macroscopic properties of β-lg stabilized O/W emulsions, such as the droplet distribution was almost not affected by the membrane hydrophilization. It was found that the β-lg adsorption assimilated the surface energy regardless of the membrane hydrophilization as predominantly hydrophilic residues are exposed on the surface regardless of the interaction with the membrane. On a molecular level, the membrane hydrophilization influenced the layer thickness and conformational state affecting the stress-residence time behavior. The trends for the wettability and membrane hydrophilization were also found for the enzyme lipase. Additionally for lipase, the impact of the lipase concentration on the membrane was investigated. The results showed that the higher the concentration the higher the adsorbed mass of lipase on the membrane resulting in higher fouling and higher pore blocking. These are the aspect to be considered for the membrane performance in general, the duration until needed cleaning or replacement. In addition, we found that the adsorption did not impact the secondary structure or functionality of protein in the PME permeate. This further promotes the premix-membrane emulsification as a mild emulsification process, especially for sensitive biomolecules.

The results of the adsorption to liquid/liquid interfaces demonstrated a significant correlation between adsorption behavior and interfacial preoccupation depending on conformational state and charge of β-lg. Molecular dynamic calculations confirmed the results by highlighting the role of electrostatic forces. Compared to the native β-lg, the pressure-treated β-lg formed more and stronger intermolecular interactions (e. g. van der Waals interactions, disulfide bonds) between the adjacent adsorbed proteins at the oil/water interface and with the oil phase (e.g., hydrophobic associations). These interactions stabilized the protein film against shear deformation and small dilatational deformation. On contrary, the high hydrodynamic radius of pressure-treated β-lg reduced the diffusional exchange, and thus the stability against high dilatational deformation. The recombinant β-lg variants showed the impact of the intermolecular disulfide bonds onto the monomer/dimer ratio and thus on the adsorption kinetics. Molecular dynamics further showed that an unfolded secondary structure due to missing disulfide bonds led to a strong adsorption to the interface.

The results promote a more holistic understanding of the stress-residence time behavior of biomolecules in premix membrane emulsification. On the one hand, the droplet breakage mechanisms inside membrane pores were further clarified, and on the other hand, the importance of controlling the protein adsorption to the membrane wall was pointed out. The latter highly impacts the wettability of the membrane, which in turn, affects the droplet breakage and thus, the emulsification performance. The impact of the structural state and confirmation of the proteins on the interfacial adsorption, and by that, on the stabilization of droplets in emulsions, is of importance for controlling and predicting the emulsification process. The better understanding of the premix membrane emulsification process will enlighten the industrial implementation.

References

1. Piacentini E, Drioli E, Giorno L (2014) Membrane emulsification technology: Twenty-five years of inventions and research through patent survey. J Memb Sci 468:410–422
2. Schröder V, Schubert H (1999) Production of emulsions using microporous, ceramin membranes. Colloids Surf Physicochem Eng Asp 152:103–109
3. Piacentini E, Mazzei R, Giorno L (2021) Comparison between lipase performance distributed at the O/W interface by membrane emulsification and by mechanical stirring. Membranes (Basel) 11(137)
4. Nazir A, Schroën K, Boom R (2010) Premix emulsification: a review. J Memb Sci 362(1–2):1–11
5. Piacentini E, Imbrogno A, Drioli E, Giorno L (2014) Membranes with tailored wettability properties for the generation of uniform emulsion droplets with high efficiency. J Memb Sci 459:96–103
6. de Aguiar IB, Schroën K (2020) Microfluidics used as a tool to understand and optimize membrane filtration processes. Membranes (Basel) 10(11):1–21
7. Trentin A, De Lamo S, Güell C, López F, Ferrando M (2011) Protein-stabilized emulsions containing beta-carotene produced by premix membrane emulsification. J Food Eng 106(4):267–274
8. Suelter CH, DeLuca M (1983) How to prevent losses of protein by adsorption to glass and plastic. Anal Biochem 135(1):112–119
9. Van den Oetelaar PJM, Mentink IM, Brinks GJ (1989) Loss of peptides and proteins upon sterile filtration due to adsorption to membrane filters. Drug Dev Ind Pharm 15(1):97–106
10. Derewenda ZS, Sharp AM (1993) News from the interface: the molecular structures of triacyglyceride lipases. Trends Biochem Sci 18(1):20–25
11. Giorno L, Drioli E (2000) Biocatalytic membrane reactors: applications and perspectives. Trends Biotechnol 18(8):339–349
12. Bowen WR, Gan Q (1992) Properties of Microfiltration Membranes: The Effects of Adsorption and Shear on the Recovery of an Enzyme. Biotechnol Bioeng 40:491–497
13. Beverung CJJ, Radke CJJ, Blanch HWWU (1999) Protein adsorption at the oil/water interface: characterization of adsorption kinetics by dynamic interfacial tension measurements. Biophys Chem 81(1):59–80
14. Baldursdottir SG, Fullerton MS, Nielsen SH, Jorgensen L (2010) Colloids and surfaces B: biointerfaces adsorption of proteins at the oil/water interface—Observation of protein adsorption by interfacial shear stress measurements. Colloids Surf B Biointerfaces 79(1):41–46
15. Berry JD, Neeson MJ, Dagastine RR, Chan DYC, Tabor RF (2015) Measurement of surface and interfacial tension using pendant drop tensiometry. J Colloid Interface Sci 454:226–237
16. Tamm F, Sauer G, Scampicchio M, Drusch S (2012) Pendant drop tensiometry for the evaluation of the foaming properties of milk-derived proteins. Food Hydrocoll 27(2):371–377
17. Roenby J, Bredmose H, Jasak H (2019) IsoAdvector: geometric VOF on. In: OpenFOAM. Springer, pp 281–296
18. Gründing D, Smuda M, Antritter T, Fricke M, Rettenmaier D, Kummer F, Stephan P, Marschall H, Bothe D (2020) A comparative study of transient capillary rise using direct numerical simulations. Appl Math Model 86:142–165
19. Raeini AQ, Blunt MJ, Bijeljic B (2012) Modelling two-phase flow in porous media at the pore scale using the volume-of-fluid method. J Comput Phys 231(17):5653–5668
20. Stephan FK (1993) Hydrodynamics of wetting. In: Be JC (ed) Wettability. Marcel Dekker, Inc., New York, USA, pp 311–430
21. Hoffman RL (1983) A study of the advancing interface: II. Theoretical prediction of the dynamic contact angle in liquid-gas systems. J Colloid Interface Sci 94(2):470–486
22. Hundschell CS, Bäther S, Drusch S, Wagemans AM (2020) Osmometric and viscometric study of levan, β-lactoglobulin and their mixtures. Food Hydrocoll 101:105580
23. Zhao H, Li W, Chu W, Bollard M, Adão R, Schuck P (2020) Quantitative analysis of protein self-association by sedimentation velocity. Curr Protoc Protein Sci 101(1)

24. Uttinger MJ, Hundschell CS, Lautenbach V, Pusara S, Bäther S, Heyn TR, Keppler JK, Wenzel W, Walter J, Kozlowska M, Wagemans AM, Peukert W (2022) Determination of specific and non-specific protein–protein interactions for beta-lactoglobulin by analytical ultracentrifugation and membrane osmometry experiments. Soft Matter
25. Kar SR, Kingsbury JS, Lewis MS, Laue TM, Schuck P (2000) Analysis of transport experiments using pseudo-absorbance data. Anal Biochem 285(1):135–142
26. Schuck P (2000) Size-distribution analysis of macromolecules by sedimentation velocity ultracentrifugation and Lamm equation modeling. Biophys J 78(3):1606–1619
27. Schuck P, Zhao H, Brautigam CA, Ghirlando R (2016) Basic principles of analytical ulatacentrifugation. CRC Press, Boca Raton
28. Hayes D, Laue T, Philo J (1995) Program Sednterp: sedimentation interpretation program. Alliance Protein Laboratories, Thousand Oaks
29. Pusara S, Yamin P, Wenzel W, Krstić M, Kozlowska M (2021) A coarse-grained xDLVO model for colloidal protein-protein interactions. Phys Chem Chem Phys 23(22):12780–12794
30. Abraham MJ, Murtola T, Schulz R, Páll S, Smith JC, Hess B, Lindah E (2015) Gromacs: high performance molecular simulations through multi-level parallelism from laptops to supercomputers. SoftwareX 1–2:19–25
31. Martínez L, Andrade R, Birgin EG, Martínez JM (2012) Software news and update PACKMOL: a package for building initial configurations for molecular dynamics simulations. J Comput Chem 30(13):2157–2164
32. Anandakrishnan R, Aguilar B, Onufriev AV (2012) H++ 3.0: Automating pK prediction and the preparation of biomolecular structures for atomistic molecular modeling and simulations. Nucleic Acids Res 40(W1):537–541
33. Cole DJ, Payne MC, Csányi G, Spearing SM, Ciacchi LC (2007) Development of a classical force field for the oxidized Si surface: application to hydrophilic wafer bonding. J Chem Phys 127(20):1–38
34. Butenuth A, Moras G, Schneider J, Koleini M, Köppen S, Meißner R, Wright LB, Walsh TR, Ciacchi LC (2012) Ab initio derived force-field parameters for molecular dynamics simulations of deprotonated amorphous-SiO2/water interfaces. Phys Status Solidi Basic Res 249(2):292–305
35. Heyse A, Gößmann MM, Giefer P, Gäbel P, Fritsching U, Wagemans AM, Drusch S (2023) Lipase adsorption during premix-membrane emulsification affects membrane surface properties and structural conformation of lipase. Colloids Surf B Biointerfaces, submitted
36. Giefer P, Bäther S, Kieserling H, Heyse A, Kaufmes N, Wagemans W, Barthel L, Mayer V, Schneck E, Drusch S, Fritsching U, Wagemans AM (2023) Characterization of β-lactoglobuling adsorption on silicate surfaces as used in the premix-membrane-emulsification process. J Colloid Interface Sci 652:1074–1084
37. Keppler JK et al (2020) Towards recombinantly produced milk proteins: physicochemical and emulsifying properties of engineered whey protein beta-lactoglobulin variants. Food Hydrocoll 110(March):2021
38. Brune SE, Hoppenreijs LJG, Kühl T, Lautenbach V, Walter J, Peukert W, Schwarz K, Imhof D, Boom RM, Krull R, Keppler JK, Biedendieck R (2022) Cysteine-modified whey protein β-lactoglobulin: recombinant substitution versus (partial) chemical reduction of disulfide bonds, submitted
39. Schestkowa H, Wollborn T, Westphal A, Wagemans AM, Fritsching U, Drusch S, Wagemans AM, Fritsching U, Drusch S (2019) Conformational state and charge determine the interfacial stabilization process of beta-lactoglobulin at preoccupied interfaces. J Colloid Interface Sci 536:300–309
40. Kieserling H, Giefer P, Uttinger MJ, Lautenbach V, Nguyen T, Sevenich R, Lübbert C, Rauh C, Peukert W, Fritsching U, Drusch S, Wagemans AM (2021) Structure and adsorption behavior of high hydrostatic pressure-treated β-lactoglobulin. J Colloid Interface Sci 596:173–183
41. Kieserling H, Pankow A, Keppler JK, Wagemans AM, Drusch S (2021) Conformational state and charge determine the interfacial film formation and film stability of β-lactoglobulin. Food Hydrocoll 114:106561

42. Kieserling H, Alsmeier IM, Steffen-Heins A, Keppler JK, Sevenich R, Rauh C, Wagemans AM, Drusch S (2021) Interfacial film formation and film stability of high hydrostatic pressure-treated β-lactoglobulin. Food Hydrocoll, 106746
43. Hanwell MD, Curtis DE, Lonie DC, Vandermeerschd T, Zurek E, Hutchison GR (2012) Avogadro: an advanced semantic chemical editor, visualization, and analysis platform. J Cheminform 4(8)
44. Wang J, Wang W, Kollman PA, Case DA (2006) Automatic atom type and bond type perception in molecular mechanical calculations. J Mol Graph Model 25(2):247–260
45. Wollborn T, Luhede L, Fritsching U (2019) Evaluating interfacial shear and strain stress during droplet deformation in micro-pores. Phys Fluids 31(1)
46. Giefer P, Fritsching U (2023) Impact of wettability on interface deformation and droplet breakup in mirco-capillaries. Phys Fluids 35:042110
47. Wollborn T, Giefer P, Kieserling H, Wagemans AM, Drusch S, Fritsching U (2022) Investigation of local and temporal interfacial shear stress distribution during membrane emulsification. Can J Chem Eng 100(5):1061–1078
48. Uttinger MJ, Wawra SE, Guckeisen T, Walter J, Bear A, Thajudeen T, Sherwood PJ, Smith A, Wagemans AM, Stafford WF, Peukert W (2019) A Comprehensive brownian dynamics approach for the determination of non-ideality parameters from analytical ultracentrifugation. Langmuir 35(35):11491–11502
49. Schaink HM, Smit JAM (2000) Determination of the osmotic second virial coefficient and the dimerization of beta-lactoglobulin in aqueous solutions with added salt at the isoelectric point. Phys Chem Chem Phys 2(7):1537–1541
50. Sackett DL, Wolff J (1987) Nile red as a polarity-sensitive fluorescent probe of hydrophobic protein surfaces. Anal Biochem 167(2):228–234
51. Stein WH (1949) Amino acid composition of β-Lactoglobulin and bovine serum albumin. J Biol Chem 178(1):79–91
52. Kape A, Ruick B, Drusch S (2016) Characterisation of the work of adhesion of food grade coating materials on a maltodextrin model surface. Chem Eng Res Des 110:152–159
53. Wollborn T, Michaelis M, Ciacchi LC, Fritsching U (2022) Protein conformational changes at the oil/water-interface induced by premix membrane emulsification. J Colloid Interface Sci 828:72–81
54. Clark WM, Bansal A, Sontakke M, Ma YH (1991) Protein adsorption and fouling in ceramic ultrafiltration membranes. J Memb Sci 55(1–2):21–38
55. Huisman IH, Prádanos P, Hernández A (2000) The effect of protein–protein and protein–membrane interactions on membrane fouling in ultrafiltration. J Memb Sci 179:79–90
56. Lee J, Jeong S, Ye Y, Chen V, Vigneswaran S, Leiknes TO, Liu Z (2017) Protein fouling in carbon nanotubes enhanced ultrafiltration membrane: fouling mechanism as a function of pH and ionic strength. Sep Purif Technol 176:323–334
57. Rivero CW, Palomo JM (2016) Covalent immobilization of Candida rugosa lipase at Alkaline pH and their application in the regioselective deprotection of Per-O-acetylated thymidine. Catalysts 6(8)
58. Militano F, Poerio T, Mazzei R, Piacentini E, Gugliuzza A, Giorno L (2016) Biointerfaces Influence of protein bulk properties on membrane surface coverage during immobilization. Colloids Surf B Biointerfaces 143:309–317
59. Reis P, Miller R, Krägel J, Leser M, Fainerman V, Watzke H, Holmberg K (2008) Lipases at interfaces: unique interfacial properties as globular proteins. Langmuir 24(13):6812–6819
60. Martinez-Monteagudo SI, Saldaña MDA (2014) Chemical reactions in food systems at high hydrostatic pressure. Food Eng Rev 6(4):105–127
61. Ravera F, Ferrari M, Santini E, Liggieri L (2005) Influence of surface processes on the dilational visco-elasticity of surfactant solutions. Adv Colloid Interface Sci 117(1–3):75–100
62. Young ME, Carroad PA, Bell RL (1980) Estimation of diffusion coefficients of proteins. Biotechnol Bioeng XXII:947–955
63. Tyn MT, Gusek TW (1990) Prediction of diffusion coefficients of proteins. Biotechnol Bioeng 35(4):327–338

64. Dickinson E, Rolfe SE, Dalgleish DG (1989) Competitive adsorption in oil-in-water emulsions containing α-lactalbumin and β-lactoglobulin. Top Catal 3(3):193–203
65. Zhai J, Hoffmann SV, Day L, Lee TH, Augustin MA, Aguilar MI, Wooster TJ (2012) Conformational changes of α-lactalbumin adsorbed at oil-water interfaces: interplay between protein structure and emulsion stability. Langmuir 28(5):2357–2367
66. Murray BS (1998) Interfacial rheology of mixed food protein and surfactant adsorption layers with respect to emulsion and foam stability. Stud Interface Sci 7(C):179–220
67. Rühs PA, Scheuble N, Windhab EJ, Mezzenga R, Fischer P (2012) Simultaneous control of pH and ionic strength during interfacial rheology of β-lactoglobulin fibrils adsorbed at liquid/liquid interfaces. Langmuir 28(34):12536–12543
68. Ziegler GR, Foegeding EA (1990) The gelation of proteins. Adv Food Nutr Res 34(C):203–298
69. Roth CM, Lenhoff AM (1993) Electrostatic and van der Waals contributions to protein adsorption: computation of equilibrium constants. Langmuir 9(4):962–972
70. Sengupta T, Damodaran S (1998) Role of dispersion interactions in the adsorption of proteins at oil-water and air-water interfaces. Langmuir 14(22):6457–6469
71. Damodaran S, Anand K (1997) Sulfhydryl-disulfide interchange-induced interparticle protein polymerization in whey protein-stabilized emulsions and its relation to emulsion stability. J Agric Food Chem 45(10):3813–3820
72. Shimada K, Cheftel JC (1989) Sulfhydryl group/disulfide bond interchange reactions during heat-induced gelation of whey protein isolate. J Agric Food Chem 37(1):161–168
73. Fang Y, Dalgleish DG (1997) Conformation of β-lactoglobulin studied by FTIR: effect of pH, temperature, and adsorption to the oil-water interface. J Colloid Interface Sci 196(2):292–298
74. Engel MFM, Van Mierlo CPM, Visser AJWG (2002) Kinetic and structural characterization of adsorption-induced unfolding of bovine α-lactalbumin. J Biol Chem 277(13):10922–10930
75. Dickinson E, Matsumura Y (1994) Proteins at liquid interfaces: role of the molten globule state. Colloids Surf B Biointerfaces 3(1–2):1–17
76. Lopez-Fandino R (2007) Functional improvement of milk whey proteins induced by high hydrostatic pressure. Crit Rev Food Sci Nutr 351–363
77. Aouzelleg A (2014) High pressure control of protein structure and functionality. Nutr Food Sci 44(1):41–46
78. Dickinson E, Matsumura Y (1991) Time-dependent polymerization of beta-lactoglobulin through disulphide bonds at the oil-water interface in emulsions. Int J Biol Macromol 13(1):26–30
79. Dumay EM, Kalichevsky MT, Cheftel JC (1994) High-pressure unfolding and aggregation of β-Lactoglobulin and the Baroprotective effects of sucrose. J Agric Food Chem 42(9):1861–1868
80. Galazka VB, Dickinson E, Ledward DA (1996) Effect of high pressure on the emulsifying behaviour of. Food Hydrocoll 10(2):213–219
81. Considine T, Patel HA, Anema SG, Singh H, Creamer LK (2007) Interactions of milk proteins during heat and high hydrostatic pressure treatments—A review. Innov Food Sci Emerg Technol 8(1):1–23
82. Loch JI, Bonarek P, Tworzydło M, Polit A, Hawro B, Łach A, Ludwin E, Lewiński K (2016) Engineered β-Lactoglobulin produced in E. coli: purification, biophysical and structural characterisation. Mol Biotechnol 58(10):605–618

Influence of Thermomechanical Stresses on Structural and Functional Changes of Highly Concentrated Protein Systems in Extrusion Processing

Maria Gräfenhahn, Felix Ellwanger, Heike P. Karbstein, and M. Azad Emin

1 General Introduction

The functional properties of proteins depend on both, their amino acid composition and sequence, and the conformation of the protein chains [1]. In the native state, proteins fold to lower the free energy and their conformation depends among others on intra- and intermolecular interactions such as hydrophobic and van der Waals interactions, and disulfide bonds [2]. Therefore, proteins are very sensitive to their environment. Changes in the native protein structure may result in loss of biological function or catalytic activity [3, 4], and, in many cases, leads to the formation of protein aggregates with different properties [5–7]. For certain applications, such

*Joint first authorship—Maria Gräfenhahn, Felix Ellwanger; **Corresponding authors—Heike P. Karbstein, M. Azad Emin.

M. Gräfenhahn
Institute of Life Sciences, University of Applied Sciences and Arts Western Switzerland Valais-Wallis (HES-SO VS), Sion, Switzerland
e-mail: maria.graefenhahn@hevs.ch

F. Ellwanger · H. P. Karbstein (✉) · M. A. Emin (✉)
Institute of Process Engineering in Life Sciences, Chair of Food Process Engineering, Karlsruhe Institute of Technology, Karlsruhe, Germany
e-mail: heike.karbstein@kit.edu

M. A. Emin
e-mail: azad.emin@kit.edu

F. Ellwanger
e-mail: felix.ellwanger@kit.edu

M. A. Emin
nexnoa GmbH, Durmersheimertraße 188A, 76189 Karlsruhe, Germany

A. Kwade and I. Kampen (eds.), *Dispersity, Structure and Phase Changes of Proteins and Bio Agglomerates in Biotechnological Processes*,
https://doi.org/10.1007/978-3-031-63164-1_10

as nutritional and biopharmaceutical applications, it is important that the proteins preserve their native state in order to exert the desired functionality [8]. On the other hand, when properly controlled protein denaturation and aggregation can result in innovative materials with novel functionalities [8]. Consequently, in many technical applications, protein denaturation and aggregation and their kinetics have to be properly understood in order to achieve the desired product properties and performance. Examples of products resulting from such targeted modification of protein structure are e.g., protein-based surfactants, stabilizers, coatings, biodegradable films, or meat analogues [8].

Past research has gained a vast amount of useful information on heat-induced structural changes in proteins in dilute model solutions under a wide range of experimental conditions including pH, concentration, and ionic strength. The type of protein and its aggregation pathway play a decisive role in the final structure and might result in an increase in viscosity of protein solutions, decrease of solubility due to exposure of hydrophobic groups, increase of reactivity of side groups, altered sensitivity to enzymatic proteolysis and altered interfacial properties [9]. Moreover, environmental conditions and types of applied stresses, including chemical, thermal, and mechanical stresses, influence the aggregation pathway of a protein and its kinetics [10, 11]. Mechanical treatment is relevant since almost all protein-related processes, from production to formulation, involve a certain kind of flow and consequently mechanical treatment at least to a certain degree [12]. Shear stresses can destabilize the native structure of proteins and therefore can modify them at their structural level [13, 14]. For example, under favorable solution conditions, various forms of agitation including shaking and stirring have been successfully applied to accelerate the aggregation of various proteins including insulin [15, 16], β-lactoglobulin [17], and albumin [18]. Moreover, the resulting protein aggregates can be soluble, insoluble, and might have an amorphous or fibrillar structure. Protein aggregates might be formed through weak forces such as hydrogen bonding, and electrostatic and hydrophobic interactions [19] or may be a result of disulfide bond linkages. However, these findings are based on dilute protein systems. In contrast, in processes, where thermomechanical treatment is applied, such as in extruders or scraped surface heat exchangers, shear stress is explicitly applied to proteins at higher concentrations to achieve a certain colloidal structure like microparticles, fibrils, foams or films [10, 20–23].

Extrusion processing using a cooling die leads to products with a fibrous anisotropic structure, as required for e.g., mozzarella-type cheese or meat analogues [13, 20]. Besides protein source, the conformational changes, chemical reactions, and the resulting functionalities of extruded proteins can vary considerably depending on the processing conditions such as barrel temperature, screw speed, screw configuration, extruder type, die geometry, feed rate, feed composition, and moisture content [24, 25]. The increase in pressure and temperature as a result of transfer of heat from the heated barrel and the conversion of mechanical energy into heat energy accompanied with the shearing and mixing of the extruder screw might lead to protein denaturation. The forces which stabilize the tertiary and quaternary structures of the proteins are weakened by a combination of elevated temperature and shear [26], bonds and interactions between amino acid side chains are disrupted, reactive sites become available, and new protein–protein interactions are formed, resulting in protein aggregation. Despite its technological and biological importance, little is known about the structural change of proteins under simultaneous thermal and mechanical stresses. The effects of extrusion on the molecular changes of proteins are difficult to isolate as high protein concentrations are exposed to several stresses, i.e., thermal and mechanical stresses, simultaneously in an extruder. Only few investigators [27–29] have addressed the mechanisms and kinetics of protein reactivity during extrusion, especially at relevant protein concentrations. Studies on structuring of soy, wheat, pea, and whey proteins have attributed them to a combination of fragmentation and aggregation, non-covalent associations, and covalent cross-linking [30–34]. However, the influence of combined thermal and mechanical treatment during extrusion processing on the protein reaction behavior is still unclear. Moreover, it is unknown how the reaction behavior of proteins at high-concentration differs from the mostly known behavior in low-concentration systems and no unified denaturation or aggregation model including relevant protein–protein interactions has been proposed to date. Therefore, it is still difficult to predict structures, textures, and functionality resulting from an extrusion process [25, 26, 35]. Product and process development thus is still driven by trial-and-error, and therefore based on experiences. Consequently, in the scope of this project, the influence of a well-defined thermomechanical treatment on the structural changes and the resulting functionality of a model protein system was investigated. Extrusion processing was chosen as a model process since extrusion has a wide application area from food to pharmaceuticals and has a growing potential with the development of novel products/structures [8, 35–38].

2 Influence of Defined Extrusion-Like Conditions on the Denaturation Behavior and Functionality of Highly Concentrated Whey Proteins

Whey proteins are widely used in the food industry because of their nutritional, technological, and functional properties. Depending on their native structure, proteins show diverse functionalities, such as the ability to form gels, increase the viscosity of solutions or dispersions, and stabilize emulsions [39]. The molecular structure of proteins and their functionality, can be modified by thermal, shear, and/or pressure treatment [40–43].

The reaction behavior of dilute whey protein solutions (protein concentration < 10%) especially under thermal stress is well summarized in e.g., Brodkorb et al. and Nicolai et al. [19, 44]. At such conditions, whey protein denaturation can be described in a simplified way by two successive steps, the unfolding of the protein globular structure and the subsequent aggregation of reactive molecules. The main whey protein fractions are β-lactoglobulin (βLG) and α-lactalbumin (αLA). βLG is predominantly present as a dimer (D) in solutions at 25 °C and neutral pH [45, 46]. Increasing the temperature leads to various conformational changes including a shift in the equilibrium towards the monomeric form [47] and exposure of free thiol groups and hydrophobic patches at around 40 °C [48]. At temperatures above 65 °C, the exposed thiol groups of the βLG monomers react with those accessible through the unfolding disulfide bonds, which results in a thiol–disulfide exchange reaction [49] and eventually in protein aggregation. In contrast to βLG, αLA in solution is very heat-stable with respect to aggregation, due to its four stabilizing disulfide bonds and no free thiol group [50]. Although thermal denaturation of αLA begins at lower temperatures compared to βLG (approx. 35 °C), it is completely reversible up to a high temperature range [51]. Only treatment at temperatures $\geq$ 100 °C, is known to result in disulfide bond loss leading to irreversible aggregation of αLa via the thiol–disulfide exchange reaction [52, 53]. In contrast, in presence of other fractions, αLA participates in the aggregation with βLG monomers via the described thiol-disulfide exchange at lower temperatures (75 °C) as shown in Schokker et al. [54]. Consequently, whey proteins used in commercial applications like concentrates (WPC) and isolates (WPI) are expected to show different reaction behavior compared to single isolated fractions as they are multi-component systems including various protein fractions (i.e., βLG and αLA) as well as non-protein components such as lactose.

Extrusion processing has been applied to design protein-based products such as meat and dairy analogues, as well as protein-based emulsifiers and thickeners with specific functionalities [21, 26, 28–30, 55–57]. During extrusion processing, highly concentrated proteins are subjected to thermal and mechanical stresses simultaneously through the heated barrel and the rotating screws. Onwulata et al. [58] extruded WPC, WPI, and whey albumin in a twin-screw extruder and showed that the gel strength of extruded isolates increased initially up to barrel temperatures of 50 °C, but their gelling and foaming properties were lost at higher temperatures. These changes in functionality were related to extrusion-induced insolubility (denaturation), which increased with barrel temperature, and aggregation via disulfide bonds [58]. The properties of aggregates (e.g., size, form, and intermolecular interactions) play a crucial role in the performance of the resulting products. Results from previous studies indicate that changes in processing conditions influence the molecular structure of whey proteins at concentrations below 40% and consequently their functionality. Therefore, to be able to produce protein products with the desired functionality, it is necessary to understand and control the influence of extrusion process parameters on the denaturation and functionality of proteins. However, the determination of the influence of extrusion processing conditions on the denaturation behavior of whey proteins in a mechanistic manner is not an easy task due to the multivariate and complex nature of the process, i.e., thermal and mechanical stresses, which are coupled to each other, change along the extruder, and are thus not defined. Therefore, their influence on the local processing conditions is still unknown and it remains difficult to determine the influencing factors on the denaturation reaction during extrusion processing.

The objective of this study was to systematically investigate the effect of extrusion-like treatment conditions, especially the effect of combined thermal and mechanical treatment, on the denaturation kinetics of a highly concentrated multi-component protein system such as whey protein isolate. Extrusion-like processing conditions were simulated using a special rheometer, the closed cavity rheometer (CCR), since it allows treatments at high temperatures, high protein concentrations, and short treatment times. Additionally, it allows the variation of mechanical stresses independently of temperature. The influence of temperature, protein concentration, and shear rate on the denaturation behavior of whey proteins was described as a function of time and was evaluated using rheological, numerical, and protein chemistry analysis. Moreover, to gain better insights about the denaturation mechanisms and kinetics at high protein concentrations in some way representative of extrusion processing of multi-component proteins for the creation of functional ingredients, the role of the main whey protein fractions (i.e., βLG and αLA) and their mixtures were also investigated separately. For this, the influence of the shear rate on the reaction onset temperature for βLG, αLA, their mixtures, and whey protein isolate at various protein concentrations was investigated. Furthermore, the degree of denaturation for the chosen whey protein systems was analyzed and described by a mathematical model to determine the kinetic parameters as a function of temperature and shear rate. Lastly. the changes in functionality of whey proteins as a consequence of different treatment conditions are briefly discussed.

2.1 Material and Methods

2.1.1 Material

Pure bovine βLG was produced at the Chair of Food and Bioprocess Engineering of the Technical University Munich at Freising-Weihenstephan, Germany. Two different charges of βLG were used during this study. The extraction methods are described in Toro-Sierra, Tolkach, Kulozik and Haller, Kulozik [59, 60]. The βLG was highly pure (< 0.05% lactose, minerals < 0.13%), the moisture content was < 4%, the degree of nativity was > 99%, with 96.8% protein on dry matter, the content of βLG in the dry matter was > 99%. Alpha-lactalbumin (αLA), BiPro Alpha 9000, was kindly provided by Agropur Inc. (Agropur, Wisconsin, USA). According to the supplier, the αLA used in this study was highly native and pure (< 0.2% lactose), the moisture content was < 5.6%, with 97.6% protein on dry matter, and the content of αLA in the dry matter was 92.2%. Whey protein isolate (WPI), German Prot 9000, with a content of 90% protein on dry matter was kindly provided by Sachsenmilch Leppersdorf GmbH (Sachsenmilch Leppersdorf, Leppersdorf, Germany). According to the supplier, the contents of minor ingredients are ≤ 3% lactose, ≤ 1% fat. Total ash content is 4% and the content of moisture is ≤ 5%.

2.1.2 Sample Preparation

Whey proteins were mixed with deionized water (Millipore Sigma, Burlington, USA) to achieve protein concentrations between 17 and 70% (w/w) in a Thermomix (Vorwerk, Wuppertal, Germany) for 3 min. To ensure a homogenous distribution of water, the samples were stored at 8 °C at least for two days.

The influence of mixing ratio of βLG and αLA on the denaturation of highly concentrated whey proteins, was investigated. For this, mixing ratios of 5:2 and 1:1 were chosen. The mixing ratio of 5:2 was chosen since this is the natural occurrence of each fraction in whey. A mixing ratio of 1:1 was applied to investigate the effect of a lower relative amount of βLG on the reactions between βLG and αLA.

2.1.3 Thermomechanical Treatment and Inline Rheological Analyses

Thermomechanical treatment and rheological measurements were performed by a closed-cavity rheometer CCR (RPA flex, TA Instruments, New Castle, DE, USA) as previously described in detail in Quevedo et al. [61, 62].

2.1.4 Degree of Denaturation

The degree of denaturation D_D was calculated as the protein concentration ratio before and after thermomechanical treatment and the applied method is described in detail in e.g., Quevedo et al. [62]. The obtained values for treatment times up to 90 s were used for the determination of the denaturation kinetic parameters.

2.1.5 Analysis of Protein–Protein Interactions Leading to Aggregation

Protein–protein interactions responsible for the aggregation of βLG, αLA, WPI, and mixtures of βLG and αLA molecules after thermal and mechanical treatment were determined by the solubility of samples in various buffer solutions cleaving specific bonds as previously described in detail in Quevedo et al. [63].

2.1.6 Kinetic Modeling

The effect of the protein concentration, presence and mixing ratio of protein fractions, and heat and shear treatment on the denaturation kinetics was described using formal reaction kinetics. The kinetic parameters were determined using a non-linear regression (NLR) method as described by Quevedo et al. [64].

2.1.7 Functional Properties

The functionality of βLG such as solubility in water, rheological properties of water dispersions, and interfacial properties was analyzed as a function of treatment conditions such as temperature, shear rate and protein concentration and the respective methods are described in detail in Quevedo et al. [63].

2.2 Results and Discussion

2.2.1 Influence of Defined Thermal and Mechanical Treatment on the Reaction Behavior and Onset Temperatures of Whey Proteins

The complex modulus |G*| of a system containing 70% (w/w) WPI is depicted in Fig. 1 as a function of temperature for three different shear rates. The complex modulus in Fig. 1a. shows two distinct regions: initial decrease followed by an increase. The decrease in the complex modulus is related to the increased mobility of the protein molecules due to increasing temperature.

As the rheological properties are a direct function of the structure of (bio-) -polymers [66], the results indicate that at temperatures, where the slope of the

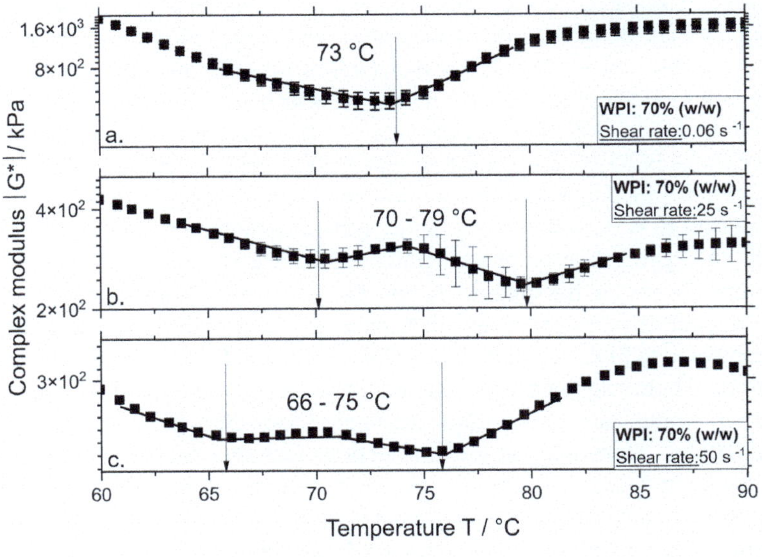

Fig. 1 Complex modulus |G*| as a function of temperature for systems with a whey protein isolate (WPI) concentration of 70% (w/w) and a shear rate of 0.06 (**a**), 25 (**b**), and 50 s^{-1} (**c**). Measurements were performed at a constant heating rate of 5 K·min^{-1}. Figure taken from [65], licensed under the Creative Commons Attribution 4.0 International license (CC BY 4.0)

complex modulus changes, structural changes are taking place. Thus, these temperatures can be defined as reaction onset temperatures. A distinct change in the slope happened at approximately 73 °C, thereafter |G*| increases. As an increase in molecular size or mass of polymers increases the viscosity [66], the increase in |G*| can be associated with structure-forming reactions (i.e., aggregation) [67–69]. Therefore, the increase in the complex modulus at this temperature is assumed to show the onset temperature of aggregation [61]. Although the aggregation might have started at lower temperatures, this is the temperature where the aggregation outweighs the effect of temperature on the molecular mobility. By increasing the shear rate to 25 and 50 s^{-1}, the course of the complex modulus changed compared to Fig. 1a. The complex modulus in Fig. 1b,c shows four distinct regions: initial decrease (Region I), followed by an increase (Region II), a further decrease (Region III), and final increase (Region IV), which led to the determination of two onset temperatures. At a shear rate of 25 s^{-1}, the aggregation onsets at 70 and 79 °C. By increasing the shear rate to 50 s^{-1}, these temperatures decreased to 66 and 75 °C, respectively. The occurrence of two onset temperatures when the samples were sheared might suggest shear induced phase separation of proteins. At a shear rate of 0.06 s^{-1}, the protein molecules of different fractions (i.e., βLG and αLA) were homogenously distributed, and the measured onset temperature implied that they are involved in the aggregation reaction simultaneously. At higher shear rates, it is possible that these protein fractions form separated phases due to their incompatibility, which might lead to variations of the molecular mobility, interactions, and aggregation at the local level.

The results for βLG and αLA samples depicted in Figs. 2a and 3a show that there were two temperatures at which the course of the complex modulus changed. These temperatures are defined as the denaturation and aggregation onset temperature, respectively. As shown in Fig. 2a, the denaturation onset temperature for βLG with a concentration of 70% was approximately 72 °C, which is in accordance with previous results in Quevedo et al., [61]. In the case of αLA (Fig. 3a), a lower denaturation of approximately 65 °C was measured. For both proteins, increasing the shear rate (depicted in b, c) led to a decrease in the aggregation onset temperature.

In contrast to WPI, which is a complex system containing more than one protein fraction, the aggregation onset temperature at a shear rate of 0.06 s^{-1} of systems containing the single fractions αLA and βLG increases from 73 to 82 °C and 80 °C, respectively. As expected, αLA shows higher aggregation onset temperatures compared to βLG due to its four stabilizing disulfide bonds and lacking a free thiol group. Higher aggregation temperatures were observed for αLA compared to βLG from various authors at various treatment conditions [52, 53]. The results of WPI at 0.06 s^{-1} show that the protein fractions including βLG have lower aggregation temperatures in this complex system. Therefore, an increase in reaction rate was expected as in WPI systems all protein fractions present could participate in the occurring reactions and form aggregates, leading to higher reaction rates, which can explain the lower onset temperature. Although no denaturation onset temperature was observed for WPI at a shear rate of 0.06 s^{-1}, it was expected that denaturation onsets between 60 and 70 °C.

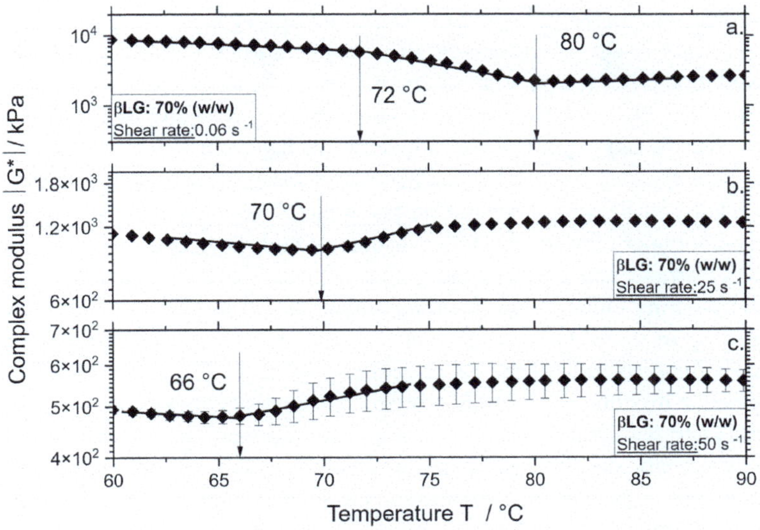

Fig. 2 Complex modulus |G*| as a function of the temperature for systems with a βLG concentration of 70% (w/w) and a shear rate of 0.06 (**a**), 25 (**b**), and 50 s^{-1} (**c**). Measurements were performed at a constant heating rate of 5 K·min^{-1}. Figure taken from [65], licensed under the Creative Commons Attribution 4.0 International license (CC BY 4.0)

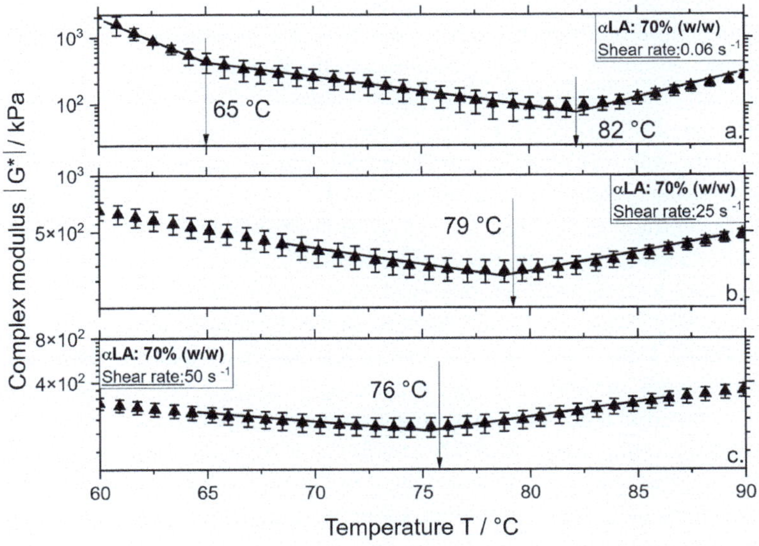

Fig. 3 Complex modulus |G*| as a function of temperature for systems with a αLA concentration of 70% (w/w) and a shear rate of 0.06 (**a**), 25 (**b**), and 50 s^{-1} (**c**). Measurements were performed at a constant heating rate of 5 K·min^{-1}. Figure taken from [65], licensed under the Creative Commons Attribution 4.0 International license (CC BY 4.0)

At a shear rate of 25 s^{-1}, the βLG shows an aggregation onset temperature of 70 °C and αLA of 79 °C. Increasing the shear rate to 50 s^{-1} led to a decrease in the aggregation onset temperature to 66 and 76 °C, for βLG and αLA, respectively. This effect was also observed in previous studies on WPI and βLG [62, 65, 70]. It seems that the aggregation onset temperatures determined for the sheared WPI systems corresponded to the onset temperatures of the single protein fractions. This can be related to the phase separation phenomena, which has been observed for various biopolymer mixtures [71]. For instance, Ould Eleya et al., [72] reported for βLG and κ-carrageenan (κ-Car) systems two slopes of the elastic modulus G' during a heating/shearing and cooling down treatment step. The first slope corresponded to the setting of the βLG network, and the second to the further protein network affected by the formation of the κ-Car gel in the system [73]. Although the authors investigated a mixture of globular proteins and polysaccharide (i.e., βLG and κ-car), similar to the results observed in Fig. 1 during heating and shearing of WPI systems, two markedly different rheological behaviors (i.e., a bi-continuous profile) were observed suggesting the formation of phase-separated gels. Similarly, the first onset temperature suggests the setting of βLG aggregates and the second one, the aggregation of αLA affecting the previously formed βLG network. Although phase separation of protein solutions has not been often reported, it is indeed expected to occur with time [74]. Since proteins have, like all polymers or all solid materials, a limited solubility, at a protein concentration above their solubility, separation into liquid different phases occur [75].

The concentration of each fraction present in the mixture was also expected to influence denaturation and aggregation reactions. In contrast to the results of WPI at a concentration of 70%, phase separation occurs in 5:2 systems at a lower protein concentration of 60% (Fig. 4). In Fig. 4a, the system with concentration of 60% (w/w) without the influence of shear rate, shows only one aggregation onset temperature. Compared to the system with same βLG and αLA mixing ratio, i.e., WPI, at higher protein concentration (Fig. 1a), the aggregation onset temperature is very similar since for both system it is approx. 73 °C. Similarly, at a shear rate of 25 s^{-1}, two different onset temperatures are observed at approx. 73 and 89 °C. Increasing the shear rate to 50 s^{-1}, leads to a slight decrease in the aggregation onset temperatures to 72.8 and 86 °C. For 1:1 system, two onset temperatures were also determined when the systems were sheared [62], and a decrease in protein concentration from 70 and 60% resulted in a decrease in aggregation onset temperatures (data not shown). Although, the mixing ratio of WPI is 5:2 for βLG and αLA, the discussed results point out differences in reaction behavior, which might be due to the presence of minor protein fractions in WPI such as bovine serum albumin (BSA) or low concentrations of other components such as lactose and fat.

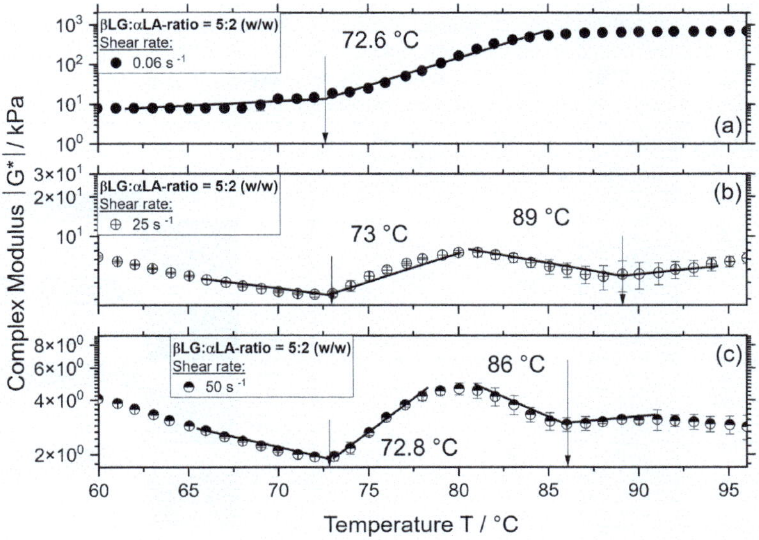

Fig. 4 Complex modulus |G*| as a function of temperature for 5:2 systems with a concentration of 60% (w/w) and a shear rate of 0.06 s^{-1} (**a**), 25 s^{-1} (**b**), and 50 s^{-1} (**c**). Measurements were performed at a constant heating rate of 5 K·min^{-1}, bars represent standard deviation (SD). Figure originally published in [70]. Permission for reuse was granted by Elsevier

2.2.2 Kinetics of Denaturation of Whey Proteins at Extrusion-Like Conditions

The kinetic parameters for the denaturation reaction for αLA and βLG after thermal and thermomechanical treatment at a concentration of 70% (w/w) are shown in Table 1. For the investigated range of treatment conditions, a denaturation reaction order of 2.262 ± 0.007 and 2.151 ± 0.025 was identified. The reaction rate constant (k) increased with temperature and a combination of thermal and mechanical treatment led to even higher values of k, and consequently the calculated values of activation energy decreased.

In contrast, the reaction order for WPI at the same concentration was 1.865 ± 0.055 as depicted in Table 2. Additionally, the mixed systems containing βLG and αLA at a mixing ratio of 5:2 and 1:1 showed even lower reaction orders of 1.815 ± 0.321 and 1.399 ± 0.061, respectively. There is clearly a dependence of protein fractions present in the protein system on the reaction order. The reaction order for single protein fractions follows a fractional second order, whereas for complex systems such as WPI or mixed systems, where multiple protein fractions are present, a fractional first order was observed. In theory, a second order is observed when the reactions depended on the concentration of two reactants or one concentration squared. For βLG it was hypothesized that a reactive monomer is needed to start the reaction, consequently starting from dimers, then native monomers are formed, which are activated and react further with other monomers to form non-native dimers, trimers,

Table 1 Values of activation energy without shear stress, reduction of activation energy due to shear stress and their ratio from the denaturation kinetic parameters for αLA and βLG samples with a protein concentration of 70% (w/w) at different thermomechanical treatment conditions. Data from Table originally published in [64, 70]. Permission for reuse was granted by Elsevier

Temperature range / °C	Shear rate / s^{-1}	Reaction order n / -	Protein concentration / %	E_t / kJ mol^{-1}	E_0 / kJ mol^{-1}	ΔE / kJ mol^{-1}	$(\Delta E/E_0)$ / %
80 - 100	0.06	2.262 ± 0.007	αLA70	117.239	117.239		
	25			112.381		4.859	4%
	50			109.441		7.799	7%
60 - 80	0.06			233.653	233.653		
	25			190.844		42.808	18%
	50			189.808		43.845	19%
80 - 100	0.05	2.151 ± 0.025	βLA70	133.646	133.646		
	25			117.200		16.446	12%
	50			107.745		25.901	19%
60 - 80	0.05			283.135	283.135		
	25			266.592		16.543	6%
	50			250.329		32.806	12%

and eventually oligomers [76]. Consequently, the denaturation reaction depends on the concentrations of dimers and reactive monomers available. Denaturation of αLA is expected to follow a similar mechanism, starting from monomers, which need a monomer with a free thiol group to react. When the thermal treatment is high enough the disulfide bonds stabilizing the αLA molecule are broken down. As shown in Schnack and Klostermeyer [53], thermal treatment at 100 °C for 10–30 min results in up to a 25% loss of disulfide bonds, which could lead to the formation of reactive groups similar to thiols [53]. Similar to βLG, the reactive αLA monomers react then with monomers and form disulfide bonded oligomers through thiol-disulfide exchange reactions.

For the WPI systems, either the free thiol group of the βLG can react with βLG monomers or αLA monomers, therefore the reactions depend mostly on the concentration of the thiol groups available resulting in a fractional first order. The reaction order of the multi-component systems is different than most of the findings of previous studies for lower protein concentrations at neutral pH. Most of the kinetic data for the denaturation reaction of whey proteins takes only into consideration the thermal denaturation of βLG in the system and a reaction order of 1.5 is frequently observed. The difference in the reaction order at the investigated conditions could arise either from the increased protein concentration or from the additional effect of shear stresses on the reactions. Furthermore, as shown in Figs. 1 and 4, for systems containing multiple protein fractions, a combination of thermal and mechanical treatment led to a more complex reaction behavior as two aggregation onset temperatures were observed. The resulting reaction orders of below 2 could then result from a combination of a lower reaction order for non-sheared systems and higher reaction order for sheared systems as phase separation occurs, leading to similar reaction

Table 2 Values of activation energy without shear stress, reduction of activation energy due to shear stress and their ratio from the denaturation kinetic parameters for WPI and samples with a defined mixing ratio of βLG to αLA with a protein concentration of 70% (w/w) at different thermomechanical treatment conditions. Table taken from [65], licensed under the Creative Commons Attribution 4.0 International license (CC BY 4.0)

Temperature range / °C	Shear rate / s^{-1}	Reaction order n / -	βLG:αLA mixing ratio	E_t / kJ mol^{-1}	E_0 / kJ mol^{-1}	ΔE / kJ mol^{-1}	$(\Delta E/E_0)$ / %
80 - 100	0.06	1.865 ± 0.055	WPI (5:2)	100.361	100.361		
	25			85.434		14.927	15%
	50			80.411		19.950	20%
60 - 80	0.06			267.176	267.176		
	25			244.372		22.804	9%
	50			234.644		32.532	12%
70 - 90	0.06	1.815 ± 0.321	5:2	81.457	81.457		
	25			67.873		13.584	17%
	50			66.625		14.832	18%
50 - 90	0.06			303.400	303.400		
	25			283.404		19.996	7%
	50			260.904		42.496	14%
70 - 90	0.06	1.399 ± 0.061	1:1	112.430	112.430		
	25			78.733		33.697	30%
	50			78.139		34.291	30%
50 - 90	0.06			200.077	200.077		
	25			183.465		16.611	8%
	50			175.183		24.894	12%

orders as for the single protein fractions. Moreover, it was shown that the denaturation reaction rate after thermomechanical treatment of complex systems containing various protein fractions also depends on the ratio and concentration of each fraction. The calculated values of the activation energy for single whey protein fractions and multi-component systems decreased with increasing thermal and mechanical treatment. In the temperature range of 80–100 °C or 70–90 °C the activation energy was highest for βLG (107–133 kJ mol^{-1}), then for αLA (109–117 kJ mol^{-1}) followed by 1:1 systems (78–112 kJ mol^{-1}) and WPI (80–100 kJ mol^{-1}), and lastly 5:2 systems (66–81 kJ mol^{-1}). Since at temperatures above 70 °C both protein fractions (αLA and βLG) are expected to aggregate, the energy input in the system due to the treatment was high enough to induce aggregation, consequently the activation energy was lowest for multi-component systems. In the temperature range between 60 and 80 °C or 50 and 70 °C, the activation energy was lowest for the αLA systems and systems containing the same ratio of αLA and βLG. Since αLA denatured at approximately 65 °C, treatment in this temperature range positively affected the reactions leading to lower values of activation energies. Similar results were observed in Kazmierski and Corredig [77] since the denaturation of αLA in solutions was found to be initially even faster than that of βLG at 65 °C [77]. The maximum reduction of the activation energy at a shear rate of 50 s^{-1} was both 19% for αLA and βLG. However, the shear rate influenced the reactions differently, since this maximum reduction was observed in different temperatures ranges. For αLA the maximum reduction was observed between 60 and 80 °C, whereas for βLG was between 80 and 100 °C. The higher reduction of activation energy when shear stress was applied is also observed at higher temperatures for al multi-component systems, implying that in these systems the maximum reduction occurs due to its influence on the βLG reaction. The reduction of activation energy for αLA seems to indicate, as expected, that shear treatment can lead to the formation of a reactive monomer, which plays a significant role in the overall aggregation pathway. Moreover, it is observed that the two regions in which the activation energy was markedly different, changes for the systems where αLA and βLG were mixed at defined ratios. The temperature range changes from 60–80 °C and 80–100 °C to 50–70 °C and 70–90 °C, respectively and the bend temperature decreases from 80 to 70 °C. The decrease in the bend temperature suggests that there is a change in the temperature dependence of the rate limiting reactions. Aggregation seems to be the limiting step at temperatures above 70 °C and no longer 80 °C. The decrease in the bend temperature suggests a more reactive system, as shown in the lower values of activation energy specially at temperatures above the expected denaturation temperature of the protein fractions. Furthermore, for αLA systems and systems containing the same ratio of αLA and βLG, increasing the shear rate from 25 to 50 s^{-1}, does not change the reduction of activation energy significantly, which could imply that shear stress indeed influences the reactions, but only up to a limit and the maximum reduction is achieved at the lower shear rate. The activation energy determined for the βLG systems was overall higher than for αLA and WPI. As already shown in the results from temperature sweeps, the complex modulus of the βLG systems was higher than for αLA and WPI systems, indicating a decreased molecular mobility. This was also the case for 5:2 mixed systems but only at the lower temperature range. Overall, the calculated activation energies for the multi-component systems and the effect of the shear rate on their reduction point out the complex behavior of such systems, which represent the combined effect observed in the main fractions, βLG and αLA.

As already shown for less concentrated systems, changes of βLG and αLA ratios in WPI systems result in aggregates stabilized by different protein–protein interactions [78]. Accordingly, as shown in Fig. 5 for both multi-component systems the proportions of newly formed non-covalent bonds outweigh that of the covalent bonds. Furthermore, as shown in Fig. 5a, b more aggregation is taking place via disulfide bonds for the 5:2 systems, compared to that for the 1:1 system (Fig. 5c, d). In contrast, the 1:1 system seems to aggregate mainly via non-covalent bonds, which appears to be meaningful, since the lower relative amount of βLG provides fewer available thiol groups. Increasing the treatment temperature leads to more aggregation, firstly non-covalent bonds are formed, then when the temperature is high enough (100 °C), there is a shift in the formation of interactions, and aggregation takes place via disulfide bonds. Comparing the samples treated at 90 and 100 °C, the degree of non-covalent bond aggregation decreases, and the disulfide bond aggregation increases further. Moreover, since non-covalent bonds are weak bonds, their activation energy is lower than e.g., disulfide bonds. Therefore, this could explain the overall lower values of activation energy for the 1:1 system in Table 2 at temperatures below 70 °C.

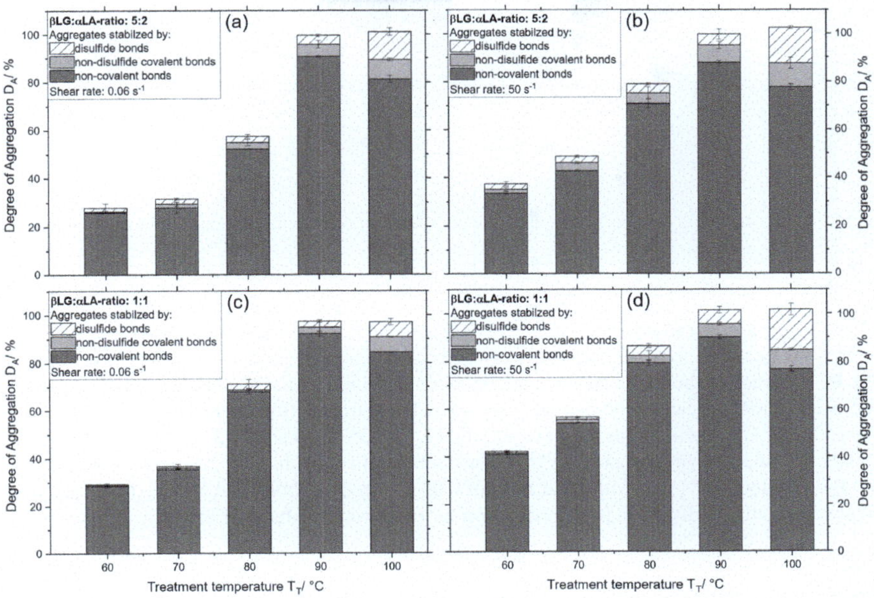

Fig. 5 Degree of aggregation as a function of temperature for 5:2 systems at a shear rate of 0.06 s^{-1} (**a**), 50 s^{-1} (**b**), and 1:1 systems at a shear rate of 0.06 s^{-1} (**c**) and 50 s^{-1} (**d**) with a protein concentration of 70% (w/w) and a treatment time of 30 s. The solid dark grey portion of the bar gives the proportion of stabilization by non-covalent bonds, that by non-disulfide covalent bonds is given by the solid light grey portion of the bar, and that by disulfide bonds is given by the hatched portion of the bar, bars represent standard deviation (SD). Figure originally published in [70]. Permission for reuse was granted by Elsevier

2.2.3 Influence of Thermal and Mechanical Treatment on the Functional Properties of Whey Proteins

The changes in the functional properties induced by extrusion-like treatments were determined for βLG systems [63]. The functionality was characterized by the water solubility, interfacial properties, and viscoelasticity and the results of thermomechanically treated samples were compared to an untreated control sample.

As shown in the previously discussed results, the denaturation reaction is favored when the intensity of treatment increases (thermomechanical compared to only thermal treatment), leading to the formation of aggregates stabilized by non-covalent or covalent bonds, whereas formation of covalent-bonds usually leads to insoluble molecules. Denaturation implies that the globular structure and thus the conformation of whey proteins changes, which can result in changes in the surface hydrophobicity. Consequently, the decreased solubility shown in Quevedo et al., [63] might result from either an increase in particle size, changes of intermolecular interactions or changes of surface hydrophobicity [43, 63, 79]. Since the surface hydrophobicity influences the interfacial properties, the interfacial tension of βLG at an oil-water

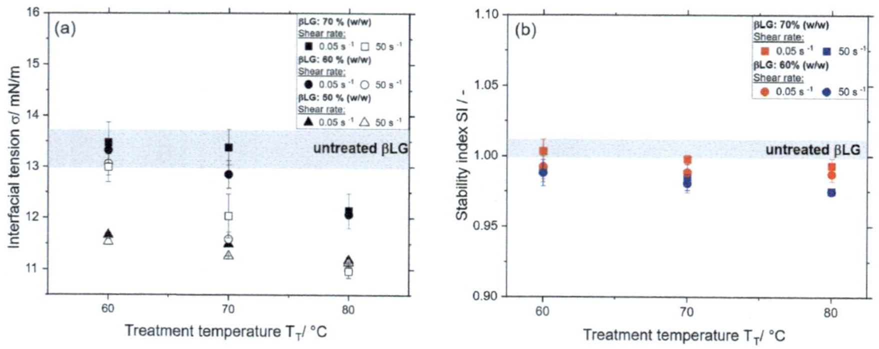

Fig. 6 Influence of thermomechanical treatment on **a** the interfacial tension and **b** stability index of o/w emulsions prepared with untreated (grey band) and treated (treatment time of 30 s) samples. In panel a, βLG concentrations were (w/w) 50% (7, 8), 60% („ –) and 70% (!, ∀), with shear rates of 0.05 s^{-1} (closed symbols and 50 s^{-1} (open symbols). In panel b, βLG concentrations were (w/w) 60% (, ,) and 70% (!, !), with shear rates of 0.05 s^{-1} (red symbols) and 50 s^{-1} (blue symbols). Figure originally published in [63]. Permission for reuse was granted by Elsevier

interface was determined. As expected, it decreased with increasing thermomechanical treatment. By minimizing the interfacial tension, the rate of phase separation or de-emulsification during the emulsification process can be reduced [80] (Fig. 6).

Additionally, the capillary pressure decreases with decreasing interfacial tension leading to less drainage of the liquid film that separates the dispersed oil droplets, preventing the coalescence of droplets close to each other. Consequently, when the samples were used to prepare o/w emulsions, the smallest Sauter mean diameter of the oil droplets and therefore more stable emulsions were observed for the samples with the lowest interfacial tension. It can be therefore stated, that by increasing the intensity of the thermomechanical treatment, the interfacial and consequently the emulsifying properties of small aggregated, non-sedimentable proteins are enhanced due to a decrease in the interfacial tension. However, it must be pointed out, that since only the soluble aggregates showed good interfacial properties, the beneficial effect after treatment was attributed to an interfacial effect of residual native proteins as proposed in Unterhaslberger et al. [81].

The viscosity of dispersions containing 10% untreated and treated βLG samples shown in Fig. 7 confirmed that the size of aggregates increased with thermomechanical treatment since they showed the highest values of viscosity and an increase in molecular size of polymers leads to an increase in viscosity [66]. A 600-fold increase in viscosity was observed for samples treated at 100 °C and 50 s^{-1} compared to untreated samples. Furthermore, compared to thermal treatment alone, combined thermal and mechanical treatment resulted in higher viscosities, due to the previously mentioned increase in denaturation and aggregation with intensity of treatment [43, 82] and possible increase in the water-holding capacity with increasing aggregation [83]. No further increase in viscosity was achieved by application of shear for samples that were already completely aggregated after thermal treatments. Although many studies have shown that shearing can also lead to de-aggregation, the observed decrease in viscosity with shear rate for partially native samples (Fig. 7c, 60 °C and shear rates 0.06 and 25 s^{-1}) is expected to arise from a change in secondary structure since the aggregation degree was below 50%.

As expected, the change in functionality after thermomechanical treatment confirm that the functionality changes as a consequence of denaturation and aggregation. Furthermore, the degree of aggregation and especially the protein–protein interactions, which stabilize the aggregates, play an important role in the resulting functionality. Although this was not investigated in detail in this study from the previous results it can be expected that since the denaturation and aggregation reaction

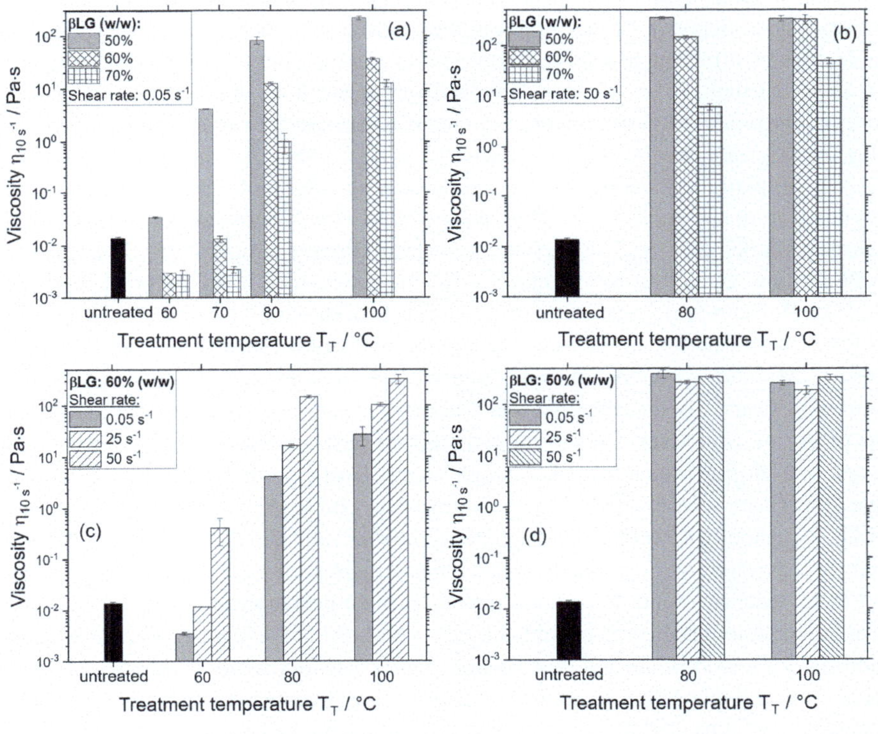

Fig. 7 Shear viscosity at a constant shear rate of 10 s^{-1} of 10% dispersions prepared with samples treated at different conditions. Panels a and b show the influence of the βLG concentration (, 50%; , 60%; , 70%, w/w) and treatment temperature at a shear rate of **a** 0.05 s^{-1} and **b** 50 s^{-1} for a treatment time of 30 s on the resulting viscosity of dispersions prepared with treated samples; panels c and d show the influence of the shear rate (, 0.05 s^{-1}; , 0.25 s^{-1}; , 50 s^{-1}) and treatment temperature for a βLG concentration, **c** 60% (w/w) and **d** 50% (w/w) for a treatment time of 90 s on the resulting viscosity of dispersions prepared with treated samples. Untreated samples are shown by the solid black bars. Figure originally published in [63]. Permission for reuse was granted by Elsevier

behavior changes, the effect of mechanical treatment is specific for single- or multi-component systems and depends on the ratio of the protein fractions present. Therefore, different functionalities are expected for e.g., αLA, WPI or systems different mixing ratios of βLG and αLA, respectively.

2.3 Conclusions

In scope of this part, it was shown that the denaturation and aggregation temperature for single protein fractions, i.e., βLG and αLA, decreases with increasing shear rate. Additionally, in multi-component protein systems such as WPI, the reaction behavior depends on the applied shear stress. For non-sheared systems, the molecules of different protein fractions present in the system seem to be homogeneously distributed and might be involved in the reactions simultaneously. Therefore, lower onset reaction temperatures were determined for WPI compared to the single main fractions. At higher shear rates, it is possible that they separate due to their incompatibility, leading to changes in the molecular mobility, interactions, and aggregation at the local level. This leads to an overall changed denaturation and aggregation behavior and influences the measured onset-temperatures. For sheared WPI systems, two onset-temperatures could be determined, which corresponded to each single main protein fraction, respectively. In addition to the shear rate, the ratio of the protein fractions influenced the reaction behavior. Compared to WPI, higher onset temperatures were determined (approx. 10 K higher) for systems with the same ratio of βLG and αLA, suggesting that other minor components might also influence the reactions. Moreover, the determined reaction order depended on the presence of single or multiple protein fractions. Single-component systems had fractional second reaction orders, whereas multi-component systems showed fractional first order. For βLG, the fractional second order probably arises from the dependency of dimers and reactive monomers on the denaturation and aggregation reaction. In contrast, in WPI, the reactions seem to depend mostly on the concentration of the thiol groups available. Although shear stress enhances the reactions of the investigated single- and multiple-protein systems and thus leads to a decrease in activation energy, it affects them differently. For αLA it seems that the shear stress enhances the unfolding of the molecules, whereas for βLG its aggregation. Additionally, the observed decrease in the bend temperature for systems with defined ratios of βLG and αLA suggests that there is a change in the temperature dependence of the rate limiting reactions. Aggregation seems to be the limiting step at temperatures above 70 °C and no longer 80 °C as observed for βLG, αLA, and WPI, which can be related to the nature of protein interactions participating in the reactions. In the mixed systems, mostly non-covalent bonds were formed, which are weak bonds with low activation energies. In contrast, for e.g., βLG mostly covalent bonds including disulfide bonds were determined.

The functionality of βLG depended on the treatment it was subjected to. An increase in intensity of the thermomechanical treatment, led to enhanced interfacial and emulsifying properties of small aggregated, non-sedimentable proteins.

However, this was due to an interfacial effect, i.e., decrease in the interfacial tension of residual native proteins. In contrast, combined thermal and mechanical treatment resulted in samples showing higher viscosities, caused by an increase in denaturation and aggregation with intensity of treatment.

Overall, processes such as extrusion processing, where proteins are subjected to high temperatures and shear rates, are favorable to induce their aggregation leading to enhanced thickening properties. In contrast, milder processing conditions, i.e., lower temperatures and shear rates, lead only to partial denaturation and aggregation. Thus, the interfacial properties of proteins are modified resulting in enhanced emulsifying properties. However, it must be considered that the effect of thermal and mechanical treatment on the denaturation and aggregation reaction behavior is specific for single- or multi-component systems and depends on the ratio of the protein fractions present. Consequently, different functionalities can be expected for e.g., αLA, WPI or systems with different mixing ratios of βLG and αLA.

3 Analysis of the Thermal and Mechanical Stress Profile in Extrusion Processing

Knowledge of processing conditions is essential to understand and control the reactions taking place in the extruder. However, this task is very difficult due to the interactions between mass, energy and momentum transfer combined with a rotating, intermeshing complex geometry and diverse physicochemical transformations of the material treated. To simplify the process, many studies assume the extruder to be a black box and describe it only by the specific mechanical energy input (SME) and the specific thermal energy input (STE) [84–86]. The SME is calculated from the torque of the motor and the STE from the temperature of the product at the outlet die of the extruder. Both are mean values, characterizing the overall energy input, but giving no information at all on the stress or thermal profiles the material is subjected to. These two process variables thus do not help to determine and control the reactions taking place even for simple homogeneous materials. Corresponding changes in complex food systems cannot be described a fortiori [87–89]. Local thermal and mechanical stresses vary greatly along the length of the extruder. Such local maxima and distributions are often the main reason behind structural changes (e.g. denaturation and aggregation) taking place during extrusion processing [90–92]. The analysis of a thermomechanical stress profile in an extruder is very challenging due to the difficulty of the determination of local process conditions. Nevertheless, for a mechanistic analysis and a better understanding, knowledge of the local processing conditions and their effects on the reaction kinetics is required.

Experimental measurements have the advantage that they reflect real product behavior. However, the measurement of the distribution of thermal and mechanical stress profile in extruders is a very difficult task, and to the best of our knowledge, experimental methods alone are not capable of monitoring these parameters. In order to obtain information about the distribution of thermal and mechanical stress distribution, computational fluid dynamics simulation has been proven to be promising tool, especially when coupled to measured material properties [93–97]. A solved flow field can be further investigated using particle tracking analyses to obtain additional information about flow patterns and thermomechanical stress distribution in the extruder [93, 96]. Information on thermal and mechanical stress gained in this way could be used in models for predicting the degree of protein denaturation. The required models for the degree of protein denaturation can be obtained offline, for example by using a closed cavity rheometer, as described in the Sect. 2.

The goal of this study is to show how processing conditions in the extruder relevant for protein reactions can be investigated using analytical and numerical methods. As reactions are time-dependent, the residence time distribution in the extruder was determined. In order to perform numerical simulations with accurate material data, two approaches were followed to determine the local viscosity values of the protein mixture in the respective section of the extruder under consideration. Moreover, numerical simulations were performed to obtain information about local thermomechanical stress and stress distributions in the screw section of the extruder.

3.1 Material and Methods

3.1.1 Material

Commercial soy protein isolate (SPI) "Supro ST" from Solae LLC (St. Louis, MO, USA) was used for extrusion trials, rheological measurements, and fitting for computational fluid dynamics. A protein content of at least 90% on a dry basis is specified by the manufacturer. A moisture content of 3.4% (w/w) was determined gravimetrically.

3.1.2 Sample Preparation

SPI was mixed with tap water to achieve protein concentrations between 40 and 60% (w/w) in a Thermomix (Vorwerk, Wuppertal, Germany) for 3 min. To ensure a homogenous distribution of water, the samples were stored at 8 °C at least for two days.

3.1.3 Extrusion Trials

The extrusion trials were performed on a lab-scale co-rotating twin-screw extruder "Process 11" (ThermoFisher Scientific Inc., Waltham, MA, USA). The length to diameter (L/D) ratio of the extruder is 40 and the screw diameter is 11 mm. At the end of the extruder, a round die with a length of 5 mm and a diameter of 3 mm was mounted in the die adapter. The protein powder was dosed via a gravimetrically controlled feeder of Brabender Technology GmbH (Duisburg, Germany) in the first barrel element. The water feed was pumped via a peristaltic pump "Masterflex L/S" of the Cole Parmer (Vernon Hills, IL, USA) in the third barrel element.

Four different screw configurations were investigated. One screw configuration was built with forward conveying elements only. The other configurations additionally include a 1.75 L/D kneading blocks with a distance of 3 L/D from the die adapter. The kneading blocks consisted of seven 0.25 L/D kneading disks with a 30° offset forward, 30° offset backward, and 90° offset neutral. The same temperature profile of the barrel was chosen for all extrusion trials whereby the temperature was 40 °C in the first zone, 60 °C in the second zone, 80 °C in the third zone and 100 °C in the following 3 zones. Screw speed and mass flow rate were varied during the experiments. The water content was kept constant at 52% (w/w).

3.1.4 Rheological Investigation of the Protein Mixture

The complex viscosity $|\eta^*(\omega)|$ was determined by performing frequency sweep experiments with a closed cavity rheometer (CCR), also called Rubber Process Analyzer flex (RPA), from TA Instruments, Inc (New Castle, DE, USA). The experiments were carried out within the linear viscoelastic regime with a strain of $\gamma = 3\%$ in an angular frequency range between $\omega = 0.05$ and 315 rad·s^{-1}. Prior to each measurement, the material was held at the zero position for three minutes to ensure a homogenous temperature profile. Measurements were performed in triplicate to determine measurement errors.

The dependence of the complex viscosity $|\eta^*(\omega)|$ on the angular frequency ω was described with a Power-law approach where K is the consistency index and n the Power-law index.

$$|\eta^*(\omega)| = K * \omega^n \tag{1}$$

The temperature dependence of the protein melt was described with an Arrhenius approach, where T is the temperature, T_a is a reference temperature, $|\eta^*(\omega, T_a)|$ the complex viscosity at the reference temperature, and a is the relationship between activation energy and the universal gas constant.

$$|\eta^*(\omega, T)| = |\eta^*(\omega, T_a)| * \exp\left(\frac{a}{T} - \frac{a}{T_a}\right) \tag{2}$$

By defining a reference temperature and combining the Power-law approach with the Arrhenius approach, the complex viscosity can be expresses as a function of angular frequency and temperature, where the coefficient k combines K and $\exp\left(-\frac{a}{T_a}\right)$.

$$|\eta^*(\omega, T)| = k * \exp\left(\frac{a}{T}\right) * \omega^n \tag{3}$$

The apparent viscosity $\eta_{app}(\dot{\gamma}_{app})$ was determined during extrusion trials using an inline slit die rheometer attached at the end of the extruder. This rheometer has a channel length of 125 mm, a width of 19 mm and a height of 4 mm. Moreover, three pressure sensors were placed along the slit die. The procedure for the determination of the apparent viscosity and further information on the method can be found in the work of Horvat et al. [98].

3.1.5 Residence Time Distribution

The residence time distribution (RTD) was determined by a tracer method coupled with image processing. For this, an ExtruVis3 instrument with associated software of ExtruVis (MeltPrep GmbH, Graz, Austria) was used. Measurement setup includes a camera pointed at the extruder outlet, a LED ring to illuminate the extrudate and a laptop computer. A small amount of red dye (~10 μg) was fed to the first section of the extruder, and the color change of the extrudate was recorded using the camera and software. The software automatically outputs the RTD with characteristic values. Here, the mean residence time (t_{mean}) is used as a comparison value. For each process condition, the RTD was measured three times.

3.1.6 Numerical Investigation

For the numerical investigations Finite Element Method (FEM) simulations were performed. The FEM solver of the commercial software ANSYS POLYFLOW® 2020 R2 (Ansys Inc., Canonsburg, PA, USA) was used, Post processing was carried out with ANSYS CFX Post 2020 R2, ANSYS Polystat, and PyCharm (JetBrains, Prague, Czech). More information about the simulation method can be found in previous works [96, 99]. A soy protein mixture was investigated with at water content of 52% (w/w). Rheological measurements as well as fitting for computational fluid dynamic were performed as previously described in detail [100]. To prevent inconvergence due to high viscosity values at low shear rates, the viscosity was implemented using a Bird-Carreau model coupled with an Arrhenius approach [101].

$$\eta(\dot{\gamma}, T) = \exp\left(1383 * \left(\frac{1}{T} - \frac{1}{333}\right)\right) \\ *\left(27 + \left(1.014 * 10^7 - 27\right) * \left(1 + 100^2 * \dot{\gamma}^2\right)^{\frac{0.099-1}{2}}\right) Pa * s \tag{4}$$

Simulation geometries correspond to the 11 mm twin screw extruder used in the experimental set up. The investigated screw elements (1 L/D conveying elements, 1.25 L/D, 45° offset forward kneading blocks, 1.25 L/D, 90° offset neutral kneading blocks, 1.25 L/D, 45° offset backward kneading blocks) and according meshes are depicted in Fig. 8. The mass flow rate was kept constant at 1.5 kg/h and the screw

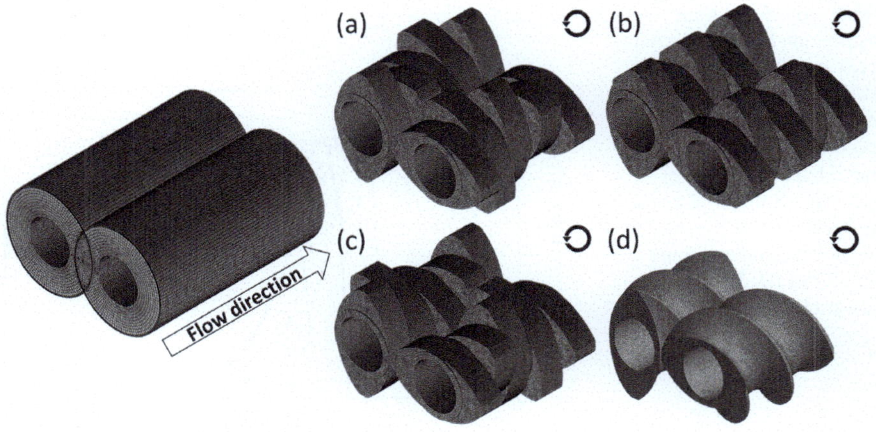

Fig. 8 Geometries with computational meshes of the barrel and the superimposed screw elements for **a** 1.25 L/D, 45° offset forward kneading blocks, **b** 1.25 L/D, 90° offset neutral kneading blocks, **c** 1.25 L/D, 45° offset backward kneading blocks, and **d** 1 L/D conveying elements

speed was kept constant at 600 rpm. The barrel temperature and the inlet temperature is set to 373 K.

3.2 Results and Discussion

3.2.1 Residence Time Distribution

To analyze the influence of thermomechanical stress proteins are subjected to during extrusion processing on the possible occurring reactions, their residence time in the extruder has to be known. In Fig. 9 the tracer intensity E(t) is plotted as a function of the residence time t_{RT} for different screw speeds and mass flow rates.

A skewed right, unimodal RTD can be seen for all process parameters with residence times between 10 and 55 s. As expected, increasing the mass flow at constant screw speed shifts the RTD towards lower residence times. The width of the RTD additionally decreases. If the throughput number (ratio of total mass flow to screw speed) is kept constant, the shifting towards lower residence time is even more pronounced. An increase in screw speed at constant mass flow mainly leads to a shift of the RTD towards lower residence times. Changes in RTD under variation of screw speed and mass flow have often been investigated for classical polymers and food systems. The decrease in residence time by increasing the total mass flow rate and the reduced width of the RTD by increasing the screw speed can be found, for example, for the extrusion of low-density polyethylene (LDPE) in the work of Pauax et al. and for wheat flour in the work of Yeh et al. [102, 103]. It can be concluded that known relationships between process parameters and residence time distribution can

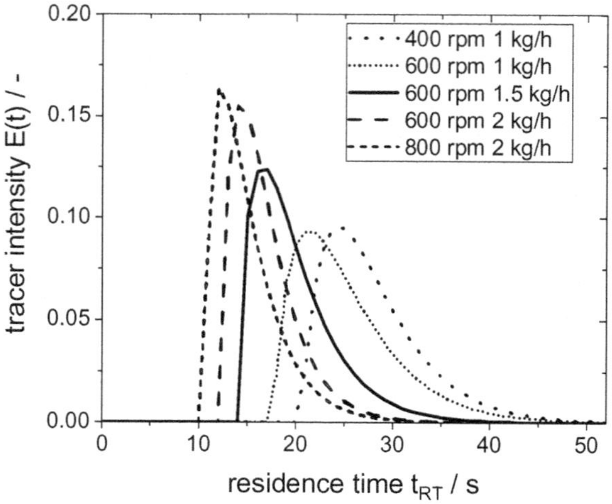

Fig. 9 Residence time distribution in an 11mm twin screw extruder, using a 52% (w/w) soy protein isolate mixture at different screw speeds and mass flows. Barrel temperature increasing from 40 °C at the feed zone to 100 °C at the die. Screw geometry build with 1.75 L/D, 30° offset forward kneading blocks

also be applied to the extrusion of highly concentrated protein systems. It should be of interest for future research how the rheological properties of the protein matrix affect the RTD. Namely, during the extrusion of reactive proteins, it is expected that the rheological properties change along the extruder, which is expected to have an influence on the RTD.

The influence of process parameters on the residence time were compared using the mean residence time t_{mean}. The influence of changes in screw configurations on t_{mean} is listed for various process parameters in Table 3. It should be mentioned that the influence of mass flow and screw speed on the RTD reported in Fig. 9 can be found for all screw configurations used. The highest t_{mean} of 34.4 s was found for the combination of lowest mass flow and lowest screw speed, for the screw with backward kneading blocks. As expected, the highest mass flow and screw speed led to the shortest t_{mean} of 14.7 s when only conveying elements were used. Therefore, an increase in t_{mean} can be achieved by either lowering the mass flow, lowering the screw speed, or reducing the conveying capacity of the screw. Conveying elements have better conveying capacities than kneading blocks. The conveying capacity of kneading blocks decreases from forward offset, to neutral offset, and to backward offset.

3.2.2 Viscosity of Protein Mixture

In Fig. 10a the complex viscosity $|\eta^*(\omega)|$ of a SPI mixture is given as a function of the angular frequency ω for different temperatures. In the considered range of $\omega = 0.05\text{–}315$ rad s^{-1}, a shear-thinning behavior can be observed for all temperatures, which has been reported for high concentrated protein systems [104–106]. The SPI used does not show any polymerization reaction or degradation reaction in the applied temperature range, hence the decreasing complex viscosity with increasing

Table 3 Mean residence time distribution in an 11mm twin screw extruder, using a soy protein isolate mixture with 52% (w/w) water content at different screw speeds and mass flows. Barrel temperature increasing from 40 °C at the feed zone to 100 °C at the die. Screw geometries built with only conveying elements, conveying elements and 1.75 L/D, 30° offset forward kneading blocks, 1.75 L/D, 90° offset neutral kneading blocks, and 1.75 L/D, 30° offset backward kneading blocks

	t_{mean}/s			
Processing conditions	Conveying elements	Forward kneading elements	Neutral kneading elements	Backward kneading elements
400 rpm, 1 kg/h	27.2 ± 1.2	28.5 ± 0.9	33.2 ± 1.0	34.4 ± 0.7
600 rpm, 1 kg/h	23.8 ± 0.4	25.6 ± 0.3	28.8 ± 0.5	29.9 ± 0.4
600 rpm, 1.5 kg/h	18.6 ± 1.2	20.4 ± 0.2	21.7 ± 0–5	21.9 ± 0.6
600 rpm, 2 kg/h	15.9 ± 0.5	17.4 ± 1.3	18.1 ± 0.3	18.8 ± 0.5
800 rpm, 2 kg/h	14.7 ± 0.3	16.7 ± 1.2	16.7 ± 0.2	17.7 ± 0.7

temperature is due to the increased mobility of molecules. Therefore, the magnitude of the complex viscosity can be described using a Power-law approach combined with an Arrhenius approach according to Eq. (3). For a reference temperature of 298 K, the following dependence of the complex viscosity on the angular frequency and temperature is obtained.

$$\left|\eta^*(\omega, T)\right| = 97 * \exp\left(\frac{2950}{T}\right) * \omega^{-0.9} \tag{5}$$

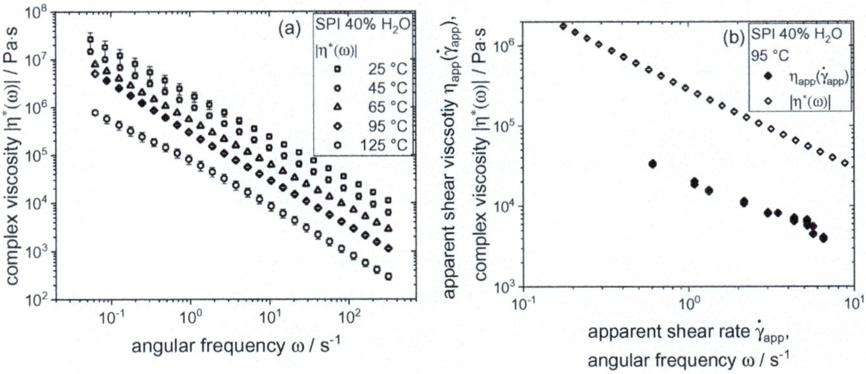

Fig. 10 **a** Complex viscosity for SPI with 40% (w/w) at different temperatures measured by frequency sweeps within the linear viscoelastic region. **b** Complex viscosity for SPI with 40% (w/w) at 95 °C measured by frequency sweeps within the linear viscoelastic region and the apparent shear viscosity measured with an inline slit rheometer

Determining the dependence of the complex viscosity on temperature and angular frequency for a reactive protein such as WPI is a more challenging task, as the complex viscosity may also changes with time (depending on the temperature). A possible approach to determine the complex viscosity is to stop the reactions by strong cooling, followed by measuring the complex viscosity at temperatures below the reaction temperatures. Subsequently, the complex viscosity must be extrapolated back to the temperature of actual interest.

In Fig. 10b the complex viscosity $|\eta^*(\omega)|$ is plotted as a function of the angular frequency ω, and the apparent shear viscosity $\eta_{app}(\dot{\gamma}_{app})$ is plotted as a function of the apparent shear rate $\dot{\gamma}_{app}$ for SPI at a temperature of T = 95 °C. As already described for the complex viscosity in Fig. 10a, b shear thinning behavior is also seen for the apparent shear viscosity. The shear thinning behavior of the different viscosities is in similar orders of magnitude, but the absolute values differ by more than a decade. Although the empirical Cox-Merz rule ($|\eta^*(\omega)| = \eta(\dot{\gamma})$ for $\dot{\gamma} = \omega$) is known to be valid for many classical polymer melts, it is assumed that the rule is not applicable to highly concentrated protein systems. Reasons for this include interactions between protein molecules, the gel-like structure and the complex composition from different protein fractions [107]. In addition, the data of the inline capillary rheometer only allow the determination of the apparent viscosity since possible wall slippage instabilities were not corrected.

However, section of dominating shear flow is present in the screw section and die section of the extruder, which is why it is relevant to know or be able to estimate the shear viscosity. Accordingly, it is important to improve existing methods, or develop new methods to determine shear viscosity under extrusion-like conditions. Recently, a method was developed for the closed cavity rheometer to determine the shear viscosity. This method has already been confirmed for classical polymers and will also play a role in the investigation of protein melts in the future [108].

3.2.3 Thermomechanical Stress Profile and Mixing Characteristics Obtained by Computational Fluid Dynamics Simulation

Figure 11 shows counters of the temperature distribution in a plane through the rotation axes of the screw elements for different screw elements at a quasi-stationary time step. It can be seen that a variation in screw elements lead to different temperature distribution. For all elements, the material flows into the simulated region with an inlet temperature of 373 K. In direction of flow, the temperature increases 39 K for the forward kneading blocks (Fig. 11a), 41 K for the neutral kneading blocks (Fig. 11b), 44 K for the backward kneading blocks (Fig. 11c), and 33 K for the conveying elements (Fig. 11d). It should be emphasized that for the neutral kneading blocks, the temperature is increased in steps in the middle of a kneading disk. In contrast, the temperature for the other elements increases continuously over the element length. This observation can be attributed to the different flow and mixing conditions in the extrusion processes. However, Fig. 11a–d can only give an idea of the temperature

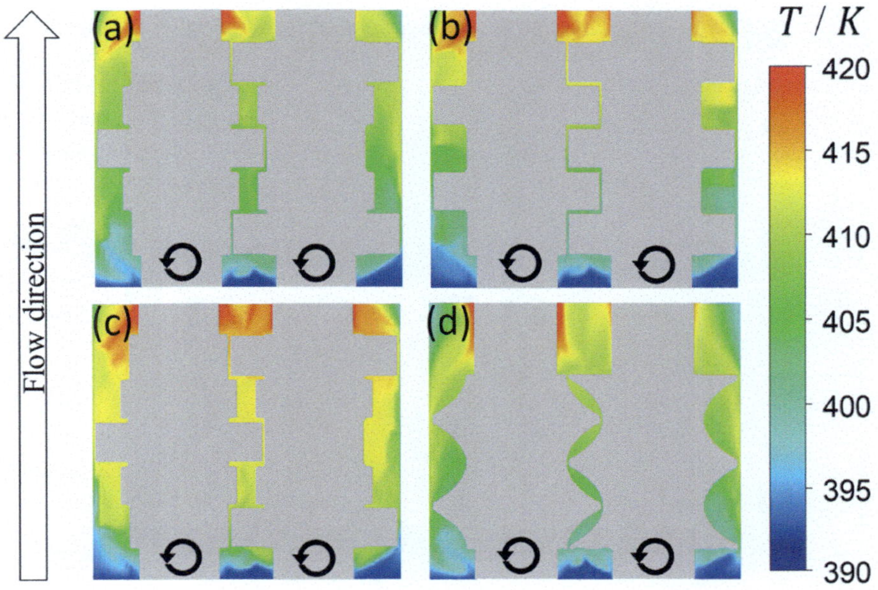

Fig. 11 Counters of the temperature distribution in a plane through the rotation axes of the screw elements in a simulated, fully filled section of a twin screw extruder for different screw elements. Mass flow rate is set to 1.5 kg/h, screw speed to 600 rpm, wall and inlet temperature is 373K. **a** 1.25 L/D, 45° offset forward kneading blocks, **b** 1.25 L/D, 90° offset neutral kneading blocks, **c** 1.25 L/D, 45° offset backward kneading blocks, and **d** 1 L/D conveying elements

stress experienced by the proteins. For more detailed conclusions, analysis of the thermal stress profile is necessary.

The shear rate distribution can be used as an indicator of the mechanical stress in the screw section. Analogous to the temperature distribution in Figs. 11 and 12 shows counters of the shear rate distribution in a plane through the rotation axes of the screw elements for different screw elements. For all four elements, the highest shear rates are observed in the gap between the screw elements, or in the gap between the screw tip and the barrel wall. In these areas, the circumferential velocity is the highest and the gap the smallest. Thereby, shear rates up to 6,000 s^{-1} are achieved, which in turn means that shear rate distribution in the extruder is very broad. It should be emphasized here, that the shear rates in this order of magnitude cannot be investigated in a model process like the closed cavity rheometer. Accordingly, it is still unclear to what extent high shear rates may influence the reaction behavior of highly concentrated proteins.

When comparing the screw elements, the low shear rates in the channel of the conveying elements (Fig. 12d) are notable. Further, it can be seen that the shear rate distribution in this counters is almost the same for the kneading blocks with forward and backward offset. For the neutral kneading blocks, the gap length between the screws where the highest shear rates occur is the largest. Depending on the rotation

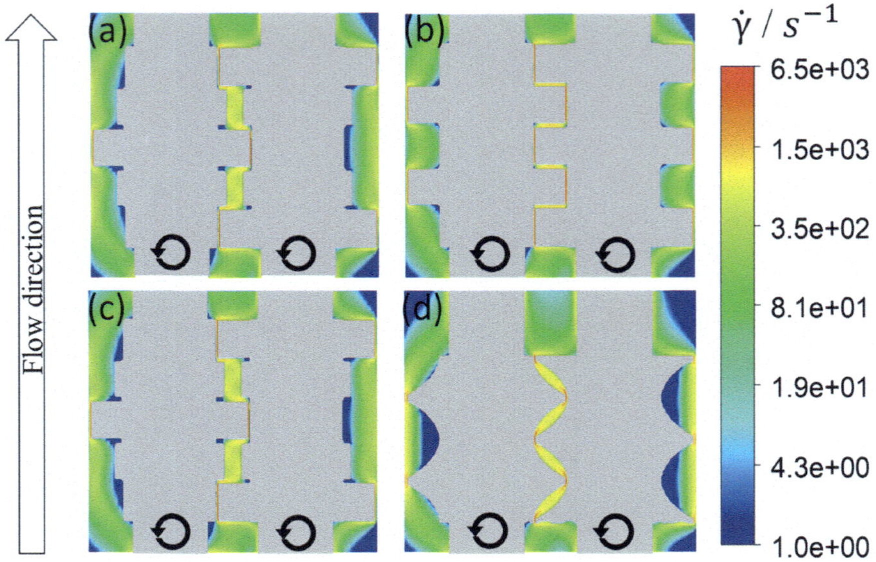

Fig. 12 Counters of the shear rate distribution in a plane through the rotation axes of the screw elements in a simulated, fully filled section of a twin screw extruder for different screw elements. Mass flow rate is set to 1.5 kg/h, screw speed to 600 rpm, wall and inlet temperature is 373K. **a** 1.25 L/D, 45° offset forward kneading blocks, **b** 1.25 L/D, 90° offset neutral kneading blocks, **c** 1.25 L/D, 45° offset backward kneading blocks, and **d** 1 L/D conveying elements

position of the screws, however, the gap becomes wider again. In summary, the maximum shear rate for all screw elements is of similar magnitude since the gap width is the same and the circumferential velocity is constant. Consequently, it cannot be concluded that different screw elements can necessarily lead to different mechanical stress histories. For this purpose, stress profile of the material must be analyzed.

To obtain this information, particle tracking analysis were performed. A set of 1,000 particles were randomly distributed in the inlet plane of the simulated elements and subsequently move through the elements based on the solved velocity fields. Meanwhile, information about local process conditions like temperature and shear rate are recorded for each particle trajectory. For instance, Fig. 13 shows an example of the temperature profile and shear rate profile for a single, randomly selected particle for the simulated conveying elements (Figs. 11d and 12d). It can be observed that the particle flowed through the section within 1.06 s. During the flow, a stepwise increase in temperature from 377 K at the beginning to 403 K at the exit can be seen. The results of the shear rate show, that the particle remains mainly in low shear regime $\dot{\gamma} < 10\ \text{s}^{-1}$, but repeating peaks with shear rates above $100\ \text{s}^{-1}$ up to $6000\ \text{s}^{-1}$ occur. Concerning the temperature, it should be noted that a gradual temperature increase is mainly found after a shear rate peak. Coupled with the regular repetition of the shear rate peaks, this observation suggests that the particle is conveyed in the axial direction by the screw in the event of a shear rate peak (axial displacement).

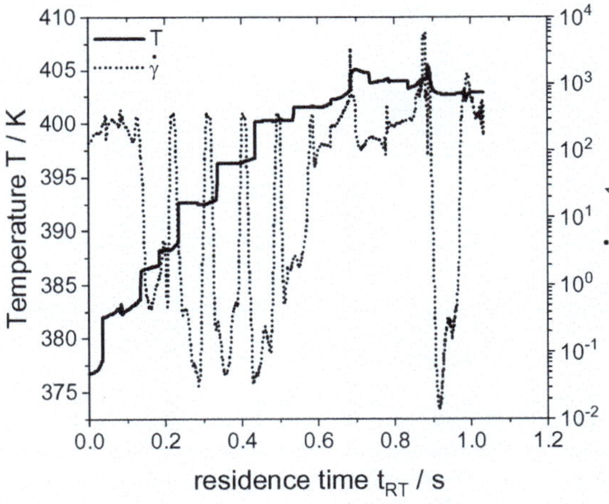

Fig. 13 Temperature history and shear rate history of a randomly picked single particle in the flow through the conveying elements

Results from particle tracking analysis can be subjected to statistical analysis, which allows deducing information about the thermomechanical stress history, stress distributions, and the comparison to other processing conditions. The following is an example of the simulated results under variation of the screw configuration. Figure 14a shows information for the residence time distribution for particles during their flow through different screw elements. Using kneading blocks in comparison to conveying elements leads to an increase in the width of the RTD and the residence time is shifted towards longer times. However, it should be mentioned that the geometry of the conveying elements has a shorter L/D compared to kneading blocks, but the residence time is still relatively longer. When comparing the kneading blocks, it can be observed that the backward kneading blocks lead to the widest RTD with the largest median ($Q_0 = 0.5$) of 3.3 s. The median of the neutral kneading blocks is at slightly shorter residence times of 3 s, but the RTD is narrower. The forward kneading blocks has the smallest median of the kneading blocks with a value of 2.3 s. The residence times differ significantly for the different kneading blocks, although the flowable volume and the mass flow rate are identical. It is assumed that not the entire available volume is involved in the axial flow and therefore there are theoretically different flowable volumes.

Considering the experienced shear rates in the extruder, a broad shear rate distribution was found in the counters (Fig. 12) and recurring shear rate peaks were observed in a particle trajectory (Fig. 13). However, observing the shear rate distribution in Fig. 12, it is noticeable that shear rates above 850 s^{-1} occur only in the gap between the two screws and between the screw tip and the barrel wall. It is known that high shear stress can result in fragmentation of biopolymer molecules like amylopectin, the main molecule fraction of starch, therefore the number of passes for these experienced high shear rates was evaluated [109]. Figure 14b shows the cumulative frequency Q_0 plotted against the number of passes n for experienced shear rates above 850 s^{-1}. The number of passes for the different screw elements varies from zero up to 78 passes. Thereby, an analogously behavior of the number

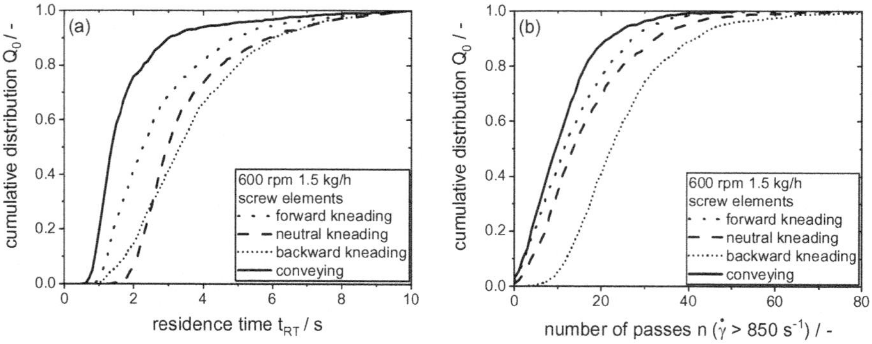

Fig. 14 **a** Residence time distribution of particle collectives flowing through different screw elements with constant screw speed of 600 rpm and constant mass flow of 1.5 kg/h. **b** Number of passes for experienced shear rates above a value of 850 s^{-1} (Gap between two screws or gap between screw and barrel wall)

of passes and the RTD can be observed. With increasing residence time, the probability of being exposed to high shear rates increases. In addition, the flow and mixing processes in the various screw elements differ. It is worth highlighting that for the backward kneading elements, the particles are stressed more frequently overall and all particles experience shear rates above of 850 s^{-1} at least twice. Finally, it was found, that regardless of the screw elements and mass flow rate, for all investigated processing conditions an inhomogeneous stress distribution can be found. The most efficient method to control the number of passes is to control the residence time.

Further information about stress history can be obtained by a cumulative consideration of the mechanical stress over the dimensionless axial length of the elements. For this purpose, the cumulative mechanical stress σ_{mech} is built of the product shear stress at a discrete time $\sigma_i(t_i)$ times time increment $\Delta t = t_{i+1} - t_i$.

$$\sigma_{mech} = \sum_{t_i = t_{entrance}}^{t_{exit}} \sigma_i(t_i) * \Delta t$$

Likewise, the cumulative thermal stress σ_{therm} can be built of the product temperature at a discrete time $T_i(t_i)$ times time increment $\Delta t = t_{i+1} - t_i$.

$$\sigma_{therm} = \sum_{t_i = t_{entrance}}^{t_{exit}} T_i(t_i) * \Delta t$$

Figure 15a plots the cumulative mechanical stress σ_{mech} (mean value of the particle collective) over the dimensionless axial length of the elements $l/l_{elements}$ for different screw elements. Particles can be at the same axial position of the extruder several times; here σ_{mech} is shown for the first pass of the axial position. The cumulative thermal stress σ_{therm} is shown analogously in Fig. 15b.

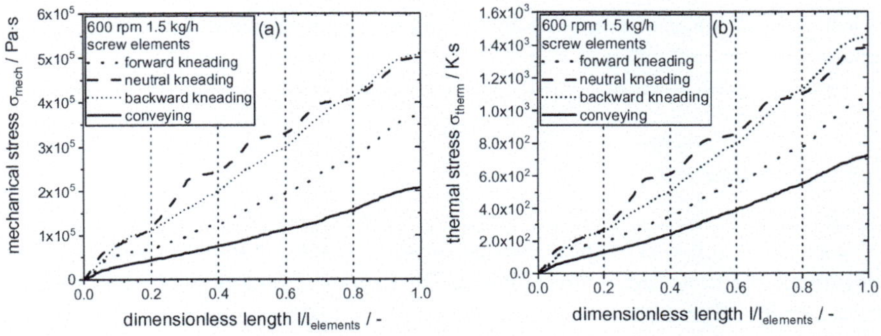

Fig. 15 **a** cumulative mechanical stress of the particle collective versus dimensionless length of element (first crossing), **b** cumulative thermal stress of the particle collective versus dimensionless length of element (first crossing). Extrusion screw speed is set constant at 600 rpm and constant mass flow rate of 1.5 kg/h. Dotted line indicate the end of a kneading disk

If the cumulative mechanical stress is considered, differences in the courses with corresponding differences in the total mechanical stress at the end of the screw elements can be observed. The lowest total cumulative mechanical stress with a value of 208 kPa s can be found for the conveying elements, followed by the forward kneading blocks with a total cumulative mechanical stress of 374 kPa s. The neutral kneading blocks with a value of 504 kPa s is slightly above the cumulative mechanical stress of the backward kneading blocks with a value of 511 kPa s. Since the cumulative mechanical stress is also influenced by the residence time, it should be taken into account that the resident time varies for the different elements. Relating the total cumulative mechanical stress e.g. to the median of the RTD (Fig. 14a), the difference is no longer as pronounced. The related cumulative mechanical stress is 148 kPa for the conveying elements, 162 kPa for the forward kneading blocks, 168 kPa for the neutral kneading blocks, and 155 kPa for the backward kneading blocks. Accordingly, it can be stated that the different residence times in different screw elements lead to strong variations in the mechanical stress.

The different stress frequencies at high shear rates shown in Fig. 14b are not reflected in the profile of cumulative mechanical stresses. The reason for this is that the viscosity model describing the concentrated protein is highly shear thinning ($n \approx 0.9$). Thus, viscosity is reduced at high shear rates, and the product of viscosity times shear rate (shear stress) increases only slightly. Additionally, the temperature of the material is higher for the elements with higher numbers of passes, which in turn decreases the viscosity for not reactive systems.

The cumulative thermal stress, shown in Fig. 15b, reveals similar trends as the cumulative mechanical stress in Fig. 15a. Accordingly, it can be concluded that the residence time also has the greatest influence on thermal stress. It should not be neglected, that the mechanical stress depends on the thermal stress and vice versa. With increasing mechanical stress, the viscous heating increases and thus the temperature. As the temperature rises, the viscosity in turn decreases, resulting in lower mechanical stresses. Therefore, these quantities should not be discussed separately from each other.

3.3 Conclusions

In scope Sect. 3, it was to show how processing conditions in the extruder relevant for protein reactions can be investigated. The residence time in the extruder, which is relevant for all reactions, was determined as the residence time distribution with conventional methods. It was shown that the process parameters screw speed, mass flow, and screw geometry influence the residence time and the width of the residence time distribution. The correlations found between process parameters and residence time could be confirmed with literature data of other polymer type raw materials. Measurements of the complex viscosity at different temperatures and shear rates provide viscosity models as a function of angular frequency and temperature, which can further be used as input data for numerical simulations. Measurements with an inline capillary rheometer made it possible to determine the apparent shear viscosity as a function of apparent shear rate. To obtain information on local thermomechanical stress and stress distributions, finite element simulations were performed. The influence of screw configuration on the local temperature was analyzed and it could be shown that higher local temperatures were related to using kneading blocks. Considering the mechanical stress, it was found that local maxima of the shear rate for different screw elements are in similar order of magnitude. Therefore, it seemed that it is not the pure value that plays a decisive role for the experienced shear stress, but how long the protein mixture remains in the screw element and which path it passes through during the flow. To investigate this, a particle tracking method was used to observe the residence time distribution and the number of passes for experienced high shear rates. Thereby, it could be shown that the choice of screw elements has an important influence on the residence time and the stress frequency. In general, as expected by general knowledge, it was shown that at longer residence times, the material is also more frequently exposed to intensive shear stress. Moreover, a change in screw elements led to higher shear stresses due to a changed flow profile. The simulation data thus confirm experience based general extrusion knowledge. In addition, the simulation allows for the first time to have concrete absolute values for the stress history the material experiences when flowing through an extruder. Examination of the cumulative mechanical and thermal stress revealed that the residence time plays a decisive role for the total experienced stress. Although the stresses at the end of the elements differ, the stress is of similar magnitude, if these are weighted with the residence time. The reason for this is the strong shear thinning behavior of proteins.

Once, concrete stress profiles in the extrusion process (Sect. 3) and stress-dependent reaction kinetic data (Sect. 2) are available, prediction of the degree of denaturation of proteins and thus functional properties could be possible. This can save a great deal of time and money in product development. Additionally, it is conceivable that the approach shown here could be applied to other raw materials such as fiber-rich products, and different processes such as cooling scrapers or mixers.

4 General Conclusions and Outlook

In the first part, the reaction behavior of whey proteins (as model biopolymer) and their main components under extrusion like conditions was determined and discussed from a mechanistic view. Among others, the influence of water content, treatment temperature and composition on denaturation and aggregation was investigated. In the second part of the study, the local process conditions in the actual extrusion process were investigated in more detail. It was possible to obtain information about residence time, local temperatures, and mechanical stresses as well as information about stress distributions for a non-reactive model system. Despite the complexity of the process as well as the protein system, the methods developed and applied allowed to gain mechanistic understanding on the extrusion processing of highly concentrated protein systems. The future work in this field should focus on the prediction of the denaturation and aggregation rate of model protein systems, by using the developed methods and models and gained knowledge. First, the shear viscosity of protein systems should be determined at extrusion-relevant conditions. Additionally, the reactivity should be considered in numerical simulations, and they should not only be focused on fully-filled sections. Lastly, the thermomechanical stress profiles in the extruder should be coupled to the results of the model protein system obtained with the closed-cavity rheometer.

References

1. Nakai S (1983) Structure-function relationships of food proteins: with an emphasis on the importance of protein hydrophobicity. J Agric Food Chem 31:676–683
2. Kannan A, Hettiarachchy NS, Marshall M (2016) Chemistry and structure. In: Hettiarachchy NS (ed) Food proteins and peptides: chemistry, functionality, interactions, and commercialization. CRC Press, Boca Raton
3. Dobson CM (1999) Protein misfolding, evolution and disease. Trends Biochem Sci 24:329–332
4. Stefani M, Dobson CM (2003) Protein aggregation and aggregate toxicity: new insights into protein folding, misfolding diseases and biological evolution. J Mol Med 81:678–699
5. Chiti F, Dobson CM (2006) Protein misfolding, functional amyloid, and human disease. Annu Rev Biochem 75:333–366
6. Calamai M, Chiti F, Dobson CM (2005) Amyloid fibril formation can proceed from different conformations of a partially unfolded protein. Biophys J 89:4201–4210
7. Dobson CM (2004) Principles of protein folding, misfolding and aggregation. Semin Cell Dev Biol 15:3–16
8. de Graaf LA (2000) Denaturation of proteins from a non-food perspective. J Biotechnol 79:299–306

9. Amin S, Barnett GV, Pathak JA et al (2014) Protein aggregation, particle formation, characterization & rheology. Curr Opin Colloid Interface Sci 19:438–449
10. Spiegel T, Huss M (2002) Whey protein aggregation under shear conditions—Effects of pH-value and removal of calcium. Int J Food Sci Technol 37:559–568
11. Murphy RM, Kendrick BS (2007) Protein misfolding and aggregation. Biotechnol Prog 23:548–552
12. Bekard IB, Asimakis P, Bertolini J et al (2011) The effects of shear flow on protein structure and function. Biopolymers 95:733–745
13. Li P-C, Makarov DE (2003) Theoretical studies of the mechanical unfolding of the muscle protein titin: Bridging the time-scale gap between simulation and experiment. J Chem Phys 119:9260–9268
14. Rohs R, Etchebest C, Lavery R (1999) Unraveling proteins: a molecular mechanics study. Biophys J 76:2760–2768
15. Nielsen L, Khurana R, Coats A et al (2001) Effect of environmental factors on the kinetics of insulin fibril formation: elucidation of the molecular mechanism. Biochemistry 40:6036–6046
16. Sluzky V, Tamada JA, Klibanov AM et al (1991) Kinetics of insulin aggregation in aqueous solutions upon agitation in the presence of hydrophobic surfaces. Proc Natl Acad Sci U S A 88:9377–9381
17. Hill EK, Krebs B, Goodall DG et al (2006) Shear flow induces amyloid fibril formation. Biomacromolecules 7:10–13
18. Oliva A, Santoveña A, Fariña J et al (2003) Effect of high shear rate on stability of proteins: kinetic study. J Pharm Biomed Anal 33:145–155
19. Nicolai T, Britten M, Schmitt C (2011) β-Lactoglobulin and WPI aggregates: formation, structure and applications. Food Hydrocolloids 25:1945–1962
20. Ruttarattanamongkol K (2012) Functionalization of whey proteins by reactive supercritical fluid extrusion. Songklanakarin J Sci Technol 34:395–402
21. Queguiner C, Dumay E, Salou-Cavalier C et al (1992) Microcoagulation of a whey protein isolate by extrusion cooking at acid pH. J Food Sci 57:610–616
22. Onwulata CI, Tunick MH, Qi PX (2011) Extrusion texturized dairy proteins: processing and application. Adv Food Nutr Res 62:173–200
23. Liu KS, Hsieh F-H (2007) Protein-protein interactions in high moisture-extruded meat analogs and heat-induced soy protein gels. J Am Oil Chem Soc 84:741–748
24. Manoi K, Rizvi SSH (2009) Physicochemical changes in whey protein concentrate texturized by reactive supercritical fluid extrusion. J Food Eng 95:627–635
25. Purwanti N, van der Goot AJ, Boom R et al (2010) New directions towards structure formation and stability of protein-rich foods from globular proteins. Trends Food Sci Technol 21:85–94
26. Camire ME (1991) Protein functionality modification by extrusion cooking. J Am Oil Chem Soc 68:200–205
27. Liu K, Hsieh F-H (2008) Protein-protein interactions during high-moisture extrusion for fibrous meat analogues and comparison of protein solubility methods using different solvent systems. J Agric Food Chem 56:2681–2687
28. Qi PX, Onwulata CI (2011) Physical properties, molecular structures, and protein quality of texturized whey protein isolate: effect of extrusion moisture content. J Dairy Sci 94:2231–2244
29. Qi PX, Onwulata CI (2011) Physical properties, molecular structures, and protein quality of texturized whey protein isolate: effect of extrusion temperature. J Agric Food Chem 59:4668–4675
30. Osen R, Toelstede S, Wild F et al (2014) High moisture extrusion cooking of pea protein isolates: raw material characteristics, extruder responses, and texture properties. J Food Eng 127:67–74
31. Akdogan H (1999) High moisture food extrusion. Int J Food Sci Technol 34:195–207
32. Arêas JA (1992) Extrusion of food proteins. Crit Rev Food Sci Nutr 32:365–392
33. Maurice TJ, Burgess LD, Stanley DW (1976) Texture-structure relationships in texturized soy protein. III. Textural evaluation of extruded products. Can Inst Food Sci Technol J 9:173–176

34. Rebello CA, Schaich KM (1999) Extrusion chemistry of wheat flour proteins: II. Sulfhydryl-disulfide content and protein structural changes. Cereal Chem J 76:756–763
35. Verbeek CJR, van den Berg LE (2010) Extrusion processing and properties of protein-based thermoplastics. Macromol Mater Eng 295:10–21
36. Guy R (ed) (2001) Extrusion cooking: technologies and applications. Woodhead Publishing in food science and technology, Woodhead, CRC Press, Cambridge, Boca Raton
37. Leuner C, Dressman J (2000) Improving drug solubility for oral delivery using solid dispersions. Eur J Pharm Biopharm 50:47–60
38. Bouvier J-M (2014) Extrusion processing technology: food and non-food biomaterials. Wiley Blackwell, Chichester, West Sussex, UK
39. Dissanayake M, Vasiljevic T (2009) Functional properties of whey proteins affected by heat treatment and hydrodynamic high-pressure shearing. J Dairy Sci 92:1387–1397
40. Bouaouina H, Desrumaux A, Loisel C et al (2006) Functional properties of whey proteins as affected by dynamic high-pressure treatment. Int Dairy J 16:275–284
41. Dissanayake M, Liyanaarachchi S, Vasiljevic T (2012) Functional properties of whey proteins microparticulated at low pH. J Dairy Sci 95:1667–1679
42. Morr CV, Ha EYW (1993) Whey protein concentrates and isolates: processing and functional properties. Crit Rev Food Sci Nutr 33:431–476
43. Nor Afizah M, Rizvi SSH (2014) Functional properties of whey protein concentrate texturized at acidic pH: effect of extrusion temperature. LWT Food Sci Technol 57:290–298
44. Brodkorb A, Croguennec T, Bouhallab S et al (2016) Heat-induced denaturation, aggregation and gelation of whey proteins. In: McSweeney PLH, O'Mahony JA (eds) Advanced dairy chemistry, vol 5. Springer, New York, New York, NY, pp 155–178
45. McKenzie HA, Sawyer WH (1967) Effect of pH on β-Lactoglobulins. Nature 214:1101–1104
46. Verheul M, Roefs SPFM, de Kruif KG (1998) Kinetics of heat-induced aggregation of β-Lactoglobulin. J Agric Food Chem 46:896–903
47. Georges C, Guinand S, Tonnelat J (1962) Etude thermodynamique de la dissociation réversible de la β-Lactoglobuline B pour des pH supérieurs à 5,5. Biochem Biophys Acta 59:737–739
48. Loveday SM (2016) β-Lactoglobulin heat denaturation: a critical assessment of kinetic modelling. Int Dairy J 52:92–100
49. Hoffmann MAM, van Mil PJJM (1997) Heat-induced aggregation of β-Lactoglobulin: role of the free thiol group and disulfide bonds. J Agric Food Chem 45:2942–2948
50. Sienkiewicz T (1981) Nomenklatur und einige Eigenschaften der Molkenproteine. 2. Mitt. α-Lactalbumin, Immunoglobuline, Proteose-Peptone, Minorproteine und Enzyme. Nahrung 25:335–343
51. Apenten RKO (1995) Thermodynamic parameters for 3-state thermal denaturation of human and bovine α-lactalbumin. Thermochim Acta 262:1–12
52. Chaplin LC, Lyster RLJ (1986) Irreversible heat denaturation of bovine α-lactalbumin. J Dairy Res 53:249–258
53. Schnack U, Klostermeyer H (1980) Thermal decomposition of alpha-lactalbumin: I. Destruction of cystine residues. Milchwissenschaft 35
54. Schokker EP, Singh H, Creamer LK (2000) Heat-induced aggregation of β-Lactoglobulin A and B with α-Lactalbumin. Int Dairy J 10:843–853
55. Cheftel JC, Kitagawa M, Quéguiner C (1992) New protein texturization processes by extrusion cooking at high moisture levels. Food Rev Intl 8:235–275
56. Holay SH, Harper JM (1982) Influence of the extrusion shear environment on plant protein texturization. J Food Sci 47:1869–1874
57. Wolz M, Kastenhuber S, Kulozik U (2016) High moisture extrusion for microparticulation of whey proteins—Influence of process parameters. J Food Eng 185:56–61
58. Onwulata CI, Konstance RP, Cooke PH et al (2003) Functionality of extrusion—Texturized whey proteins. J Dairy Sci 86:3775–3782
59. Toro-Sierra J, Tolkach A, Kulozik U (2013) Fractionation of α-Lactalbumin and β-Lactoglobulin from whey protein isolate using selective thermal aggregation, an optimized membrane separation procedure and resolubilization techniques at pilot plant scale. Food Bioprocess Technol 6:1032–1043

60. Haller N, Kulozik U (2019) Separation of whey protein aggregates by means of continuous centrifugation. Food Bioprocess Technol 12:1052–1067
61. Quevedo M, Jandt U, Kulozik U et al (2019) Investigation on the influence of high protein concentrations on the thermal reaction behaviour of β-Lactoglobulin by experimental and numerical analyses. Int Dairy J 97:99–110
62. Quevedo M, Kulozik U, Karbstein HP et al. (2020) Influence of thermomechanical treatment and ratio of β-Lactoglobulin and α-Lactalbumin on the denaturation and aggregation of highly concentrated whey protein systems. Foods 9
63. Quevedo M, Kulozik U, Karbstein HP et al (2020) Effect of thermomechanical treatment on the aggregation behaviour and colloidal functionality of β-Lactoglobulin at high concentrations. Int Dairy J 104:104654
64. Quevedo M, Kulozik U, Karbstein HP et al (2020) Kinetics of denaturation and aggregation of highly concentrated β-Lactoglobulin under defined thermomechanical treatment. J Food Eng 274:109825
65. Quevedo M, Karbstein HP, Emin MA (2020) Denaturation behavior and kinetics of single- and multi-component protein systems at extrusion-like conditions. Polymers (Basel) 12
66. Berry GC, Fox TG (1968) The viscosity of polymers and their concentrated solutions. Adv Polym Sci 5:261–357
67. Emin MA, Quevedo M, Wilhelm M et al (2017) Analysis of the reaction behavior of highly concentrated plant proteins in extrusion-like conditions. Innov Food Sci Emerg Technol 44:15–20
68. Koch L, Emin MA, Schuchmann HP (2017) Reaction behaviour of highly concentrated whey protein isolate under defined heat treatments. Int Dairy J 71:114–121
69. Madeka H, Kokini JL (1994) Changes in rheological properties of gliadin as a function of temperature and moisture: development of a state diagram. J Food Eng 22:241–252
70. Quevedo M, Karbstein HP, Emin MA (2021) Influence of thermomechanical treatment and pH on the denaturation kinetics of highly concentrated whey protein isolate. J Food Eng 292:110294
71. Tolstoguzov VB (2008) Food polymers. In: Aguilera JM, Lillford PJ, Aguilera-Lillford (eds) Food materials science: principles and practice. Springer, New York, NY, pp 21–44
72. Ould Eleya MM, Leng XJ, Turgeon SL (2006) Shear effects on the rheology of β-lactoglobulin/β-carrageenan mixed gels. Food Hydrocolloids 20:946–951
73. Ould Eleya M (2000) The effects of pH on the rheology of β-lactoglobulin/κ-carrageenan mixed gels. Food Hydrocolloids 14:245–251
74. Clark AH (1996) Biopolymer gels. Curr Opin Colloid Interface Sci 1:712–717
75. Walstra P (2003) Physical chemistry of foods. Food science and technology, vol 121. Marcel Dekker, New York
76. Roefs SPFM, Kruif KG (1994) A Model for the denaturation and aggregation of beta-Lactoglobulin. Eur J Biochem 226:883–889
77. Kazmierski M, Corredig M (2003) Characterization of soluble aggregates from whey protein isolate. Food Hydrocolloids 17:685–692
78. Dalgleish DG, Senaratne V, Francois S (1997) Interactions between α-Lactalbumin and β-Lactoglobulin in the early stages of heat denaturation. J Agric Food Chem 45:3459–3464
79. Leeb E, Haller N, Kulozik U (2018) Effect of pH on the reaction mechanism of thermal denaturation and aggregation of bovine β-lactoglobulin. Int Dairy J 78:103–111
80. Damodaran S (2005) Protein stabilization of emulsions and foams. J Food Sci 70:R54–R66
81. Unterhaslberger G, Schmitt C, Shojaei-Rami S et al (2007) β-lactoglobulin aggregates from heating with charged cosolutes: formation, characterization and foaming. In: Dickinson E, Leser ME (eds) Food colloids: self-assembly and material science; [the proceedings of the Food Colloids 2006: self-assembly and material science conference held on 23–26. April 2006 in Montreux, Switzerland. Royal Society of Chemistry, Cambridge, pp 175–192
82. Manoi K, Rizvi SSH (2008) Rheological characterizations of texturized whey protein concentrate-based powders produced by reactive supercritical fluid extrusion. Food Res Int 41:786–796

83. Resch JJ, Daubert CR (2002) Rheological and physcochemical properties of derivatized whey protein concentrate powders. Int J Food Prop 5:419–434
84. Klingler RW, Meuser F, Niediek EA (1986) Einfluß der Art der Energieübertragung auf strukturelle und funktionelle Merkmale von Stärke. Starch/Stärke 38:40–44
85. Meuser F, Lengerich BV, Köhler F (1982) Einfluß der Extrusionsparameter auf funktionelle Eigenschaften von Weizenstärke. Starch-Stärke 34:366–372
86. van Lengerich B (1990) Influence of extrusion processing on in-line rheological behavior, structure, and function of wheat starch. In: Faridi H, Faubion JM (eds) Dough rheology and baked product texture, vol 32. Springer. US, Boston, MA, pp 421–471
87. Abeykoon C, Martin PJ, Kelly AL et al (2012) A review and evaluation of melt temperature sensors for polymer extrusion. Sens Actuat A 182:16–27
88. Alig I, Steinhoff B, Lellinger D (2010) Monitoring of polymer melt processing. Meas Sci Technol 21:62001
89. Barnes SE, Sibley MG, Edwards HGM et al (2007) Process monitoring of polymer melts using in-line spectroscopy. Trans Inst Meas Control 29:453–465
90. Emin MA, Schuchmann HP (2013) Droplet breakup and coalescence in a twin-screw extrusion processing of starch based matrix. J Food Eng 116:118–129
91. van der Goot AJ, Peighambardoust SH, Akkermans C et al (2008) Creating novel structures in food materials: the role of well-defined shear flow. Food Biophys 3:120–125
92. Liu W-C, Halley PJ, Gilbert RG (2010) Mechanism of degradation of starch, a highly branched polymer, during extrusion. Macromolecules 43:2855–2864
93. Alsteens B, Legat V, Avalosse T (2004) Parametric study of the mixing efficiency in a kneading block section of a twin-screw extruder. Int Polym Proc 19:207–217
94. Bravo VL, Hrymak AN, Wright JD (2000) Numerical simulation of pressure and velocity profiles in kneading elements of a co-rotating twin screw extruder. Polym Eng Sci 40:525–541
95. Cheng H, Manas-Zloczower I (1997) Study of mixing efficiency in kneading discs of co-rotating twin-screw extruders. Polym Eng Sci 37:1082–1090
96. Emin MA, Schuchmann HP (2013) Analysis of the dispersive mixing efficiency in a twin-screw extrusion processing of starch based matrix. J Food Eng 115:132–143
97. Ishikawa T, Kihara S-i, Funatsu K (2001) 3-D non-isothermal flow field analysis and mixing performance evaluation of kneading blocks in a co-rotating twin Srew extruder. Polym Eng Sci 41:840–849
98. Horvat M, Azad Emin M, Hochstein B et al (2013) A multiple-step slit die rheometer for rheological characterization of extruded starch melts. J Food Eng 116:398–403
99. Emin MA, Wittek P, Schwegler Y (2021) Numerical analysis of thermal and mechanical stress profile during the extrusion processing of plasticized starch by non-isothermal flow simulation. J Food Eng 294:110407
100. Wittek P, Ellwanger F, Karbstein HP et al (2021) Morphology development and flow characteristics during high moisture extrusion of a plant-based meat analogue. Foods 10
101. Byron Bird R, Carreau PJ (1968) A nonlinear viscoelastic model for polymer solutions and melts—I. Chem Eng Sci 23:427–434
102. Puaux JP, Bozga G, Ainser A (2000) Residence time distribution in a corotating twin-screw extruder. Chem Eng Sci 55:1641–1651
103. Yeh A-I, Hwang S-J, Guo J-J (1992) Effects of screw speed and feed rate on residence time distribution and axial mixing of wheat flour in a twin-screw extruder. J Food Eng 17:1–13
104. Wittek P, Zeiler N, Karbstein HP et al (2020) Analysis of the complex rheological properties of highly concentrated proteins with a closed cavity rheometer. Appl Rheol 30:64–76
105. van der Sman RGM, Chakraborty P, Hua NP et al (2023) Scaling relations in rheology of proteins present in meat analogs. Food Hydrocolloids 135:108195
106. Pietsch VL, Bühler JM, Karbstein HP et al (2019) High moisture extrusion of soy protein concentrate: influence of thermomechanical treatment on protein-protein interactions and rheological properties. J Food Eng 251:11–18
107. Mezger T (2020) The rheology handbook. Vincentz Network

108. Burhin HG, Rauschmann T, Graf H-J (2022) New and highly efficient method to measure steady shear viscosity and wall slip of rubber compounds: closed-boundary rheometer (RPA). Rubber Chem Technol 95:413–424
109. van den Einde RM, Akkermans C, van der Goot AJ et al (2004) Molecular breakdown of corn starch by thermal and mechanical effects. Carbohyd Polym 56:415–422

Innovative Pharmaceutical Stabilization and Formulation Techniques for Protein Drugs and Their Influence on Sequence and Structure—Part: Process

Katharina Dauer, Jan Kozak, Alf Lamprecht, and Karl Gerhard Wagner

Abbreviations

BSA	Bovine serum albumin
BSE	Backscattered electron
DIA	Dynamic image analysis
DSC	Differential Scanning Calorimetry
EDX	Energy-dispersive X-ray detector
FT-IR	Fourier transformation-infrared spectroscopy
HME	Hot-melt extrusion
HMWS	High molecular weight species
MRT	Mean residence time
M_w	Molecular weight
PEG	Polyethylene glycol
PEO	Polyethylene oxide
PLGA	Poly Lactic-co-Glycolic Acid
PVP	Polyvinyl-pyrrolidone
RTD	Residence time distribution
SEC	Size-exclusion chromatography

K. Dauer · J. Kozak · A. Lamprecht · K. G. Wagner (✉)
Pharmaceutical Institute, Department of Pharmaceutics, University of Bonn, Bonn, Germany
e-mail: karl.wagner@uni-bonn.de

K. Dauer
e-mail: katharina.dauer@uni-bonn.de

J. Kozak
e-mail: s6jakoza@uni-bonn.de

A. Lamprecht
e-mail: alf.lamprecht@uni-bonn.de

A. Kwade and I. Kampen (eds.), *Dispersity, Structure and Phase Changes of Proteins and Bio Agglomerates in Biotechnological Processes*,
https://doi.org/10.1007/978-3-031-63164-1_11

SEM	Scanning electron microscopy
SFD	Spray-freeze drying
SME	Specific mechanical energy
TSE	Twin-screw extrusion

1 Introduction

The biopharmaceutical sector is steadily increasing, focusing on the development of peptide and protein drugs into stable biopharmaceutical drug products for the medical treatment of especially severe diseases [1, 2]. However, the formulation and administration of pharmacologically relevant proteins and peptides has hitherto been limited to the parenteral administration of solutions or liquid dosage forms. This is often due to a lack of chemical and physical stability or a lack of tissue specificity of the proteins. Many biologics are not stable in liquid formulations, and/or require time- and resources-wasting formulation development studies to achieve stable protein formulations [3]. In addition, the formation of aggregates or subvisible particles during formulation development programs are considered as major challenge for the successful development of a biopharmaceutical drug product [4]. Apart from bioavailability, these aspects also need to be considered when evaluating alternative formulations and modes of administration. In particular, solids for peroral, nasal or parenteral administration would significantly expand or improve the therapeutic use and the compliance of protein-based drugs. Moreover, solid protein formulations offer the potential of increased stability [2, 3]. Thus, there is a growing interest in the development of alternative solid protein formulations (e.g., protein-loaded extrudates or implants) as well as encapsulation procedures that employ dehydrated protein powders due to improved protein stability in the solid-state [5]. In addition to stability, the evaluation of composition and process parameters on the microscale and their scalability in the laboratory and pilot scale is of crucial importance, as due to the cost of proteins conventional developments on a gram or kilogram scale are financially not feasible. Therefore, the aim is the establishment of scalable formulation technologies for proteins and peptides (hot-melt extrusion (HME) and spray-freeze drying (SFD)) for the production of solid dosage forms [5, 15, 6–9]. The stabilization of the proteins takes place by embedding in eroding or dissolving polymer matrices (i.e., polyethylene glycol (PEG), polyethylene oxides (PEO), and Poly Lactic-co-Glycolic Acid (PLGA)). Target is to enable a good and early starting point for the preparation of protein-loaded extrudates with sufficient protein integrity and prior protein formulation development of long-term release systems.

In addition, we investigated by the formulation of highly porous lyophilizate particles of the substances, whether the rapid freezing step of the spray-freeze drying can give decisive advantages for protein stability (Fig. 1).

Various peptides/proteins (for example lysozyme, bovine serum albumin, and insulin) are available as models, so that from low-dose to high-dose and from oral

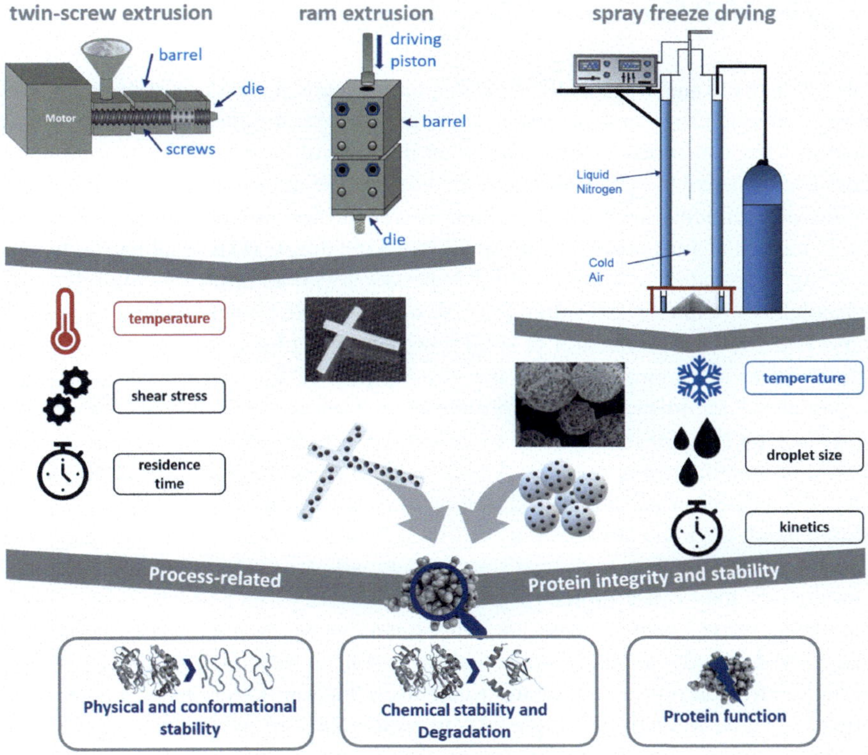

Fig. 1 Overview of the established scalable formulation technologies (twin-screw and ram extrusion and spray-freeze drying) for the production of solid dosage forms, process-related impact factors and investigation of process-related protein integrity and stability

targets to implants the entire range of formulation finding can be represented. Statements regarding stability, activity and in vitro release kinetics are considered on each scale. In view of protein analysis, a combination of methods is reported for a comprehensive protein analysis and provided information on the stability of the formulated proteins: RP-HPLC (chemical stability and protein concentration), SEC (protein fragment and aggregation analysis), SEM–EDX (protein particle distribution), DSC (unfolding temperature of the protein), FTIR spectroscopy (conformational stability), and activity assay (only for proteins with enzymatic activity, e.g., lysozyme) [6]. The second part of our studies especially focused on further analyses of lysozyme-loaded PEG-extrudates produced by ram extrusion. In the chapter "Innovative pharmaceutical stabilization and formulation processes for protein drugs and their impact on sequence and structure—Part: Analytics" lysozyme was thoroughly examined for its structural and functional properties before and after ram extrusion [7].

1.1 Hot-Melt Extrusion (HME)

Hot melt extrusion (HME) is used in the pharmaceutical sector as continuous and robust manufacturing tool for the production of solid dosage forms. Applying a HME process can result in the embedment of drugs at high loadings in polymeric matrices, is a solvent-free technique (e.g., without the use of organic solvents), and additional excipients such as surfactants or cryoprotectants are not required [10, 11]. In particular for poorly soluble small molecule drugs, HME enables the manufacturing of so-called amorphous solid dispersions with an improved solubility and bioavailability behavior of the drug [12]. Nowadays, there is an increasing awareness to use HME for the production of solid protein formulations, for instance as solid-state stabilization approach or for the development of long-term release systems for parenteral administration (e.g., protein-loaded implants). The main concern for solid-state protein formulations is to ensure protein stability and activity during the whole manufacturing process. Additionally, understanding the mechanisms of protein stabilization is critical to accelerate protein formulation development. On the one hand, the main stress factors during hot-melt extrusion processes (i.e., heat exposure in combination with shear stress) potentially affect the protein integrity and consequently may cause protein degradation or activity loss and has to be considered during the process chain. On the other hand, the use of polymers could help to ensure the thermodynamically stable and native folded structure of a protein during the melt extrusion process. In addition, the used polymer functions as a heat dissipation source between protein molecules to protect them from over-heating and thus can limit protein aggregation. The polymer can also act as energy absorber directly reducing shear stress from screw to delicate protein drugs [13, 14]. The evaluation of process parameters on the microscale and process scalability is of crucial importance, as due to the cost of protein candidates are not financially feasible. We used microscale HME for the production of highly-loaded protein extrudates, since this microscale approach is accompanied with processing small batch sizes (e.g., less than 2 g of a protein formulation), high yield, and short process times (e.g., shorter than 1–3 min for 2 g of a powdered protein formulation). In our study, protein-loaded extrudates were either produced in a ram extruder (lower shear stress) or a microscale 5 mm-TSE (higher shear stress) (Fig. 2), which enabled the control and monitoring of relevant process parameters. Protein integrity is critically affected not only by the extent of heating but also the residence time inside the barrel. The present work aimed to include all relevant extrusion process parameters to facilitate potential correlations of process parameters (e.g., screw configuration, screw speed, torque, residence time) with experimental data on polymer, and protein characteristics (e.g., unfolding temperature shifts, chemical and physical stability, biological protein activity).

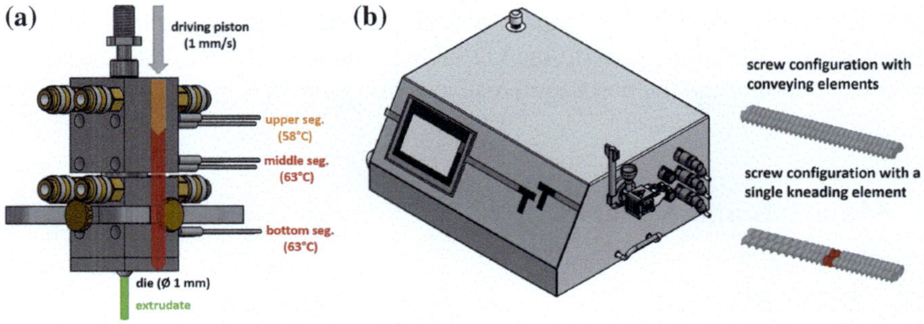

Fig. 2 Schematic drawings of **a** self-built ram extruder and **b** 5 mm twin-screw extruder (ZE5) for the production of protein-loaded extrudates; screw configurations of the 5 mm TSE with: conveying elements (5 mm pitch), a single kneading element (7.5, 5 mm pitch, 90° kneading element (dark colored), 5 mm pitch); Figure taken from [6], licensed under the Creative Commons Attribution 4.0 International license (CC BY 4.0)

1.2 Spray-Freeze Drying

Since the vast majority of biopharmaceuticals is more stable in the dry solid state, many of these products are marketed in the lyophilized form. The spray-freeze drying (SFD) is a process that combines the advantages of spray-drying (production of powder from solution) with freeze-drying (i.e., heat-gentle drying, formation of highly porous rapidly dissolving product). Compared with spray-drying, the SFD leads to porous and spherical particles with good flowability. In contrast to the bulk volume lyophilizates in vials that require re-dissolution in water before use (reconstitution), the SFD microspheres can be used as a delivery platform per se—e.g., as a free-flowing rapidly-dissolving powder for nasal administration. The multiparticulate nature enables flexible dosing. The combination of spraying of the solution to individual droplets, freezing in fall, and freeze-drying gives the SFD product—microspheres in an achievable size range between 50 and 500 μm—some exceptional characteristics. Feather-light for their high porosity and low solid content when the ice is removed by sublimation; yet freely flowing for their perfectly spherical shape, as they retain the original shape of the droplets. When in contact with water, dissolving rapidly and effortlessly, for their mutual attraction; an inherent characteristic feature all lyophilizates are known for and were named after. There are many variants of the SFD process between the different research groups; varying in all of the three involved steps: the spraying, the freezing as well as the drying. The spraying varies due to the different nozzles employed. This step is crucial especially for the particle size, size distribution and droplet agglomeration before freezing. We use a droplet generator, which provides a single stream of monodisperse droplets and hence narrow size distribution of the final SFD particles. The majority of the other research groups spray the droplets directly to the liquid nitrogen in order for the liquid to freeze rapidly. However, this leads to irregular non-spherical shapes of the particles as the droplets deform upon hitting the liquid nitrogen, to agglomeration and generally to broad particle size distribution. We use instead a pre-cooled tower-like apparatus. The fine individual droplets generated at the top of the apparatus adopt their typical spherical shape, freeze in this shape while falling, and reach the bottom as frozen microspheres. The subsequent drying via sublimation in the frozen state retains the spherical contours and the porous microspheres are formed.

2 Materials and Methods

2.1 Materials

Lysozyme from chicken egg-white, lyophilized powder was purchased from AppliChem GmbH (Darmstadt, Germany). BSA, sodium chloride, di-sodium hydrogen phosphate dihydrate, sodium dihydrogen phosphate dihydrate, acetonitrile, trifluoroacetic acid, and hexafluoroisopropanol (HPLC gradient grade) were purchased from Merck KGaA (Darmstadt, Germany). Ultrapure water was made with a Milli-Q® Lab Water Purification System (Millipore Corp., Billerica, MA, USA). Human insulin was kindly donated by Sanofi-Aventis Deutschland GmbH (Frankfurt, Germany) and was used for the production of extrudates. PEG 20,000 was obtained from Carl Roth GmbH & Co. KG (Karlsruhe, Germany). D(−)-Mannitol [Ph. Eur.] was purchased from VWR International (Amsterdam, The Netherlands). Liquid nitrogen was purchased from Linde (Pullach, Germany).

2.2 Methods

2.2.1 Preparation of Protein-Loaded Extrudates by Ram and Twin-Screw Extrusion

For the preparation of protein-loaded extrudates, a physical mixture (3 g) composed of 60% polymer and 40% protein powder (lysozyme, BSA, or human insulin) was blended for 10 min at 50 rpm using a turbula mixer (Willy A. Bachofen AG, Muttenz, Switzerland). The physical mixtures of polymer and protein powder were either ram extruded (Fig. 2a) or hot-melt extruded by using TSE (Fig. 2b).

We used a self-built ram extruder for the preparation of protein-loaded extrudates with lower shear stress (Fig. 2a). A barrel of 10 cm length and an inner cylindrical hole of 10 mm diameter consisted of three heating zones and was equipped with a cylindrical die (1 mm in diameter). Extrudates were prepared by feeding 3 g of the physical mixture into the preheated 10 mm-hole of the ram extruder. The temperature of the upper segment (filling-zone) was set to 58 °C, and the following two segments were set to 63 °C. The mixture was heated for three minutes and the molten blends were subsequently extruded through the 1 mm-die by a driving piston with a speed of 1 mm/s.

TSE was performed using a 5 mm co-rotating twin-screw extruder ZE 5 (Three-Tec GmbH, Seon, Switzerland, Fig. 2b) with a functional length of 15:1 L/D and two heating zones, equipped with a 1 mm die and either a conveying screw configuration, or a screw configuration with a single kneading element, which are presented in Fig. 2b. The physical mixtures (3 g) were fed into the extruder by using a powder belt conveyor (GUF-P Mini AD/475/75, mk Technology Group, Troisdorf, Germany) at constantly kept feeding rates of 0.4 g/min or 0.8 g/min ± 5% for lysozyme, human insulin, or BSA, respectively.

2.2.2 RP-HPLC and Size-Exclusion Chromatography (SEC)

Chemical stability and protein concentration of protein-loaded extrudates were determined by HPLC (Agilent 1100 series equipped with a diode array detector, Agilent Technologies, Waldbronn, Germany)) using a C18 reversed phase column (NUCLEOSIL 120–3, C18, 5 μm, 125 × 4 mm). Samples were dissolved in Milli-Q® water under mild agitation. 1.5 mL of the supernatant was used for analysis. The mobile phases consisted of A: acetonitrile and B: water both containing 0.1% trifluoroacetic acid (TFA). A linear gradient method was applied (0–5–7 min., 5–95–5% mobile phase B) at a flow rate of 1 mL/min for 9 min and a column temperature of 30 °C. Samples (10 μl) were injected, and chromatograms were quantified at 214 and 280 nm.

SEC was performed with a Superdex 75 Increase 10/300 GL column, 10 × 300 mm (Cytiva, Marlborough, MA, USA), where the mobile phase was 50 mM pH 7.0 phosphate buffer containing 400 mM sodium chloride at a flow rate of 0.8 mL/min and a column temperature of 30 °C. Samples were prepared as triplicates by dissolving the samples (i.e., pure protein (reference 1), the physical mixture (reference 2), or the protein-loaded extrudates) to 4 mg/mL protein concentration in the mobile phase. The UV detection detector wavelength was 214 nm. Samples vials were cooled at 5 °C in the autosampler. 10 μL of each sample was injected and potential aggregate and/or fragment formation after extrusion was analyzed in comparison with a chromatogram of the freshly prepared reference solutions.

2.2.3 Particle Size Distribution Analysis of Protein Powder

Protein powder were analyzed in triplicate in terms of particle size distribution and particle shape via dynamic image analysis (DIA) (Camsizer® X2, Retsch Technology, Haan, Germany). The continuous minimum diameter ($d_{xc\ min}$) of the analyzed particles was used for the particle size determination.

2.2.4 Scanning Electron Microscopy Coupled with Energy-Dispersive X-ray Spectroscopy (SEM–EDX) and Image Analysis of SEM–EDX-Maps

SEM images were taken using a scanning electron microscope (SU 3500, Hitachi High Technologies, Krefeld, Germany) coupled with a backscattered electron detector (BSE) and energy-dispersive X-ray detector (EDX) (EDAX Element-C2B, Ametek, Weiterstadt, Germany). The BSE detector was used to investigate the surfaces and cross sections of the prepared protein-loaded extrudates as well as the morphology of spray-freeze dried particles. The images were collected at an acceleration voltage of 5 kV at a variable pressure mode of 30 Pa. The cut extrudates were placed on an aluminum stub and sputtered with a thin platinum layer (Sputter Coater, Cressington Scientific Instruments, Watford, England). The elemental distributions were investigated by SEM–EDX. Protein particle distribution and parameterization of the distribution in the polymer matrix was examined by elemental mapping of the cross sections of extrudates for the characteristic X-ray peak of nitrogen. The percentage of detected nitrogen was evaluated by the TEAM software (Version 4.4.1, Ametek, Weiterstadt, Germany).

2.2.5 Solid-State Differential Scanning Calorimetry (Ss-DSC)

DSC studies of protein powder and protein-loaded extrudates were performed with a DSC 2 (Mettler Toledo, Gießen, Germany). ~10 mg of each sample (triplicates) were accurately weighed in 40 μL aluminum crucibles with a pierced lid. The crucibles were then heated at a rate of 10 K/min from 25 to 230 °C under a constant flow of nitrogen (30 mL/min). The calculation of the protein unfolding temperature was performed using the STAR^e version 8.01 software (Mettler Toledo, Gießen, Germany).

2.2.6 Solid-State Fourier Transform Infrared Spectroscopy (Ss-FT-IR)

All samples (protein powder, protein-loaded extrudates, and spray-freeze dried particles) were measured as triplicates using a Spectrum Two™ FT-IR spectrophotometer (PerkinElmer Inc., Waltham, USA). Spectra were recorded against an air background between 4000 and 400 cm^{-1} with an average of four scans with a resolution of 4 cm^{-1} (absorption mode). First and second derivative analysis of the generated spectra were considered in the amide-I region (1700–1600 cm^{-1}) of the protein and performed with GraphPad PRISM Software (San Diego, CA, USA).

2.2.7 Biological Lysozyme Activity Assay

The biological activity of lysozyme was determined using a EnzChek® Lysozyme Assay Kit (Molecular Probes Europe BV, Leiden, The Netherlands). The lysozyme activity is relying on its lytic activity against the cell wall of fluorescein-labeled *Micrococcus lysodeikticus* as a substrate. The resulted fluorescence increase depending on the remaining enzymatic activity of processed lysozyme was measured using a microplate reader with a fluorescein filter and OptiPlate™-96 F microwell plates (VICTOR3™ Multilabel Plate Reader; 96-Well plates black, PerkinElmer Life and Analytical Sciences, Shelton, Germany). Preparation of the lysozyme substrate stock suspension, lysozyme standard curves, as well as the general procedure were conducted according to the protocol of the manufacturer. The reaction mixtures were incubated at 37.0 ± 0.5 °C for 30 min, protected from light. The fluorescence intensity of each reaction in a microwell plate was measured at an absorption and emission maximum of 494 nm and 518 nm, respectively. The fluorescence values obtained from the control without lysozyme were subtracted.

2.2.8 Spray-Freeze Drying

The spray-freezing step was performed in a double jacket spray freeze drying tower developed earlier in our group [8]. Liquid nitrogen was used for cooling the tower from the surrounding outer layer to −100 °C (Fig. 3). The model protein (either lysozyme, BSA, or human insulin) together with the corresponding excipient(s) were dissolved in distilled water and the solution was filtered through 1.2 μm membrane filter. In the case of human insulin, the solution was acidified to pH 4 with 0.1 N hydrochloric acid. The lysozyme, BSA, and human insulin were spray-freeze dried in concentration of 5% (m/v) protein together with 5% (m/v) mannitol. The spray-step was performed using a monodisperse droplet generator (FMP, Erlangen, Germany) with a single pinhole orifice of a diameter of either 50 or 100 μm. The frozen particles collected in the glass container at the bottom of the spray-tower were transferred to a vacuum freeze dryer Alpha 1–4 (Martin Christ Gefriertrocknungsanlagen GmbH, Osterode am Harz, Germany) and freeze-dried at 0.05 mbar and condenser temperature of −50 °C.

Fig. 3 Schematic set-up for spray-freezing: the control panel is used to adjust the stimulation frequency of the drop jet nozzle, the feed solution is transported by pressurized air, the spray tower is cooled by liquid nitrogen, and particles are collected in a cooled beaker underneath the spray tower; Figure taken from [8], licensed under the Creative Commons Attribution 4.0 International license (CC BY 4.0)

3 Results and Discussions

3.1 Hot-Melt Extrusion

During melt extrusion mainly the stress factors heat exposure in combination with shear stress potentially affect the protein stability and consequently can lead to protein degradation or in case of enzymatic active proteins to their inactivation. We evaluated the protein stability and integrity after ram extrusion and TSE. The stability of lysozyme, BSA, and human insulin after extrusion at low temperatures (<70 °C) was evaluated by RP-HPLC (chemical stability and protein concentration), SEC (protein fragment and aggregation analysis), SEM–EDX (protein particle distribution over extrudate cross section), DSC (melting temperature of the protein), FT-IR spectroscopy (conformational stability), and in the case of lysozyme the biological activity (lysozyme activity assay).

3.2 General Characteristics of the Protein-Loaded Melt Extrudates

Microscale HME processing was used for embedding protein particles in a polymeric carrier. The particle size distribution analysis of the used protein powder particles for the preparation of highly-loaded protein-extrudates showed a narrow particle size distribution for lysozyme ($d_{50} = 133 \pm 7$ μm) and human insulin ($d_{50} = 142 \pm 9$ μm), whereas BSA ($d_{50} = 444 \pm 70$ μm) showed a very broad particle size distribution (Fig. 4). The degree of protein powder particle embedment in the polymeric matrix and protein particle distribution over the cross section of extrudates produced by ram extrusion or TSE was analyzed by SEM–EDX (Fig. 5). The surface of the extrudates prepared by TSE appeared slightly smoother compared to extrudates produced by ram extrusion. The cross-section cut of the extrudates produced by ram extrusion (Fig. 5) showed no pores or cracks inside of the extrudates, whereas the extrudates prepared by TSE showed only few pores. Elemental mapping of the cross-sections showed an overall homogeneous distribution of protein particles in extrudates prepared by ram extrusion and TSE. Furthermore, the protein particles were only dispersed or embedded in the PEG matrix and were not dissolved or notably grinded during the extrusion process. This observation was confirmed by particle size distribution analysis (DIA) (Fig. 4) and SEM-BSE images (Fig. 5) of the pure protein powder used for the production of protein-loaded extrudates. In the case of TSE with conveying screw configuration, BSA particles seemed to be slightly grinded. However, the used BSA powder showed a very broad particle size distribution and a d_{50} value with a high standard deviation (Fig. 4). That arouses the suspicion that the BSA-PEG powder mixture for the production of BSA-loaded extrudates by TSE with only conveying screw elements contained smaller BSA particles compared to the mixtures used for ram extrusion or TSE (kneading).

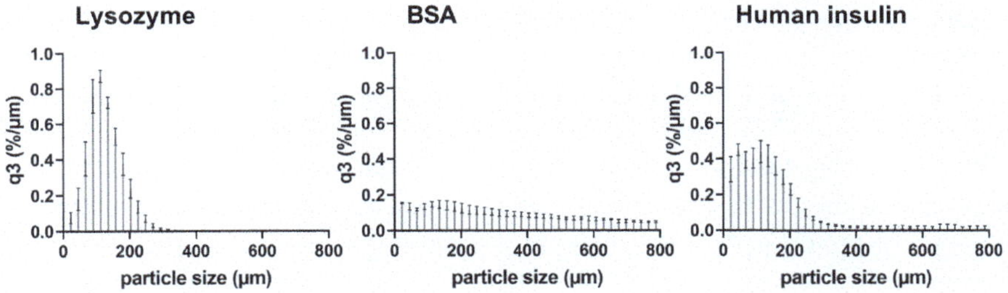

Fig. 4 Particle size distribution of protein powder (lysozyme, BSA, and human insulin); q3 is the cumulative distribution referred to the percentage (%) of particles below that micron size (μm); error bars represent the standard deviation of four measurements by DIA; Figure taken from [15], licensed under the Creative Commons Attribution 4.0 International license (CC BY 4.0)

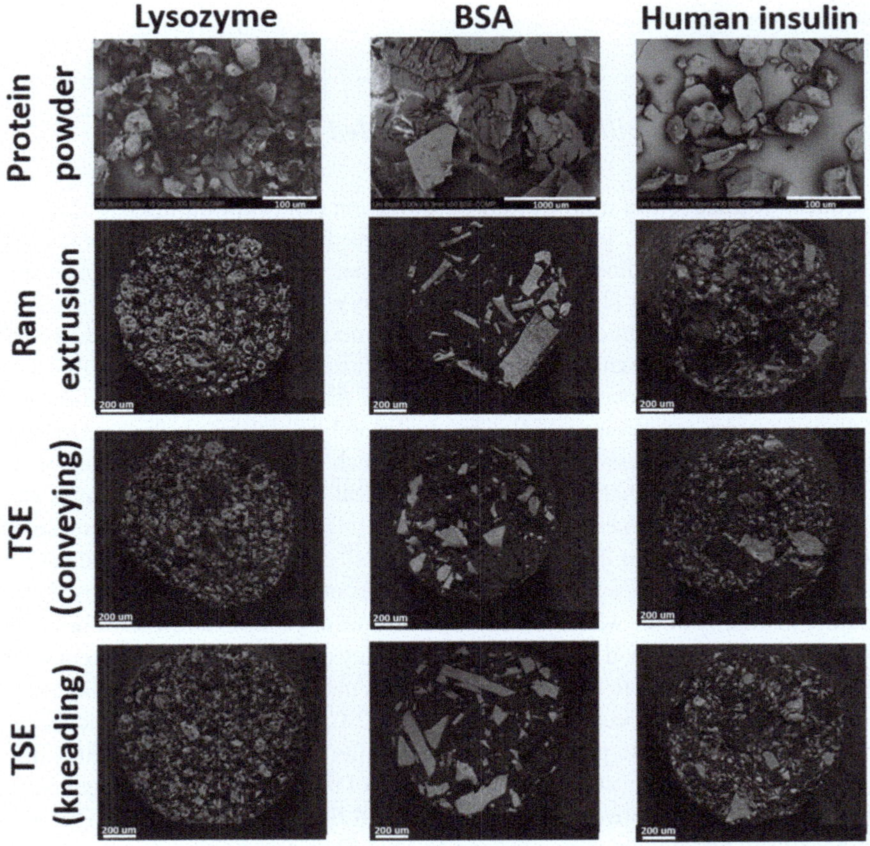

Fig. 5 Protein powder (SEM-BSE) and protein particle distribution (EDX-SEM) of lysozyme, BSA, human insulin; EDX-SEM images: elemental mapping of nitrogen (bright spots) on cross-section of 40% protein-loaded extrudates prepared by ram extrusion or TSE with only conveying screw configuration, or screws containing a single 90° kneading element, at 63 °C

Ram extrusion provided also homogenously distributed protein-loaded extrudates when small-sized protein particles (<200 μm) with a narrow particle size distribution will be used (i.e., lysozyme, human insulin). In contrast to TSE, ram extrusion offers the potential to produce highly-loaded solid protein formulations at high yield (i.e., >95%), high drug load (e.g., >40%) and at low shear stress (e.g., <100 s^{-1}). The results also highlight, that the use of conveying screws was satisfactory in terms of homogenous protein particle distribution. In comparison, the use of a single kneading element was not superior, since the protein particle distribution was not improved. In contrast to ram extrusion, TSE uses co-rotating screws to press the molten material through a cylindrical die. Once the protein-polymer powder mixture has passed through the feed system to the screws, it is conveyed by the turning motion of the screws along the extruder barrel, which introduces mixing and heat into the material through both external heaters and friction between the screws, barrel and within the material (viscous dissipation).

3.3 Chemical Stability of Protein-Loaded Melt Extrudates

Chemical degradation of a protein due to oxidation can occur during any stage of production, purification, formulation and storage. Especially proteins containing the amino acids: Histidine, Methionine, Cysteine, Tyrosine and Tryptophan are susceptible to oxidation [16]. The protein recovery rate and degradation behavior (i.e., appearance of chemically and structurally modified degradation products) of lysozyme, BSA, and human insulin after melt extrusion was monitored by RP-HPLC. The protein recovery rate was over 98% for all protein-loaded extrudates produced by ram extrusion or TSE. Chemical stability analysis showed no indication for extrusion process-related degradation such as oxidation or hydrolysis of the proteins. Protein denaturation triggered by heat exposure and shear stress can also lead to protein fragmentation or aggregation (e.g., formation of high molecular weight species (HMWS)). The protein can be partially or completely unfolded during protein denaturation, since more hydrophobic parts of the molecule are exposed and a higher flexibility of the whole protein molecule occurs. However, proteins in solid-state formulations can tend to aggregate without any detectable change in secondary structure and are therefore indistinguishable from native protein species using techniques such as FT-IR spectroscopy. Therefore, SEC was used for the detection of protein fragment formation and/or aggregation after ram extrusion and TSE. SEC analysis revealed no indication for the formation of protein fragments or aggregates and highlight that the native state of lysozyme and BSA embed in PEG 20,000 was not negatively affected by ram extrusion or TSE. However, human insulin showed the formation of <0.2% HMWS in combination with PEG 20,000. This effect was also observed for the unprocessed mixture of human insulin and PEG 20,000 and thus was independent of the extrusion process. The interaction of human insulin molecules with hydrophilic PEG 20,000 molecules is a probable explanation and may be due to a PEG-induced alteration of the local peptide structure [17]. A further analytical investigation by HPLC–MS and AUC revealed that the observed second species are soluble aggregates of human insulin molecules.

3.4 Physical Stability of Protein-Loaded Melt Extrudates

Besides chemical protein stability and biological activity, physical protein stability is also a crucial aspect for the successful development and formulation of protein pharmaceuticals. The process of unfolding and denaturation of a protein is mostly accompanied by a loss of the globular protein structure or the loss of secondary or tertiary structure elements that most proteins adopt. DSC was here used to measure the protein unfolding temperature in solid-state. The process of protein unfolding during the DSC analysis is characterized by the occurrence of a melting peak in the thermogram. The reported unfolding temperatures for solid proteins are predominately high and were analyzed as 204.7, 218.2, and 213.1 °C for native lysozyme, BSA, and human insulin, respectively and served as references (Fig. 6). A shift of the peak towards higher protein unfolding temperatures of a formulated protein compared to the protein reference (e.g., pure protein powder or physical mixture composed of the protein and polymer powder) reflects an increase in conformational protein stability. The physical mixtures composed of 40% protein powder and 60% PEG 20,000 served also as reference and showed slightly reduced unfolding temperatures for all the used proteins compared to the pure protein powder. This observation is may be due to the formation of an eutectic mixture and was further confirmed by DSC thermograms of physical mixtures composed of 20, 40, and 60% BSA. In general, the melting temperature of BSA was reduced in the presence of PEG 20,000 and depending on the percentage amount of polymer (80, 60, and 40% PEG 20,000, ranking order: 216.3, 217.3, and 218.6 °C, respectively). The unfolding temperature of processed proteins by melt extrusion was investigated since the most common stress that can cause a loss of protein structure is elevated temperature and designated as thermally induced protein denaturation. The thermal unfolding mechanisms of lysozyme, BSA, and human insulin include the appearance of a single melting peak. The proteins exhibit an irreversible unfolding mechanism as was seen in the second DSC scan, since the single peak of the first scan was completely disappeared [18]. The presence of a melting peak >200 °C in the DSC scan can indicate conservation of protein conformation after extrusion at 63 °C. Lysozyme showed a significantly reduced unfolding temperature after TSE with conveying screws (202.1 °C) and a screw configuration containing a single kneading element (201.8 °C) at 63 °C. The ram extrusion process for the production of lysozyme-loaded extrudates demonstrated no indication for a significantly reduced unfolding temperature (203.9 °C) of lysozyme. For BSA and human insulin the applied formulation processes significantly affected the unfolding temperature towards lower unfolding temperatures compared to the reference samples (i.e., pure protein powder and physical mixture). However, the total reduction of unfolding temperatures compared to the pure protein powder were <1.5 °C and <4.5 °C for BSA- and human insulin-loaded extrudates produced by TSE and a screw conformation with a kneading element, respectively. DSC was also used to evaluate extrusion processing and effects of the used extrusion process on the thermal stability of proteins. Ram extrusion resulted in the lowest, whereas TSE with a kneading screw element led to the largest reduction of unfolding temperatures of lysozyme, BSA, and human insulin.

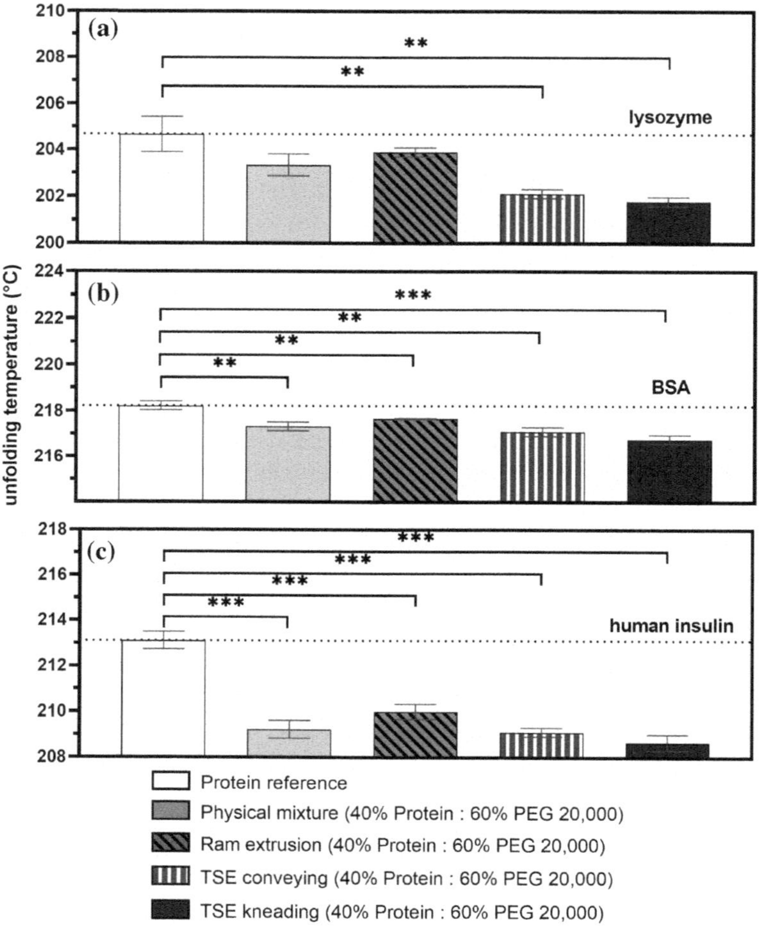

Fig. 6 Unfolding temperature of protein powders, physical mixtures, and 40% protein-loaded extrudates prepared by ram extrusion or TSE with only conveying screw configuration, or screws containing a single 90° kneading element, at 63 °C; **a** Lysozyme-loaded extrudates, **b** BSA-loaded extrudates, and **c** human insulin-loaded extrudates; the dotted line shows the unfolding temperature of the unprocessed protein (reference). Error bars represent the standard deviation of three measurements for the melting temperature by DSC; unpaired t-test (two-sample assuming equal variances) was used and statistical significance was depicted by asterisks (*): *$p < 0.05$, **$p < 0.01$, ***$p < 0.001$, ****$p < 0.0001$

FT-IR spectroscopy was used to detect changes in the molecular conformation and secondary structure elements of the processed proteins compared to the pure protein powder. The spectra and secondary structure information of powdered lysozyme, BSA, and human insulin and their physical mixtures were used as reference (Fig. 7). The model protein lysozyme is a small globular protein having a molecular weight of 14.4 kDa [19]. The secondary structure elements of lysozyme as obtained from the Protein Data Bank mainly include α-helices (34–42%) and β-sheets (7–12%) [19, 20]. The conformational stability of lysozyme in PEG-extrudates was analyzed by FT-IR spectroscopy and was not negatively affected by ram extrusion or TSE. The spectra of 40% lysozyme-loaded extrudates produced by ram extrusion or TSE were comparable to the native lysozyme powder used to prepare the physical mixture and thus the extrudates (Fig. 7). There was also no indication for denaturation (shifts or distortion of bands) or aggregation (intermolecular β-sheet formation) as a consequence of the heat exposure and shear stress during ram extrusion or TSE at 63 °C. Previous studies using CD and FT-IR spectroscopy have reported the secondary structure of BSA (M_w = 66.5 kDa): α-helical content: 52–68%, turns: 10%, but no β-sheets [21–23]. Some shifts occurred in the FT-IR spectra of BSA-loaded extrudates. A probable explanation for the observed differences especially at wavenumber of 1652 cm^{-1} may be the extrusion procedure, since melt extrusion (heat exposure and shear stress) can cause reversible changes in protein conformation and is associated with a reduction in the α-helical content and an increase in the random chain content [24, 25]. Human insulin was also studied for changes in its secondary structure as a result of heat exposure and shear stress during extrusion processing. Human insulin (M_w = 5.8 kDa) is a peptide and composed of two peptide chains (A- and B-chain) including three α-helices and a short β-sheet segment (Fig. 7) [26]. FT-IR spectra of human insulin-loaded extrudates prepared by TSE showed clear shifts and changes in secondary structure elements. The second derivative spectra of the Amide I band suggested a red-shift of the bands around 1650–1665 cm^{-1}, assigned to turns and β-sheet, which can be caused by small loosening of the turn structures. These changes indicated the transformation of α-helix content to β-sheets or a disordered structure which enhances the tendency of human insulin molecules to form aggregates. In particular, the band at 1652 cm^{-1} which corresponds to α-helical structures [24], was obviously affected by TSE with screws containing a kneading element due to higher shear stress and a longer residence time (heat exposure is increased). As can be seen in the FT-IR spectra, ram extrusion showed the lowest impact on secondary structure changes of human insulin (Fig. 7).

The biological lysozyme activity was investigated immediately after melt extrusion processing at 63 °C (i.e., initial) and after defined storage conditions (e.g., 5, 25, and 40 °C) of the lysozyme-loaded extrudates. Lysozyme-loaded extrudates were stored immediately after their production at 5, 25, and 40 °C for 3 months. Lysozyme powder was also stored at the same conditions and served as a reference. In general, proteins can rapidly denature and lose their specific function due to applied stresses during processing. Despite lysozyme is known as one of the most stable proteins, it is an accepted model for the investigation of the relationship between protein structure and enzymatic function. I.e., if a process results in a significant loss in activity of

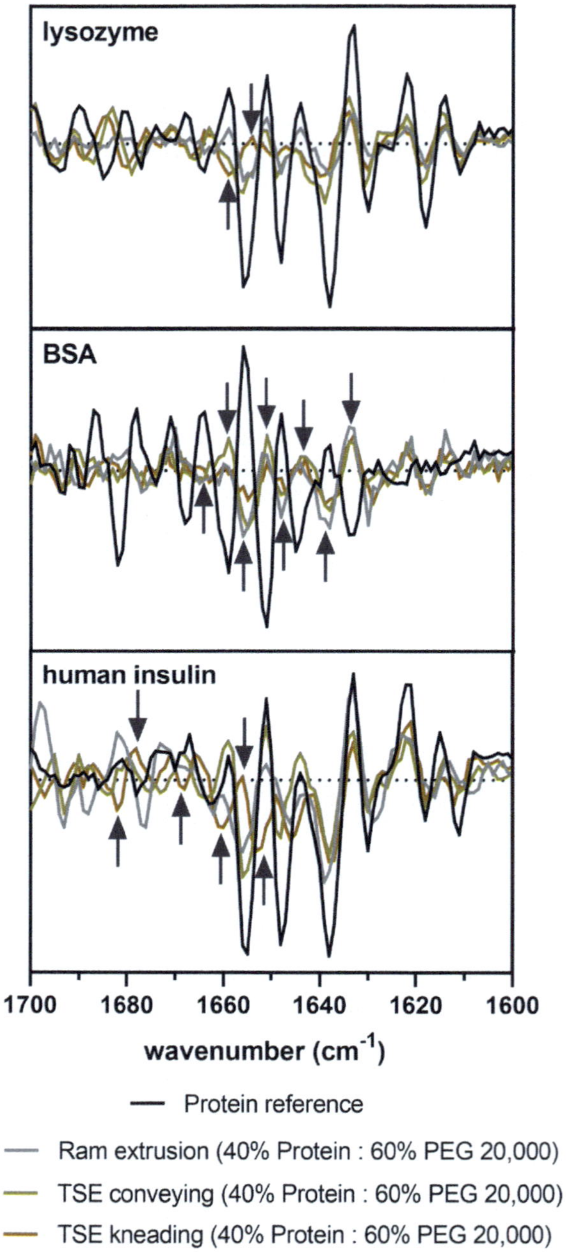

Fig. 7 Second derivative of FT-IR spectra in the Amide I region (1600–1700 cm^{-1}) of protein powder (black), and 40% protein-loaded extrudates (top: lysozyme, middle: BSA, and bottom: human insulin) produced by ram extrusion or TSE with only conveying screw configuration, or screws containing a single 90° kneading element, at 63 °C; gray arrows highlight changes in secondary structure elements

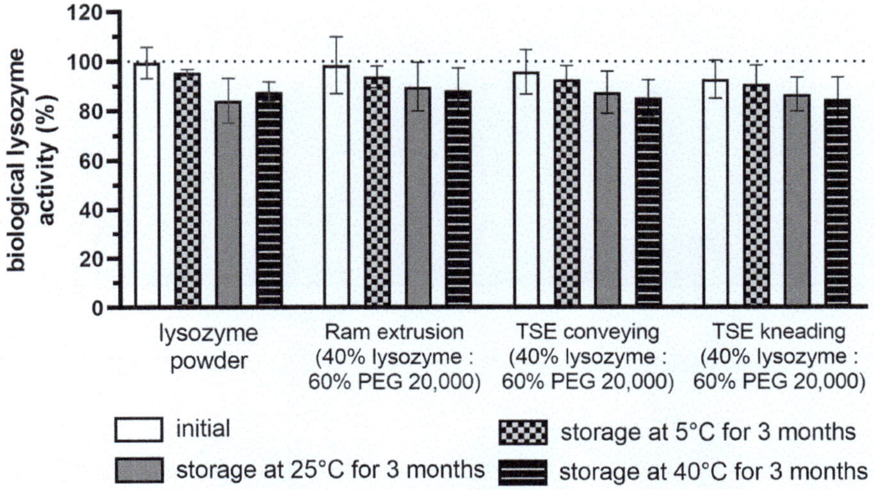

Fig. 8 Biological activity of lysozyme powder, and 40% lysozyme-loaded extrudates processed by ram extrusion or TSE with only conveying screw configuration, or screws containing a single 90° kneading element, at 63 °C; the bars represent the different storage conditions (5, 25, 40 °C) for 3 months; error bars represent the standard deviation of three measurements

lysozyme this process would disqualify itself for processing of other model proteins or relevant therapeutic proteins. Lysozyme embedded in melt extrudates by ram extrusion or TSE at 63 °C and after storage at 5, 25, and 40 °C for 3 months maintained full enzymatic activity when compared to the unprocessed lysozyme powder (t-test: $p < 0.05$) as expected (Fig. 8). The assay showed a satisfactory activity of lysozyme after ram extrusion and TSE and highlights the extreme stability of lysozyme against heat exposure and shear stress. This observation was confirmed by the results of other applied analytical methods (i.e., RP-HPLC, SEC, DSC, FT-IR spectroscopy).

3.5 *Spray-Freeze Drying*

3.5.1 General Characteristics of the SFD Microspheres

The mean particle size was in a range between 200 to 300 μm depending on the substance type and concentration, as the viscosity and surface tension are the major factors determining the droplet size and correspondingly the size of the final particles. The SFD particles had the typical nearly ideally spherical shape as well as the typical high porosity (Fig. 9). All SFD formulations containing the model proteins dissolved in water completely and very rapidly within several seconds. The rapid and complete dissolution is one of the most important quality attributes of a lyophilized product since it is critical for the reconstitution.

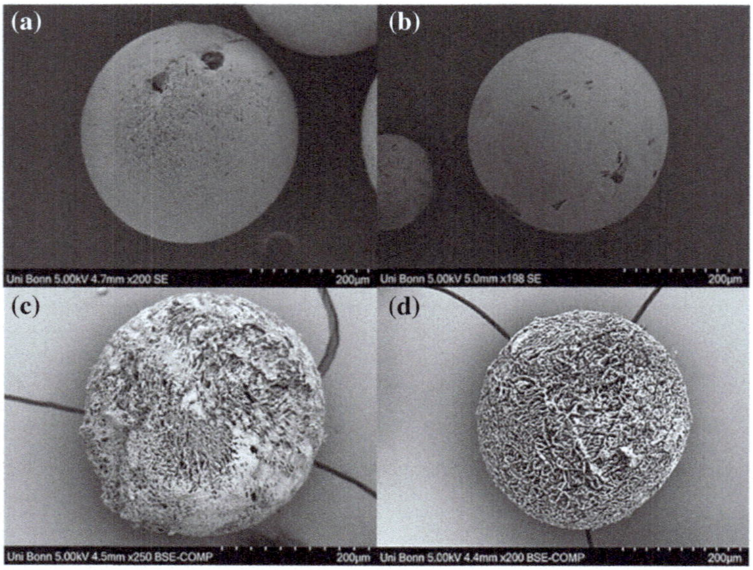

Fig. 9 SEM images of a single spray-freeze dried particle composed of either, **a** 50% BSA and 50% mannitol; **b** 50% lysozyme and 50% mannitol; **c** 50% human insulin and 50% mannitol; or **d** 50% human insulin and 50% PEG 20,000; the scale bar corresponds to 200 μm

3.5.2 Protein Stability Upon SFD

The model proteins BSA, lysozyme or human insulin formulated with mannitol or PEG were stable upon the SFD process. In all cases, more than 95% of intact protein was found by SEC compared with the original solution before SFD. The comparable peak areas as well as the unchanged retention times indicated no formation of agglomerates as well as preservation of the secondary structure of the model proteins upon SFD.

3.5.3 Nasal Delivery of Macromolecules

We have demonstrated in a recently published study that the SFD microspheres are suitable for nasal delivery of macromolecules [9]. Using different formulations of SFD microspheres with FITC-labelled dextrans of different molecular weights (4, 10, 20, 40, and 70 kDa) administered nasally to rats, we have shown that the bioavailability of macromolecules formulated in form of an SFD product across the nasal mucosa is molecular-weight dependent [9].

In a further experiment, a significant drop in the blood glucose level was detected upon nasal administration of human insulin in form of SFD microspheres in formulations containing penetration enhancers (sodium taurocholate or beta-cyclodextrin);

but no blood glucose level decrease was observed in the case of SFD insulin formulation without permeation enhancers [9]. These results confirm that the SFD formulations offer a suitable platform for nasal delivery of proteins, while retaining the enhanced storage stability at the dry solid state in contrast to the nasal liquid sprays.

4 Conclusions

The growing interest in solid-state protein formulations and protein stabilization by melt extrusion processing or spray-freeze drying and thus preservation of the native protein structure is accompanied with the need of appropriate methods to characterize the process-related protein stability. In this work, we addressed the investigation either of highly-loaded protein extrudates produced by ram or twin-screw extrusion and protein microspheres prepared by spray-freeze drying with regard to the main challenges, i.e., protein instability due to heat exposure and shear stress during extrusion, or cold exposure and freeze stress, respectively. The applied methods, including (i) RP-HPLC (chemical stability and protein concentration), (ii) SEC (protein fragment and aggregation analysis), (iii) SEM–EDX (protein particle distribution), (iv) DSC (protein's unfolding temperature), (v) FT-IR spectroscopy (conformational stability), and (vi) activity assay, enabled the successful characterization of protein stability in extrudates or microspheres (mg-scale) and proved to be very useful techniques to study process-related effects on protein stability and changes in model proteins induced by processing. PEG 20,000 was used as a simple polymer model as polymeric matrix for melt extrudates due to its low processing temperature (i.e., < 70 °C) and hydrophilicity (i.e., simplifying the sample preparation and analytical investigations of protein-loaded extrudates). We showed the successful and homogenous embedment of proteins in PEG 20,000 matrices without dissolution of the protein particles in the polymer for solid-state stabilization. Nearly complete recovery of active lysozyme highlighted that ram extrusion and TSE did not negatively affect the protein integrity. However, differences shown between lysozyme- and BSA- or human insulin-loaded extrudates indicated that melt extrusion processing could have an impact on the conformational protein stability. In particular, BSA and human insulin seemed to be more susceptible to shear stress compared to lysozyme, but ram extrusion led to less conformational changes compared to TSE due to lower shear stress. The use of conveying screws in TSE was satisfactory in terms of homogenous protein particle distribution and the use of a single kneading element did not improve the distribution. Ram extrusion showed also a good protein particle distribution and is especially for small-sized protein particles (<200 μm) the preferred method to prepare highly-loaded solid protein formulations. This work offered also the potential to compare different melt extrusion processes (i.e., ram extrusion vs. TSE (conveying) vs. TSE (kneading)) and extrusion parameters (e.g., shear stress levels, screw configurations, residence times). The spray-freeze drying proved to be suitable for protein formulation. It yields a freely flowable powder that can be used either as a lyophilized intermediate for reconstitution or as a solid dosage form per se, e.g., for nasal administration. We propose the implementation of a characterization pathway of protein stability and investigation of the impact of process-related stress on protein stability in highly-loaded solid protein formulations from small-scale melt extrusion or protein microspheres obtained from spray-freeze drying. Our work highlighted further that melt extrusion processing offers the potential for the production of sustained release protein formulations and the applied workflow provides a fundamental basis for future work, including other formulation techniques (e.g., vacuum compression molding) [27], as well as the use of sustained release polymers or other formulation excipients and stabilizing agents (e.g., pH-modifier, surfactants, plasticizer).

References

1. Kaplon H, Muralidharan M, Schneider Z, Reichert JM (2020) Antibodies to watch in 2020. mAbs 12(1):1703531. https://doi.org/10.1080/19420862.2019.1703531.
2. Lau JL, Dunn MK (2018) Therapeutic peptides: historical perspectives, current development trends, and future directions. Bioorg Med Chem 26(10):2700–2707. https://doi.org/10.1016/j.bmc.2017.06.052.
3. Mahler H-C, Friess W, Grauschopf U, Kiese S (2009) Protein aggregation: pathways, induction factors and analysis. J Pharm Sci 98(9):2909–2934. https://doi.org/10.1002/jps.21566
4. Dauer K, Kamm W, Wagner KG, Pfeiffer-Marek S (2021) High-throughput screening for colloidal stability of peptide formulations using dynamic and static light scattering. Mol Pharm 17
5. Cossé A, König C, Lamprecht A, Wagner KG (2017) Hot melt extrusion for sustained protein release: matrix erosion and in vitro release of PLGA-based implants. AAPS PharmSciTech 18(1):15–26. https://doi.org/10.1208/s12249-016-0548-5
6. Dauer K, Werner C, Lindenblatt D, Wagner KG (2022) Impact of process stress on protein stability in highly-loaded solid protein/PEG formulations from small-scale melt extrusion. Int J Pharm X 100154. https://doi.org/10.1016/j.ijpx.2022.100154.
7. Elsayed YY, Kühl T, Dauer K, Sayin A, Wagner KG, Imhof D (2022) Bioanalytical workflow for qualitative and quantitative assessment of hot-melt extruded lysozyme formulations. ACS Omega 7(45):40836–40843. https://doi.org/10.1021/acsomega.2c03559
8. Eggerstedt SN, Dietzel M, Sommerfeld M, Süverkrüp R, Lamprecht A (2012) Protein spheres prepared by drop jet freeze drying. Int J Pharm 438(1–2):160–166. https://doi.org/10.1016/j.ijpharm.2012.08.035
9. Serim TM, Kožák J, Rautenberg A, Özdemir AN, Pellequer Y, Lamprecht A (2021) Spray freeze dried lyospheres® for nasal administration of insulin. Pharmaceutics 13(6):852. https://doi.org/10.3390/pharmaceutics13060852
10. Gajda M, Nartowski KP, Pluta J, Karolewicz B (2018) The role of the polymer matrix in solvent-free hot melt extrusion continuous process for mechanochemical synthesis of pharmaceutical cocrystal. Eur J Pharm Biopharm 131:48–59. https://doi.org/10.1016/j.ejpb.2018.07.002
11. Madan S, Madan S (2012) Hot melt extrusion and its pharmaceutical applications. Asian J Pharm Sci 7(2):123–133
12. Patil H, Tiwari RV, Repka MA (2016) Hot-melt extrusion: from theory to application in pharmaceutical formulation. AAPS PharmSciTech 17(1):20–42. https://doi.org/10.1208/s12249-015-0360-7
13. Zheng Y, Pokorski JK (2021) Hot melt extrusion: an emerging manufacturing method for slow and sustained protein delivery. WIREs Nanomedicine Nanobiotechnology 13(5). https://doi.org/10.1002/wnan.1712

14. Zhang H, Li X, Tang W, Deng X, Reynolds AP, Sutton MA (2015) Heat transfer modeling of the friction extrusion process. J Mater Process Technol 221:21–30. https://doi.org/10.1016/j.jmatprotec.2015.01.032
15. Dauer K, Kayser K, Ellwanger F, Overbeck A, Kwade A, Karbstein HP, Wagner KG (2023) Highly protein-loaded melt extrudates produced by small-scale ram and twin-screw extrusion - evaluation of extrusion process design on protein stability by experimental and numerical approaches. Int J Pharm X 6:100196. https://doi.org/10.1016/j.ijpx.2023.100196
16. Manning MC, Chou DK, Murphy BM, Payne RW, Katayama DS (2010) Stability of protein pharmaceuticals: an update. Pharm Res 27(4):544–575. https://doi.org/10.1007/s11095-009-0045-6
17. Chao S-H, Matthews SS, Paxman R, Aksimentiev A, Gruebele M, Price JL (2014) Two structural scenarios for protein stabilization by PEG. J Phys Chem B 118(28):8388–8395. https://doi.org/10.1021/jp502234s
18. Mohammad MA (2016) Effect of mechanical denaturation on surface free energy of protein powders 8
19. Sheng L, Wang J, Huang M, Xu Q, Ma M (2016) The changes of secondary structures and properties of lysozyme along with the egg storage. Int J Biol Macromol 92:600–606. https://doi.org/10.1016/j.ijbiomac.2016.07.068
20. White FH Studies on secondary structure in chicken egg-white lysozyme after reductive cleavage of disulfide bonds 7
21. Chen Y, Ling J, Li M, Su Y, Arte KS, Mutukuri TT, Taylor LS, Munson EJ, Topp EM, Zhou QT (2021) Understanding the impact of protein–excipient interactions on physical stability of spray-dried protein solids. Mol. Pharm. 18(7):2657–2668. https://doi.org/10.1021/acs.molpharmaceut.1c00189
22. Servagent-Noinville S, Revault M, Quiquampoix H, Baron M-H (2000) Conformational changes of bovine serum albumin induced by adsorption on different clay surfaces: FTIR analysis. J Colloid Interface Sci 221(2):273–283. https://doi.org/10.1006/jcis.1999.6576
23. Lu R, Li W-W, Katzir A, Raichlin Y, Yu H-Q, Mizaikoff B (2015) Probing the secondary structure of bovine serum albumin during heat-induced denaturation using mid-infrared fiberoptic sensors. Analyst 140(3):765–770. https://doi.org/10.1039/C4AN01495B
24. Bolje A, Gobec S (2021) Analytical techniques for structural characterization of proteins in solid pharmaceutical forms: an overview. Pharmaceutics 13(4):534. https://doi.org/10.3390/pharmaceutics13040534
25. Bagińska K, Makowska J, Wiczk W, Kasprzykowski F, ChmurzyńSKI L (2008) Conformational studies of alanine-rich peptide using CD and FTIR spectroscopy. J Pept Sci 14(3):283–289. https://doi.org/10.1002/psc.923
26. Ciszak E, Beals JM, Frank BH, Baker JC, Carter ND, Smith GD (1995) Role of C-terminal B-chain residues in insulin assembly: the structure of hexameric LysB28ProB29-human insulin. Structure 3(6):615–622. https://doi.org/10.1016/S0969-2126(01)00195-2
27. Dauer K, Wagner KG (2023) Micro-scale vacuum compression molding as a predictive screening tool of protein integrity for potential hot-melt extrusion processes. Pharmaceutics 15(3):723. https://doi.org/10.3390/pharmaceutics15030723

Proteins Under Stress—Benefits of Simultaneous Homogenization and Mixing During the Cell Disruption Process

E. Rütten, V. Gall, and H. P. Karbstein

1 Introduction

Biotechnologically extracted enzymes or protein isolates from plant sources are often obtained via cell disruption [1–6]. High-pressure homogenizers (HPH), among others, are technically used for this purpose. The cell-disrupting stresses are generated in the flow by an extremely strong constriction of the cross-sectional area flowed through [7]. As a result, high cell disruption rates are obtained [5]. A side effect is often cavitation [7], which damages the disruption unit and contaminates the product by abrasion [8]. The literature frequently reports also problems to product quality. These include, in particular, a loss of enzyme activity and the denaturation and aggregation of proteins [9]. Reasons given are shear-induced damage to the three-dimensional protein structure in the cell disruption process itself and/or downstream. This damage occurs particularly when gas/liquid and/or solid/liquid interfaces are simultaneously present in the process [1, 10, 11]. This results from the intentional or unintentional inflow of gas bubbles or from the inflow of solid nanoparticles via abrasion [10, 11]. Both are the result of cavitating flows.

The HPH process is based on the fact that the medium is compressed to pressures of several hundred to more than a thousand bar and is then decompressed by a disruption unit [12]. Radial-flow units, so-called flat valves, are mainly used in large-scale processes such as the homogenization of milk or fruit puree. Axial-flow,

E. Rütten · V. Gall · H. P. Karbstein (✉)
Institute of Process Engineering in Life Sciences, Food Process Engineering, Karlsruhe Institute of Technology KIT, Karlsruhe, Germany
e-mail: heike.karbstein@kit.edu

E. Rütten
e-mail: eva.ruetten@kit.edu

V. Gall
e-mail: vanessa.gall@partner.kit.edu

A. Kwade and I. Kampen (eds.), *Dispersity, Structure and Phase Changes of Proteins and Bio Agglomerates in Biotechnological Processes*,
https://doi.org/10.1007/978-3-031-63164-1_12

orifice based disruption units, are more commonly found today in the production of special products with lower capacity but higher requirements, especially for the pharmaceutical industry [5, 13].

The central common feature of each disruption unit is a significant reduction of the cross-section of the flow over a very short distance, which leads to a strong increase in the flow velocity u in and downstream of the orifice. A jet of liquid forms in the orifice outlet area, which shows instabilities in the edge areas shortly after leaving the narrowest gap, starting from the boundary areas (free jet). As it continues, the jet fans out and attaches itself to the wall [14]. Radially around the exiting jet, pronounced circulation areas occur, especially in the case of sharp-edged orifices. The flow in the smallest gap can be characterized by means of the Reynolds number $Re_{orifice}$ [15]:

$$Re_{orifice} = \frac{\rho_l \cdot d_{orifice} \cdot \overline{u}_{orifice}}{\eta_l} \tag{1}$$

where ρ_l and η_l describe the density and the dynamic viscosity of the liquid phase. The diameter of the orifice is named $d_{orifice}$ and $\overline{u}_{orifice}$ describes the mean velocity inside the narrowest gap. It can be calculated as follows taking the mass flow of the high-pressure stream $\dot{M}_{HS}$ into account:

$$\overline{u}_{orifice} = \frac{\dot{M}_{HS}}{\pi \cdot \frac{d^2_{orifice}}{4} \cdot \rho_l} \tag{2}$$

At low Reynolds numbers in the narrowest cross-section, the flow downstream the orifice is also laminar and thus stationary. As the Reynolds number increases, the above-mentioned instabilities occur in the flow (transitional region) up to the formation of a turbulent flow, which is characterized by three-dimensional vortices and unsteady behavior. Previous works, summarized in [16], also show that molecules, droplets, or particles in the medium passing through such an orifice undergo a history of different stresses: Strain stresses dominate when entering the disruption unit, in the narrowest cross-section shear stresses, and in the outlet area, depending on the flow condition shear and Reynolds stresses are found. When simulating the flow in HPH valves, the turbulent kinetic energy k is often used as a measure of the Reynolds stresses [17].

Due to the strong flow constriction, cavitation may also occur in the disruption unit of HPH [18, 19], as the flow is strongly accelerated in the narrowest cross-section and the pressure decreases accordingly. The tendency of a flow to form hydrodynamic cavitation is described by the cavitation number σ. In previous works it could be shown that cavitation in orifices occurs when σ, defined according to [20, 21], falls below a value < 1:

$$\sigma = \frac{p_{out} - p_v(T)}{0.5 \cdot \rho_l \cdot \overline{u}^2_{orifice}} \tag{3}$$

with the vapor pressure p_v of the flowing liquid at process conditions. At constant vapor pressure, the cavitation number describes the ratio of static pressure downstream the orifice p_{out}, which tends to prevent the cavitation, and the hydrodynamic pressure, which promotes cavitation. The cavitation bubbles formed at the *vena contracta* are carried along with the flow far into the outflow space [22, 23]. At high pressure drops (> 400 bar), the formation of a standing cavitation cloud far beyond the narrowest flow cross-section has been observed for water. This creates a large gas–liquid interface for a short period of time, on which surface-active substances can adsorb [22–25]. When the gas bubbles implode, the gas–liquid interfaces disappear, any adsorbed molecules desorb and go back into solution. At the same time, very high local accelerations, so-called jet streams, occur with significant temperature jumps and very high stresses.

The occurrence of cavitation in orifices as disruption unit could also be determined via sono-chemoluminescence [26] and acoustic recordings [22]. Besides the inlet pressure, the pressure in the outflow chamber (also called backpressure) and the diameter ratio β between the orifice channel $d_{orifice}$ and outflow chamber d_{out} influence the occurrence of cavitation:

$$\beta = \frac{d_{orifice}}{d_{out}} \tag{4}$$

McKillop et al. [27] and Gothsch et al. [28]were able to show for convergent-divergent nozzles (Laval nozzles) and rectangular step-jump orifices, respectively, that cavitation can be completely suppressed at sufficiently high backpressure. The ratio of the pressures downstream (p_{out}) and upstream (p_{in}) of the narrowest cross-section is described as the Thoma number *Th* [29]:

$$Th = \frac{p_{out}}{p_{in}} \tag{5}$$

Schlender [22] was able to show that no cavitation occurs in orifices from $Th > 0.5$. He also classified the appearance of cavitation downstream the orifice as cavitation jet, choked cavitation, and hydraulic flip [23]. Jet cavitation refers to the state, when single cavitation bubbles exit the orifice and form a cavitation jet. When the first coherent cavitation bubble exits the gap, the cavitation pattern is named choked cavitation. By decreasing *Th* or increasing β, the choked cavitation can merge into a hydraulic flip, which refers to the state, when almost the whole outlet channel is filled with one coherent vapor bubble.

Cavitation causes, among other things, the strong abrasion of material from the disruption unit, especially in the area of the inlet and outlet of the narrowest cross-section [8]. The resulting very fine particles are carried into the liquid and form new solid–liquid interfaces.

In high-pressure homogenization of microorganisms and plant cells, cell disruption rates depend on local flow conditions and temperatures [4, 8, 30, 31]. In literature it was found that the increasing loss of enzyme activity correlates with increasing applied pressure difference [31]. This observation was attributed to higher shear rates. A multiple passage at the same pressure did not show the same impact on enzyme activity as an increase in homogenizing pressure at single passage [13]. The hypothesis that HPH exposes hydrophobic regions of proteins and enzymes and that this alters their functionality and activity has been published, for example, in [32–34].

Indications of the involvement of gas–liquid interfaces in turbulent flows have been published, for example, in [35, 36]. The authors presented the hypothesis that denaturation at the interface was the actual cause of the shear instability of proteins in solution. Hamilton-Brown et al. [37] also found that aggregation of solubilized proteins increased with stirring intensity and attributed this to an increase in the g/l interface and the fraction of protein adsorbed at that interface. The initial destabilization due to unfolding of proteins and enzymes at hydrophobic g/l and s/l interfaces is attributed to the interfacial force by [38–41]. Once unfolded, enzymes can aggregate [42–45]. Aggregation rates in flows are higher than in resting media [46]. All authors of the studies mentioned so far assume that unfolding and thus the hydrophobic groups at interfaces is the initial stage for the subsequent aggregation of the molecules.

A requirement for collisions is a relative motion between particles. This can be achieved by thermal diffusion/molecular motion or directed motion, e.g. in flows [47]. Aggregates of reactive (unfolded) enzyme or protein monomers are formed via hydrophobic and electrostatic interactions and/or newly formed covalent bonds (disulfide bonds or peptide bonds) and are therefore very stable. Further collisions lead to the formation of increasingly larger aggregates. For β-lactoglobulin, it has been shown that this can lead to the formation of insoluble aggregates [48–50].

A promising process with regard to the behavior of proteins and enzymes within the HPH process is the Simultaneous homogenizing and mixing (SHM). It represents a further development of conventional high-pressure homogenizing. As in traditional high-pressure homogenizing, a stream, referred to as the high-pressure stream (HS), is compressed to high pressures and accelerated throughout the disruption unit. The disruption unit, however, is modified in such a way that a mixing stream (MS) is fed into the outlet chamber downstream of the narrowest cross-section [51]. By varying the pressure and volume flow in the high-pressure and mixing streams, and by geometric parameters of the disruption unit, the local flow and mixing conditions in the orifice and in the outflow region of the high-pressure orifice can be changed [52].

This specific SHM design depicts many features, that could be of advantage for keeping enzyme and protein functionality in cell disruption via HPH: The direct dilution of the high-pressure stream will decrease the collision rate of unfolded molecules. By tempering the mixing stream, the product can additionally be cooled instantaneously, as already shown for another application by [52, 53]. The occurrence of cavitation in the SHM process, however, has not yet been investigated. Based on the preliminary investigations on the flow velocities, it must be assumed that the pressure distribution in the outlet chamber is significantly influenced by the mixing flow and that the findings from the HPH process cannot be completely transferred to the SHM process.

Overall, the SHM method offers several starting points to significantly improve the process of cell disruption by preserving enzyme activity and protein structure.

It is assumed that, although the stresses or shear gradients are locally very high, there is no significant flow-induced unfolding of the proteins or enzymes in the HPH process, since critical stress times are undercut. On the contrary, it is assumed that the unfolding rate increases significantly with cavitation-induced occurrence of interfaces. Based on these assumptions and the above mentioned possibilities, the following research hypotheses were defined:

- The process can be designed in such a way that the stresses in the flow required for cell breakup are not affected negatively.
- The occurring cavitation can be controlled and suppressed by process and geometry parameters, so that the SHM process can be operated cavitation-free to prevent the occurrence of g/l and s/l interfaces.
- The mixing quality in the process is sufficient to ensure rapid mixing after the orifice, thus limiting the collision rate.
- It is possible to keep the thermal stress very short in time by tempering the mixing stream and to decrease the liquid temperature already in the mixing chamber to a required level, which prevents thermal denaturation.

These hypotheses are discussed below. The methods used are summarized in the following chapter. Subsequently, the above research questions are discussed with the help of results and literature. Finally, a recommendation for improving the process based on the lessons learned is proposed.

2 Material and Methods

2.1 Plant Setup

For all experiments, the same plant setup was used. The setup is presented in Fig. 1. The liquid, which is supplied as high-pressure stream, is provided in a storage tank (a). It is fed by a two-piston pump (b). A pulsation damper (c) serves to compensate for the pulsations of the pump. The mixing stream is stored in a pressure vessel (e), which can be pressurized via a nitrogen gas cylinder (f). A filter (g) holds back contaminants. The flow rate of the high-pressure stream is controlled by the pressure of the pump, while the flow rate of the mixing stream is controlled by a needle valve (h). Both streams are mixed in the disruption unit (d). Downstream of the disruption unit, Setup A or B were used. Setup A is used to apply a back pressure within a second homogenizing step. It includes a needle valve (i) to adjust the back pressure and a collecting tank (j). Setup B is used for trials without applied back pressure, for which the needle valve is removed. The pressure and the mass flow are recorded by the use of pressure gauges and a scale, respectively.

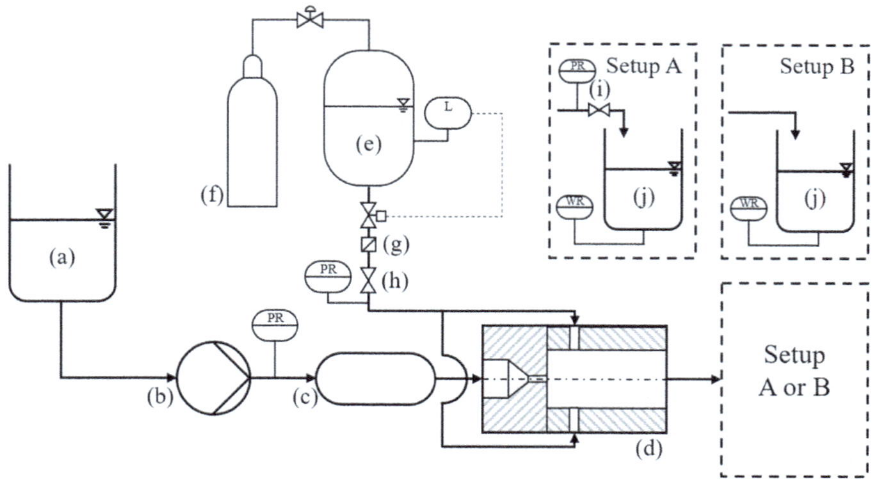

Fig. 1 Plant setup for experimental trails. Setup A is used to apply a back pressure within a second homogenizing step, Setup B is used for trials without applied back pressure

2.2 *Geometry of Disruption Unit*

The disruption unit is of modular design. This allows individual geometry parameters and materials to be adapted according to the experiments. In general, the orifice itself is drilled into a disc of stainless steel. A sketch of the orifice is shown in Fig. 2. In this study, the geometry of the orifice was kept constant. Geometrical parameters can be taken from Table 1.

The geometry of the mixing chamber may be either T- or circular shaped T-shaped or circular. The chamber was made of PMMA to allow optical measurements. As

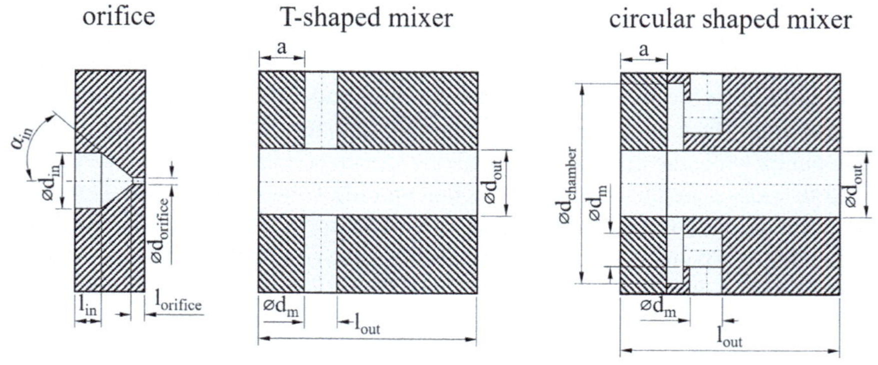

Fig. 2 Schematic drawing of the disruption unit consisting of the orifice and a mixer, which can be T-shaped or circular shaped in design

Table 1 Geometrical parameters of disruption unit

Parameter	Value
d_{in}	1.5 mm
$d_{orifice}$	0.15 mm
$l_{orifice}$	0.4 mm
α_{in}	59°
d_{out}	2 mm, 4 mm
l_{out}	40 mm
d_m	1 mm
a	0 mm, 4.5 mm, 9 mm
$d_{chamber}$	10 mm
b	0.5 mm

geometry parameters, the distance a and the outlet diameter d_{out}, respectively the diameter ratio β, were varied.

2.3 Optical Investigation of Visible Cavitation

The appearance of cavitation downstream the orifice was investigated optically. A LED lamp was placed as a light source on the backside of the optically accessible mixing chamber. A frosted glass between LED and mixing chamber ensured an even distribution of the light. With the help of a high-speed-camera in combination with a 100 mm macro objective, shadowgraphic images were recorded. The frequency was set to 10 000 fps. The exposure time varied between 50 and 70 μs. For each trial, a set of 2 000 images was recorded. During image processing, the images were converted to grey scale. For discussions of cavitation patterns, single representative images are chosen. In general, regions of vapor appear dark, while fluid regions appear light. A detailed description of the method and setup can be found in [54].

Investigations were carried out to determine the influence of process and geometrical parameters. The pressure difference $\Delta p = p_{in} - p_{out}$ was set between 100 and 500 bar. The Thoma number Th was varied between 0.002 and 0.258, and the mixing ratio of the two streams $r = \dot{V}_{MS}/\dot{V}_{HS}$ was set between 0 and 20. In addition, the geometry of the mixing chamber was varied. Both shapes, the T-mixer and the circular mixer, were used in the trails. Additionally, the distance a was studied between 0, 4.5 and 9 mm.

2.4 Mixing Efficiency in SHM-Valves

The mixing efficiency of the process was investigated using the Villermaux-Dushman reaction (VD). The iodide iodate reaction is a widely used method to determine the micromixing, which is based on two competing reactions [55, 56]. The Villermaux-Dushman reaction has been successfully used to investigate mixing efficiency in SHM processes before [57]. At the present time, however, the influence of cavitation and the influence of β have not yet been investigated.

The first reaction is an instantaneous neutralization reaction:

$$H_2BO_3^- + H^+ \Longleftrightarrow H_3BO_3 \tag{6}$$

The second reaction, which is some orders of magnitude slower, takes only place in imperfect mixing conditions. The Dushman reaction is:

$$5I^- + IO_3 + 6H^+ \Longleftrightarrow 3I_2 + 3H_2O \tag{7}$$

Iodine reacts in a quasi-instantaneous equilibrium reaction to triiodide:

$$I_2 + I^- \Longleftrightarrow I_3^- \tag{8}$$

The amount of triiodide can be analyzed by UV–visible spectroscopy. Low absorption values thereby represent high mixing efficiencies.

The chemical equilibrium constant K_B of reaction (8) depends on temperature following Eq. 10 [58]:

$$K_B = \frac{[I_3^-]}{[I_2][I^-]} \tag{9}$$

$$\log_{10} K_B = \frac{555}{T} + 7.355 - 2.575\log_{10} T \tag{10}$$

The triiodate concentration after the reaction can be quantified using Beer-Lamberts law using the measured light absorption A at 353 nm, the extinction coefficient of triiodide $\varepsilon_{353} = 26\,047$ L/mol/cm and the optical path length l [59].

$$c_{I_3^-} = \frac{A}{\varepsilon_{353} \cdot l}, \text{ valid for } 0.1 < A < 2.5 \tag{11}$$

The mixing efficiency is often decimensioned in literature using the segregation index X_S whose value lies between $X_S = 0$ indicating perfect micromixing and $X_S = 1$ indicating total segregation [60]. For partial segregation, X_S is calculated by

$$X_S = \frac{Y}{Y_{ST}} \tag{12}$$

Table 2 Reactant concentration used in trials

Chemical	Concentration/mol/l
KI	0.02320
KIO_3	0.00466
H_3BO_3	0.09090
NaOH	0.04500
H_2SO_4	0.02000

Y is the number of acid molecules consumed by reaction (7) to the total number of acid moleculs and Y_{ST} denotes the value of the ratio Y in total segregation when mixing is infinitely slow. They are given by Eqs. 13 and 14 [59].

$$Y = \frac{2 \cdot (n_{I_2} + n_{I_3^-})}{n_{H_0^+}} \tag{13}$$

$$Y_{ST} = \frac{6c_{IO_3^-,0}}{6c_{IO_3^-,0} + c_{H_2BO_3^-,0}} \tag{14}$$

In the beginning the VD reaction was mainly used in stirred vessels, where a small amount of acid is added to a defined volume of buffer solution such as in [60, 61]. In the meantime, the application in continuous micro mixers is also established, in which the buffer is mixed with the acid in a ratio of one to one [59, 62]. In that case, flow rates and Re numbers are typically much lower compared to SHM process [59, 63]. By combining a micro mixer with a high frequency ultrasound, mixing efficiency could be improved due to the formation of cavitation up to 36 % [63].

All reactant concentrations are listed in Table 2. To prepare the buffer, boric acid and NaOH were dissolved in water, resulting in a pH of 9.14. Afterwards, iodide and iodate were added.

For these investigations, the orifice diameter $d_{orifice}$ was set to 0.15 mm. The outlet diameter d_{out} varied between 2 and 4 mm. The distance between the T-shaped mixing and the orifice outlet was 4.5 mm. The high-pressure stream consisted of the buffer, the mixing stream of diluted sulfuric acid. The mixing ratio r was kept constant at a value of 1. Pressure differences of 100 and 200 bar were evaluated. Th was varied between 0 and 0.3. The absorption measurements were carried out at a wavelength of 353 nm [56]. All experiments were performed in triplicate.

2.5 *Investigation of Mixing Conditions on Oil Droplet Size*

A model system of oil droplets in a water phase was used to measure the stress-induced changes in in-flow compounds, as this model system is much easier to characterize and more sensitive to flow changes compared to protein molecules. To investigate the influence of the mixing stream on the oil droplet size, a pre-emulsion was prepared in a first step using a rotor–stator-machine. The pre-emulsion contained 1 wt-% rapeseed oil, 0.5 wt-% sodium lauryl sulfate (SDS) and 98.5 wt-% distilled water. The pre-emulsion was fed as high-pressure stream, the mixing stream consisted of distilled water. The mixing ratio r was varied between 1 and 5 and the pressure difference ranged between 100 and 300 bar. Both mixing chamber geometries were used in combination with the different distances, β was kept constant at 0.075. All trials were performed in triplicate.

To measure the oil droplet size distribution, a laser light diffraction spectrometer including polarization intensity differential scattering was used. Each sample was measured in triplicate.

2.6 Numerical Investigation of Single Phase Flow Inside the Disruption Unit

Since the investigation of the stresses occurring in the flow field is not possible experimentally, numerical flow simulations were used. The simulations were performed for HPH and SHM geometries at different process pressures.

The geometries used correspond to the T-mixer and the HPH geometry listed above (compare Fig. 2 and Table 1). The diameter of the mixing chamber was 2 mm in all cases. The distance a was set to 4.5 mm. The orifice inlet was shortened to 1 mm for the simulations, and the inlets of the mixing stream were 5 mm in each case. The mesh consisted of tetrahedra with a base length of 0.05 mm. It was refined in the area of the two inlets, the orifice and the free jet. This resulted in final meshes with 20.7 million cells for the HPH, and 27.9 million cells for the SHM geometry, respectively.

Material property data from distilled water, as given in the Ansys Fluent 2022 R2 database were used, corresponding to the continuous phase of the dispersions treated. The material properties are given in Table 3.

The boundaries were set to pressure inlet and outlet for HPH simulations with a turbulent intensity of 5% and a viscosity ratio of 10. For the SHM simulations, the inlets were set to mass flow inlets instead, using the mass flow calculated in the HPH simulations. At the walls, no slip conditions were set up. The process parameters used in the simulations are listed in Table 4.

Table 3 Material parameters for distilled water used in simulations

Parameter	Value
Density/kg/m^3	997.0
Specific heat capacity/J/(kg K)	4184.0
Thermal conductivity/W/(m K)	0.61
Viscosity/kg/(m s)	0.00089

Table 4 Overview of process parameters used in simulations

Simulation	Parameter	Inlet HS	Inlet MS	Outlet	Wall
1	**Type**	Pressure	–	Pressure	No slip
	Value	150 bar	–	50 bar	–
	Temperature	293 K	–	300 K	300 K
2	**Type**	Pressure	–	Pressure	No slip
	Value	1500 bar	–	500 bar	–
	Temperature	293 K	–	300 K	300 K
3	**Type**	Mass flow	Mass flow	Pressure	No slip
	Value	0.00178	0.00178	50 bar	–
	Temperature	293 K	293 K	300 K	300 K
	A	–	4.5 mm	–	–
4	**Type**	Mass flow	Mass flow	Pressure	No slip
	Value	0.00721	0.00721	500 bar	–
	Temperature	293 K	293 K	300 K	300 K
	A	–	4.5 mm	–	–

The RNG-k-ε model with enhanced wall treatment and Kato-Lauder production limiter was used to simulate the single-phase flow. In order to show the changes in temperature in the process, the energy equation was additionally solved.

The coupled solver was chosen to model the pressure velocity coupling. For the discretization of the gradients, the least square cell based method was applied. Turbulent kinetic energy and turbulent dissipation rate were discretized with the 1st order upwind scheme, momentum and energy with the 2nd order upwind scheme. For the discretization of the pressure, the 2nd order method was used. The simulation was set up with 30 000 iterations.

In order to analyze the critical stresses downstream the orifice, eight evenly distributed points were set up on a radius of 0.07 mm on the outlet of the orifice. Each point served as a starting point for a pathline on which the turbulent kinetic energy k and shear rate $\dot{\gamma}$ were evaluated. The disrupting stresses were assumed to have decayed when the local value on the pathline has fallen below the maximum value of the pathline by a factor of 10 in the x-direction for the last time.

To compare the local stresses and the temperatures between the HPH and the SHM setup, the distributions were plotted on planes perpendicular to the flow direction. This allows a qualitative comparison of the parameters.

3 Results and Discussions

In the following chapter, the results for the investigation of the listed working hypothesis are presented. First, the cavitation occurring in the SHM process is investigated, followed by the investigation of the mixing efficiency. The influence of the mixing stream on the oil droplet size is presented afterwards. The experimental investigations are complemented by CFD simulations.

3.1 *Influence of Process Parameters on Visible Cavitation*

The results in this chapter show the influence of process parameters in SHM geometries on visible ca-vitation in the mixing chamber. First, shadowgraphs were taken with variation of the Thoma number. The second part shows the influence of the mixing stream on the cavitation.

In [54], it could be shown, that the appearance of visible cavitation in the mixing chamber can be influenced by applying a backpressure as second homogenization stage. Figure 3 shows the cavitation for two different inlet pressures and variation of *Th*. The cavitation pattern for the lowest *Th* values depends on the inlet pressure. While jet cavitation occurs at 100 bar inlet pressure, hydraulic flip is present for 200 bar inlet pressure. With increasing *Th*, the cavitating region decreases and the flow shape does not depend any more on the inlet pressure applied. At *Th* = 0.258 and 0.254 respectively, the visible cavitation in the mixing chamber is completely suppressed. These findings are in good agreement with the findings for the conventional HPH process [19]. A more detailed discussion of the results can be found in [54].

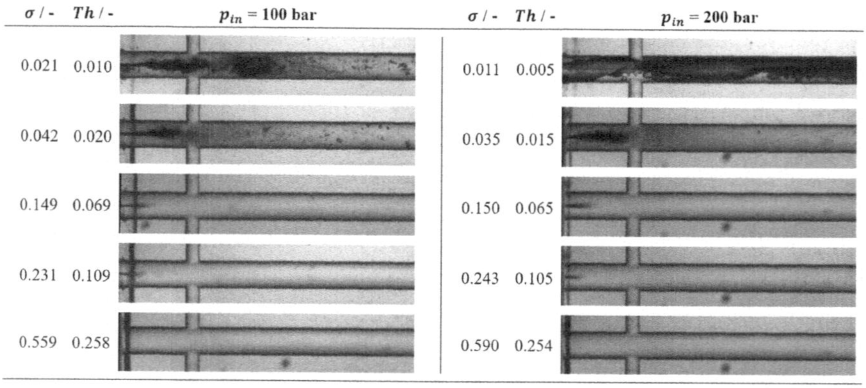

Fig. 3 Influence of backpressure on supercavitation as function of inlet pressure p_{in} and *Th* for $\beta = 0.075$ and $r = 1$. Figure taken from [54], licensed under the Creative Commons Attribution 4.0 International license (CC BY 4.0)

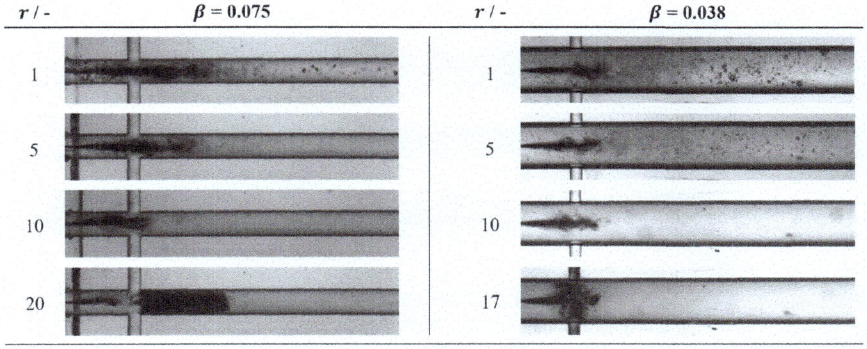

Fig. 4 Influence of increasing mixing ratio r on jet cavitation as function of diameter ratio β for a single stage process (p_{out} = 1.01325 bar) with p_{in} = 100 bar. Figure taken from [54], licensed under the Creative Commons Attribution 4.0 International license (CC BY 4.0)

The influence of the mixing ratio on the cavitation jet is shown in Fig. 4. On the left handed side, the outlet diameter was kept constant as in the previous results. On the right handed side, the outlet diameter was doubled, resulting in ratios of $\beta =$ 0.075 and 0.038 respectively. The inlet pressure was kept constant at 100 bar in both cases. The influence of the mixing stream is limited for $\beta = 0.075$, until a second cavitation region is formed, probably caused by the high total mass flow in the mixing chamber. For $\beta = 0.038$, the influence of the mixing stream is even smaller. Only the width of the cavitating region changes at $r = 17$. All results are published and discussed in detail in [54]. This publication also includes additional results dealing with the influence of the mixing stream on choked cavitation and hydraulic flip.

3.2 *Influence of Geometrical Parameters on Visible Supercavitation*

This chapter deals with the influence of geometrical parameters on the occurrence of visible cavitation in the mixing chamber of a SHM geometry. First, the influence of the diameter ratio β is discussed, followed by the influence of shape and distance of the mixing geometry.

Figure 5 shows the appearance of cavitation in the mixing chamber with variation of β (left to right) and increasing inlet pressure (top to bottom). It is visible, that the cavitation pattern changes with increasing p_{in} dependent on the diameter ratio. At an inlet pressure of 100 bar, in both cases jet cavitation is present. While the pattern changes for $\beta = 0.075$ from jet cavitation to hydraulic flip between 100 and 200 bar, jet cavitation is still present at 200 bar for $\beta = 0.038$. In this case, the pattern changes to choked cavitation for an inlet pressure of 500 bar. In conclusion, the diameter ratio

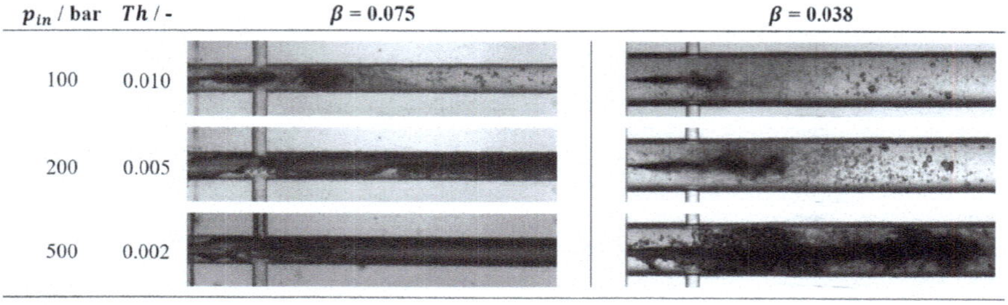

Fig. 5 Influence of inlet pressure p_{in} on supercavitation as function of outlet diameter d_{out} for a single stage process ($p_{out} = 1.01325$ bar). Figure rearranged from [54], licensed under the Creative Commons Attribution 4.0 International license (CC BY 4.0)

can influence the cavitation pattern, independent of flow parameters. These results are partly published in [54].

To further study the influence of the geometry of the mixing chamber, the shape was varied between a T-mixer and a circular mixer. Additionally, the distance of the mixing stream was varied, see Fig. 6. At the top, the cavitation in the T-mixer geometry is shown with the distance varying between 4.5 and 9 mm and two different inlet pressures (100 and 200 bar). At the bottom, the cavitation is plotted for the circular mixer for two distances (0 and 4.5 mm) and the above mentioned inlet pressures. The cavitation patterns differ dependent on both, distance and shape of the geometry. Comparing the T-mixer and the circular mixer with distances of 4.5 mm and inlet pressure of 100 bar, it becomes visible, that the length of the cavitation region deviates by a factor of almost 1.5. In addition, the cavitating region extends into the mixing disk of the circular mixer. Comparing the same geometry at an inlet pressure of 200 bar, one can observe that the cavitation regime varies between hydraulic flip (T-mixer) and choked cavitation (circular mixer). The distance of the mixing stream also influences the appearance of cavitation. A detailed description of the results can be found in [64].

3.3 *Experimental Investigation on the Mixing Efficiency in the Investigated SHM Orifices*

Since the local dilution of the concentration of cell suspensions in the process has a direct influence on the collision rate, the mixing efficiency was chosen as a related parameter to be analyzed in the mixing chamber. In the first step, it was investigated how the inlet pressure influences local mixing in cavitating flow. To ensure cavitation, no backpressure was applied and the diameter ration was set to $\beta = 0.075$. As shown in Fig. 7, higher absorption values were measured when the homogenizing pressure was increased from 100 to 200 bar. A further increase of p_{in} to 300 bar resulted in a considerable increase in the divergence of the measured values.

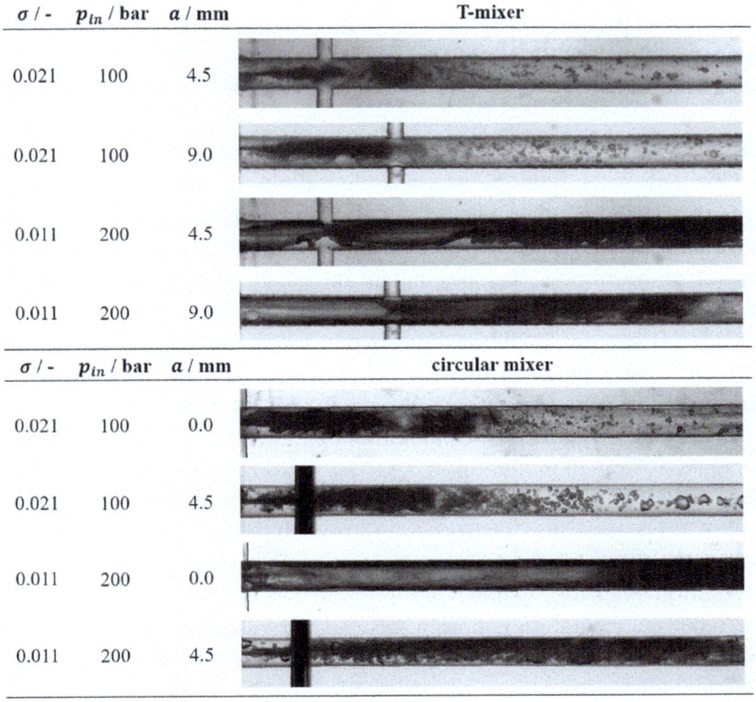

Fig. 6 Influence of mixing chamber geometry on visible cavitation for a single stage process (p_{out} = 1.01325 bar). Figure taken from [64], licensed under the Creative Commons Attribution 4.0 International license (CC BY 4.0)

Fig. 7 Absorption measured at different homogenization pressures for $\beta = 0.075$ and $p_{out} = 1.013$ bar. Compare to Fig. 5 and [54] for corresponding cavitation patterns

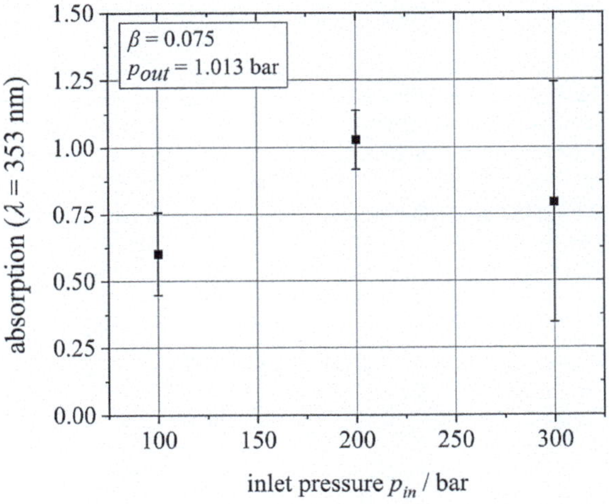

The increase of the absorption value when p_{in} increases from 100 to 200 bar contradicts the expectation that the mixing efficiency improves with increasing homogenizing pressure and thus energy input. The measured values can be attributed to the cavitation patterns observed at the same parameters, see Fig. 5 and [54]. While jet cavitation was still observed at $p_{in} = 100$ bar, at 200 bar and above, hydraulic flip occurs. One part of the liquid flow (probably mainly the HS) is located as a spray in the gas core while another part (probably mainly the MS) flows around the gas core in a ring shape. It is therefore not surprising that the hydraulic flip leads to a deterioration of the mixing efficiency and the information provided by the absorption value can be classified as credible in terms of quality.

At $p_{in} = 300$ bar, the length of the coherent cavitation bubble further increases and its collapse takes place downstream the mixing chamber. In this case, a strong scattering of the measured values was observed. Therefore, the measured absorption values in strongly cavitating flow cannot be classified as reliable and the investigated pressure range is limited in the following to $p_{in} = 200$ bar.

Following, we discuss how halving β in cavitating flow affects mixing efficiency. As shown in Fig. 8, the reduction from $\beta = 0.075$ to 0.038 causes the absorption value to drop at both 100 and 200 bar. The reduction of β is accompanied by a reduction of the cavitation intensity, as shown in Fig. 5 and described in [54]. At $p_{in} = 200$ bar and $\beta = 0.038$ jet cavitation is present instead of hydraulic flip. This explains the lower absorption values at 200 bar. The reduction of the length of the cavitating jet at 100 bar already seems to have a positive effect on the mixing efficiency.

The halving of β is also accompanied by a doubling of d_{out} and thus an enlargement of the mixing chamber. Whether this increase is also accompanied by a deterioration in mixing efficiency cannot be assessed at this point due to the overlap with the reduction in cavitation intensity, which predominates. Therefore, in the influence of β in cavitation-free flow is considered separately.

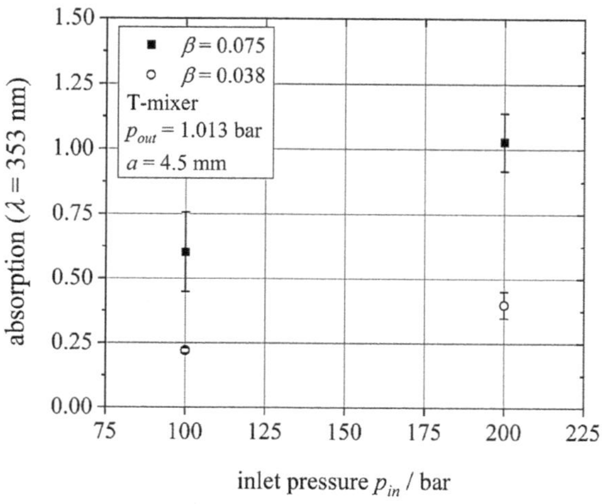

Fig. 8 Absorption measured for varying homogenization pressures and β. Compare to Fig. 5 and [54] for corresponding cavitation patterns

Since the previous results showed that cavitation has a negative effect on the mixing efficiency, we now compare the influence at *Th* numbers between 0 and 0.3 and constant Δp. A previous work shows that when the back-pressure is increased to $Th = 0.1$, only a very small cavitation jet was visible in the optically accessible mixing chamber, which disappeared completely at $Th = 0.3$ [54].

Figure 9 displays how the increase of the *Th* affects the absorption value for $\Delta p = 100$ bar and 200 bar. Since the pressure resistance of the system does not allow the backpressure required for 200 bar and $Th = 0.31$ to be applied, this experiment could not performed. It is clear that the reduction of cavitation at $Th \approx 0.1$ results in a strong improvement of the mixing efficiency at both homogenizing pressures compared to $Th \approx 0$. On average at $Th \approx 0.1$, the measured absorption values for 200 bar are below the values for 100 bar, which indicates the influence of the increased energy input. Nevertheless, the error bars in this area are so large compared to the measured values that no absolute statement can be made. This indicates that the mixing efficiency of the process in this area is already so good that small inaccuracies in the control of the process flows are clearly visible. If the *Th* number is further increased to 0.3, no further improvement in mixing efficiency can be observed. It must also be taken into account that the values can only be detected with the selected reactant concentrations and that differences can no longer be resolved.

Overall, it is clear that the occurrence of cavitation significantly changes the measurement signal and thus cavitation cannot be neglected when determining the mixing efficiency. All experiments presented so far showed a correlation between increasing cavitation intensity and higher measured absorption or worse mixing efficiency. Thus, the positive influence of cavitation on mixing efficiency observed in micro mixers with a high frequency ultrasound [63] is not evident for the SHM process. This may be due to the fact that the SHM process has by far higher *Re* numbers and thus turbulent instead of laminar flow.

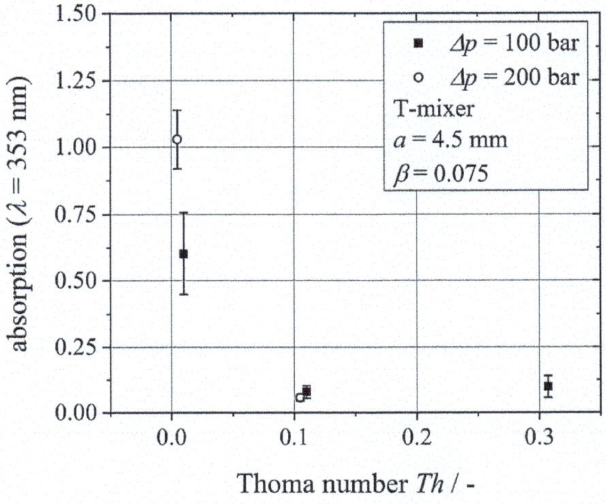

Fig. 9 Absorption measured for varying *Th* number. Compare to Fig. 5 and [54] for corresponding cavitation patterns

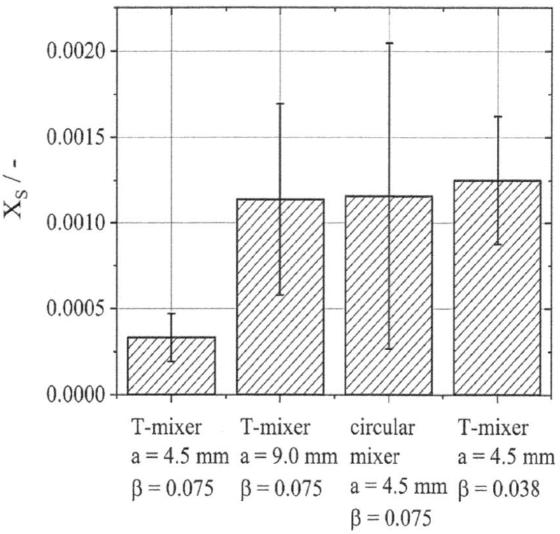

Fig. 10 Segregation indices for varying mixer geometries in cavitation free flow

Figure 10 illustrates the calculated segregation indices X_s for varying mixer geometries. The averaged segregation indizes range between 0.0003 and 0.0012. However, the scattering in multiple trials is so high that only one mixer geometry stands out from the others through better mixing efficiency. The deviation of the measured values could come from the fact that minor irregularities in the process control, such as fluctuations in the process streams and thus also in the mixing ratios, affect the already very good mixing efficiency. The segregation index in the SHM process is only about 0.1–0.5 % of the segregation index achieved in stirred tanks [60] and about 1–3 % of that achieved in micro mixers [63]. Thus, the mixing efficiency for all SHM geometries can be classified as exceptionally good compared to conventional processes. Nevertheless, it tends to show that the geometry with a T-mixer, a = 4.5 mm and $\beta = 0.075$ has the best mixing efficiency. Both an increase in the distance between orifice and insertion of the mixing stream a as well as the outlet diameter leads to a deterioration in mixing efficiency. The use of a circular mixer instead of a T-mixer also reduced the mixing efficiency. This can probably be attributed to the fact that the velocity of the mixing flow is significantly higher for the same volume flow with a T-mixer on entering the mixing chamber.

3.4 Influence of Mixing Conditions on Oil Droplet Size

Since structural changes of proteins and inactivation of enzymes occur by interaction with interfaces, it is important to prevent the formation of these interfaces. For this reason, high back pressures must be applied in experiments in order to suppress cavitation. This significantly increases the inlet pressures that would be necessary to

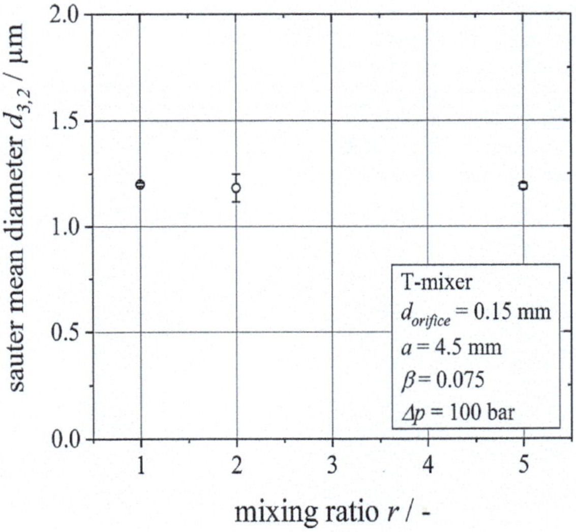

Fig. 11 Influence of mixing ratio r on mean oil droplet size for $\Delta p = 100$ bar. Figure taken from [64], licensed under the Creative Commons Attribution 4.0 International license (CC BY 4.0)

study structural changes of proteins and enzymes. These high pressures cannot be realized in the experimental setup, which is why the shear-sensitive model system of the oil droplet, as in the project of the working group of Professor Kraume (TU Berlin), see Chap. 16, was chosen. It was studied the influence of the mixing conditions on the oil droplet size instead. First, the influence of the mixing ratio r was studied, followed by the influence of geometrical parameters.

The influence of the mixing ratio on the oil droplet size is plotted in Fig. 11. The trials were carried out at a pressure difference of 100 bar. The diameter ratio was set to $\beta = 0.075$ and the distance was set to 4.5 mm. The mixing ratio r was varied between 1 and 5. For those values, no influence on the droplet size could be observed, indicating that the mixing stream does not affect significantly the disrupting forces.

In order to compare the influence of the geometry, the sauter mean diameter is plotted over the homogenization pressure for the T-mixer and the circular mixer with different distances in Fig. 12: Influence of geometry of the mixing chamber on oil droplet size [64]. Figure 12. At $\Delta p = 100$ bar, all mixing geometries result in the same droplet size. With higher pressure differences, the circular mixer, which is located directly after the gap (a = 0 mm), results in slightly smaller droplets, compared to the other geometries. However, standard deviations imply a low statistical robustness of the statement. It is possible, that this geometry prevents dead zones in the corner of the mixing chamber. In general, the differences are very small, so that it can be assumed that the mixer configuration at least does not influence the distribution of hydrodynamic stresses in the mixing zone very significantly. However, this assumption would have to be verified by e.g. flow simulation. For further results, refer to [64].

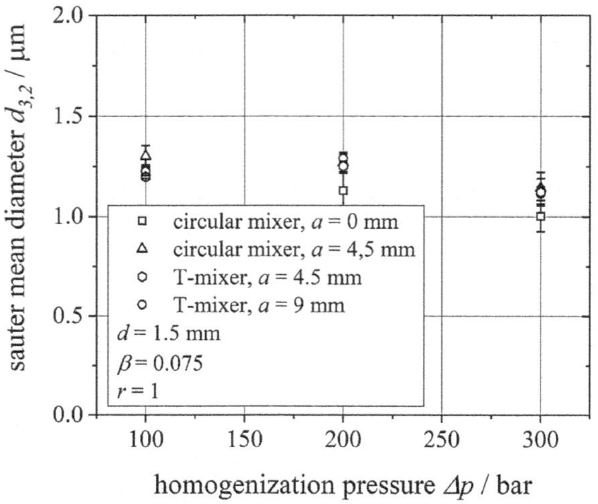

Fig. 12 Influence of geometry of the mixing chamber on oil droplet size. Figure taken from [64], licensed under the Creative Commons Attribution 4.0 International license (CC BY 4.0)

3.5 *Influence of Mixing Stream on Local Distribution of Hydrodynamic Stresses in the Mixing Chamber*

CFD simulations were performed to investigate to what extent the stresses in the mixing chamber of the SHM geometry differ from those of the HPH geometry. The parameters chosen were the shear rate and turbulent kinetic energy. As a first step, the stresses in the HPH geometry were analyzed for a pressure difference of 100 bar. Figure 13 shows the x-position for the last exceedance of 10 % of the maximum value of the shear rate and the turbulent kinetic energy for eight different pathlines. The mean value of the eight positions is plotted as dashed line. On average, the maximum shear rate has already decayed by 90 % after 0.75 mm downstream the orifice. For the turbulent kinetic energy, the x-positions are more evenly distributed around the average value of 4.8 mm. It is assumed that it is sufficient to add the mixing stream after 90 % of the maximum stresses have already decayed, so that the admixture does not negatively affect these. In the following, the center axis of the mixing stream is set to 5 mm (a = 4.5 mm).

Figure 14 shows the local distribution of turbulent kinetic energy and shear rate for the HPH and SHM geometry at a pressure difference of 100 bar. The turbulent kinetic energy is distributed mainly in the center of the plane. The maximum values occur directly after the orifice outlet. In flow direction, the area widens radially. No differences in distribution can be identified between HPH and SHM geometry. The maximum values of the shear rate also occur in the first plane shown. In this case, the flow region is characterized by turbulence around the center axis, so that the local distributions are not exactly the same, but qualitatively comparable.

In order to investigate the effect of the mixing stream at higher pressures on the stresses, the flow at a pressure difference of 1000 bar is also studied. Figure 15 shows the x-position for the last exceedance of 10 % of the maximum value of the shear

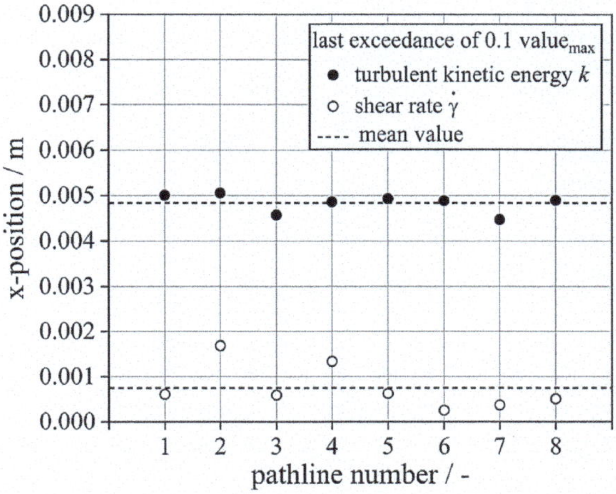

Fig. 13 Last exceedance of 10 % of the maximum of shear rate and turbulent kinetic energy on different pathlines evenly circular distributed on the outlet of the orifice with a radius of 0.7 mm for the HPH geometry at Δp =100 bar

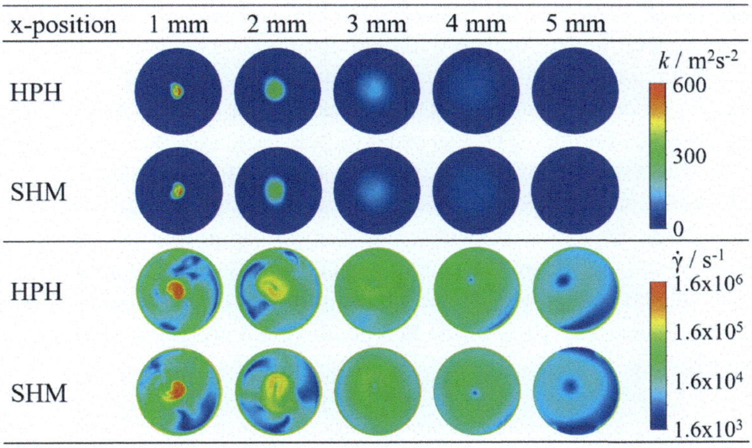

Fig. 14 Local distribution of turbulent kinetic energy and shear rate on planes perpendicular to the flow direction inside the mixing chamber for the HPH and SHM geometry at a pressure difference Δp = 100 bar

rate and the turbulent kinetic energy for eight different pathlines at Δp = 1000 bar for the HPH geometry. In comparison to Fig. 13, the x-positions are more widely distributed. For the turbulent kinetic energy, only a slight shift of the mean value to 5 mm is visible. The mean x-position of the shear rate is more than doubled compared to Δp = 100 bar, resulting in 1.9 mm. Based on these results, the centerline of the mixing stream is also set to 5 mm at 1000 bar.

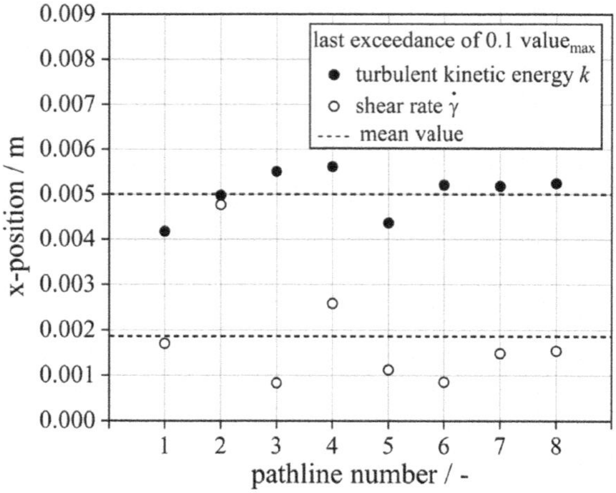

Fig. 15 Last exceedance of 10 % of the maximum of shear rate and turbulent kinetic energy on different path-lines evenly circular distributed on the outlet of the orifice with a radius of 0.7 mm for the HPH geometry at $\Delta p = 1000$ bar

Comparing the local stresses at 1000 bar on the planes perpendicular to the flow direction in Fig. 16, it becomes clear that the flow differs more between HPH and SHM geometry compared to 100 bar. The turbulent kinetic energy has its maximum at a x-position of 1 mm in both cases. For the SHM geometry the area with higher values is in general larger compared to the HPH geometry. The turbulence decays further downstream of the flow, so the mixing stream has increased the turbulence in the mixing chamber. This can be directly attributed to the doubled mass flow and the resulting Reynolds number. The increased turbulence should not negatively affect breakup mechanisms inside the flow, but rather support it. The shear rate shows a noticeable difference between the two geometries starting at a distance of 2 mm. The area of high shear rates is swirled and reduced significantly earlier and more strongly by the mixing stream. The fanning out is already more pronounced at 2 mm compared to the HPH geometry. However, the influence on cell breakup should play a limited role, since only 10 % of the maximum values are still present in this area (compare Fig. 15).

3.6 Influence of the Mixing Stream on the Temperature Distribution Inside the Mixing Chamber

To investigate the ability of the process to be cooled inline by the mixing stream, the temperature distribution in the mixing chamber is discussed below for both pressure differences. In Fig. 17, local temperatures are plotted on planes perpendicular to the flow direction for the HPH and SHM geometry. The temperature distributions in the HPH geometry look almost identical for both pressures, 100 and 1000 bar. The temperature in the center of the free jet is slightly lower compared to the surrounding flow (0.25 and 0.75 mm). At a distance of 20 mm, the influence of the wall, which

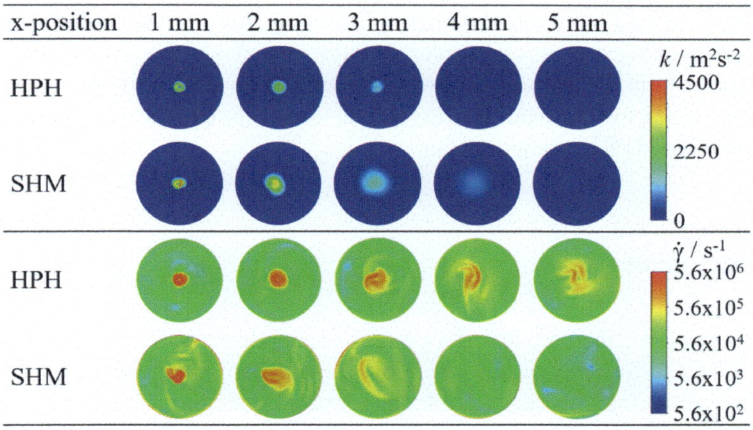

Fig. 16 Local distribution of turbulent kinetic energy and shear rate on planes perpendicular to the flow direction inside the mixing chamber for the HPH and SHM geometry at a pressure difference $\Delta p = 1000$ bar

has a temperature of 300 K, becomes apparent. This effect is much more pronounced at the lower pressure, as a result of the higher residence time due to the velocity.

By adding the mixing stream with a temperature of 283 K, the overall temperature distribution is reduced to lower levels. Already at a distance to the orifice of 0.25 mm, the temperature in the mixing chamber is in the range of the inlet temperature of the high-pressure stream (293 K). At a distance of 5 mm, a region with very low temperatures is visible in the area of the T-shaped mixer. This region is completely mixed into the flow after 0.5 mm for a pressure of 100 bar, and at 1000 bar the region has partially decayed. Again, in this geometry, the flow near the wall heats up

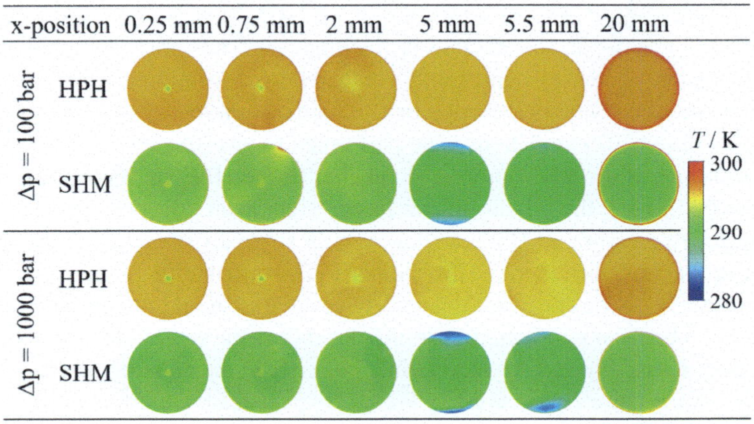

Fig. 17 Local distribution of the temperature on planes perpendicular to the flow direction inside the mixing chamber for the HPH and SHM geometry at pressure differences of 100 and 1000 bar

with increasing distance from the orifice. Due to the high mixing efficiency in the process, the mixing stream is recirculated all the way back to the orifice outlet. Thus it is possible to bring the temperature to a required level directly after the orifice without changing the mechanical stresses. The temperature of the mixing stream can be adjusted based on the required data.

4 Conclusions

The results shown indicate that the process of simultaneous homogenization and mixing (SHM) offers approaches to improve the cell disruption process, even though the influences of the flow could not be investigated directly on cell suspensions or model proteins.

Cavitation as a typical side effect of high pressure homogenization is not only harmful for the plant itself, but also for the quality of enzymes and proteins, and should therefore always be avoided in cell disruption processes. To investigate the cavitation occurring in SHM geometries, optically accessible mixing chambers were fabricated. The cavitation was characterized using shadowgraphs. It could be shown that the mixing stream has an influence on the cavitation patterns, as well as the diameter ratio β and the mass flow ratio r. In addition, the regions can also be affected by the type of mixing. Cavitation can be suppressed by applying a back pressure, as in the conventional HPH process. This makes it possible to prevent the formation of cavitation-induced new interfaces and to avoid the structural changes induced by the interaction of proteins or enzymes with these interfaces.

High mixing and dilution in the SHM process allow to decrease the collision rate, which is directly related to the aggregation behavior of the proteins. The Villermaux-Dushmann reaction was used to study the mixing efficiency depending on process parameters. It was demonstrated that the mixing quality increases with decreasing cavitation intensity. It was also shown that the SHM process mixes significantly better than, for example, stirred tanks or micromixers. Thus, it can be expected that the collision rate decreases significantly compared to conventional HPH, which will result in reduced enzyme inactivation or protein denaturation.

Single-phase CFD simulations were performed to investigate how the disruptive stresses change in the SHM process compared to the HPH. It was shown that at pressures of 100 bar, the mixing stream has no significant effect on the local stress distribution. At pressures of 1000 bar, however, the stress distributions are changed by the mixing stream added in the SHM geometry. The distribution of turbulent kinetic energy increases in the considered range. However, this should not negatively affect the process: The shear rate decreases from a distance to the orifice of 2 mm compared to the conventional process. However, this is already in the non-critical range, since the shear rate has already decayed by 90 %. To investigate the effect of the mixing geometry and various process parameters on the stress distribution, oil droplets were used as model. Their size distribution was measured as an indirect parameter of stress-induced reactions (as in Chap. 16). It was shown, that neither the geometry, nor the mass flow ratio have a relevant influence on the size disruption of the droplets.

Local high temperatures are critical to enzyme and protein quality in stress-induced cell disruption. It could be shown by CFD simulations, that the mixing stream of the SHM geometry can significantly reduce the temperature in the mixing chamber immediately after the orifice. Thermal denaturation of proteins can thus be prevented, especially by selective adjustment of the mixing flow temperature.

In summary, the SHM process offers major advantages in terms of cell breakup, if the process is operated cavitation free. For this purpose, the location of the mixing stream should be chosen as close as possible to the orifice outlet, at the moment when the interfering disruptive stresses have decayed In addition, the temperature must be controlled. To ensure that critical maximum values are not exceeded at any point in the process, the temperature of the mixing stream must be set depending on the inlet temperature and the material processed. It could be demonstrated, that CFD is a powerful tool to predict required process parameters.

Acknowledgements The authors would like to thank the DFG for financial support within the DFG priority program SPP 1934 (project number 425332181) "Proteins under stress". We also like to thank Markus Fischer for his help in conducting the experiments and Nico Leister for proofreading the manuscript.

References

1. Bekard IB, Asimakis P, Bertolini J et al (2011) The effects of shear flow on protein structure and function. Biopolymers 95:733–745. https://doi.org/10.1002/bip.21646
2. Dong X, Zhao M, Shi J et al (2011) Effects of combined high-pressure homogenization and enzymatic treatment on extraction yield, hydrolysis and function properties of peanut proteins. Innov Food Sci Emerg Technol 12:478–483. https://doi.org/10.1016/j.ifset.2011.07.002
3. Donsì F, Ferrari G, Lenza E et al (2009) Main factors regulating microbial inactivation by high-pressure homogenization: operating parameters and scale of operation. Chem Eng Sci 64:520–532. https://doi.org/10.1016/j.ces.2008.10.002
4. Follows M, Heterington PJ, Dunnill P et al (1971) Release of enzymes from bakers' yeast by disruption in an industrial homogenizer. Biotechnol Bioeng 13:549–560. https://doi.org/10.1002/bit.260130408
5. Geciova J, Bury D, Jelen P (2002) Methods for disruption of microbial cells for potential use in the dairy industry—a review. Int Dairy J 12:541–553. https://doi.org/10.1016/S0958-6946(02)00038-9
6. Hayes MG, Kelly AL (2003) High pressure homogenisation of milk (b) effects on indigenous enzymatic activity. J Dairy Res 70:307–313. https://doi.org/10.1017/S0022029903006319
7. Floury J, Bellettre J, Legrand J et al (2004) Analysis of a new type of high pressure homogeniser. A study of the flow pattern. Chem Eng Sci 59:843–853. https://doi.org/10.1016/j.ces.2003.11.017
8. Hecht LL, Schlender M, Köhler K et al (2012) Abrasion in high-pressure homogenization orifices: a new method to quantify the erosivity of particle loaded liquids. Wear 289:138–144. https://doi.org/10.1016/j.wear.2012.04.003
9. Wang W (2005) Protein aggregation and its inhibition in biopharmaceutics. Int J Pharm 289:1–30. https://doi.org/10.1016/j.ijpharm.2004.11.014

10. Talboys BL, Dunnill P (1985) Effect of shear on membrane-associated enzymes: studies of isolated progesterone 11alpha-hydroxylase complex from Rhizopus nigricans. Biotechnol Bioeng 27:1726–1729. https://doi.org/10.1002/bit.260271215
11. Thomas CR, Geer D (2011) Effects of shear on proteins in solution. Biotechnol Lett 33:443–456. https://doi.org/10.1007/s10529-010-0469-4
12. Köhler K, Schuchmann HP (2015) High-pressure homogenization with microstructured systems. In: Rayner M, Dejmek P (eds) Engineering aspects of food emulsification and homogenization. CRC Press, pp 169–194
13. Choi H, Laleye L, Amantea GF et al (1997) Release of aminopeptidase from Lactobacillus casei sp. casei by cell disruption in a Microfluidizer. Biotechnol Tech 11:451–453. https://doi.org/10.1023/A:1018489327675
14. Ball CG, Fellouah H, Pollard A (2012) The flow field in turbulent round free jets. Prog Aerosp Sci 50:1–26. https://doi.org/10.1016/j.paerosci.2011.10.002
15. Reynolds O (1883) XXIX. An experimental investigation of the circumstances which determine whether the motion of water shall be direct or sinuous, and of the law of resistance in parallel channels. Phil Trans R Soc 174:935–982. https://doi.org/10.1098/rstl.1883.0029
16. Schuchmann HP (2016) Advances in hydrodynamic pressure processing for enhancing emulsification and dispersion. In: Innovative food processing technologies, vol 10. Elsevier, pp 387–412
17. Laurien E (2013) Numerische Strömungsmechanik: Grundgleichungen und Modelle—Lösungsmethoden - Qualität und Genauigkeit, 5., überarb. u. erw. Aufl. 2013. Springer eBook Collection Computer Science and Engineering. Springer Vieweg, Wiesbaden
18. Kurzhals H-A (1977) Untersuchungen über die physikalisch-technischen Vorgänge beim Homogenisieren von Milch in Hochdruck-Homogenisiermaschinen. Dissertation, Technische Universität Hannover
19. Schlender M, Minke K, Schuchmann HP (2016) Sono-chemiluminescence (SCL) in a high-pressure double stage homogenization processes. Chem Eng Sci 142:1–11. https://doi.org/10.1016/j.ces.2015.11.028
20. Brennen CE (2014) Fundamentals of multiphase flow. Cambridge University Press
21. Yan Y, Thorpe RB (1990) Flow regime transitions due to cavitation in the flow through an orifice. Int J Multiph Flow 16:1023–1045. https://doi.org/10.1016/0301-9322(90)90105-R
22. Schlender M Untersuchungen zum Einfluss von strömungsinduzierter Kavitation auf den Emulgierprozess bei der Hochdruckhomogenisation. Dissertation
23. Schlender M, Spengler A, Schuchmann HP (2015) High-pressure emulsion formation in cylindrical coaxial orifices: influence of cavitation induced pattern on oil drop size. Int J Multiph Flow 74:84–95. https://doi.org/10.1016/j.ijmultiphaseflow.2015.04.004
24. Lee J, Kentish S, Ashokkumar M (2005) Effect of surfactants on the rate of growth of an air bubble by rectified diffusion. J Phys Chem B 109:14595–14598. https://doi.org/10.1021/jp051758d
25. Sunartio D, Ashokkumar M, Grieser F (2007) Study of the coalescence of acoustic bubbles as a function of frequency, power, and water-soluble additives. J Am Chem Soc 129:6031–6036. https://doi.org/10.1021/ja068980w
26. Miyoshi N, Hatanaka S-i, Yasui K et al (2001) Effects of pH and surfactant on the ultrasound-induced chemiluminescence of luminol. Jpn J Appl Phys 40:4097–4100. https://doi.org/10.1143/JJAP.40.4097
27. McKillop AA, Dunkley WL, Brockmeyer RL et al (1955) The cavitation theory of homogenization. J Dairy Sci 38:273–283. https://doi.org/10.3168/jds.S0022-0302(55)94971-7
28. Gothsch T, Schilcher C, Richter C et al (2015) High-pressure microfluidic systems (HPMS): flow and cavitation measurements in supported silicon microsystems. Microfluid Nanofluid 18:121–130. https://doi.org/10.1007/s10404-014-1419-6
29. Martínez-Monteagudo SI, Yan B, Balasubramaniam VM (2017) Engineering process characterization of high-pressure homogenization—from laboratory to industrial scale. Food Eng Rev 9:143–169. https://doi.org/10.1007/s12393-016-9151-5

30. Bisten A, Hirn S, Dietrich H et al. (2016) Analyse des Sedimentationsverhalten von Pfirsichnektar in Abhängigkeit von dem vorherigen Homogenisationsschritt. Dispersion Lett 1–4
31. Carle R, Jauss A, Fuesser H (1998) Cloud stability of pulp-containing tropical fruit nectars. Fruit Process 266–268:270–272
32. Guerzoni ME, Vannini L, Chaves Lopez C et al (1999) Effect of high pressure homogenization on microbial and chemico-physical characteristics of goat cheeses. J Dairy Sci 82:851–862. https://doi.org/10.3168/jds.S0022-0302(99)75303-8
33. Kheadr EE, Vachon JF, Paquin P et al (2002) Effect of dynamic high pressure on microbiological, rheological and microstructural quality of Cheddar cheese. Int Dairy J 12:435–446. https://doi.org/10.1016/S0958-6946(01)00104-2
34. Vannini L, Lanciotti R, Baldi D et al (2004) Interactions between high pressure homogenization and antimicrobial activity of lysozyme and lactoperoxidase. Int J Food Microbiol 94:123–135. https://doi.org/10.1016/j.ijfoodmicro.2004.01.005
35. Hoare M, Khan MR, Dunnill P (1993) Interfacial damage to proteins during intensive mixing in fermentation and downstream processing. In: Stability and stabilization of enzymes, vol 47. Elsevier, pp 21–27
36. Lee YK, Choo CL (1989) The kinetics and mechanism of shear inactivation of lipase from Candida cylindracea. Biotechnol Bioeng 33:183–190. https://doi.org/10.1002/bit.260330207
37. Hamilton-Brown P, Bekard I, Ducker WA et al (2008) How does shear affect Abeta fibrillogenesis? J Phys Chem B 112:16249–16252. https://doi.org/10.1021/jp805257n
38. Maa Y-F, Hsu CC (1997) Protein denaturation by combined effect of shear and air-liquid interface. Biotechnol Bioeng 54:503–512.0. https://doi.org/10.1002/(SICI)1097-0290(19970620)54:6%3C503:AID-BIT1%3E3.0.CO;2-N
39. Thomas CR, Dunnill P (1979) Action of shear on enzymes: studies with catalase and urease. Biotechnol Bioeng 21:2279–2302. https://doi.org/10.1002/bit.260211209
40. Thomas CR, Nienow AW, Dunnill P (1979) Action of shear on enzymes: studies with alcohol dehydrogenase. Biotechnol Bioeng 21:2263–2278. https://doi.org/10.1002/bit.260211208
41. Alexander-Katz A, Netz RR (2007) Surface-enhanced unfolding of collapsed polymers in shear flow. Europhys Lett 80:18001. https://doi.org/10.1209/0295-5075/80/18001
42. Ahmad A, Uversky VN, Hong D et al (2005) Early events in the fibrillation of monomeric insulin. J Biol Chem 280:42669–42675. https://doi.org/10.1074/jbc.M504298200
43. Nielsen L, Khurana R, Coats A et al (2001) Effect of environmental factors on the kinetics of insulin fibril formation: elucidation of the molecular mechanism. Biochemistry 40:6036–6046. https://doi.org/10.1021/bi002555c
44. Ramirez-Alvarado M, Merkel JS, Regan L (2000) A systematic exploration of the influence of the protein stability on amyloid fibril formation in vitro. Proc Natl Acad Sci USA 97:8979–8984. https://doi.org/10.1073/pnas.150091797
45. Sluzky V, Tamada JA, Klibanov AM et al (1991) Kinetics of insulin aggregation in aqueous solutions upon agitation in the presence of hydrophobic surfaces. Proc Natl Acad Sci USA 88:9377–9381. https://doi.org/10.1073/pnas.88.21.9377
46. Akkermans C, van der Goot AJ, Venema P et al (2008) Formation of fibrillar whey protein aggregates: influence of heat and shear treatment, and resulting rheology. Food Hydrocoll 22:1315–1325. https://doi.org/10.1016/j.foodhyd.2007.07.001
47. Mv S (1918) Versuch einer mathematischen Theorie der Koagulationskinetik kolloider Lösungen. Z Phys Chem 92U:129–168. https://doi.org/10.1515/zpch-1918-9209
48. Gough P, Jenness R (1962) Heat denaturation of β-lactoglobulins A and B. J Dairy Sci 45:1033–1039. https://doi.org/10.3168/jds.S0022-0302(62)89556-3
49. Park KH, Lund DB (1984) Calorimetric study of thermal denaturation of β-lactoglobulin. J Dairy Sci 67:1699–1706. https://doi.org/10.3168/jds.S0022-0302(84)81495-2
50. Dannenberg F, Kessler H-G (1988) Reaction kinetics of the denaturation of whey proteins in milk. J Food Sci 53:258–263. https://doi.org/10.1111/j.1365-2621.1988.tb10223.x
51. Köhler K, Aguilar F, Hensel A et al (2007) Design of a microstructured system for homogenization of dairy products with high fat content. Chem Eng Technol 30:1590–1595. https://doi.org/10.1002/ceat.200700266

52. Köhler K (2010) Simultanes Emulgieren und Mischen. Karlsruher Institut für Technologie (KIT)
53. Köhler K, Hensel A, Kraut M et al (2011) Melt emulsification—is there a chance to produce particles without additives? Particuology 9:506–509. https://doi.org/10.1016/j.partic.2011.03.009
54. Gall V, Rütten E, Karbstein HP (2022) Cavitation patterns in high-pressure homogenization nozzles with cylindrical orifices: influence of mixing stream in Simultaneous Homogenization and mixing. Exp Comput Multiph Flow 4:156–164. https://doi.org/10.1007/s42757-020-0088-9
55. Bałdyga J, Pohorecki R (1995) Turbulent micromixing in chemical reactors—a review. Chem Eng J Biochem Eng J 58:183–195. https://doi.org/10.1016/0923-0467(95)02982-6
56. Fournier M-C, Falk L, Villermaux J (1996) A new parallel competing reaction system for assessing micromixing efficiency—experimental approach. Chem Eng Sci 51:5053–5064. https://doi.org/10.1016/0009-2509(96)00270-9
57. Winkelmann M, Schuler T, Uzunogullari P et al (2012) Influence of mixing on the precipitation of zinc oxide nanoparticles with the miniemulsion technique. Chem Eng Sci 81:209–219. https://doi.org/10.1016/j.ces.2012.06.036
58. Palmer DA, Ramette RW, Mesmer RE (1984) Potentiometric studies of the thermodynamics of iodine disproportionation from 4 to 209 °C. J Solution Chem 13:685–697. https://doi.org/10.1007/BF00649009
59. Commenge J-M, Falk L (2011) Villermaux-Dushman protocol for experimental characterization of micromixers. Chem Eng Process 50:979–990. https://doi.org/10.1016/j.cep.2011.06.006
60. Guichardon P, Falk L (2000) Characterisation of micromixing efficiency by the iodide–iodate reaction system. Part I: experimental procedure. Chem Eng Sci 55:4233–4243. https://doi.org/10.1016/S0009-2509(00)00068-3
61. Guichardon P, Falk L, Villermaux J (2000) Characterisation of micromixing efficiency by the iodide–iodate reaction system. Part II: kinetic study. Chem Eng Sci 55:4245–4253. https://doi.org/10.1016/S0009-2509(00)00069-5
62. Kölbl A, Kraut M, Schubert K (2008) The iodide iodate method to characterize microstructured mixing devices. AIChE J 54:639–645. https://doi.org/10.1002/aic.11408
63. Faryadi M, Rahimi M, Safari S et al (2014) Effect of high frequency ultrasound on micromixing efficiency in microchannels. Chem Eng Process 77:13–21. https://doi.org/10.1016/j.cep.2014.01.001
64. Gall V, Karbstein HP (2020) Influence of cavitation and mixing conditions on oil droplet size in simultaneous homogenization and mixing (SHM). ChemEng 4:64. https://doi.org/10.3390/chemengineering4040064

In-Situ Investigation of Structural Changes of Biomolecules at Interfaces and under the Influence of Fluid Mechanical Stress

Vanessa Lautenbach, Maximilian J. Uttinger, Tobias Guckeisen, Christian Lübbert, Georgy Onishchukov, Johannes Walter, Saman Hosseinpour, and Wolfgang Peukert

Abstract The integrity of proteins is a decisive factor in their functionality. However, during production and processing, the flexible biomolecules are frequently exposed to external mechanical stress and interact with different surfaces, both might affect their folding. Structural changes promote aggregation induced by hydrophobic interactions. Stress-treated, genetically modified beta-lactoglobulin (BLG) proteins have been investigated in the bulk phase by analytical ultracentrifugation (AUC) and directly at the water/air and solid/water interfaces by sum-frequency generation

V. Lautenbach · M. J. Uttinger · T. Guckeisen · C. Lübbert · G. Onishchukov · J. Walter · S. Hosseinpour · W. Peukert (✉)
Institute of Particle Technology (LFG), Friedrich-Alexander-Universität Erlangen-Nürnberg (FAU), Cauerstraße 4, 91058 Erlangen, Germany
e-mail: wolfgang.peukert@fau.de

V. Lautenbach
e-mail: vanessa.lautenbach@fau.de

M. J. Uttinger
e-mail: max@uttinger.com

T. Guckeisen
e-mail: tobias.guckeisen@fau.de

C. Lübbert
e-mail: christian.luebbert@fau.de

G. Onishchukov
e-mail: georgy.onishchukov@fau.de

J. Walter
e-mail: johannes.walter@fau.de

S. Hosseinpour
e-mail: saman.hosseinpour@fau.de

C. Lübbert · J. Walter · W. Peukert
Interdisciplinary Center for Functional Particle Systems (FPS), Friedrich-Alexander-Universität Erlangen-Nürnberg (FAU), Haberstraße 9a, 91058 Erlangen, Germany

A. Kwade and I. Kampen (eds.), *Dispersity, Structure and Phase Changes of Proteins and Bio Agglomerates in Biotechnological Processes*,
https://doi.org/10.1007/978-3-031-63164-1_13

(SFG) spectroscopy. The influence of the solvent pH, salt concentration or addition of denaturating agents on the protein structural changes and the resulting particle-particle interactions have been investigated revealing the important role of electrostatic interactions. Interfacial isoelectric points (IEP) of various proteins such as BLG, bovine serum albumin (BSA), hemoglobin (Hem), lysozyme (Lys), insulin, and a number of others have been measured by surface-sensitive SFG spectroscopy and found to be almost identical to that in the bulk phase. Chiral SFG measurements indicate that proteins at the interface can be less susceptible to structural changes than in the liquid phase. Electrospray ionization mobility–classified mass spectrometry (ESI-DMA-MS) has been additionally used to characterize protein size, shape, and density in the gas phase via single-molecule pycnometry.

1 Introduction

Biomolecules such as proteins, nucleic acids or carbohydrates are flexible molecules, which are essential for all biological processes including cell division or morphogenesis [1]. Biomolecules typically react extremely sensitive to their environmental conditions [2]. During their production and processing, biomolecules may experience chemical, thermal, and fluid mechanical stresses and are exposed to external surfaces, which all can cause structural modifications and changes in dispersity (Fig. 1). The term dispersity summarizes size, shape, structure, and surface properties and their respective distributions. Typical examples are mechanically induced degradation processes, which can occur due to high shear and elongational stresses in many stages of downstream processing including stirred tanks, filtration stages, and pumping [3]. Interactions with surfaces are unavoidable in bubble and chromatographic columns, microfluidic devices, and other processes involving adsorption. Degradation causes on the one hand a reduction in the yield of the target product and lowers on the other hand their stability and functionality. When the acting stresses overcome the internal strength of the biomolecules, reversible and often irreversible structural changes are induced, which makes the product, for instance proteins, ineffective, as they lose their native conformation and therefore cannot perform properly.

Reversible protein degradation is used during targeted protein degradation to study the role of certain proteins in cellular processes. The addition of small molecules, which specifically modulate the protein structure reversibly or irreversibly, provide information about diseases and human physiology and hence enable the development of effective therapeutics [4, 5].

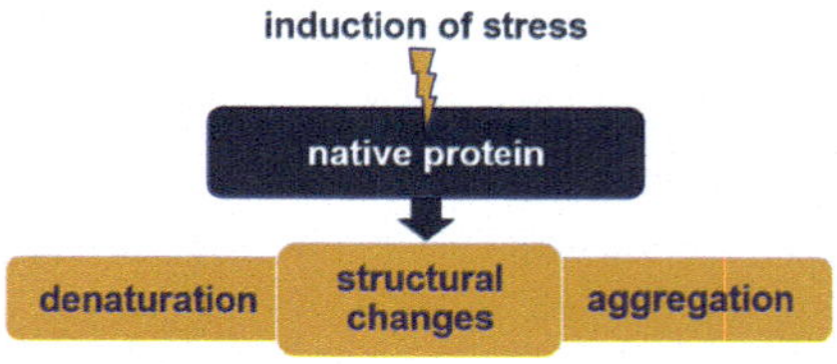

Fig. 1 Schematic representation of possible stress-induced effects on proteins

However, in most cases denaturation is undesired and must be avoided during processing. Process design and operational conditions define the distribution of stress, which acts on the proteins. Knowledge on critical, acceptable stress levels for proteins, which do not cause any denaturation, are therefore important boundary conditions for choosing operational parameters along the downstream process chain. In other words, denaturation can be avoided by mapping well-proven structure-property relationships of the proteins with process-structure correlations during their handling. A grand challenge for knowledge-based design is the determination of the structure-property functions on protein stability and functionality in dependence on different types and magnitudes of acting stresses.

In an aqueous bulk environment, proteins preferably fold in such a way that their hydrophobic parts are located in the interior, while hydrophilic fractions are arranged at the surface of the protein. Irreversible structural changes usually lead to partial or complete unfolding of a protein causing changes in its original, native conformation. The unfolding of the protein thus influences its size, shape, and surface properties. As a result, internal, in particular hydrophobic parts of the protein are exposed to the surface and trigger strong hydrophobic interactions. Hydrophobic interactions depend on the restructuring of water around hydrophobic patches at the protein surface and are thus entropic effects. The resulting attractive interactions easily induce unwanted aggregation [6]. During adsorption chromatography, for instance, hydrophobic protein patches may bind irreversibly to hydrophobic absorbents [7].

In addition to hydrophobic interactions, electrostatic interactions play a major role in intermolecular protein-protein interactions. Sufficiently strong repulsive forces are the precondition for stable colloidal solutions, which is affected by the pH and the salt concentration, i.e. the ionic strength. Depending on pH, amino-acid residues are protonated or deprotonated and thus generate local charges, which affect the surface properties and therefore influence the protein stability and via aggregation the dispersity. Interparticle interactions can be described by the classical DLVO theory and its extensions. In brief, attractive van der Waals forces, repulsive electrostatic forces, and possibly also hydrophobic interactions are summarized to determine the overall interaction potential [8, 9]. Aggregation therefore depends on the prevailing interactions and collision rates in the solution, which can be driven either by purely Brownian motion or in laminar and/or turbulent flows by the respective shear stresses [10, 11].

The isoelectric point (IEP) of a protein is a significant parameter, which corresponds to the pH at which the respective protein is net uncharged. Accordingly, electrostatic repulsive interactions vanish and attractive forces prevail at the IEP. This phenomenon is commonly applied in the field of wastewater treatment, where the pH is deliberately set to the IEP to increase flocculation and facilitate protein removal [12, 13]. Also other separation and purification techniques such as ion-exchange chromatography or electrophoresis rely on the protein charge state, which in turn is dependent on the IEP [14]. Furthermore, the IEP plays an important role in foam stability. During foam formation, proteins adsorb at the gas/liquid interface. The properties of the interfacial thin films determine bubble coalescence, foam stability, and essentially all macroscopic foam properties [15]. Elastic and inelastic properties of the film depend on whether the proteins remain intact or denature at the interface, an issue, which is intensely discussed in literature [16–18]. A further important aspect is colloidal stability, respectively, aggregation at the interface. At pH values close to the IEP, the protein film is particularly stress-resistant due to increased interfacial dilatational elasticity. The absence of repulsive forces between the individual protein molecules at IEP leads to a thicker protein layer in comparison to that at pH values away from the IEP. This multilayer protein film at the interface significantly influences foam stability and leads to improved rheological properties in comparison to protein foams with protein monolayers [15].

The adsorption to different kinds of surfaces like, for instance, pipes during industrial processes or implants in biological systems can initiate structural and conformational changes [19, 20]. In bubble columns, biomolecules adsorb to gas bubbles and denature there, whereby adsorption at the fluid/gas interface can additionally lead to undesirable foam formation. Furthermore, the adsorption to hydrophobic cell membranes is assumed to be a potential starting point of interfacial aggregation and fibril formation. The latter can trigger several neuronal disorders like Alzheimer's or Parkinson's diseases [21, 22].

While the formation of aggregates or even amyloid fibrils is considered a detrimental effect, intentional fibril formation has gained more and more attention in the recent years. Amyloid aggregates find applications as foaming, gelation or emulsification agents in food engineering, biosensors, and nanocomposites in biomaterials [23, 24]. Therefore, it is of great interest to establish a rapid, universal analysis and characterization procedure of protein aggregates. The quantification of stressed proteins is very important in order to identify possible correlations between stress induction and structural changes.

The objective of our work was to apply and further develop advanced characterization methodologies well-established in particle science and technology to biomolecules. This way, we obtain valuable insights into the behavior of proteins in the bulk phase and directly at interfaces. Both native and genetically modified proteins are considered. In particular, protein size, shape, and interactions are carefully characterized. For the analysis of proteins in the bulk phase, the focus was set to analytical ultracentrifugation (AUC), which is a versatile and highly reliable method to directly investigate proteins in their liquid environment. A relatively new method is electrospray ionization combined with mobility-classified mass spectrometry (ESI-DMA-MS) [25, 26]. In this method, the sample is first transferred into the gas phase using electrospray ionization (ESI), then classified according to its electrical mobility by differential mobility analysis (DMA) and finally detected by a mass spectrometer (MS). ESI-DMA-MS provides molecular information on the mass, density, and even shape of the analyte [27]. Protein assemblies and adducts can be characterized as well.

Proteins have been intensely studied at the surface, investigating both the arrangement and structure of the proteins at the water/air and solid/water interface. Many studies deal with the adsorption of proteins at interfaces and their influence on the protein structure. However, direct molecular detection of biomolecules at interfaces is only possible with very few techniques. The reason is that the number of molecules at an interface is usually by many orders of magnitude smaller than the number of molecules in the bulk phase. Nonlinear optical sum-frequency generation (SFG) spectroscopy can provide information on molecules directly at the interface, while signal from bulk is often suppressed by the selection rules in nonlinear optics. In brief, comprehensive infrared spectra of molecules at the surface can be measured with high sensitivity and selectivity. In particular, we investigated the influence of solvent properties, such as pH, salt concentration or the addition of denaturing substances on the protein structure at the interface.

2 Material and Methods

2.1 Analytical Ultracentrifugation (AUC)

Analytical ultracentrifugation is an established method to measure sedimentation and diffusion properties of proteins and to obtain information about their size, shape or density. In comparison to other techniques, the unique advantage of an AUC is the separation of analytes in strong centrifugal fields coupled to optical characterization. AUC provides outstanding resolution, accuracy and reproducibility. Contrary to the common particle systems, stable protein solutions show relatively narrow sedimentation distributions. With regard to the investigation of protein aggregation, it is possible to distinguish between different species based on the difference in their weight and size. For instance, monomers and dimers can be distinguished easily, which is not possible by classical particle size analysis such as dynamic light scattering. In addition, binding constants can be determined. Several reviews deal with the comprehensive analysis of biomolecules [28–32]. Moreover, hydrodynamic and thermodynamic non-ideality effects can be studied as function of analyte concentration [33, 34].

AUC operates in vacuum at rotor revolutions up to 60,000 rpm providing centrifugal accelerations equivalent to 290,000 g. Sedimentation properties of samples can be detected by multiwavelength absorption, emission, and interference techniques. The sample is filled into a special, light-transparent sector-shaped cell. Depending on the rotor design, it is possible to measure up to eight cells simultaneously. During rotation, the particles experience centrifugal forces and settle towards the bottom of the cell with a sedimentation velocity, depending on their size, shape, and density as well as on the density and viscosity of the surrounding medium.

The ratio of sedimentation velocity u and centrifugal acceleration $r \cdot \omega^2$ provides the sedimentation coefficient s, where r denotes the distance of the particle from the centre of rotation and ω the angular velocity. Considering all forces acting on the sedimenting particle (centrifugal, buoyancy, and frictional forces), the sedimentation coefficient can be calculated as a function of the particle mass m, the friction coefficient f, and the densities of the solvent ρ_s and of the particle ρ_p.

$$s = \frac{u}{r \cdot \omega^2} = \frac{m \cdot (1 - \frac{\rho_s}{\rho_p})}{f} = \frac{m \cdot (1 - \frac{\rho_s}{\rho_p})}{3 \cdot \pi \cdot \eta \cdot x_H} \tag{1}$$

In addition to classical sedimentation, flotation can also occur in the centrifugal field if the density of the particle is lower than that of the surrounding medium and the buoyancy force thus dominates over the centrifugal force. In such cases, the sedimentation coefficient is negative.

During an AUC sedimentation velocity measurement, radial scans of the sample in the measuring cell are performed periodically, whereby the evolution of radial change in extinction is recorded over time and so-called sedimentation profiles are collected. An illustration of radial scans is shown in Fig. 2a. The meniscus of the sample is located at the liquid/air interface. By simultaneous recording of the reference chamber, the absorbance of the sample can be calculated. Under sufficiently dilute conditions, Lambert Beer's law is valid. The progress of the recorded sedimentation boundary profiles (Fig. 2b) depends on the sedimentation coefficient, while their broadness is influenced by polydispersity and diffusive effects.

Small particles are subject to Brownian motion, which also comes into play as diffusive effects during centrifugation and influences respectively the sedimentation. The diffusion coefficient can be calculated according to the Stokes–Einstein relation using the temperature T, the Boltzmann constant k, the viscosity of the medium η, and the hydrodynamic diameter of the particle x_H [38].

$$D = \frac{k \cdot T}{f} = \frac{k \cdot T}{3 \cdot \pi \cdot \eta \cdot x_H} \tag{2}$$

Both the sedimentation and diffusion are affected by the friction coefficient f. The frictional ratio f/f_0 describes the ratio between the frictional coefficient of the particle and the friction coefficient f_0 of a sphere of the same volume [39]. f/f_0 is one for

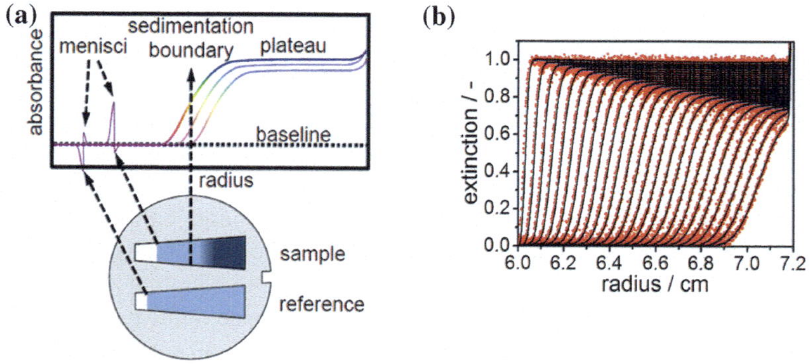

Fig. 2 **a** Measurement procedure for AUC sedimentation velocity experiments. Reprinted with permission from [35]. Copyright (2014) American Chemical Society. **b** Time-dependent sedimentation profiles. Data have been generated with SEDFIT [36] and displayed with GUSSI [37]

spherical non-solvated particles whereas deviations from one indicate an increasing asymmetry and/or solvation [40].

The Lamm equation is derived from a mass balance by taking the fluxes due to sedimentation and diffusion into account (see Eq. (3)) and describes the time-dependent change in concentration c in a sector-shaped cell [41]. The recorded sedimentation profiles can be fitted to numerical solutions of the Lamm equation to obtain information about the sedimentation and diffusion coefficients of the settling species.

$$\frac{\partial c}{\partial t} = D\left(\frac{\partial^2 c}{\partial r^2} + \frac{1 \partial c}{r \partial r}\right) - s \cdot \omega^2 \left(r \cdot \frac{\partial c}{\partial r} + 2 \cdot c\right) \tag{3}$$

With known s and D, the molar mass M_p of the sample can be calculated using the Svedberg equation [42]

$$M_p = \frac{s \cdot R \cdot T}{D \cdot (1 - \overline{v} \cdot \rho_s)} \tag{4}$$

with R being the ideal gas constant and $\overline{v}$ the partial specific volume of the sample, which describes the change in solution volume upon immersion of a defined mass of the analyte. The partial specific volume can be linked to the inverse of the effective density of the particle, in which case Mp relates to the effective mass of the particle.

Of particular interest for the study of macromolecules are elevated concentrations, where hydrodynamic and thermodynamic non-ideality effects become important. For analytes with sufficient diffusion, it is possible to determine the second virial coefficient B_{22} as measure for thermodynamic non-ideality [43, 44]. B_{22} follows from the virial expansion of the concentration-dependent reduced osmotic pressure [43, 45]. A concentration-dependent diffusion coefficient $D(c)$ can be derived from Eq. (5) [43, 46], where D_0 denotes the diffusion coefficient in the infinite dilution limit and k_s the Gralen coefficient, which is an empirically introduced constant describing concentration dependency of the friction coefficient. $O(c^2)$ refers to higher order terms, taking second order and higher terms into account [46].

$$D(c) = D_0 \cdot \frac{1 + 2 \cdot B_{22} \cdot M \cdot c - \overline{v} \cdot c + O(c^2)}{1 + k_s \cdot c} \tag{5}$$

Negative values of B_{22} indicate globally attractive interactions, while positive values indicate repulsive behavior [47]. In the context of protein–protein interactions, B_{22} is determined to characterize the crystallization behavior of proteins [48, 49] and to improve the understanding of protein aggregation and fibril formation [50].

An AUC experiment can be carried out in three different measurement modes. In a sedimentation equilibrium experiment, the rotational speed is moderate to establish an equilibrium between the contributions of sedimentation and diffusion to the mass transport in the cell. This type of measurement originates from the early days of AUC development, requires relatively long measurement times and is mainly used to determine the molecular weight and intermolecular particle–particle interactions

[51, 52]. A much more common and faster method is the sedimentation velocity experiment, in which, as described above, the cell is scanned at a constant rotational speed over radius and time. Particle size distributions can be determined with high resolution, excellent reproducibility and accuracy. For wide particle size distributions, it is recommended to measure in the gravitational sweep mode. Here, the rotor speed is varied at a constant radius position so that a wider range of particle sizes can be covered [53, 54].

Analytical ultracentrifugation provides invaluable insight into the composition of a protein sample in solution. It is a highly suitable method to study proteins in their native quaternary structure and provides information on density, hydration, and their interactions including dimerization and aggregation in the liquid phase.

The presented results were obtained at 20 °C using ultracentrifuges from Beckman Coulter (Krefeld, Germany)—either a commercial analytical centrifuge type Optima AUC or a modified preparative centrifuge type Optima L-90 K (details can be found elsewhere [35, 55]). Prior to an AUC measurement, the samples were checked for their concentration and absorption using UV–Vis spectroscopy (Varian Cary 100 (Waldbronn, Germany) or Specord 2110 plus (Analytic Jena GmbH)) and diluted accordingly. Depending on the analyte concentration, measurements were performed using either 12 mm (~400 μl sample) or 3 mm (~100 μl sample) titanium centrepieces. For AUC sedimentation velocity experiments, a rotor speed of 40,000 or 50,000 rpm was selected and the sample was centrifuged for about 10 h until complete sedimentation was observed. Unless otherwise stated, the measured data were pseudo-referenced following Kar et al. [56]. The data analysis of the sedimentation velocity experiments was performed with the continuous c(s) model implemented in SEDFIT (version 16.1) [57]. The $\overline{v}$ for beta-lactoglobulin was set to 0.751 ml/g and the solvent properties were set to those of water at 20 °C.

For the determination of the dimer dissociation constant, the procedure described by Zaho et al. was followed and an isotherm was fitted to the concentration-dependent weight-averaged sedimentation coefficients using SEDPHAT [58]. Detailed information on the experiments can be found in the respective publications.

2.2 *Electrospray Ionization Differential Mobility-Classified Mass Spectroscopy (ESI-DMA-MS)*

Electrospray ionization (ESI) coupled to differential mobility analyzer (DMA) in combination with time-of-flight mass spectroscopy (MS) is a three-step analytical process, see Fig. 3a, consisting of

1. ESI—transfer of the analyte from the liquid phase to the gas phase by electrospray,
2. DMA—separation of the species in the gas phase according to their electrical mobility in a differential mobility analyzer and
3. TOF–MS—time-of-flight mass spectrometry.

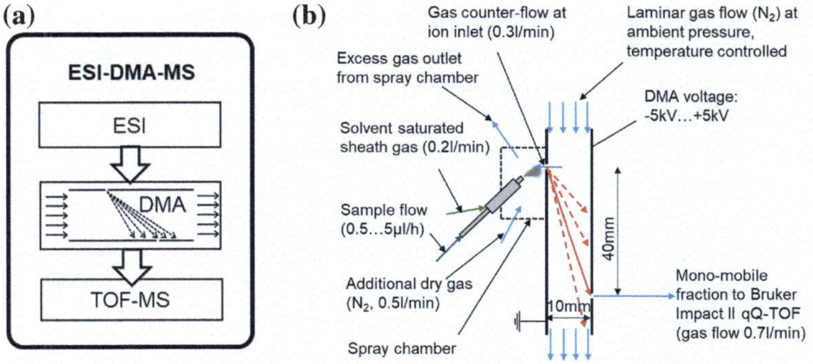

Fig. 3 **a** Principle of the electrospray ionization differential mobility-classified mass spectroscopy set-up (ESI-DMA-MS). **b** Details of the electrospray coupling to the differential mobility analyzer

2.2.1 Electrospray Ionization

Since even the best ion-mobility mass spectrometer can only analyze the ions as produced by the ion source, spray performance is a key point to generate interpretable samples, when it comes to quantification. A home-built electrospray ionization source is used for ion generation. The spray source allows operation at small flow rates (0.3–6 µl/h) in a highly stable cone-jet mode in positive and negative polarity [59]. A key feature of the source technology is a sheath gas flow, which surrounds the sprayer tip [25]. This sheath gas is saturated by solvent to prevent liquid evaporation from the Taylor cone and allows to tune the liquid composition. Under such conditions, the spray emits droplets of 50–150 nm diameter. The specific droplet size can be tuned accurately by the liquid conductivity and the flow rate. The droplet size distributions are narrow with a typical relative standard deviation of <15% and a mode diameter, which does not change by more than 1 nm during a 30 min mobility scan. Continuous visual access by a microscope camera and measurement of the spray current provide information on the spray stability. Moreover, the source is equipped with a current coupling technology, which prevents any interference of electrochemical processes on the sprayed solution [60]. The combination of such a source and the DMA-MS system allows exceptionally accurate investigations of droplet size-related problems.

During solvent evaporation, the charged drops emit mass and charge in Coulomb fission events. These events occur when the electrostatic stresses compensate the capillary pressure. A droplet loses around 1–2% of its mass and around 15% of its charge in such events, whereby 10–20 droplets of 1/8th of the parent drop are emitted, as a widely accepted assumption [61]. Many of these events occur successively during solvent evaporation. As result of this process, there are ions produced from progeny drops and that from the parent drops. The distribution of the analyte between these two types of drops is a complex function of the analyte, the solvent, and the electrolyte properties and their evolution during evaporation. Dependent on the specific goal, the spray parameters can be tuned in different directions.

Chemical conditions for evaluation of the exact analyte mass (conventional approach): If determination of the exact protein mass is the goal of the analysis, one typically combines a weak and volatile acid (acetic acid, formic acid) with solvent mixtures from water and an organic component (acetonitrile, methanol, and ethanol) and performs the spray in the positive polarity. The resulting denaturated proteins are emitted in Coulomb fission events with high probability. Due to the high-charge state of the partly deconvoluted protein ions, declustering of attached solvent in the collision cell of the mass spectrometer is simple. This approach is largely independent of the spray technology and provides adduct-free ions. However, it is not suitable for sample quantification. Due to the denaturating solvent conditions, structural information on the protein—specifically the collision cross-section as accessible from the mobility classification—is not related to the native state of the protein, which is a key point of interest in structural biology [62].

Chemical conditions for evaluation of biomolecule structure (native conditions): Conditions to gain insight into protein structure require solvents similar as close as possible to the protein natural environment. Consequently, water is used as only solvent and the liquid conductivity is optimized by pH-adjusted ammonium acetate in concentrations of typically 10–100 mmol/l. Under such conditions, structural features of native proteins are maintained (at least to some extent) after transition to the gas phase [63]. However, due to the small droplet size generated under these conditions, the number of Coulomb fission events is small and the majority of proteins is not emitted from the parent droplets. The ions observed under these conditions are to a large extent evaporation residues of the parent droplets. In case of more than a single non-volatile molecule in a parent drop, these residues are cluster ions consisting of non-volatiles in the drop. Since these clusters are not related to any clustering state in the solvent prior to the spray process, these clusters are referred to as non-specific clusters. These non-specific cluster ions can be formed as analyte-analyte clusters and clusters of analyte molecules with non-volatile contamination. If the goal of the analysis is to gain insight into specific clustering (clusters that do exist in liquid phase prior to the electrospray process), non-specific ion clustering must be excluded by careful adjustment of spray conditions. Due to the more compact structure of the gas phase ions, the comparably low vapor pressure of ammonium acetate and the high sensitivity to non-volatile contamination, it is more challenging to generate adduct-free ions suitable for exact mass analysis under these conditions [64].

Boundary conditions for the flow rate: The spray flow rate has a significant impact on the generated ion ensemble. Since in a classical electrospray, high flow rates of a few 100 μl per hour generate large droplets (typically some μm in diameter), the evaporation residues of the primary droplets are too large for detection. Consequently, only the ions that are produced in the fission process are detected. This is only an unknown fraction of all analyte species, thus quantification requires careful calibration. Furthermore, the spray is typically performed under elevated temperature to promote sufficiently fast solvent evaporation of the large drops. A large distance between the sprayer tip and the mass-spectrometer inlet results in a low transmission efficiency of the ions into the mass spectrometer. Nano electrosprays operated at low flow rates (~1 μl/h) are capable to generate very small

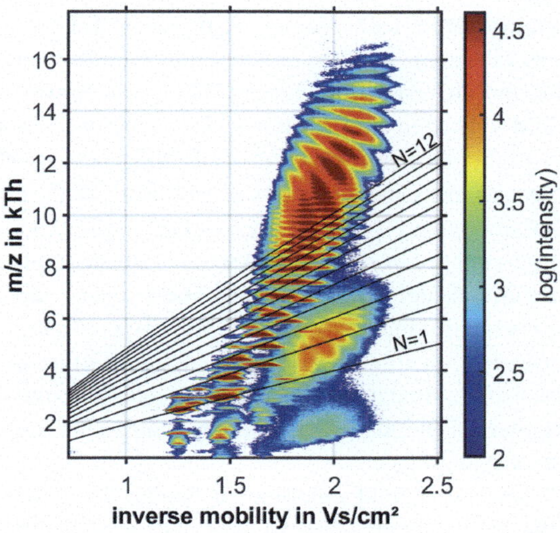

Fig. 4 Example of a mass spectrum of lysozyme clusters recorded in native ion mode. The black lines indicate expected locations of lysozyme clusters of N = 1–12 based on mobility theory

droplets (< 50 nm). Water droplets of such size evaporate in a very short time. The solvent is completely evaporated at a distance of a few millimeters from the spray tip even at ambient temperature. Consequently, the spray can be located very close to the mass-spectrometer inlet, which results is a high transfer efficiency. The small droplet volume allows for singularization of species already in the primary drops. Emission of analyte molecules is thus not limited to species emitted from the primary drops in Coulomb fission events [65]. While low flow rates are in general preferred, they are not easy to handle and are in many cases not compatible to separation processes (e.g. HPLC) employed before mass analysis. Thus, high-flow electrosprays with flow rates >200 μl/h and nanosprays with flow rates in the order of 1 μl/h are used for different analytical processes. Most important boundary conditions are shown in Fig. 3b.

Spray conditions in-between of these two extreme situations can be realized as well. For example, the spectrum in Fig. 4 was recorded spraying lysozyme from a water/formic acid solution. Using formic acid as electrolyte results in a much better desolvation of large protein clusters (here up to around 12 nm) due to the higher volatility of formic acid compared to ammonium acetate. To prevent unfolded proteins in the gas phase, the mass spectrum was recorded in the negative ion mode. As lysozyme is charged positively in solution at acidic pH, it is emitted in low quantity from the sprayed drops, and the low charge level (charge states -5 and -6 for the monomer) ensure compact clusters, which result in good mobility resolution. This specific situation is ideal for studying statistics of non-specific ion clustering.

2.2.2 Mobility Classification

In the second step, the ions are separated in the gas phase by their respective electrical drift velocity in an electric field. Ion mobility separation can be performed with different technologies. Drift tubes (DT), traveling-wave mobility spectrometers (T-wave), field-asymmetric ion mobility spectrometry (FAIMS), and trapped ion mobility spectrometers (TIMS) exist beside the differential mobility analyzer (DMA) used in our instrument. These instruments are typically combined with time-of-flight mass spectrometers (TOF–MS), but can in principle be combined with other types of mass spectrometers as well [66]. With respect to data evaluation, mobility-classified mass spectrometry has a relevantly higher peak capacity, as the overlap of different species on the mass scale is substantially reduced by pre-separation on the mobility scale. Other than mass spectrometry alone, mobility-classified mass spectrometry is capable to analyze polydisperse samples with hundreds of ion species, provided ion signals are strong enough. Fitting mobility theory to acquired data, it is even possible to attribute charge levels and ion masses to signals that are resolved neither on the mass nor mobility scale. This property allows the analysis of large protein aggregates.

Our instrument combines a Seadm P4 differential mobility analyzer (resolution 60 for single charged ions and up to 100 for multiple charged ions) with a Bruker Impact II qQ-TOF (resolution 40,000). This instrument combination allows analyzing protein clusters of masses up to 900 kDa. To promote efficient transition of large ions in the rear part of the instrument (collision cell and transfer to the TOF mass analyzer), helium is used as collision gas, elsewise the instrument is used under standard conditions. While TOF–MS is a standard instrument for mass analysis, DMAs are seldom applied for mobility analysis. A DMA consists of an electrical condenser. Between its two plates, gas flows at laminar conditions with velocities in the order of 50 m/s. Narrow slits in the condenser plates allow entering and exiting the DMA for the ions, whereby the exit slit is displaced in the downstream direction of the gas flow. With a constant gas flow velocity in the DMA, ion trajectories in the DMA depend on their drift velocity in the electric field (red lines in Fig. 3b). Only ions of specific drift velocity reach exactly the location of the outlet slit and are transmitted to the mass spectrometer. To scan the ion mobility spectrum, the voltage on the exit plate of the DMA is scanned in a linear ramp. Since the ion drift velocity is proportional to their electrical mobility Z, the transmitted ion mobility is inversely proportional to the DMA voltage U. Since Z and U are constant in the DMA for a constant gas velocity, specific values of the electrical mobility can be attributed to the DMA voltage by a one-point mobility calibration. For this calibration, an ion of known electrical mobility Z_{cal} is scanned by the DMA and the voltage U_{cal}, at which the mobility reference is transmitted, is noted. The mobility transmitted through the DMA at a voltage U is correlated to the reference mobility and voltage by:

$$Z(U) = \frac{U_{cal} Z_{cal}}{U} \tag{6}$$

Since the DMA is scanned in a linear voltage ramp, it is common to plot the mass data not over the mobility Z, but over the inverse mobility 1/Z. For data interpretation, the measured mobility of an ion can be compared to theoretically calculated values of the mobility. Beside numerical simulations, the most common model for mobility

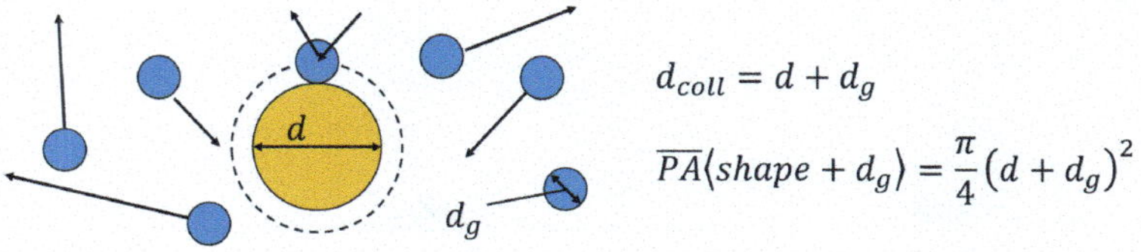

Fig. 5 Schematic representation of the collision of a particle with gas molecules

calculation is the Mason-Schamp equation:

$$Z = \frac{ze}{CCS}\frac{3}{16N}\sqrt{\frac{2\pi}{kT\mu_r}} \tag{7}$$

The Mason-Schamp equation correlates the ion mobility Z to the number z of elementary charges e, the number of gas molecules per volume N, the gas temperature T, the reduced mass of the ion $\mu_r = m_g m_i/(m_g + m_i)$, whereby the m_g and m_i are the mass of a drift gas molecule and the mass of the ion, respectively, and the Boltzmann constant k. The information on the ion size is represented by the collision cross-section CCS. For species larger than 1.5 nm in diameter that drift in air or nitrogen, the collision cross-section can be approximated by $CCS = \xi\overline{PA}\langle shape + d_g\rangle$ with a good accuracy.

$\overline{PA}\langle shape + d_g\rangle$ is the orientation-averaged projected area, extended by the gas molecule collision diameter d_g (see Fig. 5). This is straightforward for a spherical particle. For complex shaped molecules like proteins, values of $\overline{PA}\langle shape + d_g\rangle$ can be calculated by software tools. ξ is the ratio of the CCS and the mean projected area $\overline{PA}\langle shape + d_g\rangle$. Since its value is related to the momentum transfer efficiency in a collision, it is called the momentum accommodation coefficient and can theoretically have values between 1 and 1.4. Its value for organic molecules in nitrogen or air is 1.36–1.39, independent of size down to a particle diameter of 1.5 nm. Comparing calculated collision cross-sections to values derived from measured mobilities by Eq. (7) is frequently employed in structural biology for confirmation of predicted protein structures [62, 67, 68].

2.3 *Sum-Frequency Generation Spectroscopy (SFG)*

SFG is a surface-sensitive molecular spectroscopic method based on a non-linear optical second order effect. For the generation of SFG, two lasers with high pulse peak power are used, one with a fixed frequency (ω_{vis}) in the visible (vis) spectral range and a second one with an adjustable infrared (IR) frequency (ω_{IR}). When the laser beams overlap temporally and spatially at the interface, their frequencies sum up and thereby generate the SFG signal $\omega_{SFG} = \omega_{vis} + \omega_{IR}$ (see Fig. 6). The

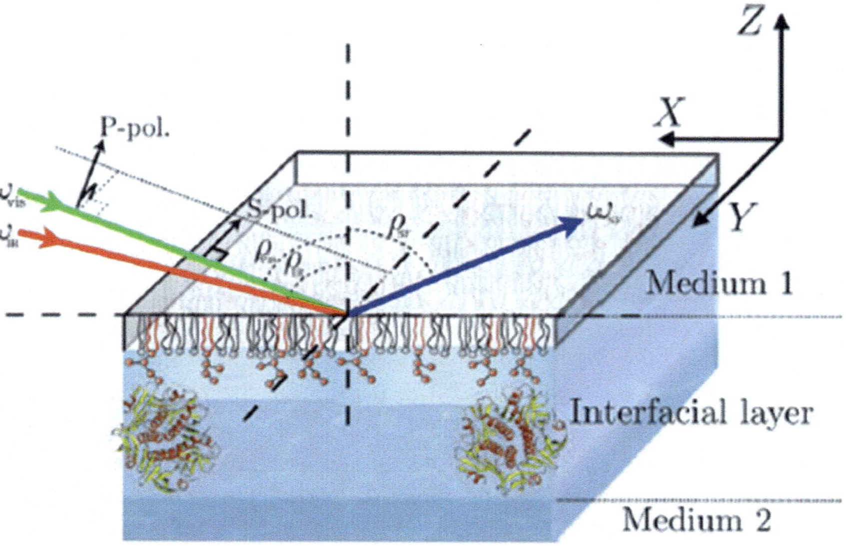

Fig. 6 Schematic representation of a protein SFG experiment, indicating definitions of the beam polarizations, incoming and outgoing angles ρ, and the lab-frame axes [71]. Reproduced from [72]. Copyright (2013) American Chemical Society, licensed under the Creative Commons Attribution 4.0 International license (CC BY 4.0)

spectroscopic SFG setup includes polarizers, which polarize the incoming visible and IR beams as well as the outgoing SFG signal. The polarization can be either S or P corresponding to the electrical field vector being perpendicular (S) or parallel (P) to the plane of incidence, resulting in eight different polarization combinations for the incident beams and the resulting SFG signal. Achiral measurements at the liquid/air interface are typically measured in the SSP (S-polarized SFG, S-polarized vis, and P-polarized IR) configuration, whereas for chiral SFG measurements other polarization modes like SPP, PSP or PPS are used [69, 70].

SFG spectroscopy can only be conducted in accordance with certain requirements. The investigated molecules must have a particular degree of order at the interface and should reside in a non-centrosymmetric environment. In addition, simultaneous IR and Raman activities of the sample are required. These selection rules make it possible to investigate molecules adsorbed on the surface without any contribution from bulk molecules and make SFG spectroscopy an inherent surface-sensitive measurement technique.

As Eq. 8 shows, the measured signal intensity of SFG spectroscopy is proportional to the intensities of the incident visible I_{vis} and infrared I_{IR} light and is a function of the non-resonant (NR) and resonant (R) parts of the second-order nonlinear susceptibility $\chi^{(2)}$ [20, 73–75].

$$\mathrm{I_{SFG}} \propto \left|\chi_{NR}^{(2)} + \chi_{R}^{(2)}\right|^2 .I_{vis}.I_{IR} \propto \left|\chi_{NR}^{(2)} + \sum_n \frac{A_n}{\omega_n - \omega_{IR} + i\Gamma_n}\right|^2 .I_{vis}.I_{IR} \qquad (8)$$

The resonant sub-term is dependent on the infrared frequency ω_{IR} and can be expressed as a function of the amplitude A_n, and the damping factor Γ_n, of the nth vibrational mode with frequency ω_n [20, 74]. As Eq. (9) shows, the oscillator strength A_n is composed of the product between the number N of ordered interfacial molecules and their tensor mode strength (the angular brackets refer to an orientation average over all interfacial molecules), which demonstrates that the SFG signal intensity is further strongly dependent on the number of molecules at the interface [20, 74]. Furthermore, on electrified surfaces, not only second-order $\chi^{(2)}$ but also third-order susceptibility $\chi^{(3)}$ contributes to the interfacial surface potential [20].

$$A_n = N\langle a_n \rangle \tag{9}$$

Proteins can be directly investigated at the interface using the described above powerful technique of surface-sensitive sum-frequency generation spectroscopy. It provides information about the adsorption behavior and can be used to characterize the charge state and molecular structure of biomolecules at fluid interfaces. Detailed information about the broadband SFG spectrometer used for the following investigations can be found elsewhere [76].

Figure 7 shows exemplary achiral SFG spectra recorded with a broadband SFG spectrometer in SSP polarization mode at the liquid/gas interface. For this purpose, individual measurements in the different spectral regions were combined to provide a wide frequency range and to give an overview of the individual characteristic regions. Region (a) demonstrates the fingerprint bands below 1350 cm^{-1}. The carboxylate (R-COO$^-$) symmetric stretching vibrations around 1400 cm^{-1} are located in area (b). (c) corresponds to the amide I bands followed by area (d), the pure non-resonant $\chi^{(2)}_{NR}$ contribution. The CH stretching region between 2800 cm^{-1} and 3000 cm^{-1} can be found in region (e) followed by the typical water vibrational mode from 3000 cm^{-1}–3800 cm^{-1} [77].

For measurements at the water/air interface, the protein sample was placed in a petri dish and measured from the top while rotating the sample. In contrast, investigations at the solid/liquid interfaces were measured from the bottom. Accordingly the beams were directed through a CaF_2 prism (Ma Teck GmbH, Jülich, Germany) to obtain the total internal reflection, which increases the collected protein signal. Prior to the measurements, the clean prism was spin coated by deuterated polystyrene (dPS) to investigate the proteins adsorbed at a hydrophobic interface. Details of sample preparation can be found in [20].

Since chiral measurements as well as the amid I region in achiral measurements correspond to the secondary structure of the proteins, it is possible to obtain information on potential structural changes of the interfacial proteins. The CH vibrations refer to the proteins backbone, and the OH vibrations reflect the protein surface charge, inducing orientation of the surrounding water molecules and therefore increase the SFG signal. Conclusively, SFG provides information about protein conformation at the interface, its arrangement, and charge state.

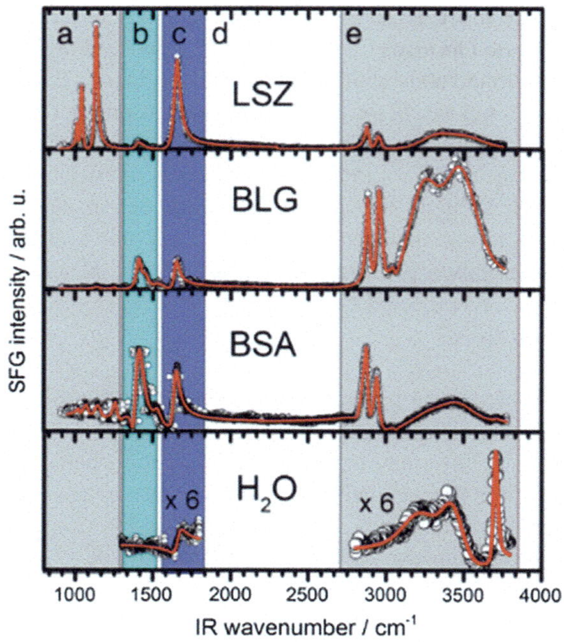

Fig. 7 Illustration of characteristic ranges and specific protein features within a typical protein SFG spectrum. For comparison, the broad frequency range (950–4000 cm^{-1}), assembled from partial measurements over shorter spectral ranges, is presented for lysozyme (LSZ), beta-lactoglobulin (BLG), bovine serum albumin (BSA), and water at the water/air interface. Reprinted from [77], Copyright (2014), with permission from Elsevier

2.4 *Proteins and Buffers*

Unless otherwise described, all proteins, salts, and solvents used were commercially obtained. Detailed sample preparation can be found in respective publications referred to in the following text. Bovine serum albumin (BSA; A7030), beta-lactoglobulin (BLG B; L8005), cytochrome C (Cyt C; C2506), α-lactalbumin (LALBA; L6010), ribonuclease A (RNase A; R5503), catalase (SRE 0041), ovalbumin (A5503), chymotrypsinogen A (Chym A; C4879), insulin (I2643), and hemoglobin (Hem; H7379) were purchased from Sigma-Aldrich (Steinheim, Germany). Antifreeze protein type III (AFP III) was purchased from A/F Protein Inc., and lysozyme (Lys; 8259.2) was obtained from Carl Roth.

The most commonly used protein was beta-lactoglobulin (BLG), which is a globular whey protein consisting of 178 amino acids available in isoforms A and B. In the context of cooperation projects, BLG was isolated from cow milk or recombinantly produced and provided by the group of Ulrich Kulozik (Technical University Munich, Germany), the group of Anja Wagemans (Technical University Berlin), or the group of Julia Keppler (Wageningen University, Netherlands). For a detailed description, please refer to the respective publications.

Sodium phosphate tribasic (pH 12.3, Alfa Aesar (11,590)), sodium phosphate monobasic (pH 4.5, Sigma Aldrich (71,505)) and NaCl (≥99.8%, Carl Roth GmbH, Karlsruhe, Germany) were used in different salt concentrations as buffer and provided a constant ionic strength. For pH adjustments, 0.1 M phosphoric acid (pH 1.6, Carl

Roth (6366)), HCl (0.1 M Honeywell™ Fluka™, Thermo Fisher Scientific Inc., Schwerte Germany), and NaOH (0.1 M Merck KGaA, Darmstadt, Germany) were added to the respective solutions.

3 Results and Discussion

In the following, the results and contributions of the findings obtained within the framework of SPP 1934 are presented. The bulk behavior of proteins in the liquid and gas phase as well as at the interface have been examined in great detail by the three main analytical methods AUC, ESI-DMA-MS and SFG.

3.1 Comprehensive Protein Characterization by AUC

Analytical ultracentrifugation is widely used to characterize the composition of several samples. Within collaborative projects of this SPP 1934, AUC analysis has been extensively used to characterize specifically modified or stress-treated proteins. Keppler et al. studied the comparability of beta-lactoglobulin (BLG) types with respect to their emulsification properties, comparing native variants of the whey protein extracted from cow milk (bovine BLG B) with recombinantly produced variants (rBLG and sBLG) [78]. As shown in Fig. 8a, the bovine and recombinant variants could be well compared by AUC analysis. No significant differences have been observed between the individual variants. The sedimentation coefficient distributions are in good agreement with each other and delivered three maxima localized around 2 S, 2.6 S, and > 3 S, correlating accordingly with BLGs monomeric (18.4 kDa), dimeric (36 kDa) and multimeric (>50 kDa) assemblies. AUC analysis confirmed the predominant presence of dimers at a salt concentration of 10 mM NaCl and neutral pH.

In a further study, eight different BLG variants were examined, which differ in terms of their cysteine (cys) content [80]. Cysteine forms disulfide bridges in proteins and thus affects the folding and structure of the respective proteins. Bovine as well as recombinant BLG variants were compared. The cysteines were modified either by chemical treatment of the bovine BLG variants extracted from milk or recombinantly by selective replacement of individual cysteines. For investigations of the monomer and dimer distribution, all variants were diluted in 10 mM sodium phosphate buffer to approximately equal optical density and measured via AUC. In most samples, the dimer dominates, as it could be expected according to the given conditions and findings of the previous study (Fig. 8a). Two remarkable samples, the completely chemically reduced one and the recombinant one without outer disulfide bond between cys66 and cys160, however, show a clear abundance of monomers in the solution. Obviously, the absence of the outer disulphide bridge significantly

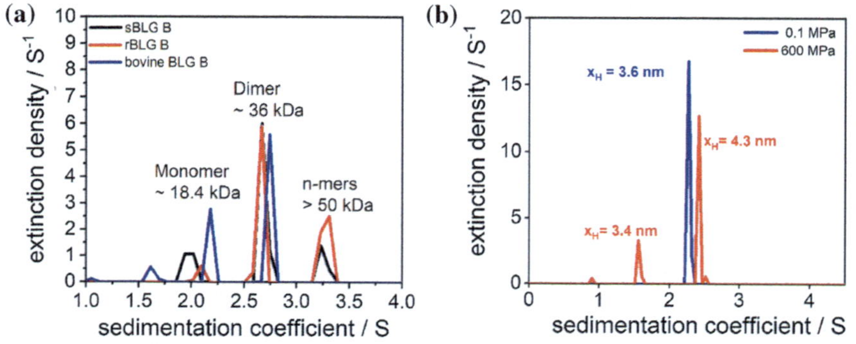

Fig. 8 **a** Sedimentation coefficient distributions for BLG B, sBLG B and rBLG B in water at pH 7 measured at a concentration of 1 g/l in 10 mM NaCl and rotor speed of 50,000 rpm. Reprinted from [78], Copyright (2020), with permission from Elsevier, licensed under the Creative Commons Attribution 4.0 International license (CC BY 4.0). **b** Comparison of sedimentation analysis of high hydrostatic pressure treated and non-treated BLG solution. Reprinted from [79], Copyright (2021), with permission from Elsevier

influences the dimerization behavior and indicates, according to the AUC findings, a clear structure–property relationship of the dimerization behavior.

Besides the characterization of BLG variants, AUC analysis provides insights into stress induced aggregate formation. Kieserling et al. investigated the structure and adsorption behaviour of high pressure treated BLG [79]. Hydrostatic pressure treatment may lead to structural changes in the BLG, which affect the intra- and intermolecular interactions. For these studies, whey protein was treated with high pressure of up to 600 MPa for 10 min at 20 °C in the bulk phase and characterized using a variety of methods. For insights into the quaternary monomer–dimer structure, AUC measurements were carried out and subsequently correlated with results from the gas phase via ESI-DMA-MS.

Figure 8b compares the sedimentation coefficient distribution of an untreated and a pressure-treated BLG sample. The sample treated with 600 MPa clearly shows two species at 1.6 S and 2.4 S correlating with the monomer–dimer structure of the protein. In contrast, the untreated sample shows only one species at 2.2 S, referring to BLG monomeric conformation and a molar mass of 18.1 kDa. The difference in the sedimentation rates between the two samples can be attributed to significant changes in protein shape that affect the frictional coefficients and thus the sedimentation behaviour. The pressure-treated sample showed slightly higher f/f_0 values, indicating changes of the outer shape as a result of the adapted intramolecular structure of the protein. The calculated hydrodynamic diameter indicates an increase in the predominant molecular size from 3.6 nm to 4.3 nm, suggesting a molten globular state of the pressure-treated protein. In addition, the AUC analysis shows a change in the monomer–dimer ratio in the sample, confirming the high pressure induced structural changes of the BLG.

3.2 Analysis of Non-Spherical Protein Aggregate Structures

Particle size distributions (PSDs) of protein solutions are often obtained using dynamic light scattering (DLS). Since the scattering intensity in the Rayleigh regime (particle size << wavelength) scales with the sixth power of particle size, the obtained PSDs are shifted to the coarse side. DLS measures scattered light from the whole protein ensemble and therefore requires inversion techniques based on regularization to extract the PSD. Due the underlying ill-posed problem, only mean values and standard deviations of the PSD are typically obtained. In addition, DLS is limited, when it comes to characterization of non-spherical particles. Consequently, the traditional, subjective and tedious particle counting using microscope images is used. A large number of manually evaluated particles are needed to make a statistically relevant statement about the actual distribution of the non-spherical sample.

In the following, a straightforward method is presented that allows the measurement of particle size distributions of BLG amyloid fibrils based on the AUC technique. The protein structures are measured directly in the aqueous environment and therefore exclude changes in size or unwanted aggregate formation as a result of sample preparation and drying phenomena. Fibril dispersions prepared at pH 2 form straight amyloid aggregates, which exhibit great potential in various applications. During biotechnological processing, they are subjected to high stresses, which can lead to fragmentation of the long fibrils. To investigate these effects in more detail, shear stress was applied using a rotor–stator setup of an ultra turrax homogenizer to obtain smaller fragments of the long-fibril BLG aggregates. From AFM images (Fig. 9a), a height of 3.8 ± 0.5 nm was obtained matching very well the BLG fibril size of 4 nm reported in literature [81]. The fibril length was found to be between 80 and 810 nm with a mean value of 39 nm. The length of fibril fragments was characterized via AUC assuming a cylindrical fibril shape and fixed fibril diameter of 4.7 nm (corresponding to the fibril size including two surrounding water layers) and compared with the AFM measurements of the samples (Fig. 9b).

Equation (1) for the sedimentation coefficient can be rearranged for cylinder-like structures with height h and length l, see Eq. (10) with the aspect ratio $q = \frac{l}{h}$ and the effective fibril density $\rho_{F,eff}$ [39, 82].

$$\frac{1}{s} \cdot \left(\frac{3 \cdot h^3}{2 \cdot q^2}\right)^{\frac{2}{3}} \cdot \left(\rho_{F,eff} - \rho_s\right) - 18 \cdot \eta \cdot \frac{f}{f_0} = 0 \tag{10}$$

For cylinders, Hansen showed that the frictional ratio $\frac{f}{f_0}$ can be described as a function of the logarithm of the aspect ratio $k = \ln(q)$ [83]:

$$\frac{f}{f_0} = 1.0304 + 0.0193 \cdot k + 0.06229 \cdot k^2 + 0.0047 \cdot k^3 + 0.00166 \cdot k^4 + 2.66 \cdot 10^{-6} \cdot k^7 \tag{11}$$

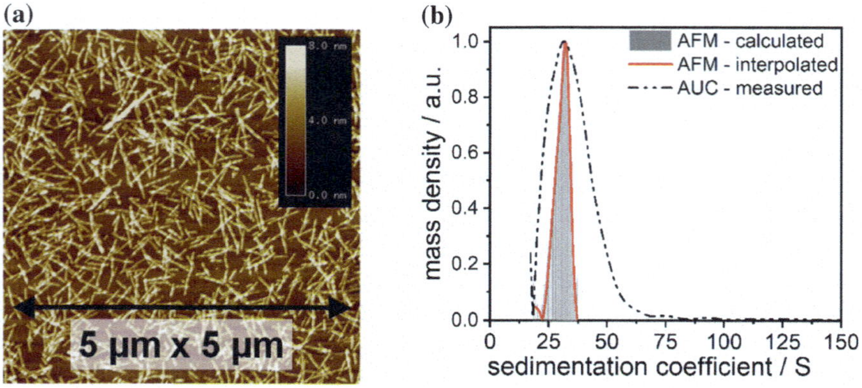

Fig. 9 **a** AFM image of BLG fibrils (processed for 60 s at 19,000 rpm), drop-coated onto mica (5 µm × 5 µm waver). **b** Comparison between the mass density sedimentation coefficient distributions obtained by AUC and AFM. A) and B) Reprinted from [39], Copyrigth (2020), licensed under the Creative Commons Attribution 4.0 International license (CC BY 4.0)

Hence, the sedimentation coefficient distribution can be calculated from the obtained AFM results, assuming the effective fibril density to be equal to $1331.6\ \frac{kg}{m^3}$. Figure 9b compares the mass density sedimentation coefficient distributions obtained by AFM and AUC. Although the widths of the distributions differ, the modal sedimentation coefficients are in excellent agreement. The difference in the distribution width can be explained due to the much better sample statistics of the AUC measurement and clearly underlines the strength of the fast sedimentation analysis over the tedious and time-consuming AFM evaluation [39].

The presented approach of non-spherical protein aggregates characterization via AUC analysis has been applied subsequently within a study by Heyn et al. and contributed to the understanding of stress-induced fibril fragmentation [24]. Additionally a relationship between the energy input of the rotor–stator system and the kinetics of the fibril size reduction was established using the modal values obtained from the AUC sedimentation coefficients. The fibril length l scales with the applied energy per volume $l \sim E_v^{-b}$ and with the rotor revolution n of the Ultra Turrax homogenizer according to $l \sim n^{-3b}$. The exponent b in the scaling law for the fibril size reduction kinetics is in the range of ~0.5, which is surprisingly in agreement with the results of other particulate materials such as graphene oxide [39, 84, 85].

3.3 Characterization of Protein–Protein Interactions

Besides information on size, mass, and composition, specific protein–protein interactions were determined quantitatively via the dissociation constant K_D obtained by AUC, which is the ratio of the dimerization and dissociation reaction constants k or respectively the dimer-to-monomer ratio:

$$K_D = \frac{k_{diss}}{k_{dim}} = \frac{[A]^2}{[A_2]} = \frac{[Monomer]^2}{[Dimer]} \tag{12}$$

BLG as a well-characterized model protein occurs in a dynamic monomer–dimer equilibrium depending on concentration and pH [86]. For the measurement of K_D, concentration series were prepared at different pH values and analyzed by AUC. Figure 10a compares again bovine and recombinant BLG variants and confirm their similarity. A clear minimum at pH values between 5 and 7 is shown for both, native and recombinant variants, which correlates with the pH of the BLG isoelectric point (IEP) around pH 5.4. Due to balanced surface charges, the aggregation effects increase near the IEP and decrease for more acidic or basic pH values, which consequently leads to an opposite behavior for the development of the dissociation.

To further investigate the influence of the solvent properties (pH and ionic strength) on the dimerization behavior of BLG, concentration studies were performed at pH 3 and 7 at salt concentrations of 10 mM and 100 mM NaCl. Figure 10b shows the concentration-dependent sedimentation coefficient distributions of a BLG sample. In agreement with previous studies, presented at the beginning of this section (Fig. 8), this sample predominantly shows dimers at pH 7, which shifts towards monomers depending on the concentration. Based on the measured concentration series, the dimerization constant can be derived from AUC analysis (Fig. 10c). In contrast to neutral pH, in acidic pH, BLG is preferentially present in its monomeric structure, which is confirmed by the increased value of the dissociation constant at pH 3 and 100 mM. With increasing salt concentration, the dissociation constant decreases, as the surface charges of the protein are more strongly screened.

Moreover, in combination with osmometry, non-specific interactions were determined based on the second virial coefficient (according to Eq. (5)) and compared with theoretically calculated data. With the dissociation constant known from AUC,

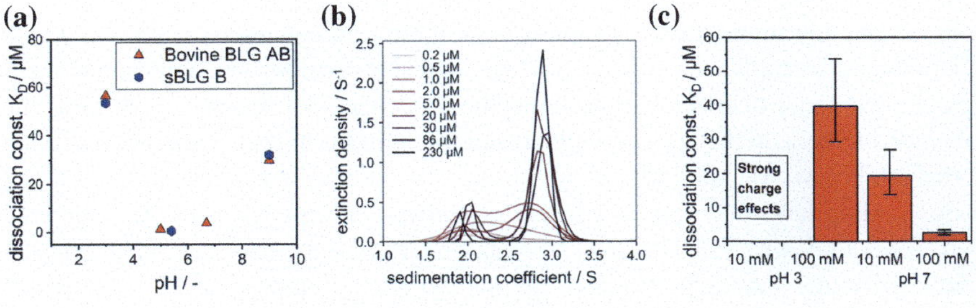

Fig. 10 **a** Dimer dissociation constant (K_D) from monomer–dimer equilibrium for bovine and recombinantly produced BLG set to pH of 3, 5, 7 and 9. Reprinted from [78], Copyright (2020), with permission from Elsevier, licensed under the Creative Commons Attribution 4.0 International license (CC BY 4.0). **b** Exemplary sedimentation coefficient distributions for various BLG loading concentration at pH 7 and 100 mM NaCl. Adopted from [46], Copyright (2022), with permission from Royal Society of Chemistry. **c** Dimer dissociation constant for different solvent conditions determined from AUC experiments. AUC analysis of the data obtained for pH 3 and 10 mM NaCl was impossible, due to strong charge effects occurring during the measurements

the second virial coefficient was determined from the molecular weight-corrected osmotic pressure measured by membrane osmometry by the group of Anja Wagemans (Technical University Berlin) within this SPP 1934 [46]. The experimental data are in very good agreement with the theoretical calculations from an extended DLVO coarse grained model (xDLVO-CG) approach provided by the group of Wolfgang Wenzel (Karlsruhe Institute of Technology KIT) within this SPP 1934 [46]. Consequently, the study confirmed that the particle–particle interactions are mainly driven by electrostatics and provided a protocol to simultaneously determine specific and non-specific protein–protein interactions.

3.4 Gas-Phase Protein Characterization by ESI-DMA-MS

Protein clusters or protein–ligand aggregates as observed by ESI-DMA-MS can have two origins: They can be generated as purely stochastic product of the electrospray process (non-specific clusters), or they can already be clustered in the liquid phase and transferred as such to the gas phase. Distinguishing of these two origins is challenging [64]. A principle solution to this problem is to analyze a dilution series of the sample. Non-specific aggregates would become smaller with increasing dilution of the sample and would eventually vanish from the mass spectrum, while specific aggregates would be detected independently of the dilution. Considering that the largest non-specific aggregates will be formed as evaporation residues of the original drops formed by the electrospray source, small droplets are needed to avoid non-specific clustering at detectable analyte concentration. Furthermore, the outcome of dilution series can only be related reliably to the clustering origin if the electrospray drop size is not affected by the dilution and if one can definitely state that the observed species are indeed evaporation residues of the original drops and species resulting from Coulomb fission events.

From an experimental point of view, this requires an electrospray source, which generates small droplets of well-defined size and size distribution. While small sizes are obtained under nanospray conditions, the drop size does not only depend on the flow rate, but also on the applied voltage and the liquid properties. Various emission modes are known to exist [87]. The best-understood emission mode is the cone-jet mode. Scaling laws that correlate liquid properties and flow rates to the generated drop size are known [59]. The drop size can be tuned by the flow rate and the liquid conductivity in a size range between 50 nm and a few micrometers. The generated drop size distributions are narrowly distributed with a relative standard deviation around 15% and the drop size modal value is highly stable and does not vary by more than 1% in consecutive size measurements. The cone-jet mode is thus an optimum choice for analysis of non-specific aggregation of analyte molecules.

Addressing the above-mentioned key points for analysis of non-specific aggregates, we have confirmed by comparing mobility-classified mass spectra of non-specifically formed lysozyme clusters to complementary measurements of these cluster ions by a scanning mobility particle sizer (SMPS) that species observed in a mobility-classified mass spectrum, include indeed both, the evaporation residues of the original drops and those which result from Coulomb fission events, see Fig. 11. A SMPS can measure particles in the size range between 3 and 1000 nm and thus the results obtained confirm that no species larger than those as observed in the mobility-classified mass spectrum are generated in the spray process. At the same time, the modal value of evaporation residues was found at equal cluster size in measurements based on ESI-DMA-MS and SMPS [26].

These measurement results prove that we record the complete outcome of the electrospray process. To assess whether a certain observed level of ion clustering is specific or non-specific, more information is required. Precise information on the drop size distribution is required to make theoretical predictions on the expected level of ion clustering for a given analyte concentration in solution. To compare these theoretical results to experimental data, it is required to know the detection efficiency as a function of the cluster mass. Since the detection efficiency of a mass spectrometer varies strongly over the m/z range, this is prerequisite to a quantitative comparison. We show in [88] that the drop size distribution generated by the electrospray can accurately be derived based on mobility scan for a dilution series of sprayed raffinose solution. Knowing that the drop generation frequency is constant in the cone-jet mode for equal liquid properties, the detection efficiency as function m/z can be determined from the data set at the same time. Having access to both, it is shown in [88] that non-specific ion clustering can be predicted in excellent agreement to experimental data. Specifically, the method is accurate enough to identify errors in protein number concentration in solution, that result from hygroscopic water take up by the protein, when the calculation of number concentration is based on the dissolved protein mass and its dry molar mass. The error in protein number concentration in the solution as derived from the measured non-specific cluster spectra is less than 5%. The accurate prediction of the clustering state is not limited to protein–protein clusters. It applies to clustering of proteins with non-volatile contaminants as well.

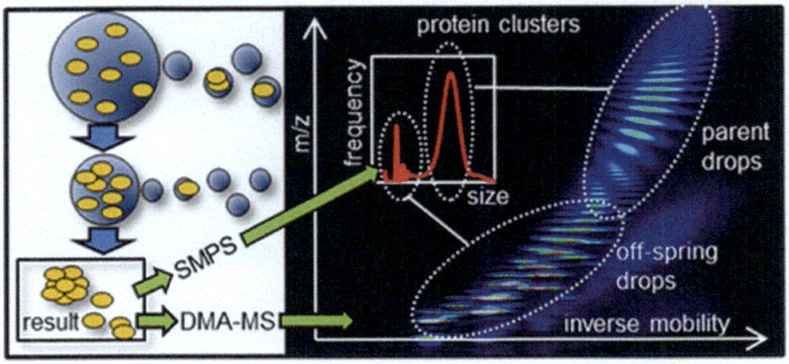

Fig. 11 Principle of combined SMPS and DMA measurements to analyze the complete outcome of an electrospray. Reprinted with permission from. [26], Copyright (2020), American Chemical Society

While a more accurate statistical prediction on the impact of contamination is possible [88], if the drop size distribution is considered, the assumption of monodisperse drops is sufficient to give good estimates. Since the probability for a non-volatile species to escape from a droplet under native spray conditions is small (in the order of 5%), statistical treatment of the problem can be conducted based on the non-volatile distribution in the primary drops. Considering an average number $\overline{N} = c_c/V_{drop}$ of contamination molecules in a drop, the probability p to find N molecules in a drop is given by Poisson statistics as:

$$p(N) = \frac{\bar{N}^N}{N!} exp\{-\bar{N}\} \tag{12}$$

The fraction of droplets, which do not contain a contaminant, is $p(N) = 0 = exp\{-c_c/V_{drop}\}$. Clean analyte molecules will be found only in this fraction of primary drops. For orientation, $\overline{N}$ is 1 for a 100 nm droplet at a concentration of 3.2 μmol/l, i.e. 37% of the drops would not contain contaminants. This analysis can be performed for the drop size distribution as well. It is also neglected that some analyte molecules are emitted in Coulomb fission events. Since the tiny droplets emitted in such events have only around 1/500th of the original drop volume, these molecules will be typically free of adducts. In Fig. 12, the BLG emitted from the primary drops is represented by sharp bottom lines of the protein monomer signal areas. The signal area above these lines represents BLG monomers that have not been emitted from the original drops. These are clustered to a relevant amount of non-volatile contamination. Note further that no such sharp bottom lines exist for the dimers and trimers. These clusters are non-specific aggregates formed as evaporation residues from the original drops only.

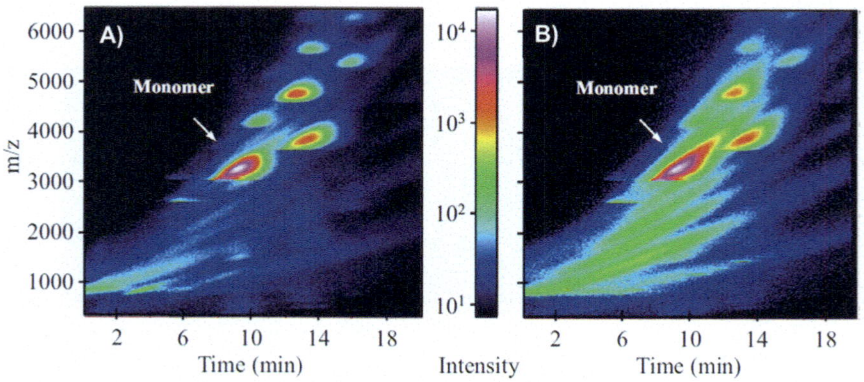

Fig. 12 Mobility-classified mass spectra of BLG **a** before and **b** after pressure treatment. Reprinted from [79], Copyright (2021), with permission from Elsevier

The signal below the BLG monomers belongs to the evaporation residues of original drops that do not contain a BLG molecule, but only non-volatile contamination. This is the case of around 80% of the original drops in both samples shown in Fig. 12.

The impact of contamination must be judged based on the analytical task. For evaluation of the clustering state, protein peaks need to be separated in the scan. In this case, the signals of the monomers, dimers, trimers, etc. can be quantified. This is excellently fulfilled in Fig. 12a and is just sufficient in Fig. 12b). The tolerable level of contamination is relatively high and can roughly be estimated to 20% of the protein volume concentration in the sample.

Clean analyte species are required for analysis of the analyte mass and mobility. At high contamination levels, one has to rely on the Coulomb fission products. Since the probability for this transfer is small for nanosprays and is material dependent [89], a simple precise statement is not possible. If the molar concentration of non-volatile contaminant molecules is not more than 10 times the analyte molar concentration, this is certainly clean enough. Applying the Poisson statistics and assuming a protein in every 10th drop, one would find 37% of the proteins to be clean. A clean protein from every 30th drop is sufficient for analysis by nano-electrospray. To improve sample cleanliness, the samples can be diluted with clean solvent. However, the protein concentration is diluted in this case as well, and the overall signal intensity decreases accordingly.

3.5 *Combining the Results from the Liquid and Gas Phases*

Mobility-classified mass spectrometry was used for analysis of the aggregation state. To allow for a statistical analysis, the samples of pressure-treated BLG and the original BLG sample (without pressure treatment) were analyzed by mobility-classified mass spectrometry. Due to a rather high contamination level of the pressure-treated sample (Fig. 12b), the samples were diluted to a level at which one finds on average only a single BLG molecule in around every 6th drop. Figure 12a) shows the much cleaner original BLG under the same conditions. With respect to quantification, note the logarithmically scaled color bar. The monomer peak maxima are one order of magnitude higher compared to the dimer peaks, which are represented by the three peaks above the indicated monomer series. Beside the three dimer peaks, there are two trimer peaks located above the dimer peaks. These trimer peaks have intensities more than two orders of magnitude lower compared to the monomer peaks. The monomers and dimers observed here are non-specific aggregates. Their relative abundances are in line with statistical expectations. Furthermore, both mass spectra show—as discussed above—clean species, which must be attributed to those proteins escaped in Coulomb fission events, only for the monomers. If dimers were tightly (e.g. covalently) bound, such clean escaped species would be expected for the dimers as well. For a qualitative comparison, the original protein concentration was adjusted close to that of the pressure-treated sample. The analysis of both, the original and the pressure-treated samples show very similar mass spectra under this condition. Since the original sample does, as found by complementary methods, contain monomers only, dimers and trimers are non-specific in this case. The non-specific aggregate nature of the pressure-treated sample is thus stated based on (a) the relative abundances of monomers, dimers, and trimers within the corresponding spectrum, (b) the contamination adduct distribution, which indicates that only monomers exit parent drops in Coulomb fission events and (c) a comparison to the original BLG spectrum recorded under identical conditions, whereby the original sample does not contain n-mers. While AUC results at higher concentration indicate that around 70% of the pressure-treated BLG is found in dimers, ESI-DMA-MS finds > 80% of the observed species to be the monomers and the dimers and trimers are to be observed due to non-specific aggregation. It is thus concluded that n-mers as observed by complementary methods such as AUC, are weakly bound. At the rather low concentration used in the ESI-DMA-MS measurements (200 nmol/l), the equilibrium cluster distribution is likely on the monomer side. While covalently bound n-mers survive the electrospray process, loosely bound n-mers might also not withstand the stresses, they are exposed to in the electrospray process.

For evaluation of altered protein mass in recombinant protein samples, we followed the classical electrospray approach. Proteins were dispersed in water with formic acid electrolyte and sprayed in the positive ion mode. The resulting ions are easily declustered from remaining solvent adducts in the mass spectrometer. Adduct-free spectra of the proteins are obtained. The mass difference between recombinant and original bovine BLG is determined to be 68 Da, based on the isotopic patterns. Figure 13 shows exemplarily mass spectra for beta-lactoglobulin isoform B with modified N-terminus (sample, red upper spectrum in Fig. 13) and the reference (black lower spectrum). Due to the relevant width of the isotopic pattern, we have included the data from the reference spectrum (upper spectrum) shifted by 68 Da in the lower spectrum as well. This shifted spectrum is represented by the red points in the lower spectrum and matches the spectrum for the reference bovine BLG. Consequently, the exact mass difference for manipulated proteins can unambiguously be identified by high-resolution mass spectrometry.

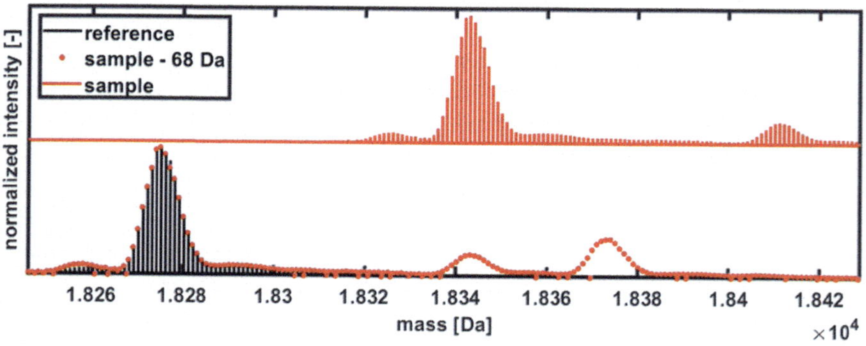

Fig. 13 Evaluation of the exact mass difference of a recombinant beta-lactoglobulin (BLG) (upper spectrum) and the original BLG sample (lower spectrum). Shifting the upper spectrum by 68 Da (red dots in the lower spectrum) matches the isotopic pattern of the original BLG

3.6 *Density and Shape Measurements*

The analysis of proteins by mobility-classified mass spectrometry provides information on their mobility in the drift gas and their mass. Mass-equivalent and mobility-equivalent diameters of globular proteins can be obtained. Prerequisite to such analysis is, however, the assumption of a globular analyte shape, which then allows determination of protein density with high precision, independent of any assumptions on the analyte shape, by careful analysis of analyte polyethylene glycole (PEG)-cluster ions following the concept of classical pycnometry. In particular, the protein is entrapped a globular PEG-protein droplet. Based on the independently derived mobility diameter and volume-equivalent diameter, it is possible to attribute two parameters to the protein—size and shape, respectively. We demonstrate the approach for lysozyme, cyano-cobalamin (vitamin B12), and glucose, which cover two orders of magnitude in analyte mass (180–14,300 Da). The derived densities for these molecules are highly accurate, i.e. they deviate less than 1% from literature values. Our method can be applied to newly isolated biomolecules and molecular clusters, where only minor amount of material is available. The obtained shape parameters of lysozyme and cyano-cobalamin agree well with the expected molecular shapes. Data evaluation relies only on locations of the species in the mass-mobility plane and is in principle independent of any mobility theory. Our approach is thus robust with respect to experimental uncertainties and produces identical results irrespective of the type of mobility classification and drift gas [27].

3.7 *SFG for Surface-Sensitive Protein Investigations at Interfaces*

In the third and last part, we focus on the characterization of protein properties at gas/liquid and solid/liquid interfaces based on the powerful surface-sensitive sum-frequency generation spectroscopy (SFG). Proteins preferentially adsorb at hydrophobic or oppositely charged surfaces. Adsorption is often accompanied by conformational alterations of the protein and can lead to changes in the protein properties. Various proteins have been investigated under different pH conditions and analyzed both at the interface and in the bulk phase in order to better understand adsorption and to clarify the role of electrostatic charges. This way, novel insights into the relationship between the charge states of the proteins, their adsorption and interfacial conformation are gained.

Proteins were first examined at the water/air interface with a particular focus to their isoelectric points directly at the surface. These investigations were later transferred to a hydrophobic polystyrene/water interface. Finally, chiral SFG measurements were carried out at the water/air interface, which provide insights into the structure of the proteins at the surface.

3.7.1 Isoelectric Point of Proteins at Hydrophobic Interfaces

The isoelectric point (IEP) is an important parameter of an individual protein and corresponds to the pH at which the respective protein is net uncharged. The charge of a protein influences its structural and colloidal stability, which is consequently a function of pH, salt type, and concentration. IEPs at the interface are seldom known but are highly relevant for protein aggregation and film formation due to the absence of repulsive forces.

In an aqueous medium, the charge of adsorbed proteins influences the orientation of the surrounding water molecules at the interface, which affect the vibrational spectrum of the SFG spectroscopy. Hence, the IEP for proteins adsorbed at the interface can be determined by measuring the OH vibration amplitude of the water molecules, as shown by Engelhardt et al. for BSA, BLG and Lys [15]. In the following, the interpretation of an exemplary pH-dependent SFG experiment, based on a protein solution measured at the water/air interface, is explained in more detail.

Figure 14 illustrates an SFG spectrum of BLG in a relatively wide spectral range of 2800–3800 cm^{-1}. The region between 2800 cm^{-1} and 3100 cm^{-1} covers the CH vibrations, which include symmetrical and asymmetrical CH_2 and CH_3 vibrations as well as Fermi resonances. The OH oscillations of the interfacial water molecules are in the adjacent region between 3200 cm^{-1} and 3500 cm^{-1}. A special feature around 3050 cm^{-1} (see Fig. 14b) is the ring mode attributed to the CH stretching vibrations of proteins, which originate from the aromatic rings of amino acids such as tyrosine, phenylalanine, and tryptophan.

The amplitudes of the water vibrations increase with increasing charge of the interface at higher and lower pH values. Consequently, the amplitude of the OH

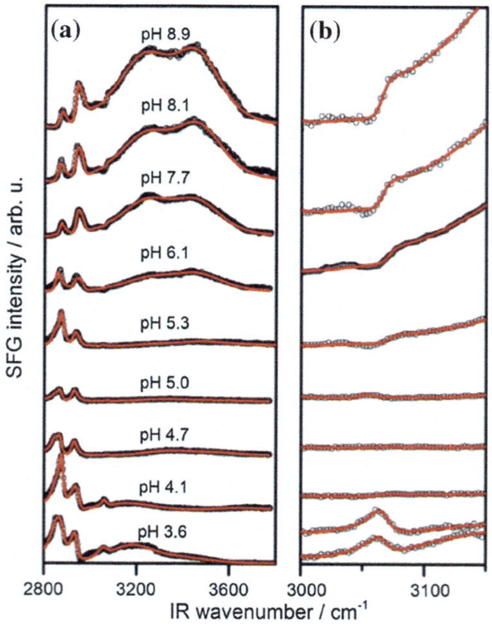

Fig. 14 **a** pH-dependent vibrational SFG spectra of BLG adsorbed at the water/air interface showing the CH region (2800 – 3100 cm^{-1}) and OH stretching vibrations (3000 – 3800 cm^{-1}). **b** Visualization of aromatic CH stretching band at 3050 cm^{-1} showing the changes in polarity in more detail. Adapted with permission from [15], Copyright (2013) American Chemical Society

vibrational band passes at the IEP through a pronounced minimum (see Fig. 14). Crossing the point of zero charge at the interface leads to a change in the sign of the electric field and consequently to a change in the net orientation of the H_2O at the interface by 180°. Changes in the net orientation of interfacial water lead to constructive interference of the 3050 cm^{-1} band with the broad OH stretching band at pH < 5, while destructive interference is observed at pH > 5. Consequently, the interfacial IEP is additionally indicated by the changes in the polarity of the vibration band around 3050 cm^{-1}, which is positive at a pH < IEP and negative at a pH > IEP (see Fig. 14b). In this particular example of BLG in Fig. 14, the interfacial IEP is located between pH 5.3 and pH 5.0. The amplitude of the OH oscillations passes through a minimum there, and the vibration amplitude at the ring mode (3050 cm^{-1}) undergoes a sign change. From zeta-potential measurements, it is known that the IEP is around pH 5.1 for BLG in the bulk phase and is therefore in surprisingly good agreement with the determined interfacial IEP at the water/air interface [15, 77].

Guckeisen et al. studied twelve proteins with different structural features such as size, number of disulphide bridges, composition of secondary structural elements and hydrophobicity at the water/air interface using SFG [14]. The surface IEP was determined from the integral SFG signal intensity of the OH stretching region (3100 – 3750 cm^{-1}), which passes through a pH-dependent minimum (red lines in Fig. 16), that is assigned to the IEP. As shown in Fig. 15, the analysis of the twelve proteins at the surface showed very good agreement between the results in the bulk phase, the pH of maximum layer thickness, and the IEP at the interface [14]. The error bars indicate a rather broad minimum of the SFG signal intensities for Lys and AFP III, which makes the accurate determination of their interfacial IEP rather difficult. For all other proteins, the minimum is well pronounced and the error bars being small are omitted. Taking into account all the measurement uncertainties, the IEP at the interface could be determined by SFG spectroscopy with a good accuracy of one pH unit.

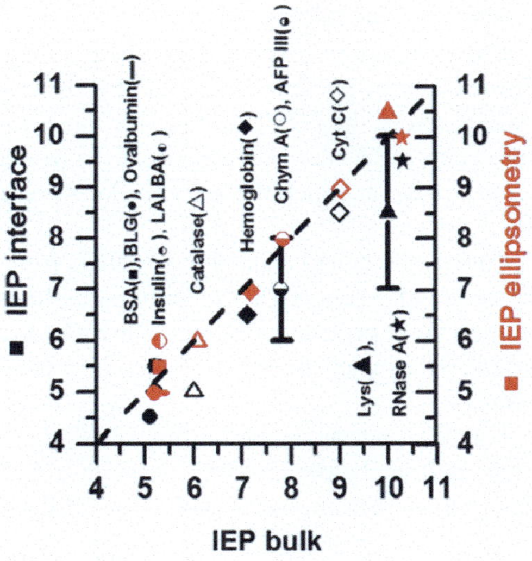

Fig. 15 Overview of measured IEPs of twelve proteins in the bulk (IEP bulk) and at the water/air surface (IEP interface) as well as the pH of maximum layer thickness (IEP ellipsometry). Due to overlapping IEPs for insulin, LALBA, BSA, BLG, and ovalbumin, their corresponding symbols cannot be distinguished in this figure. Reprinted with permission from [14], Copyright (2019), American Chemical Society

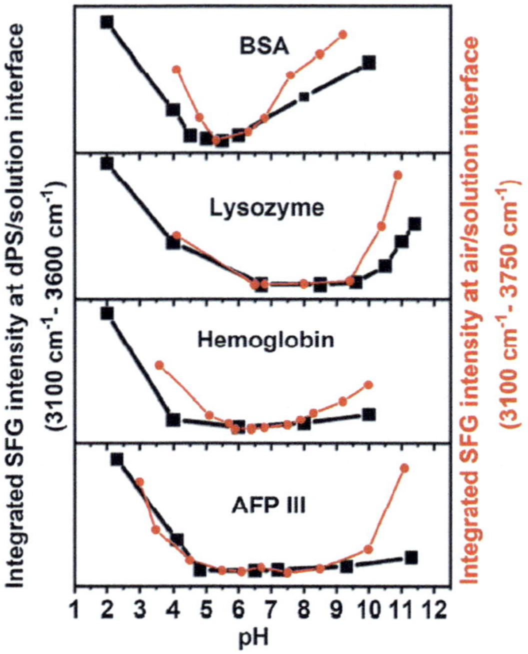

Fig. 16 Integrated SFG signal intensity at the buried dPS/solution interface in black and at the solution/air interface in red (data reproduced with permission from [14], Copyright (2019), American Chemical Society) for four different proteins. Figure reprinted with permission from [20], Copyright (2021), Frontiers, licensed under the Creative Commons Attribution 4.0 International license (CC BY 4.0)

The pH-dependent layer thickness was examined by ellipsometry. The measured layer thickness was never less than the hydrodynamic diameter of a single native protein at the interface. This indicates that the proteins are present at the interface at least in a monolayer and seem to retain their native structure. At least, a complete unfolding can be excluded for the examined proteins. The thickness of the layer at the interface is an important parameter that reflects the electrostatic properties of the proteins at the surface. Aggregation at the interface in the vicinity of the IEP leads to increased film thickness.

Charged proteins experience repulsive forces and hence arrange in a monolayer. However, as the pH approaches the IEP, the net protein charge is reduced and finally reaches zero at the IEP. The attractive forces then dominate, and protein agglomeration occurs at the interface, which leads to an increased layer thickness. Consequently, the pH value, at which the layer thickness reaches its maximum, coincide with the IEP.

To strengthen and expand the findings from the water/air interface, further work has been carried out at a hydrophobic solid/liquid interface. In this case, the adsorbed proteins are expected to undergo stronger hydrophobic interactions than at the liquid/gas interface, where the proteins can orient themselves flexibly on the surface or even rotate.

For SFG measurements at the solid/liquid interface, the two laser beams were directed from below and pass through a CaF_2 prism. The special prism geometry enables measurements within the total internal reflection area (TIR) and thus enhances the weak protein signal at the interface by increasing the signal-to-noise ratio. The flat top plane of the prism was coated with polystyrene. To avoid interference vibrations originating from CH compounds in the polymer and the protein, deuterated polystyrene (dPS) was used.

For the studies at the hydrophobic solid/liquid interface, four model proteins (BSA, lysozyme, hemoglobin, AFP III) were investigated, which mainly differ in terms of their bulk IEP and hydrophobicity. These parameters have a significant influence on the electrostatic and hydrophobic interactions and thus on the adsorption and reorganization of the protein at the hydrophobic interface. Hydrophobic parts are usually hidden in the inner part of a folded protein and only reach the surface during refolding or denaturation, where the hydrophobic intermolecular forces promote aggregation. The hydrophobicity index is a parameter that can be calculated from the hydrophobicity of the respective amino-acid sequence and can be either positive for increased hydrophobicity or negative, indicating less hydrophobic properties.

Figure 16 shows the comparison between the SFG measurements at the liquid/gas and solid/liquid interfaces. Striking is that the curves of the two integral OH vibrations hardly differ and are congruent in the area of the minimum. This indicates that the interfacial IEP is identical for both conditions and that the electrostatic interactions of the proteins are similar regardless of their interfacial environment or whether the protein is adsorbed to a solid or gas interface.

Other details, such as the symmetry of the curves or the width of the minimum correlate with the hydrophobicity of the respective protein and thus with the respective amino-acid content. The more hydrophobic the protein is, the flatter the curves are, since it is assumed that the strongest interactions with the hydrophobic dPS layer occur there.

We believe that the amino-acid sequence of proteins plays a deterministic role here. The pH-dependent charge of proteins is determined by the amino-acid composition of the proteins and their respective pK values. As a first approximation, the acidic amino-acid residues that are dominant in the acidic proteins (i.e., those with high IEPs) are aspartic acid (optimized pK_a: 3.87) and glutamic acid (optimized pK_a: 4.41). For basic proteins (i.e., those with low IEPs), the respective dominant amino-acid residues are arginine (optimized pK_a: 11.84) and lysine (optimized pKa: 9.05). The pK_a values of the acid residues are much closer to each other than the pK_a values of the basic residues. This distribution of amino acids is expected to result in a more sudden charge change for acidic proteins and a more gradual charge change for arginine- and lysine-dominated proteins [14].

As already established for the water/air interface, the minima of the integral intensities of the water oscillations (3100–3600 cm^{-1}) at the solid/liquid, dPS/water interface corresponds to the minimum charge state and can thus be identified as the IEP at the interface. The comparison with the measurements of the IEP at the water/air and that of the IEP in the bulk phase by measuring the zeta potential show good agreements for all four proteins (see Fig. 16).

Irrespective of the individual characteristic parameters and structural features, the same trend was observed for all investigated proteins. Contrary to expectations, no significant difference was observed between the IEP of the proteins at the solid/

liquid interface and in the bulk phase, although it is known that after adsorption, the protein changes its conformation due to the interaction of the hydrophobic parts of the protein with the hydrophobic interface.

The IEP is mainly determined by the hydrophilic parts of the protein, which are supposed to be directed into the water phase [14, 90]. The hydrophobic regions of the protein therefore rarely contribute to the charge of the protein and consequently barely influence the electrostatic properties. Denaturation at the interface should change the environment of the dissociable groups and thus the IEP.

None of the measured IEPs measured at the interface differ by more than one pH unit. Therefore, we suggest that this is a widely applicable feature for many proteins. To expand the above list of twelve proteins to further species and to explain possible deviations from this rule due to a specific protein structure is a highly interesting task for the future.

3.7.2 Proteins Secondary Structure at the Water/Air Interface

In addition to the consideration of the interface influence on the electrostatic properties of protein, chiral SFG measurements were carried out, which provide information on the preservation of the protein secondary structure at the interface. The preservation of an intact structure plays an enormous role for proteins, as their functionality and performance in industrial processes significantly depend on the protein conformation. Three selected proteins (alpha-chymotropsinogen A (Chym A), catalase (Cat) and beta-lactoglobulin (BLG)) were investigated using circular dichroism (CD) in the bulk phase and SFG at the water/air interface. By means of SPP-polarization modes during the interface-specific measurements, additional structure-sensitive chiral SFG measurements could be performed. The proteins were exposed to different pH conditions and were initiated to denaturation by the addition of 8 M urea.

The CD spectra of the bulk samples show the characteristic features of the respective proteins (exemplarily shown for Chym A and BLG at pH 4.5 in Fig. 17 (black curve)), which match well with published results. No major changes were observed in the CD spectra of Chym A across the pH range from pH 1.6 to pH 11.

Due to differences in the measured amplitudes of the characteristic minima in the CD spectra of BLG and Cat, a pH-dependent structural change could not be excluded. In the acidic bulk environment, no complete loss of the secondary structure of these two proteins could be detected. Under strongly alkaline conditions (pH > 11), however, the CD spectra of BLG and Chym show significant differences from the profiles at low pH values, indicating a major loss of secondary structure.

For more insights into the interfacial protein structure, SFG measurements were performed to validate the findings from the bulk phase at the interface. Conventional achiral SFG measurements in the SSP polarization mode, as used for the previous studies presented above, are not applicable for these considerations. Although the amide I band (see 1600–1800 cm^{-1} in Fig. 7) correlates with the secondary structures of proteins, this signal is dominated by the overlapping bending mode of the OH water

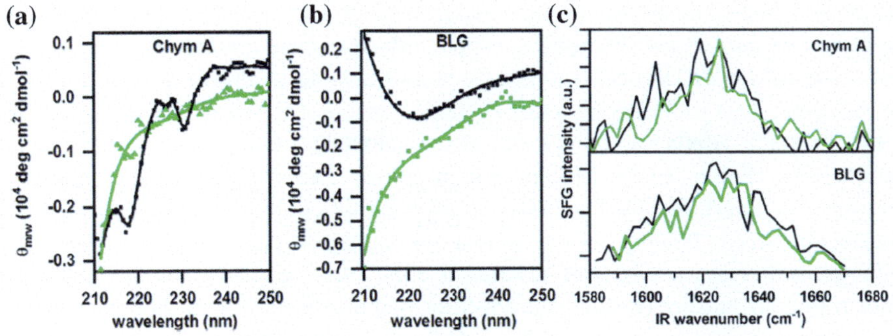

Fig. 17 **a** and **b** Circular dichroism measurements at pH 4.5 without (black line) and with 8 M urea (green line) for Chym A and BLG. **c** Respective chiral SFG spectra in the amide region in SPP polarization mode. **a**, **b**, and **c** reprinted from [69], Copyright (2021), with permission from Elsevier

vibrations. To eliminate this phenomenon, chiral SFG measurements were carried out in the SPP polarization mode. The detected amide I bands and N–H stretching correlate with the alpha-helix and beta-sheet structures of the proteins.

For all three proteins, it was observed that the amplitude of the chiral N–H stretching region remained relatively constant for the proteins in acidic conditions, whereas it decreases in alkaline conditions depending on the IEP (Chym from pH 9, BLG from pH 5.5 and Cat from pH 8). The amid I vibrations, which according to literature [69, 91] correlate with parallel and anti-parallel beta-sheet structures, showed almost no changes. Only for BLG, no SFG signal was detected at pH > 11, which correlates well with the observations in the bulk phase and indicates a complete loss of the secondary structure.

In summary, for the pH-dependent investigations, the observations in the bulk phase correspond well to those at the interface for the acidic milieu—no complete structural loss could be observed. However, in alkaline conditions, the protein behaviour at the interface seems to deviate from the bulk. The investigated proteins showed a change in secondary structure and thus a stronger denaturation at the interface than in the bulk, particularly at lower pH values.

Furthermore, the protein structure after the addition of urea was compared in the bulk phase and at the interface. The structure of the protein is expected to be completely unfolded after the addition of the denaturant. The detailed mechanism is not yet elucidated in literature, but it is evident that it has a major effect on the hydrophobic part of the protein [14, 92]. The distribution of the hydrophobic parts of the protein at the interface may differ from that in the bulk phase. Whereas they tend to be located in the interior of the protein in the bulk, they are oriented towards the hydrophobic interface at the phase boundary. Therefore, it can be assumed that the denaturation with urea will be differently pronounced in both cases.

Figures 17a and 17b show exemplarily the CD spectra of BLG before (black curve) and after (green curve) addition of urea. A clear change in the CD spectrum can be observed, which confirms the strong structural changes in the bulk. Figure 17c shows the corresponding chiral SFG spectra. Apparently, these two spectra are very similar, and no significant differences between the protein samples with and without urea can be observed. This indicates an intact protein structure and absence of denaturation at the interface. It is assumed that this is related to the limited accessibility of the hydrophobic regions of the protein, which are directed towards the gas phase at the water/air interface. Another possible explanation could be the phenomenon of interface-induced "healing" of the structure, i.e. the protein, denatured in the bulk, assumes its native structure after adsorption at the interface. The hydrophobic forces at the liquid/gas surface facilitate the refolding of alpha-helix or beta-sheet structures.

In summary, the interface can have both, a positive or negative effect on the conformational stability of the protein, depending on the underlying denaturation mechanism. The change in the secondary structure of the proteins, adsorbed on the surface in alkaline solutions, occurs at pH values lower than in the bulk solution [69]. In contrast, 8 M urea completely denatures the studied proteins in the bulk solution (Figs. 17a and 17b), while the liquid/air interface prevents the urea-induced denaturation of the surface-adsorbed proteins by limiting the access of urea to the hydrophobic side-chains of proteins protruding to air (Fig. 17c).

4 Conclusion

The three outstanding measurement techniques analytical ultracentrifugation (AUC), electrospray ionization differential mobility-classified mass spectroscopy (ESI-DMA-MS), and sum frequency generation spectroscopy (SFG) have contributed significantly to the understanding of structural changes in stress-treated and genetically modified protein variants. It has been successfully demonstrated that these measurement techniques, common in particle technology, can be effectively applied to biological systems.

Within the framework of this SPP 1934 project, a large number of different samples have been analyzed using AUC, identifying changes within the quaternary structure. In addition to the characterization of spherical, globular proteins, a straightforward method was developed to quantitatively analyze non-spherical particles such as BLG amyloid fibrils via AUC. With the knowledge of the structure and effective density of cylinder-shaped fibrils, mean sedimentation properties can be derived directly from dispersions of amyloid aggregates. This allows a comprehensive, fast and efficient analysis of fragmented cylindrical fibrils by AUC and provides insight into the size-reduction kinetics considering the modal sedimentation coefficient as a function of stress energy. The amount of monomers and dimers measured by AUC provides information on protein–protein interactions. Depending on the salt concentration or the pH value, the influence of solvent properties on the particle–particle interactions was investigated. The important role of electrostatics was revealed.

The methodology of ESI-DMA-MS was extended and further developed for the analysis of protein aggregates. Detection limits have been determined, and protein aggregates in the gas phase were characterized and compared with those in the liquid phase.

The particular interface-sensitive method of SFG spectroscopy allowed investigations on the interfacial IEP. Surprisingly, analogies were found between the interfacial IEP of a protein and the IEP in the bulk phase. It could be shown that the electrostatic properties of a protein are independent of its location at the interface and that, contrary to initial assumptions, there is no difference to the bulk behavior of a dissolved protein after adsorption at solid or gaseous interfaces.

Furthermore, insight into the secondary structure of interfacial proteins was gained by means of chiral SFG measurements. A positive effect of the interface on the preservation of the secondary structure was found. Proteins that denature in the bulk phase under the same solvent conditions (ionic strength and urea content) remain stable at the interface in terms of their secondary structure. However, an earlier loss of secondary structure was observed for the effect of the alkaline pH at the interface compared to the bulk phase.

The extensive characterization of proteins in the bulk and at the interface considerably contribute to the understanding of protein properties. The comprehensive usage of highly specific measurement techniques such as AUC, ESI-DMA-MS, and SFG contributed to the elucidation of numerous scientific questions within the priority program DiSPBiotech (SPP1934) and provided novel insights into protein properties and their structure.

Acknowledgements All authors gratefully thank the German Research Foundation (DFG) for funding of the priority program DiSPBiotech (SPP1934), projects PE427/32-1 and PE427/32-2. MJU, JW and WP gratefully acknowledge funding by the DFG through the project INST 90-1123-1 FUGG "Analytical ultracentrifuge with absorption and interference detector". CL and WP also acknowledge the DFG for funding of the project INST 90/913-1 FUGG "Modular ion mobility mass spectrometer".

References

1. Bunge M (1979) Treatise on basic philosophy. Ontology II: a world of systems. Springer, Berlin. https://doi.org/10.1007/978-94-009-9392-1
2. Bao G (2002) Mechanics of biomolecules. J Mech Phys Solids 50(11):2237–2274. https://doi.org/10.1016/S0022-5096(02)00035-2
3. Hawe A, Wiggenhorn M, van de Weert M, Garbe JHO, Mahler H-C et al (2012) Forced degradation of therapeutic proteins. J Pharm Sci 101(3):895–913. https://doi.org/10.1002/jps.22812
4. Wu T, Yoon H, Xiong Y, Dixon-Clarke SE, Nowak RP et al (2020) Targeted protein degradation as a powerful research tool in basic biology and drug target discovery. Nat Struct Mol Biol 27(7):605–614. https://doi.org/10.1038/s41594-020-0438-0
5. Kiely-Collins H, Winter GE, Bernardes GJL (2021) The role of reversible and irreversible covalent chemistry in targeted protein degradation. Cell Chem Biol 28(7):952–968. https://doi.org/10.1016/j.chembiol.2021.03.005
6. Wang W, Nema S, Teagarden D (2010) Protein aggregation–pathways and influencing factors. Int J Pharm 390(2):89–99. https://doi.org/10.1016/j.ijpharm.2010.02.025

7. Millitzer M, Wenzig E, Peukert W (2005) Adsorption isotherms and irreversible binding of proteins on commercially available hydrophobic adsorbents. Chem Eng Technol 28(7):756–761. https://doi.org/10.1002/ceat.200500072
8. Verwey EJW (1947) Theory of the stability of lyophobic colloids. J Phys Colloid Chem 51(3):631–636. https://doi.org/10.1021/j150453a001
9. Oliveira R (1997) Understanding adhesion: a means for preventing fouling. Exp Thermal Fluid Sci 14(4):316–322. https://doi.org/10.1016/S0894-1777(96)00134-3
10. Kyriakidis AS, Yiantsios SG, Karabelas AJ (1997) A study of colloidal particle brownian aggregation by light scattering techniques. J Colloid Interface Sci 195(2):299–306. https://doi.org/10.1006/jcis.1997.5150
11. Sugimoto T, Kobayashi M, Adachi Y (2014) The effect of double layer repulsion on the rate of turbulent and Brownian aggregation: experimental consideration. Colloids Surf A 443:418–424. https://doi.org/10.1016/j.colsurfa.2013.12.002
12. Mohebrad B, Rezaee A, Sohrabi B (2019) Effect of isoelectric point on cheese whey wastewater treatment using a microbial electrochemical system. Bioelectrochemistry 130:107200. https://doi.org/10.1016/j.bioelechem.2018.08.004
13. Jiang J-Q (2015) The role of coagulation in water treatment. Curr Opin Chem Eng 8:36–44. https://doi.org/10.1016/j.coche.2015.01.008
14. Guckeisen T, Hosseinpour S, Peukert W (2019) Isoelectric points of proteins at the air/liquid interface and in solution. Langmuir 35(14):5004–5012. https://doi.org/10.1021/acs.langmuir.9b00311
15. Engelhardt K, Lexis M, Gochev G, Konnerth C, Miller R et al (2013) PH effects on the molecular structure of β-lactoglobulin modified air-water interfaces and its impact on foam rheology. Langmuir 29(37):11646–11655. https://doi.org/10.1021/la402729g
16. D'Imprima E, Floris D, Joppe M, Sánchez R, Grininger M, et al (2018) The deadly touch: protein denaturation at the water-air interface and how to prevent it. Biorxiv.org. https://doi.org/10.1101/400432.
17. Graham D, Phillips M (1979) Proteins at liquid interfaces. J Colloid Interface Sci 70(3):427–439. https://doi.org/10.1016/0021-9797(79)90050-X
18. Kim SH, Kinsella JE (1985) Surface activity of food proteins: relationships between surface pressure development, viscoelasticity of interfacial films and foam stability of bovine serum albumin. J Food Sci 50(6):1526–1530. https://doi.org/10.1111/j.1365-2621.1985.tb10525.x
19. Mitra SP (2020) Protein adsorption on biomaterial surfaces: subsequent conformational and biological consequences—a review. J Surf Sci Technol. https://doi.org/10.18311/jsst/2020/23282
20. Lautenbach V, Hosseinpour S, Peukert W (2021) Isoelectric point of proteins at hydrophobic interfaces. Front Chem 9:712978. https://doi.org/10.3389/fchem.2021.712978
21. Saghir AE, Farrugia G, Vassallo N (2021) The human islet amyloid polypeptide in protein misfolding disorders: mechanisms of aggregation and interaction with biomembranes. Chem Phys Lipid 234:105010. https://doi.org/10.1016/j.chemphyslip.2020.105010
22. Gonzalez-Garcia M, Fusco G, de Simone A (2021) Membrane interactions and toxicity by misfolded protein oligomers. Front Cell Dev Biol 9:642623. https://doi.org/10.3389/fcell.2021.642623
23. Heyn TR, Garamus VM, Neumann HR, Uttinger MJ, Guckeisen T et al (2019) Influence of the polydispersity of pH 2 and pH 3.5 beta-lactoglobulin amyloid fibril solutions on analytical methods. Eur Polym J 120:109211. https://doi.org/10.1016/j.eurpolymj.2019.08.038
24. Heyn TR, Uttinger MJ, Kwade A, Peukert W, Keppler JK et al (2021) Whey protein (amyloid)-aggregates in oil-water systems: The process-related comminution effect. J Food Eng 311:110730. https://doi.org/10.1016/j.jfoodeng.2021.110730
25. Lübbert C, Peukert W (2018) The mass transfer at Taylor cones. J Aerosol Sci 123:39–51. https://doi.org/10.1016/j.jaerosci.2018.05.014
26. Lübbert C, Peukert W (2020) Mobility-classified mass spectrometry reveals a complete picture of the electrospray outcome. J Phys Chem A 124(42):8842–8852. https://doi.org/10.1021/acs.jpca.0c05076

27. Lübbert C, Supper M, Kaspereit M, Walter J, Peukert W (2023) Single molecule pycnometry and shape analysis of ions in the gas phase. Anal Chem 95(35):13010–13017. https://doi.org/10.1021/acs.analchem.3c00625
28. Cole JL, Lary JWP, Moody T, Laue TM (2008) Analytical ultracentrifugation: sedimentation velocity and sedimentation equilibrium. Methods Cell Biol 84:143–179. https://doi.org/10.1016/S0091-679X(07)84006-4
29. Laue TM, Stafford WF (1999) Modern applications of analytical ultracentrifugation. Annu Rev Biophys Biomol Struct 28:75–100. https://doi.org/10.1146/annurev.biophys.28.1.75
30. Houde DJ, Berkowitz SA (eds) (2024) Biophysical characterization of proteins in developing biopharmaceuticals. Elsevier. https://doi.org/10.1016/B978-0-444-59573-7.00013-0
31. Badmalia MD, Siddiqui MQ, Mrozowich T, Gemmill DL, Patel TR (2020) Analytical ultracentrifuge: an ideal tool for characterization of non-coding RNAs. Eur Biophys J: EBJ 49(8):809–818. https://doi.org/10.1007/s00249-020-01470-9
32. Berkowitz SA, Philo JS (2020) Characterizing biopharmaceuticals using analytical ultracentrifugation. In: Biophysical characterization of proteins in developing biopharmaceuticals. Elsevier, pp 225–283. https://doi.org/10.1016/B978-0-444-64173-1.00009-3
33. Uttinger MJ, Wawra SE, Guckeisen T, Walter J, Bear A et al (2019) A comprehensive Brownian dynamics approach for the determination of non-ideality parameters from analytical ultracentrifugation. Langmuir 35(35):11491–11502. https://doi.org/10.1021/acs.langmuir.9b01916
34. Yang D, Correia JJ, Stafford Iii WF, Roberts CJ, Singh S et al (2018) Weak IgG self- and hetero-association characterized by fluorescence analytical ultracentrifugation. Protein Sci: Publ Protein Soc 27(7):1334–1348. https://doi.org/10.1002/pro.3422
35. Walter J, Löhr K, Karabudak E, Reis W, Mikhael J et al (2014) Multidimensional analysis of nanoparticles with highly disperse properties using multiwavelength analytical ultracentrifugation. ACS Nano 8(9):8871–8886. https://doi.org/10.1021/nn503205k
36. Schuck P (1998) Sedimentation analysis of noninteracting and self-associating solutes using numerical solutions to the Lamm equation. Biophys J 75(3):1503–1512. https://doi.org/10.1016/S0006-3495(98)74069-X
37. Brautigam CA (2015) Calculations and publication-quality illustrations for analytical ultracentrifugation data. Methods Enzymol 562:109–133. https://doi.org/10.1016/bs.mie.2015.05.001
38. Miller CC (1924) The Stokes-Einstein law for diffusion in solution. In: Proceedings of the royal society of London. Series A, containing papers of a mathematical and physical character, vol 106, No. 740, pp 724–749. https://doi.org/10.1098/rspa.1924.0100
39. Uttinger MJ, Heyn TR, Jandt U, Wawra SE, Winzer B et al (2020) Measurement of length distribution of beta-lactoglobulin fibrils by multiwavelength analytical ultracentrifugation. Eur Biophys J 49(8):745–760. https://doi.org/10.1007/s00249-020-01421-4
40. Smith CA (1988) Estimation of sedimentation coefficients and frictional ratios of globular proteins. Biochem Educ 16(2):104–106. https://doi.org/10.1016/0307-4412(88)90075-1
41. Schuck P (2000) Size-distribution analysis of macromolecules by sedimentation velocity ultracentrifugation and lamm equation modeling. Biophys J 78(3):1606–1619. https://doi.org/10.1016/S0006-3495(00)76713-0
42. Svedberg T, Fåhraeus R (1926) A new method for the determination of the molecular weight of the proteins. J Am Chem Soc 48(2):430–438. https://doi.org/10.1021/ja01413a019
43. Harding SE, Johnson P (1985) The concentration-dependence of macromolecular parameters. Biochem J 231(3):543–547. https://doi.org/10.1042/bj2310543
44. Harding SE, Johnson P (1985) Physicochemical studies on turnip-yellow-mosaic virus. homogeneity, relative molecular masses, hydrodynamic radii and concentration-dependence of parameters in non-dissociating solvents. Biochem J 231(3):549–555. https://doi.org/10.1042/bj2310549
45. Uttinger MJ, Walter J, Thajudeen T, Wawra SE, Peukert W (2017) Brownian dynamics simulations of analytical ultracentrifugation experiments exhibiting hydrodynamic and thermodynamic non-ideality. Nanoscale 9(45):17770–17780. https://doi.org/10.1039/C7NR06583C

46. Uttinger MJ, Hundschell CS, Lautenbach V, Pusara S, Bäther S et al (2022) Determination of specific and non-specific protein-protein interactions for beta-lactoglobulin by analytical ultracentrifugation and membrane osmometry experiments. Soft Matter 18(35):6739–6756. https://doi.org/10.1039/d2sm00908k
47. Neal B, Asthagiri D, Velev O, Lenhoff A, Kaler E (1999) Why is the osmotic second virial coefficient related to protein crystallization? J Cryst Growth 196(2–4):377–387. https://doi.org/10.1016/S0022-0248(98)00855-0
48. Deszczynski M, Harding SE, Winzor DJ (2006) Negative second virial coefficients as predictors of protein crystal growth: evidence from sedimentation equilibrium studies that refutes the designation of those light scattering parameters as osmotic virial coefficients. Biophys Chem 120(2):106–113. https://doi.org/10.1016/j.bpc.2005.10.003
49. Wanka J, Peukert W (2011) Optimized production of protein crystals: from 1D crystallization slot towards 2D supersaturation B22 diagram. Chem Eng Technol 34(4):510–516. https://doi.org/10.1002/ceat.201000445
50. Saluja A, Fesinmeyer RM, Hogan S, Brems DN, Gokarn YR (2010) Diffusion and sedimentation interaction parameters for measuring the second virial coefficient and their utility as predictors of protein aggregation. Biophys J 99(8):2657–2665. https://doi.org/10.1016/j.bpj.2010.08.020
51. Vistica J, Dam J, Balbo A, Yikilmaz E, Mariuzza RA et al (2004) Sedimentation equilibrium analysis of protein interactions with global implicit mass conservation constraints and systematic noise decomposition. Anal Biochem 326(2):234–256. https://doi.org/10.1016/j.ab.2003.12.014
52. Zhao H, Brautigam CA, Ghirlando R, Schuck P (2013) Overview of current methods in sedimentation velocity and sedimentation equilibrium analytical ultracentrifugation. Curr Protoc Protein Sci, Unit20.12. https://doi.org/10.1002/0471140864.ps2012s71
53. Ma J, Zhao H, Sandmaier J, Alexander Liddle J, Schuck P (2016) Variable field analytical ultracentrifugation: II. Gravitational sweep sedimentation velocity. Biophys J 110(1):103–112. https://doi.org/10.1016/j.bpj.2015.11.027
54. Stafford WF, Braswell EH (2004) Sedimentation velocity, multi-speed method for analyzing polydisperse solutions. Biophys Chem 108(1–3):273–279. https://doi.org/10.1016/j.bpc.2003.10.027
55. Pearson J, Walter J, Peukert W, Cölfen H (2018) Advanced multiwavelength detection in analytical ultracentrifugation. Anal Chem 90(2):1280–1291. https://doi.org/10.1021/acs.analchem.7b04056
56. Kar SR, Kingsbury JS, Lewis MS, Laue TM, Schuck P (2000) Analysis of transport experiments using pseudo-absorbance data. Anal Biochem 285(1):135–142. https://doi.org/10.1006/abio.2000.4748
57. Schuck P, Rossmanith P (2000) Determination of the sedimentation coefficient distribution by least-squares boundary modeling. Biopolymers 54(5):328–341. https://doi.org/10.1002/1097-0282(20001015)54:5%3c328::AID-BIP40%3e3.0.CO;2-P
58. Zhao H, Li W, Chu W, Bollard M, Adão R et al (2020) Quantitative analysis of protein self-association by sedimentation velocity. Curr Protoc Protein Sci 101(1):e109. https://doi.org/10.1002/cpps.109
59. Chen D-R, Pui DYH (1997) Experimental investigation of scaling laws for electrospraying: dielectric constant effect. Aerosol Sci Technol 27(3):367–380. https://doi.org/10.1080/02786829708965479
60. Lübbert C, Peukert W (2019) How to avoid interfering electrochemical reactions in ESI-MS analysis. J Mass Spectrom 54(4):301–310. https://doi.org/10.1002/jms.4315
61. Hogan CJ, Carroll JA, Rohrs HW, Biswas P, Gross ML (2009) Combined charged residue-field emission model of macromolecular electrospray ionization. Anal Chem 81(1):369–377. https://doi.org/10.1021/ac8016532
62. Konijnenberg A, Butterer A, Sobott F (2013) Native ion mobility-mass spectrometry and related methods in structural biology. Biochem Biophys Acta 1834(6):1239–1256. https://doi.org/10.1016/j.bbapap.2012.11.013

63. Leney AC, Heck AJR (2017) Native mass spectrometry: what is in the name? J Am Soc Mass Spectrom 28(1):5–13. https://doi.org/10.1007/s13361-016-1545-3
64. Kitova EN, El-Hawiet A, Schnier PD, Klassen JS (2012) Reliable determinations of protein-ligand interactions by direct ESI-MS measurements. are we there yet? J Am Soc Mass Spectrom 23(3):431–441. https://doi.org/10.1007/s13361-011-0311-9
65. Karas M, Bahr U, Dülcks T (2000) Nano-electrospray ionization mass spectrometry: addressing analytical problems beyond routine. Fresenius J Anal Chem 366(6–7):669–676. https://doi.org/10.1007/s002160051561
66. Dodds JN, Baker ES (2019) Ion mobility spectrometry: fundamental concepts, instrumentation, applications, and the road ahead. J Am Soc Mass Spectrom 30(11):2185–2195. https://doi.org/10.1007/s13361-019-02288-2
67. Gabelica V, Marklund E (2018) Fundamentals of ion mobility spectrometry. Curr Opin Chem Biol 42:51–59. https://doi.org/10.1016/j.cbpa.2017.10.022
68. Bohrer BC, Merenbloom SI, Koeniger SL, Hilderbrand AE, Clemmer DE (2008) Biomolecule analysis by ion mobility spectrometry. Annu Rev Anal Chem 1:293–327. https://doi.org/10.1146/annurev.anchem.1.031207.113001
69. Guckeisen T, Hosseinpour S, Peukert W (2021) Effect of pH and urea on the proteins secondary structure at the water/air interface and in solution. J Colloid Interface Sci 590:38–49. https://doi.org/10.1016/j.jcis.2021.01.015
70. Yan ECY, Fu L, Wang Z, Liu W (2014) Biological macromolecules at interfaces probed by chiral vibrational sum frequency generation spectroscopy. Chem Rev 114(17):8471–8498. https://doi.org/10.1021/cr4006044
71. Hosseinpour S, Roeters SJ, Bonn M, Peukert W, Woutersen S et al (2020) Structure and dynamics of interfacial peptides and proteins from vibrational sum-frequency generation spectroscopy. Chem Rev 120(7):3420–3465. https://doi.org/10.1021/acs.chemrev.9b00410
72. Roeters SJ, van Dijk CN, Torres-Knoop A, Backus EHG, Campen RK et al (2013) Determining in situ protein conformation and orientation from the amide-I sum-frequency generation spectrum: theory and experiment. J Phys Chem A 117(29):6311–6322. https://doi.org/10.1021/jp401159r
73. Richmond GL (2002) Molecular bonding and interactions at aqueous surfaces as probed by vibrational sum frequency spectroscopy. Chem Rev 102(8):2693–2724. https://doi.org/10.1021/cr0006876
74. Shen YR, Ostroverkhov V (2006) Sum-frequency vibrational spectroscopy on water interfaces: polar orientation of water molecules at interfaces. Chem Rev 106(4):1140–1154. https://doi.org/10.1021/cr040377d
75. Zhang C, Myers J, Chen Z (2013) Elucidation of molecular structures at buried polymer interfaces and biological interfaces using sum frequency generation vibrational spectroscopy. Soft Matter 9:4738–4761. https://doi.org/10.1039/C3SM27710K
76. Beierlein FR, Clark T, Braunschweig B, Engelhardt K, Glas L et al (2015) Carboxylate ion pairing with alkali-metal ions for β-lactoglobulin and its role on aggregation and interfacial adsorption. J Phys Chem B 119(17):5505–5517. https://doi.org/10.1021/acs.jpcb.5b01944
77. Engelhardt K, Peukert W, Braunschweig B (2014) Vibrational sum-frequency generation at protein modified air–water interfaces: effects of molecular structure and surface charging. Curr Opin Colloid Interface Sci 19(3):207–215. https://doi.org/10.1016/j.cocis.2014.03.008
78. Keppler JK, Heyse A, Scheidler E, Uttinger MJ, Fitzner L et al (2021) Towards recombinantly produced milk proteins: physicochemical and emulsifying properties of engineered whey protein beta-lactoglobulin variants. Food Hydrocoll 110:106132. https://doi.org/10.1016/j.foodhyd.2020.106132
79. Kieserling H, Giefer P, Uttinger MJ, Lautenbach V, Nguyen T et al (2021) Structure and adsorption behavior of high hydrostatic pressure-treated β-lactoglobulin. J Colloid Interface Sci 596:173–183. https://doi.org/10.1016/j.jcis.2021.03.051
80. Brune SE, Hoppenreijs LJG, Kühl T, Lautenbach V, Walter J et al (2023) Precision fermentation as a route to modify β-lactoglobulin structure through substitution of specific cysteine residues. Int Dairy J 147:105772. https://doi.org/10.1016/j.idairyj.2023.105772

81. Adamcik J, Jung J-M, Flakowski J, de Los Rios P, Dietler G et al (2010) Understanding amyloid aggregation by statistical analysis of atomic force microscopy images. Nat Nanotechnol 5(6):423–428. https://doi.org/10.1038/nnano.2010.59
82. Walter J, Nacken TJ, Damm C, Thajudeen T, Eigler S et al (2015) Determination of the lateral dimension of graphene oxide nanosheets using analytical ultracentrifugation. Small 11(7):814–825. https://doi.org/10.1002/smll.201401940
83. Hansen S (2004) Translational friction coefficients for cylinders of arbitrary axial ratios estimated by Monte Carlo simulation. J Chem Phys 121(18):9111–9115. https://doi.org/10.1063/1.1803533
84. Halbig CE, Nacken TJ, Walter J, Damm C, Eigler S et al (2016) Quantitative investigation of the fragmentation process and defect density evolution of oxo-functionalized graphene due to ultrasonication and milling. Carbon 96:897–903. https://doi.org/10.1016/j.carbon.2015.10.021
85. Peukert W, Segets D, Pflug L, Leugering G (2015) Unified design strategies for particulate products, mesoscale modeling in chemical engineering part I. Elsevier, pp 1–81. https://doi.org/10.1016/bs.ache.2015.10.004
86. Khan S, Ipsen R, Almdal K, Svensson B, Harris P (2018) Revealing the dimeric crystal and solution structure of β-lactoglobulin at pH 4 and its pH and salt dependent monomer-dimer equilibrium. Biomacromol 19(7):2905–2912. https://doi.org/10.1021/acs.biomac.8b00471
87. Cloupeau M, Prunet-Foch B (1994) Electrohydrodynamic spraying functioning modes: a critical review. J Aerosol Sci 25(6):1021–1036. https://doi.org/10.1016/0021-8502(94)90199-6
88. Lübbert C, Peukert W (2021) Characterization of electrospray drop size distributions by mobility-classified mass spectrometry: implications for ion clustering in solution and ion formation pathways. Anal Chem 93(38):12862–12871. https://doi.org/10.1021/acs.analchem.1c00727
89. Cech NB, Enke CG (2000) Relating electrospray ionization response to nonpolar character of small peptides. Anal Chem 72(13):2717–2723. https://doi.org/10.1021/ac9914869
90. Kozlowski LP (2016) IPC—isoelectric point calculator. Biol Direct 11(1):55. https://doi.org/10.1186/s13062-016-0159-9
91. Yan ECY, Wang Z, Fu L (2015) Proteins at interfaces probed by chiral vibrational sum frequency generation spectroscopy. J Phys Chem B 119(7):2769–2785. https://doi.org/10.1021/jp508926e
92. Ahmadi S, Farokhi M, Padidar P, Falahati M (2015) Combined spectroscopic and calorimetric studies to reveal absorption mechanisms and conformational changes of protein on nanoporous biomaterials. Int J Mol Sci 16(8):17289–17302. https://doi.org/10.3390/ijms160817289

Generalized Morphology Modeling of Aggregating, Filamentous Microorganisms

Henri Müller, Stefan Schmideder, and Heiko Briesen

Abstract Widely used in industrial biotechnology, filamentous fungi produce chemicals, enzymes, acids, antibiotics, and further lifesaving pharmaceuticals. Due to a strong relation between fungal morphology, process conditions, and productivity, targeted morphology engineering techniques aim to improve fungal morphology in terms of process behavior and productivity. However, the exact impact of such genetic and process modification methods are often unknown. The reason for this is the lack of appropriate analytical methods and models that correlate the fungal morphology to substrate supply. In this chapter, we present a method based on X-ray micro-computed tomography and image processing for detailed analysis of three-dimensional fungal morphology. This method has been extended in terms of sample throughput and, in combination with an additionally developed 2D image analysis method, allows the description of the micro- and macro-morphological development of complete fungal cultures. Furthermore, we used Monte Carlo growth simulations to expand the database of experimentally 3D-imaged fungal pellets by 3125 in-silico generated, morphologically different fungal structures. This database was used to perform diffusion computations through the 3D images, revealing a universal correlation between the structure and diffusivity through hyphal networks. The developed methods and results eventually contribute to targeted morphology engineering and improved control of filamentous fungal bioprocesses, leading to increased productivity.

H. Müller · S. Schmideder · H. Briesen (✉)
Process Systems Engineering, School of Life Sciences, Technical University of Munich, 85350 Freising, Germany
e-mail: heiko.briesen@tum.de

H. Müller
e-mail: henri.mueller@tum.de

S. Schmideder
e-mail: ga56baz@mytum.de

A. Kwade and I. Kampen (eds.), *Dispersity, Structure and Phase Changes of Proteins and Bio Agglomerates in Biotechnological Processes*,
https://doi.org/10.1007/978-3-031-63164-1_14

1 Introduction

Filamentous fungi are an integral part of industrial biotechnology. They are masters of biosynthesis, producing chemicals, enzymes, acids, antibiotics, and further lifesaving pharmaceuticals in bioreactors with several hundred cubic meters capacity [1–5]. Features such as their potential to build various bioactive molecules [6–8], their highly expanded protein secretion apparatus [9], and their ability to perform complex post-translational modifications [10] make them these efficient cell factories. Filamentous fungi are also capable of completely degrading lignocellulosic biomass into a variety of useful and sustainable products, making them unique as microorganisms [11].

When cultivated under industrially relevant submerged conditions, filamentous fungi adopt different morphologies. The developed morphology strongly influences the productivity of the bioprocess [12–14]. In general, filamentous fungi consist of branched hyphae that may be loosely packed, which corresponds to the morphology of dispersed mycelia, or very dense and highly branched, which is referred to as pellets [12, 15]. These two forms of morphology each bring advantages and disadvantages. Dispersed mycelia grow rapidly, but cause a non-Newtonian behavior and high viscosity of the cultivation broth, and thus, reduce the nutrient supply due to insufficient mixing [13, 16]. Contrary, cultivation broths with pellets as predominant macromorphology show low viscosities, but the transport of nutrients and oxygen into pellets is diffusion-limited by the dense structure [17, 18], reducing growth and growth-associated product formation [12, 13, 19]. This clearly indicates that the morphology and inner structure of pellets, e.g., the spatial distribution of tips and hyphal material, affects their productivity. Thus, various strategies have been developed to modify fungal morphology. These strategies include the variation of titre, vitality and hydrophobicity of the spore inoculum [20, 21] and cultivation conditions, including medium composition, temperature, shear stress [15, 22], and the addition of microparticles [14, 23–25], as well as genetic modifications of the fungus [19, 26, 27]. However, the mode of action of such strategies is often unknown, limiting targeted morphology engineering of filamentous fungi for increased productivity. These knowledge gaps are due the lack of experimental methods and mathematical models to (1) study the micromorphology within whole intact pellets [28], (2) correlate the three-dimensional (3D) structure and diffusive mass transport of oxygen and nutrients within pellets [29, 30], and (3) analyze the morphological evolution of complete cultivations holistically with a statistically representative number of samples, starting from single spores, through spore agglomeration, to final heterogeneous pellet populations.

In context of the priority program DiSPBiotech of the German Science Foundation (DFG SPP 1934), 3D image analysis methods based on X-ray micro-computed tomography (μ-CT) were developed to determine the micromorphology of whole pellets [28]. Furthermore, diffusion computations through 3D images of pellets from μ-CT measurements and in-silico generated filamentous fungi from 3D Monte Carlo growth simulations revealed a universal correlation between the structure and diffusivity through hyphal networks [29, 30]. Moreover, high-throughput 2D and 3D image and data processing methods based on stereomicroscopy and synchrotron radiation based micro-computed tomography (SR-μ-CT) were developed to track the macro- and micro-morphological development of complete cultivations [31, 32]. Such methods and models can serve to generate a database providing mechanistic insights into the development of fungal morphologies including growth, breakage, and aggregation processes. In addition, modeling approaches can be validated/developed to predict the evolution of fungal morphology, for which suitable experimental data on the internal structure and diffusivity of pellets have been lacking [33–36]. Eventually, the results generated during the research project DFG SPP 1934 may contribute to targeted morphology engineering and increased control of filamentous fungal bioprocesses resulting in enhanced productivities.

Since important parameters for model development are missing in the literature, e.g. local hyphal densities of pellets, we set ourselves the task of detailed characterization of fungal morphology by 2D and 3D image analysis. This topic has received an unexpected higher momentum in the project due to the collaborative opportunities within the SPP and successful exploratory tests to characterize pellet structures with X-ray μ-CT in an instrument that became available in our group (Major Research Instrumentation Grant DFG INST 95/1111-1). Therefore, our overall project title of this report could be misleading with respect to the weighting between modeling and image analysis content.

2 Material and Methods

2.1 Cultivation, Sampling, and Preparation of Pellets

To obtain pellet structures for method development and analysis, the following industrially relevant fungal species/strains were cultivated: *Penicillium chrysogenum* strain MUM17.85 (Micoteca da Universidade do Minho, Braga, Portugal), *Aspergillus niger* strain MF22.4 [27] and its parental strain MF19.5 [37], *Rhizopus stolonifer* (isolated organisms), *Rhizopus oryzae* CBS 607.68 (unknown source). *Aspergillus niger* and *Rhizopus stolonifer* were cultivated at the TU Berlin in cooperation with the SPP working group of Prof. Meyer. Sporangiospores (*Rhizopus* spp.) and conidiospores (*A. niger* and *P. chrysogenum*) of all strains were received from agar plate cultures [38]. Erlenmeyer flasks with liquid media were inoculated with the spores and pellets were grown by submerged cultivation [28, 30, 31].

For μ-CT measurements, a stable and dried state of the pellet samples is required. Therefore, we developed a protocol to generate freeze-dried fungal pellets [28]. Following our protocol, a culture sample with pellets was carefully pipetted into an Eppendorf tube (Eppendorf SE), washed three times with sterile tap water to remove impurities from the medium, and deep frozen in liquid nitrogen while pellets were floating in water. Subsequently, the frozen pellet sample was freeze-dried, resulting in one Eppendorf tube containing several dried pellets with preserved structure. Figure 1 indicates the preserved structure.

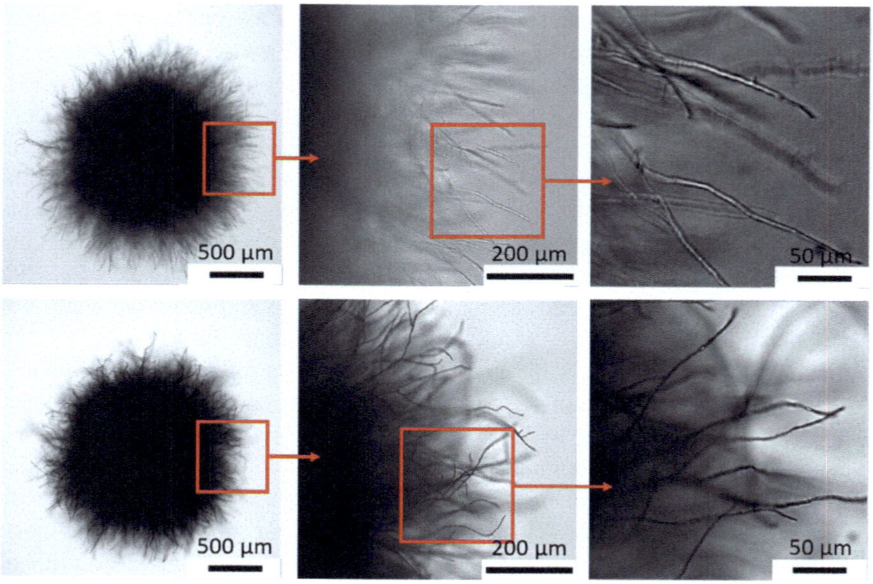

Fig. 1 Light microscope images of a *P. chrysogenum* pellet. The upper row shows the "wet" state of the pellet, whereas the lower row shows the freeze-dried state. (Reprinted with permission from [28], Copyright (2019) John Wiley and Sons)

2.2 *Micro-computed Tomography*

To generate 3D images of the freeze-dried filamentous pellets using laboratory-scale µ-CT [28–30] and SR-µ-CT [32], the freeze-dried pellets were placed on specially designed sample holders which were mounted on a sample pin that fit into the sample rotation stage between the X-ray source and the camera. The generated 2D projections from different angles taken during µ-CT measurements were reconstructed into a 3D image by the filtered back projection algorithm.

In case of the laboratory-scale µ-CT (XCT-1600HR; Matrix Technologies, Feldkirchen, Germany), measurements were performed inhouse with a cone X-ray beam generated at an energy of 60 kV. The exposure time was between 5–6 s per projection. At least 2000 projections were acquired per CT scan using a 10.8 MP CCD camera system. A custom-designed software (Matrix Technologies, Feldkirchen, Germany) that uses CERA (Siemens, Munich, Germany) was used for 3D image reconstruction. The final reconstructed 3D gray level image had a size of $2000 \times 2000 \times 1225$ voxel with a voxel size of 1 µm.

SR-µ-CT was realized at the Imaging Beamline P05 [39–41] at the storage ring PETRA III (Deutsches Elektronen-Synchrotron—DESY, Hamburg, Germany). Imaging was performed at a photon energy of 18 keV and 30 ms exposure time per projection. Per tomographic scan, 2401 projections were recorded using a 20 MP CMOS camera system. Two scans per measured sample holder were performed to increase the field of view in the vertical direction. The volumes were reconstructed with a custom reconstruction pipeline [42] based on MATLAB (MathWorks) and the Astra Toolbox [43–45]. The reconstructed bottom and upper volumes of a sample were stitched together, resulting in a final 3D gray level image with 2595 $\times$ 2595 $\times$ 3718 voxel with a voxel size of 1.28 μm. It must be mentioned, that the high brilliance of synchrotron radiation enables a higher signal to noise radiation and thus better image contrast compared to a laboratory μ-CT.

2.3 3D Image Segmentation of Multiple Pellets

Within the project, several steps were developed and iteratively refined for processing the 3D gray level images showing pellets, parts of the sample holder, and background. During this development period, parts of the image processing were performed with different software, e.g., MAVI (version 1.4.1; Fraunhofer ITWM, Kaiserslautern, Germany), VGSTUDIO MAX (version 3.2, Volume Graphics GmbH, Heidelberg, Germany), and FIJI/ImageJ. These processing parts were finally transferred completely and identically to MATLAB (MathWorks). The following description therefore concentrates on the image processing steps that were established in MATLAB. Furthermore, for the first μ-CT based study, fungal pellets were segmented manually from the 3D grayscale images and subsequently post-processed for detailed and accurate determination of micromorphological parameters, e.g., the locally resolved branch and tip density of pellets [28]. The manual segmentation of pellets was replaced by an automated method to increase sample throughput in μ-CT measurements and image processing for statistically meaningful results, and to cope with the large amount of 3D image data from SR-μ-CT measurements [30, 32]. The developed automated method enabled the segmentation and processing of up to 579 pellets from a single SR-μ-CT image. In the following, this method is presented.

For μCT measurements, several freeze-dried fungal pellets were transferred into a specially designed sample holder consisting of a polyether ether ketone (PEEK) base, a Kapton tube, and a PEEK cap. PEEK and Kapton ensure high X-ray resistance and transparency (Fig. 2a). Figure 2b shows a microtomography from SR-μ-CT measurements depicting a sample holder filled with several pellets. To enable the analysis of the micromorphology of individual pellets described in the next section, pellets need to be segmented out of this pellet stack. Before segmentation, the complete gray level tomography was binarized with Otsu's method [46]. Subsequently, the sample holder was removed based on a normalized cross-correlation [47, 48] between a template of the Kapton tube and the tomography. Labeling and segmentation into individual pellets were established by a marker-controlled watershed segmentation. Here, pellets were dilated with a spherical structured element and remaining holes within the dilated pellets were filled. Based on the dilated and filled pellets a distance transformation was performed. Using the resulting distance image, markers representing the inner part of the pellet spheres were determined by masking voxels that were a certain distance from the background with zero. For each marker, a pellet

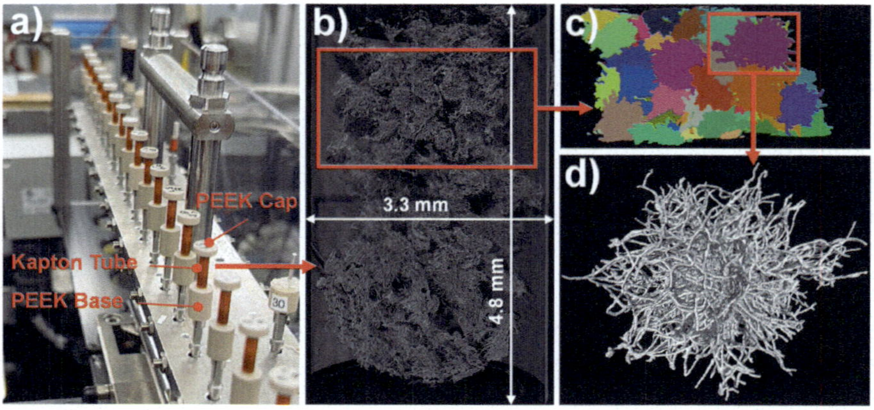

Fig. 2 Sample preparation, microtomography, and image processing. **a** Sample holders with freeze dried pellets at the P05 beamline; **b** synchrotron radiation based X-ray microtomography with a voxel size of 1.28 μm; **c** segmented and labeled pellets; **d** segmented and binarized pellet with an equivalent diameter of 557 μm. (Reprinted with permission from [32], Copyright (2023) John Wiley and Sons)

was labeled and segmented by the watershed segmentation (Fig. 2c). Each grayscale image, representing only a single pellet, was obtained by masking all voxels on the original grayscale tomography with zero except the corresponding segmented pellet. The resulting gray level pellet images were binarized (Fig. 2d) and further processed for tip and branching point detection as well as for diffusion computations of the filamentous structures.

2.4 *Post-processing of Segmented 3D Pellets*

Following the developed automatic segmentation method (previous section) [30, 32], the segmented 3D gray level images depicting only the pellet were binarized by setting a global threshold using Otsu's method [46]. Subsequently, remaining small connected objects that were not part of the pellet were deleted (Fig. 3a, b). Using the processed and binarized image, the skeleton of the pellet representing the centerline of hyphae was calculated by an iterative thinning algorithm [49, 50] (Fig. 3c). Based on the skeleton, tips, and branching point voxels were computed (Fig. 3d). Branching point voxels are characterized by at least three neighboring voxels. Contrary, tips have only one neighbor and represent the endpoints of the skeleton. "Normal" hyphae voxels, are those with two neighboring voxels. When detecting tips and branching points based on the skeleton, errors may occur that lead to artefactual detection of individual tips and branching points. Thus, postprocessing steps were performed to correct the skeleton. The first step eliminates falsely detected tips due to short branches caused by surface roughness of hyphae (Fig. 4a, b). Here, the Euclidean length of each branch was calculated. If this length was shorter than 4 μm, which is close to the hyphal diameter of the used strains, the branch and thus the tip was eliminated. In the second step, wrongly detected branching points, caused by overlapping hyphae were corrected (Fig. 4c, d). Since true hyphal branching points connect only three branches, the number of branches was determined for each branching point. If the number was greater than three, the branching point was falsely detected and consequently deleted. In the third postprocessing step, the branching points were corrected, which result from parallel and closely spaced hyphae due to small bridges between hyphae (Fig. 4e, f). These connections can result from the limited spatial resolution, making it impossible to distinguish between background and hyphae if the distance between two hyphae is too small. To correct this issue, the distance between two branching points must be at least 6.5 μm, otherwise the branching points are deleted and recognized as "normal" voxels of the skeleton. For a more detailed description of the developed postprocessing steps, we refer to [28].

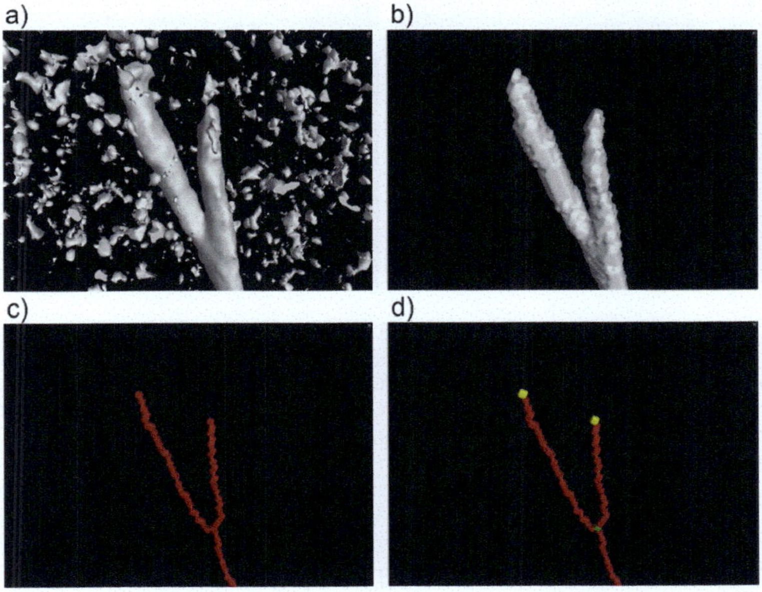

Fig. 3 Image processing steps for a small region of a *P. chrysogenum* pellet; images are rendered with VGSTUDIO MAX. **a** Raw image: CT data with remaining image noise and small connected objects. **b** Preprocessed and binarized image with deleted small connected objects. **c** Skeletonized image. **d** Analyzed image: analysis of the skeleton; tips are marked yellow and branches green. (Reprinted with permission from [28], Copyright (2019) John Wiley and Sons)

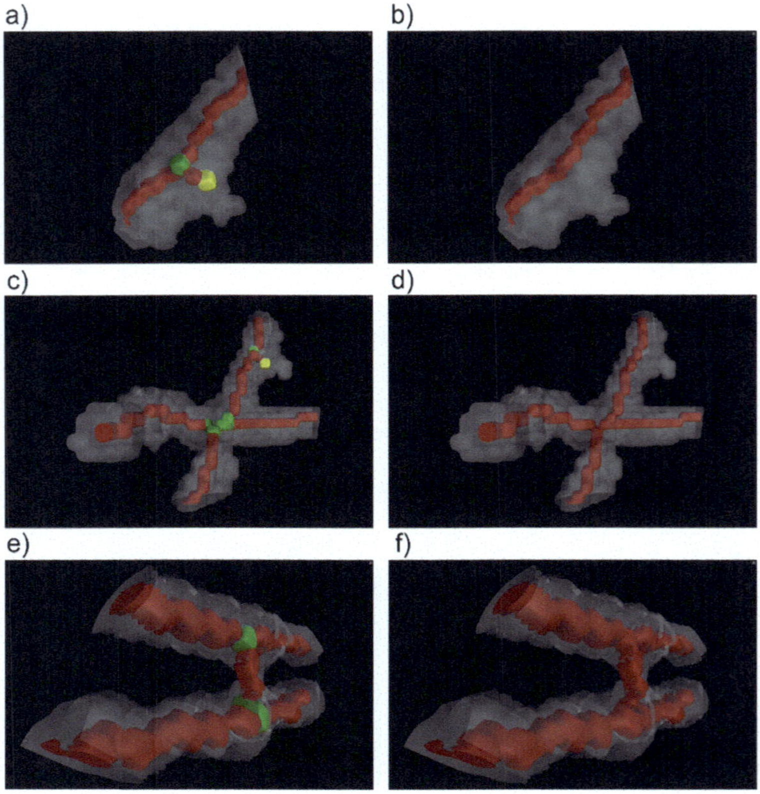

Fig. 4 Post-processing steps for small regions of *P. chrysogenum* pellet; transparent grey objects illustrate the binarized 3D data, red objects the skeleton, yellow markers the tips, and green markers the branches. **a** Rough surface of binarized hyphae result in a short junction. **b** Post-processing: short junction from rough surface is deleted. **c** Overlapping hyphae result in a branch. **d** Post-processing: incorrect branch is deleted. **e** Close parallel hyphae result in bridges imitating branches. **f** Post-processing: incorrect branches are deleted. (Reprinted with permission from [28], Copyright (2019) John Wiley and Sons)

2.5 Global and Local Morphology Properties of μ-CT Pellets

Based on the full 3D pellet structure from μ-CT measurements and image processing the following global morphology features were determined within the studies [28, 32]: total tip number, total branch number, total hyphal volume, total hyphal fraction, pellet equivalent diameter, total hyphal length, average hyphal diameter, hyphal growth unit (HGU), and average branch length.

The total number of tips N_{Tips} and branching points $N_{Branches}$, were obtained by counting the corresponding segmented object voxels of the pellet. To calculate the total hyphal volume V_{Hyphae}, all hyphal voxel of the binarized pellet were summarized and multiplied with the voxel volume. The convex hull, extending over the outermost tips of the pellet, was used to calculate the total volume of the pellet V_{Pellet}. The ratio between V_{Hyphae} and V_{Pellet} defined the total hyphal fraction of a pellet. The total pellet volume was further used for the calculation of the equivalent spherical diameter of the pellet as $d_{pellet} = \sqrt[3]{\frac{6 \cdot V_{pellet}}{\pi}}$ [28]. Similarly, the equivalent spherical diameter of the spore core, which is the spore agglomerate within a pellet, was calculated [32]. During the project, different calculation methods for the average hyphal diameter and total hyphal length were applied. To process a high number of fungal pellets without further input from different software, we calculated the total hyphal length as $L_{Hyphae} = \frac{4 \cdot V_{Hyphae}}{\pi \cdot d_{Hyphae}{}^2}$, where d_{Hyphae} is the average hyphal diameter. d_{Hyphae} was calculated by masking all voxels of the distance image of the binarized pellet with zero, except for those representing the skeleton of the pellet and thus the regional radii of the pellet. The regional radii were used to calculate the arithmetic mean of the hyphal diameter [32]. Furthermore, the HGU was calculated as $\frac{L_{Hyphae}}{N_{Tips}}$ [51]. Complementary, the average branch length L_{Branch} for pellets, which is the length between two branching points or a branching point and a tip, was estimated as $L_{Branch} = \frac{L_{Hyphae}}{N_{Tips}+N_{Branches}}$ [32].

The analyzed pellets had a spherical structure, which allowed quantifying the morphological properties distributed along their radii. For this, the number densities of tips, branches, and the hyphal fractions of a pellet were determined in shells along its distance from the pellet center. The width of the shells was at least 15 and 50 μm for the central sphere. The pellet center was defined as the mass center of the pellet [28] and in a later study as the mass center of the determined spore agglomerate within pellets, the spore core of a pellet [32]. The hyphal fraction of a shell was determined by dividing the hyphal volume of the shell by the total volume of the shell. Accordingly, to calculate the number densities, the number of tips and branches for the shell was divided by the volume of the corresponding shell.

2.6 *Three-Dimensional Growth Simulation of Filamentous Fungal Pellets*

To generate nature-like 3D images of filamentous fungal pellets in-silico, a microscopic modeling approach was adopted. This allowed the calculation of the entire micromorphological 3D structure of pellets including the exact positions of hyphae, branches, and tips. Furthermore, the modeling approach enabled the voxel-based generation of 3125 morphologically different fungal structures covering a broad range of fungal morphologies. The simulated pellets increased the database of pellet structures to be analyzed with respect to diffusion properties as will be presented and discussed in the following section.

The morphological simulations were implemented using MATLAB and based on stochastic growth models for filamentous microorganisms [34, 52]. Compared to Celler et al. [34], we neglected oxygen limitation and cross wall-formation during growth. Furthermore, we introduced the pellet formation from spore agglomerates to generate coagulative type pellet structures in addition to non-coagulative ones. A voxel-based output of the pellet structures was introduced to compare the simulated data with the μCT data. The developed model includes 3 main parts: (1) spore aggregation, (2) growth, and (3) generation of voxel-based 3D structure (Fig. 5). In order to simulate morphologically different pellets, the following parameters were varied in the corresponding main steps: number of initial spores, branching interval, branching angle, maximum growth angle, and the hyphal diameter. For each parameter fives values were varied to generate $5^5 = 3125$ morphologically different structures. Figure 6 summarizes the varied parameters which were estimated based on

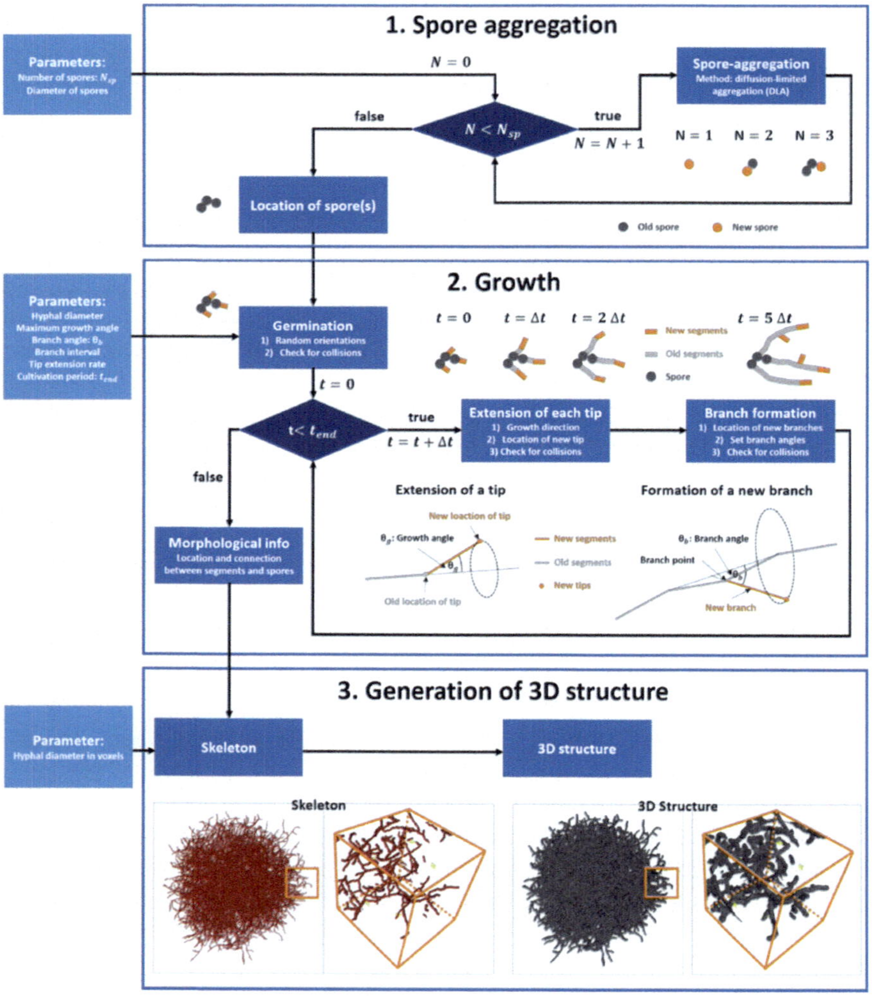

Fig. 5 Overview of the algorithm of morphological growth simulations. (Reprinted with permission from [30], Copyright (2020) John Wiley and Sons)

values for filamentous fungi from literature [11, 53, 54]. Spore aggregation to produce the initial spore agglomerates that grow into the pellet structure was estimated using the diffusion-limited-limited aggregation (DLA) method [55]. The spore agglomerate was initialized with a first seed spore and iteratively enlarged by further spores, which performed a random walk until they came into contact with the first spore or the already formed spore agglomerate. This process was terminated, if the desired number of spores was reached (Fig. 5, top section). The second main step, growth, began with germination, where random orientated segments of the length l_{germ} developed from the spores. In this process, collisions of the formed segments with other germlings were prevented. After germination, tip extension and branching of the hyphae began. These two operations were repeated at each time step Δt until the end of the growth simulation. For tip extension, a segment with the length $l_{segment}$, which is the product of Δt and the tip extension rate α_t, was attached to the previous segment at a random growth angle θ_g between 0° and the desired maximum growth angle (Fig. 5, middle section). Overlapping and collisions between new segments and already simulated hyphal material was prevented with closest-point computations [56]. New possible branching points (node between two segments) were computed in subapical regions of hyphae, which were defined by the length $l_{subapical}$ beginning from the tip, if three conditions were fulfilled: (1) location of branching point in subapical region, (2) no branches in the neighboring four nodes, (3) current number of branches of the hyphae was lower than the possible number of branches, calculated by dividing the length of the hyphae by the minimum mean distance between two branches $d_{b,min}$. If these three conditions were met, a new possible branch with the desired branching angle θ_b was calculated, checked for collisions with already simulated hyphal material [56], and rejected if a collision was detected (Fig. 5, middle section).

The output of the growth section includes the morphological data of the locations and connections of nodes, hyphal segments, and spores, which were converted to the final 3D voxel-based output. Within this last main step (Fig. 5, bottom section), the locations were scaled with the factor $f_{scale} = \frac{d_{h,vx}}{d_h}$ to represent the hyphae with the desired hyphal diameter in μm d_h and resolution depending on the hyphal diameter in voxels $d_{h,vx}$. Subsequently, the connections between nodes were discretized with Bresenham's line algorithm [57] to compute skeletons of pellets. Finally, the skeletons were dilated with a spherical structuring element resulting in voxel-based 3D pellet images with the desired hyphal diameter. For a more detailed description of the Monte Carlo growth model, we refer to [30].

2.7 Diffusion Computations Within Fungal Pellets

Dense hyphal networks can limit mass transport resulting in oxygen/nutrient starvation [17, 18, 58, 59] and morphological heterogeneity [60] inside fungal pellets. To predict the concentration of substrates inside pellets, it is necessary to know the

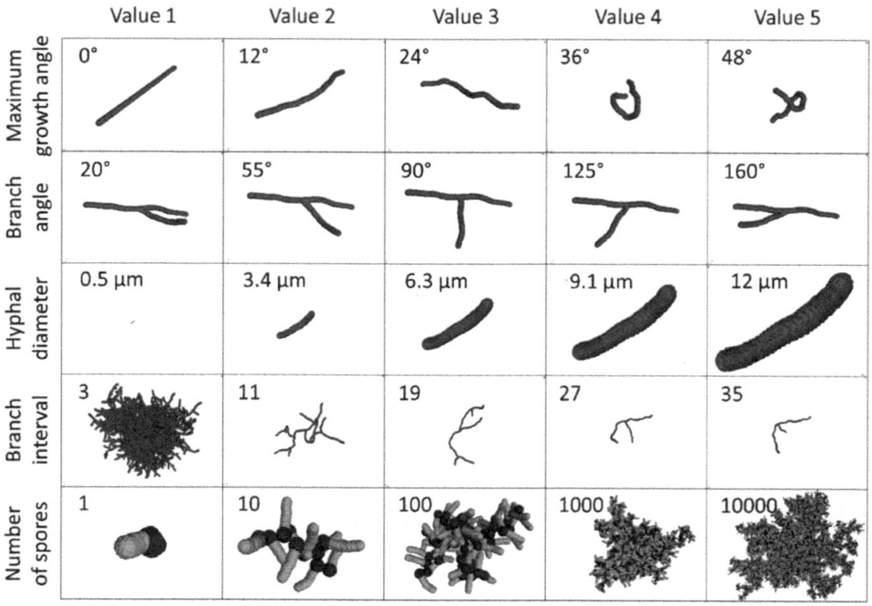

Fig. 6 Applied morphological parameters and values for simulations of filamentous structures. Each of these parameters was varied to five values resulting in a total of $5^5 = 3125$ simulations performed with the Monte Carlo method. (Reprinted with permission from [30], Copyright (2020) John Wiley and Sons)

metabolic rate and diffusive mass transfer through hyphal networks, which is considered the main transport phenomenon inside submerged cultivated pellets [34, 36, 61, 62]. To calculate the diffusive mass transport, the effective diffusion coefficient $D_{i,eff}$ of the diffusing component i (e.g. oxygen or a nutrient) is required, which is defined as [63]

$$D_{i,eff} = D_{i,bulk} \cdot k_{eff}, \tag{1}$$

where $D_{i,bulk}$ is the diffusion coefficient of the diffusing component i in the bulk medium and k_{eff} the effective diffusion factor. k_{eff} should be independent of the diffusing substance and describes the geometrical diffusion hindrance due to the pore geometry. Thus, it reduces the free bulk diffusion $D_{i,bulk}$ to the effective diffusion coefficient $D_{i,eff}$ in the pore network. Since $D_{i,bulk}$ can be estimated from the medium condition [64], k_{eff} is the only unknown variable to determine $D_{i,eff}$ and thus the effective diffusivity inside pellets. The diffusion factor k_{eff} is defined as [65]

$$k_{eff} = \frac{\epsilon}{\tau^2}, \tag{2}$$

where ϵ is the porosity and τ the tortuosity of the porous medium. The tortuosity τ can be described as the ratio of the average pore length to the length of the porous medium along the major flow or diffusion axis. The definition of the effective diffusion factor and the tortuosity implies that the detailed 3D micro-structure of the analyzed porous medium has to be known to determine these parameters. In the absence of suitable models and experimental methods for predicting the effective diffusion factor and thus the effective diffusivity inside whole non-destructive fungal pellets [29], we harnessed μ-CT and 3D image analysis to resolve the 3D micro-structure of whole pellets and computed locally resolved effective diffusion factors.

To compute the spatially resolved effective diffusivity inside filamentous fungal pellet structures gained from μ-CT measurements as well as from Monte Carlo growth simulations, several representative cubic sub-volumes were cropped out of each voxel-based 3D pellet image. Within each representative elementary volume (REV), diffusion computation were performed with the module DiffuDict of the commercial software GeoDict (Math2Market GmbH) [63, 66, 67].

Using MATLAB, the REVs were extracted along the three main axis originating from the calculated mass center of the 3D pellet structures (Fig. 7). The distance between the centers of cubic REVs along the main axis was set at 25 μm, resulting in 50–150 single REVs per pellet depending on the pellet size. In our studies [29, 30], we proved that an edge length of at least six times the hyphal diameter of the analyzed fungal species and strains was sufficient as REV to investigate the hyphal fraction and diffusivity.

Since DiffuDict enables the voxel-based solution of transport equations, the cubic REVs could directly be processed. Diffusion computations were performed in the radial direction, pointing to the mass center of the pellet, and two tangential directions

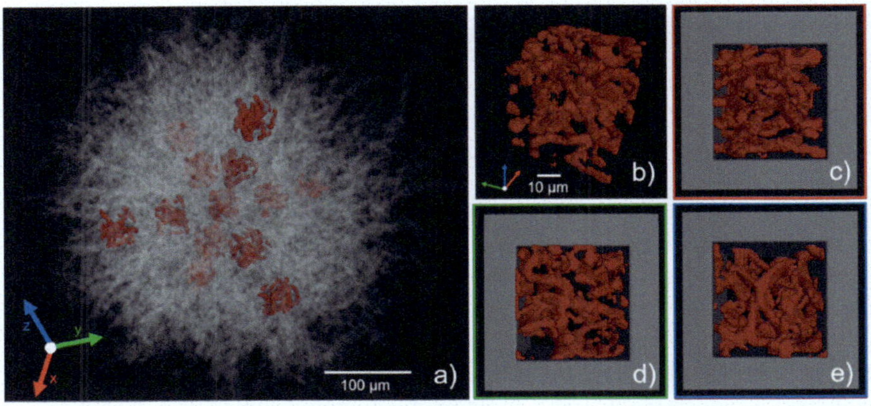

Fig. 7 Binarized and processed 3D μ-CT image of an *Aspergillus niger* pellet and cropped cubes for the diffusion computations: **a** transparent: projection of a whole pellet; red: exemplary cubes used for diffusion computations. **b–e** Morphology of a single cube from different viewing directions; the gray boundaries in (**c–e**) illustrate the boundaries parallel to the diffusion computation. (Reprinted with permission from [29], Copyright (2019) John Wiley and Sons)

(perpendicular to the radial direction) for each cube to compute the effective diffusion factor for the corresponding direction. The diffusion was computed in the space/liquid between the hyphae, the porous medium. We applied pure bulk diffusion which was modeled by the Laplace equation with Neumann boundary conditions on the pores-to-solids boundaries and a concentration drop of the diffusing component in the diffusion direction [63]. For the in- and outlet, we used Dirichlet boundary conditions and symmetric boundary conditions on the other four faces. Pure bulk diffusion is the predominant diffusion regime in liquids [63, 68] and was therefore the method of choice. Surface effects on the solid–liquid interface that could affect diffusion were neglected [29]. In addition to the diffusion computations, the porosity ϵ and the hyphal fraction $c_h = 1 - \epsilon$ were determined of each REV.

3 Results and Discussion

3.1 *Morphological Development of Pellets Based on 2D and 3D Imaging*

The μ-CT measurements and developed image analysis revealed micromorphological properties of whole filamentous fungal pellets [28, 32]. Due to the lack of measurement techniques, most of these properties had not been determined for pellets before [69]. Moreover, properties such as the number density of tips, branches, and the hyphal fraction were locally resolved enabling a comprehensive analysis of the pellets' morphology [28]. Furthermore, the development of a high-throughput 2D image analysis method enabled the tracking of macromorphological pellet growth and population heterogeneity based on size distributions of spore agglomerates and pellets. For the sake of brevity we did not include any information on the 2D image analysis in the in Method Section but instead refer to [31]. In addition, SR-μ-CT allowed the three-dimensional resolution of large volumes of pellet samples from multiple time steps revealing the micromorphological development during pellet growth [32].

Figure 8 shows the size evolution of spore agglomerates evolving into spore cores and of pellets in an *A. niger* cultivation. The data show good agreement between the developed 2D [31] and 3D [32] image analysis and thus a correct segmentation of single pellets from a large volume with multiple pellets. With the 2D analysis it is possible to track the spore agglomeration and pellet growth with one reproducible method. Contrary, the developed 3D image analysis does not provide information on spore agglomeration. However, it provides much more detailed data, which can be locally resolved as exemplarily indicated in Fig. 9. Here, an increase of the hyphal fraction in outer pellet regions can be observed during the cultivation (Fig. 9a). The hyphal fraction up to 50 μm ranges between 20.5% (18 h) to 22% (32 h) representing mainly the solid fraction of the spores/spore agglomerate which are embedded in the pellet center. Since branching points remain constant at their place of origin, the branch density increases in outer pellet parts but remains constant in inner parts with cultivation time (Fig. 9b). For a more detailed interpretation of these results we refer to our publications [31, 32].

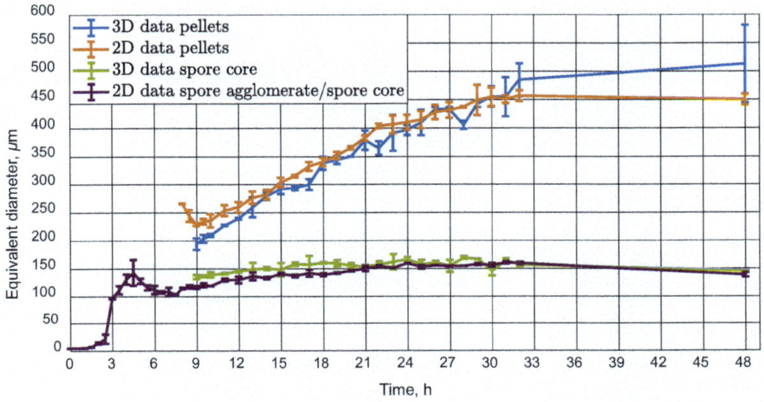

Fig. 8 Mean equivalent diameters of pellets and spore cores over cultivation time using 2D and 3D μ-CT image data. Spore cores are detected spore agglomerates within pellets that are detected from 8 h. The error bars indicate the standard deviation between replicated (flask A and B) cultivations. (Adapted and reprinted with permission from [32], Copyright (2023) John Wiley and Sons)

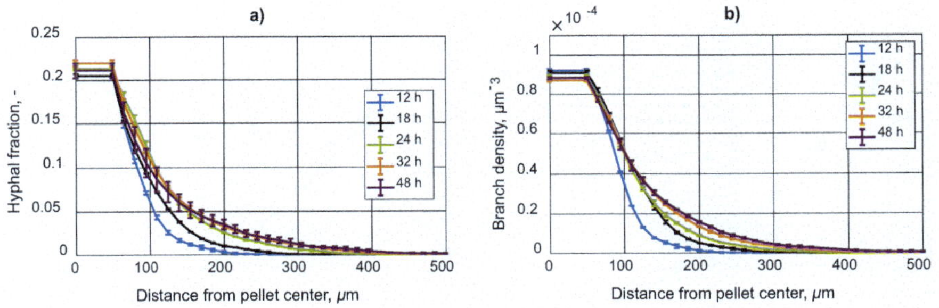

Fig. 9 Hyphal fractions and branch densities over the distance from the pellet center. The error bars indicate the standard deviation between replicated cultures. **a** Mean hyphal fractions; **b** mean branch densities. (Adapted and reprinted with permission from [32], Copyright (2023) John Wiley and Sons)

3.2 Effective Diffusivity Through Mycelial Networks

As a main result of the project, a law for predicting the effective diffusion factor as a function of the hyphal fraction was found [29]. This law was shown to be applicable to all experimentally measured and in-silico generated filamentous fungi [30].

In a first diffusion study [29], the radial effective diffusion factors from five *A. niger* pellets of μ-CT measurements were plotted over the hyphal fractions of the corresponding representative elementary volumes (REVs, Fig. 10). Furthermore, existing models from the literature predicting the diffusion factor depending on the hyphal fraction/porosity ($\epsilon = 1 - c_h$) of filamentous microorganisms (Fig. 10a) and fibers (Fig. 10b) were compared to the experimental diffusion data of *A. niger*. The linear assumption $k_{eff} = \epsilon$ applied in simulation studies for filamentous microorganisms by Celler et al. [34], Cui et al. [61], and Buschulte [33] overestimates the effective diffusion factors compared to the experimental data. This model ($k_{eff} = \epsilon$) is also known as the "law of mixtures" [70], which was developed for flow along parallel fibers and does not take into account the tortuosity of the filamentous network resulting in an increased diffusion factor. In contrast, Lejeune and Baron [36] assumed a constant tortuosity of two $\left(k_{eff} = \epsilon/2\right)$. Thus, their model would result in a reduced, geometrically hindered diffusion for a hyphal fraction of zero making their model not applicable. The non-linear model of Meyerhoff et al. [62] satisfies the mandatory condition of an effective diffusion factor of one and zero for a hyphal fraction of zero and one, respectively. However, the proposed model overestimates the experimental effective diffusion factors. Buschulte [33] used a second model ($k_{eff} = e^{-2.8c_h}$) that best fits the experimental data compared to other models for filamentous fungi in

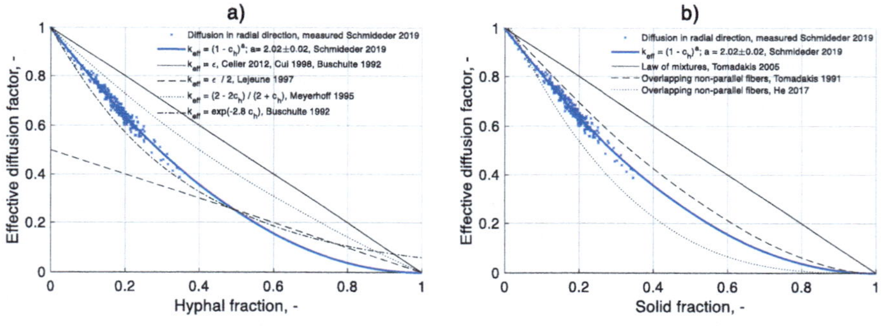

Fig. 10 Correlations between hyphal fraction c_h/porosity ($\epsilon = 1 - c_h$) and effective diffusion factor k_{eff}. The blue data points correspond to the computed effective diffusion factors of five *A. niger* pellets in the radial direction. The solid bold blue line shows the new correlation between the hyphal fraction and the effective diffusion factor; the black lines represent existing correlations in the literature for **a** filamentous microorganisms and **b** 3D random distributed overlapping fibers. (Reprinted with permission from [29], Copyright (2019) John Wiley and Sons)

the literature. This model underestimates our experimental data and also does not meet the condition of an effective diffusion factor of zero for a hyphal fraction of one. Figure 10b shows the "law of mixtures" and correlations for 3D random distributed overlapping fibers. Compared to the models formulated for filamentous microorganisms (Fig. 10a), our experimental data were fitted better by the non-linear models of Tomadakis and Sotirchos [71] and He et al. [72] for overlapping non-parallel fibers (Fig. 10b). Both models include a power function of porosity, which is often used to estimate the effective diffusion factor of isotropic structures [73]. However, additional parameters are needed in these models: the percolation porosity [71] or a further fitting parameter [72]. Since the existing assumptions in the literature for modeling diffusion in filamentous microorganisms inaccurately correlated the effective diffusion factor with the porosity/hyphal fraction, and in order to obtain a relation with only one fitting parameter, we proposed a new non-linear correlation:

$$k_{eff} = (1 - c_h)^a, \tag{3}$$

with a fitting parameter a of 2.02 ± 0.02 within 95% confidence bounds, this model fitted the experimental diffusion factors of the five investigated *A. niger* pellets very well. Furthermore, our simple equation is theory-consistent with the condition of a diffusion factor of one and zero for a hyphal fraction of zero and one, respectively [29].

To get closer to a generally applicable correlation for filamentous fungi, we performed diffusion computations with REVs of additional strains of industrial-relevant fungal species including *P. chrysogenum* (penicillin producer), *R. stolonifer* (fumaric acid), and *R. oryzae* (enzymes, tempeh) [30]. Subsequently, we fitted the parameter a of the developed correlation (Eq. (3)) based on the computed diffusion factors and hyphal fractions of 66 μCT analyzed Pellets (Fig. 11). Although the pellet morphologies of the strains studied differed significantly, the fitting parameter a was similar for all strains.

To cover the wide variety of fungal morphologies, we performed further diffusion computation within 3125 morphological different simulated filamentous structures using a Monte Carlo growth model (see Sect. 2) and fitted the parameter a (Fig. 12) [30]. Since we observed a dependence between image resolution and the fitted parameter a, we simulated pellets with an image resolution representing hyphae with 5 (Fig. 12a) and 13 voxels (Fig. 12b) in diameter. By the variation of growth parameters, the Monte Carlo growth simulations also computed very loose structures with a mean hyphal fraction of the respective cubes less than 0.05 and pellet structures with very unlikely morphological parameters, e.g., straight hyphae or extreme branch angles (20 and 180°). We discarded these simulated structures and fitted the parameter a for the remaining 898 and 712 pellets with hyphae represented by 5 and

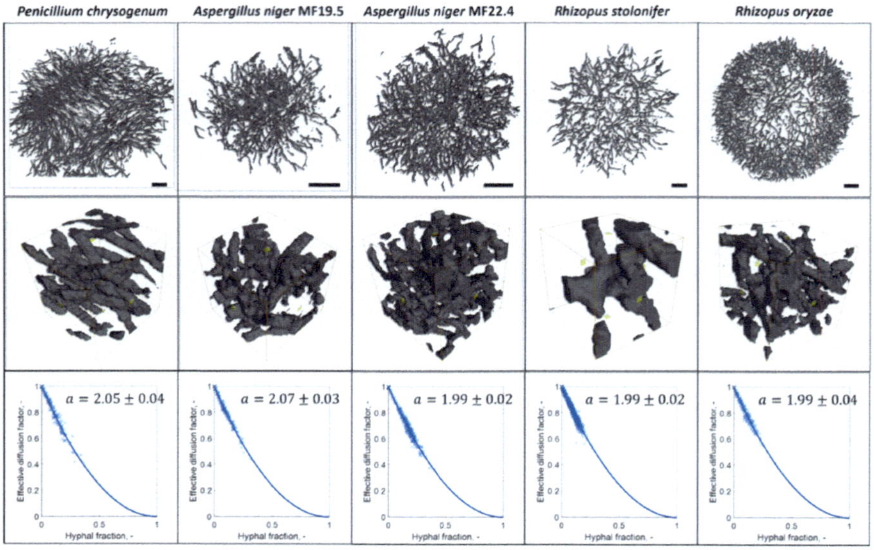

Fig. 11 Morphology and diffusivity of experimentally investigated pellets. Upper two rows: projections of processed μ-CT images. Cubes are 50 × 50 × 50 μm. Bottom row: correlation between hyphal fraction (c_h) and effective diffusion factor (k_{eff}). Each blue data point corresponds to one cube applied for diffusion computations. The solid blue line shows the new correlation $k_{eff} = (1 - c_h)^a$, resulting in the fitted exponent a. ± specifies the 95% confidence interval. (Reprinted with permission from [30], Copyright (2019) John Wiley and Sons)

13 voxels, respectively. As a result, a mean correlation factor with a narrow distribution of $a = 2.072 \pm 0.025$ and $a = 1.760 \pm 0.023$ was obtained. In Fig. 12a, μ-CT pellets were also included, indicating a fitting parameter a in the range of simulated pellets for the comparable resolution, where hyphae are represented with a diameter of 5 voxels. The narrow range of the fitting parameter a indicated a universal application of the diffusion law. Since we showed, that the fitting parameter a converges to $1.6 < a < 1.8$ with increasing resolution, we proposed the law for high resolutions as universal applicable law to calculate the diffusivity through fungal pellets [30]:

$$k_{eff} = (1 - c_h)^{1.76}, \tag{4}$$

where the fitting parameter a was calculated based on 712 morphologically different pellet structures and hyphae were represented with a resolution of 13 voxel in diameter.

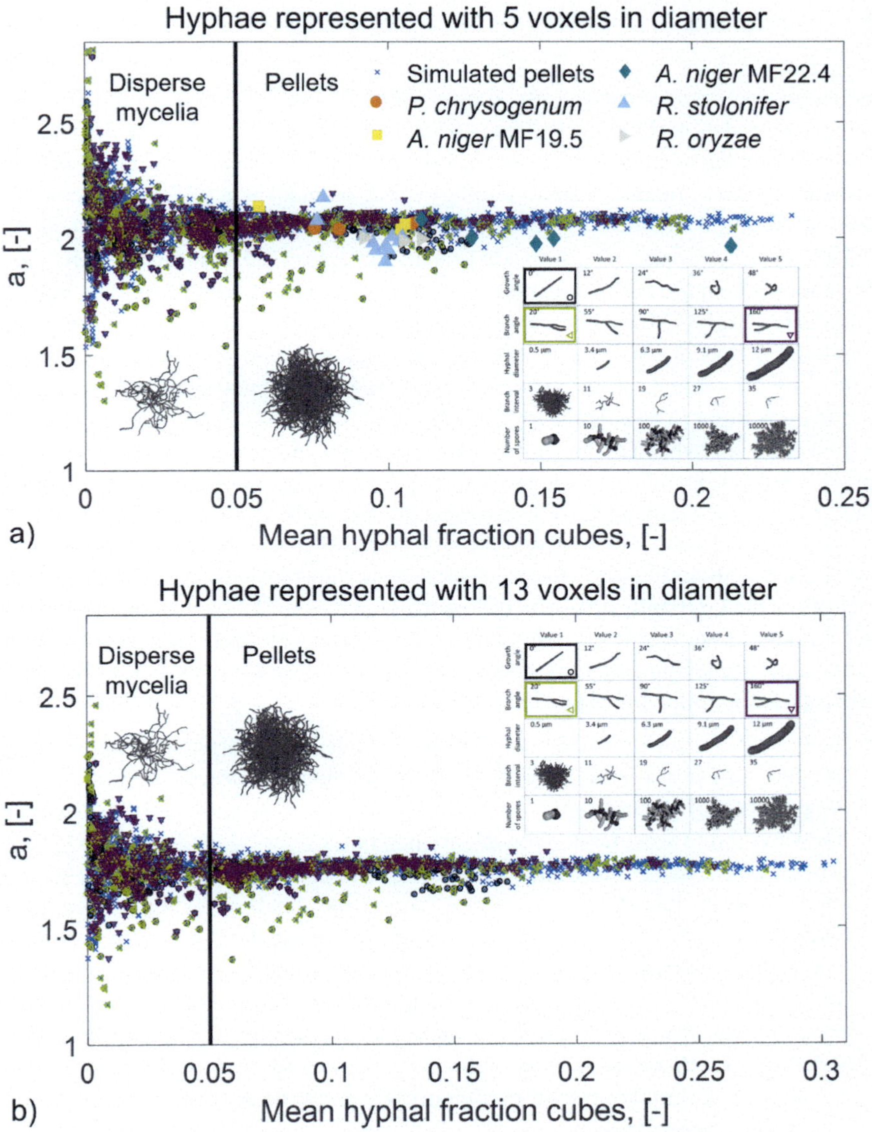

Fig. 12 Correlation factor a for simulated and experimentally determined pellets. Each data point corresponds to the correlation ($k_{eff} = (1 - c_h)^a$) between the hyphal fraction c_h and the effective diffusion factor k_{eff} of one pellet, resulting in the fitted exponent a. . For each pellet, the fit is based on diffusion computations through several cubes. Hyphae of simulated pellets are represented with a "resolution" of five (top) and 13 (bottom) voxels in diameter. (Reprinted with permission from [30], Copyright (2020) John Wiley and Sons)

4 Conclusion

In our project within the priority program DiSPBiotech funded by the German Science Foundation (DFG SPP 1934), three main methods/results were elaborated: (1) a method based on μ-CT measurements and image analysis was developed to quantify the three-dimensional morphology of whole intact pellets [28]; (2) the developed 3D analysis method was significantly improved in terms of sample throughput and allows in combination with an additionally developed 2D image analysis method the description of the micro- and macro-morphological evolution of complete fungal cultures [31, 32]; (3) based on diffusion computations through μ-CT measured and simulated pellets, a universal correlation between micromorphology and effective diffusivity was found, allowing the diffusive transport of nutrients and oxygen in filamentous fungi to be calculated [29, 30].

It is well known, that the supply of oxygen and nutrients strongly affects the growth and productivity of fungal pellets [12, 13, 19]. Morphology engineering therefore aims to create idealized structures that allow the desired nutrient supply of the whole pellet [26]. The concentration of nutrients inside pellets depends on the metabolic consumption rate of the fungus and the transport through the filamentous network, which is mainly driven by diffusion [34, 36, 61, 62]. Our generated methods and findings make it now feasible to compute the concentration profile of nutrients within a pellet making targeted morphology engineering possible. For this objective, we propose the following procedure:

1. Using μ-CT measurements and 3D image analysis, a morphology database of fungal cultivations with different strains or at modified process conditions can be generated. Furthermore, growth simulations allow to extend this database by the variation of reasonable growth parameters.
2. The diffusive mass transport within each pellet can be solved using the generated diffusion law ($k_{eff} = (1 - c_h)^{1.76}$) and the database including the local hyphal fractions. The concentration profile of nutrients inside pellets can be calculated with modeling approaches based on partial differential diffusion–reaction equations [33, 34, 74]. The micromorphology required for these models, e.g., number of tips and branches, can in turn be derived from the morphology database. Furthermore, the reaction term within these models requires the substrate consumption rate of the fungus which can be acquired from literature or additional experiments.
3. The calculated nutrient concentration profiles enable to determine substrate-limited and substrate-saturated areas within each pellet. Based on the requirements for the product of interest, e.g., limitation or saturation needed, a suitable pellet structure can be selected from the database.
4. Using process and/or genetic engineering approaches, the upscaling of the cultivation with the desired pellet structure to the size of industrially relevant bioreactors could be realized.

Some steps of this proposed procedure are still lacking detailed elaboration. However, in addition to modeling/simulation approaches even the pure data from synchrotron radiation based μ-computed tomography and subsequent high-throughput 3D image analysis will lead to a holistic understanding of the morphological development of pellets. This will open new opportunities for morphological engineering towards enhanced process control and productivity.

References

1. Hoffmeister D, Keller NP (2007) Natural products of filamentous fungi: enzymes, genes, and their regulation. Nat Prod Rep. https://doi.org/10.1039/b603084j
2. Meyer V (2008) Genetic engineering of filamentous fungi—progress, obstacles and future trends. Biotechnol Adv 26:177–185
3. Punt PJ, van Biezen N, Conesa A, Albers A, Mangnus J, van den Hondel C (2002) Filamentous fungi as cell factories for heterologous protein production. Trends Biotechnol. https://doi.org/10.1016/s0167-7799(02)01933-9
4. Cairns TC, Nai C, Meyer V (2018) How a fungus shapes biotechnology: 100 years of *Aspergillus niger* research. Fungal Biol Biotechnol. https://doi.org/10.1186/s40694-018-0054-5
5. Füting P, Barthel L, Cairns TC, Briesen H, Schmideder S (2021) Filamentous fungal applications in biotechnology: a combined bibliometric and patentometric assessment. Fungal Biol Biotechnol. https://doi.org/10.1186/s40694-021-00131-6
6. Brakhage AA (2013) Regulation of fungal secondary metabolism. Nat Rev Microbiol. https://doi.org/10.1038/nrmicro2916
7. Keller NP (2019) Fungal secondary metabolism: regulation, function and drug discovery. Nat Rev Microbiol. https://doi.org/10.1038/s41579-018-0121-1
8. Nielsen JC, Grijseels S, Prigent S, Ji B, Dainat J, Nielsen KF, Frisvad JC, Workman M, Nielsen J (2017) Global analysis of biosynthetic gene clusters reveals vast potential of secondary metabolite production in *Penicillium* species. Nat Microbiol. https://doi.org/10.1038/nmicrobiol.2017.44
9. Ward OP (2012) Production of recombinant proteins by filamentous fungi. Biotechnol Adv. https://doi.org/10.1016/j.biotechadv.2011.09.012
10. Wang Q, Zhong C, Xiao H (2020) Genetic engineering of filamentous fungi for efficient protein expression and secretion. Front Bioeng Biotechnol. https://doi.org/10.3389/fbioe.2020.00293
11. Meyer V, Basenko EY, Benz JP, Braus GH, Caddick MX, Csukai M, de Vries RP, Endy D, Frisvad JC, Gunde-Cimerman N, Haarmann T, Hadar Y, Hansen K, Johnson RI, Keller NP, Kraševec N, Mortensen UH, Perez R, Ram AFJ, Record E, Ross P, Shapaval V, Steiniger C, van den Brink H, van Munster J, Yarden O, Wösten HAB (2020) Growing a circular economy with fungal biotechnology: a white paper. Fungal Biol Biotechnol. https://doi.org/10.1186/s40694-020-00095-z
12. Veiter L, Rajamanickam V, Herwig C (2018) The filamentous fungal pellet-relationship between morphology and productivity. Appl Microbiol Biotechnol 102:2997–3006
13. Krull R, Wucherpfennig T, Esfandabadi ME, Walisko R, Melzer G, Hempel DC, Kampen I, Kwade A, Wittmann C (2013) Characterization and control of fungal morphology for improved production performance in biotechnology. J Biotechnol 163:112–123
14. Böl M, Schrinner K, Tesche S, Krull R (2021) Challenges of influencing cellular morphology by morphology engineering techniques and mechanical induced stress on filamentous pellet systems—a critical review. Eng Life Sci. https://doi.org/10.1002/elsc.202000060
15. Papagianni M (2004) Fungal morphology and metabolite production in submerged mycelial processes. Biotechnol Adv 22:189–259
16. Krull R, Cordes C, Horn H, Kampen I, Kwade A, Neu TR, Nörtemann B (2010) Morphology of filamentous fungi: linking cellular biology to process engineering using Aspergillus niger. In: Wittmann C, Krull R (eds) Biosystems engineering II: linking cellular networks and bioprocesses. Springer Berlin Heidelberg, Berlin, Heidelberg, pp 1–21

17. Hille A, Neu TR, Hempel DC, Horn H (2005) Oxygen profiles and biomass distribution in biopellets of Aspergillus niger. Biotechnol Bioeng 92:614–623
18. Hille A, Neu TR, Hempel DC, Horn H (2009) Effective diffusivities and mass fluxes in fungal biopellets. Biotechnol Bioeng 103:1202–1213
19. Cairns TC, Zheng X, Zheng P, Sun J, Meyer V (2019) Moulding the mould: understanding and reprogramming filamentous fungal growth and morphogenesis for next generation cell factories. Biotechnol Biofuels. https://doi.org/10.1186/s13068-019-1400-4
20. Papagianni M, Mattey M (2006) Morphological development of Aspergillus niger in submerged citric acid fermentation as a function of the spore inoculum level. Application of neural network and cluster analysis for characterization of mycelial morphology. Microb Cell Factories 5
21. Dynesen J, Nielsen J (2003) Surface hydrophobicity of Aspergillus nidulans conidiospores and its role in pellet formation. Biotechnol Prog. https://doi.org/10.1021/bp0340032
22. Wucherpfennig T, Kiep KA, Driouch H, Wittmann C, Krull R (2010) Morphology and rheology in filamentous cultivations. Adv Appl Microbiol. https://doi.org/10.1016/S0065-2164(10)72004-9
23. Laible AR, Dinius A, Schrader M, Krull R, Kwade A, Briesen H, Schmideder S (2021) Effects and interactions of metal oxides in microparticle-enhanced cultivation of filamentous fungi and bacteria. Eng Life Sci. https://doi.org/10.1002/elsc.202100075
24. Driouch H, Sommer B, Wittmann C (2010) Morphology engineering of Aspergillus niger for improved enzyme production. Biotechnol Bioeng 105:1058–1068
25. Kaup J-A, Ehrich K, Pescheck M, Schrader J (2008) Microparticle-enhanced cultivation of filamentous microorganisms: increased chloroperoxidase formation by Caldariomyces fumago as an example. Biotechnol Bioeng 99:491–498
26. Meyer V, Cairns T, Barthel L, King R, Kunz P, Schmideder S, Müller H, Briesen H, Dinius A, Krull R (2021) Understanding and controlling filamentous growth of fungal cell factories: novel tools and opportunities for targeted morphology engineering. Fungal Biol Biotechnol. https://doi.org/10.1186/s40694-021-00115-6
27. Fiedler MRM, Barthel L, Kubisch C, Nai C, Meyer V (2018) Construction of an improved Aspergillus niger platform for enhanced glucoamylase secretion. Microb Cell Fact. https://doi.org/10.1186/s12934-018-0941-8
28. Schmideder S, Barthel L, Friedrich T, Thalhammer M, Kovačević T, Niessen L, Meyer V, Briesen H (2019) An X-ray microtomography-based method for detailed analysis of the three-dimensional morphology of fungal pellets. Biotechnol Bioeng. https://doi.org/10.1002/bit.26956
29. Schmideder S, Barthel L, Müller H, Meyer V, Briesen H (2019) From three-dimensional morphology to effective diffusivity in filamentous fungal pellets. Biotechnol Bioeng. https://doi.org/10.1002/bit.27166
30. Schmideder S, Müller H, Barthel L, Friedrich T, Niessen L, Meyer V, Briesen H (2020) Universal law for diffusive mass transport through mycelial networks. Biotechnol Bioeng. https://doi.org/10.1002/bit.27622
31. Müller H, Barthel L, Schmideder S, Schütze T, Meyer V, Briesen H (2022) From spores to fungal pellets: a new high-throughput image analysis highlights the structural development of Aspergillus niger. Biotechnol Bioeng. https://doi.org/10.1002/bit.28124
32. Müller H, Deffur C, Schmideder S, Barthel L, Friedrich T, Mirlach L, Hammel JU, Meyer V, Briesen H (2023) Synchrotron radiation-based microcomputed tomography for three-dimensional growth analysis of Aspergillus niger pellets. Biotechnol Bioeng. https://doi.org/10.1002/bit.28506
33. Buschulte TC (1992) Mathematische Modellbildung und Simulation von Zellwachstum, Stofftransport und Stoffwechsel in Pellets aus Streptomyceten. PhD thesis, Fakultät Verfahrenstechnik der Universität Stuttgart, Stuttgart, Germany
34. Celler K, Picioreanu C, van Loosdrecht MCM, van Wezel GP (2012) Structured morphological modeling as a framework for rational strain design of Streptomyces species. Antonie Leeuwenhoek Int J Gen Mol Microbiol 102:409–423

35. King R (2015) A framework for an organelle-based mathematical modeling of hyphae. Fungal Biol Biotechnol. https://doi.org/10.1186/s40694-015-0014-2
36. Lejeune R, Baron GV (1997) Simulation of growth of a filamentous fungus in 3 dimensions. Biotechnol Bioeng 53:139–150
37. Kwon MJ, Arentshorst M, Roos ED, van den Hondel CAMJJ, Meyer V, Ram AFJ (2011) Functional characterization of Rho GTPases in Aspergillus niger uncovers conserved and diverged roles of Rho proteins within filamentous fungi. Mol Microbiol. https://doi.org/10.1111/j.1365-2958.2010.07524.x
38. Bennet JW, Lasure LL (1991) More gene manipulations in fungi. Academic Press
39. Greving I, Wilde F, Ogurreck M, Herzen J, Hammel JU, Hipp A, Friedrich F, Lottermoser L, Dose T, Burmester H, Müller M, Beckmann F (2014) P05 imaging beamline at PETRA III: first results. In: Stock SR (ed) Developments in X-ray tomography IX. SPIE Optical Engineering + Applications, San Diego, California, United States, Sunday 17 August 2014, 92120O. SPIE. https://doi.org/10.1117/12.2061768
40. Haibel A, Ogurreck M, Beckmann F, Dose T, Wilde F, Herzen J, Müller M, Schreyer A, Nazmov V, Simon M, Last A, Mohr J (2010) Micro- and nano-tomography at the GKSS imaging beamline at PETRA III. In: Stock SR (ed) Developments in X-ray tomography VII. SPIE Optical Engineering + Applications, San Diego, California, USA, Sunday 1 August 2010, 78040B. SPIE. https://doi.org/10.1117/12.860852
41. Wilde F, Ogurreck M, Greving I, Hammel JU, Beckmann F, Hipp A, Lottermoser L, Khokhriakov I, Lytaev P, Dose T, Burmester H, Müller M, Schreyer A (2016) Micro-CT at the imaging beamline P05 at PETRA III. AIP Conf Proc 1741:030035
42. Moosmann J, Ershov A, Weinhardt V, Baumbach T, Prasad MS, LaBonne C, Xiao X, Kashef J, Hofmann R (2014) Time-lapse X-ray phase-contrast microtomography for in vivo imaging and analysis of morphogenesis. Nat Protoc. https://doi.org/10.1038/nprot.2014.033
43. Palenstijn WJ, Batenburg KJ, Sijbers J (2011) Performance improvements for iterative electron tomography reconstruction using graphics processing units (GPUs). J Struct Biol. https://doi.org/10.1016/j.jsb.2011.07.017
44. van Aarle W, Palenstijn WJ, de Beenhouwer J, Altantzis T, Bals S, Batenburg KJ, Sijbers J (2015) The ASTRA toolbox: a platform for advanced algorithm development in electron tomography. Ultramicroscopy. https://doi.org/10.1016/j.ultramic.2015.05.002
45. van Aarle W, Palenstijn WJ, Cant J, Janssens E, Bleichrodt F, Dabravolski A, de Beenhouwer J, Joost Batenburg K, Sijbers J (2016) Fast and flexible X-ray tomography using the ASTRA toolbox. Opt Express. https://doi.org/10.1364/OE.24.025129
46. Otsu N (1979) A threshold selection method from gray-level histograms. IEEE Trans Syst, Man Cybern 9:62–66
47. Haralick RM, Shapiro LG (1992) Computer and robot vision. Addison-Wesley, Boston, MA
48. Lewis JP (1995) Fast normalized cross-correlation. Industrial Light & Magic. http://scribblethink.org/Work/nvisionInterface/nip.pdf
49. Kerschnitzki M, Kollmannsberger P, Burghammer M, Duda GN, Weinkamer R, Wagermaier W, Fratzl P (2013) Architecture of the osteocyte network correlates with bone material quality. J Bone Miner Res: Off J Am Soc Bone Miner Res. https://doi.org/10.1002/jbmr.1927
50. Lee TC, Kashyap RL, Chu CN (1994) Building skeleton models via 3-D medial surface axis thinning algorithms. CVGIP: Graph Model Image Process. https://doi.org/10.1006/cgip.1994.1042
51. Caldwell IY, Trinci APJ (1973) The growth unit of the mould Geotrichum candidum. Arch Mikrobiol 88:1–10
52. Yang H, Reichl U, King R, Gilles ED (1992) Measurement and simulation of the morphological development of filamentous microorganisms. Biotechnol Bioeng. https://doi.org/10.1002/bit.260390108
53. Zacchetti B, Andrianos A, van Dissel D, de Ruiter E, van Wezel GP, Claessen D (2018) Microencapsulation extends mycelial viability of Streptomyces lividans 66 and increases enzyme production. BMC Biotechnol. https://doi.org/10.1186/s12896-018-0425-2

54. Lehmann A, Zheng W, Soutschek K, Roy J, Yurkov AM, Rillig MC (2019) Tradeoffs in hyphal traits determine mycelium architecture in saprobic fungi. Sci Rep. https://doi.org/10.1038/s41598-019-50565-7
55. Witten TA, Sander LM (1981) Diffusion-limited aggregation, a kinetic critical phenomenon. Phys Rev Lett. https://doi.org/10.1103/PhysRevLett.47.1400
56. Ericson C (2004) Real-time collision detection. CRC Press
57. Bresenham JE (1965) Algorithm for computer control of a digital plotter. IBM Syst J. https://doi.org/10.1147/sj.41.0025
58. Veiter L, Kubicek M, Hutter H, Pittenauer E, Herwig C, Slouka C (2020) Study of metabolism and identification of productive regions in filamentous fungi via spatially resolved time-of-flight secondary ion mass spectrometry. Anal Bioanal Chem. https://doi.org/10.1007/s00216-019-01980-2
59. Wittier R, Baumgartl H, Lübbers DW, Schügerl K (1986) Investigations of oxygen transfer into Penicillium chrysogenum pellets by microprobe measurements. Biotechnol Bioeng. https://doi.org/10.1002/bit.260280713
60. Zacchetti B, Wösten HAB, Claessen D (2018) Multiscale heterogeneity in filamentous microbes. Biotechnol Adv. https://doi.org/10.1016/j.biotechadv.2018.10.002
61. Cui YQ, van der Lans R, Luyben K (1998) Effects of dissolved oxygen tension and mechanical forces on fungal morphology in submerged fermentation. Biotechnol Bioeng 57:409–419
62. Meyerhoff J, Tiller V, Bellgardt K-H (1995) Two mathematical models for the development of a single microbial pellet. Bioprocess Eng. https://doi.org/10.1007/BF00369507
63. Becker J, Wieser C, Fell S, Steiner K (2011) A multi-scale approach to material modeling of fuel cell diffusion media. Int J Heat Mass Transf. https://doi.org/10.1016/j.ijheatmasstransfer.2010.12.003
64. Yaws CL (2014) Transport properties of chemicals and hydrocarbons. William Andrew
65. Epstein N (1989) On tortuosity and the tortuosity factor in flow and diffusion through porous media. Chem Eng Sci 44:777–779
66. Velichko A, Wiegmann A, Mücklich F (2009) Estimation of the effective conductivities of complex cast iron microstructures using FIB-tomographic analysis. Acta Mater. https://doi.org/10.1016/j.actamat.2009.07.004
67. Wiegmann A, Zemitis A (2006) EJ-HEAT: a fast explicit jump harmonic averaging solver for the effective heat conductivity of composite materials. Berichte des Fraunhofer-Instituts für Techno- und Wirtschaftsmathematik (ITWM Report), vol 94. Fraunhofer-Institut für Techno- und Wirtschaftsmathematik, Fraunhofer (ITWM), Kaiserslautern
68. Panerai F, Ferguson JC, Lachaud J, Martin A, Gasch MJ, Mansour NN (2017) Micro-tomography based analysis of thermal conductivity, diffusivity and oxidation behavior of rigid and flexible fibrous insulators. Int J Heat Mass Transf. https://doi.org/10.1016/j.ijheatmasstransfer.2016.12.048
69. Schmideder S (2022) Following fungal features—micromorphology and diffusivity of filamentous fungal pellets revealed by three-dimensional imaging and simulation. Universitätsbibliothek der TU München, München
70. Tomadakis MM, Robertson TJ (2005) Viscous permeability of random fiber structures: comparison of electrical and diffusional estimates with experimental and analytical results. J Compos Mater. https://doi.org/10.1177/0021998305046438
71. Tomadakis MM, Sotirchos SV (1991) Effects of fiber orientation and overlapping on Knudsen, transition, and ordinary regime diffusion in fibrous substrates. MRS Proc. https://doi.org/10.1557/PROC-250-221
72. He X, Guo Y, Li M, Pan N, Wang M (2017) Effective gas diffusion coefficient in fibrous materials by mesoscopic modeling. Int J Heat Mass Transf. https://doi.org/10.1016/j.ijheatmasstransfer.2016.11.097

73. Nam JH, Kaviany M (2003) Effective diffusivity and water-saturation distribution in single- and two-layer PEMFC diffusion medium. Int J Heat Mass Transf. https://doi.org/10.1016/S0017-9310(03)00305-3
74. King R (1998) Mathematical modelling of the morphology of Streptomyces species. In: Schügerl K (ed) Relation between morphology and process performances. Springer Berlin Heidelberg, Berlin, Heidelberg, pp 95–124

Harnessing Genetic and Microfluidic Approaches to Model Shear Stress Response in Cell Wall Mutants of the Filamentous Cell Factory *Aspergillus niger*

Lars Barthel, Philipp Kunz, Rudibert King, and Vera Meyer

Abstract *Aspergillus niger* is an efficient biotechnological cell factory harnessed for the production of proteins, enzymes, platform chemicals and secondary metabolites. Critical for productivity of filamentous fungi such as *A. niger* is their hyphal growth and the ability of their cell walls to withstand shear stress conditions during submerged bioreactor cultivations. As chitin is thought to be fundamental for the rigidity of fungal hyphae, we generated a chitin synthase knock-out library for *A. niger* and studied the impact of these genetic modifications on its ability to withstand shear stress during submerged bioreactor cultivations. We isolated a double deletion mutant in which the genes *chsE* and *chsD* were inactivated and observed a reduction in chitin content by 60%, cell wall thickness by 20%, hyphal diameter by 60% and maximum growth rate by 25%. We furthermore developed a parallelized microfluidic growth chamber system that enabled us to track the growth of the double mutant and the respective control strain using a confocal laser scanning microscope, and, additionally, to compare their growth responses when subjected to different shear stress conditions. A newly developed mathematical model allowed us to compute and evaluate lag-phase, maximum growth rate and shear stress response based on

L. Barthel · V. Meyer (✉)
Chair of Applied and Molecular Microbiology, Institute of Biotechnology, Technische Universität Berlin, Berlin, Germany
e-mail: vera.meyer@tu-berlin.de

L. Barthel
e-mail: lars.barthel@posteo.de

P. Kunz · R. King
Chair of Measurement and Control, Institute of Chemical and Process Engineering, Technische Universität Berlin, Berlin, Germany
e-mail: p.kunz@posteo.de

R. King
e-mail: rudibert.king@tu-berlin.de

A. Kwade and I. Kampen (eds.), *Dispersity, Structure and Phase Changes of Proteins and Bio Agglomerates in Biotechnological Processes*,
https://doi.org/10.1007/978-3-031-63164-1_15

the microscopic images obtained. Overall, this microfluidic approach clearly identified differences in shear stress response between both strains and has high potential as scale-down screening tool to predict growth characteristics under submerged bioreactor cultivations.

1 Introduction

The filamentous fungus *Aspergillus niger* is one of the main cell factories used in microbial biotechnology for more than 100 years. It is harnessed for the production of organic acids, proteins, enzymes and secondary metabolites because of its extraordinary versatile metabolism, potent secretion routes and the availability of efficient genetic and metabolic engineering tools [6, 7, 30]. However, much room is left for innovation since not all molecules become efficiently secreted and mutant screening approaches for hyper-secretors are very laborious because high-throughput screening tools are still lacking [5, 8]. *A. niger* belongs to the group of *fungi imperfecti*, which means that its reproduction is purely based on an asexual life cycle (Fig. 1a). In brief, the life cycle begins with dormant spores that start to germinate and form young hyphae (cells that are separated by septa) once they sense favorable nutrients in a moist environment. The hyphae elongate and eventually start branching to create a complex interwoven three-dimensional network called mycelium. Once nutrients become limiting, hyphae conquer the airspace and conidiophores are produced out of the mycelium to form new spores at their ends. During the mycelial development of *A. niger*, vegetative growth is limited to the hyphal tip (polar growth) and strictly coupled to protein secretion ([5, 8], Fig. 1b). Proteins destined for incorporation into the growing hyphal tip (i.e. plasma membrane and cell wall components) or destined for secretion into the environment become packed into secretory vesicles and transported from the endoplasmic reticulum through the *trans*-Golgi network towards the hyphal tip along the microtubule and actin cytoskeleton. Once the vesicles reach the hyphal tip region, they fuse with the plasma membrane thus building up a constantly growing plasma membrane and cell wall whilst also releasing protein cargo into the environment. Most of the secreted proteins belong to hydrolytic enzymes (e.g. amylases including glucoamylase, proteases, lipases) whose function is to degrade polymeric organic material in the external environment into monomeric components, which eventually are taken up by specific transporters into the fungal cytoplasm. Fungi thus first digest then ingest their nutrients. Remarkably, many more vesicles are transported to the hyphal tip than can be accommodated by the growing plasma membrane, hence excess vesicle membrane material becomes subsequently taken up by endocytosis. Importantly, this vesicle-based secretion also ensures that proteins become transported via cytoskeletal elements to septa which divide hyphae into compartments. For more detailed information on the cell biological and molecular mechanisms driving and linking hyphal growth, septation and protein secretion, the readers are directed to [7, 44]. In agreement, we could recently show that the enzyme glucoamylase, which is the most abundant protein secreted by *A. niger*, is not only

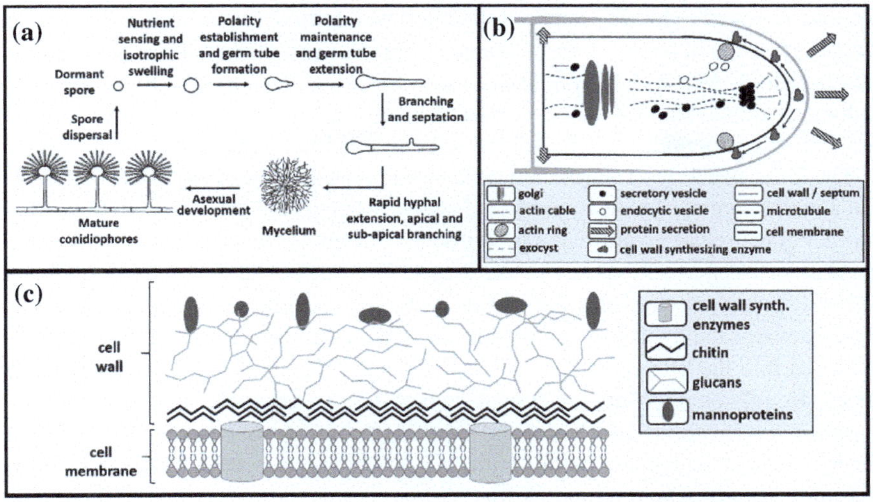

Fig. 1 A very simplified view on how the life cycle, hyphal tip growth, protein secretion and cell wall composition are linked in *A. niger*. For more detailed information the readers are directed to the papers cited in the main text. **a** The asexual life cycle starts with spores that germinate and form a three-dimensional mycelium, which eventually produces new spores under nutrient-limited conditions. **b** Vesicular transport in hyphal tips is crucial for both fungal growth and protein secretion. **c** Plasma membrane and cell wall form the outer barrier of the hyphae. Cell wall synthesizing enzymes are anchored in the cell membrane to produce cell wall polymers. Figures are adapted from [7], licensed under the Creative Commons Attribution 4.0 International license (CC BY 4.0)

secreted at the hyphal tip but most likely also via septal junctions. Through overexpression of the secretion-related regulatory protein ArfA, we were able to show that more glucoamylase localizes at septa and the total amount of secreted protein increases [14].

Plasma membranes and cell walls of fungi thus shield and protect the intracellular fungal space from the environment and concomitantly orchestrate hyphal growth and protein secretion. Whereas the composition and function of fungal plasma membranes show many similarities to the plasma membrane of higher eukaryotes, the composition of fungal cell walls is unique to the fungal kingdom [17]. In general, fungal cell walls are very complex networks of interlinked polymers including chitin, α- and β-glucans, galactomannans, and glycosylated proteins such as mannoproteins (Fig. 3c, [17, 18]). Intriguingly, the architecture and composition of fungal cell walls change throughout the fungal life cycle, and continuously adapt to changing physical and biochemical conditions in the external environment. Hence, fungal cell walls should not be considered as rigid structures with defined composition, but as a very dynamic viscoelastic surface meshwork that allows communication and vesicular traffic in both outward and inward direction [10]. This dynamic behavior is ensured by different signaling pathways that sense physical or biochemical stress conditions via surface-embedded protein receptors (mechanosensors) and which transmit potential life-threatening conditions to the cell nucleus. Here, different transcription factors become activated to modify expression of cell wall remodeling enzymes that fortify fungal cell walls in a very dynamic manner. For *A. niger*, at least three transcription factors (RlmA, MsnA, CrzA) linked to three signaling pathways (cell wall integrity pathway, osmotic stress response, calcium signaling) have been proposed to ensure cell wall integrity under cell wall stress conditions [13, 36, 40, 49].

The inner structure of fungal cell walls is composed of chitin and alkali-insoluble glucans that largely define strength and shape, whereas the outer cell wall consists of other glycans and glycosylated proteins and determines hydrophobicity and adhesiveness of hyphae and spores to surfaces [17, 18]. For the biosynthesis of inner cell wall polymers, enzymes like chitin synthases and glucan synthases are transported towards the hyphal tip via vesicle-mediated transport (see above) and become anchored into the plasma membrane. Interestingly, the number of chitin synthases varies strongly among different fungal species. For example, the unicellular fungus *Saccharomyces cerevisiae* relies on three different chitin synthase genes (Chs1, Chs2, Chs3), whereas many more chitin synthase genes are predicted for filamentous fungi (e.g. nine *chs* genes in *A. niger*, [37]).

From a biotechnological perspective, it is of utmost importance to understand the genetic basis that governs the cell wall architecture in filamentous fungal cell factories during submerged bioreactor cultivation. On the one hand, because spores and hyphae tend to self-assemble during submerged bioreactor cultivation into macromorphologies that range from disperse mycelia over loose clumps to dense pellets and on the other hand, because medium composition and stirrer speed can cause substantial shear stress and thus cell wall stress due to hyphal damage or fragmentation. All these factors in turn impact product formation and secretion [31]. If the genetic link between a dynamically changing cell wall architecture, spore outgrowth, hyphal elongation, hyphal branching, vesicular transport, and product secretion are mechanistically understood, fungal cell factories can be rationally designed for various industrial applications [49].

As mentioned above, chitin is critical for defining rigidity of hyphal cell walls, making it an interesting target for rational genetic modifications. The impact of deleting or silencing chitin synthase genes have therefore been studied in several industrial fungi. For example, the downregulation of *chsC* in *A. niger* by RNAi silencing reduced sporulation, decreased chitin levels in the fungal cell wall and altered macromorphological phenotypes during submerged cultivation [24, 46]. Disruption of *csmA* and *chsB* in *A. oryzae* decreased tip extension rate in both cases, whereby disruption of *csmA* improved lipid production [23, 35].

To pave the way for mechanistic understanding of hyphal growth, vesicular transport and protein secretion and its link to shear stress response and cell wall architecture, we have recently developed a mathematical model that describes intracellular transport via secretory vesicles and the overall growth speed inside of *A. niger* hyphae, when cultivated in a microscopic growth chamber without shear stress [27]. This model abstracted individual secretory vesicles as a concentration and linked local vesicle consumption with tip expansion. This initial model showed a good fit to confocal microscopy data of vesicle-tagged *A. niger* germlings and was then extended and fitted to a larger sample size of young hyphae, which led to the identification of two subpopulations that differed in their growth behavior after germination, i.e., actively growing germlings and germlings for which a stagnation of growth occurred after some time [28]. As a next step, growth of *A. niger* was investigated in an environment where cells were exposed to shear stress. In order to mimic mechanical stress encountered in a bioreactor on a microscopic scale, time-variant shear fluctuations were reproduced inside a self-developed flow chamber device [26]. Culture medium flowing over a backward-facing step resulted in a phenomenon known as Kelvin–Helmholtz instability [38]. This instability gave rise to turbulent flow patterns downstream, i.e., in the wake of the step and, more importantly, to adjustable levels of cell wall shear stress on the upper side of microscope slides where spores germinated. The stress reaction on the macromorphological appearance of the growing hyphae could thus be observed through the microscope. Mycelia of *A. niger* adapted to the differences in shear stress with changing from a homogeneous mycelia growth in regions of low shear stress to cloud-like patterns in high stress regions [26]. Moreover, both the amount of available secretory vesicles inside the hyphae and of the considered target protein glucoamylase were affected by the wall shear stress.

Whereas the mathematical model mentioned above focused on the description of subcellular processes for a specific *A. niger* strain based on data from experiments with the backward-facing step, in the present study, we model cell growth from a more macroscopic point of view exploiting a new flow chamber detailed below. Most notably, here we consider the effects of cell wall mutants of *A. niger* from various perspectives. The mutants were created by genetic inactivation of chitin biosynthetic genes, in which at least one chitin synthase gene was knocked out. Our aim was to obtain strains with substantially reduced chitin levels in their cell walls and septa. On the one hand, we assumed that these modifications could make cell walls of *A. niger* leakier, thus allowing molecules like proteins to easier pass through. On the other hand, we speculated that more vesicles might potentially be available for protein secretion if less secretory vesicles are needed to transport cell wall building material to the hyphal tip and septa. Thus, we wished to generate a mutant strain collection, in which the impact of cell wall thickness on growth and protein secretion capacities of *A. niger* can be systematically studied. Here, we describe and phenotypically characterize the growth of one selected double chitin synthase knockout strain compared to its progenitor control strain when exposed to different levels of shear stress in a microfluidic growth chamber as well as in a 5 L submerged bioreactor. As the mutants do not possess additional tags for GlaA and a protein attached to secretory vesicles, the detailed mathematical model described in [27] could not be adapted for the mutants. Instead, a simpler mathematical growth model was built based on data obtained in the microfluidic growth chamber with a parental strain and a mutant. This model can be used to identify parameters that serve as indicators for shear-stress dependent variation of the growth dynamics.

2 Materials and Methods

2.1 Strains, Molecular Techniques and Growth Conditions

The uridine-auxotrophic strain *A. niger* MA169.4 [9] served as parental strain. Chitin synthase genes were deleted in this strain using uridine-prototrophy for selection (*pyrG* gene of *A. oryzae* served as a marker). To allow for direct comparison, control strain MJK17.25 was established, in which uridine-auxotrophy was restored to prototrophy via integration of an intact *pyrG* copy. For all genetic modifications, well-established molecular standard procedures, gene-inactivation approaches and transformation protocols for *A. niger* were followed [2]. Genomic DNA extraction, diagnostic PCR and Southern blot analyses to confirm correct genetic modifications were performed as described elsewhere [19]. FOA counterselection for reuse of the *pyrG* marker to generate chitin synthase double mutants followed a protocol described earlier [2]. Details on cloning procedures, primers used, plasmids and strains will be published elsewhere or can be requested from the authors. All strains generated and used in this study are listed in Table 1.

All strains were grown on minimal medium (MM) plates as described in [33] or complete medium (MM + 1% (w/v) yeast extract + 0.5% (w/v) casamino acids) agar plates. If required, 10 mM uridine was added.

Table 1 List of all *A. niger* strains used in this study, their genetic background, their uridine requirement, and the chitin synthase (*chs*) genes deleted. Nomenclature of *chs* genes and open reading frame (ORF) code is according to [37]

Name	Background	Uridine-requirement	Deleted *chs* gene (ORF code)
MA196.4	–	Auxotroph	–
MJK17.25	MA169.4	Prototroph	–
LB7.1	MA169.4	Prototroph	Δ*csmB* (An02g02340)
LB8.18	MA169.4	Prototroph	Δ*csmA* (An02g02360)
LB9.1	MA169.4	Prototroph	Δ*chsF* (An03g06360)
LB10.1	MA169.4	Prototroph	Δ*chsA* (An07g05570)
LB11.1	MA169.4	Prototroph	Δ*chsG* (An08g05290)
LB12.14	MA169.4	Prototroph	Δ*chsD* (An09g02290)
LB13.3	MA169.4	Prototroph	Δ*chsB* (An09g04010)
LB14.1	MA169.4	Prototroph	Δ*chsE* (An12g10380)
LB19.2	LB14.1	Auxotroph	Δ*chsE* (An12g10380)
ME1.4	LB19.2	Prototroph	Δ*chsE* (An12g10380) + Δ*chsD* (An09g02290)

2.2 Bioreactor Cultivation

Bioreactor cultivation of *A. niger* strains followed established protocols [34]. In brief, glucose-limited batch cultivation in 5 L BioStat Sartorius bioreactors were performed, for which a final concentration of 1×10^6 spores/mL were used for inoculation. Four liters minimal medium (0.8% (w/v) glucose) were set to 30 °C, pH3, stirrer speed 250 rpm and an air flow of 0 L/min. After 6 h cultivation time, stirrer speed was increased to 750 rpm and air flow was set to 2 L/min. 15 mL samples were taken approximately every 2 h throughout cultivation. Biomass and culture supernatant were separated using a vacuum filter device and biomass was freeze-dried.

2.3 Microscopy

Spores of *A. niger* strains were dotted on minimal medium agar plates in different concentrations and incubated at 20 °C for 48 h. Colonies that were visible to the naked eye were cut out and treated with calcofluor white to stain fungal cell walls. Here, one drop of calcofluor white staining solution (Sigma Aldrich), which binds specifically to chitin [4, 20], was added to the mycelium followed by one drop of 10% (w/v) potassium hydroxide. After 1 min incubation at room temperature, the sample was investigated using differential interference contrast (DIC) microscopy. Biomass samples from bioreactor cultivation were microscopically investigated without prior staining. A DIC Leica DM6000 microscope equipped with a 40x oil immersion objective (HC PL APO 40x/1.30 Oil C52) and a Leica DFC365FX camera were used. For fluorescence imaging of calcofluor white stained samples, a Leica CFP filter cube was applied. All images were acquired using the software LasX version 3.5.7.23225 (Leica microsystems). For measuring of the hyphal diameter, the inbuilt measuring tool was used manually.

For transmission electron microscopy, 50 mL MM in a 250 mL shake flask were inoculated with 1×10^5 spores/mL and cultivated for 72 h at 37 °C and 120 rpm. 4320 μL culture broth were transferred into a 15 mL reaction tube and mixed with 480 μL 25% glutaraldehyde solution and 1200 μL 25% paraformaldehyde solution and incubated over night at room temperature. Samples were washed twice with 0.1 M EM-HEPES buffer (HEPES 0.1, 0.09 M sucrose, 10 mM $CaCl_2$, 10 mM $MgCl_2$, pH 6.9), and exposed to osmium tetroxide (1% in HEPES buffer) for 1 h at room temperature, followed by a washing step with HEPES buffer. Dehydration took place in a graded series of ethanol (10, 30, 50, 70, and 90%) on ice, and two steps in 100% ethanol at room temperature, each step for 30 min. The 70% ethanol step was performed over night at 4 °C with the addition of 2% uranyl acetate. Infiltration with the resin LR White (LRW) was performed subsequently (LRW:EtOH: 1:1, 2:1, 2× 100%), each incubating step for approx. 8 h respectively overnight. After addition of a starter, samples were polymerized at 50 °C for 2 days. Ultrathin sections of approx. 50–70 nm thickness were cut with an Ultramicrotome (Ultracut, Reichert) and counterstained with 4% aqueous uranyl acetate for 3 min and lead citrate for 15 s. Images were acquired with a Libra 120 Plus (Zeiss, Germany) with an acceleration voltage of 120 kV and at calibrated magnifications. Contrast and brightness were adjusted using the WinTEM software.

2.4 Chitin Quantification

For quantification of microscopic fluorescence images, the quantification tool of software LasX version 3.5.7.23225 (Leica microsystems) was used. Per strain, sections in three separate images showing a septum and its surrounding hyphae that were totally in focus were chosen. In each of these sections, three lines were drawn in the hyphal lumen close to the septum. Then the 'Mean Value' of these three lines was taken to calculate the average background signal intensity in this section. Next, three lines were drawn through the outer cell wall close to the septum and the peak values ('Max Amplitude') were used to calculate the average peak cell wall signal. The average peak septum signal strength was determined similarly. Finally, the average background signal was subtracted from average cell wall and septum peak values.

2.5 Flow Chamber Design, Cultivation and Microscopy

The designed flow chamber setup features a closed-circuit pumping system [26]. Minimal medium consisting of 1% (w/v) glucose at 37 °C was pumped between two bottle reservoirs. One bottle served as a supply and the other as a receiver tank and vice versa. A system of valves guaranteed that the flow direction through different flow chambers attached, hosting growing mycelia, was always maintained. A controlled volume flow speed was obtained by compressing the air overlaying the medium inside the supply bottle and releasing the pressure in the other one. When the medium level inside a bottle reached a lower threshold, the system of valves switched to release pressure in the former supply bottle and to pressurize the now full, former receiver bottle. By this, the function of the bottles changed. The setup of the valves guaranteed that the flow direction through the experimental flow chambers remained unchanged, even when a switch between the reservoirs occurred. Details of this general setup can be found in [26]. Here, we only describe the newly designed manifold setup in greater detail that replaced the backward-facing step chamber in the aforementioned work.

For this work, a specific design of a manifold (Fig. 2) was attached to the general setup. Instead of producing different turbulent shear stresses behind a backward-facing step, a commercially available laminar flow device was exploited for simplicity that featured six parallel plate flow chambers (Ibidi VI 0.4 channel slides, ibidi GmbH, Germany). It was precoated with Poly-L-Lysine and filled with a spore suspension of 5.8×10^5 spores/mL. Such precoating generally improves adherence of swelling spores to the plastic slide thus preventing washout due to medium flow. Minimal medium, coming from the supply bottle, reaching the manifold was directed into three different pathways (see Fig. 2). With a volumetric flow $\dot{V}$ set, it passed a first channel, channel no. 2, with a related shear stress level. The shear stress based on $\dot{V}$ was calculated using an average formula by the manufacturer, assuming fully laminar flow characteristics [22]. After leaving channel no. 2, the flow was split in half and redirected to two further channels, no. 1 and 3. This resulted in a reduced volume flow $0.5\dot{V}$ in each channel and thereby reduced shear stress. The experiments were conducted separately for two different flow levels of $\dot{V}_1 = 59.15$ mL min^{-1} and $\dot{V}_2 = 78.87$ mL min^{-1}, resulting in the average wall shear stresses $\tau_1 = 7.5$ Pa and $\tau_2 = 10$ Pa in channel no. 2 and, accordingly, in $\tau_1/2 = 3.75$ Pa and $\tau_2/2 = 5$ Pa in both channels no. 1 and 3.

In addition to that, medium was pumped through a so-called reference channel (called *MM exchange* in Fig. 2). Here, a very low flow rate (3–4 mL min^{-1}) resulted in neglectable shear stress, but still forwarded oxygen-rich, fresh medium. As a third flow option, medium bypassed the chambers completely. During this third, intermediate configuration, high volume-flow medium was filtered using two parallel round filters (Macherey-Nagel, Chromafil Xtra PES, 5 μm) in order to remove potential spores and hyphae that had detached from the channels from the circulating medium.

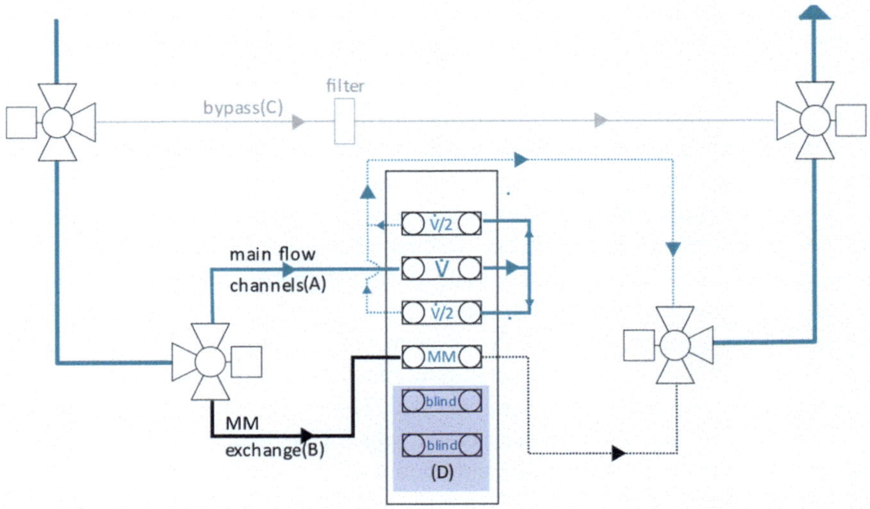

Fig. 2 Flow circuit and valve system of the manifold part used for the experiments: the incoming medium entered the first valve (top left). The medium was automatically directed either through the bypass for filtration (**c**) or to the manifold. There, the flow was directed to the three top main channels (**a**), or to the MM exchange channel (**b**). In the former case, the flow entered channel 2 at full volume flow speed and was reduced to half of the volumetric flow speed through channels 1 and 3. During the experiments, no medium was directed through the blind channels (**d**)

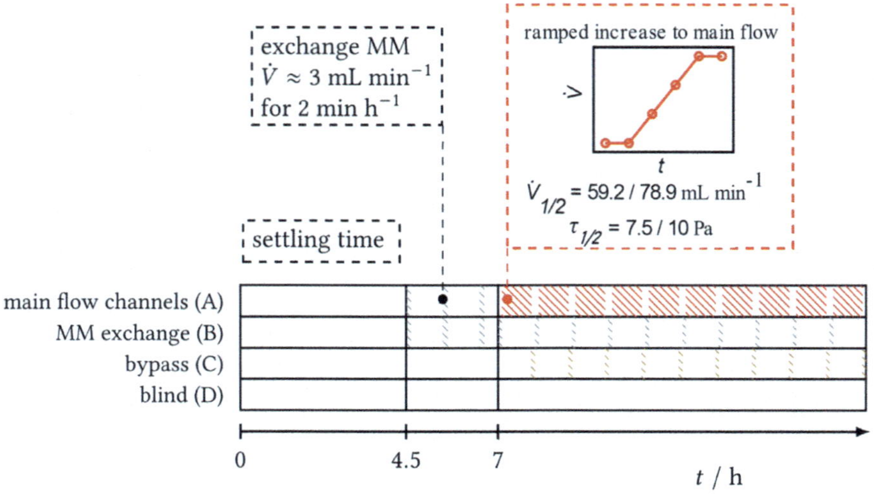

Fig. 3 Time course of the fluid flow experiments conducted: after a settling time of no fluid movement, during which spores started to sediment and to swell in the culture medium, the medium was slowly exchanged in the shear flow channels 1–3 (**a**) and the MM exchange channel 4 (**b**) between 4.5 and 7 h for sufficient oxygen and nutrient supply. Starting at a cultivation time of 7 h, one-hour cycles began: first, MM was exchanged in (**b**) for 2 min. Second, continuous flow in the main shear channels was set for 56 min, where the target flow was obtained via ramped increases over a period of 30 s. Third, the medium was pumped at high flow rate through the bypass filter to remove spores and mycelia from the medium (**c**). No minimal medium was pumped through the blind channels (**d**)

As a full blind comparison without shear and without continued oxygen supply, channels no. 5 and 6 were not attached to the flow circuit.

During the experiments, the sequence of flow through the chambers was coordinated assuring different goals (see Fig. 3). First, after a spore solution was filled into all six channels, a settling time was applied to allow the spores to sink to the chamber bottoms and to swell. This was done in an external incubator at 37 °C. After 4.5 h, the manifold was connected to the flow system and medium was slowly exchanged at $3 - 4$ mL min^{-1} in all four connected flow channels for 2 min every hour in order to guarantee that enough oxygen was provided. The temperature was maintained at 37 °C. After 7 h, the main program started, which consisted of 2 min of medium exchange in channel no. 4 (Fig. 3b) and 56 min of the main flow at $\dot{V}$ and at $0.5\dot{V}$ in the split channels 1–3 (Fig. 3a). As spores and mycelia might detach during the experiments and as the medium was recycled in the chosen closed-circuit pumping system, a final 2 min period was used to remove this residual biomass through a filter in the bypass (Fig. 3c).

The growth behavior of *A. niger* under shear stress, in the medium exchange channel and in the full blind channels was tracked using a confocal laser-scanning microscope (CLSM) SP8, Leica, Germany, using an OPS laser at $\lambda = 552$ nm wavelength and a dry objective HC PL APO 10x/0.40. Laser transmission was detected using a transmission detector. Per channel, eight regions of interest in the middle area of the chamber were predefined, resulting in images of size 1024 × 1024 pixels and in N_t time frames.

2.6 Image Analysis for CLSM Data

Light absorption of *A. niger* in the transmission images (see Sect. 3.5) was quantified using the open-source program Fiji [41]. For an image stack in three spatial, and one time dimension, thus, consisting of M pixels in x, N pixels in y, O pixels in z direction and N_t time frames, a transmission dataset of the form $\mathcal{J}[m, n, o, t_i]$ was constructed. The following calculations were done to quantify light extinction, $\mathcal{J}^{-1}$, by the biomass and to determine the initial spore number.

2.6.1 Quantification of biomass

The change of biomass concentration could be calculated from how much light the sample absorbed. Due to the power conservatism of the transmission radiation, an arbitrary height plane of the sample image, e.g. $o = (O - 1)/2$, could be chosen to determine the amount of overall biomass [50]. The images were inverted to obtain the intensity of absorbed light (extinction) as positive intensity values

$$\mathcal{J}^{-1}[m, n, t_i] = N_{\text{gray}} - \mathcal{J}[m, n, t_i], \tag{1}$$

with the maximal number of gray values of the digital image, $N_{\text{gray}} = 255$. The spatial local mean in an area of the size $M \times N$pixels for time $t = t_i$ was calculated by

$$\langle \mathcal{J}^{-1}[t_i] \rangle = \frac{1}{M \cdot N} \sum_{m=1}^{M} \sum_{n=1}^{N} \mathcal{J}^{-1}[m, n, t_i] \tag{2}$$

In the time course of the experiments, the mean average extinction $\langle \mathcal{J}^{-1}[t_i] \rangle$ increased with growing biomass, $X(t)$, starting from a minimal light extinction, $\mathcal{J}^{-1}_{min}$. The initial value $\mathcal{J}^{-1}_{min}$ could not be linked to the number of spores due to their small size and, therefore, small impact on the mean average extinction in the beginning of the experiments. Thus, using f_e as a proportionality factor, the following relation had to be stated

$$\langle \mathcal{J}^{-1}[t_i] \rangle = \mathcal{J}^{-1}_{min} + (X(t_i) - X_0) \cdot f_e, \tag{3}$$

linking the increasing extinction $\langle \mathcal{J}^{-1}[t_i] \rangle$ with time t_i with the increasing and initial biomass $X(t_i)$ and X_0, respectively. Rewriting this equation leads to

$$(X(t_i) - X_0) \cdot f_e = \langle \mathcal{J}^{-1}[t_i] \rangle - \mathcal{J}_{min}^{-1}. \tag{4}$$

As terms on the right-hand side of this equations were measured, the equation

$$Y(t_i) - Y_0 = \langle \mathcal{J}^{-1}[t_i] \rangle - \mathcal{J}_{min}^{-1} \tag{5}$$

served as a measurement equation with $Y = X \cdot f_e$. The proportionality factor f_e, linking real biomass X in the channels with Y is unknown und remains unidentifiable with the experiments described above. Therefore, we will hereafter deal with biomass growth in the form of change in light extinction and refer to Y as the 'relative biomass'.

To account for intensity fluctuations of the laser, a macro for quantile normalization was used that calculated the 99.8th percentiles of the image histograms, divides all pixels by that value and thus normalized the images to a standard range of [0,1] [47]. By doing so, the comparability of the data sets was ensured, and the impact of the laser intensity fluctuations was minimized. Due to this transformation and due to Eq. (6), the measurements $Y(t_i) - Y_0$, indicating the increase of biomass from its initial value, started at 0 and increased during the experiments, see Fig. 4.

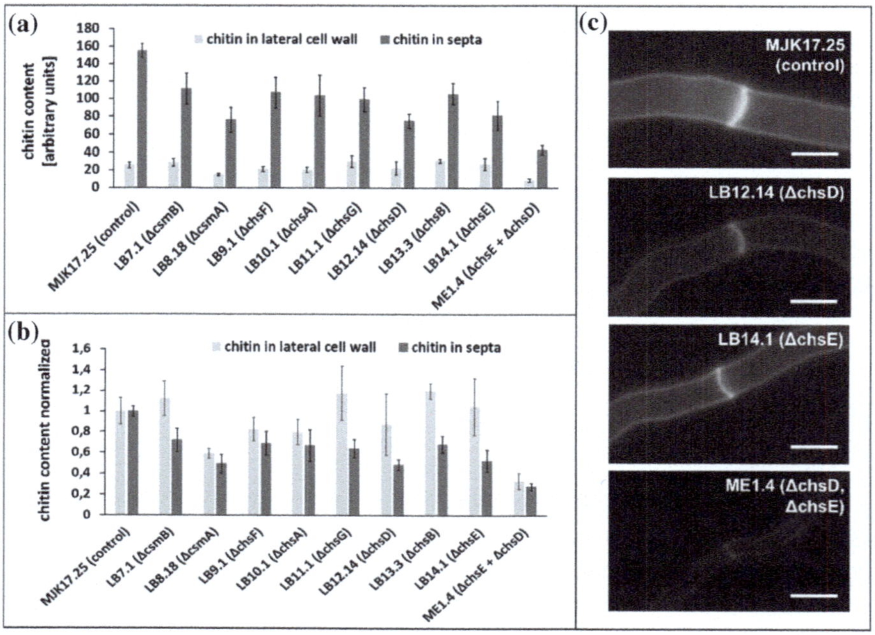

Fig. 4 Relative chitin content in lateral cell wall and septa of *A. niger* chitin synthase knock-out and control strains. Strains were cultivated for 2 day at 20 °C on minimal medium agar plates. A 1 cm^2 square piece was cut out and stained with calcofluor white. The fluorescence signal intensity across the cell wall and septa are shown as total values in arbitrary units (**a**) as well as normalized against the chitin value of the control strain MJK17.25 (**b**). Exemplary microscopic images of calcofluor white stained septal areas are shown for four strains in (**c**). Scale bar: 5 μm

2.6.2 Spore Counting for Initial Biomass

For a quantification of fungal growth, the initial biomass, X_0, or more appropriately, $Y_0 = X_0 f_e$, needed to be calculated. As light extinction by the spores at the beginning of the experiment was too low to be quantified using the spatial local mean (see above), it was calculated by counting the number of spores before biomass growth started. The feature "Analyze Particles" in Fiji war used for this purpose: After preprocessing the image time stacks (background subtraction, thresholding using "Otsu"-method, closing and filling of the resulting binary picture), the number of circular objects per frame was counted for each image and averaged over all pictures taken per flow channel. As the initially injected spore concentration decreased after plugging the flow channels into the circuit pumping system and starting the programmed flow pattern (see Fig. 3), the number of spores shortly after the start of the flow at $t = 7$ h was taken as the reference frame to calculate initial biomass, see the following paragraph.

2.7 *Mathematical Model for Flow Chamber Experiments*

In order to quantify the growth speed of the specific strains under varying shear stress conditions, a simple exponential growth model approach was used, exploiting a transitional function $f(t)$ [21] to account for an observed lag time, Δt, at the beginning of the growth period:

$$\frac{dX}{dt} = \mu X f(t), X(t = 0) = X_0 \tag{6}$$

with

$$f(t) = \frac{1}{1 + e^{-\alpha(t - \Delta t)}} \tag{7}$$

Here, X denotes the biomass, and μ an average growth rate, see as well Table 2. The initial biomass is X_0. The transitional function, with the heuristic parameter α, allows for a smooth transition between lag time and beginning of growth, which rendered it beneficial for numerical reasons during parameter identification compared to the introduction of a strict lag. For identification of the estimated average growth rate, the parameter α is a heuristic value and was chosen to be 25 h^{-1}, in accordance with [21]. Estimates are denoted here by the ^-symbol, i.e., the estimated growth rate is given with $\hat{\mu}$. Estimating $\hat{\mu}$ allows to compare the average growth rate between different strains or under different shear stress conditions. Moreover, the second parameter to be estimated, $\widehat{\Delta t}$, shows the influence of the lag time if it exists in a specific experiment.

Table 2 Overview of variables and parameters used

X		Biomass	τ	[Pa]	Wall shear stress
$\hat{\mu}$	$[h^{-1}]$	Estimated average growth rate			
Δt	[h]	Lag time	$\mathcal{J}$	[Pxl]	Image value
α	$[h^{-1}]$	Smoothing parameter	$\mathcal{J}^{-1}$	[Pxl]	Inverted image value
Y		Relative biomass	N_{gray}		Image value range

2.8 Identification of Model Parameters

The spore counting procedure allowed to calculate the drop of spores per image due to wash-out after volume flow was started. As only spores were present in the chamber at the start of the experiment but not mycelia, the number of spores was directly proportional to the initial biomass at this stage. Initial biomass was calculated for all channels in the following way: First, an initial relative biomass used as a reference, $Y_{0,\text{ref}}$, was identified by fitting the above described growth model to a reference data set allowing all parameters $\hat{\mu}$, $\widehat{\Delta t}$ and $\hat{Y}_0$ to be estimated. This reference set referred to a blind channel where no wash-out occurred. Therefrom, the reference initial relative biomass, $Y_{0,\text{ref}} = \hat{Y}_0 = 0.0038$, was obtained from pictures, containing on average $N_{\text{ref}} = 438.8$ spores/image. In this way, a fixed linkage between spore count and initial relative biomass could be provided that can be used for other channels with different initial numbers of spores per image N_i. For channels $i = \{\tau, 0.5\tau, \text{ MMEx}\}$, where medium washed out spores in the beginning and would, therefore, decrease the initial biomass, the N_i/N_{ref}-th fraction of the reference biomass allowed to calculate the initial relative biomass $Y_{0,i}$ aforehand instead of estimating it numerically:

$$Y_{0,i} = Y_{0,\text{ref}} \cdot \frac{N_\text{i}}{N_{\text{ref}}} \quad \text{with } i = \{\tau, 0.5\tau, \text{MMEx}\}. \tag{8}$$

This procedure was necessary, as mentioned already, because the initial biomass of the spore solution could not be measured directly.

Simulation of the basic model and parameter inference were conducted using the package AMICI [15] in Matlab R2017a (MathWorks Inc, US), which allowed for automatic generation of analytical derivatives up to second order and fast execution due to conversion to mex-files.

For the initial optimization-based parameter adjustment, a combination of global and local methods was chosen. For this purpose, a global parameter search was first performed (multi-start value optimization, PESTO-toolkit, [43]). Parameter confidence intervals were approximated using the Markov Chain Monte Carlo (MCMC) method Parallel Tempering from the toolbox of [43].

3 Results and Discussions

A total of nine putative chitin synthase encoding genes have been predicted to be present in the genome of *A. niger* [37]. To study the impact of all these genes on the amount of chitin in its lateral and septal cell walls, we aimed to delete each of the nine predicted genes. For eight chitin synthases, we were successful to generate single knock-out strains (Table 1) however, several attempts to knock-out *chsC* (An14g00660) remained unsuccessful for reasons unknown.

Chitin was quantified in the control and deletion strains using microscopic imaging of calcofluor white stained hyphae. As summarized in Fig. 4, the signal intensity in septa substantially surpasses the signal in the lateral cell wall in all strains, indicating a higher chitin content in septal regions. As the smaller values for chitin in lateral cell wall made it difficult to analyze inter-strain differences, both the fluorescence signal for chitin in septa as well as in lateral cell wall were normalized against values measured for the control strain *A. niger* MJK17.25 (Fig. 4b). When comparing chitin levels in the lateral cell wall, the single chitin synthase knock-out strains LB7.1 (Δ*csmB*), LB11.1 (Δ*chsG*), LB13.3 (Δ*chsB*) and LB14.1 (Δ*chsE*) showed a signal increase of 5–20% compared to the control. For other mutant strains LB8.18 (Δ*csmA*), LB9.1 (Δ*chsF*), LB10.1 (Δ*chsA*) and LB12.14 (Δ*chsD*) a decrease of fluorescence signal intensity of 12–41% was observed. At septa, the calcofluor white signal decreased for all single chitin synthase knock-out strains in a range from 28% in case of LB7.1 (Δ*csmB*) and 52% for LB12.14 (Δ*chsD*).

As we have previously shown that the *A. niger* secreted enzyme glucoamylase localizes not only at hyphal tips but considerably at septa [14], we aimed to generate an additional chitin synthase double knock-out strain, in which the amount of chitin in septa should be even more reduced when compared to the single knock-out strains. With such a strain in hand, we wished to increase shear sensitivity and thus to improve protein secretion as more glucoamylase could potentially become secreted through thinner septal walls. The combination of Δ*csmA* (LB8.18) and Δ*chsD* (LB12.14), which led to a reduction of chitin in the septa of 51% and 52%, respectively, was thus of potential interest. However, the knock-out strain LB8.18 displayed considerably reduced sporulation (data not shown), a phenotype which is very inconvenient in large-scale industrial processes. We thus decided to delete *chsD* in a strain deleted for *chsE* (LB14.1, 48% less chitin in septa) and obtained a double knock-out strain ME1.4. As expected, the chitin content in this strain reduced in both lateral cell walls and septa down to 33 and 28% compared to the control strain MJK17.25, respectively, when cultivated on minimal medium agar plates (Fig. 4a, b).

Remarkably, the hyphal diameter of 3.6 μm in the control strain MJK17.25 decreased drastically to 1.5 μm in the *chs* double mutant ME1.4 (Figs. 4c and 5a). While a hyphal diameter of 3.6 μm corresponds well to diameters reported in literature for hyphae of *A. niger* [42], the thinner hyphae in ME1.4 suggested that deletion of both chitin synthases *chsD* and *chsE* reduces the elasticity of the cell wall, which is key to withstand the internal hydrostatic pressure (turgor). Earlier reports in plants and algae (whose cell walls consists of cellulose) have shown that the cell volume

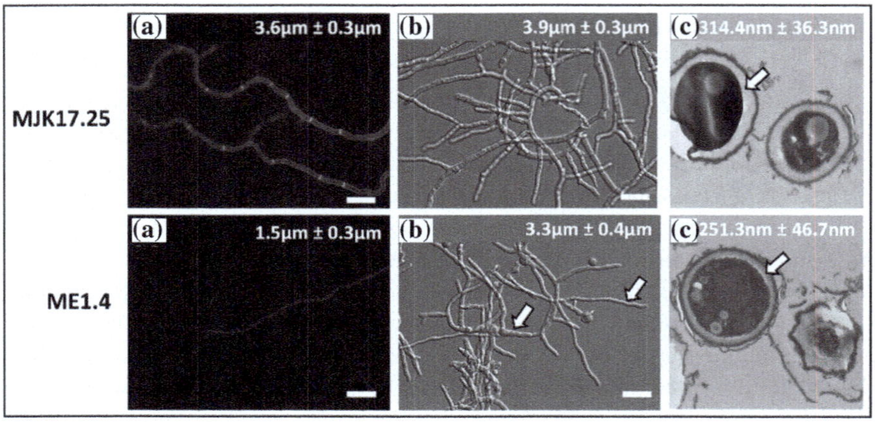

Fig. 5 Morphological comparison of *A. niger* MJK17.25 and ME1.4. Exemplary images of **a** calcofluor white stained hyphae grown on minimal medium agar plate for 2 day at 20 °C. Average hyphal diameter at ten septa is given in upper right. **b** Hyphae grown under shear stress in stirred tank bioreactor in minimal medium at 30 °C for 18 h. Average hyphal diameter at ten septa is given in upper right. Arrows indicate thinner and thicker hyphal subpopulations in ME1.4. **c** The thickness of hyphal cell walls from shake flask samples were investigated using transmission electron microscopy. Average cell wall thickness of eight cuts is given in the upper right. Arrows indicate cell wall region. Scale bar for **a** and **b**: 25 µm

increases with turgor pressure when placed in medium containing water and that high turgor pressure is key to initiate cell wall stretching and thus extension growth [45]. Similarly, hyphal growth in fungi is driven by internal turgor pressure which itself is dependent on vesicular transport, calcium gradients at the hyphal tip and intracellular de-novo osmolyte biosynthesis to sequester water as reviewed previously [29]. In a recent study, it was demonstrated for several filamentous fungi including *A. nidulans* and *A. oryzae* that the lower the elastic modulus of the cell wall (i.e. the force that balances the turgor pressure in the opposite direction) the lower the hyphal elongation rate [16].

We have therefore studied chitin content, cell wall thickness and growth rate in both MJK17.25 and ME1.4 strains when cultivated in minimal medium under stirred tank bioreactor conditions, i.e., when hyphae were subjected to external shear forces and thus cell wall stress. Under these conditions, the hyphal diameter of MJK17.25 and ME1.4 were on average 3.9 µm and 3.3 µm, respectively (Fig. 5b). Thus, the switch from low to high shear stress induced an 8% increase in hyphal diameter in MJK17.25 but a 120% increase in ME1.4, suggesting a stronger cell wall stress response in the double chitin synthase knock-out mutant and thus substantial increase in its hyphal diameters. Such an exaggerated response is a well-described phenomenon in filamentous fungal cell wall mutants. If cell wall integrity is disturbed due to the inactivation of some cell wall synthesizing genes (chitin synthases, glucan synthases) or due to the presence of antifungal compounds inhibiting cell wall integrity, filamentous fungi including *A. niger* respond with fortification of their cell walls due to activation of several signaling pathways (e.g. cell wall integrity pathway, calcium signaling and osmotic stress response) which in turn upregulate expression of genes encoding other cell wall synthetizing enzymes [12, 13, 17, 32, 39]. It was most recently reported for the yeast *Candida albicans* (which can grow as unicellular or filamentous forms) that this overexpression response, i.e. high chitin production, comes at a fitness cost [11]. Therefore, activation of both calcium and cell wall integrity pathways in response to cell wall stress generates a subpopulation of *C. albicans* cells that have supra-high chitin levels but are non-growing, supposedly due to loss of wall elasticity required for cell wall expansion. As mutations in the calcium signaling pathway can prevent such supra-high chitin levels, the authors proposed that calcium signaling may down-regulate the cell wall integrity pathway in some cells [11]. If a similar biological response is also occurring in *A. niger*, is unknown but it is tempting to speculate that this might be the case. As can be seen in Fig. 5b, two hyphal populations can be observed for the double chitin synthase mutant ME1.4, but not for the control strain MJK17.25. One hyphal population of MA1.4 is very thin and mainly belongs to young growing hyphae, whereas the other hyphal population is characterized by thicker and irregular hyphal compartments in the older mycelium.

To determine the thickness of cell walls in both *A. niger* strains, biomass samples were harvested from submerged cultivation in shake flasks (which cause less cell wall stress than bioreactor cultivation) and single hyphae were examined using transmission electron microscopy. As summarized in Fig. 5c, cell wall thickness in strain MJK17.25 was about 300 nm and decreased to about 240 nm in strain ME1.4. Such a reduction by about 20% confirmed that deletion of *chsD* and *chsE* genes resulted in significantly thinner cell walls, thus rendering strain ME1.4 more susceptible to cell wall stress.

According to the considerations discussed above (the thinner the cell wall, the lower its elastic modulus and the lower the hyphal elongation rate), one would expect a longer lag-phase and a lower maximum growth rate in ME1.4 when compared to the control strain MJK17.25. Indeed, data obtained for stirred tank bioreactor cultivations demonstrate an unusual long lag phase (20 h instead of 10 h) and a considerably reduced maximum growth rate μ_{max} (0.18 h^{-1} instead of 0.26 h^{-1}) for the double chitin synthase knock-out strain in comparison to MJK17.25 (Fig. 6a). Growth defects became also obvious during microscopic analyses of biomass samples taken throughout bioreactor runs (Fig. 6b). After 16 h of cultivation, MJK17.25 cultures have already entered exponential growth, whereas spores of ME1.4 where just about to germinate. At 22 h of cultivation, MJK17.25 was about to enter the post-exponential growth phase (data not shown), whereas ME1.4 germlings where about to form young hyphae that formed first branches. At 36 h of cultivation, ME1.4 had formed less straight but hyper-branching hyphae, a phenotype that was not observed for the mycelium of MJK17.25 at the same growth stage at 22 h.

Taken together, hyphae of the mutant strain *A. niger* ME1.4 lacking both chitin synthase genes *chsE* and *chsD*, display considerably reduced diameters, thinner cell walls with less chitin and are unable to achieve maximum hyphal elongation rates

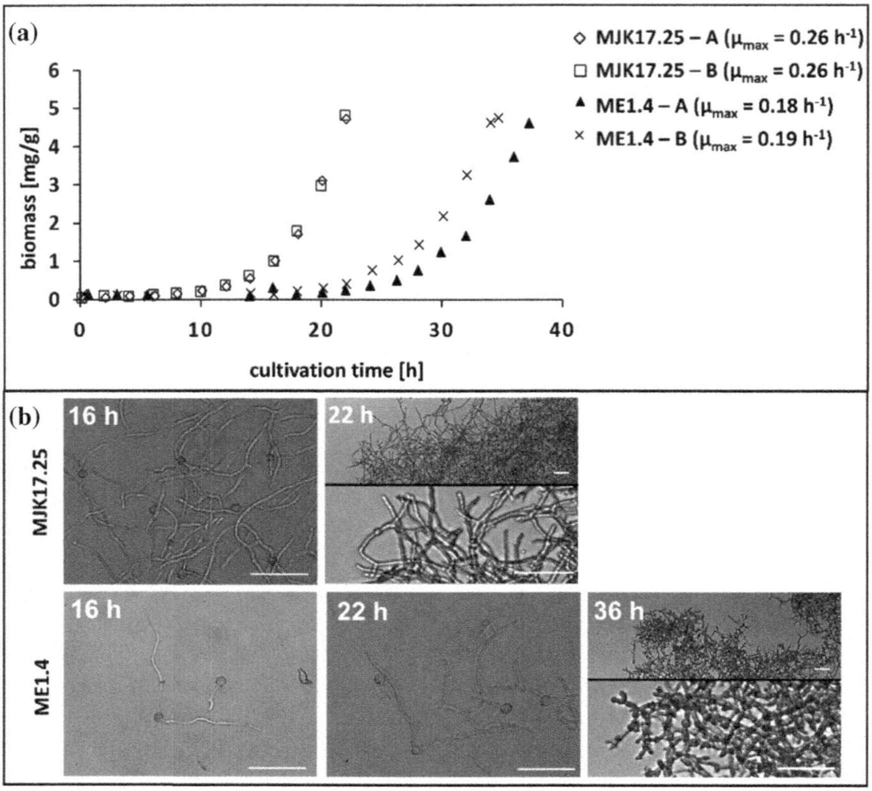

Fig. 6 Duplicate runs of *A. niger* strains MJK17.25 and ME1.4 in stirred tank bioreactor. **a** Biomass samples were taken until the end of the exponential growth phase to determine the growth curve and maximal growth rate. **b** Exemplary microscopic images of biomass samples taken at the time points indicated. Scale bar: 50 μm

when compared to the control strain, in which all nine chitin synthases can contribute to withstand shear forces during bioreactor cultivations.

To further corroborate any relationship between shear stress and hyphal elongation rate, ME1.4 and MJK17.25 were cultivated in microfluidic flow chambers. As described in the Materials and Methods section, the microfluidic design used allowed us to apply different shear stresses to both strains. First, the parameters of the mathematical model introduced in Sect. 3.7 were identified for both strains. Figure 7 displays the temporal evolution of the relative biomass for a single experiment using the control strain MJK17.25 with different mechanical stimuli obtained in different flow channels.

Clearly, the amount of relative biomass over time increases only very slowly in the blind channel and the MM Ex control, potentially because of oxygen limitation due the reduced medium exchange. When shear stress levels of 3.75 and 7.5 Pa were applied, the biomass accumulation matches the standard growth curve of hyphal

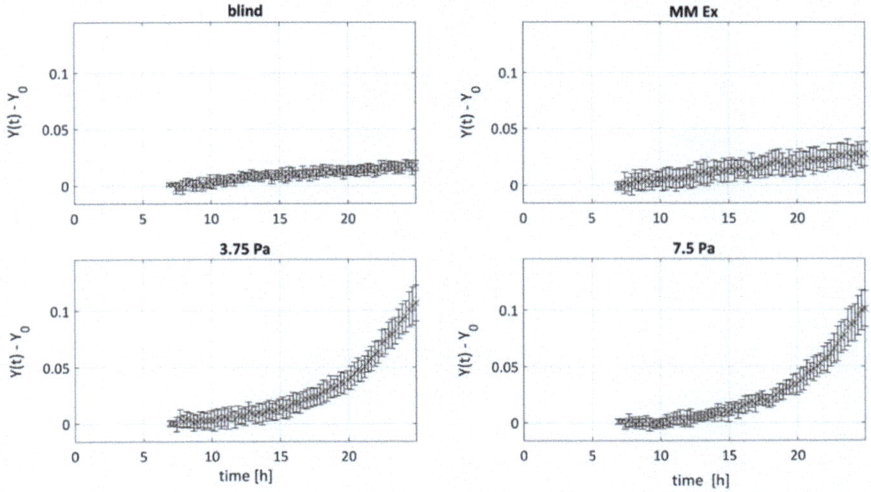

Fig. 7 Temporal evolution of the relative biomass, $Y(t) - Y_0$, for strain MJK17.25, obtained in four channels of one experimental run for shear stress levels $\tau_1 = 7.5$ Pa, $\tau_1/2 = 3.75$ Pa and for the short-term exchange of the minimal medium (MM Ex). For further comparison, a channel without medium flow was measured as a blind. In the channels with increased shear stress and better oxygen supply, biomass increased exponentially after a lag time and reached (nearly) linear growth at the end of the experimental duration

microorganisms better [25]. Here, an initial lag-phase is followed by an exponentially increasing growth speed followed by linear growth. Despite linear growth of a single hypha, mycelia can grow exponentially in total biomass due to an exponentially increase in the number of branches [3, 48].

Overall, eight flow chamber experiments were conducted. In the first setup, blind, MM Ex, 3.75 and 7.5 Pa were measured three times for both strains MJK17.25 and ME1.4. In a second setup, blind, MM Ex, 5 and 10 Pa were measured once per strain. Using this data, the parameters $\hat{\mu}$ and $\widehat{\Delta t}$ were identified for both strains and all shear stress conditions. The results are depicted in an error bar plot in Fig. 8. The initial relative biomass concentration is also shown. Zero-like error bars indicate the parameter was held constant on the precalculated value.

For the control strain MJK17.25, an increase in the average growth rate can be seen under flow compared to the blind and the pure exchange of medium. The apparent increase starts at 3.75 Pa and reaches a plateau of around 0.25 h^{-1}, which is very similar to the detected maximum growth rate obtained in stirred tank bioreactor of $\mu_{max} = 0.26\,h^{-1}$ (Fig. 6). At higher wall shear stresses (7.5 and 10 Pa), the parameter estimates start to vary between different experimental runs, which is most likely the result of varying spore washout. The detected average growth rates of the chitin synthase double knock-out strain ME1.4 increased equally with the shear stress levels but the plateau reached lower levels (about 0.2 h^{-1}) compared to the control strain MJK17.25, supporting the observations made for bioreactor cultivations (Fig. 6). A

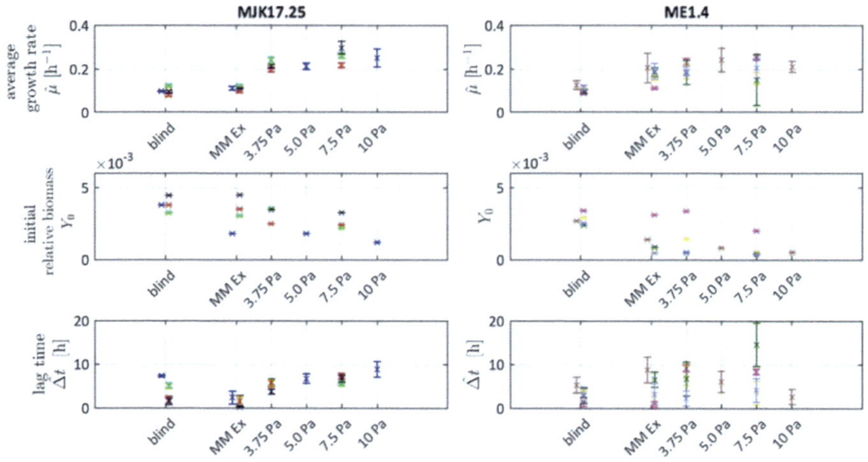

Fig. 8 Parameter estimation results of the three model parameters, Eq. (6), including estimates for μ and Δt. The depicted error bars show the estimates of different experimental runs (control strain MJK17.25, left column; double mutant strain ME1.4, right column) and for the four different shear stress intensity levels (namely $\tau_1 = 7.5$ Pa, $\tau_1/2 = 3.75$ Pa as a triplicate and $\tau_2 = 10$ Pa, $\tau_2/2 = 5$ Pa as a unicate). The error bars include 1σ confidence intervals as calculated from Markov Chain Monte Carlo based parameter sampling. As Y_0 was fixed regarding the spore count, its error bars have zero width in the plot and serve to show the flush out of spores when volumetric flow was started. Different colors indicate biological replicates

slower overall growth rate in the mutant strain was also apparent from microscopic analyses (data not shown). In general, the calculated average growth values for both strains did not change significantly for the cultivations under 3.75, 5, 7.5 and 10 Pa shear stress.

The initial relative biomass detected (Y_0) is an indicator for the amount of spores washed out of the chamber due to the volumetric flow and shows a strong correlation with the applied shear force. Despite large variations between the biological replicates at a certain condition, the overall trend for both strains shows a decrease in initial relative biomass with an increase in shear stress. In a stirred-tank bioreactor experiment, however, such situation does not occur as spores cannot be washed out. Remarkably, the mutant strain ME1.4 shows a lower initial relative biomass at all conditions compared to the control strain MJK17.25. An explanation for this could be the drastically increased lag time for this strain (see Fig. 6), potentially leaving the spores in a less adhesive stage for a longer time. This could promote washout when the volumetric flow was started after 7 h. Another possible explanation could be differences in cell wall composition in both strains (see Figs. 5 and 6). Given that the chitin synthase mutant ME1.4 hypothetically produces less hydrophobic spore surfaces, spore attachment to the chamber would be less efficient. Such an effect was observed in *A. fumigatus*, where the knock-out of the chitin synthase genes *csmA* and *csmB* led to a reduction of the hydrophobic layer of rodlet nanostructures covering wildtype conidiospores, resulting in an increased hydrophilicity of these spores [1].

Finally, the detected lag time Δt for control strain MJK17.25 correlates positively with an increase in shear forces. This could potentially suggest that higher shear stress might provoke first cell wall salvage responses in spores after which they can initiate germination. Notably, the lag times for strain ME1.4 vary strongly between biological replicates at the same conditions. As this might potentially be because of reduced spore hydrophobicity (the more spores become washed out of the system the less spores can germinate the higher the error rate will be) it is also tempting to speculate that the two hyphal subpopulations observed under bioreactor conditions might already be detectable in the microfluidic flow chamber experiment. If so, high biological variability between spores and young hyphae of ME1.4 will cause washout of spores with a larger lag time leaving only spores in the flow chamber with a smaller lag time.

4 Conclusion

Here, we have successfully demonstrated that deletion of the two chitin synthase genes *chsD* and *chsE* in *A. niger* leads to a hyper-shear sensitive mutant strain, which is phenotypically characterized by thinner hyphal diameters, thinner cell walls and reduced chitin levels. When cultivated in microfluidic growth chambers, on agar plate or in stirred-tank bioreactors, this mutant exhibits an extended lag-phase and a lower maximum growth rate.

We have furthermore designed and evaluated a microfluidic flow chamber experiment with which lag phase, maximum growth rate and washout of *A. niger* spores can be calculated under different shear stress conditions. Despite deviations between biological replicates (which were mainly observed for the mutant strain), similar trends were observed compared to bioreactor cultivations conditions. This study thus highlights the potential of microfluidic flow chamber experiments to become harnessed as scale-down models to study the impact of shear stress on hyphal growth and productivity of industrial filamentous fungi such as *A. niger*. As cellular and molecular responses to defined levels of shear stress and potentially other process parameters can be much fast analysed and microscopically followed online, we propose that this simple experimental setup can be harnessed as promising high-throughput screening tool to predict growth characteristics of industrial fungi in stirred-tank bioreactor experiments and beyond.

Acknowledgements The authors thank the Deutsche Forschungsgemeinschaft (DFG) for financial supporting the DFG priority program 1934 "DiSPBiotech" (315305620 for VM and 315457657 for RK). Furthermore, we thank Marcel Schrader and Luca Bendisch from the Institute for Particle Technology (iPAT) at the Technische Universität Braunschweig, Germany and Mathias Müsken from the Helmholtz-Centre for Infection Research Braunschweig, Germany (HZI) for performing transmission electron microscopy analyses. We furthermore acknowledge the DFG for open access funding.

References

1. Alsteens D, Aimanianda V, Hegde P, Pire S, Beau R, Bayry J, Latgé J-P, Dufrêne YF (2013) Unraveling the nanoscale surface properties of chitin synthase mutants of *Aspergillus fumigatus* and their biological implications. Biophys J 105(2):320–327. https://doi.org/10.1016/j.bpj.2013.05.040
2. Arentshorst M, Ram AFJ, Meyer V (2012) In: Bolton MD, Thomma BPHJ (eds) Using non-homologous end-joining-deficient strains for functional gene analyses in filamentous fungi, pp 133–150. https://doi.org/10.1007/978-1-61779-501-5_9
3. Bergter F (1978) Kinetic model of mycelial growth. Zeitschrift für Allgemeine Mikrobiologie 18(2):143–145. https://doi.org/10.1002/jobm.3630180209
4. Bidhendi AJ, Chebli Y, Geitmann A (2020) Fluorescence visualization of cellulose and pectin in the primary plant cell wall. J Microsc 278(3):164–181. https://doi.org/10.1111/jmi.12895
5. Cairns TC, Barthel L, Meyer V (2021) Something old, something new: challenges and developments in *Aspergillus niger* biotechnology. Essays Biochem 65(2):213–224. https://doi.org/10.1042/EBC20200139
6. Cairns TC, Nai C, Meyer V (2018) How a fungus shapes biotechnology: 100 years of *Aspergillus niger* research. Fungal Biol Biotechnol 5(1):13. https://doi.org/10.1186/s40694-018-0054-5
7. Cairns TC, Zheng X, Zheng P, Sun J, Meyer V (2019) Moulding the mould: understanding and reprogramming filamentous fungal growth and morphogenesis for next generation cell factories. Biotechnol Biofuels 12(1):77. https://doi.org/10.1186/s13068-019-1400-4
8. Cairns TC, Zheng X, Zheng P, Sun J, Meyer V (2021) Turning inside out: filamentous fungal secretion and its applications in biotechnology, agriculture, and the clinic. J Fungi 7(7):535. https://doi.org/10.3390/jof7070535
9. Carvalho NDSP, Arentshorst M, Kwon MJ, Meyer V, Ram AFJ (2010) Expanding the ku70 toolbox for filamentous fungi: establishment of complementation vectors and recipient strains for advanced gene analyses. Appl Microbiol Biotechnol 87(4):1463–1473. https://doi.org/10.1007/s00253-010-2588-1
10. Casadevall A, Gow NAR (2022) Ending the (cell) wall metaphor in microbiology. Cell Surf 8:100087. https://doi.org/10.1016/j.tcsw.2022.100087
11. da Silva Dantas A, Nogueira F, Lee KK, Walker LA, Edmondson M, Brand AC, Lenardon MD, Gow NAR (2021) Crosstalk between the calcineurin and cell wall integrity pathways prevents chitin overexpression in *Candida albicans*. J Cell Sci 134(24). https://doi.org/10.1242/jcs.258889
12. Dichtl K, Samantaray S, Wagener J (2016) Cell wall integrity signalling in human pathogenic fungi. Cell Microbiol 18(9):1228–1238. https://doi.org/10.1111/cmi.12612
13. Fiedler MRM, Lorenz A, Nitsche BM, van den Hondel CAMJJ, Ram AFJ, Meyer V (2014) The capacity of *Aspergillus niger* to sense and respond to cell wall stress requires at least three transcription factors: RlmA, MsnA and CrzA. Fungal Biol Biotechnol 1(1):5. https://doi.org/10.1186/s40694-014-0005-8
14. Fiedler MRM, Cairns TC, Koch O, Kubisch C, Meyer V (2018) Conditional expression of the small GTPase ArfA impacts secretion, morphology, growth, and actin ring position in *Aspergillus niger*. Front Microbiol 9. https://doi.org/10.3389/fmicb.2018.00878
15. Fröhlich F, Weindl D, Schälte Y, Pathirana D, Paszkowski Ł, Lines GT, Stapor P, Hasenauer J (2021) AMICI: high-performance sensitivity analysis for large ordinary differential equation models. Bioinformatics 37(20):3676–3677. https://doi.org/10.1093/bioinformatics/btab227
16. Fukuda S, Yamamoto R, Yanagisawa N, Takaya N, Sato Y, Riquelme M, Takeshita N (2021) Trade-off between plasticity and velocity in mycelial growth. MBio 12(2). https://doi.org/10.1128/mBio.03196-20
17. Gow NAR, Latgé J-P, Munro CA (2017) The fungal cell wall: structure, biosynthesis, and function. Microbiol Spectr 5(3). https://doi.org/10.1128/microbiolspec.FUNK-0035-2016
18. Gow NAR, Lenardon MD (2022) Architecture of the dynamic fungal cell wall. Nat Rev Microbiol. https://doi.org/10.1038/s41579-022-00796-9

19. Green MR, Sambrook J (2012) In: Inglis J, Boyle A, Gann A, Cuddihy J, Janssen K, Zierler M, Bubbeo K (eds) Molecular cloning: a laboratory manual. Cold Spring Harbor Laboratory Press, Cold Spring Harbor. ISBN: 978-1-936113-42-2
20. Herth W, Schnepf E (1980) The fluorochrome, calcofluor white, binds oriented to structural polysaccharide fibrils. Protoplasma 105(1–2):129–133. https://doi.org/10.1007/BF01279855
21. Huang L (2008) Growth kinetics of *Listeria monocytogenes* in broth and beef frankfurters—determination of lag phase duration and exponential growth rate under isothermal conditions. J Food Sci 73(5):E235–E242. https://doi.org/10.1111/j.1750-3841.2008.00785.x
22. Ibidi GmbH (2022) Shear stress and shear rates for ibidi μ-slides based on numerical calculations. Version 6.1. https://ibidi.com/img/cms/support/AN/AN11_Shear_stress.pdf
23. Jeennor S, Anantayanon J, Panchanawaporn S, Chutrakul C, Laoteng K (2019) Morphologically engineered strain of *Aspergillus oryzae* as a cell chassis for production development of functional lipids. Gene 718:144073. https://doi.org/10.1016/j.gene.2019.144073
24. Jiang C, Wang H, Liu M, Wang L, Yang R, Wang P, Lu Z, Zhou Y, Zheng Z, Zhao G (2022) Identification of chitin synthase activator in *Aspergillus niger* and its application in citric acid fermentation. Appl Microbiol Biotechnol 106(21):6993–7011. https://doi.org/10.1007/s00253-022-12174-9
25. King R (2015) A framework for an organelle-based mathematical modeling of hyphae. Fungal Biol Biotechnol 2(1):5. https://doi.org/10.1186/s40694-015-0014-2
26. Kunz PJ, King R (2022) Secretory vesicle and glucoamylase distribution in *Aspergillus niger* and macromorphology in regions of varying shear stress. Front Microbiol 13. https://doi.org/10.3389/fmicb.2022.842249
27. Kunz PJ, Barthel L, Meyer V, King R (2020) Vesicle transport and growth dynamics in *Aspergillus niger*: microscale modeling of secretory vesicle flow and centerline extraction from confocal fluorescent data. Biotechnol Bioeng 117(9):2875–2886. https://doi.org/10.1002/bit.27452
28. Kunz PJ (2021) Der Einfluss sekretorischer Vesikel auf fungales Wachstum, Technische Universität Berlin. https://doi.org/10.14279/depositonce-12201
29. Lew RR (2011) How does a hypha grow? The biophysics of pressurized growth in fungi. Nat Rev Microbiol 9(7):509–518. https://doi.org/10.1038/nrmicro2591
30. Meyer V (2021) Metabolic engineering of filamentous fungi. In: Nielsen J, Stephanopoulos G, Lee SY (eds) Metabolic engineering: concepts and applications. Wiley-VCH Verlag. https://doi.org/10.1002/9783527823468.ch20
31. Meyer V, Cairns TC, Barthel L, King R, Kunz PJ, Schmideder S, Müller H, Briesen H, Dinius A, Krull R (2021) Understanding and controlling filamentous growth of fungal cell factories: novel tools and opportunities for targeted morphology engineering. Fungal Biol Biotechnol 8(1):8. https://doi.org/10.1186/s40694-021-00115-6
32. Meyer V, Damveld RA, Arentshorst M, Stahl U, van den Hondel CAMJJ, Ram AFJ (2007) Survival in the presence of antifungals. Genome-wide expression profiling of *Aspergillus niger* in response to sublethal concentrations of caspofungin and fenpropimorph. J Biol Chem 282(45):32935–32948. https://doi.org/10.1074/jbc.M705856200
33. Meyer V, Ram AFJ, Punt PJ (2014) Genetics, genetic manipulation, and approaches to strain improvement of filamentous fungi. In: Bull AT, Junker B, Katz L, Lynd LR, Masurekar P, Reeves CD, Zhao H (eds) Manual of industrial microbiology and biotechnology. ASM Press, pp 318–329. https://doi.org/10.1128/9781555816827.ch22
34. Meyer V, Wanka F, van Gent J, Arentshorst M, van den Hondel CAMJJ, Ram AFJ (2011) Fungal gene expression on demand: an inducible, tunable, and metabolism-independent expression system for *Aspergillus niger*. Appl Environ Microbiol 77(9):2975–2983. https://doi.org/10.1128/AEM.02740-10
35. Müller C, McIntyre M, Hansen K, Nielsen J (2002) Metabolic engineering of the morphology of *Aspergillus oryzae* by altering chitin synthesis. Appl Environ Microbiol 68(4):1827–1836. https://doi.org/10.1128/AEM.68.4.1827-1836.2002
36. Park J, Hulsman M, Arentshorst M, Breeman M, Alazi E, Lagendijk EL, Rocha MC, Malavazi I, Nitsche BM, Hondel CAMJJ, Meyer V, Ram AFJ (2016) Transcriptomic and molecular genetic

analysis of the cell wall salvage response of *Aspergillus niger* to the absence of galactofuranose synthesis. Cell Microbiol 18(9):1268–1284. https://doi.org/10.1111/cmi.12624
37. Pel HJ, de Winde JH, Archer DB, Dyer PS, Hofmann G, Schaap PJ, Turner G, de Vries RP, Albang R, Albermann K, Andersen MR, Bendtsen JD, Benen JAE, van den Berg M, Breestraat S, Caddick MX, Contreras R, Cornell M, Coutinho PM, et al (2007) Genome sequencing and analysis of the versatile cell factory *Aspergillus niger* CBS 513.88. Nat Biotechnol 25(2):221–231. https://doi.org/10.1038/nbt1282
38. Rani HP, Sheu TWH, Tsai ESF (2007) Eddy structures in a transitional backward-facing step flow. J Fluid Mech 588:43–58. https://doi.org/10.1017/S002211200700763X
39. Ries LNA, Rocha MC, de Castro PA, Silva-Rocha R, Silva RN, Freitas FZ, de Assis LJ, Bertolini MC, Malavazi I, Goldman GH (2017) The *Aspergillus fumigatus* CrzA transcription factor activates chitin synthase gene expression during the caspofungin paradoxical effect. MBio 8(3). https://doi.org/10.1128/mBio.00705-17
40. Riquelme M, Munro C, Gow N (2020) The fungal cell wall: biology, biosynthesis and biotechnology. Cell Surf 6:100037. https://doi.org/10.1016/j.tcsw.2020.100037
41. Schindelin J, Arganda-Carreras I, Frise E, Kaynig V, Longair M, Pietzsch T, Preibisch S, Rueden C, Saalfeld S, Schmid B, Tinevez J-Y, White DJ, Hartenstein V, Eliceiri K, Tomancak P, Cardona A (2012) Fiji: an open-source platform for biological-image analysis. Nat Methods 9(7):676–682. https://doi.org/10.1038/nmeth.2019
42. Schmideder S, Barthel L, Friedrich T, Thalhammer M, Kovačević T, Niessen L, Meyer V, Briesen H (2019) An X-ray microtomography-based method for detailed analysis of the three-dimensional morphology of fungal pellets. Biotechnol Bioeng 116(6):1355–1365. https://doi.org/10.1002/bit.26956
43. Stapor P, Weindl D, Ballnus B, Hug S, Loos C, Fiedler A, Krause S, Hroß S, Fröhlich F, Hasenauer J (2018) PESTO: Parameter EStimation TOolbox. Bioinformatics 34(4):705–707. https://doi.org/10.1093/bioinformatics/btx676
44. Steinberg G, Peñalva MA, Riquelme M, Wösten HA, Harris SD (2017) Cell biology of hyphal growth. Microbiol Spectr 5(2). https://doi.org/10.1128/microbiolspec.FUNK-0034-2016
45. Steudle E, Zimmermann U, Lüttge U (1977) Effect of turgor pressure and cell size on the wall elasticity of plant cells. Plant Physiol 59(2):285–289. https://doi.org/10.1104/pp.59.2.285
46. Sun X, Wu H, Zhao G, Li Z, Wu X, Liu H, Zheng Z (2018) Morphological regulation of *Aspergillus niger* to improve citric acid production by *chsC* gene silencing. Bioprocess Biosyst Eng 41(7):1029–1038. https://doi.org/10.1007/s00449-018-1932-1
47. Tupper B (2021) Percentile normalization macro for ImageJ. https://github.com/deepimagej/imagej-macros/blob/master/PercentileNormalization.ijm
48. Yang H, King R, Reichl U, Gilles ED (1992) Mathematical model for apical growth, septation, and branching of mycelial microorganisms. Biotechnol Bioeng 39(1):49–58. https://doi.org/10.1002/bit.260390109
49. Yoshimi A, Miyazawa K, Kawauchi M, Abe K (2022) Cell wall integrity and its industrial applications in filamentous fungi. J Fungi 8(5):435. https://doi.org/10.3390/jof8050435
50. Zhang B, Zerubia J, Olivo-Marin J-C (2007) Gaussian approximations of fluorescence microscope point-spread function models. Appl Opt 46(10):1819. https://doi.org/10.1364/AO.46.001819

Interaction of Mechanical Stresses and Productivity of Biological Agglomerates in Stirred Fermenters

Chrysoula Bliatsiou, Philipp Waldherr, Lutz Böhm, and Matthias Kraume

1 Introduction

Stirred-tank reactors (STRs) are standard bioreactor configurations in the biotechnological industry. Performing a cultivation process of filamentous microorganisms in STRs is highly complicated since several requirements should be fulfilled for the growth of the biological matter and the product secretion [1]. The interplay among process parameters, i.e., biomass growth, morphology, rheology, and oxygen transfer, defines the total system's productivity [2]. Many literature studies focus on the effect of fluid dynamics in the bioreactor and specifically of the employed impeller type on various process parameters such as biomass concentration, gas dispersion, k_La, and productivity [3–6]. The decisive influence of the fungus' morphology on productivity has also been proven [7], however, the link between the fluid dynamics and the fungus' morphology is still not fully understood. The concept of "high shear" and "low shear" impellers, which dominated the discussion for many years, is nowadays widely disapproved as the criteria set behind such a categorisation were biased [8]. The works of Jüsten et al. [9], Cui et al. [10], Hardy et al. [11] with filamentous microorganisms and the studies of Henzler and Biedermann [12], Wollny [13] with model particle systems significantly enlightened the effect of impeller geometry on

C. Bliatsiou · P. Waldherr · L. Böhm (✉) · M. Kraume
Chair of Chemical and Process Engineering, Technische Universität Berlin, Ackerstraße 76, 13355 Berlin, Germany
e-mail: lutz.boehm@tu-berlin.de

C. Bliatsiou
e-mail: c.bliatsiou@tu-berlin.de

P. Waldherr
e-mail: p.waldherr@tu-berlin.de

M. Kraume
e-mail: Matthias.Kraume@tu-berlin.de

A. Kwade and I. Kampen (eds.), *Dispersity, Structure and Phase Changes of Proteins and Bio Agglomerates in Biotechnological Processes*,
https://doi.org/10.1007/978-3-031-63164-1_16

particle breakage and enabled an objective classification of impeller geometries in more- or less-stress-intensive. Still, the mechanism of breakage is not fully understood. The advanced experimental and numerical methods available today can give a new perspective on this research field.

This research study aims at achieving a deeper understanding of the influence of fluid dynamic stress on the fungus' morphology in lab-scale aerated stirred tank bioreactors, and at identifying low-stress stirring conditions beneficial for the growth of stress-sensitive microorganisms. The effect of stirring on the morphology of various strains of the filamentous fungus *Aspergillus niger* was on focus [14, 15]. Various conventional, commercial, and newly designed impellers were examined. The investigations were expanded significantly by using model particle systems [15–18]. Focus was given on the rheology of the cultivation broth [19] and the stress that particles experience in non-Newtonian flow conditions [15]. The additional stress on particles due to aeration was also investigated [15]. Different parameters, such as the maximum energy dissipation rate ε_{max} or the circulation frequency $t_c{}^{-1}$ of the particles in the reactor, were evaluated to interpret the impellers' stress-intensity and to set a base for a successful scale-up [20]. To determine such flow properties, experimental and numerical fluid dynamic analysis was carried out. In this chapter, the main findings of the research conducted within SPP 1934 are presented. The reader can access further details and extended analysis in the group's publications given as references in this chapter.

2 Cultivation of *Aspergillus niger*

The recombinant fungal strain *Aspergillus niger* SKAn 1015 was mainly employed in this work to study the influence of hydrodynamic stress on filamentous microorganisms. The strain produces β-fructofuranosidase. It was provided by the Institute of Biochemical Engineering at the Technical University Braunschweig in the frame of the priority program. Based on that group's previous works, the morphology of *A. niger* SKAn 1015 can be controlled in salt-enhanced cultivations [7]. In this study, the main focus was given to pellet-producing cultivations (see Fig. 1) to facilitate the morphological analysis of the biological system. Through the cooperation of our working group and the Institute of Biochemical Engineering at the Technical University Braunschweig within the SPP 1934, additional emphasis was given to the rheological characterisation of the cultivation process [19]. The rheology of the broth is a crucial parameter for the mixing in the bioreactor, thus for the whole cultivation process affecting the nutrient and oxygen transfer, the biomass growth, the stress on bio-agglomerates, etc. Figure 1 shows exemplarily the significant rheological changes recorded for the cultivation of *A. niger*. Independent of the bio-agglomerates' shape (pellet or freely dispersed mycelia), as the biomass increases over time, the broth evolves from a Newtonian medium with water-like viscosity to a shear-thinning system due to particle-to-particle interactions and the ability of the filaments to elongate and compress. The specific morphological properties of the fungus define the

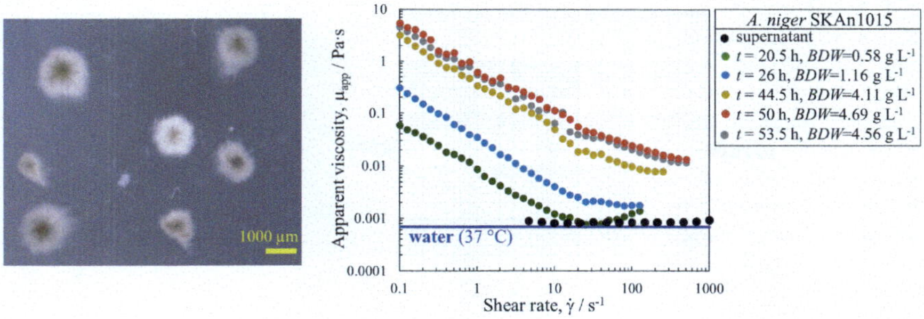

Fig. 1 Microscope image of *A. niger* SKAn 1015 and flow curves of the broth during batch cultivation

broth viscosity, i.e., lower pellet compactness and higher surface roughness (longer external hyphae) increase the viscosity. In the cooperative publication of Bliatsiou et al. [19], a deeper rheological analysis of the specific cultivation system of *A. niger* SKAn 1015, as well as of the filamentous bacteria *Lentzea aerocolonigenes* (main employed microorganism of the Institute of Biochemical Engineering at the Technical University Braunschweig within SPP 1934) and *Actinomadura namibiensis* was provided.

To verify the general validity of the results obtained with *A. niger* SKAn 1015 and to check their transferability to other cultivation systems, further cooperation within SPP 1934 was built with the Department of Applied and Molecular Microbiology at the Technische Universität Berlin. That research group genetically modified the *A. niger* strain MJK 17.25, developing the mutant ME 1.4 [21]. The synthesis of chitin for the cell walls was reduced for ME 1.4, and a lower mass transfer resistance for products through the cell wall and to the hyphal tip at the parallel cost of a more stress-sensitive organism was expected. Both the parental strain MJK 17.25 and the expectedly more stress-sensitive ME 1.4 were employed in this work to study the hydrodynamic stress induced by various impellers in stirred tank reactors.

All cultivations of the study were conducted in 3–4 L stirred tank bioreactors under optimised physicochemical conditions. The selection of agitation and aeration conditions in the bioreactor to ensure sufficient spore suspension, oxygen supply and allow the growth of bio agglomerates in a rheologically continuously changing system was challenging (see Fig. 2). Using Rushton turbines for the cultivations proved to be very robust on creating a fluid dynamic environment on which *A. niger* could grow in the required pellet form.

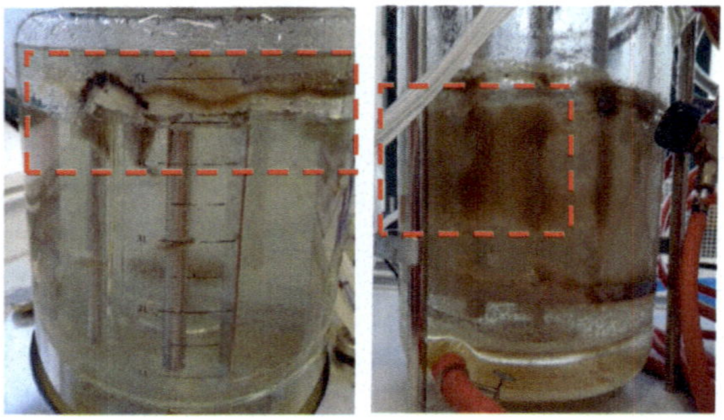

Fig. 2 Selection of unsuited agitation–aeration conditions during cultivation led to unsuccessful runs, e.g., discharge of spores on the fluid surface due to strong aeration (left) and biomass adhesion/growth on tank walls due to "wrong" choice of stirrer geometry (right)

3 Development of Experimental Methods

3.1 Characterisation of Particle Breakage

3.1.1 Fragmentation of *Aspergillus niger*

The reproducibility of any cultivation process is often a significant challenge, complicating in this study a fair and valid comparison of stirring conditions that would extend further than a qualitative morphological characterisation. Moreover, as often reported in the literature and verified in this study, the interdependency between the process parameters (stirring, aeration, rheology, morphology, etc.) was very strong during cultivation. Therefore, it was difficult to attribute the difference in the produced pellet morphologies only to stirring effects. A methodology of offline fragmentation experiments was developed to overcome these limitations and to precisely examine and evaluate the hydrodynamic stress on pellets induced by stirring.

Based on that method, the cultivations were stopped after the substrate consumption (approx. 48 h), and the broth was collected and diluted. The diluted broth was mixed in three parallel 3 L stirred tank reactors under varying controlled stirring conditions, i.e., different stirrer types and power inputs. This way, it was ensured that in all reactors, the initial pellet morphology was identical, and the differences in final pellet sizes and shapes resulted solely from the exposure of pellets to different levels of hydrodynamic stress. The level of broth dilution was set at approx. 0.1 g/L to minimise particle-to-particle interactions, which would influence the system rheology, and to provide a large broth volume for a high number of batch experiments with varying stirring conditions to be conducted. In the published work of Waldherr et al. [14], details about the development of the fragmentation methodology were provided.

Using model fluids as the continuous phase of the batches, the rheology of the actual cultivation broth was simulated, and the interdependency among rheology-stirring-pellet breakage was objectively investigated. The selection of model fluids was based on this group's published findings [19]. Therefore, the fragmentation experiments were mainly conducted with water as continuous phase to mimic the original cultivation medium and initial cultivation phases, and with 0.05% w w^{-1} and 0.1% w w^{-1} Xanthan gum solutions (XGS) to mimic the shear-thinning rheology at progressed cultivation stages. The non-Newtonian rheological behavior of Xanthan gum was characterised through the power law model with measured consistency and flow indices $K = 0.172$ Pa s^n, $n = 0.385$ for 0.05% w w^{-1} XGS and $K = 0.328$ Pa s^n, $n = 0.379$ for 0.1% w w^{-1} XGS.

3.1.2 Development and Experimentation with Model Systems

Despite the methodology of offline pellet fragmentation and the respective dilutions, the available volume for batch experimentation after each cultivation run was still limited, restricting the range of stirring variables that could be examined each time. Moreover, the cultivations were time-, material- and cost-intensive. Therefore, the development and utilisation of model particle systems that would mimic the actual biological system and facilitate the characterisation of hydrodynamic stress acting on particulate matter in stirred tank reactors was widely employed in this research. Table 1 summarises the characteristics and properties of the four employed model particle systems. The experimentation with these systems consisted of batch stirring experiments where the breakage of particles under varying stirring conditions, i.e., stirring intensity, stirrer type and number of stirrers, was quantified.

Table 1 Employed model particle systems

Model system	Composition	Description	Particle morphology
Silicon oil/TX-100/H_2O	• Silicon oil $\varphi = 1.5\%$ v v^{-1} • Surfactant Triton X-100 $\gamma = 0.15\%$ w w^{-1} • Ultrapure water	Liquid/liquid, non-coalescing, Newtonian system	Spherical
Silicon oil/H_2O	• Silicon oil $\varphi = 1.5\%$ v v^{-1} • Ultrapure water	Liquid/liquid, coalescing, Newtonian system	
Clay/Praestol 650 BC/H_2O	• Clay $c = 0.7$ g L^{-1} • Flocculation agent Praestol 650 BC $c = 0.002$ g L^{-1} (copolymer of acrylamide) • Sodium chloride $c = 1$ g L^{-1} • Water 1000 μS/cm	Solid/liquid, re-agglomerating, Newtonian system	Non-spherical
Silicon oil/TX-100/XGS (0.05% w w^{-1})	• Silicon oil $\varphi = 1.5\%$ v v^{-1} • Surfactant Triton X-100 $\gamma = 0.15\%$ w w^{-1} • 0.05% w w^{-1} Xanthan gum sol	Liquid/liquid, non-coalescing, shear thinning system	Spherical

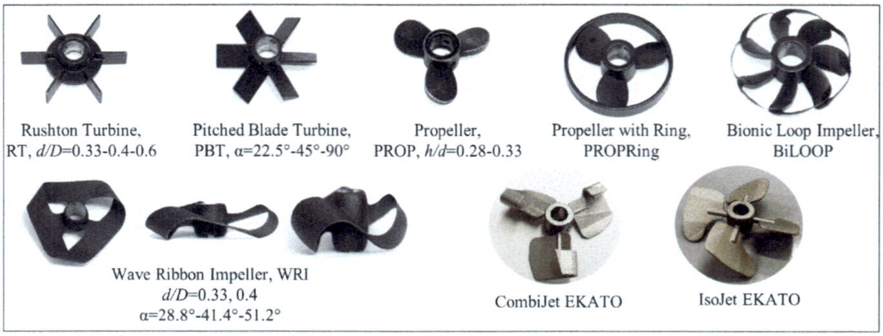

Fig. 3 Impeller geometries under investigation

3.2 *Impellers Under Investigation*

Multiple identical set-ups were designed and built for the fragmentation experiments with *A. niger* pellets and the experimentation with model systems. They consisted of a baffled stirred tank with an inner diameter of $D = 160$ mm and a torispherical-shaped bottom. The geometrical features of the tank, the shaft and baffles were as close as manufacturing possible to the actual bioreactor geometry employed for the cultivations. In the single impeller experiment with a ratio of fluid height to tank diameter $H/D = 1$, the typical impeller bottom clearance was set at $h_{\text{imp}}/H = 0.33$. For the dual-impeller experiments with $H/D = 1.5$, the spacing between the stirrers was an additional variable under investigation. In the study, in total 16 impellers were tested (see Fig. 3), including four Rushton turbines (RT), two 6-blade pitched blade turbines (PBT), two 3-blade propellers (PROP), a modified propeller with a circumferential ring (PROPRing), a bionic loop impeller (BiLoop) designed by EvoLogics GmbH, three wave-ribbon impellers with no blades (WRI) designed by EvoLogics GmbH and two commercial stirrers of EKATO HOLDING GmbH (Ekato Combijet, Ekato Isojet). Geometrical features such as impeller diameter d, blade (or ribbon) angle α, and blade height h were also variables of investigation.

3.3 *Inline Particle Size Measurement Technique and Image Analysis*

Hydrodynamic stress was quantified based on the particle breakage caused by stirring. A precise morphological particle analysis was demanded to evaluate and quantify the breakage. An inline photo-optical particle measurement technique (SOPAT GmbH) was employed. It consists of an endoscopic probe inserted in the reactor. In total, four endoscopic probes were utilised in the study, each allowing to capture particles of different size ranges (total range of interest 10 μm–7 mm). The evaluation of particle

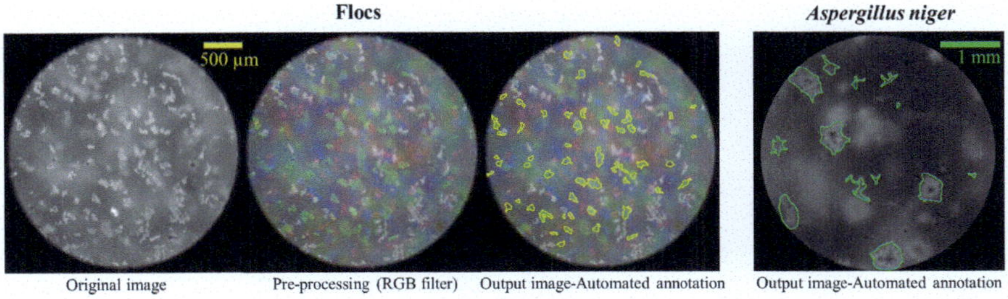

Fig. 4 An original endoscope image of flocs in the reactor, image pre-processing, automated detection of flocs through CNN developed algorithms (left), and automated CNN detection of *Aspergillus niger* pellets (right)

size distributions was conducted by the commercial software of SOPAT GmbH for spherical particles. Convolutional Neural Networks (CNN) were employed and trained for image analysis and particle detection in the case of bio-agglomerates and flocs, where the shapes are non-spherical and very complex. Specifically, the U-Net architecture implemented in the deep learning framework Caffe was utilised to create a sophisticated algorithm that allowed particle detection. Time-intensive training followed based on manually marked particles in images given as input to the computing system. Specifically, eight training rounds were conducted for the clay-floc system, and a total of 2070 manually annotated particles were given as training input. Multiple image rescalings and augmentations were additionally used to artificially increase the amount of training data. The result was an exact and robust particle detection in various depths inside the bulk fluid region that could even differentiate between moving and non-moving particles (attachments on endoscopic probe lens) by pre-processing the images applying an RGB filter. Details about this novel image analysis method were described in the publication of Panckow at al. [17]. A respective process was followed for *A. niger* with pellet and dispersed morphology. Figure 4 shows exemplarily image material obtained by the inline photo-optical technique and the particle detection achieved by the respective parametrised algorithm.

4 Experimental Investigation on Particle Stress

4.1 Fragmentation of Aspergillus niger

Figure 5a, b summarise representative results from fragmentation experiments with *A. niger* SKAn 1015 and ME 1.4 in water (continuous phase). The steady-state arithmetic mean of the maximum Feret diameter of the pellets d_P, used as characteristic pellet dimension to quantify breakage, is plotted as a function of the mean energy dissipation rate $\overline{\varepsilon}$ in the reactor for two fragmentation runs with different initial pellet sizes $d_{P,0}$ for each strain.

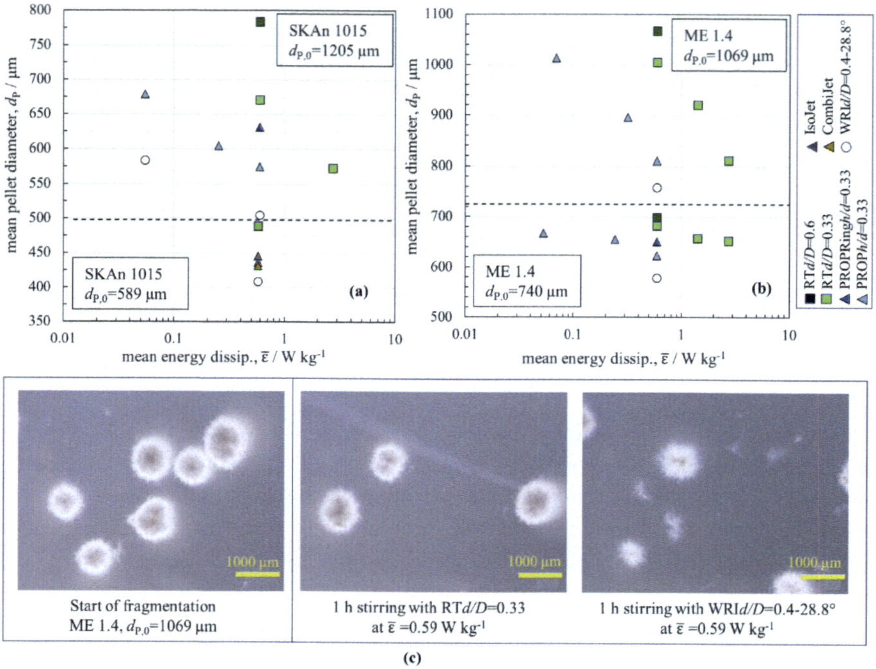

Fig. 5 The arithmetic mean of maximum Feret diameter of pellets d_P over mean energy dissipation rate $\bar{\varepsilon}$ for fragmentation experiments with varying impellers for **a** *A. niger* SKAn 1015 and **b** *A. niger* ME 1.4 (the dashed line in (**a**, **b**) separates data obtained for different fragmentation runs, i.e., different initial pellet size). **c** Exemplary microscope images of *A. niger* ME 1.4 before fragmentation (left) and after stirring with a Rushton turbine (middle) and with a wave-ribbon impeller (right)

The initial pellet size and morphology before fragmentation were identified as a significant influencing factor of the stress-sensitivity of the bio-agglomerates. A decline in particle size with increasing average specific power input was recorded for all fragmentation runs independent of the initial pellet morphology. However, it was found that initially larger particles show a greater dependence on the mean energy dissipation range. Comparing data obtained with the same impeller for the same strain but varying initial pellet sizes, it became apparent that the slope of pellet size decline was steeper for initially larger pellets. Moreover, the effect of impeller geometry became more significant for initially bigger pellets as the differences seen in the proportional pellet size decline among stirrers became wider. During a fragmentation run, pellet size reduces until the cohesive forces of the pellet, i.e., the tensile strength, come in equilibrium with the hydrodynamic stress. The tensile strength is directly connected to the pellet micromorphology (e.g., hyphae branching frequency, pellet density), but as proved in this work, it is also influenced by the macromorphological features. Finally, since the chitin is believed to be essential for the mechanical robustness of the fungus, the lower chitin concentration of the lateral cell wall of the strain ME 1.4 (1/3 of that one of the parental MJK 17.25) was expected to increase the fungus sensitivity to hydrodynamic stress. However, that was not confirmed by the experiments of this study. When pellets of ME 1.4 and MKJ 17.25 of similar initial size, i.e., $d_{P,0} \approx 745 \pm 4$ μm, were stirred with the same impeller RT$d/D = 0.33$ at equal $\bar{\varepsilon} = 2.73$ W kg^{-1}, it was found that the finally obtained arithmetic mean of the ME 1.4 pellets was 25% larger than that of MJK 17.25 pellets. Therefore, a lower chitin content does not prove to be a crucial parameter for strain stress-sensitivity. The pellet size and the micromorphology primarily define the resistance to stress instead.

Independent of the specific morphological features of *A. niger* pellets, using the arithmetic mean diameter to measure the hydrodynamic stress led to the following conclusions. When considering the equal energy dissipation rate (i.e., mass-specific power input) $\bar{\varepsilon}$ as a comparison condition, the radial-flow Rushton turbines cause minor pellet breakage, with a larger turbine diameter d further reducing the breakage. The axial-flow propellers induce substantial particle stress. However, modifying the propeller geometry with a circumferential ring showed partly beneficial effects regarding reduced pellet breakage compared to the conventional propeller. On the other hand, the wave-ribbon impeller induced the strongest hydrodynamic stress in all cases and thus decreased pellet size. The commercial EKATO impellers prove to be similarly stress-intensive as a standard 3-blade propeller design, despite the larger blade surface available for fluid mixing for the commercial IsoJet (axial-flow stirrer) and CombiJet (stirrer of combined flow in both radial and axial direction). Figure 5c shows microscopic images of ME 1.4 pellets before the fragmentation experiment and at steady-state (end of fragmentation) for experiments conducted with a Rushton turbine and a wave-ribbon impeller at the same $\bar{\varepsilon}$. The pellet breakage, chip-off of pellet hyphae, and release of fragments in the fluid became visually apparent when the wave-ribbon impeller was used to stir the broth. These conclusions were valid for all *A. niger* strains employed in the study.

4.2 Agitation-Induced Stress on Model Systems

The fragmentation experiments provided valuable information on how stress-intensive an impeller geometry is, as the interaction between the stirrer and the actual bio-agglomerates could be recorded and evaluated in a close and controlled way. However, as already mentioned they were related to time- and cost-intensive experimentation and a limited range of operation conditions that could be examined, since the produced volume of cultivation broth was enough only for a restricted number of experiments. The usage of model systems facilitated the investigation of a wider variety of stirrer features. Figure 6 summarises the breakage results obtained with the non-coalescing and coalescing liquid/liquid dispersion systems (Fig. 6a) and the re-agglomerating solid/liquid floccular system (Fig. 6b). Detailed presentation of the results is provided in [15–18]. In Fig. 6 for each system, the steady-state characteristic particle diameter, i.e., Sauter mean drop diameter d_{32} and median floc diameter d_{n50}, is plotted over the mean energy dissipation rate $\bar{\varepsilon}$. Independent of the

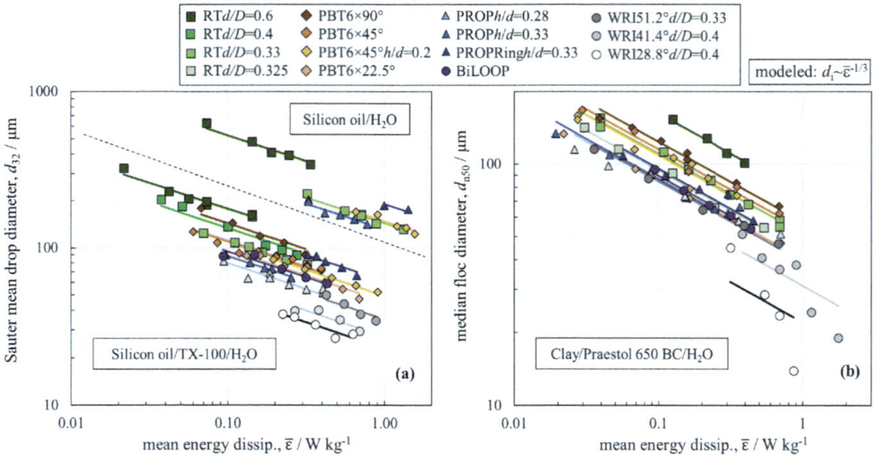

Fig. 6 Steady-state characteristic particle diameter as a function of the mean energy dissipation rate $\overline{\varepsilon}$ for the various impellers of investigation **a** for the liquid/liquid silicon oil systems and, **b** for the solid/liquid clay/polymer floc system

employed model system and thus the nature of particles experiencing stress (drops or flocs), the main conclusions remain identical.

When operated at equal $\overline{\varepsilon}$, stirring with an impeller of larger diameter d leads to formation of larger particles, i.e., to less particle breakage (see comparison among Rushton turbines in Fig. 6). At equal $\overline{\varepsilon}$, the larger stirrer operates at a significantly lower rotational frequency, without this parameter being adequate to interpret the recorded effect, since even at equal impeller tip speed v_{tip} the particles (drops and flocs) produced by the larger impeller were still bigger. To this result seems to attribute (1) the larger blade surface, which means that the same amount of energy is inserted by a more extensive area into the fluid, resulting in lower values of local energy peaks, and (2) differentiations in the flow field, as they have been determined with Particle Image Velocimetry (PIV) measurements. In Fig. 7, the variation of the flow field of the Rushton turbines with $d/D = 0.33$ (distant-wall impeller) and $d/D = 0.6$ (near-wall impeller) can be directly compared. The Rushton turbine with the smaller diameter does not generate a purely radial secondary flow, but a partly downwards-oriented flow due to an interplay among the torispherical tank bottom, the stirrer bottom clearance, and the partly baffled reactor.

The influence of the impeller blade angle α was also clearly depicted in the results of Fig. 6. At equal $\overline{\varepsilon}$, steeper (bigger) blade angles induce less particle stress. As the blade angle increases, the generated secondary flow acquires more radial components (recorded by PIV measurements), and the impeller has a higher power number. Thus, it operates at lower stirrer speeds to insert a specific power in the reactor.

The comparison among the conventional 3-blade propellers and the modified PROPRing either at equal $\overline{\varepsilon}$ or equal v_{tip} showed that adding a ring in an axial-flow agitator reduces the particle breakage. Characteristic changes in the flow field were

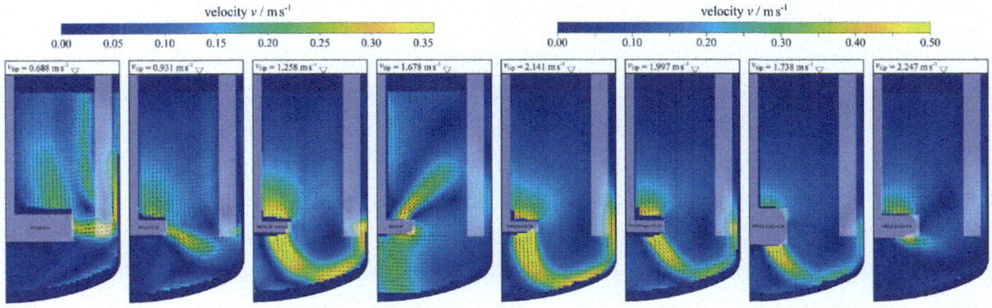

Fig. 7 Secondary flow fields experimentally determined with particle image velocimetry (PIV) for (from left to right) RT$d/D = 0.6$, RT$d/D = 0.33$, PBT6 $\times$ 45° $-$ $h/d = 0.2$, BiLOOP, PROP$h/d = 0.33$, PROPRing$h/d = 0.33$, WRI51.2°$d/D = 0.33$, WRI28.8°$d/D = 0.4$ at $\overline{\varepsilon} = 0.1$ W kg^{-1}

again determined, since through the ring addition the PROPRing generates a flow with more significant radial components (see Fig. 7). Energy is introduced to the fluid through a wider surface and not only through the blade tip, expecting lower values of local energy peaks in the trailing vortex structure build by the stirrer. The newly designed BiLoop also produces bigger particles than conventional propellers, at a lower degree than the PROPRing though, an effect mainly attributed to the higher power number of this stirrer and thus operation at lower stirrer speeds.

The wave-ribbon impellers have no blades, and the ribbon configuration induces the fluid motion. The decrease in ribbon angle causes a significant differentiation in the flow field, reducing the pumping capacity of the stirrer and leading to a very poor fluid motion that strongly circulates through the impeller swept volume for the impeller with the smallest angle $\alpha = 28.8°$ (see Fig. 7). This flow variation with decreasing ribbon angle was found to result in an extreme increase in particle breakage.

All in all, when operated at equal $\overline{\varepsilon}$, radial-flow impellers induce significantly less particle stress than axial-flow agitators. This can be partly explained by the overall higher power numbers of radial-flow agitators, which lead to operation at lower stirrer speeds to obtain a specific power input level in the reactor compared to axial-flow stirrers. However, exceptions of this interrelation were recorded, e.g., when looking at the results of the Rushton turbine RT$d/D = 0.33$ and pitched blade turbines PBT6 $\times$ 45° and PBT6 $\times$ 45°$h/d = 0.2$. These agitators produce particles of similar sizes, despite the more significant power number of RT$d/D = 0.33$ (Ne = 4) compared to pitched blade turbines (Ne = 1.4–1.6). This result is attributed to a similar blade surface area available for the fluid mixing for these three impellers and similarities in the flow field. Therefore, these factors are also crucial for the hydrodynamic stress caused by an agitator. Among the axial-flow impellers, apart from the 6-blade PBT with large blade angle, the modified propeller with the ring is also considered beneficial for stress-sensitive applications. On the contrary, wave-ribbon impellers with small ribbon angles are disadvantageous for stress-sensitive organisms.

The experimentation findings with model systems in water as continuous phase agree with the respective fragmentation results of *Aspergillus niger*. Even though the main conclusions are drawn by considering a characteristic particle diameter, i.e., a statistic percentile of the particle size distribution, a closer observation of the whole distribution releases more information about the stirring-induced breakage. Differences between the model systems became apparent under this aspect. Figure 8 shows the width of the distributions produced by each tested stirrer, defined as span $= (d_{90} - d_{10})/d_{50}$. Even though for the non-coalescing liquid/liquid system, the particle size distributions of all stirrers were self-similar, i.e., their width deviated insignificantly (except for the WRIs), that was not the case for the coalescing liquid/liquid system. For this system, the distributions formed by the Rushton turbines were narrower than those formed by the axial-flow agitators. The comparison between RT$d/D = 0.33$ and PBT6 × 45°$h/d = 0.2$ shows that even though the characteristic particle diameter of Fig. 6 is identical for the two stirrers, the total distribution formed by the pitched blade turbine is wider. The findings agreed with the results of the re-agglomerating solid/liquid floc system. For the latter, a clear differentiation in distribution properties, e.g., width and standard deviation, and in the detailed morphological analysis of the flocs, e.g., particle circularity and aspect ratio, was recorded. That led to conclusions about different breakage mechanisms of the various impellers. A deeper analysis of the findings is provided in detail in Panckow et al. and Bliatsiou et al. [17, 18].

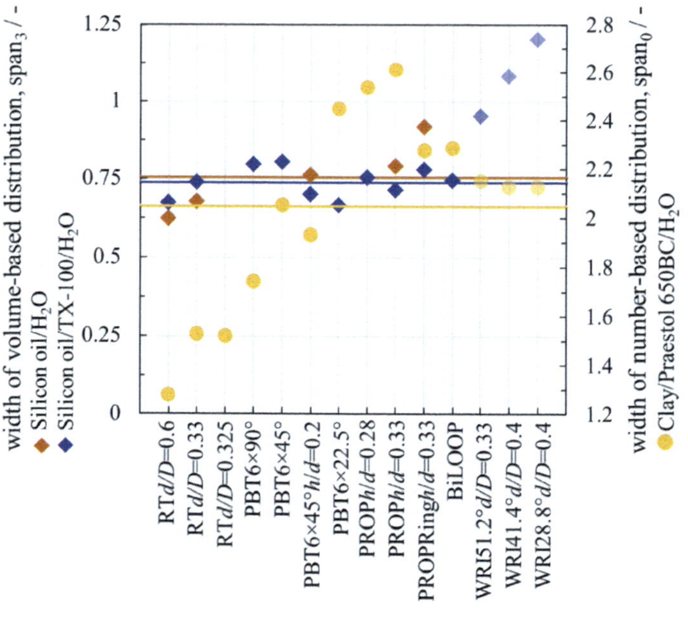

Fig. 8 Width of the particle size distribution (span) formed by the impellers under investigation for the three Newtonian model systems (data of WRIs must be treated with caution. Due to the very small particles produced by these stirrers, the measuring limits of the employed technique were reached, and the extreme percentiles of the distribution were harder to determine reliably)

The tank's maximum energy dissipation rate ε_{max} in the tank is crucial for particle breakage. The successful modelling of the experimental data with this local flow property verifies its importance in multiphase particulate systems. Figure 9 depicts such modelling when ε_{max} is approximated for each impeller using the generalised mathematical description of Grenville et al. [22]. They experimentally determined the trailing vortex structure for various agitators, which is expected to be crucial for the particle breakage. A comparison with other literature approaches on quantifying ε_{max} was also conducted (see [18]) and showed that the long-existing concept of the impeller swept volume defined as a cylinder based on the stirrer dimensions is equally helpful to achieve a rapid, rough approximation of ε_{max}. Nevertheless, ε_{max} remains a difficult to determine value and even a highly resolved experimental or numerical determination contains uncertainties. For new impeller designs, such as the WRIs, approximations of ε_{max} based on literature approaches are highly imprecise. That is the reason for the deviating data for WRIs seen in Fig. 9 for Silicon oil/TX-100/H_2O and Clay/Praestol 650BC/H_2O.

Applying the Energy Dissipation Circulation Function model, which includes the effect of both ε_{max} and the particle circulation frequency through the impeller swept volume t_c^{-1}, did not prove to correlate/model the data better [18]. It is believed that the role of the circulation time in small lab-scale tanks remains insignificant, opposite to what might be expected for pilot- and industrial-scale apparatuses. By determining experimentally through a decolourisation method, the mixing time in the reactor, i.e., the time needed to reach a certain degree of homogeneity, it was found that at an operation range of $\overline{\varepsilon} = 0.05$–$0.5$ W kg^{-1}, mixing times lie at $t_{mix} = 3$–10 s for a system with water-like viscosity. Since multiple fluid circulations are required before uniformity is achieved in the tank, the circulation time is connected to the mixing time

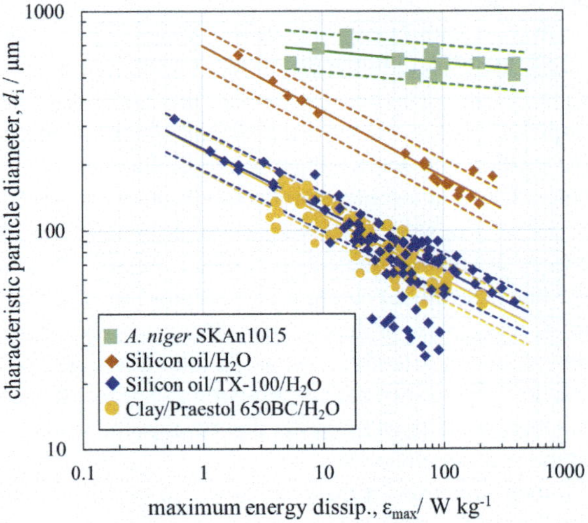

Fig. 9 Steady-state characteristic particle diameter d_i as a function of maximum energy dissipation rate ε_{max} for the stirrers under investigation for a cultivation of *A. niger* SKAn 1015 and all employed model systems (−continuous line corresponds to power function fitting, --- dotted line corresponds to ±20% deviation)

$t_{mix} \sim 3 \ldots 5\ t_c$ [23]. This rough approximation implies extremely short circulation times ($t_c < 4$ s in the tank under investigation). This time scale is not relevant for the breakage experiments of the study that last up to 2 h till a steady state is reached. At higher viscosities in shear-thinning Xanthan gum solutions, mixing times up to $t_{mix} = 200$ s were measured for WRI impellers, meaning that respectively longer circulation times of up to few minutes could be expected. Therefore, the effect of the circulation time and frequency through which particles enter the impeller region becomes more significant in more viscous media, as well as bigger working volumes and reactor scales.

The experimentation with *A. niger* showed that the interdependence of pellet breakage on mass-specific power input is influenced by the initial pellet morphology. The relative decline of pellet diameter under stirring at a certain intensity is stronger for pellets with an initial larger size (see Fig. 5). However, in general, the power input dependency of the particle diameter for the model systems, which is well-predicted by the model of Shinnar and Church [24] for the dissipation range of turbulence $d_i \sim \overline{\varepsilon}^{-1/3}$ (see Fig. 6), is always stronger than that of the actual bio-pellets. The same behavior is depicted in Fig. 9. The slope of the fitting line for the data of *A. niger* is essentially smaller than that for the model systems, showing that the damage caused under exposure at a certain ε_{max} is stronger for the drops and flocs and less important for the biological pellets of *A. niger*. These findings prove a varying breakage mechanism for the inorganic particles and the biological matter. Through morphological observation under the microscope, it was seen that the bio-pellets often do not break completely, but parts of the hyphae are separated from the retaining pellets (chip-off). This breakage mechanism does not change the diameter drastically. Moreover, the fragments have an elongated shape and thus a relatively large Feret diameter. For these reasons, an increase in stress has a limited effect on average pellet size.

In bioreactor operation, ratios of fluid height to tank diameter larger than $H/D > 1$ and, therefore, multistage impeller systems are usually employed [20]. Figure 10 shows experimentation results with the floc system and dual impellers. The relative impeller spacing *s/d* was an additional factor of investigation. Figure 10a, b depict the median floc diameters produced by dual Rushton turbines and varying *s/d*. The single Rushton turbine is plotted additionally as a reference. At equal $\overline{\varepsilon}$, the spacing of $s/d = 1$ induces the lowest particle breakage. At equal v_{tip}, an equal maximum energy dissipation rate ε_{max} could be expected. Thus, when stirring under equal v_{tip}, the floc size should be identical independent of the impeller spacing or the number of stirrers (single or dual). This is the case only for $s/d \geq 2$, when the two impellers operate independently, creating parallel flows, and producing floc sizes identical in size to a single stirrer. The formation of parallel flows in the reactor is confirmed by the increase in power number that reaches the sum of the individual power numbers of the two impellers when $s/d \geq 2$ (see Fig. 10b). For smaller spacings, as the power number decreases, a merging flow can be reasonably assumed. This variation of the flow field seems to lead to lower values of ε_{max}, which could justify the lower particle stress for specific spacings. The same results were verified with the liquid/liquid non-coalescing system, proving again the configuration $2 \times \mathrm{RT}d/D = 0.33$ –

s/d = 1 was the least stress-intensive. Figure 10c shows the comparison among dual configurations of varying impeller geometries. At equal $\overline{\varepsilon}$, a dual Rushton turbine induces less particle breakage than combinations of the Rushton turbine with an axial-flow impeller or a dual axial-flow impeller configuration since the impellers operate at lower tip speeds due to higher power numbers. The position of the Rushton turbine in the reactor affects the breakage intensity. This effect is depicted in Fig. 10c, when comparing the floc sizes produced by combinations of RT*d/D* = 0.33 with BiLOOP and combinations of RT*d/D* = 0.33 with PROP*h/d* = 0.33. Once being the bottom impeller, the Rushton turbine induces lower stress than when being the top impeller, despite the smaller power number and thus operation at higher tip speed for the configuration with the Rushton turbine as the bottom impeller. The reason behind this effect might lie in the significant variation of the flow field depending on the relative position of the Rushton turbine in the tank, see Fig. 10d.

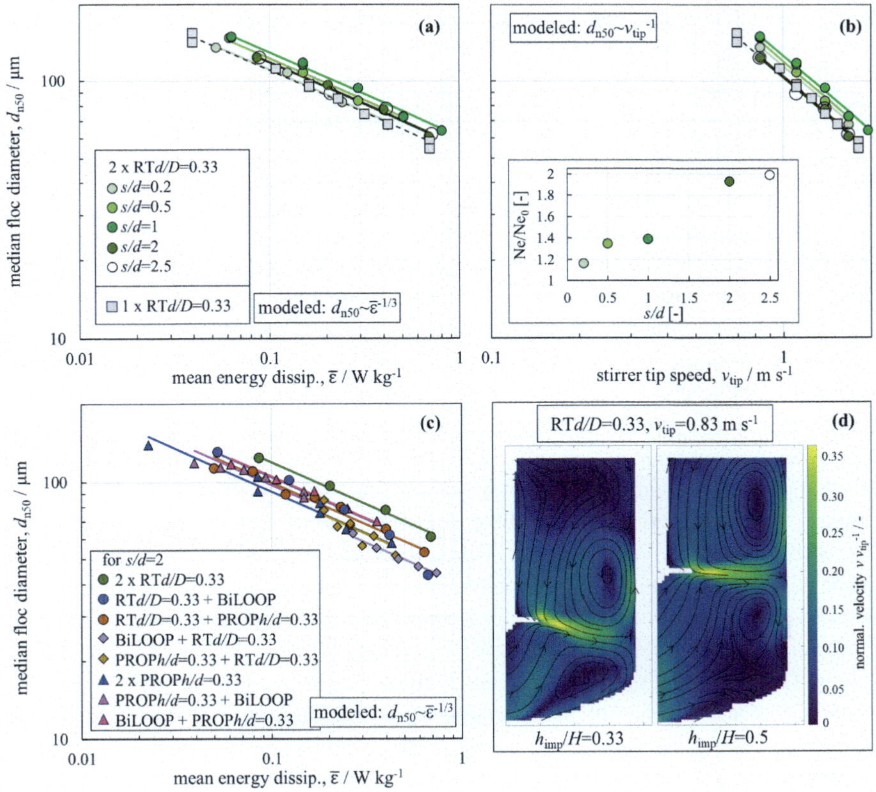

Fig. 10 Steady-state median floc diameter d_{n50} **a** as a function of the mean energy dissipation rate $\overline{\varepsilon}$ and **b** as a function of stirrer tip speed v_{tip}, for single and dual Rushton turbines at various spacings *s/d* and **c** as a function of the mean energy dissipation rate $\overline{\varepsilon}$ for dual configurations of different impeller geometries at constant spacing *s/d* = 2. **d** Experimentally determined secondary flow fields for RT*d/D* = 0.33 at bottom clearance h_{imp}/H = 0.33 and h_{imp}/H = 0.5

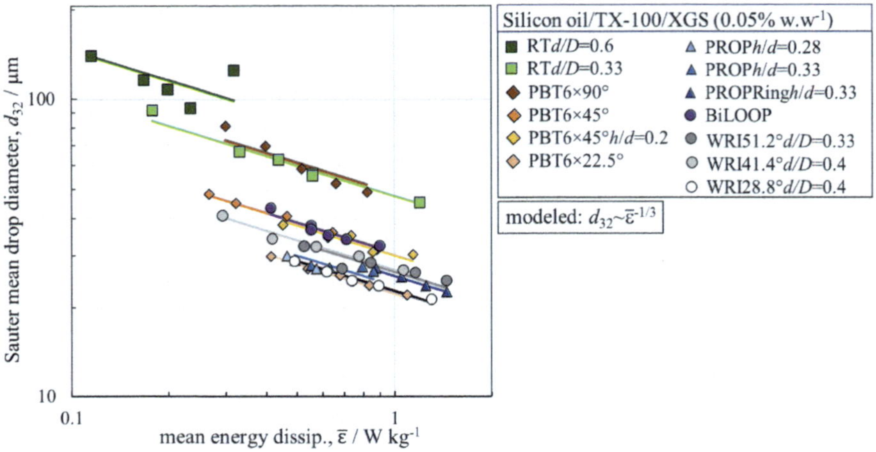

Fig. 11 Steady-state Sauter mean diameter d_{32} as a function of the mean energy dissipation rate $\bar{\varepsilon}$ for the non-Newtonian shear-thinning liquid/liquid model system

4.3 Effect of Rheology and Aeration on Particle Stress

The non-coalescing liquid/liquid system with a 0.05% w w^{-1} Xanthan gum solution as continuous phase was additionally employed to investigate the effect of rheology on the impeller-induced particle stress. Figure 11 shows that the radial-flow stirrers (Rushton turbine and PBT6 × 90°) produce significantly larger particles at equal $\bar{\varepsilon}$. The differences among axial-flow agitators become less important in the non-Newtonian medium. Fragmentation experiments with *A. niger* SKAn 1015 broth diluted in 0.05% w w^{-1} and 0.1% w w^{-1} Xanthan gum solutions agree with the findings of model systems, as reported in Waldherr et al. [15].

For the microorganisms' growth, an oxygen supply is necessary. Aeration typically happens in bioreactors through a sparger. By bubble formation at the sparger, bubbles' breakage/coalescence, bubble rise, and bursting at the liquid–gas interface, additional fluid dynamic stress occurs that is detrimental for bio agglomerates. Figure 12 shows results of particle breakage under aeration at gas flow rates typical for cultivations of filamentous microorganisms and stirring with RT$d/D$ = 0.33 and CombiJet in the case of *A. niger* SKAn 1015 and with RT$d/D$ = 0.33 for the model floc system. The EKATO CombiJet was chosen to be investigated in the study, since it is a commercial stirrer recommended for gas–liquid applications due to its blade shape that prevents the formation of gas pockets and results in no loss of power input in the presence of gas bubbles. Overall, increased aeration flow rate leads to significant particle size reduction. By quantifying the particle size decline, it was concluded that aeration-induced volumetric power input generates higher mechanical stresses than agitation-induced power input. This finding highlights that aeration is a significant parameter influencing fluid dynamic stress. Therefore, during an actual cultivation process of stress-sensitive microorganisms, it would be beneficial to meet the oxygen

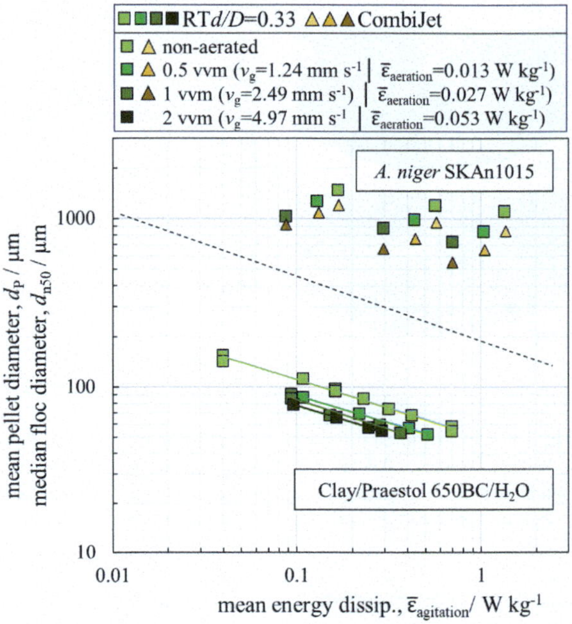

Fig. 12 Steady-state characteristic particle diameter d_i as a function of the mean energy dissipation rate $\bar{\varepsilon}$ for the aerated broth and the floc model system (the dotted line separates visually the data of the two particulate systems)

demands by increasing agitation intensity rather than the aeration rate. The radial flow RT$d/D$ = 0.33 causes less particle damage than the commercial CombiJet despite the intense power loss recorded for the radial-flow impellers in the presence of gas bubbles. Furthermore, it was experimentally determined that the Rushton turbines would avoid the flooding of the reactor in lower power inputs than the axial-flow agitators, proving that even though not designed for aeration processes, this impeller is a very robust agitator geometry.

5 CFD Based-Numerical Analysis

The results shown earlier revealed that the maximum energy dissipation rate ε_{max} defines the particle breakage in the reactor. For new impeller designs, predictions of ε_{max} are complicated; therefore in Fig. 9, the WRI data showed a substantial deviation from the general fitting, since the length scale of the trailing vortex is unknown for this impeller. CFD simulations with a commercial finite-volume solver (STAR-CCM+, Siemens PLM), employing the k–ε RANS turbulent model, were conducted to increase prediction accuracy. The turbulent energy dissipation field was extracted and processed. Typically, the average energy dissipation rate is calculated by the torque on the stirrer and shaft. However, as this work aims to examine the local dissipation rate values, the integration of the local energy dissipation must be conducted. It must be close to the experimental value so that the analysis of local

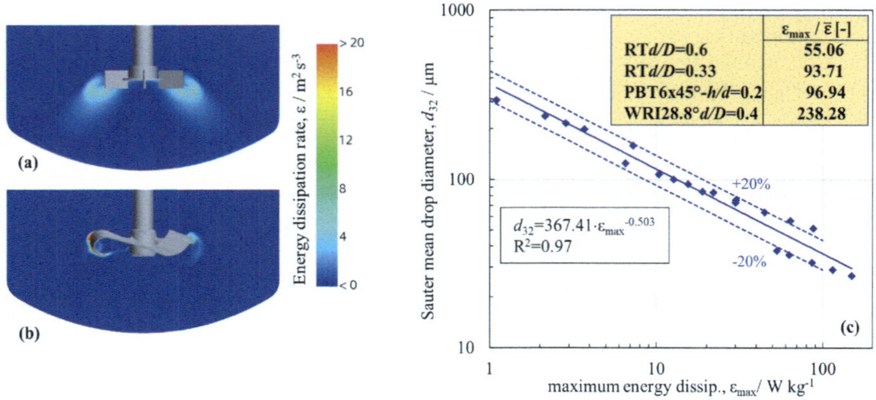

Fig. 13 CFD-simulated field of energy dissipation rate in the reactor with **a** RT*d/D* = 0.33 and **b** WRI28.8°*d/D* = 0.4 ($\overline{\varepsilon} = 0.2$ W kg^{-1}). **c** Steady-state Sauter mean drop diameter d_{32} as a function of CFD-calculated $\varepsilon_{\max}$ for four stirrers for the non-coalescing liquid/liquid model system (—continue line corresponds to power function fitting, ---doted line corresponds to ±20% deviation)

quantities can be considered reliable. Such integration of the local quantities agreed reasonably with the experimental values, allowing a further comparison of local values. Figure 13a and b show the distribution of the energy dissipation rate for the "high-stress" WRI and the "low-stress" RT*d/D* = 0.33 at equal $\overline{\varepsilon}$. Energy dissipation takes place in a significantly larger volume for the Rushton turbine, whereas for the WRI, energy dissipation is limited to an area close to the ribbon boundaries. From the local values of $\overline{\varepsilon}$, the volume-based distribution was calculated and the $\varepsilon_{\max}$ was then determined as the 99.99%-percentile. With the CFD obtained $\varepsilon_{\max}$ values, the drop steady-state diameters were re-plotted for the non-coalescing liquid/liquid model system and four impellers, showing a significantly enhanced correlation, see Fig. 13c, with good fitting achieved for the WRI28.8° − *d/D* = 0.4 as well (compare with Fig. 9). The $\varepsilon_{\max}/\overline{\varepsilon}$ ratios obtained (Fig. 13c) justify the classification of stirrers based on their stress-intensity as described earlier. In conclusion, CFD simulations significantly improved the $\varepsilon_{\max}$ prediction, suggesting that $\varepsilon_{\max}$ is the best correlator for particle stress if a precise estimation for this local value is possible. CFD simulations can be a valuable tool to predict this value for new stirrer designs, decreasing the amount of experimental work necessary.

6 Conclusions

The project "Interaction of mechanical stresses and productivity of biological agglomerates in stirred fermenters" was carried out in the funding period August 2016 to October 2023 at the Chair of Chemical and Process Engineering at the Technische Universität Berlin. The project aims were the characterisation of fluid dynamic stress in aerated stirred bioreactors and the identification of low-stress stirring conditions beneficial for the growth of stress-sensitive microorganisms.

As model microorganism, the filamentous fungus *Aspergillus niger* was employed. For the successful cultivation of *A. niger*, the choice of agitation and aeration conditions was proven to be challenging. The conditions should allow a reasonable suspension of the spores during the initial germination phase and further growth of bio agglomerates while ensuring adequate oxygen supply in a medium that undergoes significant rheological changes over time. Rushton turbines proved to be very robust agitators for the operation of the 3–4 L stirred tank bioreactors of the study. However, the interplay among all process parameters in the bioreactor did not allow a fair evaluation of the hydrodynamic stress acting on bio agglomerates exclusively due to agitation.

For these reasons, two direct methods were employed in this work to investigate the particle stress induced by various impeller geometries, i.e., offline fragmentation experiments with three strains of *A. niger* and breakage batch experiments with four model particle systems. The fragmentation experiments allowed a direct examination of the interaction between impeller and pellets in a controlled environment. The model systems allowed a significantly more extended examination of various impeller types in a more time- and cost-efficient way. The results proved that employing model systems is a valid and efficient method to quantify the breakage intensity of different stirrer types in stirred tank reactors. Each model system has various advantages and drawbacks for the experimental investigations, therefore investing in building a comprehensive toolbox of available methods and systems was a successful strategy for the study.

In single impeller stirred tanks, examining the separate geometrical features of an impeller led to the conclusion that larger impeller diameters, blade surfaces and blade angles result in less particle breakage. Connecting the impeller blades, e.g., through a ring, as in PROPRing and BiLoop designs, reduces the agitator's stress intensity. Overall, when radial-flow, high-power number agitators, such as Rushton turbines, are employed, particles experience less hydrodynamic stress in turbulent flows in Newtonian media and transient flow regimes in non-Newtonian media at the same specific power input. Combining two Rushton turbines induces the smallest particle size decline in dual impeller configurations. The impeller spacing and the position of the radial-flow agitator (bottom vs top) additionally affect the stress acting on particles in dual impeller systems. In aerated stirred tanks, the Rushton turbine delivered the best results regarding reduced particle disintegration. However, the aeration induces significantly stronger stress on particulate matter than agitation. Therefore, it is suggested to meet the oxygen needs in a bioreactor by increasing agitation intensity rather than the aeration rate.

All comparisons were based on the mean energy dissipation rate, i.e., mass-specific power input, as this criterion also considers the energy demands for the mechanical agitation. In total, it was experimentally determined that the Rushton turbine is a very robust agitator geometry. It could ensure low particle breakage. In parallel, it showed a good pumping capacity, fast mixing, and thus good fluid homogenisation (short mixing times) while providing an easy bubble loading (dispersion) in the aerated reactor. The high power loss of this turbine under aeration is indisputably a disadvantage that the effortless combination of this stirrer with axial-flow agitators in dual impeller configurations could compensate.

To interpret the stress intensity of the investigated stirrer types, the maximum energy dissipation rate ε_{max} proved to be the key parameter. Applying literature approaches to approximate this local fluid dynamic property showed a good correlation of the experimental data, i.e., steady-state particle sizes, for all systems (*A. niger* broths and model systems). For the newly designed impellers of the work, e.g., wave ribbon impellers, no literature data were available. Through CFD numerical investigations, more information could be extracted for these agitators and a more precise characterisation could also be achieved for conventional impellers for the specific geometry of the lab-scale baffled tank. The numerically determined $\varepsilon_{max}/\bar{\varepsilon}$ ratios agree with the results of particle breakage, i.e., the higher $\varepsilon_{max}/\bar{\varepsilon}$, the higher the stress intensity of an impeller and the smaller the particles formed during the stirring process. Nevertheless, even for stirrers with somewhat similar $\varepsilon_{max}/\bar{\varepsilon}$ ratios and, thus, almost identical steady-state mean particle sizes, the produced particle size distributions can still differ, as particles experience different fluctuating environments in the significantly varying flow fields generated by the impellers (e.g., see comparison between RT$d/D = 0.33$ and PBT6 $\times$ 45° $- h/d = 0.2$). Finally, the circulation time as an averaged quantity could not be identified as a defining parameter for the particle breakage at this scale of stirred tank reactors of this study.

References

1. Nienow AW (1998) Hydrodynamics of stirred bioreactors. Appl Mech Rev 51(1):3–32. https://doi.org/10.1115/1.3098990
2. Paul EL, Atiemo-Obeng VA, Kresta SM (2003) Handbook of industrial mixing: science and practice. John Wiley & Sons, New York
3. dos Santos MC, Cerri MO, Bicas JL (2022) Relation of shear stress and $K_L\alpha$ on bikaverin production by *Fusarium oxysporum* CCT7620 in a bioreactor. Bioprocess Biosyst Eng 45:733–740. https://doi.org/10.1007/s00449-022-02693-0
4. Roque T, Delettre J, Hardy N, Nienow AW, Augier F, Ben Chaabane F, Béal C (2021) The impact of fluid-dynamic stress in stirred tank bioreactors on the synthesis of cellulases by *Trichoderma reesei* at the intracellular and extracellular levels. Chem Eng Sci 232:116353. https://doi.org/10.1016/j.ces.2020.116353
5. Suhaili N, Tan JS, Mohamed M, Halim M, Ariff AB (2015) Effects of dual impeller system of Rushton turbine, concave disk turbine and their combinations on the performance of kojic acid fermentation by *Aspergillus flavus* Link 44-1. Asia-Pac J Chem Eng 10:65–74. https://doi.org/10.1002/apj
6. Nienow AW, Bujalski W (2004) The versatility of up-pumping hydrofoil agitators. Chem Eng Res Des 82(A9):1073–1081. https://doi.org/10.1205/cerd.82.9.1073.44150
7. Wucherpfennig T, Hestler T, Krull R (2011) Morphology engineering—osmolality and its effect on *Aspergillus niger* morphology and productivity. Microb Cell Fact 10:58. https://doi.org/10.1186/1475-2859-10-58

8. Nienow AW (2021) The Impact of fluid dynamic stress in stirred bioreactors—the scale of the biological entity: a personal view. Chem Ing Tech 93(1–2):1–15. https://doi.org/10.1002/cite.202000176
9. Jüsten P, Paul GC, Nienow AW, Thomas CR (1996) Dependence of mycelial morphology on impeller type and agitation intensity. Biotechnol Bioeng 52:672–684. https://doi.org/10.1002/(SICI)1097-0290(19961220)52:6%3c672::AID-BIT5%3e3.0.CO;2-L
10. Cui YQ, Van Der Lans RGJM, Luyben KCAM (1997) Effect of agitation intensities on fungal morphology of submerged fermentation. Biotechnol Bioeng 55(5):715–7126
11. Hardy N, Augier F, Nienow AW, Béal C, Ben Chaabane F (2017) Scale-up agitation criteria for *Trichoderma reesei* fermentation. Chem Eng Sci 172:158–168. https://doi.org/10.1016/j.ces.2017.06.034
12. Henzler H-J, Biedermann A (1996) Modelluntersuchungen zur Partikelbeanspruchung in Reaktoren (Model studies on particle stress in reactors). Chem Ing Tech 68:1546–1561. https://doi.org/10.1002/cite.330681205
13. Wollny S (2010) Experimentelle und numerische Untersuchungen zur Partikelbeanspruchung in gerührten (Bio-Reaktoren). PhD thesis, Technische Universität Berlin
14. Waldherr P, Bliatsiou C, Böhm L, Kraume M (2023) Fragmentation of *Aspergillus niger* pellets in stirred tank bioreactors due to hydrodynamic stress. Chem Eng Res Des 195:116–131. https://doi.org/10.1016/j.cherd.2023.05.038
15. Waldherr P, Bliatsiou C, Böhm L, Kraume M (2024) Characterizing particulate systems towards the analysis of fluid dynamic stress in aerated stirred tank bioreactors. Chem Ing Tech 96(4):471–485. https://doi.org/10.1002/cite.202300193
16. Bliatsiou C, Malik A, Böhm L, Kraume M (2019) Influence of impeller geometry on hydromechanical stress in stirred liquid/liquid dispersions. Ind Eng Chem Res 58(7):2537–2550. https://doi.org/10.1021/acs.iecr.8b03654
17. Panckow RP, Bliatsiou C, Nolte L, Böhm L, Maaß S, Kraume M (2023) Characterisation of particle stress in turbulent impeller flows utilising photo-optical measurements of a flocculation system—part 1. Chem Eng Sci 267:118333. https://doi.org/10.1016/j.ces.2022.118333
18. Bliatsiou C, Panckow RP, Nolte L, Böhm L, Maaß S, Kraume M (2024) Characterisation of particle stress in turbulent impeller flows utilising photo-optical measurements of a flocculation system—part 2. Chem Eng Sci 291:119853. https://doi.org/10.1016/j.ces.2024.119853
19. Bliatsiou C, Schrinner K, Waldherr P, Tesche S, Böhm L, Kraume M, Krull R (2020) Rheological characteristics of filamentous cultivation broths and suitable model fluids. Biochem Eng J 163:107746. https://doi.org/10.1016/j.bej.2020.107746
20. Böhm L, Hohl L, Bliatsiou C, Kraume M (2019) Multiphase stirred tank bioreactors—new geometrical concepts and scale-up approaches. Chem Ing Tech 91(12):1724–1746. https://doi.org/10.1002/cite.201900165
21. Barthel L (2023) Understanding the chitin synthases repertoire of the filamentous fungus *Aspergillus niger*. PhD thesis, Technische Universität Berlin
22. Grenville RK, Giacomelli JJ, Padron G, Brown DAR (2017) Mixing: impeller performance in stirred tanks. Chem Eng 124:42–51
23. Bisgaard J, Muldbak M, Tajsoleiman T, Rydal T, Rasmussen T, Huusom JK, Gernaey KV (2021) Characterization of mixing performance in bioreactors using flow-following sensor devices. Chem Eng Res Des 174:471–485. https://doi.org/10.1016/j.cherd.2021.08.008
24. Shinnar R, Church JM (1960) Statistical theories of turbulence in predicting particle size in agitated dispersions. Ind Eng Chem 52(3):253–256. https://doi.org/10.1021/ie50603a036

Modeling, Simulation and Experimental Characterization of Mechanical and Surface-Induced Stresses on Pellets of the Filamentous Microorganism *Lentzea aerocolonigenes*

Anna Dinius, Marcel Schrader, Kathrin Schrinner, Ingo Kampen, Rainer Krull, and Arno Kwade

Abbreviations

DSMZ	German collection of microorganisms and cell cultures
HPLC	High pressure liquid chromatography
CDW	Cell dry weight
AESD	Area-equivalent-spherical-diameter
CLSM	Confocal laser scanning microscopy
MPEC	Microparticle-enhanced cultivation
AYD	Averaged yield coefficient

A. Dinius · K. Schrinner · R. Krull (✉)
Institute of Biochemical Engineering, Technische Universität Braunschweig, Braunschweig, Germany
e-mail: r.krull@tu-braunschweig.de

A. Dinius
e-mail: a.dinius@tu-braunschweig.de

M. Schrader · I. Kampen · A. Kwade
Institute for Particle Technology, Technische Universität Braunschweig, Braunschweig, Germany
e-mail: marcel.schrader@tu-braunschweig.de

I. Kampen
e-mail: i.kampen@tu-braunschweig.de

A. Kwade
e-mail: a.kwade@tu-braunschweig.de

A. Dinius · M. Schrader · K. Schrinner · I. Kampen · R. Krull · A. Kwade
Center of Pharmaceutical Engineering, Technische Universität Braunschweig, Braunschweig, Germany

A. Kwade and I. Kampen (eds.), *Dispersity, Structure and Phase Changes of Proteins and Bio Agglomerates in Biotechnological Processes*,
https://doi.org/10.1007/978-3-031-63164-1_17

STR	Stirred tank reactor
EDCF	Energy dissipation circulation function
PTV	Particle tracking velocimetry
SAR	Stress area ratio

1 Introduction

Filamentous fungi and bacteria play a crucial role as cell factories in various biotechnological industries, including the production of organic acids and antibiotics. In fact, many of today's known antibiotics are produced by filamentous bacteria, such as actinomycetes [2–4]. Unlike other common prokaryotic producers, filamentous microorganisms in submerged cultures exhibit complex cellular morphologies that range between the two extremes of dispersed mycelial structures and dense spherical pellets (Fig. 1).

However, achieving high product yields in the cultivation of natural antibiotic producers remains a challenge. One key factor is that the optimal morphology for high productivity strictly depends on the specific biological system, posing a

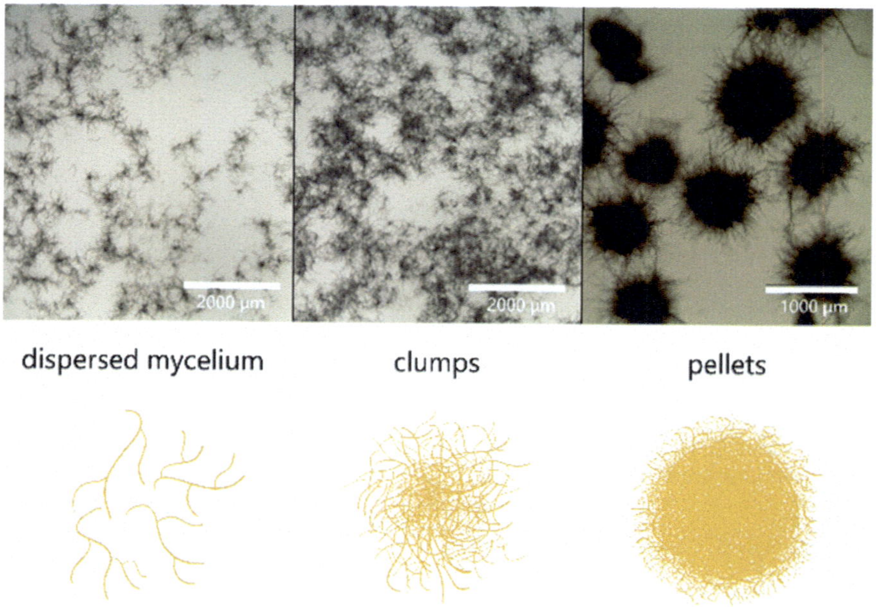

Fig. 1 The three main macro-morphological structures: microscopic images (upper row) and schematic illustrations (bottom row). Figure taken from [5], licensed under the Creative Commons Attribution 4.0 International license

recurring challenge itself [6], and the product formation (e.g. primary or secondary metabolite). Over time, several strategies for *morphology engineering* have been successfully developed to access control over the cellular morphology of filamentous microorganisms, besides changing basic process settings [5, 7].

Lentzea aerocolonigenes, a filamentous actinomycete, is a natural producer of rebeccamycin, a drug associated with antibiotic and antitumoral properties. Nevertheless, in order for rebeccamycin to be medically relevant, rather high quantities are required [8–11]. As *L. aerocolonigenes* does not naturally produce rebeccamycin in concentrations suitable for pharmaceutical applications, the supplementation of macroparticles (>200 μm), glass microparticles (<50 μm) and the emulsifier soy lecithin were investigated in regard to their impact on productivity and cellular morphology.

Glass bead supplementation (macroparticle-enhanced cultivation) has been studied as a technique for significant alterations in the cultivation of filamentous microorganisms [12–21]. Detailed information regarding these studies are elaborated in [5, 7, 22]. Overall, glass beads were proven to evoke changes in cellular morphology and affect the product synthesis in the investigated cultivation systems, particularly influencing the pellet size. Thus, in the course of this project a more systematic and sophisticated approach was conducted to experimentally and simulatively investigate different parameters for macroparticle addition, such a diameter, concentration, and density as well as the shaking frequency, for a better understanding of the particle-induced stress on *L. aerocolonigenes.* In addition, simulations of pellet compression between two parallel plates were performed for the first time in this project to investigate the micromechanical behavior as a function of morphological growth parameters as well as simulation parameters.

A further important *morphology engineering* strategy is the supplementation of microparticles (e.g., talc powder, aluminum oxide or titanate), referred to as microparticle-enhanced cultivation (MPEC). This approach has been widely discussed in the literature [5, 7, 23–25] and primely impacts the cellular micromorphology of filamentous microorganisms. In spore agglomerating systems, MPEC can disturb the early spore agglomeration process, resulting in smaller pellets. Also, microparticles can become directly incorporated into the hyphal pellet structure. These entanglements are postulated to loosen the hyphal network, thereby enhancing the supply of nutrients and oxygen (e.g., as the cell growth-limiting substrate S) towards the center of the additionally shrunken pellets. Consequently, the critical pellet radius r_{crit} for nutrient supply is shifted toward the pellet center, and the share of the active pellet Δr_{active} is relatively increased, potentially leading to enhanced product synthesis (Fig. 2) [5].

Furthermore, as discussed in Laible et al. [25], multilayered physico-chemical effects between the hyphae and particle surfaces cannot be excluded. Therefore, the particles should be well characterized for application [25]. Thus, within this work glass microparticles are introduced for MPEC in *L. aerocolonigenes* as they are associated to be a well characterizable and modifiable particle system for tailor-made surface modifications.

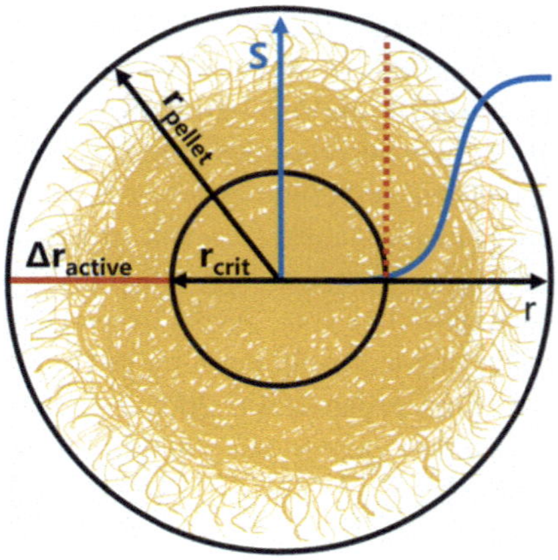

Fig. 2 Macroscopic pellet growth and substrate limitation. Figure taken from [5], licensed under the Creative Commons Attribution 4.0 International license

The supplementation of the emulsifier lecithin has also emerged as a promising supplementation strategy for enhancing the cultivation outcomes of filamentous microorganisms. Previous studies have demonstrated the beneficial effects of lecithin supplementation in various actinomycetes isolated from soil (e.g., [26–28]), further discussions can be found in [5]. Consequently, soy lecithin was investigated as a supplement for rebeccamycin production with *L. aerocolonigenes*. Additionally, for the first time, combinatorial approaches involving lecithin and other *morphology engineering* strategies were investigated to assess the potential synergistic mechanisms that reinforce each other.

2 Material and Methods

2.1 *Cultivation of* L. aerocolonigenes

Lentzea aerocolonigenes DSM 44217 was purchased from the German Collection of Microorganisms and Cell Cultures (DSMZ). Cultivation was conducted in 250 mL shaking flasks with four defined baffles containing of 50 mL GYM medium (4 g L^{-1} glucose, 4 g L^{-1} yeast extract, 10 g L^{-1} malt extract, pH 7.2). Pre-culture preparation was conducted by inoculation with 1 mL of thawed cryo-culture (in 30% glycerol (v/v)). Incubation took place in an orbital shaker (Certomat BS-1, Sartorius, Germany) with a shaking frequency of 120 min^{-1} (if not stated otherwise) and an amplitude of 50 mm at 28 °C for 2 days in darkness. For main culture inoculation pre-culture (300 μL) was transferred to new flasks and incubated for 10 days under the same conditions in triplicates for all experiments.

For rebeccamycin quantification culture broth was used for extraction with ethyl acetate in an overhead shaker (Intelli Mixer RM-2 M, LTF Labortechnik, Wasserburg, Germany). After centrifugation (4000 min^{-1}, 10 min; Heraeus Varifuge 3.0R,

Thermo Fisher Scientific, Waltham, United States), the organic phase was used for HPLC analysis as described in [29]. Furthermore, HPLC was used for substrate quantification during cultivation and was performed as described in [29]. The cell dry weight concentration (CDW) was quantified gravimetrically by filtration through dried and pre-weighted filters [29]. Macro-morphological investigations were performed accordingly to [29, 30] with a digital inverse microscope (EVOS XL, AMG, Bothell, WA, USA) and image analysis followed either by using Matlab (Version 2019a, MathWorks) [29] or ImageJ (1.52h) (National Institute of Health, USA) [30]. In order to provide comparability, the area-equivalent-spherical diameter (AESD) was determined for pellet sizes (Eq. (1)):

$$AESD = \sqrt{\frac{4 \cdot area}{\pi}} \tag{1}$$

2.2 Supplements

Glass beads were obtained from Sigmund Lindner GmbH (Warmensteinach, Germany) and included eight size types (type S and micro glass beads; soda lime glass) and one size of ceramic beads (type Z; zirconium silicate), which have a 1.5 times higher density than glass beads (see Table 1). Particles were added in varying sizes and concentrations to *L. aerocolonigenes* cultivations [22, 29, 30]. Particle size distributions and the related mean particle sizes ($\overline{x}_{3,2}$) were determined by laser diffraction (Mastersizer 3000, Malvern Panalytical Ltd, Worcestershire, United Kingdom) and are given in Table 1. For simplicity, the beads are referred to by the mean particle diameter in the following. Macroparticles were either supplemented in certain weight concentrations, indicated with the corresponding results, or at a defined bead number, calculated from bead density and size.

For experiments with microparticles soda-lime glass microparticles ($x_{50} = 7.9$ µm and $x_{50} = 30.5$ µm) were purchased from Sigmund Lindner GmbH (Warmensteinach, Germany). Varying microparticle amounts were added at certain weight concentrations and prepared as described in [1]. In some approaches soy lecithin (MP Biomedicals, Eschwege, Germany or Cargill Texturizing Solutions Deutschland GmbH and Co. KG, Hamburg, Germany) was supplemented additionally. Cultivations were prepared as described in [1, 22].

2.3 Viability Staining, Pellet Slicing and CLSM Imaging

The viability of pellets was investigated accordingly to [1, 29]. Shortly, pellets were stained with SYTO9 and propidium iodide (PI) for living areas (green) and dead areas (red), respectively. Subsequently, stained pellets were frozen in liquid cryosectioning

Table 1 Range of glass bead sizes, according to the manufacturer and the corresponding mean glass bead size $\bar{x}_{3,2}$

Bead size range (μm)	Mean bead diameter ($\bar{x}_{3,2}$) (μm)
Glass beads (ρ = 2500 kg m^{-3})	
200–300	282
400–600	540
500–750	658
750–1000	969
1000–1300	1183
1250–1650	1513
1550–1850	1746
1700–2100	1932
Ceramic beads (ρ = 3800 kg m^{-3})	
800–1000	918

medium. After slicing into cross sections of 100 μm thickness, an equatorial region section was monitored via CLSM (C2si, Nikon Instruments, Amsterdam, Netherlands). Live ratio was calculated as depicted in Eq. (2). For comparison with CLSM analytic quality, flow cytometry measurements were conducted [29].

$$Live\ ratio = \frac{Living\ area}{Total\ area} \tag{2}$$

2.4 CFD-DEM-Simulation Environments for a Shaking Flask

Multiphase simulations were carried out to numerically characterize the stresses caused by macroparticles during cultivations. For the shake flask, the computational fluid mechanics (CFD) was unresolved four-way coupled with the discrete element method (DEM). In detail, mass and momentum balance equations are applied in CFD simulations to describe the fluid flow. In addition, DEM was used to calculate the translational and angular velocity and position of discrete particles with the help of Newton's second law of motion (Eqs. (3) and (4)). In each time step, the position $\boldsymbol{x_b}$ of a particle i with the mass m_b is calculated based on the sum of all forces acting on it.

$$m_{b,i} \cdot \frac{d^2 \boldsymbol{x}_{b,i}}{dt^2} = \sum_{i=1}^{n} \boldsymbol{F}_{ij}^{c} + \boldsymbol{F}_{i}^{pf} + \boldsymbol{F}_{i}^{g} + \left(\boldsymbol{F}_{i}^{a}\right) \tag{3}$$

$$I_{b,i} \cdot \frac{d\boldsymbol{\omega}_{b,i}}{dt} = \sum_{i=1}^{n} \boldsymbol{M}_{ij}^{t} + \sum_{i=1}^{n} \boldsymbol{M}_{ij}^{r} \tag{4}$$

The Hertz-Mindlin model was used to calculate the contact forces $\boldsymbol{F}_{ij}^{c}$. Furthermore, $\boldsymbol{M}^{t}$ stands for the tangential moment due to contact forces and $\boldsymbol{M}^{r}$ for the rolling friction moment, which was calculated with the constant directional rolling friction model.

For the particle–fluid interaction force $\boldsymbol{F}_{i}^{pf}$, the sum of drag, pressure gradient, viscous, surface tension and lubrication forces were taken into account. In addition to the gravity force $\boldsymbol{F}_{i}^{g}$, a centrifugal force $\boldsymbol{F}_{i}^{a}$ was applied to the particles and the fluid to model the effect of the orbital shaker motion. The volume of fluid (VOF) approach was chosen to represent the air and liquid flow in the shake flask. For more details on all models and parameters used in the shaking flask simulations, the reader is referred to Schrader et al. [30, 31]. The addition of the macroparticles creates additional stress events between the beads or between the beads and the walls shake flask. To analyze the collisions, the concept of stress energy SE (Eq. (5)) and stress frequency SF, originally introduced for stirred media mills [32], was used.

$$SE \propto SE_b = \frac{1}{2} \cdot d_b^3 \cdot v_{b,rel}^2 \cdot \rho_b \tag{5}$$

On the one hand, SE is a measure of the maximum energy that can be transferred to product particles during a collision. The SE corresponds to the kinetic collision energy, which is dependent on the bead size d_b, bead density ρ_b and the relative speed v_b of the collision partners. For a detailed consideration of the type of contact, the SE definitions of Beinert et al. [33] for normal, shear, torsional and rolling contacts were used. On the other hand, the SF describes the number of contacts per second.

3 Macroparticle-Enhanced Cultivation

3.1 CFD-DEM Simulations of the Shake Flask for Stress Analysis

Figure 3a shows the distribution of the total SE together with proportions of the six contact types for a bead size of 0.969 mm at a shaking frequency of 120 min^{-1}. It can be seen that the highest SEs occur as a result of translational shear. In order to localize the maximum stresses, Fig. 3b shows the SE distributions for translational shear between the beads and between the beads and the wall of the shaking flask. From this it can be concluded that the highest shear stresses are found between the beads and the wall of the flask.

For the evaluation of the simulations, it was assumed that only the top ten percent of all contacts can cause a relevantly high stress. The 90% value of the SE was therefore used as a reference value to compare simulations with different parameters. For a more accurate consideration of cell and pellet damage, the knowledge of a critical SE would be necessary. As an additional step, the relative velocity used to calculate

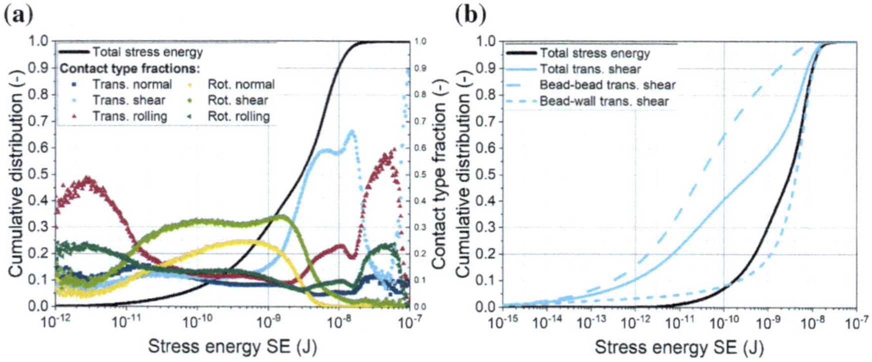

Fig. 3 **a** Sum distribution of the stress energy including the fractions of the six contact types for 0.969 mm beads at a shaking frequency of 120 min^{-1}; **b** sum distributions of the total stress energy as well as the distribution of the stress energy for the translational shear (bead-sphere and bead-wall). Figures taken from [31], licensed under the Creative Commons Attribution 4.0 International license

the translational shear SE, hereafter referred to as shear velocity, was used for the comparison. This measurement provides a direct analysis of shear because, unlike SE, it is not influenced by bead size and density.

3.2 Effect of Varying Shake Flask Cultivation Conditions

3.2.1 Variation of the Bead Concentration

To assess the impact of different bead numbers at a constant bead size, the concentration of 969 μm glass beads was varied between 25 g L^{-1} (10 mL L^{-1}) and 150 g L^{-1} (60 mL L^{-1}) in shaking flask experiments (Fig. 4). The rebeccamycin concentration and averaged yield coefficient (AYD) increased until a glass bead concentration of 100 g L^{-1} (40 m L^{-1}), reaching a concentration of 121 mg L^{-1} after 10 days of cultivation. However, a further increase in glass bead concentration up to 150 g L^{-1} (60 mL L^{-1}) resulted in a subsequent decrease in product titer. In comparison, the unsupplemented control culture only achieved a titer of 45 mg L^{-1}. Dobson and O'Shea [12] also observed a positive effect of increasing number of beads on the formation of the antibiotic. Here, the number of pellets increased while the pellet diameter decreased.

As the concentration of beads increases, the total space required at the bottom of the flask increases. The product of the number of beads and the projection area of a bead has been used in Fig. 5b as a measure of the space required by all the beads. Initially, the bead-wall stress frequency increases strongly up to a projection area of about 30 cm^2 and then tends towards a limit. The limit is due to the fact that once the planar base area of the flask of about 25 cm^2 is exceeded, only a few new contacts can

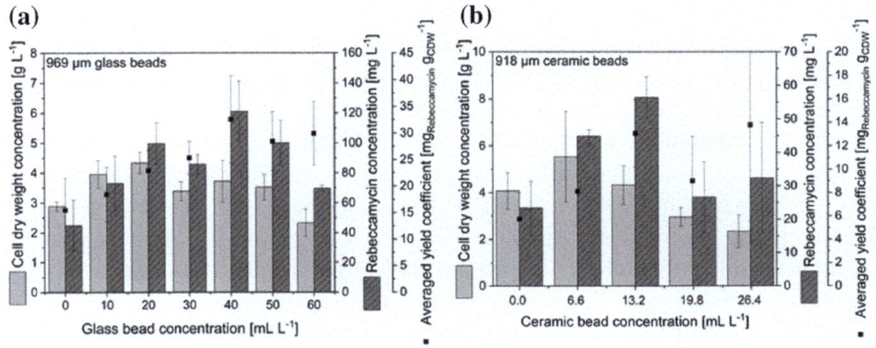

Fig. 4 CDW concentration, rebeccamycin concentration, and averaged yield coefficient of a 10 day *Lentzea aerocolonigenes* cultivation **a** with different concentrations of 969 μm glass beads; **b** parameters of **a** with addition of different concentrations of 918 μm ceramic beads; figures taken and modified from [22], licensed under the Creative Commons Attribution 4.0 International license

be formed on the side wall of the shaking flask. Vertical layers of beads are formed, which can also cause additional shear stress on the microorganisms. However, the shear velocity between the glass beads and the flask wall was consistently 2.5 times higher than between the beads themselves. The bead-wall shear velocity (Fig. 5a) increases slightly up to a glass bead concentration of around 50 mL L^{-1} and then decreases.

While the used ceramic beads have a similar size (0.918 mm) compared the glass beads with 0.969 mm, they possess an approximately 1.5 times higher density. Comparable to the experiment of varying the amount of glass beads, the concentration of ceramic beads was changed in the range of 25–100 g L^{-1} (6.6–26.4 mL L^{-1}). By comparing the supplemented mass concentrations of the ceramic beads to those

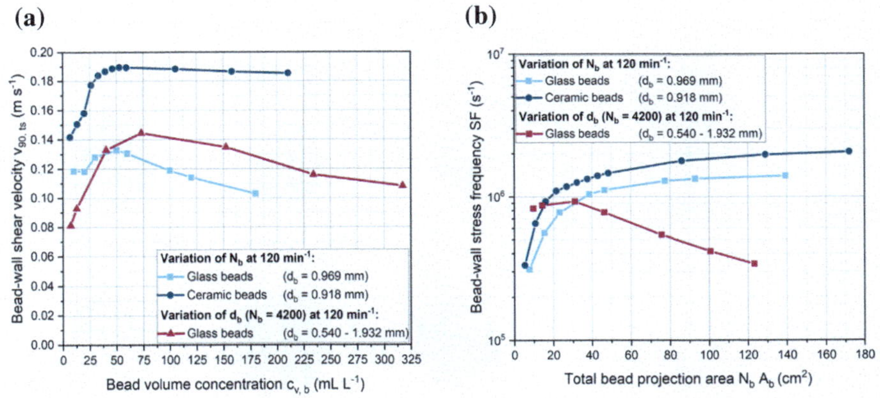

Fig. 5 **a** Bead-wall shear velocity for different volume concentrations of beads, **b** stress frequency as a function of total bead projection area. Figures taken from [31], licensed under the Creative Commons Attribution 4.0 International license

of glass beads, the optimal mass concentration for rebeccamycin production with ceramic beads was determined to be only half the value 50 g L^{-1} (13.2 mL L^{-1}). Furthermore, the maximum product concentration of about 56 mg L^{-1} is only approximately half as high as that achieved with the glass beads.

Similar to simulations with the glass beads, the ceramic beads simulations (Fig. 5b) showed a strong initial increase in SF between the beads and the wall, followed by an approach to a limit. However, the SF of the ceramic beads increases slightly faster than that of the glass beads. With increasing bead concentration, the shear velocity increases significantly more for the ceramic beads than for the glass beads and then remains almost constant. Furthermore, a comparison of simulations with similar volume concentrations (Fig. 6) shows that the ceramic beads distribute less on the bottom of the shake flask and move faster overall. According to the simulations, ceramic beads caused higher shear stress than glass beads. In addition, the higher bead mass results in higher centrifugal and gravitational forces, both leading to higher normal forces. The combination of higher shear rates and higher normal forces could be a reason for the observed cultivation results.

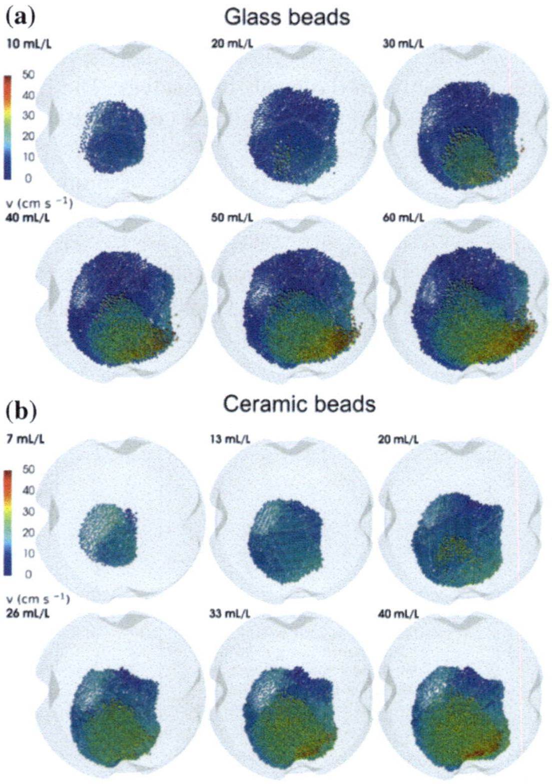

Fig. 6 Variation of volume concentration of **a** glass (0.969 mm) and **b** ceramic (0.918 mm) beads at 120 min^{-1}. Figures taken from [31], licensed under the Creative Commons Attribution 4.0 International license

3.2.2 Variation of the Shaking Frequency and Bead Size

Further experiments were conducted to investigate optimized rebeccamycin production by using different shaking frequencies (100, 120, and 160 min^{-1}) during cultivations. Moreover, different sizes of glass beads were chosen at a concentration of 100 g L^{-1} (40 mL L^{-1}) in the culture broth (Fig. 7).

Theoretically, a shaking frequency of 100 min^{-1} would reduce the bead shear velocity. In comparison to a standard cultivation at 120 min^{-1}, a shift in the maximum rebeccamycin titer was observed from 969 μm (optimal bead size at 120 min^{-1}) to 1183 μm at 100 min^{-1}. In turn, with a shaking frequency of 160 min^{-1}, shear velocity and stress frequency were increased compared to the 120 min^{-1} approach (Fig. 8a), and a shift towards smaller glass bead sizes was expected for the maximum product titer. However, due to high standard deviations at lower bead sizes, this hypothesis could not be experimentally confirmed, and a maximum rebeccamycin titer of 140 mg L^{-1} was achieved with 969 μm glass beads. A direct comparison of the strongly differing absolute values is not necessarily reasonable, as they result from different biological approaches. When the coarsest glass beads were used, relatively low CDW and rebeccamycin concentrations were obtained due to resulting high shear velocities, which are assumed to cause cell destruction or growth inhibition. As the CDW concentration for small and medium-sized glass beads remained similar for varying shear velocities and stress frequencies, it can be theorized that the effects of the glass beads are not solely related to biomass. It is more likely that other factors are relevant, such as micro-morphological changes or alterations of the inner pellet structure. These should be considered and further investigated in future studies.

In addition to the cultivations, simulations for different bead sizes were carried out for a range of shaking frequencies from 100 to 200 min^{-1}. The SF between the beads and the wall increases linearly at constant bead volume concentration for smaller bead diameters and higher shaking frequencies. As the bead diameter increases and

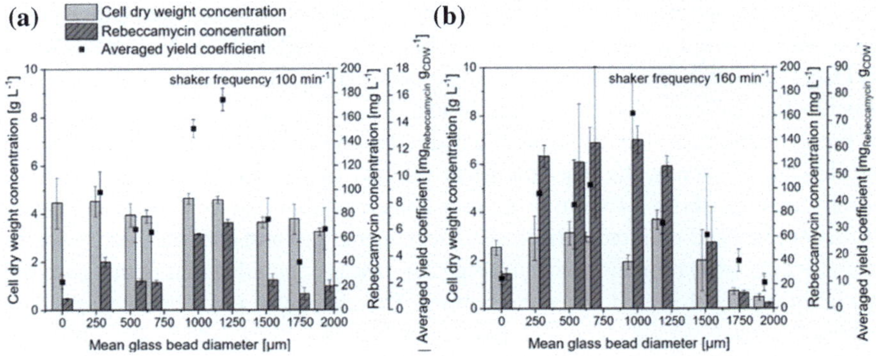

Fig. 7 Cell dry weight concentration, rebeccamycin concentration and averaged yield coefficient of a 10 day cultivation with addition of different glass bead diameters with a concentration of 100 g L^{-1} at different shaking frequencies of **a** 100 min^{-1} and **b** 160 min^{-1}. Figures taken from [22], licensed under the Creative Commons Attribution 4.0 International license

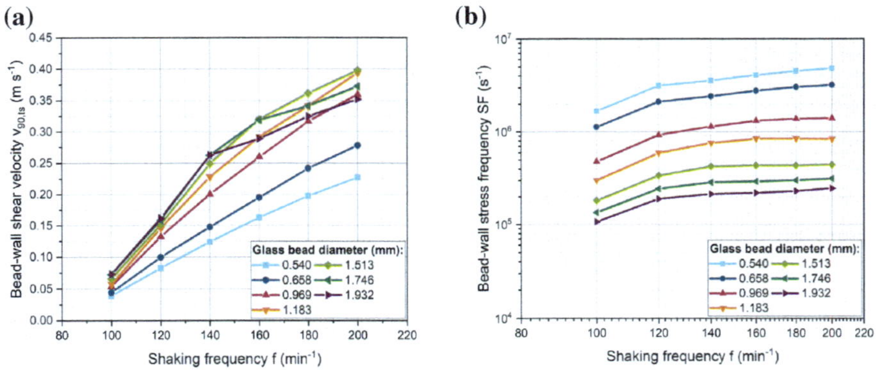

Fig. 8 **a** Bead-wall shear velocity and **b** bead-wall stress frequency as a function of the shaking frequency for different glass bead diameters. Figures taken from [31], licensed under the Creative Commons Attribution 4.0 International license

the number of beads decreases, the influence on the SF decreases. In addition, there is a noticeable increase in SF between 100 and 120 min^{-1} as the dynamics of the bead movement becomes more pronounced.

The shear velocity of the beads in contact with the wall of the shaking flask increased with the shaking frequency to different degrees depending on the bead size. The initially stronger increase in shear rate can probably be explained by the influence of centrifugal forces. The centrifugal force increases quadratically with the shaker frequency, so that the beads are pushed further out towards the edge of the shaking flask as the frequency increases. Due to the larger circular orbit, the beads rotating further out on the bottom of the shaking flask have a higher orbital velocity and thus a higher shear velocity. Above a certain frequency, most of the beads are at the edge of the bottom, so that the influence of the previously described effect decreases. Furthermore, larger beads experience higher centrifugal forces, which can lead to an intensification of the previously described effect. Interestingly, the shear velocities did not increase as much in the range of high shaking frequencies and large bead diameters, so that, for example, the shear velocity for 1.932 mm beads is below the velocity of the 1.513 mm beads.

Figure 9 illustrates the data from Fig. 8 as a function of bead diameter. The higher the shaking frequency, the more the shear velocity of the beads increases (Fig. 9a). For frequencies above 140 min^{-1}, there is a decrease in shear velocity in the range of the largest bead diameters. Instead of decreasing, the translational bead velocity (x and y direction) remains constant (Fig. 9b). It was concluded that in this case the beads roll rather than slide over the bottom. The change of the rotational speed of a sphere depends on the rotational moment of inertia as well as the rolling friction. It follows from this fact that larger beads change their rotational speed more slowly as a result of moments acting on them. Furthermore, moments act on a sphere as a result of frictional forces, which in turn depend on normal forces and the coefficient of friction. Moreover, in layers of beads, a blockade of spherical rotations can occur.

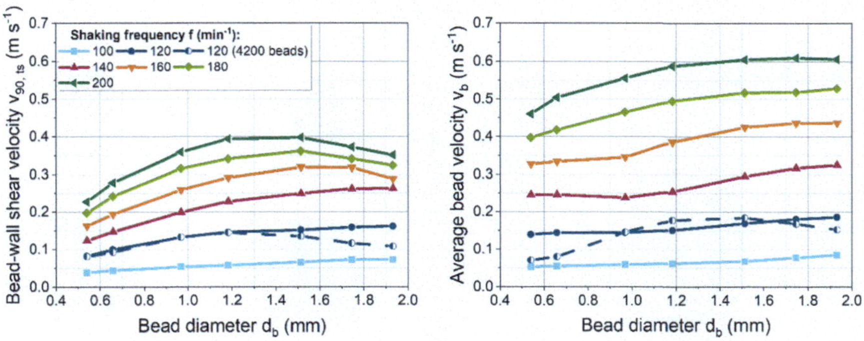

Fig. 9 **a** Bead-wall shear velocity and **b** bead-wall stress frequency as a function of the bead diameter for different shaking frequencies. Figures taken from [31], licensed under the Creative Commons Attribution 4.0 International license

Thus, the translational shear velocities actually occurring are very difficult to predict without simulations due to the complex influencing factors.

3.2.3 Variation of the Bead Diameter at Constant Shaking Frequency and Bead Number

In the previous bead size experiment both, bead size and bead number, were varied. To further investigate the optimal bead size while keeping SF constant, a different approach was employed. The number of 4200 beads (corresponding to the amount of glass beads at 100 g L^{-1} (40 mL L^{-1}), 969 μm) was kept constant for all bead sizes, which resulted in a simultaneous variation of bead size and concentration (Fig. 10a). The CDW concentrations of a 10 day cultivation fluctuated around 3.5 g L^{-1}, not showing a distinct trend. However, the rebeccamycin concentration and AYC showed a clear increase as the glass bead size increased, peaking at 969 μm glass beads. This resulted in a rebeccamycin titer of almost 100 mg L^{-1}. Beyond this optimum, a sharp decline can be observed.

Simulations with 4200 beads were performed for a better understanding of the cultivation results. The data in Fig. 9a shows that the translational shear velocity along the wall reaches a maximum shear velocity at a bead size of 1.183 mm. This maximum coincides with the maximum of the product titer of the associated cultivation. Furthermore, the bead-wall stress frequency decreases significantly above a total bead projection area of about 30 cm^2, which corresponds to a bead size of 0.969 mm. This is because larger beads have a larger projection area, so fewer beads can be in contact with the wall of the shake flask (Fig. 10b). Instead, vertical layers of beads are formed so that the shear stress between the beads increases. Furthermore, the beads extend into the curved area of the shake flask so that some of the beads interact with the baffles.

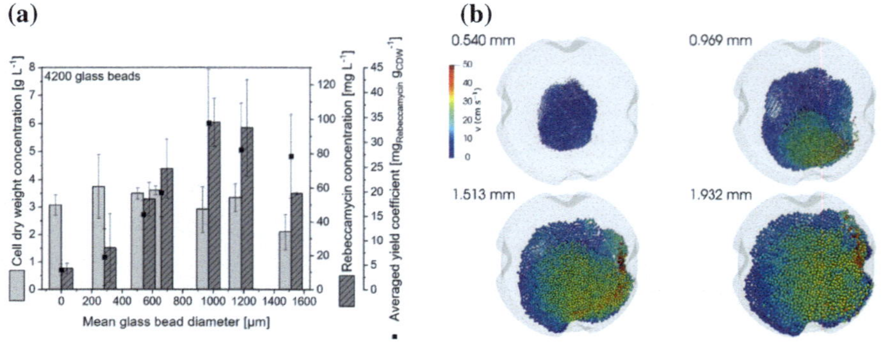

Fig. 10 **a** CDW concentration, rebeccamycin concentration, and averaged yield coefficient of a 10 day *Lentzea aerocolonigenes* cultivation with addition of different glass bead sizes with the same glass bead number of 4200; figure taken and modified from [13], licensed under the Creative Commons Attribution 4.0 International license; **b** variation of the bead diameter for a constant bead number of 4200 at a shaking frequency of 120 min^{-1}. Figures taken from [31], licensed under the Creative Commons Attribution 4.0 International license

3.3 Comparison of Growth Kinetics

Thereafter, more detailed investigations were conducted regarding the comparison between a standard cultivation without glass bead supplementation and a cultivation with the above-mentioned glass bead supplementation conditions (969 µm, 100 g L^{-1}), as it is shown in Fig. 11. For both approaches, severe biomass growth can be observed until day 2 (Fig. 11b). After reaching a peak at a similar CDW concentration, the addition of glass beads causes a slightly faster decrease in biomass content over time. Additionally, glucose consumption does not significantly differ in the course of the cultivation, as in both experiments limitations are reached after two days. This indicates that the presence of glass beads does not significantly affect the growth kinetics. From day 4 onwards, the concentration of rebeccamycin increases more rapidly in the cultivation with glass beads compared to the control cultivation without glass beads. Thus, a final rebeccamycin concentration of 95 mg L^{-1} after 10 day is obtained for the supplemented cultivation, whereas the control culture reaches a concentration of 38 mg L^{-1}. Consequently, the specific rebeccamycin productivity q_p is significantly increased from the beginning of the rebeccamycin production phase (day 2) in the supplemented cultivation, reaching value four times higher on day 5.

Furthermore, cellular macro-morphology was examined through microscopic image analysis, examining approximately 54,800 pellets from the control cultivation and 103,800 pellets from the glass bead experiment over the 10 day cultivation period. Overall, the AESD of the pellets originating from cultivations with and without glass beads was 106 and 172 µm, respectively.

Additionally, the calculated number and volume–density-distributions (q_0, q_3) showed that the glass bead cultivation resulted in slightly smaller (and more circular)

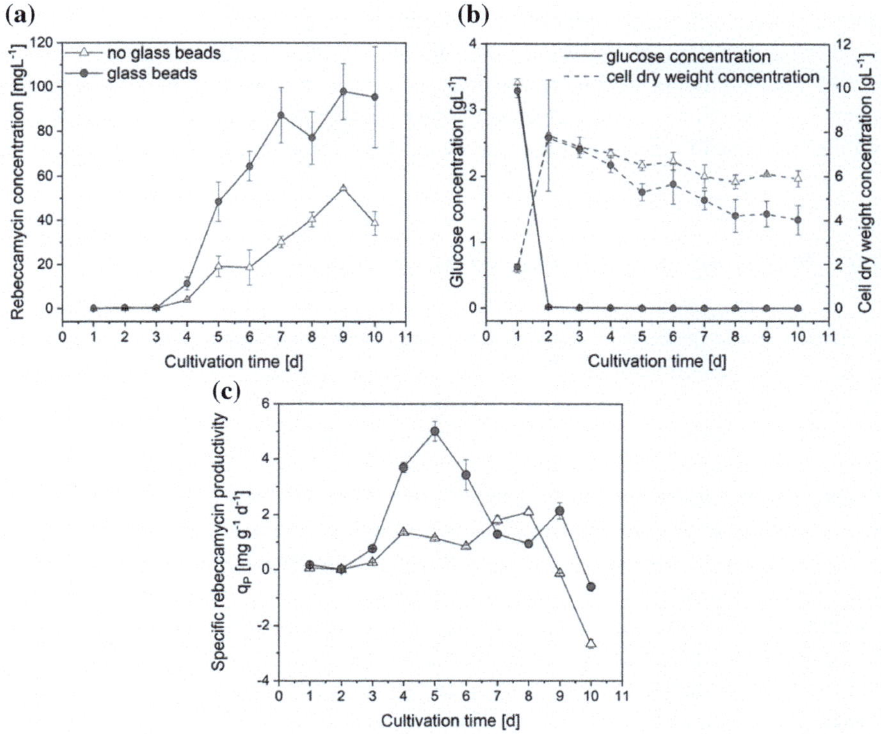

Fig. 11 **a** Rebeccamycin, **b** glucose and cell dry weight concentrations and **c** specific rebeccamycin productivity q_P of a 10 day cultivation of *L. aerocolonigenes* with (circle) and without (triangle) glass bead addition ($x_{50} = 969$ µm, 100 g L^{-1} (40 mL L^{-1})). Presented data are mean values with their respective standard deviations. Figures taken from [29], licensed under the Creative Commons Attribution 4.0 International license

pellets compared to the control cultivation. However, the findings of all glass bead experiments together do not provide conclusive evidence regarding the effect of glass beads on pellet size or circularity in relation to product formation. Therefore, it is theorized that the addition of glass beads may have a greater influence on micro-morphology rather than on macro-morphology, but this theory still requires confirmation.

Cellular viability of pellets is a crucial factor for achieving efficient productivity in microorganisms. Therefore, investigations were conducted in order to evaluate the impact of glass bead supplementation on the share of metabolically active zones. This was accomplished through CLSM and subsequent image analysis of live/dead stained pellet cross-sections. Higher live ratio values were observed for pellets supplemented with glass beads compared to those without glass bead addition, specifically during cultivation days 3–8, which is also the main period of rebeccamycin production. Thus, it can be assumed that an increased live ratio correlates with enhanced rebeccamycin production in *L. aerocolonigenes* cultivations with the addition of glass beads.

However, due to the labor-intensive technique, only one pellet per sample could be considered, resulting in statistical uncertainties. To address this, the results of the CLSM measurements were compared and validated by flow cytometry measurements [34], which were conducted on a larger number of pellets cultivated with and without glass bead supplementation, as described in Schrinner et al. [11, 29].

3.4 Modeling of Bead Induced Shear Stress

In classic stirred tanks, the turbulent flow causes temporal and spatial velocity fluctuations. The resulting relative velocities between the biomass and the fluid lead to hydrodynamic stress. Various modelling approaches exist in the literature for modelling the hydrodynamic stress of filamentous microorganisms in stirred bioreactors. Among others, the so-called *energy dissipation circulation function* (EDCF) has been successfully applied to describe the influence of hydrodynamic stress on cell morphology, product formation or the growth of filamentous microorganisms [35]. Alternatively, the ratio of the maximum to the average energy dissipation rate was used to characterise the hydrodynamic stress [36–38]. This showed that the maximum energy dissipation rate in an agitated bioreactor is up to ten times greater in relation to the mean energy dissipation rate than in a shaking flask. Macroparticles are used to increase shear stress in the shaking flask, but a suitable modeling approach is lacking in the literature. The approach described below was developed for this purpose.

For a stirred tank bioreactor, Eslahpazir et al. [39] showed by means of CFD that the normal Reynolds stresses were significantly larger compared to the shear stresses. Based on this, it was hypothesized that pellet deformation occurs first due to normal forces and size reduction of the pellets due to shear stresses. During the contact of beads with other beads or solid walls, both normal and shear forces occur. As a measure of the intensity of shear, the shear rate was defined via the ratio of the tangential relative velocity $v_{90,ts}$ of the contact partners and the size of the bioparticle (d_a), such as a pellet (Eq. (6)). For bioparticles that have a size of 400 μm, shear rates within the range of 100–1000 s^{-1} can be attained, depending on the frequency of shaking. This frequency is also within the range of the average shear rate detected in stirred bioreactors [40]. However, the size of the bioparticles is subject to a distribution that also changes over the cultivation time. For this reason, only the shear rate was considered in Eq. (7) for modeling.

$$\dot{\gamma}_{90} = \frac{v_{90,ts}}{d_a} \tag{6}$$

$$S_s \propto v_{90,ts} \tag{7}$$

As a result of the gravitational force acting on the sphere, a bioparticle located between the wall and the spheres becomes compressed. For the modeling of the normal forces as a result, the quantity S_n was therefore defined as follows (Eq. (8)):

$$S_n \propto F_g \propto m_b \cdot g \propto d_b^3 \cdot \rho_b \cdot g \text{ (N)} \quad (8)$$

To describe the total active stress area A_s, it was assumed that the area is proportional to the number of bead-wall contacts N_c and thus to the stress frequency SF (Eq. (9)). In addition, a proportionality of the stress area per bead with the projection area A_b of a bead was assumed.

$$A_s \propto N_c \cdot A_b \propto SF \cdot d_b^2 \left(\frac{\text{m}^2}{\text{s}}\right) \quad (9)$$

For the newly defined stress area ratio (SAR), the product of the shear stress S_s and normal stress S_n was related to the total stress area A_s (Eq. (10)).

$$SAR \propto \frac{S_s \cdot S_n}{A_s} = \frac{v_{90,ts} \cdot d_b^3 \cdot \rho_b \cdot g}{SF \cdot d_b^2} = \frac{v_{90,ts} \cdot d_b \cdot \rho_b \cdot g}{SF} \left(\frac{\text{N}}{\text{m}}\right) \quad (10)$$

The normal and shear stresses were multiplied, as it was expected that higher normal forces would result in greater compression of the bioparticles. The resulting stresses in the mycelium can be amplified by shear forces, so that hyphae can rupture or pellets may break.

3.5 Correlation of Bead Shear Stress with Cultivation Results

The cultivation results discussed in the previous sections are plotted in Fig. 12 as a function of the characteristic parameter SAR presented in the last section. In the cultivations of the filamentous bacterium *L. aerocolonigenes*, the size, number and density of the macroparticles were varied in addition to the shaking frequency [22, 30]. It should be noted that several pre-cultures were necessary for the cultivations, which limits the comparability of the results. Furthermore, comparatively high standard deviations for the production of secondary metabolites in shaking flasks compared to microtiter plates have been described in the literature [13, 41]. Nevertheless, a collective optimum can be identified in the data for rebeccamycin concentration (Fig. 12a). It is noticeable that too high or too low stresses do not lead to an increase in the product titer compared to the control cultures without macroparticles. In fact, a decrease in product titer was observed at the highest rotational frequencies. There is also a shift in the y-axis of the product concentration for different shaking frequencies, indicating that other shaking frequency-dependent effects appear to influence product formation. For the averaged yield coefficient (Fig. 12b), which represents the relationship between product titer and biomass concentration, an optimum curve can also be seen in the Fig. 12b. Here, a clearer trend can be observed especially for

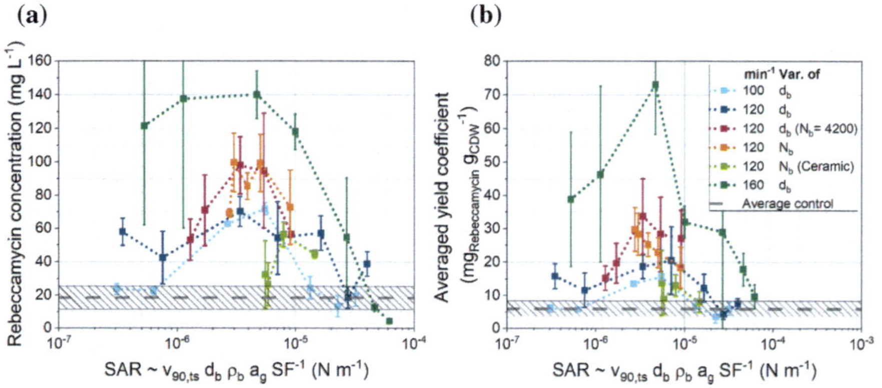

Fig. 12 Correlation of **a** rebeccamycin concentration and **b** averaged yield coefficient (from 10 day cultivations of the strain *L. aerocolonigenes*) with stress area ratio (SAR) for different macroparticle parameters (bead diameter d_b, bead number N_b, bead density, shaking frequency). Figures taken from [31], licensed under the Creative Commons Attribution 4.0 International license

the variation of the number of glass beads, so that the number of beads seems to have a particularly strong influence on the biomass concentration.

Overall, the formation of rebeccamycin seems to be linked to the ratio of shear and normal forces to the total shear stress area. It appears that there is a certain process window within which product formation is increased up to a factor of about seven. Therefore, introducing glass beads offers the chance to study the impact of shear stress on cultivation at an early stage in the process development. Simulations can deliver valuable information on the magnitude of stress prior to cultivation. By choosing appropriate parameters with the help of simulations, stress screening can be conducted with minimal need for experimental studies. Furthermore, the calculation of characteristic parameters for the bead induced stresses allows the comparison of cultivations of different strains.

3.6 *Validation of Shaking Flask CFD-DEM-Simulations*

For a qualitative as well as quantitative comparison of experiments and CFD-DEM simulations of the particle enhanced shake flask cultivation, particle tracking velocimetry (PTV) was used. In this method, the distance travelled by a particle is first determined by comparing two successive images. Subsequently, the velocity of the particle can be calculated from the recording interval and the distance. For the experimental recording of the bead movement at the bottom of the 250 mL shaking flask with 4 baffles, pictures were taken with a high-speed camera. Here, the flask was filled with 50 mL of water and beads of different sizes and then fixed on a platform mounted on the orbital shaker. In the case of the transparent glass beads, a mixture of equally sized black beads (marker) and transparent beads was chosen. This step improved the detection and tracking of individual beads in the image analysis. Before the actual PTV analysis, the marker beads were detected by an algorithm and their position was saved in a new image without disturbing reflections of the water surface.

The Fig. 13 shows an exemplary comparison of the camera recordings with the simulations for a bead size of 1.932 mm at three different shaking frequencies. At the lowest frequency of 100 min^{-1}, the beads are distributed over a large part of the shaking flask. Increasing the frequency causes the beads to be pushed to the outer edge of the flask together with the water due to higher centrifugal forces. The comparison with the bottom view of the simulations shows a very similar distribution of the beads. In addition, it is noticeable in the side view that the beads almost exclusively form a monolayer at low rotational frequencies, which develops into a multilayer at higher rotational frequencies.

Figure 14a–c shows the simulative and experimental velocity distributions for three bead sizes as a function of shaker speeds. At a speed of 100 min^{-1} the velocity distributions are almost independent of the bead size. Increasing the shaking frequency results in larger beads having a narrower velocity distribution. The maximum speed of the beads, on the one hand, is independent of the bead size. Here, the maximum velocity of the beads is limited by the diameter of the shake flask (rotation diameter) and the shaker frequency. On the other hand, the width of the distribution depends on the spatial distribution of the beads on the bottom of the shaking flask. Accordingly, larger beads move further away from the center of the flask. One reason is that the bead size was varied with a constant total mass. As a result, the total bead space requirement, which correlates with the product of the number of beads and the cross-sectional area of the beads, increases as the diameter of the beads decreases. Overall, there is good agreement between the simulation results and the experimental data. However, it is noticeable that a bimodal velocity distribution occurs with smaller beads. A possible cause is that all beads are included in the evaluation of the simulations. In contrast, the experimental distribution does not contain any information about the frequency of bead velocities in the vertical direction. This has the consequence that the frequency of a bead velocity is different for spherical layers, which can lead to a change in the velocity distribution. The greater deviations between experiment and simulation at a frequency of 160 min^{-1} arise most likely from the fact that the detection and assignment of partners between two images becomes more susceptible to error. This is due to the fact that the beads are distributed over a smaller space in several layers. In Fig. 14c, white ceramic balls with a density of 3800 kg m^{-3} were taken instead of glass beads with a density of 2500 kg m^{-3}. The lack of marker beads in a different color made it difficult to detect the beads in the images, so volume concentration of the beads was reduced and the bead detection algorithm was adjusted. Overall, the ceramic beads have lower maximum velocities at the same rotational speed compared to the glass beads.

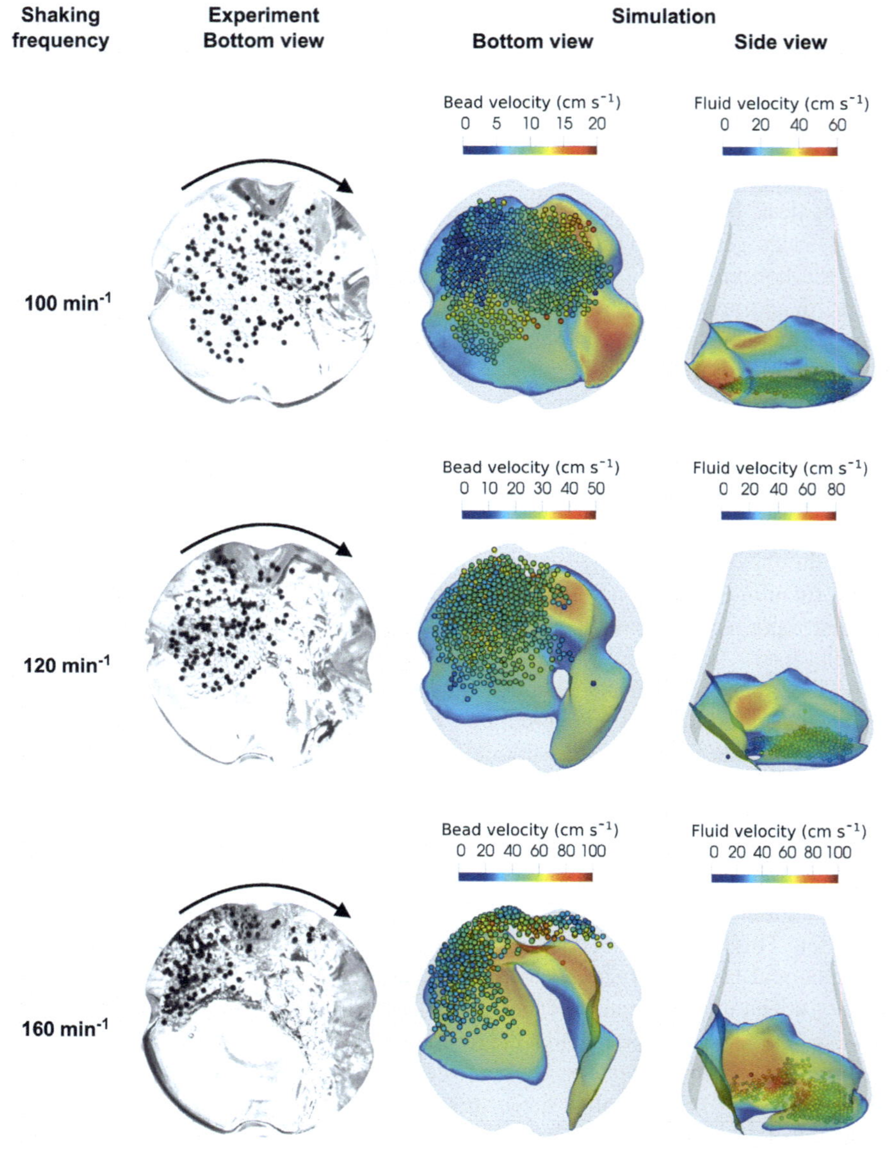

Fig. 13 High-speed images of the shaking flask bottom with transparent and black 1.932 mm glass beads at different shaking frequencies and CFD-DEM simulations of the glass beads with the contour of the water surface. Figure taken from [31], licensed under the Creative Commons Attribution 4.0 International license

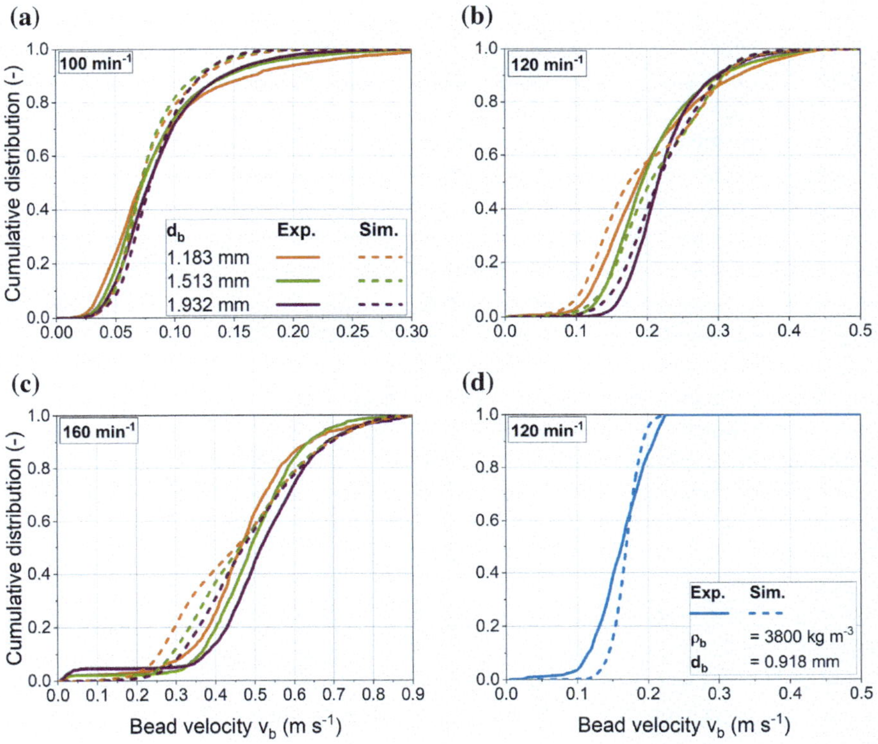

Fig. 14 Experimentally and simulatively determined cumulative distributions of bead velocities for three glass **a**–**c** bead diameters and one ceramic **d** bead size and different shaking frequencies. Due to the fact that the particle tracking method can only capture velocities in a plane parallel to the bottom of the baffled shaking flask, the velocities from the simulations are also shown only for the x- and y-directions. Figures taken from [31], licensed under the Creative Commons Attribution 4.0 International license

4 Microparticle-Enhanced Cultivation

An initial experiment demonstrated glass microparticles ($x_{50} = 7.9\ \mu m$) to generally yield a more significant improvement in productivity (+286%) in *L. aerocolonigenes* cultivations compared to conventional talc microparticles ($x_{50} = 10\ \mu m$; + 43%). Further on, a screening was conducted to assess the optimal concentration (0–10 g L^{-1}) and size ($x_{50} = 7.9$ or $30.5\ \mu m$) of glass microparticles. Two clear trends were identified regarding the rebeccamycin production. Firstly, in comparison to the unsupplemented control culture, product formation tends to increase with higher microparticle supplementation (up to +357%). Secondly, the product titers for smaller glass microparticles are consistently slightly higher than those for larger microparticles. Consequentially, a microparticle concentration of 10 g L^{-1} and a particle size of $x_{50} = 7.9\ \mu m$ were chosen as the setup for enhancing product formation.

Subsequently, light was shed on the changes in pellet morphology resulting from the supplementation of glass microparticles. In line with observations on the effects of MPEC in other cultivation systems [42–44], the MPEC approach (10 g L^{-1}) led to a significant decrease in pellet size (approximately 75 μm mean AESD) compared to the unsupplemented control culture (approximately 135 μm mean AESD). This reduction in pellet size is beneficial as smaller pellets are associated with improved product formation due to an overall reduced share of oxygen limited biomass. Moreover, the entanglement or incorporation of microparticles into the hyphal pellet structure, which has been observed previously [43–45], could be confirmed in the case for MPEC in *L. aerocolonigenes.* A broad incorporation of glass microparticles (10 g L^{-1}) was found throughout all investigated pellets, displaying distinct patterns such as encircling the pellet center or radiating towards the pellet edge. Thus, it is theorized that the incorporation is not random. Additionally, it is assumed to enhance product formation by loosening the pellet structure and reducing the biomass density, facilitating the supply of growth-limiting nutrients (e.g., oxygen). These two effects are considered key factors for enhanced rebeccamycin production. Nevertheless, additional physico-chemical effects associated with the interaction between particles and hyphae, as summarized by Laible et al. [25], should be considered in future studies.

To gain a better understanding of the MPEC effect on metabolic activity, the viability of the biomass was further examined. Therefore, fluorescence staining, cell slicing and CLSM were performed in the course of a 10 day cultivation to differentiate between live (green dye: SYTO 9) and dead (red dye: PI) shares of the biomass within single pellets (Fig. 15).

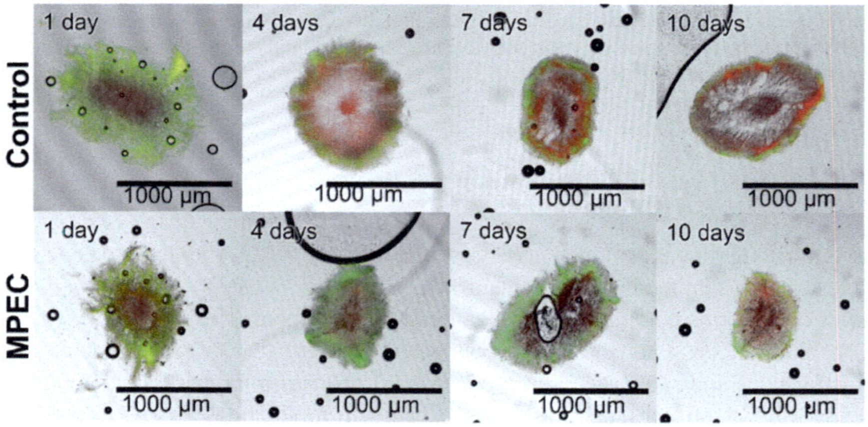

Fig. 15 Exemplary microscopic images of sliced pellets from unsupplemented cultivation (control) (upper row) and a glass microparticle-supplemented cultivation (10 g L^{-1}, $x_{50} = 7.9$ μm) (lower row) of a 10 day *L. aerocolonigenes* cultivation obtained with CLSM. Staining was performed with SYTO9 (green fluorescence, living cells) and PI (red fluorescence, dead cells). Figure taken from [1], licensed under the Creative Commons Attribution 4.0 International license

After one day of cultivation, both approaches show similar high levels of viability in the pellets. Subsequently, the viability of the unsupplemented control decays quickly over time. However, the live portions of MPEC pellets remain relatively constant, on a significantly higher level. Consequentially, the live ratio is consistently higher in MPEC pellets compared to the unsupplemented control throughout the entire cultivation, which is in line with previous studies [22, 46, 47]. Overall, there is a correlation between glass microparticles being supplemented and the formation of smaller pellets, the formation of biomass with higher porosity due to microparticle incorporation, a higher live ratio in the pellets, and enhanced synthesis of rebeccamycin.

5 Particle-Enhanced Cultivation Combined with Soy Lecithin Supplementation

Lecithin has been reported to have beneficial effects on various actinomycete cultivations [26–28]. Therefore, its potential as an additive in *L. aerocolonigenes* cultures was investigated. By varying the lecithin concentration (0–10 g L^{-1}), it was demonstrated that, consistent with observations in literature, the addition of lecithin generally had a positive impact on the cultivation system. The rebeccamycin titer reached an optimum concentration of 103 mg L^{-1} at 7.5 g L^{-1} lecithin, representing a +687% increase compared to the unsupplemented control. The increase in CDW concentration with increasing lecithin supplementation, up to threefold, suggested that lecithin could serve as an additional carbon source.

A subsequent investigation was conducted to compare the growth and synthesis kinetics of a culture supplemented with glass beads (100 g L^{-1}, $x_{50} = 969$ μm) with a culture containing both 5 g L^{-1} lecithin and glass beads. In this comparison, the trend in CDW concentration resulted to be similar for both approaches, but the lecithin supplemented one exhibited significantly higher CDW concentrations.

Furthermore, the exponential growth phase was prolonged, which delayed the synthesis of rebeccamycin. However, the enhanced rebeccamycin synthesis was confirmed with a concentration of 79 mg L^{-1} without lecithin compared to 259 mg L^{-1} for the culture with 5 g L^{-1} lecithin (Fig. 16a). Besides its role as an additional carbon source, lecithin may have other potential effects. Its emulsifying properties can facilitate the solution of rebeccamycin, which is highly hydrophobic, into the culture broth. This, in turn, could potentially lead to feedback regulatory effects that enhance intracellular product formation or it can alter the oxygen transfer rate in the culture, leading to secondary effects. Additionally, a direct comparison of cultures with and without glass beads (100 g L^{-1}, $x_{50} = 969$ μm) and lecithin (5 and 10 g L^{-1}) and a unsupplemented control culture is depicted in Fig. 16a.

It can be highlighted that a combination of 5 g L^{-1} lecithin and glass beads (100 g L^{-1}, $x_{50} = 969$ μm) resulted in significantly higher rebeccamycin concentration of 388 mg L^{-1}. Compared to the other approached, it presents the highest

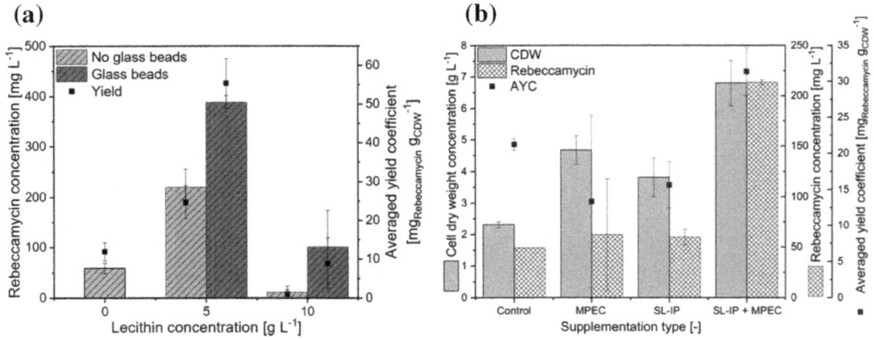

Fig. 16 **a** Rebeccamycin concentration and AYC of a 10 day cultivation with supplementation of soy lecithin in different concentrations up to 10 g L^{-1} and glass beads (100 g L^{-1} x_{50} = 969 μm); **b** CDW concentration, rebeccamycin concentration, and AYC of a 10 day cultivation of an unsupplemented control culture, MPEC (10 g L^{-1}, x_{50} = 7.9 μm), soy lecithin supplemented culture (SL-IP, 7.5 g L^{-1}) and MPEC (10 g L^{-1}, x_{50} = 7.9 μm) combined with soy lecithin addition (SL-IP, 7.5 g L^{-1}) (SL-IP + MPEC). Figures taken from [1, 22], licensed under the Creative Commons Attribution 4.0 International license

reported rebeccamycin titer in shaking flasks to date. Moreover, in contrast to the experiment with lecithin supplemention alone, the AYC was also increased in the combinatorial approach. Also, microscopic images revealed the pellets to appear of relatively low density. Thus, it can be assumed that micromorphological changes in the inner pellet structure are the potential key factor contributing to the superior synthesis of rebeccamycin in the combinatorial supplementation.

Since the combination of glass beads and soy lecithin demonstrated superior results compared to separate approaches, lecithin supplementation (EMULPUR™ IP, Cargill Texturizing Solutions Deutschland GmbH and Co. KG, Hamburg, Germany, 7.5 g L^{-1}) was matched with MPEC (10 g L^{-1}, x_{50} = 7.9 μm) for further combinatorial experiments (control culture, lecithin, MPEC, lecithin + MPEC) as shown in Fig. 16b. All supplementation approaches, in comparison to the control culture, evoked significant increases in CDW concentration, while the combinatorial approach reached a nearly threefold amount. The rebeccamycin titer slightly increased for the single supplement experiments, while the combination yielded 213 mg L^{-1}, an almost fourfold enhancement compared to the unsupplemented control and the highest titer achieved so far with MPEC. Moreover, a reduction in AESD (data not shown) and significant alterations of the inner pellet, observed with cross-sectional images, can explain the beneficial effects of lecithin and glass microparticles on enhancing rebeccamycin production. It can be theorized that lecithin, functioning as an emulsifier, influences the surface properties (e.g., hydrophobicity) of the microparticles, which in turn may affect the interaction between microparticles and hyphae. The combination of MPEC and lecithin seems to be promising for enhancing the cultivation of filamentous microorganisms in bioreactors, especially in regard to a potential scale-up. Implementing MPEC appears to be rather simple compared to larger glass beads, except for any desired separation of microparticles from biomass, despite the slight productivity advantage observed with macroparticles in shaking flasks.

6 Methods for Pellet Investigations

6.1 Oxygen Measurements in Pellets

It is assumed that targeted supplementation of glass particles affects the porosity of the pellets, resulting in an improved oxygen supply to the biomass and a shift of the critical radius r_{crit} towards the pellet center, thus increasing the proportion of viable and producing biomass. In the course of glass particle supplementation (both micro and macro-sized) the oxygen concentration within the pellets was monitored using oxygen microelectrode measurements for the first time in prokaryotic organisms. For oxygen microprofiling, the microelectrode stand described in Dinius et al. [1] was used (Fig. 17).

Shortly, pellets were fixed in a loop at the tip of a gel-loading tip which was located in the center of a flow cell. To imitate cultivation conditions, the aerated and diluted minimal medium circulated in the tube with a temperature of 28 °C. Pellet surface determination was enabled by camera (board-level camera) and detected oxygen decrease. Subsequently, the sensor was driven stepwise (10 μm each step) into the pellet at the equator area of the pellet, while the local oxygen concentration was measured. A minimum of three pellets per particle concentration was investigated,

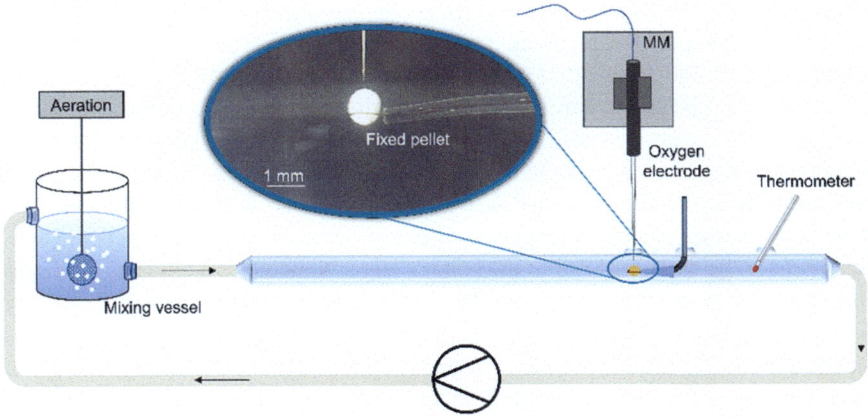

Fig. 17 Scheme of the experimental setup for oxygen microprofiling in biopellets (blue encircled close-up) of filamentous microorganisms. Pellet analysis occurs in a glass tube (laminar flow conditions) with the help of oxygen micro-sensors fixed on a software-controlled micromanipulator (MM). Oxygen concentration is measured in 10 μm step sizes inside the pellet. Figure taken from [1], licensed under the Creative Commons Attribution 4.0 International license

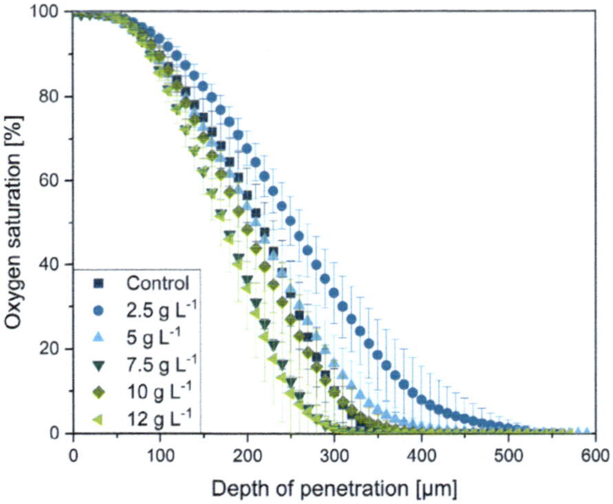

Fig. 18 Oxygen microprofiling experimentally acquired with the experimental setup described in [1]. Oxygen saturation in dependence on the relative penetration depth into the *L. aerocolonigenes* pellet for varying microparticle concentrations in the culture broth. Pellets were investigated after 64 h of cultivation and were selected within a similar size range. Figure taken from [1], licensed under the Creative Commons Attribution 4.0 International license

while at least three profiles were recorded for each pellet. Resulting profiles are depicted in Fig. 18.

Unsupplemented pellets showed the typical sigmoidal saturation pattern while the addition of 2.5 g L^{-1} glass microparticles resulted in the expected flattening of the oxygen profile curve, tantamount to an improved substrate supply to the pellet center. However, further increases in particle concentration (5, 7.5, 10, 12 g L^{-1}) led to steeper oxygen profiles, reaching a maximum saturation gradient $(dSatO_2/dr)_{max}$ with the 12 g L^{-1} supplementation. Consequently, the smallest oxygen-consuming layer Δr_{active} can be found here. Thus, contrary to previous expectations, the addition of glass microparticles does not result in an extended oxygen supply towards the pellet center. However, rebeccamycin synthesis is still improved by the addition of glass microparticles. The enhanced activity of the biomass in microparticle supplemented pellets (Fig. 15) leads to higher oxygen consumption in the outer pellet region, causing r_{crit} to be located closer to the pellet's edge. Nonetheless, oxygen microprofiling was also conducted with glass bead-supplemented pellets, which generally show higher activity compared to control cultures, too. However, the oxygen profiles of these pellets did not show a clear tendency in any specific direction. Therefore, it can be theorized that the directed influence on oxygen profiles can be exclusively attributed to the incorporation of microparticles into the hyphal network.

All in all, the measured oxygen profiles in pellets from MPEC reflect a balance between the enhanced oxygen transport into the pellets due to higher porosity and the increased consumption by the more active biomass. Depending on which effect contributed stronger, the profiles are affected accordingly. For example, the effect of higher activity may increasingly overshadow the impact of increasing microparticle concentration.

6.2 Micromechanical Behavior of Pellets

Pellets of filamentous microorganisms are subjected to high mechanical stress as a result of the power consumption of, e.g., stirring and aeration during cultivation [48, 49]. If the critical strength of the cell wall is exceeded, erosion of individual hyphae on the surface of the pellet or even breakage of the pellet alter the pellet size and morphology [50–52]. In addition to size of the pellets, the internal structure of the pellets can also be affected by external forces [48]. The internal pellet structure is connected with the mass transport into the pellet [1, 53], which in turn determines the activity and productivity of the biomass in the pellet. Furthermore, the influence of mechanical stresses on the morphology is important for the modelling of cultivation process [54–56].

In this context, much work has been done in the past on the global evolution of pellet size as a function of power input at the bioreactor level. However, there is currently no detailed understanding of the micromechanical behavior of filamentous pellets in the literature. Only Dittmann et al. [57] experimentally investigated the micromechanical behavior of filamentous pellets of the fungus *Aspergillus niger* SKAn1015 and the bacterium *Actinomadura namibiensis* using multiple compression between two plates. The results showed a plastic deformation of the pellets after the first compression. One possible reason for the plastic deformation was spatial displacement of individual hyphae in the loose hyphal network. It was also suggested that the extracellular matrix of the hyphae may cause the hyphae to adhere to each other after compression, resulting in permanent deformation. In addition, increasing maximum forces were observed with increasing pellet diameter and density. The authors concluded that, on the one hand, a sufficiently high mechanical stability of the pellets is necessary to withstand the mechanical properties of the pellets in order to achieve high levels of productivity. On the other hand, a certain porosity of the pellet structure is required to avoid nutrient and oxygen limitations in the pellet. In order to be able to obtain an optimum pellet morphology, a deep and comprehensive understanding of the extremely complex factors influencing pellet and product formation is required. As a step in this direction, a simulative method for characterizing the mechanical behavior pellets was developed in this project.

6.3 DEM-Simulation Environment for the Compression of Pellet Structures

The investigation of the structure-properties relationship for pellets of filamentous microorganisms by simulation requires the generation of three-dimensional mycelial structures. In the literature, different approaches with varying degrees of complexity have been described. In the study by Islam et al. [58], a random fiber network with a defined fiber density was generated for a representative volume element. Alternatively, Schmideder et al. [59] were the first to reconstruct the hyphal network of real pellets using X-ray micro-computed tomography (μCT). The method allowed the determination of a variety of morphological properties such as hyphal length, number of tips, number of branches, hyphal growth unit, porosity and mean hyphal diameter. However, this process is time consuming and costly, so it is limited to the analysis of a small number of pellets. The use of synchrotron radiation-based X-ray microtomography, in contrast, allows the examination of a much larger number of pellets, so that the growth of the pellets over the cultivation time can be studied [60].

In this work, Monte-Carlo simulations were used to quickly generate a digital pellet [53, 61]. In detail, three-dimensional pellet structures of different sizes can be created from a single spore or from a spore agglomerate over several iteration steps. The structures are depended on hyphal diameter, maximum growth angle, branch angle and branch interval [53]. The maximum growth angle defines the maximum angle relative to the previous tip element (Fig. 19a). At each growth step, an angle within the limits is randomly selected. Combined with the length of a hyphal element, this gives a circular path for the position of the new hyphal tip. The position of the hyphal tip is randomly selected on the circular path and checked for collisions with other elements. A branch can be formed if there are no branches in the four adjacent nodes, and the number of allowed branches is less than the maximum number allowed. In addition, it was checked that there were no collisions with other elements. The maximum number of branches is determined by the branch interval, which is the ratio between the length between two branches and the hyphal diameter. Unlike the growth angle, the branching angle is constant. Again, a circular path is defined together with the element length, on which the position of the hyphal tip at the branch is randomly selected. The branch angle is defined relative to the average of the orientations of the elements before and after the branch point. In this study, the pellet was created from a single spore with a hyphal diameter of 6.3 μm, a maximum growth angle of 24°, a branching angle of 90° and a branching interval of 19. The values used in the growth model correspond to the middle values of the parameter space investigated by Schmideder et al. [53]. A specific strain was deliberately not chosen as a reference, as the aim of this initial work was to investigate strain-independent dependencies. Pellet structures often show a surface with low hyphal density, making it difficult to clearly define a pellet size. For this reason, the hyphal (volume) fraction per spherical shell was calculated. The pellet size is derived from the diameter of the outermost spherical shell, which has at least a hyphal fraction of 5%. This threshold corresponds to the limit defined by Schmideder et al. [53] between growth as dispersed mycelium

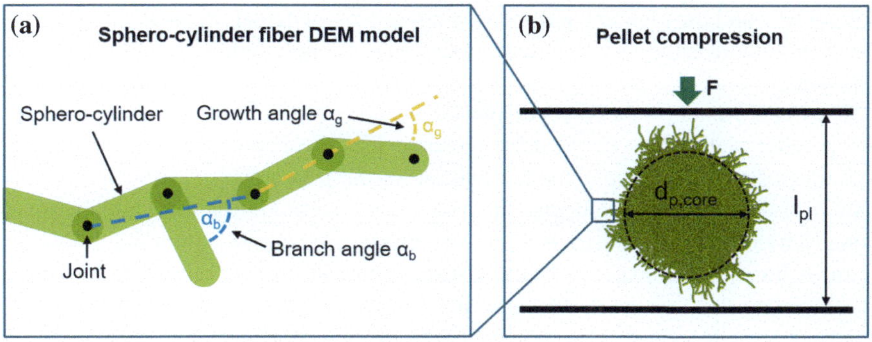

Fig. 19 **a** Schematic representation of the sphero-cylinder model used to represent hyphae in the DEM simulation, **b** pellet generated by Monte-Carlo growth simulation and positioned between two compression plates

and pellet form. The corresponding diameter is referred to as the core diameter $d_{p,core}$ in the following.

The generated hyphal network was modelled in the DEM software Ansys Rocky using a linear-elastic fiber model. In this approach, the hyphae are represented at discrete intervals by flexible connected sphero-cylinders (Fig. 19a) [62]. Forces and moments are calculated in the joints between the hyphal segments. The fiber model is based on a full cylinder when calculating forces and moments. This assumption is not correct for hyphae as only the cell wall contributes to the cell mechanics. For this reason, the stiffnesses used in the fiber model have been adapted to the cross-sectional areas and moments of inertia of a hollow cylinder. In addition to the joint model, a linear-elastic contact model was also used for the interactions between adjacent hyphae.

To investigate the micro-mechanical behavior, a single pellet was compressed between two parallel plates and the resulting force was recorded over the compression distance (Fig. 19b). The initial plate spacing l_{pl} corresponded to the maximum dimensions of the pellet in the z-direction plus a safety distance of twice the hyphal diameter. In experimental pellet compression tests by Dittmann et al. [57] a deformation rate of 1% s^{-1} was used. Depending on the pellet size, this corresponds to a speed of 1–4 $\mu m\ s^{-1}$. Also Overbeck et al. [63] compressed yeast and algae with a similar velocity of 2 $\mu m\ s^{-1}$, whereas Islam et al. [58] used compression speeds of 100–1000 $\mu m\ s^{-1}$ for dry mycelium. Thereby, they could not determine any significant influence of the strain rate in the range of the examined strains. To reduce the simulation time, the pellets were pre-compressed with an increased top plate speed of 50 mm s^{-1}. The compression distance for the initial pre-compression results from the difference of initial plate distance and the pellet core diameter. After pre-compression, all pellets were compacted with a reduced top plate speed of 20 mm s^{-1} to a compression ratio of 60% of the pellet core diameter. Due to the small hyphae diameter of 6.3 μm, a constant time step of $1.9 \cdot 10^{-9}$ s was used. For a purely elastic material behavior with no energy dissipation, the restitution coefficient was set to 1 and the element and joint damping was switched off. A Poisson's ratio of 0.5 (incompressible material) was chosen as a common value for biological cells [64–66].

The Fig. 20 displays the force–displacement curve of a pellet compressed between two parallel plates. The initial delay in force increase is attributed to the pellet's random positioning. Once a stable position on the lower plate is achieved, deformation of the outer hyphae begins, causing slow increase in the force on the upper plate. Subsequently, the entire network of hyphae is gradually compressed, leading to an exponential increase in force. Unlike an ideally elastic material, the curve for unloading does not track the curve for loading, resulting in hysteresis behavior. However, the unloading curve returns to the original starting point of the loading curve, indicating a reversible deformation. The area beneath the unloading curve corresponds to the elastic energy. On the other hand, the area between the loading and unloading curves represents hysteresis energy. Linear elastic force and moment models without adhesion were used in the simulations, so that energy dissipation during compression can be excluded. Instead, rearrangements of hyphae in the network are expected to occur so that some of the energy remains in the form of stresses in the hyphae even after the pellet is unloaded.

Currently, the focus is on systematically investigating the influence of modeling and simulation parameters to evaluate their effect on the mechanical behavior of the pellet. Additionally, the effect of the growth behavior of filamentous microorganisms on different pellet structures will be investigated in order to obtain a detailed knowledge of the structure–property relationships. Additionally, an analysis will be performed to determine the stress present in the pellet during loading to investigate possible damage of single hyphae or fracture of the pellet due to defined external forces. In the long term, the combination of numerical methods and genetic modification of the strains will allow for a targeted adaptation of the mechanical behavior of

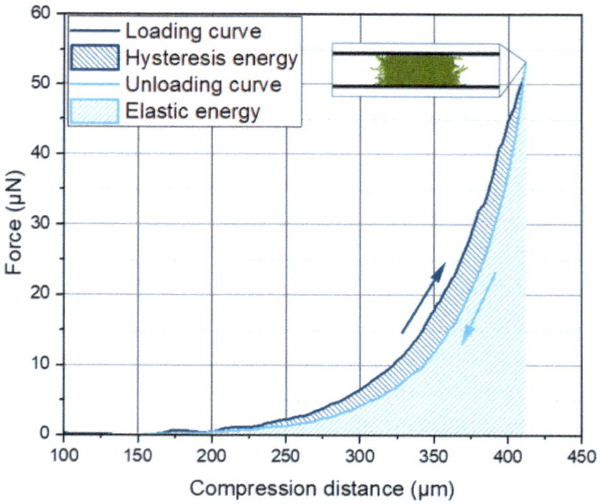

Fig. 20 Force–displacement curve for the compression and decompression of a pellet with a core diameter of 450 μm. The colored areas correspond to the elastic and hysteresis energy, respectively

the hyphal network structure depending on the field of application and the associated requirements.

7 Conclusion and Outlook

In this chapter, it was demonstrated how various morphology engineering methods were successfully implemented to enhance rebeccamycin production with the filamentous bacterium *L. aerocolonigenes*. Based on these findings, four main conclusions can be drawn:

- A positive correlation between the rebeccamycin synthesis and the addition of macro glass beads was demonstrated and linked to the induced mechanical stress by the glass beads. To gain a better understanding of the particle-induced stress, a closer look was taken at the interplay between product synthesis and pellet morphology, as well as various parameters for macroparticle addition, such as bead diameter, concentration, density and shaking frequency. Multiphase simulations were used to characterize and quantify the stresses caused by macroparticles. Based on this, shake flask scale cultivation can be optimized for ideal stress levels to achieve high product titers. The setting of optimum stress conditions depends on the strain and the type of product formation (growth-related or growth-independent). In the future, the optimal process parameters, which depend on the morphological characteristic, can be determined more systematically and efficiently with the presented experimental and simulative methods.
- Glass microparticles were found to be a superior type compared to conventional microparticles for MPEC in *L. aerocolonigenes*. Furthermore, being well characterizable particles, they provide the foundation for further investigations into the interaction between hyphae and varying particle surface properties through surface modifications.
- The oxygen microelectrode technique was implemented for measurements in actinomycetes and can be employed to correlate the MPEC-induced structural changes in pellets with the oxygen supply.
- The emulsifier soy lecithin is a highly beneficial supplement for cultivations of *L. aerocolonigenes*. Its combination with macro glass beads and glass microparticles demonstrates even more superior cultivations results.

The external and internal morphology of pellets is determined by the influence of mechanical forces on the mycelium. To investigate the mechanical behavior of pellets and their dependence on biological growth behavior, simulations of pellet compression were established for the first time. Furthermore, in a future study, the scale-up of the shear-sensitive *L. aerocolonigenes* into the stirred tank bioreactor will be presented. For this purpose, a novel membrane gassing-unit for bubble-free cultivation was designed, implemented, and characterized in 3 L scale bioreactor, which also allows cultivation accompanying CFD-DEM simulations.

References

1. Dinius A, Schrinner K, Schrader M, Kozanecka ZJ, Brauns H, Klose L, Weiß H, Kwade A, Krull R (2023) Morphology engineering for novel antibiotics: effect of glass microparticles and soy lecithin on rebeccamycin production and cellular morphology of filamentous actinomycete *Lentzea aerocolonigenes*. Front Bioeng Biotechnol. https://doi.org/10.3389/fbioe.2023.1171055
2. Meyer V, Cairns T, Barthel L, King R, Kunz P, Schmideder S, Müller H, Briesen H, Dinius A, Krull R (2021) Understanding and controlling filamentous growth of fungal cell factories: novel tools and opportunities for targeted morphology engineering. Fungal Biol Biotechnol. https://doi.org/10.1186/s40694-021-00115-6
3. Mast Y, Stegmann E (2019) Actinomycetes: the antibiotics producers. Antibiotics, Basel, Switzerland. https://doi.org/10.3390/antibiotics8030105
4. Genilloud O (2017) Actinomycetes: still a source of novel antibiotics. Nat Prod Rep. https://doi.org/10.1039/C7NP00026J
5. Dinius A, Kozanecka ZJ, Hoffmann KP, Krull R (2024) Intensification of bioprocesses with filamentous microorganisms. Phys Sci Rev 9(2):777–823. https://doi.org/10.1515/psr-2022-0112
6. Whitaker A (1992) Actinomycetes in submerged culture. Appl Biochem Biotechnol. https://doi.org/10.1007/BF02922146
7. Böl M, Schrinner K, Tesche S, Krull R (2021) Challenges of influencing cellular morphology by morphology engineering techniques and mechanical induced stress on filamentous pellet systems—a critical review. Eng Life Sci. https://doi.org/10.1002/elsc.202000060
8. Nettleton Jr DE, Bush JA, Bradner WT, Doyle TW. Process for producing rebeccamycin. Patent US4552842A
9. Bush JA, Long BH, Catino JJ, Bradner WT (1987) Production and biological activity of rebeccamycin, a novel antitumor agent. J Antibiot. https://doi.org/10.7164/antibiotics.40.668
10. Pommerehne K, Walisko J, Ebersbach A, Krull R (2019) The antitumor antibiotic rebeccamycin-challenges and advanced approaches in production processes. Appl Microbiol Biotechnol. https://doi.org/10.1007/s00253-019-09741-y
11. Schrinner K, Althof K, Ebersbach AT, Grosch J-H, Krull R (2020) XAD resins for increased rebeccamycin productivity in cultivations of *Lentzea aerocolonigenes*. Chem Ing Tec. https://doi.org/10.1002/cite.202000131
12. Dobson LF, O'Shea DG (2008) Antagonistic effect of divalent cations Ca^{2+} and Mg^{2+} on the morphological development of *Streptomyces hygroscopicus var. geldanus*. Appl Microbiol Biotechnol. https://doi.org/10.1007/s00253-008-1627-7
13. Sohoni SV, Bapat PM, Lantz AE (2012) Robust, small-scale cultivation platform for *Streptomyces coelicolor*. Microb Cell Fact. https://doi.org/10.1186/1475-2859-11-9
14. Holtmann D, Vernen F, Müller JM, Kaden D, Risse JM, Friehs K, Dähne L, Stratmann A, Schrader J (2017) Effects of particle addition to *Streptomyces* cultivations to optimize the production of actinorhodin and streptavidin. Sustain Chem Pharm. https://doi.org/10.1016/j.scp.2016.09.001
15. Hotop S, Möller J, Niehoff J, Schügerl K (1993) Influence of the preculture conditions on the pellet size distribution of *Penicillium chrysogenum* cultivations. Process Biochem. https://doi.org/10.1016/0032-9592(93)80013
16. Cai M, Zhang Y, Hu W, Shen W, Yu Z, Zhou W, Jiang T, Zhou X, Zhang Y (2014) Genetically shaping morphology of the filamentous fungus *Aspergillus glaucus* for production of antitumor polyketide aspergiolide A. Microb Cell Fact. https://doi.org/10.1186/1475-2859-13-73
17. Liang J, Xu Z, Liu T, Lin J, Cen P (2008) Effects of cultivation conditions on the production of natamycin with *Streptomyces gilvosporeus* LK-196. Enzyme Microb Technol. https://doi.org/10.1016/j.enzmictec.2007.08.012
18. Královà S, Sandoval-Powers M, Fawwal DV, Degnes KF, Lewin AS, Klinkenberg G, Nguyen G-S, Liles MR, Wentzel A (2021) *Streptomyces tardus sp. nov.*: a slow-growing *Actinobacterium*

producing candicidin, isolated from sediments of the Trondheim Fjord. Front Microbiol (2021). https://doi.org/10.3389/fmicb.2021.714233
19. Song X, Zhang Y, Xue J, Li C, Wang Z, Wang Y (2018) Enhancing nemadectin production by *Streptomyces cyaneogriseus ssp. noncyanogenus* through quantitative evaluation and optimization of dissolved oxygen and shear force. Bioresour Technol (2018). https://doi.org/10.1016/j.biortech.2017.09.033
20. Li G, Sun J, Li F, Wang Y, Li Q (2022) Macroparticle-enhanced bioleaching of uranium using *Aspergillus niger*. Miner Eng. https://doi.org/10.1016/j.mineng.2022.107493
21. Zhu W-Z, Wang S-H, Gao H-M, Ge Y-M, Dai J, Zhang X-L, Yang Q (2021) Characterization of bioactivities and biosynthesis of angucycline/angucyclinone derivatives derived from *Gephyromycinifex aptenodytis* gen. nov., sp. nov. Mar Drugs (2021). https://doi.org/10.3390/md20010034
22. Schrinner K, Schrader M, Niebusch J, Althof K, Schwarzer FA, Nowka P-F, Dinius A, Kwade A, Krull R (2021) Macroparticle-enhanced cultivation of *Lentzea aerocolonigenes*: variation of mechanical stress and combination with lecithin supplementation for a significantly increased rebeccamycin production. Biotechnol Bioeng. https://doi.org/10.1002/bit.27875
23. Krull R, Wucherpfennig T, Esfandabadi ME, Walisko R, Melzer G, Hempel DC, Kampen I, Kwade A, Wittmann C (2013) Characterization and control of fungal morphology for improved production performance in biotechnology. J Biotechnol. https://doi.org/10.1016/j.jbiotec.2012.06.024
24. Walisko R, Moench-Tegeder J, Blotenberg J, Wucherpfennig T, Krull R (2015) The taming of the shrew-controlling the morphology of filamentous eukaryotic and prokaryotic microorganisms. Adv Biochem Eng Biotechnol. https://doi.org/10.1007/10_2015_322
25. Laible AR, Dinius A, Schrader M, Krull R, Kwade A, Briesen H, Schmideder S (2021) Effects and interactions of metal oxides in microparticle-enhanced cultivation of filamentous microorganisms. Eng Life Sci. https://doi.org/10.1002/elsc.202100075
26. Adelson LM, Schatz A, Trelawny GS (1957) Metabolism of lipids and lipid derivatives by a soil actinomycete. J Bacteriol 73:148–153
27. Schatz A, Adelson LM, Trelawny GS (1956) Phospholipids and related substances as growth substrates for soil microorganisms. Appl Microbiol 4:223–228
28. Brock TD (1956) The effect of oils and fatty acids on the production of filipin. Appl Microbiol 4:131–133
29. Schrinner K, Veiter L, Schmideder S, Doppler P, Schrader M, Münch N, Althof K, Kwade A, Briesen H, Herwig C, Krull R (2020) Morphological and physiological characterization of filamentous *Lentzea aerocolonigenes*: comparison of biopellets by microscopy and flow cytometry. PLoS ONE. https://doi.org/10.1371/journal.pone.0234125
30. Schrader M, Pommerehne K, Wolf S, Finke B, Schilde C, Kampen I, Lichtenegger T, Krull R, Kwade A (2019) Design of a CFD-DEM-based method for mechanical stress calculation and its application to glass bead-enhanced cultivations of filamentous *Lentzea aerocolonigenes*. Biochem Eng J. https://doi.org/10.1016/j.bej.2019.04.014
31. Schrader M, Schrinner K, Polomsky L, Ivanov D, Kampen I, Schilde C, Krull R, Kwade A (2023) Quantification and modeling of macroparticle-induced mechanical stress for varying shake flask cultivation conditions. Front Bioeng Biotechnol. https://doi.org/10.3389/fbioe.2023.1254136
32. Kwade A (2003) A stressing model for the description and optimization of grinding processes. Chem Eng Technol. https://doi.org/10.1002/ceat.200390029
33. Beinert S, Fragnière G, Schilde C, Kwade A (2015) Analysis and modelling of bead contacts in wet-operating stirred media and planetary ball mills with CFD–DEM simulations. Chem Eng Sci. https://doi.org/10.1016/j.ces.2015.05.063
34. Veiter L, Herwig C (2019) The filamentous fungus *Penicillium chrysogenum* analysed via flow cytometry-a fast and statistically sound insight into morphology and viability. Appl Microbiol Biotechnol. https://doi.org/10.1007/s00253-019-09943-4
35. Böhm L, Hohl L, Bliatsiou C, Kraume M (2019) Multiphase stirred tank bioreactors—new geometrical concepts and scale-up approaches. Chem Ing Tec. https://doi.org/10.1002/cite.201900165

36. Büchs J, Zoels B (2001) Evaluation of maximum to specific power consumption ratio in shaking bioreactors. J Chem Eng Jpn. https://doi.org/10.1252/jcej.34.647
37. Daub A, Böhm M, Delueg S, Mühlmann M, Schneider G, Büchs J (2014) Characterization of hydromechanical stress in aerated stirred tanks up to 40 m^3 scale by measurement of maximum stable drop size. J Biol Eng. https://doi.org/10.1186/1754-1611-8-17
38. Daub A, Böhm M, Delueg S, Mühlmann M, Schneider G, Büchs J (2014) Maximum stable drop size measurements indicate turbulence attenuation by aeration in a 3 m^3 aerated stirred tank. Biochem Eng J. https://doi.org/10.1016/j.bej.2014.02.006
39. Eslahpazir Esfandabadi M, Wucherpfennig T, Krull R (2012) Agitation induced mechanical stress in stirred tank bioreactors—linking CFD simulations to fungal morphology. J Chem Eng Jpn. https://doi.org/10.1252/jcej.12we019
40. Campesi A, Cerri MO, Hokka CO, Badino AC (2009) Determination of the average shear rate in a stirred and aerated tank bioreactor. Bioprocess Biosyst Eng. https://doi.org/10.1007/s00449-008-0242-4
41. Siebenberg S, Bapat PM, Lantz AE, Gust B, Heide L (2010) Reducing the variability of antibiotic production in *Streptomyces* by cultivation in 24-square deepwell plates. J Biosci Bioeng. https://doi.org/10.1016/j.jbiosc.2009.08.479
42. Kaup B-A, Ehrich K, Pescheck M, Schrader J (2008) Microparticle-enhanced cultivation of filamentous microorganisms: increased chloroperoxidase formation by *Caldariomyces fumago* as an example. Biotechnol Bioeng. https://doi.org/10.1002/bit.21713
43. Driouch H, Hänsch R, Wucherpfennig T, Krull R, Wittmann C (2012) Improved enzyme production by bio-pellets of Aspergillus niger: targeted morphology engineering using titanate microparticles. Biotechnol Bioeng. https://doi.org/10.1002/bit.23313
44. Kuhl M, Gläser L, Rebets Y, Rückert C, Sarkar N, Hartsch T, Kalinowski J, Luzhetskyy A, Wittmann C (2020) Microparticles globally reprogram *Streptomyces albus* toward accelerated morphogenesis, streamlined carbon core metabolism, and enhanced production of the antituberculosis polyketide pamamycin. Biotechnol Bioeng. https://doi.org/10.1002/bit.27537
45. Walisko J, Vernen F, Pommerehne K, Richter G, Terfehr J, Kaden D, Dähne L, Holtmann D, Krull R (2017) Particle-based production of antibiotic rebeccamycin with *Lechevalieria aerocolonigenes*. Process Biochem. https://doi.org/10.1016/j.procbio.2016.11.017
46. Driouch H, Roth A, Dersch P, Wittmann C (2010) Optimized bioprocess for production of fructofuranosidase by recombinant *Aspergillus niger*. Appl Microbiol Biotechnol. https://doi.org/10.1007/s00253-010-2661-9
47. Driouch H, Sommer B, Wittmann C (2010) Morphology engineering of *Aspergillus niger* for improved enzyme production. Biotechnol Bioeng. https://doi.org/10.1002/bit.22614
48. Lin P-J, Scholz A, Krull R (2010) Effect of volumetric power input by aeration and agitation on pellet morphology and product formation of *Aspergillus niger*. Biochem Eng J. https://doi.org/10.1016/j.bej.2009.12.016
49. Kelly S, Grimm LH, Bendig C, Hempel DC, Krull R (2006) Effects of fluid dynamic induced shear stress on fungal growth and morphology. Process Biochem. https://doi.org/10.1016/j.procbio.2006.06.007
50. Cui YQ, van der Lans RGJM, Luyben KCAM (1997) Effect of agitation intensities on fungal morphology of submerged fermentation. Biotechnol Bioeng. https://doi.org/10.1002/(SICI)1097-0290(19970905)55:5%3c715:AID-BIT2%3e3.0.CO;2-E
51. Heydarian S, Ison A, Lilly M, Ayazi Shamlou P (2000) Turbulent breakage of filamentous bacteria in mechanically agitated batch culture. Chem Eng Sci. https://doi.org/10.1016/S0009-2509(99)00449-2
52. Waldherr P, Bliatsiou C, Böhm L, Kraume M (2023) Fragmentation of *Aspergillus niger* pellets in stirred tank bioreactors due to hydrodynamic stress. Chem Eng Res Des. https://doi.org/10.1016/j.cherd.2023.05.038
53. Schmideder S, Müller H, Barthel L, Friedrich T, Niessen L, Meyer V, Briesen H (2020) Universal law for diffusive mass transport through mycelial networks. Biotechnol Bioeng. https://doi.org/10.1002/bit.27622

54. Edelstein L, Hadar Y (1983) A model for pellet size distributions in submerged mycelial cultures. J Theor Biol. https://doi.org/10.1016/0022-5193(83)90185-6
55. Papagianni M (2004) Fungal morphology and metabolite production in submerged mycelial processes. Biotechnol Adv. https://doi.org/10.1016/j.biotechadv.2003.09.005
56. Krabben P, Nielsen J, Michelsen ML (1997) Analysis of single hyphal growth and fragmentation in submerged cultures using a population model. Chem Eng Sci. https://doi.org/10.1016/S0009-2509(97)00079-1
57. Dittmann J, Tesche S, Krull R, Böl M (2019) The influence of salt-enhanced cultivation on the micromechanical behaviour of filamentous pellets. Biochem Eng J. https://doi.org/10.1016/j.bej.2019.04.023
58. Islam MR, Tudryn G, Bucinell R, Schadler L, Picu RC (2017) Morphology and mechanics of fungal mycelium. Sci Rep. https://doi.org/10.1038/s41598-017-13295-2
59. Schmideder S, Barthel L, Friedrich T, Thalhammer M, Kovačević T, Niessen L, Meyer V, Briesen H (2019) An X-ray microtomography-based method for detailed analysis of the three-dimensional morphology of fungal pellets. Biotechnol Bioeng. https://doi.org/10.1002/bit.26956
60. Müller H, Deffur C, Schmideder S, Barthel L, Friedrich T, Mirlach L, Hammel JU, Meyer V, Briesen H (2023) Synchrotron radiation-based microcomputed tomography for three-dimensional growth analysis of *Aspergillus niger* pellets. Biotechnol Bioeng. https://doi.org/10.1002/bit.28506
61. Celler K, Picioreanu C, van Loosdrecht MCM, van Wezel GP (2012) Structured morphological modeling as a framework for rational strain design of *Streptomyces* species. Antonie Van Leeuwenhoek. https://doi.org/10.1007/s10482-012-9760-9
62. Guo Y, Wassgren C, Curtis JS, Xu D (2018) A bonded sphero-cylinder model for the discrete element simulation of elasto-plastic fibers. Chem Eng Sci. https://doi.org/10.1016/j.ces.2017.09.029
63. Overbeck A, Günther S, Kampen I, Kwade A (2017) Compression testing and modeling of spherical cells—comparison of yeast and algae. Chem Eng Technol. https://doi.org/10.1002/ceat.201600145
64. Zhao L, Schaefer D, Xu H, Modi SJ, LaCourse WR, Marten MR (2005) Elastic properties of the cell wall of *Aspergillus nidulans* studied with atomic force microscopy. Biotechnol Prog. https://doi.org/10.1021/bp0497233
65. A-Hassan E, Heinz WF, Antonik MD, D'Costa NP, Nageswaran S, Schoenenberger CA, Hoh JH (1998) Relative microelastic mapping of living cells by atomic force microscopy. Biophys J. https://doi.org/10.1016/S0006-3495(98)77868-3
66. Touhami A, Nysten B, Dufrêne YF (2003) Nanoscale mapping of the elasticity of microbial cells by atomic force microscopy. Langmuir. https://doi.org/10.1021/la034136x

SPRINGER NATURE

GPSR Compliance

The European Union's (EU) General Product Safety Regulation (GPSR) is a set of rules that requires consumer products to be safe and our obligations to ensure this.

If you have any concerns about our products, you can contact us on ProductSafety@springernature.com

In case Publisher is established outside the EU, the EU authorized representative is:

Springer Nature Customer Service Center GmbH
Europaplatz 3
69115 Heidelberg, Germany

The manufacturer's authorised representative in the EU is Springer
Nature Customer Service Centre GmbH, Europaplatz 3, 69115 Heidelberg,
Germany. If you have any concerns regarding our products, please
contact ProductSafety@springernature.com

Printed and bound by CPI Group (UK) Ltd, Croydon, CR0 4YY
01/10/2026
02225965-0016